Continuum Mechanics of Solids

Lallit Anand

Department of Mechanical Engineering, Massachusetts Institute of Technology, Cambridge, MA 02139, USA

Sanjay Govindjee

Department of Civil and Environmental Engineering, University of California, Berkeley, Berkeley, CA 94720, USA

OXFORD UNIVERSITY PRESS

OXFORD
UNIVERSITY PRESS

Great Clarendon Street, Oxford, OX2 6DP,
United Kingdom

Oxford University Press is a department of the University of Oxford. It furthers the University's objective of excellence in research, scholarship, and education by publishing worldwide. Oxford is a registered trade mark of Oxford University Press in the UK and in certain other countries

First Edition published in 2020
Reprinted with corrections in 2023

Published in the United States of America by Oxford University Press
198 Madison Avenue, New York, NY 10016, United States of America

British Library Cataloguing in Publication Data
Data available

Library of Congress Control Number: 2020938849

ISBN 978-0-19-886472-1

DOI: 10.1093/oso/9780198864721.001.0001

Printed and bound by
CPI Group (UK) Ltd, Croydon, CR0 4YY

Continuum Mechanics of Solids

Contents

PART I

Vectors and tensors

1 Vectors and tensors: Algebra

The theories of Solid Mechanics are mathematical in nature. Thus in this chapter and the next, we begin with a discussion of the essential mathematical definitions and operations regarding vectors and tensors needed to discuss the major topics covered in this book.[1]

The approach taken is a pragmatic one, without an overly formal emphasis upon mathematical proofs. The end-goal is to develop, in a reasonably systematic way, the essential mathematics of vectors and tensors needed to discuss the major aspects of mechanics of solids. The physical quantities that we will encounter in this text will be of three basic types: *scalars*, *vectors*, and *tensors*. Scalars are quantities that can be described simply by numbers, for example things like temperature and density. Vectors are quantities like velocity and displacement—quantities that have both magnitude and direction. Tensors will be used to describe more complex quantities like stress and strain, as well as material properties. In this chapter we deal with the algebra of vectors and tensors. In the next chapter we deal with their calculus.

1.1 Cartesian coordinate frames. Kronecker delta. Alternating symbol

Throughout this text *lower case Latin subscripts range over the integers*

$$\{1, 2, 3\}.$$

A **Cartesian coordinate frame** for the Euclidean point space $\mathcal{E}$ consists of a reference point $\mathbf{o}$ called **the origin** together with a positively oriented orthonormal basis $\{\mathbf{e}_1, \mathbf{e}_2, \mathbf{e}_3\}$ for the associated vector space $\mathcal{V}$. Being positively oriented and orthonormal, the basis vectors obey

$$\mathbf{e}_i \cdot \mathbf{e}_j = \delta_{ij}, \qquad \mathbf{e}_i \cdot (\mathbf{e}_j \times \mathbf{e}_k) = e_{ijk}. \tag{1.1.1}$$

Here δ_{ij}, the Kronecker delta, is defined by

$$\delta_{ij} = \begin{cases} 1, & \text{if } i = j, \\ 0, & \text{if } i \neq j, \end{cases} \tag{1.1.2}$$

[1] The mathematical preliminaries in this chapter and the next are largely based on the clear expositions by Gurtin in Gurtin (1981) and Gurtin et al. (2010).

Continuum Mechanics of Solids. Lallit Anand and Sanjay Govindjee, Oxford University Press (2020).

DOI: 10.1093/oso/9780198864721.001.0001

while e_{ijk}, the alternating symbol,[2] is defined by

$$e_{ijk} = \begin{cases} 1, & \text{if } \{i,j,k\} = \{1,2,3\}, \{2,3,1\}, \text{ or } \{3,1,2\}, \\ -1, & \text{if } \{i,j,k\} = \{2,1,3\}, \{1,3,2\}, \text{ or } \{3,2,1\}, \\ 0, & \text{if an index is repeated,} \end{cases} \tag{1.1.3}$$

and hence has the value $+1$, -1, or 0 when $\{i, j, k\}$ is an even permutation, an odd permutation, or not a permutation of $\{1, 2, 3\}$, respectively. Where convenient, we will use the compact notation

$$\{\mathbf{e}_i\} \stackrel{\text{def}}{=} \{\mathbf{e}_1, \mathbf{e}_2, \mathbf{e}_3\},$$

to denote our positively oriented orthonormal basis.

1.1.1 **Summation convention**

For convenience, we employ the Einstein *summation convention* according to which summation over the range $1, 2, 3$ is implied for any index that is repeated twice in any term, so that, for instance,

$$u_i v_i = u_1 v_1 + u_2 v_2 + u_3 v_3,$$

$$S_{ij} u_j = S_{i1} u_1 + S_{i2} u_2 + S_{i3} u_3,$$

$$S_{ik} T_{kj} = S_{i1} T_{1j} + S_{i2} T_{2j} + S_{i3} T_{3j}.$$

In the expression $S_{ij} u_j$ the subscript i is *free*, because it is not summed over, while j is a *dummy* subscript, since

$$S_{ij} u_j = S_{ik} u_k = S_{im} u_m.$$

As a rule when using the Einstein summation convention, one can not repeat an index more than twice. Expressions with such indices indicate that a computational error has been made. In the rare case where three (or more) repeated indices must appear in a single term, an explicit summation symbol, for example $\sum_{i=1}^{3}$, must be used.

Note that the Kronecker delta δ_{ij} modifies (or contracts) the subscripts in the coefficients of an expression in which it appears:

$$a_i \delta_{ij} = a_j, \qquad a_i b_j \delta_{ij} = a_i b_i = a_j b_j, \qquad \delta_{ij} \delta_{ik} = \delta_{jk}, \qquad \delta_{ik} S_{kj} = S_{ij}.$$

This property of the Kronecker delta is sometimes termed the substitution property.

Further, when an expression in which an index is repeated twice *but summation is not to be performed* we state so explicitly. For example,

$$u_i v_i \qquad \text{(no sum)}$$

signifies that the subscript i is not to be summed over.

[2] The alternating symbol is sometimes called the permutation symbol.

Because $\{\mathbf{e}_i\}$ is a basis, every vector $\mathbf{u}$ admits the unique expansion

$$\mathbf{u} = u_j \mathbf{e}_j; \tag{1.1.4}$$

the scalars u_i are called the (Cartesian) **components of u** (relative to this basis). If we take the inner product of (1.1.4) with $\mathbf{e}_i$ we find that, since $\mathbf{e}_i \cdot \mathbf{e}_j = \delta_{ij}$,

$$u_i = \mathbf{u} \cdot \mathbf{e}_i. \tag{1.1.5}$$

Guided by this relation, we define the coordinates of a point $\mathbf{x}$ with respect to the origin $\mathbf{o}$ by

$$x_i = (\mathbf{x} - \mathbf{o}) \cdot \mathbf{e}_i. \tag{1.1.6}$$

In view of (1.1.4), the inner product of vectors $\mathbf{u}$ and $\mathbf{v}$ may be expressed as

$$\begin{aligned} \mathbf{u} \cdot \mathbf{v} &= (u_i \mathbf{e}_i) \cdot (v_j \mathbf{e}_j) \\ &= u_i v_j \delta_{ij} \\ &= u_i v_i = u_1 v_1 + u_2 v_2 + u_3 v_3. \end{aligned} \tag{1.1.7}$$

It should also be recalled that this relation is equivalent to $\mathbf{u} \cdot \mathbf{v} = |\mathbf{u}|\,|\mathbf{v}| \cos \theta_{uv}$, where $|\cdot|$ denotes the length of a vector (its magnitude or norm) and θ_{uv} is the angle between vectors $\mathbf{u}$ and $\mathbf{v}$. Equation (1.1.4) further implies that the cross product of two vectors

$$\begin{aligned} \mathbf{u} \times \mathbf{v} &= (u_j \mathbf{e}_j) \times (v_k \mathbf{e}_k) \\ &= u_j v_k (\mathbf{e}_j \times \mathbf{e}_k) \\ &= e_{ijk} u_j v_k \mathbf{e}_i. \end{aligned} \tag{1.1.8}$$

In particular, (1.1.8) implies that the vector $\mathbf{u} \times \mathbf{v}$ has the component form

$$(\mathbf{u} \times \mathbf{v})_i = e_{ijk} u_j v_k. \tag{1.1.9}$$

This definition of the cross product is fully compatible with the relation $|\mathbf{u} \times \mathbf{v}| = |\mathbf{u}|\,|\mathbf{v}| \sin \theta_{uv}$, which indicates that the magnitude of the cross product of two vectors is equal to the area of the parallelogram whose edges are defined by $\mathbf{u}$ and $\mathbf{v}$.

It can be shown that the permutation symbol is related to the Kronecker delta by

$$e_{ijk} = \det \begin{bmatrix} \delta_{i1} & \delta_{i2} & \delta_{i3} \\ \delta_{j1} & \delta_{j2} & \delta_{j3} \\ \delta_{k1} & \delta_{k2} & \delta_{k3} \end{bmatrix} = \det \begin{bmatrix} \delta_{i1} & \delta_{j1} & \delta_{k1} \\ \delta_{i2} & \delta_{j2} & \delta_{k2} \\ \delta_{i3} & \delta_{j3} & \delta_{k3} \end{bmatrix}, \tag{1.1.10}$$

where $\det[\cdot]$ represents the determinant of a 3×3 matrix. A consequence of (1.1.10) is that

$$\begin{aligned} e_{ijk} e_{pqr} &= \det \begin{bmatrix} \delta_{ip} & \delta_{iq} & \delta_{ir} \\ \delta_{jp} & \delta_{jq} & \delta_{jr} \\ \delta_{kp} & \delta_{kq} & \delta_{kr} \end{bmatrix}, \\ &= \delta_{ip}(\delta_{jq}\delta_{kr} - \delta_{jr}\delta_{kq}) - \delta_{iq}(\delta_{jp}\delta_{kr} - \delta_{jr}\delta_{kp}) + \delta_{ir}(\delta_{jp}\delta_{kq} - \delta_{jq}\delta_{kp}). \end{aligned} \tag{1.1.11}$$

Some useful identities which follow from successive contractions of (1.1.11), the substitution property of the Kronecker delta and the identity $\delta_{ii} = 3$ are

$$e_{ijk}e_{ipq} = \delta_{jp}\delta_{kq} - \delta_{jq}\delta_{kp}, \tag{1.1.12}$$

and hence also that

$$\begin{aligned} e_{ijk}e_{ijl} &= 2\delta_{kl}, \\ e_{ijk}e_{ijk} &= 6. \end{aligned} \tag{1.1.13}$$

Also useful is the identity

$$\mathbf{e}_i = \tfrac{1}{2} e_{ijk} \mathbf{e}_j \times \mathbf{e}_k. \tag{1.1.14}$$

Let $\mathbf{u}$, $\mathbf{v}$, and $\mathbf{w}$ be vectors. Useful relations involving the inner and cross products then include,

$$\begin{gathered} \mathbf{u} \cdot (\mathbf{v} \times \mathbf{w}) = \mathbf{v} \cdot (\mathbf{w} \times \mathbf{u}) = \mathbf{w} \cdot (\mathbf{u} \times \mathbf{v}), \\ \mathbf{u} \cdot (\mathbf{u} \times \mathbf{v}) = 0, \\ \mathbf{u} \times \mathbf{v} = -\mathbf{v} \times \mathbf{u}, \\ \mathbf{u} \times (\mathbf{v} \times \mathbf{w}) = (\mathbf{u} \cdot \mathbf{w})\mathbf{v} - (\mathbf{u} \cdot \mathbf{v})\mathbf{w}. \end{gathered} \tag{1.1.15}$$

The operation

$$[\mathbf{u}, \mathbf{v}, \mathbf{w}] \stackrel{\text{def}}{=} \mathbf{u} \cdot (\mathbf{v} \times \mathbf{w}), \tag{1.1.16}$$

is known as the *scalar triple product* and is physically equivalent to the (signed) volume of the parallelepiped whose edges are defined by any three linearly independent vectors $\mathbf{u}$, $\mathbf{v}$, and $\mathbf{w}$.

1.2 Tensors

1.2.1 What is a tensor?

In mathematics the term **tensor** is a synonym for the phrase "a linear transformation which maps a vector into a vector." A tensor $\mathbf{S}$ is therefore a *linear* mapping that assigns to each vector $\mathbf{u}$ a vector

$$\mathbf{v} = \mathbf{S}\mathbf{u}. \tag{1.2.1}$$

One might think of a tensor $\mathbf{S}$ as a machine with an input and an output: if a vector $\mathbf{u}$ is the input, then the vector $\mathbf{v} = \mathbf{S}\mathbf{u}$ is the output. **Linearity** of a tensor $\mathbf{S}$ is the requirement that

$$\begin{aligned} \mathbf{S}(\mathbf{u} + \mathbf{v}) &= \mathbf{S}\mathbf{u} + \mathbf{S}\mathbf{v} \qquad \text{for all vectors } \mathbf{u} \text{ and } \mathbf{v}, \\ \mathbf{S}(\alpha\mathbf{u}) &= \alpha\mathbf{S}\mathbf{u} \qquad \text{for all vectors } \mathbf{u} \text{ and scalars } \alpha. \end{aligned} \tag{1.2.2}$$

1.2.2 Zero and identity tensors

Two basic tensors are the **zero tensor** $\mathbf{0}$ and the **identity tensor** $\mathbf{1}$:

$$\mathbf{0v} = \mathbf{0}, \qquad \mathbf{1v} = \mathbf{v}$$

for all vectors $\mathbf{v}$.

1.2.3 Tensor product

Another example of a tensor is the **tensor product**[3] $\mathbf{u} \otimes \mathbf{v}$, of two vectors $\mathbf{u}$ and $\mathbf{v}$, defined by

$$(\mathbf{u} \otimes \mathbf{v})\mathbf{w} = (\mathbf{v} \cdot \mathbf{w})\mathbf{u} \tag{1.2.3}$$

for all $\mathbf{w}$; the tensor $\mathbf{u} \otimes \mathbf{v}$ maps any vector $\mathbf{w}$ onto a scalar multiple of $\mathbf{u}$.

EXAMPLE 1.1 Projection tensor

As a concrete example of the tensor product consider the tensor

$$\mathbf{P} = \mathbf{1} - \mathbf{n} \otimes \mathbf{n},$$

where $\mathbf{n}$ is a vector of unit length. The action of $\mathbf{P}$ on an arbitrary vector $\mathbf{u}$ is then given by

$$\begin{aligned}\mathbf{Pu} &= (\mathbf{1} - \mathbf{n} \otimes \mathbf{n})\mathbf{u},\\ &= \mathbf{1u} - (\mathbf{n} \otimes \mathbf{n})\mathbf{u},\\ &= \mathbf{u} - (\mathbf{u} \cdot \mathbf{n})\mathbf{n}.\end{aligned} \tag{1.2.4}$$

Equation (1.2.4) can be interpreted to be the vector $\mathbf{u}$ minus the projection of $\mathbf{u}$ in the direction of $\mathbf{n}$. Hence $\mathbf{Pu}$ is equal to the projection of the vector $\mathbf{u}$ into the plane orthogonal to $\mathbf{n}$. Note that in this example we have exploited the fact that the dot product of a vector with a unit vector results in its projection onto the direction defined by the unit vector. Further, we have exploited the additive algebraic structure of tensors, whereby for tensors $\mathbf{S}$ and $\mathbf{T}$, $(\mathbf{S} + \mathbf{T})\mathbf{u} = \mathbf{Su} + \mathbf{Tu}$ for all vectors $\mathbf{u}$.

1.2.4 Components of a tensor

Given a tensor $\mathbf{S}$, the quantity $\mathbf{Se}_j$ is a vector. The components of a tensor $\mathbf{S}$ with respect to the basis $\{\mathbf{e}_1, \mathbf{e}_2, \mathbf{e}_3\}$ are defined by

$$S_{ij} \stackrel{\text{def}}{=} \mathbf{e}_i \cdot \mathbf{Se}_j. \tag{1.2.5}$$

[3] Other common terminologies for a *tensor product* are *tensor outer product* or simply *outer product*, as well as, *dyadic product*.

With this definition of the components of a tensor, the Cartesian component representation of the relation $\mathbf{v} = \mathbf{S}\mathbf{u}$ is

$$\begin{aligned} v_i = \mathbf{e}_i \cdot \mathbf{v} = \mathbf{e}_i \cdot \mathbf{S}\mathbf{u} \\ &= \mathbf{e}_i \cdot \mathbf{S}(u_j \mathbf{e}_j) \\ &= (\mathbf{e}_i \cdot \mathbf{S}\mathbf{e}_j) u_j \\ &= S_{ij} u_j. \end{aligned} \tag{1.2.6}$$

It should be observed that (1.2.6) indicates that each component of $\mathbf{v}$ is a linear combination of the components of $\mathbf{u}$, where the weights are given by the components of $\mathbf{S}$.

Further, a tensor $\mathbf{S}$ has the representation

$$\mathbf{S} = S_{ij}\, \mathbf{e}_i \otimes \mathbf{e}_j, \tag{1.2.7}$$

in terms of its components S_{ij} and the basis tensors $\mathbf{e}_i \otimes \mathbf{e}_j$.

EXAMPLE 1.2 Identity tensor components

Using (1.2.5), the components of the identity tensor are given by

$$(\mathbf{1})_{ij} = \mathbf{e}_i \cdot \mathbf{1}\mathbf{e}_j = \mathbf{e}_i \cdot \mathbf{e}_j = \delta_{ij}.$$

Thus the identity tensor can be expressed as $\mathbf{1} = \delta_{ij}\mathbf{e}_i \otimes \mathbf{e}_j$, equivalently as $\mathbf{1} = \mathbf{e}_i \otimes \mathbf{e}_i$.

EXAMPLE 1.3 Zero tensor components

Using (1.2.5), the components of the zero tensor are given by

$$(\mathbf{0})_{ij} = \mathbf{e}_i \cdot \mathbf{0}\mathbf{e}_j = \mathbf{e}_i \cdot \mathbf{0} = 0.$$

Thus the zero tensor can be expressed as $\mathbf{0} = \sum_{i=1}^{3} \sum_{j=1}^{3} 0\, \mathbf{e}_i \otimes \mathbf{e}_j$.

1.2.5 Transpose of a tensor

The **transpose** $\mathbf{S}^\top$ of a tensor $\mathbf{S}$ is the unique tensor with the property that

$$\mathbf{u} \cdot \mathbf{S}\mathbf{v} = \mathbf{v} \cdot \mathbf{S}^\top \mathbf{u} \tag{1.2.8}$$

for all vectors $\mathbf{u}$ and $\mathbf{v}$. The components of the transpose are determined via (1.2.5) as

$$\mathbf{e}_i \cdot \mathbf{S}^\top \mathbf{e}_j = \mathbf{e}_j \cdot \mathbf{S}\mathbf{e}_i = S_{ji}\,;$$

thus the components of $\mathbf{S}^\top$ are

$$(\mathbf{S}^\top)_{ij} = S_{ji}. \tag{1.2.9}$$

1.2.6 **Symmetric and skew tensors**

A tensor $\mathbf{S}$ is **symmetric** if

$$\mathbf{S} = \mathbf{S}^{\top}, \qquad S_{ij} = S_{ji}, \tag{1.2.10}$$

and **skew** if

$$\mathbf{S} = -\mathbf{S}^{\top}, \qquad S_{ij} = -S_{ji}. \tag{1.2.11}$$

Clearly,

$$S_{ij} = \tfrac{1}{2}(S_{ij} + S_{ji}) + \tfrac{1}{2}(S_{ij} - S_{ji}).$$

Thus, every tensor $\mathbf{S}$ admits the decomposition

$$\mathbf{S} = \operatorname{sym}\mathbf{S} + \operatorname{skw}\mathbf{S} \tag{1.2.12}$$

into a symmetric part and a skew part, where

$$\begin{aligned} \operatorname{sym}\mathbf{S} &= \tfrac{1}{2}(\mathbf{S} + \mathbf{S}^{\top}), \\ \operatorname{skw}\mathbf{S} &= \tfrac{1}{2}(\mathbf{S} - \mathbf{S}^{\top}), \end{aligned} \tag{1.2.13}$$

with components

$$\begin{aligned} (\operatorname{sym}\mathbf{S})_{ij} &= \tfrac{1}{2}(S_{ij} + S_{ji}), \\ (\operatorname{skw}\mathbf{S})_{ij} &= \tfrac{1}{2}(S_{ij} - S_{ji}). \end{aligned} \tag{1.2.14}$$

Note that a symmetric tensor has at most six independent components, and a skew tensor has at most three independent components; the latter point follows since $(\operatorname{skw}\mathbf{S})_{ii} = 0$ (no sum).

1.2.7 **Axial vector of a skew tensor**

Since a skew tensor has only three independent components, it is possible to define its action on vectors by another vector. Given any *skew* tensor $\mathbf{\Omega}$, there is a unique vector $\boldsymbol{\omega}$—called the **axial vector** of $\mathbf{\Omega}$—such that

$$\mathbf{\Omega}\mathbf{u} = \boldsymbol{\omega} \times \mathbf{u} \tag{1.2.15}$$

for all vectors $\mathbf{u}$. Since

$$(\mathbf{\Omega}\mathbf{u})_i = \Omega_{ij}u_j \qquad \text{and} \qquad (\boldsymbol{\omega} \times \mathbf{u})_i = e_{ijk}\omega_j u_k = e_{ikj}\omega_k u_j,$$

we have that

$$\Omega_{ij} = e_{ikj}\omega_k \equiv -e_{ijl}\omega_l. \tag{1.2.16}$$

Further, operating on both sides of (1.2.16) with e_{ijk} and using the epsilon-delta identity $(1.1.13)_1$ we obtain

$$e_{ijk}\Omega_{ij} = -e_{ijk}e_{ijl}\omega_l = -2\delta_{kl}\omega_l = -2\omega_k,$$

and hence in terms of the three independent components of the skew tensor $\mathbf{\Omega}$, the three components of its axial vector $\boldsymbol{\omega}$ are given by

$$\omega_i = -\tfrac{1}{2}e_{ijk}\Omega_{jk}, \tag{1.2.17}$$

which in expanded form is

$$\omega_1 = \Omega_{32} = -\Omega_{23}, \qquad \omega_2 = \Omega_{13} = -\Omega_{31}, \qquad \omega_3 = \Omega_{21} = -\Omega_{12}. \tag{1.2.18}$$

1.2.8 **Product of tensors**

Given tensors **S** and **T**, the **product ST** is defined by composition; that is, **ST** is defined by

$$(\mathbf{ST})\mathbf{v} = \mathbf{S}(\mathbf{T}\mathbf{v}) \tag{1.2.19}$$

for all vectors **v**. By (1.2.5), $\mathbf{T}\mathbf{e}_j = T_{lj}\mathbf{e}_l$ and by (1.2.9), $\mathbf{S}^\top\mathbf{e}_i = S_{ik}\mathbf{e}_k$; thus,

$$\begin{aligned}\mathbf{e}_i \cdot \mathbf{ST}\mathbf{e}_j &= \mathbf{S}^\top\mathbf{e}_i \cdot \mathbf{T}\mathbf{e}_j \\ &= (S_{ik}\mathbf{e}_k) \cdot (T_{lj}\mathbf{e}_l) \\ &= S_{ik}T_{lj}\underbrace{(\mathbf{e}_k \cdot \mathbf{e}_l)}_{\delta_{kl}} \\ &= S_{ik}T_{kj},\end{aligned}$$

and, hence,

$$(\mathbf{ST})_{ij} = S_{ik}T_{kj}. \tag{1.2.20}$$

Generally, $\mathbf{ST} \neq \mathbf{TS}$. When $\mathbf{ST} = \mathbf{TS}$, the tensors **S** and **T** are said to **commute**.

EXAMPLE 1.4 Product with the identity tensor

The product of the identity tensor **1** with any arbitrary tensor **T**, is **T** itself:

$$(\mathbf{1T})\mathbf{v} = \mathbf{1}(\mathbf{T}\mathbf{v}) = \mathbf{T}\mathbf{v},$$

and since this holds for all vectors **v** we have

$$\mathbf{1T} = \mathbf{T}.$$

Alternatively, in components

$$\begin{aligned}(\mathbf{1T})_{ij} &= \mathbf{e}_i \cdot \mathbf{1T}\mathbf{e}_j \\ &= \mathbf{1}^\top\mathbf{e}_i \cdot \mathbf{T}\mathbf{e}_j \\ &= \mathbf{e}_i \cdot \mathbf{T}\mathbf{e}_j \\ &= T_{ij}.\end{aligned}$$

The components of **1T** are equal to the components of **T**. The two are equal to each other.

1.2.9 Trace of a tensor. Deviatoric tensors

The **trace** is the linear operation that assigns to each tensor **S** a scalar tr **S** and satisfies

$$\mathrm{tr}(\mathbf{u}\otimes\mathbf{v}) = \mathbf{u}\cdot\mathbf{v} \tag{1.2.21}$$

for any vectors **u** and **v**. Linearity is the requirement that (cf. (1.2.2))

$$\mathrm{tr}(\alpha\mathbf{S}+\beta\mathbf{T}) = \alpha\,\mathrm{tr}(\mathbf{S}) + \beta\,\mathrm{tr}(\mathbf{T})$$

for all tensors **S** and **T** and all scalars α and β. Thus, by (1.2.7),

$$\begin{aligned}\mathrm{tr}\,\mathbf{S} &= \mathrm{tr}(S_{ij}\mathbf{e}_i\otimes\mathbf{e}_j)\\ &= S_{ij}\,\mathrm{tr}(\mathbf{e}_i\otimes\mathbf{e}_j)\\ &= S_{ij}(\mathbf{e}_i\cdot\mathbf{e}_j)\\ &= S_{ii},\end{aligned} \tag{1.2.22}$$

and the trace is well-defined for all tensors. Some useful properties of the trace are

$$\left.\begin{aligned}\mathrm{tr}(\mathbf{Su}\otimes\mathbf{v}) &= \mathbf{v}\cdot\mathbf{Su},\\ \mathrm{tr}(\mathbf{S}^{\top}) &= \mathrm{tr}\,\mathbf{S},\\ \mathrm{tr}(\mathbf{ST}) &= \mathrm{tr}(\mathbf{TS}),\\ \mathrm{tr}\,\mathbf{1} &= 3.\end{aligned}\right\} \tag{1.2.23}$$

As a consequence of $(1.2.23)_2$,

$$\mathrm{tr}\,\mathbf{S} = 0 \quad \text{whenever } \mathbf{S} \text{ is skew.} \tag{1.2.24}$$

A tensor **S** is **deviatoric** (or traceless) if

$$\mathrm{tr}\,\mathbf{S} = 0, \tag{1.2.25}$$

and we refer to

$$\begin{aligned}\mathbf{S}' &\equiv \mathrm{dev}\,\mathbf{S}\\ &\stackrel{\text{def}}{=} \mathbf{S} - \tfrac{1}{3}(\mathrm{tr}\,\mathbf{S})\mathbf{1}\end{aligned} \tag{1.2.26}$$

as the **deviatoric part** of **S**,[4] and to

$$\tfrac{1}{3}(\mathrm{tr}\,\mathbf{S})\mathbf{1} \tag{1.2.27}$$

as the **spherical part** of **S**. Trivially,

$$\mathbf{S} = \underbrace{\mathbf{S} - \tfrac{1}{3}(\mathrm{tr}\,\mathbf{S})\mathbf{1}}_{\mathbf{S}'} + \underbrace{\tfrac{1}{3}(\mathrm{tr}\,\mathbf{S})\mathbf{1}}_{s\mathbf{1}},$$

[4] The notation "dev" is useful for denoting the deviatoric part of the product of many tensors; e.g., dev($\mathbf{ST}\cdots\mathbf{M}$).

where $s = \frac{1}{3}\operatorname{tr}\mathbf{S}$, and it is noted that $\operatorname{tr}\operatorname{dev}\mathbf{S} = 0$. Thus every tensor $\mathbf{S}$ admits the decomposition

$$\mathbf{S} = \mathbf{S}' + s\mathbf{1} \tag{1.2.28}$$

into a deviatoric tensor and a spherical tensor.

EXAMPLE 1.5 Deviatoric and spherical parts of a tensor

The deviatoric part of a tensor

$$\mathbf{T} = 3\mathbf{e}_1 \otimes \mathbf{e}_1 + 2\mathbf{e}_1 \otimes \mathbf{e}_3$$

is determined as

$$\begin{aligned} \mathbf{T}' &= \mathbf{T} - \tfrac{1}{3}(\operatorname{tr}\mathbf{T})\mathbf{1} \\ &= (3\mathbf{e}_1 \otimes \mathbf{e}_1 + 2\mathbf{e}_1 \otimes \mathbf{e}_3) - \tfrac{1}{3} \cdot 3\mathbf{1} \\ &= 2\mathbf{e}_1 \otimes \mathbf{e}_1 + 2\mathbf{e}_1 \otimes \mathbf{e}_3 - 1\mathbf{e}_2 \otimes \mathbf{e}_2 - 1\mathbf{e}_3 \otimes \mathbf{e}_3. \end{aligned}$$

The spherical part of $\mathbf{T}$ is given by $\frac{1}{3}\operatorname{tr}\mathbf{T}\mathbf{1} = \mathbf{1}$.

1.2.10 **Positive definite tensors**

A tensor $\mathbf{C}$ is **positive definite** if and only if

$$\mathbf{u} \cdot \mathbf{C}\mathbf{u} > 0, \qquad u_i C_{ij} u_j > 0 \tag{1.2.29}$$

for all vectors $\mathbf{u} \neq \mathbf{0}$.

1.2.11 **Inner product of tensors. Magnitude of a tensor**

The inner product of two vectors $\mathbf{u}$ and $\mathbf{v}$ is defined by

$$\mathbf{u} \cdot \mathbf{v} = \mathbf{v} \cdot \mathbf{u} = u_i v_i.$$

Analogously, the **inner product** of two tensors $\mathbf{S}$ and $\mathbf{T}$ is defined by

$$\mathbf{S} : \mathbf{T} = \mathbf{T} : \mathbf{S} = S_{ij} T_{ij}.$$

The symbol : is known as the double contraction symbol.

By analogy to the notion of the magnitude (or norm) of a vector $\mathbf{u}$,

$$|\mathbf{u}| = \sqrt{\mathbf{u} \cdot \mathbf{u}} = \sqrt{u_i u_i},$$

the magnitude (or norm) $|\mathbf{S}|$ of a tensor $\mathbf{S}$ is defined by

$$|\mathbf{S}| = \sqrt{\mathbf{S} : \mathbf{S}} = \sqrt{S_{ij} S_{ij}}.$$

EXAMPLE 1.6 Contraction of a skew tensor with a symmetric tensor

Consider a symmetric tensor $\mathbf{S}$ ($S_{ij} = S_{ji}$) and a skew tensor $\boldsymbol{\Omega}$ ($\Omega_{ij} = -\Omega_{ji}$). Their *double contraction* $\mathbf{S}:\boldsymbol{\Omega} = S_{ij}\Omega_{ij}$ will always be zero:

$$S_{ij}\Omega_{ij} = -S_{ij}\Omega_{ji} = -S_{ji}\Omega_{ji} = -S_{ij}\Omega_{ij}.$$

Since the only scalar with the property that it is equal to its negative is zero, we conclude that $S_{ij}\Omega_{ij} = 0$ whenever $\mathbf{S}$ is symmetric and $\boldsymbol{\Omega}$ is skew.

1.2.12 **Matrix representation of tensors and vectors**

Tensors and vectors of the type presented have related matrix representations. We write $[\mathbf{u}]$ and $[\mathbf{S}]$ for the matrix representations of a vector $\mathbf{u}$ and a tensor $\mathbf{S}$ with respect to the basis $\{\mathbf{e}_i\}$:

$$[\mathbf{u}] = \begin{bmatrix} u_1 \\ u_2 \\ u_3 \end{bmatrix}, \qquad [\mathbf{S}] = \begin{bmatrix} S_{11} & S_{12} & S_{13} \\ S_{21} & S_{22} & S_{23} \\ S_{31} & S_{32} & S_{33} \end{bmatrix}.$$

- All the operations and operators we have introduced for vectors and tensors are in one-to-one correspondence to the same operations and operators for matrices.

For example,

$$\begin{aligned} [\mathbf{S}][\mathbf{u}] &= \begin{bmatrix} S_{11} & S_{12} & S_{13} \\ S_{21} & S_{22} & S_{23} \\ S_{31} & S_{32} & S_{33} \end{bmatrix} \begin{bmatrix} u_1 \\ u_2 \\ u_3 \end{bmatrix} \\ &= \begin{bmatrix} S_{11}u_1 + S_{12}u_2 + S_{13}u_3 \\ S_{21}u_1 + S_{22}u_2 + S_{23}u_3 \\ S_{31}u_1 + S_{32}u_2 + S_{33}u_3 \end{bmatrix} \\ &= \begin{bmatrix} \sum_i S_{1i}u_i \\ \sum_i S_{2i}u_i \\ \sum_i S_{3i}u_i \end{bmatrix} \\ &= [\mathbf{S}\mathbf{u}], \end{aligned}$$

so that the action of a tensor on a vector is consistent with that of a 3×3 matrix on a 3×1 matrix. Further, the matrix $[\mathbf{S}^\top]$ of the transpose $\mathbf{S}^\top$ of $\mathbf{S}$ is identical to the transposition of the matrix $[\mathbf{S}]$:

$$[\mathbf{S}^\top] \equiv [\mathbf{S}]^\top = \begin{bmatrix} S_{11} & S_{21} & S_{31} \\ S_{12} & S_{22} & S_{32} \\ S_{13} & S_{23} & S_{33} \end{bmatrix}.$$

Similarly, the trace of a tensor **S** is equivalent to the conventional definition of this quantity from matrix algebra

$$\operatorname{tr}\mathbf{S} \equiv \operatorname{tr}[\mathbf{S}] = S_{11} + S_{22} + S_{33} = S_{kk}.$$

The inner product of two vectors

$$\mathbf{u}\cdot\mathbf{v} \equiv [\mathbf{u}]^{\top}[\mathbf{v}] = \begin{bmatrix} u_1 & u_2 & u_3 \end{bmatrix}\begin{bmatrix} v_1 \\ v_2 \\ v_3 \end{bmatrix} = u_1v_1 + u_2v_2 + u_3v_3 \equiv u_iv_i,$$

and the tensor product

$$[\mathbf{u}\otimes\mathbf{v}] \equiv [\mathbf{u}][\mathbf{v}]^{\top} = \begin{bmatrix} u_1 \\ u_2 \\ u_3 \end{bmatrix}\begin{bmatrix} v_1 & v_2 & v_3 \end{bmatrix} = \begin{bmatrix} u_1v_1 & u_1v_2 & u_1v_3 \\ u_2v_1 & u_2v_2 & u_2v_3 \\ u_3v_1 & u_3v_2 & u_3v_3 \end{bmatrix}.$$

If one chooses to perform computations using matrix representations of vectors and tensors, care must be taken to use a single basis in all calculations, as the matrix representations are basis dependent.

1.2.13 **Determinant of a tensor**

Tensors, like square matrices, have determinants. The general definition of the **determinant** is an operation that assigns to each tensor **S** a scalar $\det\mathbf{S}$ defined by

$$\det\mathbf{S} = \frac{\mathbf{S}\mathbf{u}\cdot(\mathbf{S}\mathbf{v}\times\mathbf{S}\mathbf{w})}{\mathbf{u}\cdot(\mathbf{v}\times\mathbf{w})} \tag{1.2.30}$$

for any three non-coplanar vectors $\{\mathbf{u},\mathbf{v},\mathbf{w}\}$. Thus, $\det\mathbf{S}$ is the ratio of the volume of the parallelepiped defined by the vectors **Su**, **Sv**, and **Sw** to the volume of the parallelepiped defined by the vectors **u**, **v**, and **w**.

Definition (1.2.30) is fully equivalent to the conventional one given for square matrices; in particular

$$\begin{aligned}\det\mathbf{S} \equiv \det[\mathbf{S}] &= \begin{vmatrix} S_{11} & S_{12} & S_{13} \\ S_{21} & S_{22} & S_{23} \\ S_{31} & S_{32} & S_{33} \end{vmatrix} \\ &= S_{11}(S_{22}S_{33} - S_{23}S_{32}) - S_{12}(S_{21}S_{33} - S_{23}S_{31}) + S_{13}(S_{21}S_{32} - S_{22}S_{31}) \\ &= e_{ijk}S_{i1}S_{j2}S_{k3} = \tfrac{1}{6}e_{ijk}e_{pqr}S_{ip}S_{jq}S_{kr}.\end{aligned} \tag{1.2.31}$$

It is useful also to observe that

$$\det(\mathbf{S}\mathbf{T}) = \det\mathbf{S}\,\det\mathbf{T}, \tag{1.2.32}$$

and that

$$\det\mathbf{S}^{\top} = \det\mathbf{S}. \tag{1.2.33}$$

1.2.14 **Invertible tensors**

A tensor $\mathbf{S}$ is **invertible** if there is a tensor $\mathbf{S}^{-1}$, called the **inverse of S**, such that

$$\mathbf{S}\mathbf{S}^{-1} = \mathbf{S}^{-1}\mathbf{S} = \mathbf{1}. \tag{1.2.34}$$

Further,

$$\mathbf{S} \text{ is invertible} \quad \text{if and only if} \quad \det \mathbf{S} \neq 0. \tag{1.2.35}$$

Note also that $\det \mathbf{S}^{-1} = 1/\det \mathbf{S}$.

1.2.15 **Cofactor of a tensor**

Let $\mathbf{S}$ be an invertible tensor, then the tensor

$$\operatorname{cof} \mathbf{S} \stackrel{\text{def}}{=} (\det \mathbf{S})\mathbf{S}^{-\top}, \tag{1.2.36}$$

is called the **cofactor** of $\mathbf{S}$. A straight forward but slightly involved calculation shows that for all linearly independent vectors $\mathbf{u}$ and $\mathbf{v}$,

$$\operatorname{cof} \mathbf{S}(\mathbf{u} \times \mathbf{v}) = \mathbf{S}\mathbf{u} \times \mathbf{S}\mathbf{v}. \tag{1.2.37}$$

That is, $\operatorname{cof} \mathbf{S}$ transforms the *area vector* $\mathbf{u} \times \mathbf{v}$ of the parallelogram defined by $\mathbf{u}$ and $\mathbf{v}$ into the *area vector* $\mathbf{S}\mathbf{u} \times \mathbf{S}\mathbf{v}$ of the parallelogram defined by $\mathbf{S}\mathbf{u}$ and $\mathbf{S}\mathbf{v}$.

1.2.16 **Orthogonal tensors**

A tensor $\mathbf{Q}$ is **orthogonal** if and only if

$$\mathbf{Q}^{\top}\mathbf{Q} = \mathbf{Q}\mathbf{Q}^{\top} = \mathbf{1}. \tag{1.2.38}$$

If $\mathbf{Q}$ is orthogonal, then

$$\det \mathbf{Q} = \pm 1.$$

An orthogonal tensor is a **rotation** (or a proper rotation) if $\det \mathbf{Q} = 1$, and a **reflection** (or an improper rotation) if $\det \mathbf{Q} = -1$.

1.2.17 **Transformation relations for components of a vector and a tensor under a change in basis**

Given a Cartesian coordinate system with a right-handed orthonormal basis $\{\mathbf{e}_i\}$, a vector $\mathbf{v}$ and a tensor $\mathbf{S}$ may be represented by their components

$$v_i = \mathbf{e}_i \cdot \mathbf{v} \qquad \text{and} \qquad S_{ij} = \mathbf{e}_i \cdot \mathbf{S}\mathbf{e}_j,$$

respectively. In another coordinate system, cf. Fig. 1.1, with a right-handed orthonormal basis $\{\mathbf{e}_i^*\}$, $\mathbf{v}$ and $\mathbf{S}$ have the component representations

$$v_i^* = \mathbf{e}_i^* \cdot \mathbf{v} \qquad \text{and} \qquad S_{ij}^* = \mathbf{e}_i^* \cdot \mathbf{S}\mathbf{e}_j^*.$$

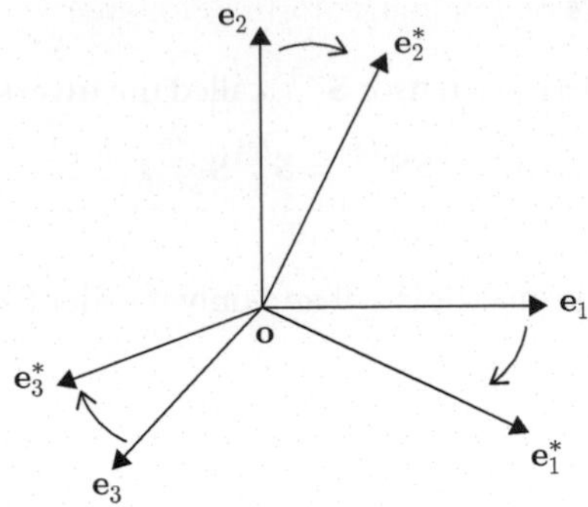

Fig. 1.1 Change in basis.

We now consider how the two representations (v_i, S_{ij}) and (v_i^*, S_{ij}^*) in the two coordinate systems are related to each other.

Let $\mathbf{Q}$ be the rotation tensor defined by

$$\mathbf{Q} \stackrel{\text{def}}{=} \mathbf{e}_1 \otimes \mathbf{e}_1^* + \mathbf{e}_2 \otimes \mathbf{e}_2^* + \mathbf{e}_3 \otimes \mathbf{e}_3^* \equiv \mathbf{e}_k \otimes \mathbf{e}_k^*, \tag{1.2.39}$$

so that

$$\mathbf{e}_i^* = \mathbf{Q}^\top \mathbf{e}_i \qquad i = 1, 2, 3. \tag{1.2.40}$$

That $\mathbf{Q}$ is a rotation follows from $\mathbf{Q}\mathbf{Q}^\top = (\mathbf{e}_i \otimes \mathbf{e}_i^*)(\mathbf{e}_j^* \otimes \mathbf{e}_j) = \delta_{ij} \mathbf{e}_i \otimes \mathbf{e}_j = \mathbf{1}$ and similarly for $\mathbf{Q}^\top\mathbf{Q}$, along with the right-handedness of the bases to ensure $\det \mathbf{Q} = +1$. The components of $\mathbf{Q}$ with respect to $\{\mathbf{e}_i\}$ are given by

$$Q_{ij} = \mathbf{e}_i \cdot (\mathbf{e}_k \otimes \mathbf{e}_k^*)\mathbf{e}_j = \delta_{ik} \mathbf{e}_k^* \cdot \mathbf{e}_j = \mathbf{e}_i^* \cdot \mathbf{e}_j.$$

Thus, a base vector $\mathbf{e}_i^*$ may be expressed in terms of the basis $\{\mathbf{e}_j\}$, with Q_{ij} serving as components of the vector $\mathbf{e}_i^*$:

$$\mathbf{e}_i^* = Q_{ij} \mathbf{e}_j, \tag{1.2.41}$$

and the matrix representation of the components of $\mathbf{Q}$ with respect to the basis $\{\mathbf{e}_i\}$ is

$$[\mathbf{Q}] = \begin{bmatrix} Q_{11} & Q_{12} & Q_{13} \\ Q_{21} & Q_{22} & Q_{23} \\ Q_{31} & Q_{32} & Q_{33} \end{bmatrix}.$$

Transformation relation for vectors

For a vector $\mathbf{v}$ the components with respect to the basis $\{\mathbf{e}_i^*\}$ are

$$v_i^* = \mathbf{e}_i^* \cdot \mathbf{v} = (Q_{ij}\mathbf{e}_j) \cdot \mathbf{v} = Q_{ij}(\mathbf{e}_j \cdot \mathbf{v}) = Q_{ij} v_j.$$

Thus, the **transformation relation** for the Cartesian components of a vector $\mathbf{v}$ under a change of basis are:

$$\boxed{v_i^* = Q_{ij} v_j.} \tag{1.2.42}$$

These relations may be expressed in matrix form as follows:

$$\begin{bmatrix} v_1^* \\ v_2^* \\ v_3^* \end{bmatrix} = \begin{bmatrix} Q_{11} & Q_{12} & Q_{13} \\ Q_{21} & Q_{22} & Q_{23} \\ Q_{31} & Q_{32} & Q_{33} \end{bmatrix} \begin{bmatrix} v_1 \\ v_2 \\ v_3 \end{bmatrix},$$

where the components of $\mathbf{Q}$ are given in the $\{\mathbf{e}_i\}$ basis.

Transformation relation for tensors

For a tensor $\mathbf{S}$ the components with respect to the basis $\{\mathbf{e}_i^*\}$ are

$$S_{ij}^* = \mathbf{e}_i^* \cdot \mathbf{S}\mathbf{e}_j^* = \Big(Q_{ik}\mathbf{e}_k\Big) \cdot \mathbf{S}\Big(Q_{jl}\mathbf{e}_l\Big) = Q_{ik}Q_{jl}\Big(\mathbf{e}_k \cdot \mathbf{S}\mathbf{e}_l\Big) = Q_{ik}Q_{jl}S_{kl}.$$

Thus, the **transformation relation** for the Cartesian components of a tensor $\mathbf{S}$ under a change of basis are:

$$\boxed{S_{ij}^* = Q_{ik}Q_{jl}S_{kl}.} \tag{1.2.43}$$

These relations may be expressed in matrix form as follows:

$$\begin{bmatrix} S_{11}^* & S_{12}^* & S_{13}^* \\ S_{21}^* & S_{22}^* & S_{23}^* \\ S_{31}^* & S_{32}^* & S_{33}^* \end{bmatrix} = \begin{bmatrix} Q_{11} & Q_{12} & Q_{13} \\ Q_{21} & Q_{22} & Q_{23} \\ Q_{31} & Q_{32} & Q_{33} \end{bmatrix} \begin{bmatrix} S_{11} & S_{12} & S_{13} \\ S_{21} & S_{22} & S_{23} \\ S_{31} & S_{32} & S_{33} \end{bmatrix} \begin{bmatrix} Q_{11} & Q_{12} & Q_{13} \\ Q_{21} & Q_{22} & Q_{23} \\ Q_{31} & Q_{32} & Q_{33} \end{bmatrix}^{\top},$$

where the components of $\mathbf{Q}$ are given in the $\{\mathbf{e}_i\}$ basis.

1.2.18 Eigenvalues and eigenvectors of a tensor. Spectral theorem

A scalar ω is an **eigenvalue** of a tensor $\mathbf{S}$, if there is a unit vector $\boldsymbol{\mu}$ such that

$$\mathbf{S}\boldsymbol{\mu} = \omega\boldsymbol{\mu}, \tag{1.2.44}$$

in which case $\boldsymbol{\mu}$ is an **eigenvector** of $\mathbf{S}$ corresponding to the eigenvalue ω. Physically, eigenvectors of $\mathbf{S}$ are those vectors which remain parallel to themselves when mapped by the tensor $\mathbf{S}$, Fig. 1.2.

If ω and $\boldsymbol{\mu}$ are an eigenvalue and corresponding eigenvector for a tensor $\mathbf{S}$, then (1.2.44) implies that

$$(\mathbf{S} - \omega\mathbf{1})\boldsymbol{\mu} = \mathbf{0}. \tag{1.2.45}$$

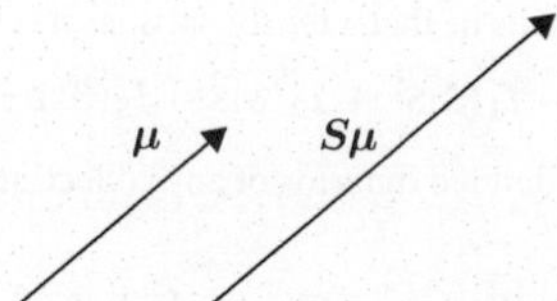

Fig. 1.2 Eigenvector $\boldsymbol{\mu}$ of a tensor $\mathbf{S}$.

For (1.2.45) to possess a non-trivial ($\boldsymbol{\mu} \neq \mathbf{0}$) solution, the tensor $(\mathbf{S}-\omega\mathbf{1})$ can not be invertible, so that, by (1.2.35),

$$\det(\mathbf{S}-\omega\mathbf{1})=0.$$

Thus, the determinant of the 3×3 matrix $[\mathbf{S}]-\omega[\mathbf{1}]$ must vanish,

$$\det\big([\mathbf{S}]-\omega[\mathbf{1}]\big)=0,$$

which is simply the classic requirement for a system of homogeneous equations to have a non-trivial solution. Each eigenvalue ω of a tensor $\mathbf{S}$ is, therefore, a solution of a polynomial equation of the form $\omega^3 - a_1\omega^2 + a_2\omega - a_3 = 0$, where the coefficients are functions of $\mathbf{S}$. A tedious computation shows that this **characteristic equation**[5] has the explicit form

$$\omega^3 - I_1(\mathbf{S})\omega^2 + I_2(\mathbf{S})\omega - I_3(\mathbf{S}) = 0, \tag{1.2.47}$$

where $I_1(\mathbf{S})$, $I_2(\mathbf{S})$, and $I_3(\mathbf{S})$, called the **principal invariants**[6] of $\mathbf{S}$, are given by

$$\left.\begin{aligned} I_1(\mathbf{S}) &= \operatorname{tr}\mathbf{S},\\ I_2(\mathbf{S}) &= \tfrac{1}{2}\big[\big(\operatorname{tr}(\mathbf{S})\big)^2 - \operatorname{tr}(\mathbf{S}^2)\big],\\ I_3(\mathbf{S}) &= \det\mathbf{S}. \end{aligned}\right\} \tag{1.2.48}$$

$I_k(\mathbf{S})$ are called invariants because of the way they transform under the group of orthogonal tensors:

$$I_k(\mathbf{Q}\mathbf{S}\mathbf{Q}^{\top}) = I_k(\mathbf{S}) \quad \text{for any orthogonal tensor } \mathbf{Q}. \tag{1.2.49}$$

The solutions of the characteristic equation (1.2.47), cubic in ω, are the eigenvalues ω_i, $i = 1, 2, 3$. Since the principal invariants $I_k(\mathbf{S})$ are always real, the theory of polynomials tells us that

- *The characteristic equation has (i) either three real roots (not necessarily distinct), or (ii) one real and two complex conjugate roots.*

Once the eigenvalues have been determined, the eigenvectors corresponding to each eigenvalue can be found by substituting back into (1.2.45).

---

[5] The Cayley–Hamilton theorem also tells us that a tensor satisfies its characteristic equation; that is

$$\mathbf{S}^3 - I_1(\mathbf{S})\mathbf{S}^2 + I_2(\mathbf{S})\mathbf{S} - I_3(\mathbf{S})\mathbf{1} = \mathbf{0}. \tag{1.2.46}$$

[6] The principal invariants can also be defined in terms of any collection of three vectors $\mathbf{a}, \mathbf{b}, \mathbf{c}$, with $[\mathbf{a}, \mathbf{b}, \mathbf{c}] \neq 0$, via the relations

$$\begin{aligned} I_1(\mathbf{S}) &= \big([\mathbf{Sa}, \mathbf{b}, \mathbf{c}] + [\mathbf{a}, \mathbf{Sb}, \mathbf{c}] + [\mathbf{a}, \mathbf{b}, \mathbf{Sc}]\big)/[\mathbf{a}, \mathbf{b}, \mathbf{c}],\\ I_2(\mathbf{S}) &= \big([\mathbf{Sa}, \mathbf{Sb}, \mathbf{c}] + [\mathbf{a}, \mathbf{Sb}, \mathbf{Sc}] + [\mathbf{Sa}, \mathbf{b}, \mathbf{Sc}]\big)/[\mathbf{a}, \mathbf{b}, \mathbf{c}],\\ I_3(\mathbf{S}) &= \big([\mathbf{Sa}, \mathbf{Sb}, \mathbf{Sc}]\big)/[\mathbf{a}, \mathbf{b}, \mathbf{c}]. \end{aligned}$$

Eigenvalues of symmetric tensors

It turns out that for a **symmetric** tensor the eigenvalues are all **real**, and the corresponding eigenvectors $\{\boldsymbol{\mu}_i\}$ are **mutually orthogonal**. Suppose that **S** is *symmetric* and that $\omega_1 \neq \omega_2$ are real eigenvalues of **S** with corresponding eigenvectors $\boldsymbol{\mu}_1$ and $\boldsymbol{\mu}_2$. Then, since **S** is symmetric,

$$\begin{aligned}
\mathbf{0} &= (\mathbf{S}\boldsymbol{\mu}_1 - \omega_1\boldsymbol{\mu}_1) \\
0 &= \boldsymbol{\mu}_2 \cdot (\mathbf{S}\boldsymbol{\mu}_1 - \omega_1\boldsymbol{\mu}_1) \\
&= \mathbf{S}\boldsymbol{\mu}_2 \cdot \boldsymbol{\mu}_1 - \omega_1\boldsymbol{\mu}_2 \cdot \boldsymbol{\mu}_1 \\
&= \omega_2\boldsymbol{\mu}_2 \cdot \boldsymbol{\mu}_1 - \omega_1\boldsymbol{\mu}_2 \cdot \boldsymbol{\mu}_1 \\
&= (\omega_2 - \omega_1)\boldsymbol{\mu}_2 \cdot \boldsymbol{\mu}_1
\end{aligned}$$

Since $\omega_1 \neq \omega_2$, we must have

$$\boldsymbol{\mu}_2 \cdot \boldsymbol{\mu}_1 = 0.$$

Thus, for a symmetric tensor, eigenvectors corresponding to distinct eigenvalues are orthogonal. When the eigenvalues are repeated, the situation is a bit more complex and is treated below. Notwithstanding, if **S** is **symmetric**, with distinct eigenvalues $\{\omega_1 > \omega_2 > \omega_3\}$, then they are **real**, and there exists an **orthonormal** set of corresponding eigenvectors $\{\boldsymbol{\mu}_1, \boldsymbol{\mu}_2, \boldsymbol{\mu}_3\}$ such that $\mathbf{S}\boldsymbol{\mu}_i = \omega_i\boldsymbol{\mu}_i$ (no sum).

SPECTRAL THEOREM *Let **S** be symmetric with distinct eigenvalues* ω_i. Then there is a corresponding orthonormal set $\{\boldsymbol{\mu}_i\}$ of eigenvectors of **S** and, what is most important, one uniquely has that

$$\mathbf{S} = \sum_{i=1}^{3} \omega_i \boldsymbol{\mu}_i \otimes \boldsymbol{\mu}_i. \tag{1.2.50}$$

The relation (1.2.50), which is called a **spectral decomposition** of **S**, gives **S** as a linear combination of projections, with each $\boldsymbol{\mu}_i \otimes \boldsymbol{\mu}_i$ (no sum) a projection tensor onto the eigenvector $\boldsymbol{\mu}_i$.

Since the eigenvectors $\{\boldsymbol{\mu}_i\}$ are orthonormal, they can also function as a basis, and in this basis the matrix representation **S** is diagonal:

$$[\mathbf{S}] = \begin{bmatrix} \omega_1 & 0 & 0 \\ 0 & \omega_2 & 0 \\ 0 & 0 & \omega_3 \end{bmatrix}.$$

When the eigenvalues of **S** are repeated, the spectral theorem needs to be modified. The remaining relevant cases are when two of the eigenvalues are the same and when all three eigenvalues are the same. Respectively, in these cases one has

(a) If $\omega_1 \neq \omega_2 = \omega_3$, then **S** admits the representation

$$\mathbf{S} = \omega_1\boldsymbol{\mu}_1 \otimes \boldsymbol{\mu}_1 + \omega_2(\mathbf{1} - \boldsymbol{\mu}_1 \otimes \boldsymbol{\mu}_1), \tag{1.2.51}$$

which indicates that $\boldsymbol{\mu}_1$, as well as any vector orthogonal to $\boldsymbol{\mu}_1$, is an eigenvector of **S**.

(b) If $\omega \equiv \omega_1 = \omega_2 = \omega_3$, then **S** admits the representation

$$\mathbf{S} = \omega\mathbf{1}, \tag{1.2.52}$$

which indicates that any vector will qualify as an eigenvector of **S** and that **S** is spherical.

Other invariants and eigenvalues

The principal invariants (1.2.48) of a tensor are completely characterized by the eigenvalues $\{\omega_1, \omega_2, \omega_3\}$:

$$\left.\begin{aligned} I_1(\mathbf{S}) &= \omega_1 + \omega_2 + \omega_3, \\ I_2(\mathbf{S}) &= \omega_1\omega_2 + \omega_2\omega_3 + \omega_3\omega_1, \\ I_3(\mathbf{S}) &= \omega_1\omega_2\omega_3. \end{aligned}\right\} \tag{1.2.53}$$

Invariants play an important role in continuum mechanics. As is clear from (1.2.53) for a tensor $\mathbf{S}$ the eigenvalues $\{\omega_1, \omega_2, \omega_3\}$ are the basic invariants in the sense that any invariant of $\mathbf{S}$ can be expressed in terms of them. In many applications, as an alternate to (1.2.53) it is often more convenient to choose as invariants the following three *symmetric functions* of $\{\omega_1, \omega_2, \omega_3\}$:

$$\left.\begin{aligned} \operatorname{tr}\mathbf{S} &= \omega_1 + \omega_2 + \omega_3, \\ \operatorname{tr}\mathbf{S}^2 &= \omega_1^2 + \omega_2^2 + \omega_3^2, \\ \operatorname{tr}\mathbf{S}^3 &= \omega_1^3 + \omega_2^3 + \omega_3^3. \end{aligned}\right\} \tag{1.2.54}$$

These three quantities are clearly invariant and they are independent in the sense that no one of them can be expressed in terms of the other two.

As a further alternative set of invariants recall that a tensor $\mathbf{S}$ may always be decomposed into a deviatoric and spherical part as

$$\mathbf{S} = \underbrace{\mathbf{S}'}_{\text{deviatoric part}} + \underbrace{\frac{1}{3}(\operatorname{tr}\mathbf{S})\mathbf{1}}_{\text{spherical part}}, \qquad \text{with} \qquad \operatorname{tr}\mathbf{S}' = 0. \tag{1.2.55}$$

The decomposition (1.2.55) leads to the possibility of using the invariant $(\operatorname{tr}\mathbf{S})$ and the invariants of $\mathbf{S}'$ as an alternative set of invariants. Since $\mathbf{S}'$ has only two independent non-zero invariants

$$\{\operatorname{tr}\mathbf{S}'^2, \operatorname{tr}\mathbf{S}'^3\}, \tag{1.2.56}$$

it is sometimes convenient to adopt the list

$$\{\operatorname{tr}\mathbf{S}, \operatorname{tr}\mathbf{S}'^2, \operatorname{tr}\mathbf{S}'^3\} \tag{1.2.57}$$

as a set of alternative invariants of a tensor $\mathbf{S}$.

1.2.19 **Fourth-order tensors**

In our discussions we will also need the concept of a **fourth-order tensor**. A fourth-order tensor is defined as a linear transformation that maps a second-order tensor to a second-order tensor. A fourth-order tensor $\mathbb{C}$ is therefore a linear mapping that assigns to each second-order tensor $\mathbf{A}$ a second-order tensor

$$\mathbf{B} = \mathbb{C}\mathbf{A}. \tag{1.2.58}$$

Linearity of $\mathbb{C}$ is the requirement that

$$\mathbb{C}(\mathbf{A}+\mathbf{B}) = \mathbb{C}\mathbf{A}+\mathbb{C}\mathbf{B} \qquad \text{for all tensors } \mathbf{A} \text{ and } \mathbf{B},$$
$$\mathbb{C}(\alpha\mathbf{A}) = \alpha\mathbb{C}\mathbf{A} \qquad \text{for all tensors } \mathbf{A} \text{ and scalars } \alpha.$$

It should also be noted that common alternate notations for $\mathbb{C}\mathbf{S}$ are

$$\mathbb{C}[\mathbf{S}] \quad \text{and} \quad \mathbb{C}:\mathbf{S}.$$

The latter form, $\mathbb{C}:\mathbf{S}$, emphasizes the action of $\mathbb{C}$ via its component form, which is discussed next.

Recall that the components T_{ij} of a second-order tensor are defined as

$$T_{ij} = \mathbf{e}_i \cdot \mathbf{T}\mathbf{e}_j.$$

For discussing the component form of fourth-order tensors it is convenient to introduce the basis tensors

$$\mathbf{E}_{ij} \stackrel{\text{def}}{=} \mathbf{e}_i \otimes \mathbf{e}_j, \tag{1.2.59}$$

with the *orthonormality* property

$$\mathbf{E}_{ij} : \mathbf{E}_{kl} = \delta_{ik}\delta_{jl}.$$

Using this notation, the components C_{ijkl} of $\mathbb{C}$ are defined as

$$C_{ijkl} = \mathbf{E}_{ij} : \mathbb{C}\mathbf{E}_{kl}. \tag{1.2.60}$$

This allows one to also express a fourth-order tensor as

$$\mathbb{C} = C_{ijkl}\mathbf{E}_{ij} \otimes \mathbf{E}_{kl} \equiv C_{ijkl}\mathbf{e}_i \otimes \mathbf{e}_j \otimes \mathbf{e}_k \otimes \mathbf{e}_l. \tag{1.2.61}$$

From these expressions, one can also infer the component-wise action of a fourth-order tensor on a second-order tensor as

$$\mathbf{S} = \mathbb{C}\mathbf{T}, \qquad S_{ij} = C_{ijkl}T_{kl}. \tag{1.2.62}$$

EXAMPLE 1.7 Fourth-order identity tensor

The fourth-order tensor

$$\mathbb{I} = \delta_{ik}\delta_{jl}\mathbf{E}_{ij} \otimes \mathbf{E}_{kl}$$

is the identity tensor. Verify that $\mathbb{I}$ as defined possesses the property of an identity operator, viz. $\mathbb{I}\mathbf{T} = \mathbf{T}$ for all tensors $\mathbf{T}$.

$$\begin{aligned}
\mathbb{I}\mathbf{T} \equiv \mathbb{I}:\mathbf{T} &= (\delta_{ik}\delta_{jl}\mathbf{E}_{ij} \otimes \mathbf{E}_{kl}) : (T_{pq}\mathbf{E}_{pq}) \\
&= \delta_{ik}\delta_{jl}T_{pq}(\mathbf{E}_{kl} : \mathbf{E}_{pq})\mathbf{E}_{ij} \\
&= \delta_{ik}\delta_{jl}T_{pq}\delta_{kp}\delta_{lq}\mathbf{E}_{ij} \\
&= T_{ij}\mathbf{E}_{ij} = \mathbf{T}.
\end{aligned}$$

Since this expression was derived for an arbitrary tensor $\mathbf{T}$, $\mathbb{I}$ must be the fourth-order identity tensor.

EXAMPLE 1.8 Symmetric fourth-order identity tensor

$$\mathbb{I}^{\mathrm{sym}} = \tfrac{1}{2}(\delta_{ik}\delta_{jl} + \delta_{il}\delta_{jk})\mathbf{E}_{ij} \otimes \mathbf{E}_{kl}$$

is the symmetric identity tensor. It maps any second-order tensor to its symmetric part. Check:

$$\begin{aligned}
\mathbb{I}^{\mathrm{sym}}\mathbf{T} \equiv \mathbb{I}^{\mathrm{sym}} : \mathbf{T} &= \big(\tfrac{1}{2}(\delta_{ik}\delta_{jl} + \delta_{il}\delta_{jk})\mathbf{E}_{ij} \otimes \mathbf{E}_{kl}\big) : (T_{pq}\mathbf{E}_{pq}) \\
&= \tfrac{1}{2}(\delta_{ik}\delta_{jl} + \delta_{il}\delta_{jk})T_{pq}(\mathbf{E}_{kl} : \mathbf{E}_{pq})\mathbf{E}_{ij} \\
&= \tfrac{1}{2}(\delta_{ik}\delta_{jl} + \delta_{il}\delta_{jk})T_{pq}\delta_{kp}\delta_{lq}\mathbf{E}_{ij} \\
&= \tfrac{1}{2}(\delta_{ip}\delta_{jq} + \delta_{iq}\delta_{jp})T_{pq}\mathbf{E}_{ij} \\
&= \tfrac{1}{2}(T_{ij} + T_{ji})\mathbf{E}_{ij} \\
&= \operatorname{sym}\mathbf{T}.
\end{aligned}$$

If $\mathbf{T}$ is already symmetric, then $\mathbb{I}^{\mathrm{sym}}$ functions as an alternate identity tensor.

Transformation rules for fourth-order tensors

The transformation rule for the components of a fourth-order tensor under a change in basis from $\{\mathbf{e}_i\}$ to $\{\mathbf{e}_i^*\}$ with $\mathbf{e}_i^* = Q_{ij}\mathbf{e}_j$ can be determined in the same way as for second-order tensors by first noting that

$$\mathbf{E}_{ij}^* = \mathbf{e}_i^* \otimes \mathbf{e}_j^* = Q_{ip}Q_{jq}\mathbf{e}_p \otimes \mathbf{e}_q = Q_{ip}Q_{jq}\mathbf{E}_{pq}.$$

Then we obtain

$$\begin{aligned}
C_{ijkl}^* &= \mathbf{E}_{ij}^* : \mathbb{C}\mathbf{E}_{kl}^*, \\
&= Q_{ip}Q_{jq}Q_{kr}Q_{ls}\big(\mathbf{E}_{pq} : \mathbb{C}\mathbf{E}_{rs}\big),
\end{aligned}$$

or

$$\boxed{C_{ijkl}^* = Q_{ip}Q_{jq}Q_{kr}Q_{ls}C_{pqrs}.} \tag{1.2.63}$$

2 Vectors and tensors: Analysis

In continuum mechanics, the scalars, vectors, and tensors whose properties that we discussed in Chapter 1 are used to represent physical quantities such as temperature, displacement, stress, and the like. These quantities are typically functions of position within a body. For example, the temperature at the left end of a rod $\vartheta(\mathbf{x}_{\text{left}})$ could differ from the temperature at the right end of the rod $\vartheta(\mathbf{x}_{\text{right}})$, in which case physical experience teaches us that there will be a flow of energy from the hotter end to the cooler end, and the flow rate is dependent on the gradient (rate of change with respect to position) of the temperature. In the general case, the temperature is what we term a *scalar field*, a function of position, $\vartheta(\mathbf{x})$, defined over the domain of the material body, Ω, of interest. We will also encounter *vector fields* and *tensor fields* in what follows. An example of a vector field would be the displacement field in a deformed body, $\mathbf{u}(\mathbf{x})$. An example of a tensor field would be the stresses in a body under load, $\boldsymbol{\sigma}(\mathbf{x})$. In each of these cases we will be interested in computing rates of change (from position to position) and in integral manipulations of the fields.

2.1 Directional derivatives and gradients

In order to compute rates of change with position, consider the domain Ω shown in Fig. 2.1 and an arbitrary scalar field $\varphi(\mathbf{x})$ defined over Ω. If we wish to discuss the rate of change of φ at a point $\mathbf{x}$, then we first need to specify along which direction in space we are interested in finding the rate of change. We will define this direction by the vector $\mathbf{h}$.

The rate of change of φ at the point $\mathbf{x}$ in the direction $\mathbf{h}$ is known as the *directional derivative*, and defined by,

$$D\varphi(\mathbf{x})[\mathbf{h}] \stackrel{\text{def}}{=} \lim_{\alpha\to 0}\left[\frac{\varphi(\mathbf{x}+\alpha\mathbf{h})-\varphi(\mathbf{x})}{\alpha}\right] = \frac{d}{d\alpha}\varphi(\mathbf{x}+\alpha\mathbf{h})\bigg|_{\alpha=0}. \tag{2.1.1}$$

The notation $D\varphi(\mathbf{x})[\mathbf{h}]$ stands for an operator $D\varphi(\mathbf{x})$ operating linearly on $\mathbf{h}$. The linear operator $D\varphi(\mathbf{x})$ is the *gradient* of φ at $\mathbf{x}$. Other common notations for the gradient are $\operatorname{grad}\varphi(\mathbf{x})$ and $\nabla\varphi(\mathbf{x})$. It should be noted that the computed rate of change will be per unit $\mathbf{h}$. Thus it is common to assume $|\mathbf{h}| = 1$, so that the directional derivative provides rates of change of φ per unit $\mathbf{x}$. The expression after the last equality in (2.1.1) is often the most convenient way of computing the directional derivative.

Continuum Mechanics of Solids. Lallit Anand and Sanjay Govindjee, Oxford University Press (2020).

DOI: 10.1093/oso/9780198864721.001.0001

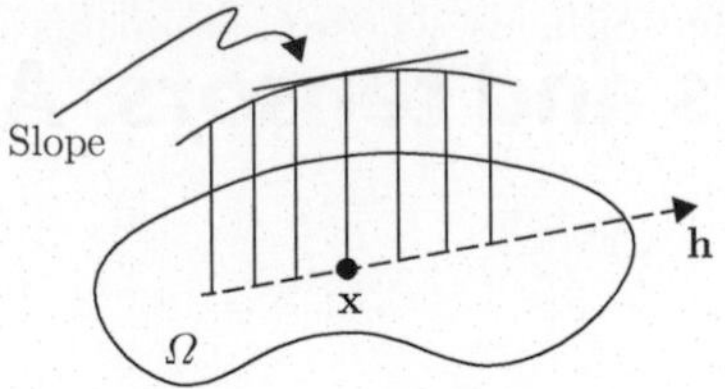

Fig. 2.1 Directional derivative construction for a generic scalar field.

EXAMPLE 2.1 Gradient of a scalar field

The gradient of $\varphi(\mathbf{x}) = \mathbf{x} \cdot \mathbf{a} + \mathbf{x} \cdot \mathbf{x}$, where $\mathbf{a}$ is a constant vector, is given by first finding the directional derivative as

$$\begin{aligned}\operatorname{grad}\varphi(\mathbf{x})[\mathbf{h}] &= \frac{d}{d\alpha}\varphi(\mathbf{x}+\alpha\mathbf{h})\Big|_{\alpha=0}\\ &= \frac{d}{d\alpha}\left[(\mathbf{x}+\alpha\mathbf{h})\cdot\mathbf{a} + (\mathbf{x}+\alpha\mathbf{h})\cdot(\mathbf{x}+\alpha\mathbf{h})\right]\Big|_{\alpha=0} = \mathbf{h}\cdot\mathbf{a} + 2\mathbf{x}\cdot\mathbf{h}\\ &= (\mathbf{a}+2\mathbf{x})\cdot\mathbf{h}.\end{aligned}$$

Since $(\mathbf{a} + 2\mathbf{x}) \cdot \mathbf{h}$ is linear in $\mathbf{h}$, the gradient of $\varphi(\mathbf{x})$ is,

$$\operatorname{grad}\varphi(\mathbf{x}) = \mathbf{a} + 2\mathbf{x},$$

which is a vector field.

The definition given also applies to vector and tensor fields. Thus for a vector field $\mathbf{v}(\mathbf{x})$, the directional derivative at a point $\mathbf{x}$ in the direction $\mathbf{h}$ is given by

$$D\mathbf{v}(\mathbf{x})[\mathbf{h}] = \lim_{\alpha\to 0}\left[\frac{\mathbf{v}(\mathbf{x}+\alpha\mathbf{h}) - \mathbf{v}(\mathbf{x})}{\alpha}\right] = \frac{d}{d\alpha}\mathbf{v}(\mathbf{x}+\alpha\mathbf{h})\Big|_{\alpha=0}, \tag{2.1.2}$$

and similarly for tensor fields. It should be noted that in the case of a vector field, the directional derivative is also a vector each of whose components gives the rate of change of the corresponding component of $\mathbf{v}$ in the direction of $\mathbf{h}$. The gradient in this case will be a tensor field (that when applied to $\mathbf{h}$ gives the directional derivative of $\mathbf{v}$ in the direction of $\mathbf{h}$).

2.2 Component expressions for differential operators

Let $\varphi(\mathbf{x})$ and $\mathbf{v}(\mathbf{x})$ be scalar and vector fields. The gradients of these fields

$$\operatorname{grad}\varphi(\mathbf{x}) \quad \text{and} \quad \operatorname{grad}\mathbf{v}(\mathbf{x}),$$

are, respectively, vector and tensor fields, whose components are given by

$$\mathbf{e}_i \cdot \operatorname{grad} \varphi(\mathbf{x}) = \left[\operatorname{grad} \varphi(\mathbf{x})\right]_i = \frac{\partial \varphi(\mathbf{x})}{\partial x_i} ,$$
$$\mathbf{e}_i \cdot \operatorname{grad} \mathbf{v}(\mathbf{x}) \mathbf{e}_j = \left[\operatorname{grad} \mathbf{v}(\mathbf{x})\right]_{ij} = \frac{\partial v_i(\mathbf{x})}{\partial x_j} . \tag{2.2.1}$$

The latter equalities for each case show that the components of the gradient expressions are simply the derivatives of the components in the coordinate directions.

EXAMPLE 2.2 Components of the gradient of a scalar field

To show the correctness of the component expression $(2.2.1)_1$ expand Definition (2.1.1):

$$\begin{aligned} \frac{d}{d\alpha} \varphi(x_1 + \alpha h_1, x_2 + \alpha h_2, x_3 + \alpha h_3)\bigg|_{\alpha=0} &= \frac{\partial \varphi}{\partial x_1} h_1 + \frac{\partial \varphi}{\partial x_2} h_2 + \frac{\partial \varphi}{\partial x_3} h_3 \\ &= \underbrace{\frac{\partial \varphi}{\partial x_i}}_{(\operatorname{grad} \varphi)_i} h_i . \end{aligned}$$

2.2.1 **Divergence, curl, and Laplacian**

Three other important differential operators are the divergence, curl, and Laplacian. The **divergence** and **curl** of a vector field $\mathbf{v}$ and a tensor field $\mathbf{T}$ may be defined as follows:

$$\begin{aligned} \operatorname{div} \mathbf{v} = \operatorname{tr}[\operatorname{grad} \mathbf{v}] &= \frac{\partial v_i}{\partial x_i} , \\ (\operatorname{curl} \mathbf{v})_i &= e_{ijk} \frac{\partial v_k}{\partial x_j} , \\ (\operatorname{div} \mathbf{T})_i &= \frac{\partial T_{ij}}{\partial x_j} , \\ (\operatorname{curl} \mathbf{T})_{ij} &= e_{ipq} \frac{\partial T_{jq}}{\partial x_p} ; \end{aligned} \tag{2.2.2}$$

thus $\operatorname{div} \mathbf{v}$ is a scalar field, $\operatorname{curl} \mathbf{v}$ and $\operatorname{div} \mathbf{T}$ are vector fields, and $\operatorname{curl} \mathbf{T}$ is a tensor field.[1]

[1] It also common to see the alternate notations $\nabla \cdot \mathbf{v} = \operatorname{div} \mathbf{v}$, $\nabla \times \mathbf{v} = \operatorname{curl} \mathbf{v}$, $\nabla \cdot \mathbf{T} = \operatorname{div} \mathbf{T}$, and $\nabla \times \mathbf{T} = \operatorname{curl} \mathbf{T}$ for the divergence and curl of vectors and tensors. Also, the definition of the curl of a tensor field is not consistent in the literature, and one needs to be aware of the definition being used by a particular author. We will employ the definition given in (2.2.2), but it is common to also see the alternate definition $(\operatorname{curl} \mathbf{T})_{ij} = e_{ipq} T_{qj,p}$.

The **Laplacian** or **Laplace operator** $\triangle$ for scalar fields φ and vector fields $\mathbf{v}$ are defined as follows:

$$\begin{aligned} \triangle\varphi &= \operatorname{div}\operatorname{grad}\varphi, \qquad & \triangle\varphi &= \frac{\partial^2\varphi}{\partial x_i \partial x_i}, \\ \triangle\mathbf{v} &= \operatorname{div}\operatorname{grad}\mathbf{v}, \qquad & \triangle v_i &= \frac{\partial^2 v_i}{\partial x_j \partial x_j}; \end{aligned} \tag{2.2.3}$$

thus, $\triangle\varphi$ is a scalar field and $\triangle\mathbf{v}$ is a vector field. Further, for a tensor field $\mathbf{T}$, $\triangle\mathbf{T}$ is the tensor field defined such that

$$(\triangle\mathbf{T})\mathbf{a} = \triangle(\mathbf{Ta}) \quad \text{for every constant vector } \mathbf{a}, \tag{2.2.4}$$

or equivalently, using components,

$$\triangle T_{ij} = \frac{\partial T_{ij}}{\partial x_k \partial x_k}. \tag{2.2.5}$$

Notation:
In indicial notation it is customary to employ a comma to denote partial differentiation with respect to a spatial coordinate, as follows:

$$\frac{\partial\varphi}{\partial x_i} = \varphi_{,i}, \qquad \frac{\partial v_i}{\partial x_j} = v_{i,j}, \qquad \frac{\partial T_{ij}}{\partial x_k} = T_{ij,k}, \qquad \text{etc.}$$

Thus, the Laplacians of φ and $\mathbf{v}$ may be written as

$$\triangle\varphi = \varphi_{,ii} \qquad \text{and} \qquad \triangle v_i = v_{i,jj},$$

and so on. We will use this shorthand notation whenever convenient.

EXAMPLE 2.3 Gradient and divergence of a vector field

The gradient of the vector field

$$\mathbf{v}(\mathbf{x}) = (\mathbf{x}\cdot\mathbf{x})\mathbf{x},$$

can be computed by first noting that $v_i = x_k x_k x_i$, then the components of the gradient are given by

$$\begin{aligned} v_{i,j} &= (x_k x_k x_i)_{,j} \\ &= \delta_{kj} x_k x_i + x_k \delta_{kj} x_i + x_k x_k \delta_{ij} \\ &= 2x_i x_j + x_k x_k \delta_{ij}. \end{aligned}$$

In the computation we have taken advantage of the fact that $\partial x_i/\partial x_j = \delta_{ij}$, since the coordinate directions are independent of each other. Thus,

$$\operatorname{grad}\mathbf{v} = 2\mathbf{x}\otimes\mathbf{x} + (\mathbf{x}\cdot\mathbf{x})\mathbf{1}.$$

The divergence follows immediately as,

$$\operatorname{div}\mathbf{v} = \operatorname{tr}[\operatorname{grad}\mathbf{v}] = 2\mathbf{x}\cdot\mathbf{x} + (\mathbf{x}\cdot\mathbf{x})3 = 5\mathbf{x}\cdot\mathbf{x}.$$

EXAMPLE 2.4 Divergence of a tensor field

The divergence of the tensor field

$$\mathbf{T} = \beta x_2(\mathbf{e}_1 \otimes \mathbf{e}_3 + \mathbf{e}_3 \otimes \mathbf{e}_1) - \beta x_1(\mathbf{e}_2 \otimes \mathbf{e}_3 + \mathbf{e}_3 \otimes \mathbf{e}_2),$$

where β is a constant scalar, can be found by applying the definition for its components $T_{ij,j}$:

$$[\operatorname{div} \mathbf{T}] = \begin{bmatrix} \beta \dfrac{\partial x_2}{\partial x_3} \\ -\beta \dfrac{\partial x_1}{\partial x_3} \\ \beta \dfrac{\partial x_2}{\partial x_1} - \beta \dfrac{\partial x_1}{\partial x_2} \end{bmatrix} = \begin{bmatrix} 0 \\ 0 \\ 0 \end{bmatrix}.$$

2.3 Generalized derivatives

It is also possible to define derivatives of functions with respect to arguments that are not simply position. For example, if one has a function of the displacement vector, it is possible to define the derivative of the function with respect to the displacement vector. The meaning is, as expected, the rate of change of the function with respect to changes in the displacement vector.

Consider a function $f(\varphi, \mathbf{v}, \mathbf{T})$ whose value is possibly scalar, vector, or tensor valued, and where φ, $\mathbf{v}$, and $\mathbf{T}$ are scalar, vector, and tensor arguments. The derivatives of f with respect to its arguments are defined as follows:

- The derivative $\partial f/\partial\varphi$ is the expression such that

$$\frac{d}{d\alpha} f(\varphi + \alpha h, \mathbf{v}, \mathbf{T})\bigg|_{\alpha=0} = \frac{\partial f}{\partial \varphi} h$$

  for all scalars h.

- The derivative $\partial f/\partial\mathbf{v}$ is the expression such that

$$\frac{d}{d\alpha} f(\varphi, \mathbf{v} + \alpha\mathbf{h}, \mathbf{T})\bigg|_{\alpha=0} = \frac{\partial f}{\partial \mathbf{v}} \cdot \mathbf{h}$$

  for all vectors $\mathbf{h}$.

- The derivative $\partial f/\partial\mathbf{T}$ is the expression such that

$$\frac{d}{d\alpha} f(\varphi, \mathbf{v}, \mathbf{T} + \alpha\mathbf{H})\bigg|_{\alpha=0} = \frac{\partial f}{\partial \mathbf{T}} : \mathbf{H}$$

  for all tensors $\mathbf{H}$.

EXAMPLE 2.5 Derivative of a scalar valued function with respect to a vector

The derivative of the scalar valued function

$$n(\mathbf{v}) = |\mathbf{v}|,$$

that is the norm of **v**, is determined as follows:

$$\begin{aligned}
n(\mathbf{v}+\alpha\mathbf{h}) &= |\mathbf{v}+\alpha\mathbf{h}| = \sqrt{(\mathbf{v}+\alpha\mathbf{h})\cdot(\mathbf{v}+\alpha\mathbf{h})} \\
\left.\frac{d}{d\alpha}n(\mathbf{v}+\alpha\mathbf{h})\right|_{\alpha=0} &= \left.\frac{d}{d\alpha}\sqrt{(\mathbf{v}+\alpha\mathbf{h})\cdot(\mathbf{v}+\alpha\mathbf{h})}\right|_{\alpha=0} \\
&= \left[\frac{1}{2}\frac{1}{\sqrt{(\mathbf{v}+\alpha\mathbf{h})\cdot(\mathbf{v}+\alpha\mathbf{h})}}\left(\mathbf{h}\cdot(\mathbf{v}+\alpha\mathbf{h})+(\mathbf{v}+\alpha\mathbf{h})\cdot\mathbf{h}\right)\right]_{\alpha=0} \\
&= \frac{1}{2}\frac{1}{|\mathbf{v}|}\left(\mathbf{h}\cdot\mathbf{v}+\mathbf{v}\cdot\mathbf{h}\right) \\
&= \frac{\mathbf{v}}{|\mathbf{v}|}\cdot\mathbf{h}.
\end{aligned}$$

Thus,

$$\frac{\partial n}{\partial \mathbf{v}} = \frac{\mathbf{v}}{|\mathbf{v}|}.$$

EXAMPLE 2.6 Derivative of a vector valued function with respect to a tensor

The derivative of the vector valued function

$$\mathbf{w}(\mathbf{T}) = \mathbf{T}\mathbf{T}\mathbf{a},$$

where **a** is a constant vector, is found as follows. Since,

$$\mathbf{w}(\mathbf{T}+\alpha\mathbf{H}) = (\mathbf{T}+\alpha\mathbf{H})(\mathbf{T}+\alpha\mathbf{H})\mathbf{a}, \qquad \text{and}$$

$$D\mathbf{w}(\mathbf{T})[\mathbf{H}] = \left.\frac{d}{d\alpha}\mathbf{w}(\mathbf{T}+\alpha\mathbf{H})\right|_{\alpha=0} = \left[\mathbf{H}(\mathbf{T}+\alpha\mathbf{H})\mathbf{a}+(\mathbf{T}+\alpha\mathbf{H})\mathbf{H}\mathbf{a}\right]_{\alpha=0} = \mathbf{H}\mathbf{T}\mathbf{a}+\mathbf{T}\mathbf{H}\mathbf{a},$$

using components we obtain,

$$\begin{aligned}
\left(D\mathbf{w}(\mathbf{T})\right)_{ijk}H_{jk} &= H_{ip}T_{pq}a_q + T_{is}H_{st}a_t, \\
&= \left(\delta_{ij}\delta_{pk}T_{pq}a_q + T_{is}\delta_{sj}\delta_{tk}a_t\right)H_{jk}, \\
&= \left(\delta_{ij}T_{kq}a_q + T_{ij}a_k\right)H_{jk},
\end{aligned}$$

which gives,

$$\left(D\mathbf{w}(\mathbf{T})\right)_{ijk} = \delta_{ij}T_{kq}a_q + T_{ij}a_k.$$

Thus in direct notation,

$$D\mathbf{w}(\mathbf{T}) \equiv \frac{\partial \mathbf{w}}{\partial \mathbf{T}} = \mathbf{1} \otimes \mathbf{Ta} + \mathbf{T} \otimes \mathbf{a},$$

which is a third-order tensor.

EXAMPLE 2.7 Derivative of a tensor with itself

The derivative of a tensor with itself can be found by considering the function

$$\mathbf{G}(\mathbf{T}) = \mathbf{T},$$

so that

$$D\mathbf{G}(\mathbf{T})[\mathbf{H}] = \frac{\partial \mathbf{T}}{\partial \mathbf{T}}\mathbf{H}.$$

Expanding gives

$$\frac{d}{d\alpha}(\mathbf{T} + \alpha\mathbf{H})\bigg|_{\alpha=0} = \mathbf{H} = \mathbb{I}\mathbf{H}.$$

Thus

$$\frac{\partial \mathbf{T}}{\partial \mathbf{T}} = \mathbb{I}, \qquad \frac{\partial T_{ij}}{\partial T_{kl}} = \delta_{ik}\delta_{jl}.$$

In the case that $\mathbf{T}$ is symmetric this result does not possess the correct (minor) symmetries upon exchange of $i \leftrightarrow j$ or $k \leftrightarrow l$. For the symmetric case, with some effort, it can be shown that the appropriate expression should be $\partial\mathbf{T}/\partial\mathbf{T} = \mathbb{I}^{\text{sym}}$, $\partial T_{ij}/\partial T_{kl} = \frac{1}{2}(\delta_{ik}\delta_{jl} + \delta_{il}\delta_{jk})$.

2.4 Integral theorems

In addition to the differential operations introduced above, we will also have need for integral operations on various fields. There are three basic theorems that we will need: (i) the localization theorem, (ii) the divergence theorem, and (iii) the Stokes theorem. Each of these is discussed below.

2.4.1 Localization theorem

LOCALIZATION THEOREM *Let $f(\mathbf{x})$ be an arbitrary scalar, vector, or tensor field, continuous at all points $\mathbf{x}$ in a domain Ω. If $\int_R f(\mathbf{x})\, dv = 0$ for all subdomains $R \subset \Omega$, then $f(\mathbf{x}) = 0$ for all $\mathbf{x} \in \Omega$.*

This theorem is central to deriving the local equations governing the behavior of bodies from global (laboratory level) observations of physical phenomena.

2.4.2 **Divergence theorem**

DIVERGENCE THEOREM *Let R be a bounded region with boundary ∂R. Assume we are given a scalar field φ, a vector field $\mathbf{v}$, and a tensor field $\mathbf{T}$, over the domain R. Let $\mathbf{n}$ denote the outward unit normal field on the boundary ∂R of R. Then*

$$
\begin{aligned}
\int_{\partial R} \varphi \mathbf{n}\, da &= \int_{R} \operatorname{grad} \varphi\, dv, \\
\int_{\partial R} \mathbf{v} \cdot \mathbf{n}\, da &= \int_{R} \operatorname{div} \mathbf{v}\, dv, \\
\int_{\partial R} \mathbf{T}\mathbf{n}\, da &= \int_{R} \operatorname{div} \mathbf{T}\, dv.
\end{aligned}
\tag{2.4.1}
$$

The identities (2.4.1) have the component forms:

$$
\begin{aligned}
\int_{\partial R} \varphi n_i\, da &= \int_{R} \frac{\partial \varphi}{\partial x_i}\, dv, \\
\int_{\partial R} v_i n_i\, da &= \int_{R} \frac{\partial v_i}{\partial x_i}\, dv, \\
\int_{\partial R} T_{ij} n_j\, da &= \int_{R} \frac{\partial T_{ij}}{\partial x_j}\, dv.
\end{aligned}
\tag{2.4.2}
$$

Note that these identities follow a general rule: the n_i in the surface integral results in the partial derivative $\partial/\partial x_i$ in the volume integral; thus, for a tensor of any order,

$$
\int_{\partial R} T_{ij\cdots k} n_r\, da = \int_{R} \frac{\partial T_{ij\cdots k}}{\partial x_r}\, dv. \tag{2.4.3}
$$

The central utility of the divergence theorem is that it allows one to convert surface expressions/information to volume expressions/information. Later we will see that the divergence theorem together with the localization theorem will allow us to derive important governing equations for many physical phenomena.

EXAMPLE 2.8 Application of the divergence theorem

Considering a domain Ω with volume $\mathrm{vol}(\Omega)$ and a *constant* tensor ε, find the value of the surface integral

$$
\frac{1}{\mathrm{vol}(\Omega)} \int_{\partial \Omega} \varepsilon_{ij} x_j n_i\, da.
$$

This integral can be computed using (2.4.3) as

$$\frac{1}{\mathrm{vol}(\Omega)}\int_{\partial\Omega}\varepsilon_{ij}x_j n_i\,da = \frac{1}{\mathrm{vol}(\Omega)}\int_{\Omega}(\varepsilon_{ij}x_j)_{,i}\,dv$$
$$= \frac{1}{\mathrm{vol}(\Omega)}\int_{\Omega}\varepsilon_{ij}x_{j,i}\,dv$$
$$= \frac{1}{\mathrm{vol}(\Omega)}\int_{\Omega}\varepsilon_{ij}\delta_{ji}\,dv$$
$$= \frac{1}{\mathrm{vol}(\Omega)}\int_{\Omega}\varepsilon_{ii}\,dv$$
$$= \frac{\mathrm{vol}(\Omega)}{\mathrm{vol}(\Omega)}\varepsilon_{ii} = \varepsilon_{ii}.$$

2.4.3 **Stokes theorem**

STOKES THEOREM *Let φ, $\mathbf{v}$, and $\mathbf{T}$ be scalar, vector, and tensor fields with common domain R. Then given any positively oriented surface S, with boundary C a closed curve, in R,*

$$\int_C \varphi\,\mathbf{dx} = \int_S \mathbf{n}\times\mathrm{grad}\,\varphi\,da$$
$$\int_C \mathbf{v}\cdot\mathbf{dx} = \int_S \mathbf{n}\cdot\mathrm{curl}\,\mathbf{v}\,da \tag{2.4.4}$$
$$\int_C \mathbf{T}\mathbf{dx} = \int_S (\mathrm{curl}\,\mathbf{T})^{\top}\mathbf{n}\,da.$$

The identities (2.4.4) have the component forms:

$$\int_C \varphi\,dx_i = \int_S e_{ijk}n_j\varphi_{,k}\,da,$$
$$\int_C v_i\,dx_i = \int_S n_i e_{ijk}\frac{\partial v_k}{\partial x_j}\,da, \tag{2.4.5}$$
$$\int_C T_{ij}\,dx_j = \int_S e_{jpq}\frac{\partial T_{iq}}{\partial x_p}n_j\,da.$$

The Stokes theorem plays a very important role in fluid mechanics and theories of electromagnetic phenomena, to name a few. In the mechanics of solid materials it plays a less important role, but we will see one important application of the theorem in the context of small deformation mechanics.

EXAMPLE 2.9 Green's theorem in a plane

Applying the Stokes theorem $(2.4.5)_2$ to a planar surface $\mathcal{S}$ oriented perpendicular to $\mathbf{e}_3$ and for

$$\mathbf{v} = f\mathbf{e}_1 + g\mathbf{e}_2$$

gives Green's theorem:

$$\int_{\mathcal{S}} \Big(\frac{\partial g}{\partial x_1} - \frac{\partial f}{\partial x_2}\Big)\, da = \int_{\mathcal{C}} (f dx_1 + g dx_2). \tag{2.4.6}$$

Further, the special choices $f = 0$ and $g = 0$, respectively, imply

$$\int_{\mathcal{S}} \frac{\partial g}{\partial x_1}\, da = \int_{\mathcal{C}} g dx_2 = \int_{\mathcal{C}} g n_1 ds, \qquad \int_{\mathcal{S}} \frac{\partial f}{\partial x_2}\, da = -\int_{\mathcal{C}} f dx_1 = \int_{\mathcal{C}} f n_2 ds, \tag{2.4.7}$$

where we have used the geometric relations

$$dx_2 = n_1\, ds, \qquad dx_1 = -n_2\, ds, \tag{2.4.8}$$

with ds an elemental arc length of the bounding curve $\mathcal{C}$.

PART II

Kinematics

3 Kinematics

Continuum mechanics, is for the most part, a study of *deforming* bodies, and *kinematics* is the mathematical description of the possible deformations that a body may undergo—it is the first step in understanding the mechanics of solids. In this chapter we will develop the necessary tools for describing general deformations and for analyzing them. We will start with descriptions of deformations of arbitrary magnitude, which leads to non-linear measures of strain. Following this, we will specialize our results to the important case of small deformations, which results in linear measures of strain.

3.1 Motion: Displacement, velocity, and acceleration

The shape of a solid body changes with time during a deformation process, and will occupy different regions of space, called configurations, at different times. To characterize a deformation process, we first adopt any convenient configuration of the body as reference, and set our clocks to measure time from zero at the moment when the body exists in this reference configuration. Typically, the reference configuration is taken as an unstressed state. We denote our chosen **reference configuration** of the body by $\mathcal{B}$ (cf. Fig. 3.1). Points $\mathbf{X}$ in $\mathcal{B}$ are called **material points**, and identified by their position vectors $(\mathbf{X}-\mathbf{o})$ which joins them to the origin $\mathbf{o}$ of a rectangular Cartesian coordinate system. The coordinates of a material point are then given by the scalar product of the vector $(\mathbf{X}-\mathbf{o})$ with each of the three mutually orthogonal unit base vectors $\{\mathbf{e}_i \mid i = 1, 2, 3\}$ of the coordinate system:

$$X_i = \mathbf{e}_i \cdot (\mathbf{X}-\mathbf{o}).$$

At some later time t the body is in the configuration $\mathcal{B}_t$, and the material point which was at $\mathbf{X}$ would have moved to some position $\mathbf{x}$ in $\mathcal{B}_t$. We call $\mathcal{B}_t$ the **deformed configuration** of the body at time t. The coordinates of the new position vector $(\mathbf{x}-\mathbf{o})$ of the material point are given by

$$x_i = \mathbf{e}_i \cdot (\mathbf{x}-\mathbf{o}).$$

The deformation of the body at each time t is described mathematically by a mapping, the **deformation map**,

$$\mathbf{x} = \chi(\mathbf{X}, t), \qquad x_i = \chi_i(X_1, X_2, X_3, t), \tag{3.1.1}$$

which gives the **place** occupied by the **material point** $\mathbf{X}$ at time t. The deformation map χ describes the **motion** of the body; it maps points in the reference configuration $\mathcal{B}$ to the deformed configuration $\mathcal{B}_t$ (Fig. 3.1). Clearly, $\chi(\mathbf{X}, 0) = \mathbf{X}$. We assume that every $\mathbf{x} \in \mathcal{B}_t$

Continuum Mechanics of Solids. Lallit Anand and Sanjay Govindjee, Oxford University Press (2020).

DOI: 10.1093/oso/9780198864721.001.0001

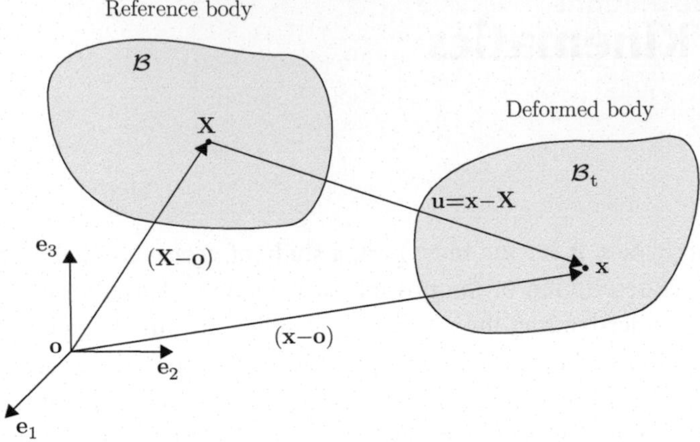

Fig. 3.1 Reference configuration and a current deformed configuration of a body.

is the unique image of some $\mathbf{X} \in \mathcal{B}$, that is the mapping is one-to-one and has a unique inverse

$$\mathbf{X} = \chi^{-1}(\mathbf{x}, t), \qquad X_i = \chi_i^{-1}(x_1, x_2, x_3, t), \tag{3.1.2}$$

for every point in the body.

The vector

$$\mathbf{u}(\mathbf{X}, t) = \chi(\mathbf{X}, t) - \mathbf{X}, \qquad u_i(X_1, X_2, X_3, t) = \chi_i(X_1, X_2, X_3, t) - X_i, \tag{3.1.3}$$

represents the **displacement** of $\mathbf{X}$ at time t.

The partial derivatives

$$\boxed{\begin{aligned} \dot{\mathbf{u}}(\mathbf{X}, t) &= \dot{\chi}(\mathbf{X}, t) = \frac{\partial}{\partial t}\chi(\mathbf{X}, t), \\ \ddot{\mathbf{u}}(\mathbf{X}, t) &= \ddot{\chi}(\mathbf{X}, t) = \frac{\partial^2}{\partial t^2}\chi(\mathbf{X}, t), \end{aligned}} \tag{3.1.4}$$

represent the **velocity** and **acceleration** (vectors) of material points, respectively. Using the inverse map (3.1.2) we can alternatively describe the velocity and acceleration as functions of $(\mathbf{x}, t)$:

$$\mathbf{v}(\mathbf{x}, t) \equiv \dot{\chi}\left(\chi^{-1}(\mathbf{x}, t), t\right), \tag{3.1.5}$$

$$\dot{\mathbf{v}}(\mathbf{x}, t) \equiv \ddot{\chi}\left(\chi^{-1}(\mathbf{x}, t), t\right). \tag{3.1.6}$$

These then provide the velocity and acceleration of the material particle currently occupying the position $\mathbf{x}$ in space.

3.1.1 Example motions

There are a number of simple motions that are useful to keep in mind as we develop the tools needed to analyze arbitrary deformations of solid bodies.

Simple shear

Simple shear is a deformation that is similar to the deformation that occurs when one slides a stack of papers or a deck of cards. The motion involves only displacement in the $\mathbf{e}_1$-direction and that motion is proportional to a parameter γ that controls the magnitude of the motion; cf. Fig. 3.2. Expressed mathematically, one has

$$\begin{aligned}\boldsymbol{\chi}(X_1, X_2, X_3, t) &= (X_1 + \gamma(t)X_2)\mathbf{e}_1 + X_2\mathbf{e}_2 + X_3\mathbf{e}_3,\\ x_1 = X_1 + \gamma(t)X_2, &\quad x_2 = X_2, \quad x_3 = X_3.\end{aligned} \tag{3.1.7}$$

Elementary elongation

Elementary elongation is a motion which involves a displacement only in the $\mathbf{e}_1$-direction and that motion is proportional to a parameter α that controls the magnitude of the motion; cf. Fig. 3.3. Mathematically, one has

$$\boldsymbol{\chi}(X_1, X_2, X_3, t) = \alpha(t)X_1\mathbf{e}_1 + X_2\mathbf{e}_2 + X_3\mathbf{e}_3, \quad x_1 = \alpha(t)X_1, \quad x_2 = X_2, \quad x_3 = X_3. \tag{3.1.8}$$

Rigid motion

A rigid motion is a motion in which the distance between any two points in the body remains constant at all times. Such a motion can be described by a translation vector $\mathbf{c}(t)$, a rotation tensor $\mathbf{Q}(t)$, and a fixed point, say, $\mathbf{o}$, as

$$\boldsymbol{\chi}(\mathbf{X}, t) = \mathbf{c}(t) + \mathbf{Q}(t)(\mathbf{X} - \mathbf{o}). \tag{3.1.9}$$

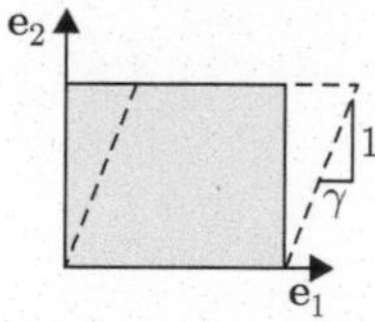

Fig. 3.2 Simple shear deformation. Dashed outline indicates the deformed body.

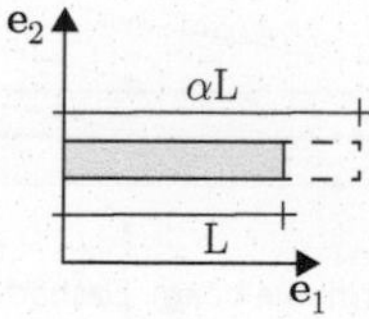

Fig. 3.3 Elementary elongation deformation. Dashed outline indicates the deformed body.

EXAMPLE 3.1 Distance between two points in a rigid motion

Verify that the distance between any two points in a rigid body remains constant during a rigid motion. Consider any two material points $\mathbf{X}_a$ and $\mathbf{X}_b$. At any time t the distance between these points is given by

$$\begin{aligned}
|\mathbf{x}_a - \mathbf{x}_b| &= |\mathbf{c}(t) + \mathbf{Q}(t)(\mathbf{X}_a - \mathbf{o}) - \mathbf{c}(t) - \mathbf{Q}(t)(\mathbf{X}_b - \mathbf{o})| \\
&= |\mathbf{Q}(t)(\mathbf{X}_a - \mathbf{X}_b)| \\
&= \sqrt{\mathbf{Q}(t)(\mathbf{X}_a - \mathbf{X}_b) \cdot \mathbf{Q}(t)(\mathbf{X}_a - \mathbf{X}_b)} \\
&= \sqrt{(\mathbf{X}_a - \mathbf{X}_b) \cdot \mathbf{Q}(t)^{\top}\mathbf{Q}(t)(\mathbf{X}_a - \mathbf{X}_b)} \\
&= \sqrt{(\mathbf{X}_a - \mathbf{X}_b) \cdot (\mathbf{X}_a - \mathbf{X}_b)} \\
&= |\mathbf{X}_a - \mathbf{X}_b|,
\end{aligned}$$

and is hence a constant for all time.

Bending deformation

The first three example deformations are relatively simple and good for testing one's intuition. This last example motion is a bit more complex, but at the same time also represents an intuitive motion. Consider a beam whose axis displaces according to a given function $v(X_1, t)$ and whose cross-sections rotate (independently of $v(X_1, t)$) according to a second given function $\theta(X_1, t)$; cf. Fig. 3.4. The resulting motion can be expressed as

$$\begin{aligned}
\chi(X_1, X_2, X_3, t) = {} & [X_1 - X_2 \sin\theta(X_1, t)]\, \mathbf{e}_1 + [X_2 + v(X_1, t) - X_2\,(1 - \cos\theta(X_1, t))]\, \mathbf{e}_2 \\
& + X_3 \mathbf{e}_3
\end{aligned}$$

$$\begin{aligned}
x_1 &= X_1 - X_2 \sin\theta(X_1, t), \\
x_2 &= X_2 + v(X_1, t) - X_2\,(1 - \cos\theta(X_1, t)), \\
x_3 &= X_3.
\end{aligned} \tag{3.1.10}$$

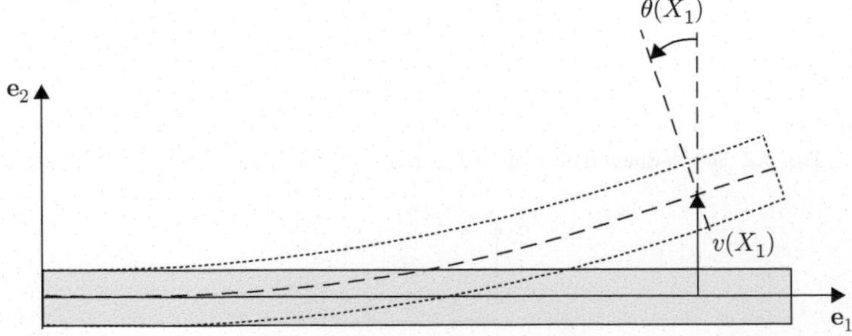

Fig. 3.4 Kinematics of a shear deformable beam. Dashed outline indicates the deformed body.

3.2 Deformation and displacement gradients

The gradient

$$\boxed{\mathbf{F}(\mathbf{X},t) = \nabla\boldsymbol{\chi} = \frac{\partial}{\partial\mathbf{X}}\boldsymbol{\chi}(\mathbf{X},t), \qquad F_{ij} = \frac{\partial}{\partial X_j}\chi_i\left(X_1, X_2, X_3, t\right),} \tag{3.2.1}$$

is called the **deformation gradient** (tensor).

The **displacement gradient** (tensor) is defined by

$$\boxed{\mathbf{H}(\mathbf{X},t) = \nabla\mathbf{u} = \frac{\partial}{\partial\mathbf{X}}\mathbf{u}(\mathbf{X},t), \qquad H_{ij}(X_1, X_2, X_3, t) = \frac{\partial}{\partial X_j}u_i(X_1, X_2, X_3, t).} \tag{3.2.2}$$

Thus, since

$$\mathbf{u}(\mathbf{X},t) = \mathbf{x} - \mathbf{X} = \boldsymbol{\chi}(\mathbf{X},t) - \mathbf{X},$$

the displacement gradient and the deformation gradient tensors are related by

$$\boxed{\mathbf{H}(\mathbf{X},t) = \mathbf{F}(\mathbf{X},t) - \mathbf{1}, \qquad H_{ij} = F_{ij} - \delta_{ij}.} \tag{3.2.3}$$

- If at a given time the deformation gradient $\mathbf{F}$, and hence the displacement gradient $\mathbf{H}$, is independent of $\mathbf{X}$, then the deformation is said to be **homogeneous** at that time.

EXAMPLE 3.2 Deformation gradient in simple shear

To find deformation gradient that corresponds to simple shear, start with (3.1.7) and apply (3.2.1):

$$[\mathbf{F}] = \begin{bmatrix} \partial\chi_1/\partial X_1 & \partial\chi_1/\partial X_2 & \partial\chi_1/\partial X_3 \\ \partial\chi_2/\partial X_1 & \partial\chi_2/\partial X_2 & \partial\chi_2/\partial X_3 \\ \partial\chi_3/\partial X_1 & \partial\chi_3/\partial X_2 & \partial\chi_3/\partial X_3 \end{bmatrix} = \begin{bmatrix} 1 & \gamma & 0 \\ 0 & 1 & 0 \\ 0 & 0 & 1 \end{bmatrix}.$$

With these basic definitions, it is now possible to discuss how lines of material particles, volumes of material, and areas of material in a body transform during deformation. These relations will then allow us to deduce sensible definitions for stretch and strain in bodies undergoing arbitrary deformations.

3.2.1 Transformation of material line elements

From the definition of the deformation gradient tensor (3.2.1) it follows that

$$\boxed{\mathbf{dx} = \mathbf{F}\mathbf{dX}, \qquad dx_i = F_{ij}dX_j.} \tag{3.2.4}$$

The vector $\mathbf{dX}$ represents an infinitesimal segment of material at $\mathbf{X}$ in $\mathcal{B}$, and the vector $\mathbf{dx}$ represents the deformed image of $\mathbf{dX}$ at $\mathbf{x}$ in $\mathcal{B}_t$; cf. Fig. 3.5.

It is important to note that the physical interpretation of $\mathbf{F}$ mapping a vector of material points from the reference configuration to the deformed configuration holds in the limit as $|d\mathbf{X}| \to 0$. This can be appreciated by observing

$$d\mathbf{x} = \chi(\mathbf{X}^{(2)}, t) - \chi(\mathbf{X}^{(1)}, t).$$

If we now replace $\chi(\mathbf{X}^{(2)}, t) = \chi(\mathbf{X}^{(1)} + d\mathbf{X}, t)$ by its Taylor series expansion in its first argument about $\mathbf{X}^{(1)}$, then

$$\begin{aligned} d\mathbf{x} &= \left[\chi(\mathbf{X}^{(1)}, t) + \mathbf{F}(\mathbf{X}^{(1)}, t)d\mathbf{X} + o(|d\mathbf{X}|)\right] - \chi(\mathbf{X}^{(1)}, t) \\ &= \mathbf{F}(\mathbf{X}^{(1)}, t)d\mathbf{X} + o(|d\mathbf{X}|) \\ &\approx \mathbf{F}(\mathbf{X}^{(1)}, t)d\mathbf{X}, \end{aligned}$$

where in the last step we have ignored the higher order terms in the Taylor series expansion. Thus in summary, the tensor $\mathbf{F}$ maps short material vectors (line elements) from the reference configuration to the deformed configuration.

Further, since at each time t the mapping $\mathbf{x} = \chi(\mathbf{X}, t)$ is one-to-one, the determinant of $\mathbf{F}$

$$J \equiv \det\left(\frac{\partial \chi}{\partial \mathbf{X}}\right) = \det \mathbf{F} \neq 0. \tag{3.2.5}$$

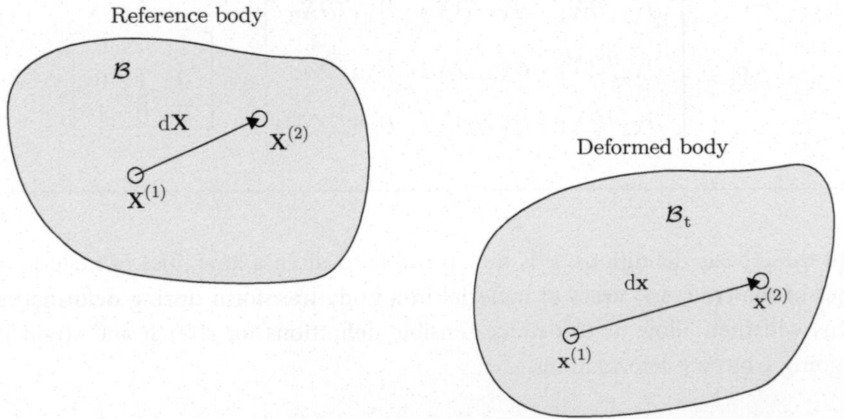

Fig. 3.5 Schematic showing how $\mathbf{F}$ maps $d\mathbf{X}$ in $\mathcal{B}$ to $d\mathbf{x}$ in $\mathcal{B}_t$, where $d\mathbf{X} = \mathbf{X}^{(2)} - \mathbf{X}^{(1)}$ and $d\mathbf{X} = \mathbf{X}^{(2)} - \mathbf{X}^{(1)}$, as $|d\mathbf{X}| \to 0$.

Since $\mathbf{F}$ is not singular, the inverse $\mathbf{F}^{-1}$ exists, and hence the following also holds true:

$$d\mathbf{X} = \mathbf{F}^{-1}d\mathbf{x}. \tag{3.2.6}$$

EXAMPLE 3.3 Relative change in length in simple shear

Vertical segments of material in simple shear have a second-order change in relative length.

In simple shear, $\mathbf{F} = \mathbf{1} + \gamma\mathbf{e}_1 \otimes \mathbf{e}_2$. Consider a short material vector in the reference body of the form $d\mathbf{X} = \zeta\mathbf{e}_2$, ($\zeta \ll 1$), its change in relative length is then given by

$$\epsilon = \frac{|\mathbf{F}d\mathbf{X}| - \zeta}{\zeta} = \frac{|\mathbf{F}(\zeta\mathbf{e}_2)| - \zeta}{\zeta} = \frac{|\zeta(\gamma\mathbf{e}_1 + \mathbf{e}_2)| - \zeta}{\zeta} = \sqrt{1+\gamma^2} - 1.$$

The quantity just computed is the change in length of the material line segment divided by its original length and is hence the (normal) strain in the vertical direction.

The quantity

$$\lambda \stackrel{\text{def}}{=} |\mathbf{F}\zeta\mathbf{e}_2|/\zeta = \sqrt{1+\gamma^2} = 1 + \epsilon,$$

defines the stretch of the line segment.

From this computation, we note that stretch and strain at a point depends on the direction of interest.

3.2.2 **Transformation of material volume elements**

Consider now a small volume of material at $\mathbf{X}$ in the reference body $\mathcal{B}$ in the shape of a parallelepiped whose edges are defined by three short, non-colinear vectors $\mathbf{a}$, $\mathbf{b}$, and $\mathbf{c}$, and denote this infinitesimal volume of material as

$$dv_{\mathrm{R}} = [\mathbf{a}, \mathbf{b}, \mathbf{c}].$$

According to (3.2.4) each edge of the parallelepiped will map to a new vector when the body is deformed, thus defining a new deformed parallelepiped with volume

$$dv = [\mathbf{Fa}, \mathbf{Fb}, \mathbf{Fc}]$$

in the deformed body $\mathcal{B}_t$; cf. Fig. 3.6.

Recalling from (1.2.30) that $[\mathbf{Fa}, \mathbf{Fb}, \mathbf{Fc}]/[\mathbf{a}, \mathbf{b}, \mathbf{c}]$ is equal to $J \equiv \det \mathbf{F}$, we find that

$$\boxed{dv = J dv_{\mathrm{R}}.} \tag{3.2.7}$$

Since an infinitesimal volume element dv_{R} cannot be deformed to zero volume or negative volume, we require that

$$J > 0. \tag{3.2.8}$$

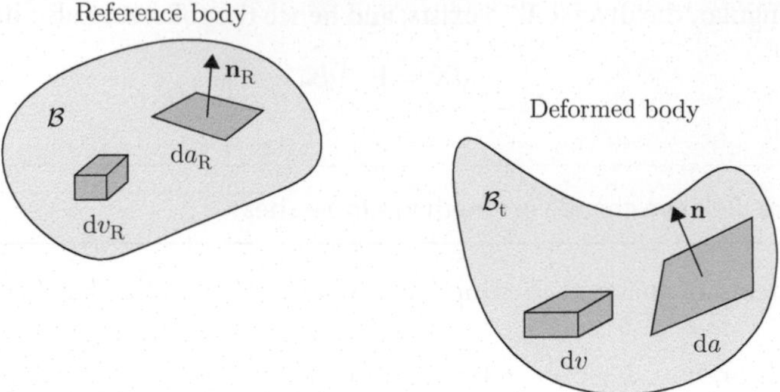

Fig. 3.6 Transformation of infinitesimal volume and area elements.

The quantity J, which was introduced in (3.2.5) as a shorthand notation for the determinant of $\mathbf{F}$, is called the *volumetric Jacobian*, or simply the *Jacobian* of the deformation. It is the appropriate mathematical quantity for describing how infinitesimal volumes of material transform from the reference to the deformed configuration. One also observes that J is the appropriate measure of volumetric stretch (ratio of current to original volume) at a point, and $e = J - 1$ is an appropriate measure of volumetric strain.

EXAMPLE 3.4 Volume stretch during rigid motion

The ratio of the volume of an infinitesimal volume element of material during rigid motion to its volume in the reference configuration is unity. To see this compute the Jacobian of the motion given in (3.1.9).

$$J = \frac{dv}{dv_\mathrm{R}} = \det \mathbf{F} = \det \mathbf{Q} = 1,$$

where the last equality follows from the property that the determinant of a rotation is $+1$; see Sec. 1.2.16.

3.2.3 **Transformation of material area elements**

The construction of an appropriate expression for the transformation of infinitesimal area elements follows a similar pattern. Consider the oriented area element $da_\mathrm{R}\mathbf{n}_\mathrm{R}$ at $\mathbf{X}$ in the reference configuration $\mathcal{B}$; cf. Fig. 3.6. Assuming this area element to be a parallelogram, it can be defined by two infinitesimal edge vectors $\mathbf{a}$ and $\mathbf{b}$, such that

$$\mathbf{n}_\mathrm{R} da_\mathrm{R} = \mathbf{a} \times \mathbf{b}.$$

Upon deformation this area element will map to a new oriented area element

$$\mathbf{n} da = \mathbf{F}\mathbf{a} \times \mathbf{F}\mathbf{b}.$$

Exploiting expression (1.2.37) for the cofactor of an invertible tensor, gives NANSON'S FORMULA:

$$\boxed{\mathbf{n}da = J\mathbf{F}^{-\top}\mathbf{n}_{\mathrm{R}}da_{\mathrm{R}}.} \tag{3.2.9}$$

Nanson's formula tells us that the appropriate mathematical quantity for describing how area elements transform from the reference to the deformed configuration is the cofactor of the deformation gradient. From this one can also define area stretch at a point, $\lambda_{\text{area}} = da/da_{\mathrm{R}} = |J\mathbf{F}^{-\top}\mathbf{n}_{\mathrm{R}}|$, and area strain $e_{\text{area}} = \lambda_{\text{area}} - 1$. Note that these quantities depend on the orientation of the area element at the point of interest.

EXAMPLE 3.5 Area stretch in simple shear

To find the area stretch for an area element with normal $\mathbf{e}_1$ in three-dimensional simple shear we need to compute the inverse transpose of the deformation gradient and its Jacobian:

$$[\mathbf{F}] = \begin{bmatrix} 1 & \gamma & 0 \\ 0 & 1 & 0 \\ 0 & 0 & 1 \end{bmatrix}, \qquad [\mathbf{F}^{-1}] = \begin{bmatrix} 1 & -\gamma & 0 \\ 0 & 1 & 0 \\ 0 & 0 & 1 \end{bmatrix}, \qquad \text{and}$$

$$[\mathbf{F}^{-\top}] = \begin{bmatrix} 1 & 0 & 0 \\ -\gamma & 1 & 0 \\ 0 & 0 & 1 \end{bmatrix}.$$

Since $J = 1$ in simple shear, it follows that

$$\lambda_{\text{area}} = |\mathbf{F}^{-\top}\mathbf{e}_1| = \sqrt{1+\gamma^2}.$$

3.3 Stretch and rotation

As is evident from the tools developed for describing various aspects of a general motion, the mapping of material vectors by the deformation gradient is a *basic* concept. If we consider that the deformation gradient is simply a tensor, then we are led to the observation that its affect on a vector can be broken down into two steps: the stretching of the vector and then the rotation of the vector, or alternately the rotation of the vector and then the stretching of the vector. This interpretation of the action of the deformation gradient is the central content of the polar decomposition theorem.

3.3.1 **Polar decomposition theorem**

Let $\mathbf{F}$ *be a tensor with* $\det \mathbf{F} > 0$. *Then there are unique, symmetric, positive definite tensors* $\mathbf{U}$ *and* $\mathbf{V}$ *and a rotation* $\mathbf{R}$ *such that*

$$\boxed{\mathbf{F} = \mathbf{R}\mathbf{U} = \mathbf{V}\mathbf{R}.} \tag{3.3.1}$$

We refer to $\mathbf{F} = \mathbf{R}\mathbf{U}$ and $\mathbf{F} = \mathbf{V}\mathbf{R}$, respectively, as the **right** and **left polar decompositions** of $\mathbf{F}$. Granted these decompositions, $\mathbf{F}$ determines $\mathbf{U}$ and $\mathbf{V}$ through the relations[1]

$$\begin{aligned} \mathbf{U} &= \sqrt{\mathbf{F}^\top\mathbf{F}}, \\ \mathbf{V} &= \sqrt{\mathbf{F}\mathbf{F}^\top}, \end{aligned} \tag{3.3.2}$$

whereby

$$\boxed{\mathbf{V} = \mathbf{R}\mathbf{U}\mathbf{R}^\top,} \tag{3.3.3}$$

and

$$\boxed{\mathbf{R} = \mathbf{F}\mathbf{U}^{-1} = \mathbf{V}^{-1}\mathbf{F}.} \tag{3.3.4}$$

- $\mathbf{U}$ and $\mathbf{V}$ are known as the **right and left stretch tensors**, respectively.
- The left and right stretch tensors are useful in theoretical discussions but are often problematic to apply because of the need to compute the square root of a tensor. For that reason, we introduce the **right and left Cauchy–Green (deformation) tensors** $\mathbf{C}$ and $\mathbf{B}$ defined by:

$$\begin{aligned} \mathbf{C} &= \mathbf{U}^2 = \mathbf{F}^\top\mathbf{F}, \qquad C_{ij} = F_{ki}F_{kj} = \frac{\partial \chi_k}{\partial X_i}\frac{\partial \chi_k}{\partial X_j}, \\ \mathbf{B} &= \mathbf{V}^2 = \mathbf{F}\mathbf{F}^\top, \qquad B_{ij} = F_{ik}F_{jk} = \frac{\partial \chi_i}{\partial X_k}\frac{\partial \chi_j}{\partial X_k}. \end{aligned} \tag{3.3.5}$$

- Note that the stretch tensors, $\mathbf{U}$ and $\mathbf{V}$, as well as the Cauchy–Green tensors, $\mathbf{C}$ and $\mathbf{B}$, *are symmetric and positive definite.*
- Being symmetric and positive definite, $\mathbf{U}$ and $\mathbf{V}$ admit spectral representations of the form

$$\begin{aligned} \mathbf{U} &= \sum_{i=1}^{3} \lambda_i \mathbf{r}_i \otimes \mathbf{r}_i, \\ \mathbf{V} &= \sum_{i=1}^{3} \lambda_i \mathbf{l}_i \otimes \mathbf{l}_i, \end{aligned} \tag{3.3.6}$$

[1] The square root of a symmetric tensor is defined via its spectral decomposition (see Sec. 1.2.18). A symmetric tensor $\mathbf{S}$ admits the spectral representation $\mathbf{S} = \sum_i s_i \mathbf{v}_i \otimes \mathbf{v}_i$, and its square root is defined as $\sqrt{\mathbf{S}} \overset{\text{def}}{=} \sum_i \sqrt{s_i}\mathbf{v}_i \otimes \mathbf{v}_i$. The square root is always taken as the positive square root in this context.

where

- $\lambda_1, \lambda_2, \lambda_3 > 0$, the **principal stretches**, are the eigenvalues of $\mathbf{U}$ and, by (3.3.3), also of $\mathbf{V}$;
- $\mathbf{r}_1$, $\mathbf{r}_2$, and $\mathbf{r}_3$, the **right principal directions**, are the eigenvectors of $\mathbf{U}$

$$\mathbf{U}\mathbf{r}_i = \lambda_i \mathbf{r}_i \qquad \text{(no sum on } i\text{)}. \tag{3.3.7}$$

- $\mathbf{l}_1$, $\mathbf{l}_2$, and $\mathbf{l}_3$, the **left principal directions**, are the eigenvectors of $\mathbf{V}$:

$$\mathbf{V}\mathbf{l}_i = \lambda_i \mathbf{l}_i \qquad \text{(no sum on } i\text{)}. \tag{3.3.8}$$

- From (3.3.3) and (3.3.6) it follows that

$$\mathbf{l}_i = \mathbf{R}\mathbf{r}_i, \qquad i = 1, 2, 3, \qquad \text{and also that} \qquad \mathbf{R} = \mathbf{l}_i \otimes \mathbf{r}_i. \tag{3.3.9}$$

- The tensors $\mathbf{C}$ and $\mathbf{B}$ have the following forms when expressed in terms of principal stretches and directions:

$$\begin{aligned} \mathbf{C} &= \sum_{i=1}^{3} \lambda_i^2 \, \mathbf{r}_i \otimes \mathbf{r}_i, \\ \mathbf{B} &= \sum_{i=1}^{3} \lambda_i^2 \, \mathbf{l}_i \otimes \mathbf{l}_i. \end{aligned} \tag{3.3.10}$$

- Further, since $\mathbf{F} = \mathbf{R}\mathbf{U}$,

$$\mathbf{F} = \sum_{i=1}^{3} \lambda_i \, \mathbf{l}_i \otimes \mathbf{r}_i. \tag{3.3.11}$$

3.3.2 Properties of the tensors U and C

Consider now infinitesimal undeformed line elements $d\mathbf{X}^{(1)}$ and $d\mathbf{X}^{(2)}$ and corresponding deformed line elements, cf. Fig. 3.7,

$$d\mathbf{x}^{(1)} = \mathbf{F}d\mathbf{X}^{(1)} \qquad \text{and} \qquad d\mathbf{x}^{(2)} = \mathbf{F}d\mathbf{X}^{(2)}. \tag{3.3.12}$$

Since $\mathbf{R}^\top\mathbf{R} = \mathbf{1}$, $\mathbf{U} = \mathbf{U}^\top$, and $\mathbf{C} = \mathbf{U}^2$, it follows that

$$\begin{aligned} d\mathbf{x}^{(1)} \cdot d\mathbf{x}^{(2)} &= (\mathbf{R}\mathbf{U}d\mathbf{X}^{(1)}) \cdot (\mathbf{R}\mathbf{U}d\mathbf{X}^{(2)}), \\ &= \mathbf{U}d\mathbf{X}^{(1)} \cdot \mathbf{U}d\mathbf{X}^{(2)}, \end{aligned} \tag{3.3.13}$$

$$\begin{aligned} &= d\mathbf{X}^{(1)} \cdot \mathbf{U}^2 d\mathbf{X}^{(2)}, \\ &= d\mathbf{X}^{(1)} \cdot \mathbf{C}d\mathbf{X}^{(2)}. \end{aligned} \tag{3.3.14}$$

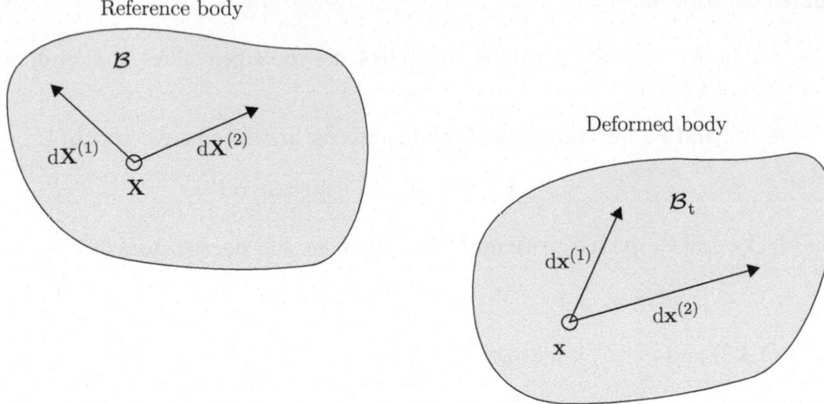

Fig. 3.7 Schematic showing how $\mathbf{F}$ maps two non-colinear, infinitesimal line elements at $\mathbf{X}$.

Fiber stretch and strain

A consequence of (3.3.13) is that

$$|d\mathbf{x}^{(1)}| = |\mathbf{U}d\mathbf{X}^{(1)}|; \tag{3.3.15}$$

the right stretch tensor $\mathbf{U}$ *therefore characterizes the deformed length of infinitesimal fibers.* Let

$$dS = |d\mathbf{X}| \qquad \text{and} \qquad \mathbf{e} = \frac{d\mathbf{X}}{|d\mathbf{X}|},$$

denote the magnitude and direction of an infinitesimal undeformed element, and

$$ds = |d\mathbf{x}| \qquad \text{and} \qquad \tilde{\mathbf{e}} = \frac{d\mathbf{x}}{|d\mathbf{x}|},$$

the magnitude and direction of the corresponding deformed element. Then, by (3.3.15)

$$\lambda(\mathbf{e}) \stackrel{\text{def}}{=} \frac{ds}{dS} = |\mathbf{U}\mathbf{e}| \tag{3.3.16}$$

represents the **stretch** in the direction $\mathbf{e}$ at $\mathbf{X}$. Bearing this in mind, we refer to fibers at $\mathbf{X}$ in the direction $\mathbf{e}$ as *stretched* or *unstretched* according as $\lambda \neq 1$ or $\lambda = 1$. Further, one also notes that

$$\lambda^2 = \mathbf{e} \cdot \mathbf{C}(\mathbf{X})\mathbf{e}. \tag{3.3.17}$$

In summary:

- *The stretch* λ *at* $\mathbf{X}$ *relative to any given material direction* $\mathbf{e}$ *is determined by the right stretch tensor* $\mathbf{U}(\mathbf{X})$ *through the relation* $\lambda(\mathbf{e}) = |\mathbf{U}(\mathbf{X})\mathbf{e}|$, *or equivalently by* $\lambda^2 = \mathbf{e} \cdot \mathbf{C}\mathbf{e}$.

- The stretch λ at $\mathbf{X}$ takes its extremal values when $\mathbf{e}$ is one of the right principal directions; thus the principal stretches are the extremal values. This result follows because the necessary equations for maximizing (3.3.16) subject to $|\mathbf{e}| = 1$ are given by the relations $\mathbf{U}\mathbf{e} = \lambda\mathbf{e}$, the eigenvalue equations for $\mathbf{U}$.

- The *engineering (normal) strain* ϵ at $\mathbf{X}$ relative to any given material direction $\mathbf{e}$ is defined as,[2]

$$\epsilon = \lambda - 1. \tag{3.3.18}$$

This result, like the one for stretch, is applicable to deformations of any magnitude.

EXAMPLE 3.6 Stretch in the coordinate directions

The stretch in the coordinate directions is characterized by

$$\begin{aligned}\lambda &= \sqrt{\mathbf{e}_i \cdot \mathbf{C}\mathbf{e}_i} && \text{(no sum)}\\ &= \sqrt{C_{ii}}. && \text{(no sum)}\end{aligned}$$

Thus the diagonal elements of the right Cauchy–Green deformation tensor fully specify the stretches in the coordinate directions. If the coordinate directions are taken to be the right principal directions, then we recover the principal stretches.

Angle change and engineering shear strain

We now determine the angle

$$\theta = \angle(d\mathbf{x}^{(1)}, d\mathbf{x}^{(2)})$$

between infinitesimal deformed fibers $d\mathbf{x}^{(1)}$ and $d\mathbf{x}^{(2)}$. By (3.3.13) and (3.3.15),

$$\cos\theta = \frac{d\mathbf{x}^{(1)} \cdot d\mathbf{x}^{(2)}}{|d\mathbf{x}^{(1)}|\,|d\mathbf{x}^{(2)}|} = \frac{\mathbf{U}d\mathbf{X}^{(1)} \cdot \mathbf{U}d\mathbf{X}^{(2)}}{|\mathbf{U}d\mathbf{X}^{(1)}|\,|\mathbf{U}d\mathbf{X}^{(2)}|}.$$

Thus,

$$\theta = \angle(d\mathbf{x}^{(1)}, d\mathbf{x}^{(2)}) = \angle(\mathbf{U}d\mathbf{X}^{(1)}, \mathbf{U}d\mathbf{X}^{(2)}), \tag{3.3.19}$$

and, hence, ***U** also characterizes the angle between deformed infinitesimal line elements.*

If we start with two material fibers which are orthogonal to each other, $d\mathbf{X}^{(1)} \cdot d\mathbf{X}^{(2)} = 0$, then the decrease in the angle between the fibers is defined to be the **engineering shear strain** with respect to the directions $\mathbf{e}^{(1)} = d\mathbf{X}^{(1)}/|d\mathbf{X}^{(1)}|$ and $\mathbf{e}^{(2)} = d\mathbf{X}^{(2)}/|d\mathbf{X}^{(2)}|$:

$$\gamma \overset{\text{def}}{=} \frac{\pi}{2} - \angle(\mathbf{U}d\mathbf{X}^{(1)}, \mathbf{U}d\mathbf{X}^{(2)}) = \sin^{-1}\left[\frac{\mathbf{U}d\mathbf{X}^{(1)} \cdot \mathbf{U}d\mathbf{X}^{(2)}}{|\mathbf{U}d\mathbf{X}^{(1)}|\,|\mathbf{U}d\mathbf{X}^{(2)}|}\right] = \sin^{-1}\left[\frac{\mathbf{e}^{(1)} \cdot \mathbf{C}\mathbf{e}^{(2)}}{\lambda(\mathbf{e}^{(1)})\lambda(\mathbf{e}^{(2)})}\right], \tag{3.3.20}$$

where we have used, $\sin\gamma = \cos\theta$.

[2] There are many other possible definitions of strain, such as $\epsilon = \ln(\lambda)$, which is called the logarithmic strain; cf. Sec. 3.4.

EXAMPLE 3.7 Engineering shear strain between right principal directions

The engineering shear strain between any two right principal directions $\mathbf{r}_i$ and $\mathbf{r}_j$, $i \neq j$, is found as

$$\gamma = \sin^{-1}\left[\frac{\mathbf{r}_i \cdot \mathbf{C}\mathbf{r}_j}{\lambda(\mathbf{r}_i)\lambda(\mathbf{r}_j)}\right] = \sin^{-1}\left[\frac{\mathbf{r}_i \cdot \lambda_j^2\mathbf{r}_j}{\lambda_i\lambda_j}\right] = \sin^{-1}(0) = 0.$$

Thus the engineering shear strain between the right principal directions is always zero.

3.4 Strain

For selected directions we have scalar definitions of both stretch and strain. We also have tensorial definitions of stretch tensors, $\mathbf{U}$ and $\mathbf{V}$, that embed all multidimensional aspects of the concept of stretch, including concepts of normal and shear strains. Similarly, we can extend our definition of strain into a tensorial one that embeds additional information. There are numerous definitions of tensorial strain, each with its own advantages and disadvantages. In what follows we introduce a few of the many known possibilities.

3.4.1 Biot strain

The most direct extension of strain to the tensorial setting is the Biot strain defined as

$$\mathbf{E}_{\text{Biot}} \stackrel{\text{def}}{=} \mathbf{U} - \mathbf{1}. \tag{3.4.1}$$

Note that $\mathbf{U} = \mathbf{1}$ for all rigid motions and thus the Biot strain vanishes for all rigid motions. This is a basic requirement for all strain tensors.

3.4.2 Green finite strain tensor

A second tensor useful in applications is the **Green strain tensor**

$$\mathbf{E} \stackrel{\text{def}}{=} \tfrac{1}{2}(\mathbf{F}^{\mathsf{T}}\mathbf{F} - \mathbf{1}), \tag{3.4.2}$$

$$= \tfrac{1}{2}(\mathbf{C} - \mathbf{1}), \tag{3.4.3}$$

$$= \tfrac{1}{2}(\mathbf{U}^2 - \mathbf{1}). \tag{3.4.4}$$

Note that $\mathbf{E}$ vanishes when $\mathbf{F}$ is a rotation, for then $\mathbf{F}^{\mathsf{T}}\mathbf{F} = \mathbf{1}$. Further, the spectral decomposition of E is

$$\mathbf{E} = \sum_{i=1}^{3} \tfrac{1}{2}(\lambda_i^2 - 1)\mathbf{r}_i \otimes \mathbf{r}_i. \tag{3.4.5}$$

3.4.3 Hencky's logarithmic strain tensors

Other strain measures found in the literature include the logarithmic strain tensors of HENCKY:

$$
\begin{aligned}
\ln \mathbf{U} &\stackrel{\text{def}}{=} \sum_{i=1}^{3} (\ln \lambda_i)\mathbf{r}_i \otimes \mathbf{r}_i, \\
\ln \mathbf{V} &\stackrel{\text{def}}{=} \sum_{i=1}^{3} (\ln \lambda_i)\mathbf{l}_i \otimes \mathbf{l}_i.
\end{aligned}
\tag{3.4.6}
$$

REMARK 3.1

Note that given $\mathbf{F}$, the Green strain measure (3.4.2) is relatively easy to compute because it only involves simple matrix operations. However, a computation of the Biot or Hencky strain measure is tedious since each involves calculating the polar decomposition of $\mathbf{F}$ to determine $\mathbf{U}$. This latter step in particular requires the determination of the spectral decomposition of $\mathbf{U}$. It is for this reason of ease of computation that the Green strain measure has been historically used in large deformation theories of mechanics. However, these days, with ever increasing computational power, this reason is no longer as compelling.

3.5 Infinitesimal deformation

In many practical circumstances the deformations encountered for solid materials are quite small, in the sense that the magnitude of the displacement gradient $\mathbf{H} = \nabla \mathbf{u}$ is small, $|\mathbf{H}| \ll 1$.[3] This situation is called the *small deformation* case, and it allows us to greatly simplify the kinematical developments of the prior sections. When the magnitude of $\mathbf{H}$ is small we can linearize each of the kinematic quantities that we have developed so far. Loosely speaking this means we ignore all effects of $\mathbf{H}$ that are higher than first order. In what follows, we will employ the results of linearization, while omitting the technical proofs. The formal process of linearization is discussed in Appendix 3.A. One of the main outcomes of linearization is that the distinction between different strain measures is seen to be immaterial.

3.5.1 Infinitesimal strain tensor

From

$$
\mathbf{E} = \tfrac{1}{2}(\mathbf{C} - \mathbf{1}) = \tfrac{1}{2}(\mathbf{F}^{\top}\mathbf{F} - \mathbf{1}) \qquad \text{and} \qquad \mathbf{H} = \mathbf{F} - \mathbf{1},
$$

---

[3] Here, and henceforth, the symbol ∇ represents the gradient with respect to position $\mathbf{X}$ in the reference body, while the gradient with respect to position $\mathbf{x}$ in the deformed body is denoted by grad.

we obtain

$$\mathbf{E} = \tfrac{1}{2}\left(\mathbf{H} + \mathbf{H}^{\top} + \mathbf{H}^{\top}\mathbf{H}.\right)$$

By **small deformations** we mean $\mathbf{F}$ is sufficiently close to $\mathbf{1}$, or equivalently that

$$|\mathbf{H}| \ll 1.$$

Under the **approximation of small displacement gradients**, we may neglect the quadratic term $(\mathbf{H}^{\top}\mathbf{H})$, and define the **infinitesimal strain tensor** by,

$$\boxed{\boldsymbol{\epsilon} = \frac{1}{2}\left[\mathbf{H} + \mathbf{H}^{\top}\right], \qquad \boldsymbol{\epsilon} = \boldsymbol{\epsilon}^{\top}.} \tag{3.5.1}$$

The matrix of the components ϵ_{ij} of $\boldsymbol{\epsilon}$ is

$$[\boldsymbol{\epsilon}] = \begin{bmatrix} \epsilon_{11} & \epsilon_{12} & \epsilon_{13} \\ \epsilon_{21} & \epsilon_{22} & \epsilon_{23} \\ \epsilon_{31} & \epsilon_{32} & \epsilon_{33} \end{bmatrix}.$$

Of the nine components listed above, only six are independent since the strain is symmetric:

$$\epsilon_{21} = \epsilon_{12}, \quad \epsilon_{31} = \epsilon_{13}, \quad \epsilon_{32} = \epsilon_{23}.$$

Written out in full, the components

$$\epsilon_{11} = \frac{\partial u_1}{\partial X_1}, \quad \epsilon_{22} = \frac{\partial u_2}{\partial X_2}, \quad \epsilon_{33} = \frac{\partial u_3}{\partial X_3},$$

are called the **normal strain components** in the coordinate directions, and the components

$$\epsilon_{12} = \frac{1}{2}\left[\frac{\partial u_1}{\partial X_2} + \frac{\partial u_2}{\partial X_1}\right] = \epsilon_{21}, \quad \epsilon_{13} = \frac{1}{2}\left[\frac{\partial u_1}{\partial X_3} + \frac{\partial u_3}{\partial X_1}\right] = \epsilon_{31},$$

$$\epsilon_{23} = \frac{1}{2}\left[\frac{\partial u_2}{\partial X_3} + \frac{\partial u_3}{\partial X_2}\right] = \epsilon_{32},$$

are called the **tensorial shear strain components** with respect to the coordinate directions.

The **engineering shear strain components** are defined as

$$\gamma_{12} = \left[\frac{\partial u_1}{\partial X_2} + \frac{\partial u_2}{\partial X_1}\right] = \gamma_{21}, \quad \gamma_{13} = \left[\frac{\partial u_1}{\partial X_3} + \frac{\partial u_3}{\partial X_1}\right] = \gamma_{31},$$

$$\gamma_{23} = \left[\frac{\partial u_2}{\partial X_3} + \frac{\partial u_3}{\partial X_2}\right] = \gamma_{32}.$$

Note that the engineering shear strains are defined as twice the magnitude of the tensorial shear strains. Beware of this difference; it is often a source of error. Note we will normally avoid using the engineering shear strain components, because they together with the normal strain components do not form the components of a tensor, and this situation is mathematically very inconvenient.

REMARK 3.2

The normal strain component in an arbitrary (unit) direction $\mathbf{n}$ can be shown to be given by

$$\epsilon_n = \mathbf{n} \cdot \boldsymbol{\epsilon}\mathbf{n}, \tag{3.5.2}$$

wherein this quantity is maximized by the eigenvector associated with the largest eigenvalue of $\boldsymbol{\epsilon}$. The (tensorial) shear strain between any two orthogonal (unit) directions $\mathbf{n}$ and $\mathbf{m}$ is given by

$$\epsilon_{nm} = \mathbf{n} \cdot \boldsymbol{\epsilon}\mathbf{m} = \mathbf{m} \cdot \boldsymbol{\epsilon}\mathbf{n} = \epsilon_{mn},$$

wherein this latter quantity is maximized when $\mathbf{n} = (\boldsymbol{\mu}_1 + \boldsymbol{\mu}_3)/\sqrt{2}$ and $\mathbf{m} = (\boldsymbol{\mu}_1 - \boldsymbol{\mu}_3)/\sqrt{2}$, where $\boldsymbol{\mu}_1$ and $\boldsymbol{\mu}_3$ are respectively, the eigenvectors associated to the algebraically largest and smallest eigenvalues of $\boldsymbol{\epsilon}$.

3.5.2 Infinitesimal rotation tensor

The displacement gradient $\mathbf{H}$ may be uniquely decomposed as

$$\mathbf{H} = \boldsymbol{\epsilon} + \boldsymbol{\omega}, \tag{3.5.3}$$

where $\boldsymbol{\epsilon}$ is the infinitesimal strain defined in (3.5.1), the **symmetric part** of $\mathbf{H}$, and

$$\boxed{\boldsymbol{\omega} = \operatorname{skw}\mathbf{H} = \frac{1}{2}\left[\mathbf{H} - \mathbf{H}^\top\right], \qquad \boldsymbol{\omega}^\top = -\boldsymbol{\omega},} \tag{3.5.4}$$

is the **skew part** of $\mathbf{H}$, and is called the **infinitesimal rotation tensor**. The component representation of the infinitesimal rotation tensor is

$$\omega_{ij} = \frac{1}{2}\left[\frac{\partial u_i}{\partial X_j} - \frac{\partial u_j}{\partial X_i}\right], \qquad \omega_{ij} = -\omega_{ji}. \tag{3.5.5}$$

The axial vector $\mathbf{w}$ of the infinitesimal rotation tensor $\boldsymbol{\omega}$ (see Sec. 1.2.7) is,

$$w_i = -\tfrac{1}{2}e_{ijk}\omega_{jk} = -\tfrac{1}{2}e_{ijk}\tfrac{1}{2}[u_{j,k} - u_{k,j}] = \frac{1}{4}[-e_{ijk}u_{j,k} + e_{ijk}u_{k,j}] = \tfrac{1}{2}e_{ijk}u_{k,j},$$

or

$$\mathbf{w} = \tfrac{1}{2}\operatorname{curl}\mathbf{u}. \tag{3.5.6}$$

The magnitude $|\mathbf{w}|$ represents the infinitesimal angle of rotation, and $\mathbf{w}/|\mathbf{w}|$ represents the axis of rotation.

Equation (3.5.3) expresses the fact that any infinitesimal deformation is the sum of an infinitesimal strain represented by $\boldsymbol{\epsilon}$ and an infinitesimal rotation represented by $\boldsymbol{\omega}$. If the strain at a point in a body in an infinitesimal deformation vanishes, then the neighborhood of that point rotates like a rigid body.

REMARK 3.3

The infinitesimal rotation tensor may also be defined by formally linearizing the expression for the rotation tensor $\mathbf{R}$ in the polar decomposition with respect to the displacement gradient, which according to (3.A.7) gives $\mathbf{R} \approx \mathbf{1} + \operatorname{skw} \mathbf{H}$.

3.6 Example infinitesimal homogeneous strain states

Next, we discuss some simple but important states of **homogeneous** strain, that is strain states ϵ which are *independent* of $\mathbf{X}$.

3.6.1 Uniaxial compression

Uniaxial compression in the $\mathbf{e}_1$-direction is defined by the displacement field

$$\mathbf{u} = -\epsilon\, X_1\, \mathbf{e}_1, \quad u_1 = -\epsilon X_1, \quad u_2 = u_3 = 0, \quad \epsilon = \text{const} \ll 1. \tag{3.6.1}$$

A sketch of the displacement field corresponding to uniaxial compression is shown in Fig. 3.8. Note that although the displacement component u_1 varies linearly with position X_1, the resulting strain is *uniform* throughout the body.

The matrix of the components of ϵ is

$$[\epsilon] = \begin{bmatrix} -\epsilon & 0 & 0 \\ 0 & 0 & 0 \\ 0 & 0 & 0 \end{bmatrix}, \tag{3.6.2}$$

while that of the infinitesimal rotation ω is

$$[\omega] = \begin{bmatrix} 0 & 0 & 0 \\ 0 & 0 & 0 \\ 0 & 0 & 0 \end{bmatrix}. \tag{3.6.3}$$

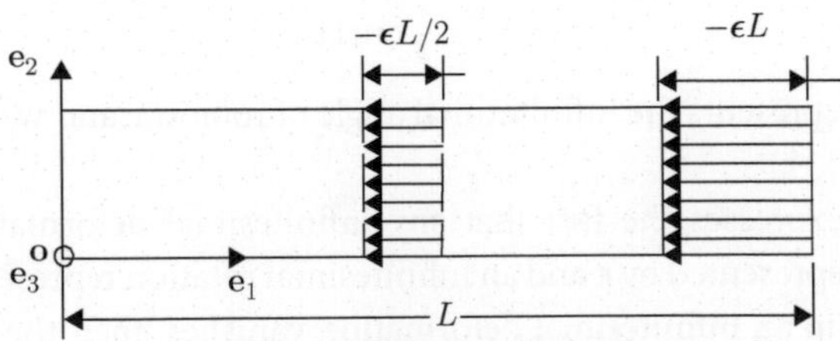

Fig. 3.8 Uniaxial compression.

EXAMPLE 3.8 Normal strain at an angle

In a state of uniaxial compression, the normal strain for material fibers oriented at 30° to the $\mathbf{e}_1$-direction in the $(\mathbf{e}_1, \mathbf{e}_2)$-plane can be found by noting that this direction is given by the unit vector

$$\mathbf{e} = \frac{\sqrt{3}}{2}\mathbf{e}_1 + \frac{1}{2}\mathbf{e}_2.$$

Thus

$$\epsilon_{30^o} = \begin{bmatrix} \sqrt{3}/2 \\ 1/2 \\ 0 \end{bmatrix}^{\top} \begin{bmatrix} -\epsilon & 0 & 0 \\ 0 & 0 & 0 \\ 0 & 0 & 0 \end{bmatrix} \begin{bmatrix} \sqrt{3}/2 \\ 1/2 \\ 0 \end{bmatrix} = -\frac{3}{4}\epsilon.$$

3.6.2 **Simple shear**

As outlined in Sec. 3.1.1, simple shear with respect to $(\mathbf{e}_1, \mathbf{e}_2)$ is defined by

$$\mathbf{u} = \gamma X_2 \mathbf{e}_1, \quad u_1 = \gamma X_2, \quad u_2 = u_3 = 0, \quad |\gamma| = \text{const} \ll 1. \tag{3.6.4}$$

The displacement field corresponding to simple shear for $\gamma > 0$ is shown in Fig. 3.9. Here, material line elements initially parallel to the $\mathbf{e}_1$-axis do not change orientation with deformation, while those parallel to the $\mathbf{e}_2$-axis rotate clockwise about the origin by an angle θ. The matrix of the components of ϵ in this case is[4]

$$[\epsilon] = \begin{bmatrix} 0 & \gamma/2 & 0 \\ \gamma/2 & 0 & 0 \\ 0 & 0 & 0 \end{bmatrix}. \tag{3.6.5}$$

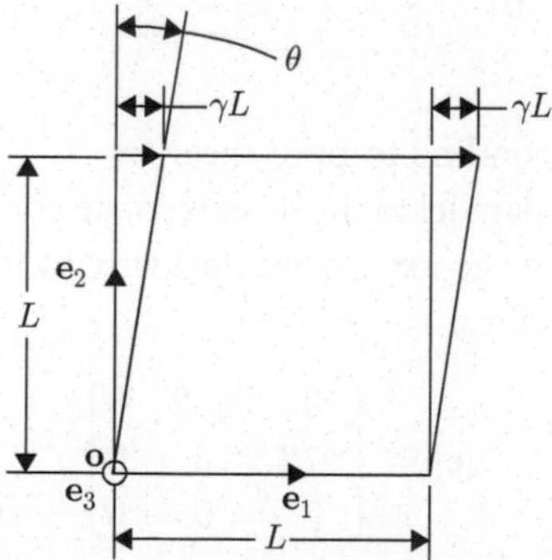

Fig. 3.9 Simple shear.

[4] It is interesting to observe that in the infinitesimal setting $\epsilon_{22} = 0$, whereas in the finite deformation setting the Green strain gives $E_{22} = (1/2)\gamma^2$, a quantity that is vanishingly small for $|\gamma| \ll 1$.

The non-zero engineering shear strain is

$$\gamma_{12} = 2 \times \epsilon_{12} = \gamma = \tan\theta.$$

In simple shear the infinitesimal rotation $\boldsymbol{\omega}$ is *non-zero*. It is given by

$$[\boldsymbol{\omega}] = \begin{bmatrix} 0 & \gamma/2 & 0 \\ -\gamma/2 & 0 & 0 \\ 0 & 0 & 0 \end{bmatrix}. \tag{3.6.6}$$

EXAMPLE 3.9 Principal strains

The value of the maximum normal strain in simple shear is found by maximizing $\epsilon(\mathbf{e}) = \mathbf{e} \cdot \boldsymbol{\epsilon}\mathbf{e}$ subject to the condition $\mathbf{e} \cdot \mathbf{e} = 1$. The necessary equations for this problem are the eigenvalue equations $\boldsymbol{\epsilon}\mathbf{e} = \lambda\mathbf{e}$, where the eigenvalues λ provide the extremal values over all orientations. Thus,

$$\det[\boldsymbol{\epsilon} - \lambda\mathbf{1}] = -\lambda^3 + \left(\frac{\gamma}{2}\right)^2 \lambda = 0 \qquad \Rightarrow \qquad \{\lambda_1, \lambda_2, \lambda_3\} = \{0, \frac{\gamma}{2}, -\frac{\gamma}{2}\},$$

yielding $\gamma/2$ as the maximal normal strain over all directions for a body in simple shear. The corresponding eigenvector gives the direction as $\mathbf{e} = (1/\sqrt{2})\mathbf{e}_1 + (1/\sqrt{2})\mathbf{e}_2$. The eigenvalues and eigenvectors of $\boldsymbol{\epsilon}$ are known as the principal strains and the principal strain directions, respectively.

3.6.3 **Pure shear**

Pure shear with respect to $(\mathbf{e}_1, \mathbf{e}_2)$ is defined by

$$\mathbf{u} = \frac{\gamma}{2} X_2 \,\mathbf{e}_1 + \frac{\gamma}{2} X_1 \,\mathbf{e}_2, \quad u_1 = \frac{\gamma}{2} X_2, \quad u_2 = \frac{\gamma}{2} X_1, \quad u_3 = 0, \quad |\gamma| = \text{const} \ll 1. \tag{3.6.7}$$

The displacement field corresponding to pure shear for $\gamma > 0$ is shown in Fig. 3.10. Here, material line elements initially parallel to the $\mathbf{e}_1$-axis rotate counter-clockwise by a small angle $(\gamma/2)$, while those parallel to the $\mathbf{e}_2$-axis rotate clockwise by a small angle $(\gamma/2)$. The matrix of the components of $\boldsymbol{\epsilon}$ in pure shear is

$$[\boldsymbol{\epsilon}] = \begin{bmatrix} 0 & \gamma/2 & 0 \\ \gamma/2 & 0 & 0 \\ 0 & 0 & 0 \end{bmatrix}, \tag{3.6.8}$$

just as in simple shear. The non-zero engineering shear strain is likewise,

$$\gamma_{12} = 2 \times \epsilon_{12} = \gamma.$$

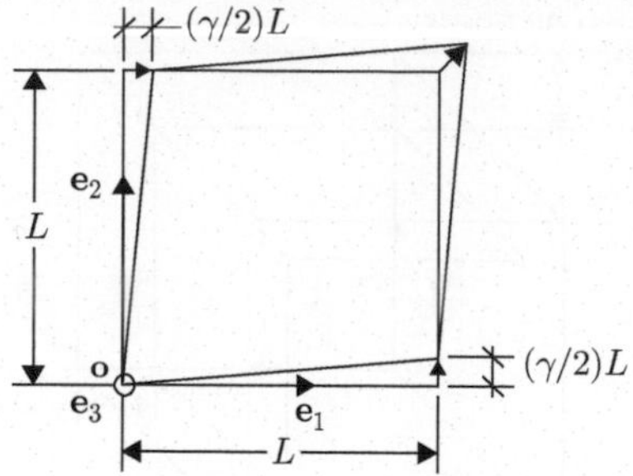

Fig. 3.10 Pure shear.

However, in pure shear the infinitesimal rotation $\boldsymbol{\omega}$ is *zero*:

$$[\boldsymbol{\omega}] = \begin{bmatrix} 0 & 0 & 0 \\ 0 & 0 & 0 \\ 0 & 0 & 0 \end{bmatrix}. \tag{3.6.9}$$

Thus, while the strains in simple shear and pure shear are the same, the infinitesimal rotation tensor in the two cases is not the same. All components of the infinitesimal rotation tensor $\boldsymbol{\omega}$ in pure shear are zero, while those for simple shear are not; they are given by $\omega_{12} = (1/2)\gamma = -\omega_{21}$, cf. (3.6.6).

3.6.4 **Uniform compaction (dilatation)**

The displacement field in uniform compaction (dilatation) is given by

$$\mathbf{u} = -\left(\frac{1}{3}\Delta\right) X_1\,\mathbf{e}_1 - \left(\frac{1}{3}\Delta\right) X_2\,\mathbf{e}_2 - \left(\frac{1}{3}\Delta\right) X_3\,\mathbf{e}_3. \tag{3.6.10}$$

The displacement field corresponding to a uniform **compaction**, $\Delta > 0$, is shown in Fig. 3.11. The matrix of the components of $\boldsymbol{\epsilon}$ is

$$[\boldsymbol{\epsilon}] = \begin{bmatrix} -\left(\frac{1}{3}\Delta\right) & 0 & 0 \\ 0 & -\left(\frac{1}{3}\Delta\right) & 0 \\ 0 & 0 & -\left(\frac{1}{3}\Delta\right) \end{bmatrix}, \tag{3.6.11}$$

and clearly for this deformation $\boldsymbol{\omega} = \mathbf{0}$. Note that in uniform compaction/dilatation the volume change per unit original volume is given by:

$$\frac{v - v_{\mathrm{R}}}{v_{\mathrm{R}}} = (1 + \epsilon_{11})(1 + \epsilon_{22})(1 + \epsilon_{33}) - 1.$$

Hence, for small strains ($|\epsilon_{ij}| \ll 1$), neglecting higher order terms, we obtain

$$\frac{v - v_{\mathrm{R}}}{v_{\mathrm{R}}} = \epsilon_{11} + \epsilon_{22} + \epsilon_{33} = \epsilon_{kk} = \operatorname{tr}\boldsymbol{\epsilon}. \tag{3.6.12}$$

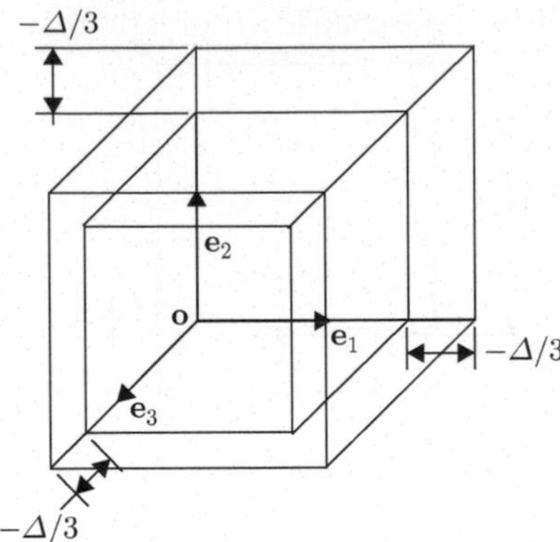

Fig. 3.11 Uniform compaction.

3.6.5 **Infinitesimal rigid displacement**

An infinitesimal rigid displacement is one whose displacement field $\mathbf{u}$ has the form

$$\mathbf{u}(\mathbf{X}) = \mathbf{u}_0 + \mathbf{W}_0(\mathbf{X} - \mathbf{X}_0), \tag{3.6.13}$$

with $\mathbf{X}_0$ a fixed point, $\mathbf{u}_0$ a constant vector, and $\mathbf{W}_0$ a **constant** *skew* tensor. In this case

$$\nabla\mathbf{u}(\mathbf{X}) = \mathbf{W}_0, \qquad \boldsymbol{\omega}(\mathbf{X}) \equiv \mathbf{W}_0, \qquad \text{and} \qquad \boldsymbol{\epsilon}(\mathbf{X}) = \mathbf{0}, \tag{3.6.14}$$

for every $\mathbf{X}$ in $\mathcal{B}$. Thus the infinitesimal strain field corresponding to the infinitesimal rigid displacement field vanishes. Note that the rigid displacement field (3.6.13) may also be written as

$$\mathbf{u}(\mathbf{X}) = \mathbf{u}_0 + \mathbf{w}_0 \times (\mathbf{X} - \mathbf{X}_0), \tag{3.6.15}$$

where $\mathbf{w}_0$ is the axial vector of $\mathbf{W}_0$.

3.7 **Volumetric deviatoric split**

In the analysis of deformation it is often advantageous to have a decomposition of the state of strain that separates volumetric changes from non-volumetric changes. We call the non-volume changing part of the deformation its deviatoric part.

3.7.1 **Volume changes**

Recall that an elemental volume dv_{R} at $\mathbf{X}$ is related to an elemental volume dv at $\mathbf{x}$ by (cf. (3.2.7))

$$dv = J dv_{\mathrm{R}}. \tag{3.7.1}$$

Thus,

$$\frac{dv - dv_{\mathrm{R}}}{dv_{\mathrm{R}}} = (J - 1)\,. \tag{3.7.2}$$

Hence $(J-1)$ represents the volume change per unit original volume. Using $\mathbf{F}=\mathbf{1}+\mathbf{H}$, we have $J=\det\mathbf{F}=\det(\mathbf{1}+\mathbf{H})$. However, for small displacement gradients it may be shown[5]

$$\det(\mathbf{1}+\mathbf{H})\approx 1+\operatorname{tr}\mathbf{H},$$

where tr is the trace operator. Since $\operatorname{tr}\mathbf{H}=\operatorname{tr}\boldsymbol{\epsilon}$, we have

$$\frac{dv-dv_{\mathrm{R}}}{dv_{\mathrm{R}}}=(J-1)\approx\operatorname{tr}\boldsymbol{\epsilon}.$$

Thus,

- *for infinitesimal deformations, the local volume change per unit original volume is given by the* **trace** *of the strain:*

$$\boxed{\frac{dv-dv_{\mathrm{R}}}{dv_{\mathrm{R}}}=\operatorname{tr}\boldsymbol{\epsilon}=\epsilon_{kk}.} \tag{3.7.3}$$

3.7.2 Strain deviator

Next we introduce the important concept of the **strain deviator**:

- *The part of a local strain state that* **deviates** *from a state of pure dilatation (or compaction) is called the* **strain deviator**, *and is defined by*

$$\boxed{\boldsymbol{\epsilon}'=\boldsymbol{\epsilon}-\frac{1}{3}(\operatorname{tr}\boldsymbol{\epsilon})\mathbf{1},\qquad \epsilon'_{ij}=\epsilon_{ij}-\frac{1}{3}(\epsilon_{kk})\,\delta_{ij}.} \tag{3.7.4}$$

Thus, the matrix of components of the strain deviator $\boldsymbol{\epsilon}'$ is given by

$$[\boldsymbol{\epsilon}']=\begin{bmatrix}\epsilon'_{11} & \epsilon'_{12} & \epsilon'_{13}\\ \epsilon'_{21} & \epsilon'_{22} & \epsilon'_{23}\\ \epsilon'_{31} & \epsilon'_{32} & \epsilon'_{33}\end{bmatrix}=\begin{bmatrix}\epsilon_{11} & \epsilon_{12} & \epsilon_{13}\\ \epsilon_{21} & \epsilon_{22} & \epsilon_{23}\\ \epsilon_{31} & \epsilon_{32} & \epsilon_{33}\end{bmatrix}-\frac{1}{3}(\epsilon_{11}+\epsilon_{22}+\epsilon_{33})\begin{bmatrix}1 & 0 & 0\\ 0 & 1 & 0\\ 0 & 0 & 1\end{bmatrix}. \tag{3.7.5}$$

Clearly

$$\operatorname{tr}[\boldsymbol{\epsilon}']=(\epsilon_{11}+\epsilon_{22}+\epsilon_{33})-\frac{1}{3}(\epsilon_{11}+\epsilon_{22}+\epsilon_{33})\times 3=0; \tag{3.7.6}$$

that is, the volume change associated with the deviatoric part of the strain is *zero*.

---

[5] Recall from our discussion of eigenvalues and eigenvectors that given a tensor $\mathbf{S}$, the determinant of $\mathbf{S}-\omega\mathbf{1}$ admits the representation

$$\det(\mathbf{S}-\omega\mathbf{1})=-\omega^3+I_1(\mathbf{S})\omega^2-I_2(\mathbf{S})\omega+I_3(\mathbf{S}).$$

Thus, with $\omega=-1$, $\det(\mathbf{1}+\mathbf{H})$ admits the representation

$$\det(\mathbf{1}+\mathbf{H})=1+I_1(\mathbf{H})+I_2(\mathbf{H})+I_3(\mathbf{H})=1+\operatorname{tr}\mathbf{H}+o(\mathbf{H})$$

as $\mathbf{H}\to\mathbf{0}$; see also Appendix 3.A.

3.8 Summary of major kinematical concepts related to infinitesimal strains

1. **Motion:**

$$\mathbf{x} = \boldsymbol{\chi}(\mathbf{X}, t), \qquad x_i = \chi_i(X_1, X_2, X_3, t). \tag{3.8.1}$$

   Here $\mathbf{X}$ is a material point in the reference configuration, and $\mathbf{x}$ is the place occupied by $\mathbf{X}$ in the deformed body at time t.

2. **Displacement:**

   The displacement vector $\mathbf{u}$ of each material point $\mathbf{X}$ at time t is given by

$$\mathbf{u}(\mathbf{X}, t) = \boldsymbol{\chi}(\mathbf{X}, t) - \mathbf{X}, \qquad u_i(X_1, X_2, X_3) = \chi_i(X_1, X_2, X_3) - X_i. \tag{3.8.2}$$

3. **Velocity and acceleration:**

   The vectors

$$\dot{\mathbf{u}}(\mathbf{X}, t) = \frac{\partial \boldsymbol{\chi}(\mathbf{X}, t)}{\partial t}, \qquad \dot{u}_i(X_1, X_2, X_3, t) = \frac{\partial \chi_i(X_1, X_2, X_3, t)}{\partial t}, \tag{3.8.3}$$

   and

$$\ddot{\mathbf{u}}(\mathbf{X}, t) = \frac{\partial^2 \boldsymbol{\chi}(\mathbf{X}, t)}{\partial t^2}, \qquad \ddot{u}_i(X_1, X_2, X_3, t) = \frac{\partial^2 \chi_i(X_1, X_2, X_3, t)}{\partial t^2}, \tag{3.8.4}$$

   represent the velocity and acceleration of the material point $\mathbf{X}$ at time t.

4. **Deformation gradient, displacement gradient:**

$$\mathbf{F}(\mathbf{X}, t) = \frac{\partial}{\partial \mathbf{X}} \boldsymbol{\chi}(\mathbf{X}, t), \quad F_{ij} = \frac{\partial}{\partial X_j} \chi_i(X_1, X_2, X_3, t), \quad \det \mathbf{F}(\mathbf{X}, t) > 0$$

$$\mathbf{H}(\mathbf{X}, t) = \frac{\partial}{\partial \mathbf{X}} \mathbf{u}(\mathbf{X}, t) \qquad H_{ij}(X_1, X_2, X_3, t) = \frac{\partial}{\partial X_j} u_i(X_1, X_2, X_3, t).$$

$$\mathbf{H}(\mathbf{X}, t) = \mathbf{F}(\mathbf{X}, t) - \mathbf{1}, \qquad H_{ij} = F_{ij} - \delta_{ij},$$

   where

$$\delta_{ij} = \begin{cases} 1 & \text{if} \quad i = j, \\ 0 & \text{if} \quad i \neq j, \end{cases}$$

   is the **Kronecker delta.**

5. **Infinitesimal strain:**

$$\boldsymbol{\epsilon} = \frac{1}{2}\left[\mathbf{H} + \mathbf{H}^{\top}\right], \qquad \boldsymbol{\epsilon} = \boldsymbol{\epsilon}^{\top}, \qquad |\mathbf{H}| \ll 1. \tag{3.8.5}$$

$$\epsilon_{ij} = \frac{1}{2}\left[\frac{\partial u_i}{\partial X_j} + \frac{\partial u_j}{\partial X_i}\right], \qquad \epsilon_{ji} = \epsilon_{ij}, \qquad \left|\frac{\partial u_i}{\partial X_j}\right| \ll 1.$$

   The components $\{\epsilon_{11}, \epsilon_{22}, \epsilon_{33}\}$ are called the **normal strain** components, while the components $\{\epsilon_{12}, \epsilon_{13}, \epsilon_{23}\}$ are called the **tensorial shear strain** components.

The **engineering shear strain** components are defined as **twice** the value of the tensorial shear strain components:

$$\gamma_{12} = 2\epsilon_{12}, \quad \gamma_{13} = 2\epsilon_{13}, \quad \gamma_{23} = 2\epsilon_{23}. \tag{3.8.6}$$

Volume change:
For finite deformations:

$$dv = (\det \mathbf{F})\, dv_{\mathrm{R}}.$$

For small strains:

$$\frac{dv - dv_{\mathrm{R}}}{dv_{\mathrm{R}}} \approx \operatorname{tr} \boldsymbol{\epsilon} = \epsilon_{kk} = \epsilon_{11} + \epsilon_{22} + \epsilon_{33}.$$

Strain deviator:

$$\boldsymbol{\epsilon}' = \boldsymbol{\epsilon} - \frac{1}{3}(\operatorname{tr} \boldsymbol{\epsilon})\, \mathbf{1}, \qquad \epsilon'_{ij} = \epsilon_{ij} - \frac{1}{3}(\epsilon_{kk})\, \delta_{ij}.$$

Appendices

3.A **Linearization**

Given any function $f(\mathbf{Y})$, where f can take on scalar, vector, or tensor values and $\mathbf{Y}$ can be a scalar, vector, or tensor argument, the linear approximation of f at a point $\mathbf{Y}_o$ is given by

$$\operatorname{Lin}_{\mathbf{Y}_o} f(\mathbf{Y}) = f(\mathbf{Y}_o) + \frac{d}{d\alpha} f\left(\mathbf{Y}_o + \alpha(\mathbf{Y} - \mathbf{Y}_o)\right)\Big|_{\alpha=0}. \tag{3.A.1}$$

This expression is nothing more than the first two terms of the Taylor series expansion for f at $\mathbf{Y}_o$.

When linearizing kinematic expressions to obtain their small displacement gradient counterparts, we apply this definition with $\mathbf{Y} = \mathbf{H}$ and $\mathbf{Y}_o = \mathbf{0}$, so that

$$\operatorname{Lin}_{\mathbf{0}} f(\mathbf{H}) = f(\mathbf{0}) + \frac{d}{d\alpha} f(\alpha\mathbf{H})\Big|_{\alpha=0}. \tag{3.A.2}$$

EXAMPLE 3.10 Linearization of the right Cauchy-Green tensor **C**

The linearization of **C** for a small displacement gradient is given by

$$\begin{aligned}\operatorname{Lin}_{\mathbf{0}} \mathbf{C}(\mathbf{H}) &= \mathbf{C}(\mathbf{0}) + \frac{d}{d\alpha}\mathbf{C}(\alpha\mathbf{H})\Big|_{\alpha=0}, \\ &= \mathbf{1} + \frac{d}{d\alpha}\left[(\mathbf{1} + \alpha\mathbf{H})^{\top}(\mathbf{1} + \alpha\mathbf{H})\right]\Big|_{\alpha=0}, \\ &= \mathbf{1} + \left[\mathbf{H}^{\top} + \mathbf{H}\right].\end{aligned}$$

EXAMPLE 3.11 Linearization of the right stretch tensor **U**

The linearization of $\mathbf{U}$ for a small displacement gradient is given by

$$\text{Lin}_{\mathbf{0}}\,\mathbf{U}(\mathbf{H}) = \mathbf{U}(\mathbf{0}) + \left.\frac{d}{d\alpha}\mathbf{U}(\alpha\mathbf{H})\right|_{\alpha=0}.$$

The easiest way to compute this is indirectly by noting that

$$\left.\frac{d}{d\alpha}\mathbf{C}(\alpha\mathbf{H})\right|_{\alpha=0} = \left.\frac{d}{d\alpha}\mathbf{U}^2(\alpha\mathbf{H})\right|_{\alpha=0} = \mathbf{U}(\mathbf{0})\left.\frac{d}{d\alpha}\mathbf{U}(\alpha\mathbf{H})\right|_{\alpha=0} + \left.\frac{d}{d\alpha}\mathbf{U}(\alpha\mathbf{H})\right|_{\alpha=0}\mathbf{U}(\mathbf{0}).$$

Thus,

$$\left.\frac{d}{d\alpha}\mathbf{U}(\alpha\mathbf{H})\right|_{\alpha=0} = \frac{1}{2}\left.\frac{d}{d\alpha}\mathbf{C}(\alpha\mathbf{H})\right|_{\alpha=0} = \underbrace{\tfrac{1}{2}(\mathbf{H}+\mathbf{H}^{\top})}_{\boldsymbol{\epsilon}},$$

and hence

$$\text{Lin}_{\mathbf{0}}\,\mathbf{U}(\mathbf{H}) = \mathbf{1} + \boldsymbol{\epsilon}.$$

A corollary to this result is that the linearized Biot strain is simply the small strain tensor (as it should be).

Additional useful linearizations are:

Green strain	$\text{Lin}_{\mathbf{0}}\,\mathbf{E}(\mathbf{H}) = \boldsymbol{\epsilon}$	(3.A.3)
Deformation gradient	$\text{Lin}_{\mathbf{0}}\,\mathbf{F}(\mathbf{H}) = \mathbf{1} + \mathbf{H}$	(3.A.4)
Inverse deformation gradient	$\text{Lin}_{\mathbf{0}}\,\mathbf{F}^{-1}(\mathbf{H}) = \mathbf{1} - \mathbf{H}$	(3.A.5)
Inverse right stretch	$\text{Lin}_{\mathbf{0}}\,\mathbf{U}^{-1}(\mathbf{H}) = \mathbf{1} - \boldsymbol{\epsilon}$	(3.A.6)
Rotation	$\text{Lin}_{\mathbf{0}}\,\mathbf{R}(\mathbf{H}) = \mathbf{1} + \boldsymbol{\omega}$	(3.A.7)
Jacobian	$\text{Lin}_{\mathbf{0}}\,J(\mathbf{H}) = 1 + \text{tr}\,\boldsymbol{\epsilon}$	(3.A.8)
Area strain	$\text{Lin}_{\mathbf{0}}\,\lambda_{\text{area}}(\mathbf{n}_{\text{R}}) - 1 = (\mathbf{1} - \mathbf{n}_{\text{R}} \otimes \mathbf{n}_{\text{R}}) : \boldsymbol{\epsilon},$	(3.A.9)

where $\boldsymbol{\epsilon} = \text{sym}\,\mathbf{H}$ and $\boldsymbol{\omega} = \text{skw}\,\mathbf{H}$; see Casey (1992) for further details.

3.B Compatibility conditions

For any given displacement field $\mathbf{u}$ describing a one-to-one mapping of a reference configuration onto a current configuration, the components ϵ_{ij} of the infinitesimal strain tensor $\boldsymbol{\epsilon}$ are easily calculated using the strain-displacement relations

$$\epsilon_{ij} = \tfrac{1}{2}(u_{i,j} + u_{j,i}). \tag{3.B.1}$$

However, sometimes one faces the problem of finding the displacements from a given strain field. When a strain field ϵ_{ij} is given, the corresponding displacement field u_i is unknown and needs to be determined by solving for the three components u_i from the six equations (3.B.1). This problem is *over-determined*, and it is possible that a continuous, single-valued displacement field may not exist. If an allowable displacement field exists, then the corresponding strain field is said to be *compatible*; otherwise, the strain field is *incompatible* and unacceptable unless some additional provisions are made (as, for instance, in the theory of fracture mechanics).

It is thus useful to have a set of necessary and sufficient conditions that can be checked to ensure that a given state of strain is compatible. Satisfaction of these *compatibility conditions* will guarantee that there is a continuous, single-valued displacement field corresponding to the given strain field.

Consider first a two-dimensional situation in which we have three strain components ϵ_{11}, ϵ_{22}, and ϵ_{12} and two displacement components u_1 and u_2 related by

$$\left.\begin{aligned} u_{1,1} &= \epsilon_{11}, \\ u_{2,2} &= \epsilon_{22}, \\ u_{1,2} + u_{2,1} &= 2\epsilon_{12}. \end{aligned}\right\} \tag{3.B.2}$$

If the given data $(\epsilon_{11}, \epsilon_{22}, \epsilon_{12})$ is compatible, then any two of the three equations should yield the same values for the displacement components u_1 and u_2. The compatibility of the data can be established as follows. Differentiate $(3.B.2)_1$ twice with respect to X_2, $(3.B.2)_2$ twice with respect to X_1, and $(3.B.2)_3$ with respect to X_1 and X_2 to obtain

$$\left.\begin{aligned} u_{1,122} &= \epsilon_{11,22}, \\ u_{2,211} &= \epsilon_{22,11}, \\ u_{1,221} + u_{2,112} &= 2\epsilon_{12,12}, \end{aligned}\right\} \tag{3.B.3}$$

from which we see that the three strain components $(\epsilon_{11}, \epsilon_{22}, \epsilon_{12})$ must satisfy

$$\epsilon_{11,22} + \epsilon_{22,11} - 2\epsilon_{12,12} = 0. \tag{3.B.4}$$

Equation (3.B.4) represents the *compatibility condition* for the three strain components $(\epsilon_{11}, \epsilon_{22}, \epsilon_{12})$ for a two-dimensional situation. As derived we have only shown that (3.B.4) is a necessary condition for compatibility. However, it turns out to also be a sufficient condition. The proof of this point is deferred to Section 3.B.1. The general result for three dimensions is given next.

Compatibility theorem:[6]
The strain field ϵ corresponding to a sufficiently smooth displacement field $\mathbf{u}$ satisfies the **compatibility condition**[7]

$$\operatorname{curl}(\operatorname{curl}\epsilon) = \mathbf{0}, \qquad e_{ipq}e_{jrs}\epsilon_{qs,rp} = 0. \tag{3.B.6}$$

---

[6] A proof of the compatibility theorem is technical, and relegated to Appendix 3.B.1.

[7] The curl of a second-order tensor $\mathbf{T}$ is a second-order tensor $\operatorname{curl}\mathbf{T}$ with components

$$(\operatorname{curl}\mathbf{T})_{ij} = e_{ipq}T_{jq,p}. \tag{3.B.5}$$

This is easy to verify by replacing ϵ_{qs} with $(1/2)(u_{q,s} + u_{s,q})$ in (3.B.6) and expanding. Conversely, let $\mathcal{B}$ be **simply connected**,[8] and let ϵ be a sufficiently smooth symmetric tensor field on $\mathcal{B}$ that satisfies the compatibility condition (3.B.6). Then there exists a field $\mathbf{u}$ on $\mathcal{B}$ such that ϵ and $\mathbf{u}$ satisfy the strain-displacement relation (3.B.1).

The compatibility condition (3.B.6) is necessary and sufficient for the existence of a corresponding displacement field in a simply connected body.

- Note that since a rigid displacement field produces no strain, the displacements may be determined by integrating the compatibility relations *to within an arbitrary rigid rotation*.

Further, if the solid is **not simply connected** then it may be possible to calculate a displacement field, but the displacement field may not be *single-valued* — one might get different solutions depending how the path of integration goes around holes and other geometric discontinuities in the body.[9]

Since expression (3.B.6) is symmetric in the indices i and j, there are only six independent compatibility conditions; in explicit form these are:

$$\boxed{\begin{aligned}\epsilon_{11,22} + \epsilon_{22,11} - 2\epsilon_{12,12} &= 0,\\ \epsilon_{22,33} + \epsilon_{33,22} - 2\epsilon_{23,23} &= 0,\\ \epsilon_{33,11} + \epsilon_{11,33} - 2\epsilon_{31,13} &= 0,\\ (\epsilon_{12,3} - \epsilon_{23,1} + \epsilon_{31,2})_{,1} - \epsilon_{11,23} &= 0,\\ (\epsilon_{23,1} - \epsilon_{31,2} + \epsilon_{12,3})_{,2} - \epsilon_{22,31} &= 0,\\ (\epsilon_{31,2} - \epsilon_{12,3} + \epsilon_{23,1})_{,3} - \epsilon_{33,12} &= 0,\end{aligned}} \tag{3.B.7}$$

which may be written in succinct form as

$$\epsilon_{ij,kl} + \epsilon_{kl,ij} - \epsilon_{ik,jl} - \epsilon_{jl,ik} = 0. \tag{3.B.8}$$

Thus, whenever, in the course of solving a problem, one arrives at an expression for the infinitesimal strain field in a body without first obtaining an expression for the corresponding displacement field, then one must ensure that the strain satisfies these compatibility conditions. However, if the strain field is derived from a given displacement field, then this requirement is trivially satisfied.

---

[8] Bodies in which every closed curve C is the boundary of at least one unbroken surface are said to be *simply connected*. In a simply connected body, every closed curve can be "shrunk" to a point without passing outside of the body. For example a hollow sphere is simply connected, however a torus (doughnut) is not.

[9] For non-simply connected bodies one requires additional conditions, which we do not discuss here. Interested readers may, for example, consult Yavari (2013).

3.B.1 **Proof of the compatibility theorem**

Compatibility theorem:[10]

The strain field ϵ corresponding to a sufficiently smooth displacement field $\mathbf{u}$ satisfies the **compatibility condition**

$$\operatorname{curl}(\operatorname{curl}\epsilon) = \mathbf{0}, \qquad e_{ipq}e_{jrs}\epsilon_{qs,rp} = 0. \tag{3.B.9}$$

Conversely, let $\mathcal{B}$ be simply connected, and let ϵ be a sufficiently smooth symmetric tensor field on $\mathcal{B}$ that satisfies the compatibility condition (3.B.9). Then there exists a field $\mathbf{u}$ on $\mathcal{B}$ such that ϵ and $\mathbf{u}$ satisfy the strain-displacement relation (3.B.1).

Necessity:
The following proposition supplies the first step in the derivation of the equation of compatibility.

The strain field ϵ and the rotation vector

$$\boldsymbol{w} = \tfrac{1}{2}\operatorname{curl}\boldsymbol{u}$$

*corresponding to a displacement field **u** satisfies*

$$\operatorname{curl}\epsilon = \nabla\mathbf{w}. \tag{3.B.10}$$

The proof of proposition (3.B.10) is as follows. The curl of a second-order tensor $\mathbf{T}$ is a second-order tensor $\operatorname{curl}\mathbf{T}$ with components

$$(\operatorname{curl}\mathbf{T})_{ij} = e_{ipq}T_{jq,p}. \tag{3.B.11}$$

We also need the following identities for vectors $\mathbf{f}$,

$$\operatorname{curl}\nabla\mathbf{f} = \mathbf{0} \qquad \text{and} \qquad \operatorname{curl}(\nabla\mathbf{f}^{\top}) = \nabla\operatorname{curl}\mathbf{f}. \tag{3.B.12}$$

The proof of these identities is as follows:

$$(\operatorname{curl}\nabla\mathbf{f})_{ij} = e_{ipq}(\nabla\mathbf{f})_{jq,p} = e_{ipq}f_{j,qp}.$$

Since $f_{j,qp} = f_{j,pq}$ and $e_{ipq} = -e_{iqp}$, we have $(3.B.12)_1$. Next,

$$(\operatorname{curl}\nabla\mathbf{f}^{\top})_{ij} = e_{ipq}(\nabla\mathbf{f}^{\top})_{jq,p} = e_{ipq}f_{q,jp} = (e_{ipq}f_{q,p})_j.$$

and since $(\operatorname{curl}\mathbf{f})_i = e_{ipq}f_{q,p}$, the second of (3.B.12) follows.

Now, apply the curl operator to the strain-displacement relation

$$\epsilon = \tfrac{1}{2}(\nabla\mathbf{u} + \nabla\mathbf{u}^{\top}), \tag{3.B.13}$$

and use the identities (3.B.12):

$$\operatorname{curl}\epsilon = \tfrac{1}{2}\operatorname{curl}(\nabla\mathbf{u} + \nabla\mathbf{u}^{\top}) = \tfrac{1}{2}\operatorname{curl}\nabla\mathbf{u} + \tfrac{1}{2}\operatorname{curl}\nabla\mathbf{u}^{\top} = \tfrac{1}{2}\nabla\operatorname{curl}\mathbf{u} = \nabla\mathbf{w}. \tag{3.B.14}$$

---

[10] The statement and proof of the compatibility theorem given in this Appendix closely follow that given by Gurtin (1972).

Next, if we take the curl of (3.B.14) and use $(3.B.12)_1$, we are immediately led to the conclusion that a *necessary* condition for the existence of a displacement field $\mathbf{u}$ is that the strain ϵ satisfies the compatibility condition (3.B.9).

Sufficiency:
Assume that the necessary condition (3.B.9) is satisfied everywhere in a region $\mathcal{R}$ of a body. If this guarantees the existence of continuous, single-valued function $\mathbf{u}$, then the necessary conditions are also sufficient. We proceed to show this in steps.

Step 1. Our first step is to show that if $\operatorname{curl}(\operatorname{curl}\epsilon) = \mathbf{0}$ in a simply connected body, then a corresponding rotation field $\mathbf{w}$ exists and can be uniquely determined.

If the infinitesimal rotation field $\mathbf{w}$ is known, then it can be expressed in terms of its gradient field as

$$\mathbf{w}(\mathbf{X}) = \mathbf{w}(\mathbf{X}_0) + \int_{\mathbf{X}_0}^{\mathbf{X}} (\nabla \mathbf{w})d\mathbf{X}, \tag{3.B.15}$$

where $\mathbf{w}(\mathbf{X}_0)$ is the infinitesimal rotation vector at some reference point $\mathbf{X}_0$, which must be specified to fix the rigid body rotation, and the line integral is along an *arbitrary* path C, completely within the body, from $\mathbf{X}_0$ to $\mathbf{X}$.

On the other hand, if an infinitesimal strain field ϵ is given then $\nabla\mathbf{w} = \operatorname{curl}\epsilon$, and (3.B.15) gives the rotation field as

$$\mathbf{w}(\mathbf{X}) = \mathbf{w}(\mathbf{X}_0) + \int_{\mathbf{X}_0}^{\mathbf{X}} (\operatorname{curl}\epsilon)d\mathbf{X} \tag{3.B.16}$$

However, this expression for the rotation vector field corresponding to the infinitesimal strain field ϵ is uniquely determined if and only if for every pair of points $(\mathbf{X}_0, \mathbf{X})$ in the body, the value of the integral in (3.B.16) is *independent of the path of integration between these two points.* This condition of path independence can be alternatively stated as

$$\int_C (\operatorname{curl}\epsilon)d\mathbf{X} = \mathbf{0} \tag{3.B.17}$$

for all *closed* curves C that lie completely within the body. This can be seen by considering the path independence of (3.B.16) when $\mathbf{X} = \mathbf{X}_0$. Next, by Stokes theorem,

$$\mathbf{0} = \int_C (\operatorname{curl}\epsilon)d\mathbf{X} = \int_S \Big(\operatorname{curl}(\operatorname{curl}\epsilon)\Big)^{\top} \mathbf{n}\,da \tag{3.B.18}$$

where S is any unbroken surface in the body with boundary C and $\mathbf{n}$ is the normal to that surface.

Thus, we have shown that if for *every* closed curve C in the body there exits *at least one* unbroken surface S that is bounded by C, then the condition $\operatorname{curl}(\operatorname{curl}\epsilon) = \mathbf{0}$ everywhere in the body guarantees path independence of the integral in (3.B.16), and therefore the existence of the infinitesimal rotation vector $\mathbf{w}$.

Bodies in which every closed curve C is the boundary of at least one unbroken surface are said to be *simply connected.* In a simply connected body, every closed curve can be "shrunk" to a point without passing outside of the body. For example a hollow sphere is simply connected, however a torus (doughnut) is not.

Thus, if a strain field ϵ is specified that satisfies $\operatorname{curl}(\operatorname{curl}\epsilon) = \mathbf{0}$ everywhere in a simply connected body, then we have shown that the corresponding rotation vector $\mathbf{w}$ is uniquely determined by (3.B.16).

Step 2. Next, with

$$\boldsymbol{\omega} \stackrel{\text{def}}{=} \tfrac{1}{2}(\nabla\mathbf{u} - \nabla\mathbf{u}^\top) \tag{3.B.19}$$

denoting the skew symmetric infinitesimal rotation tensor, and since recalling that the infinitesimal rotation vector $\mathbf{w}$ is the axial vector of the skew tensor $\boldsymbol{\omega}$,

$$\boldsymbol{\omega} = \mathbf{w}\times, \qquad \omega_{ij} = e_{ikj}w_k, \tag{3.B.20}$$

we note knowing the components w_k of the rotation vector uniquely determines the rotation tensor $\boldsymbol{\omega}$, and thus the displacement gradient field

$$\nabla\mathbf{u} = \epsilon + \boldsymbol{\omega} \tag{3.B.21}$$

is also uniquely determined.

Step 3. Next, the displacement field can be found by integrating $\nabla\mathbf{u}$:

$$\mathbf{u}(\mathbf{X}) = \mathbf{u}(\mathbf{X}_0) + \int_{\mathbf{X}_0}^{\mathbf{X}} (\epsilon + \boldsymbol{\omega})d\mathbf{X}. \tag{3.B.22}$$

The displacement field $\mathbf{u}$ is uniquely determined from (3.B.22) if and only if the integral on the right-hand side of (3.B.22) is path independent:

$$\int_{\mathcal{C}} (\epsilon + \boldsymbol{\omega})d\mathbf{X} = \mathbf{0}. \tag{3.B.23}$$

However, from Stokes theorem, assuming that the body is simply connected,

$$\int_{\mathcal{C}} (\epsilon + \boldsymbol{\omega})d\mathbf{X} = \int_{\mathcal{S}} (\operatorname{curl}(\epsilon + \boldsymbol{\omega}))^\top \mathbf{n}\,da. \tag{3.B.24}$$

Then since

$$\operatorname{curl}\boldsymbol{\omega} = \tfrac{1}{2}(\operatorname{curl}\nabla\mathbf{u} - \operatorname{curl}(\nabla\mathbf{u}^\top)) = -\tfrac{1}{2}\nabla(\operatorname{curl}\mathbf{u}) = -\nabla\mathbf{w}, \tag{3.B.25}$$

and recalling that $\operatorname{curl}\epsilon = \nabla\mathbf{w}$ we have $\operatorname{curl}(\epsilon + \boldsymbol{\omega}) = \mathbf{0}$, and therefore the displacement field is uniquely determined by (3.B.22).

Thus, given a strain field ϵ, the compatibility condition $\operatorname{curl}(\operatorname{curl}\epsilon) = \mathbf{0}$ everywhere is sufficient in a simply connected body to guarantee the existence of an infinitesimal rotation vector field $\mathbf{w}$ given by (3.B.16). The gradient of the displacement field can then be determined by (3.B.21), and finally the displacement field $\mathbf{u}$ found by evaluating (3.B.22).

PART III

Balance laws

4 Balance laws for mass, forces, and moments

The responses of all bodies which may be treated as continua, regardless of whether they be a solid or fluid, are governed by the following fundamental laws:[1]

(i) balance of mass;

(ii) balance of linear and angular momentum;

(iii) balance of energy; and

(iv) an entropy imbalance.

In this chapter we discuss the balance of mass as well as the balance of linear and angular momentum, with the latter phrased as a balance of forces and moments. We initially formulate these balance laws globally for parts $\mathcal{P}_t$ that convect with the body. Using the requirement that the underlying parts be arbitrary, we then derive local balance laws in the form of partial differential equations.

We will discuss the balance of energy and the entropy imbalance in Chapter 5.

4.1 Balance of mass

Let $\mathcal{P}_t$ be a spatial region that convects with the body, so that $\mathcal{P}_t = \chi_t(\mathcal{P})$ for some material region $\mathcal{P}$ (Fig. 4.1). We write $\rho_{\text{R}}(\mathbf{X}) > 0$ for the mass density at the material point $\mathbf{X}$ in the reference body $\mathcal{B}$, so that

$$\int_{\mathcal{P}} \rho_{\text{R}}(\mathbf{X})\, dv_{\text{R}}(\mathbf{X})$$

represents the mass of the material region $\mathcal{P}$. We refer to $\rho_{\text{R}}(\mathbf{X})$ as the **reference density**. Analogously, given a motion, we write $\rho(\mathbf{x}, t) > 0$ for the (mass) **density** at the spatial point $\mathbf{x}$ in the deformed body $\mathcal{B}_t$, so that

$$\int_{\mathcal{P}_t} \rho(\mathbf{x}, t)\, dv(\mathbf{x})$$

represents the mass of the spatial region $\mathcal{P}_t$ occupied by $\mathcal{P}$ at time t.

---

[1] A large fraction of this chapter is based on an exposition of these central topics in *The Mechanics and Thermodynamics of Continua* by Gurtin et al. (2010).

Continuum Mechanics of Solids. Lallit Anand and Sanjay Govindjee, Oxford University Press (2020).

DOI: 10.1093/oso/9780198864721.001.0001

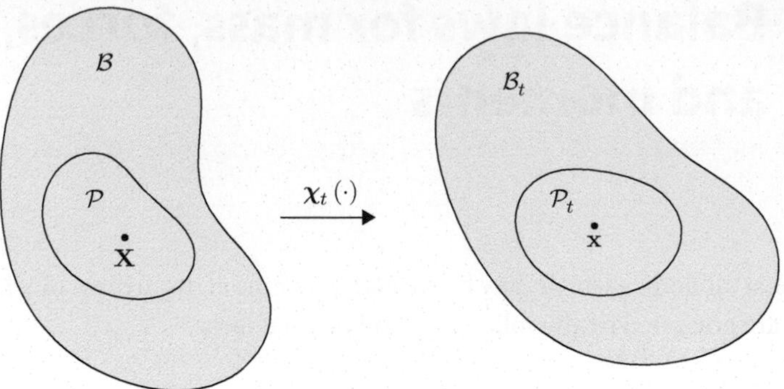

Fig. 4.1 Material region $\mathcal{P}$ which convects to $\mathcal{P}_t = \chi_t(\mathcal{P})$.

Global **balance of mass** is then the requirement that, given any motion,

$$\int_{\mathcal{P}} \rho_{\mathrm{R}}(\mathbf{X})\, dv_{\mathrm{R}}(\mathbf{X}) = \int_{\mathcal{P}_t} \rho(\mathbf{x}, t)\, dv(\mathbf{x}) \tag{4.1.1}$$

for every material region $\mathcal{P} \subset \mathcal{B}$. Recall that an elemental volume dv_{R} at $\mathbf{X}$ is related to an elemental volume dv at $\mathbf{x}$ by (cf., (3.2.7))

$$dv = J dv_{\mathrm{R}}. \tag{4.1.2}$$

Thus the integral on the right side of (4.1.1) can be written as

$$\int_{\mathcal{P}_t} \rho\, dv = \int_{\mathcal{P}} \rho J\, dv_{\mathrm{R}},$$

so that, by (4.1.1),

$$\int_{\mathcal{P}} (\rho_{\mathrm{R}} - \rho J)\, dv_{\mathrm{R}} = 0.$$

Since $\mathcal{P}$ is arbitrary, the *Localization Theorem* of Sec. 2.4.1 yields

$$\boxed{\rho = \frac{\rho_{\mathrm{R}}}{J},} \tag{4.1.3}$$

which is a local form for the balance of mass.

4.2 **Balance of forces and moments. Stress tensor. Equation of motion**

Motions are accompanied by **forces**. Classically, forces in continuum mechanics are described *spatially* (i.e., with reference to the deformed body at a given time t) by

(i) contact forces between *adjacent* spatial regions; that is, *spatial regions that intersect along their boundaries*;

(ii) contact forces exerted on the boundary of the body by its environment;

(iii) body forces exerted on the interior points of a body by the environment.

One of the most important and far reaching axioms of continuum mechanics is **Cauchy's hypothesis** concerning the form of the contact forces. Cauchy introduced a **surface traction field** $\mathbf{t}(\mathbf{n}, \mathbf{x}, t)$ — defined for each unit vector $\mathbf{n}$, each $\mathbf{x}$ in $\mathcal{B}_t$, and each t (Fig. 4.2) — assumed to have the following property:

- Given any oriented spatial surface $\mathcal{S}$ in $\mathcal{B}_t$, $\mathbf{t}(\mathbf{n}, \mathbf{x}, t)$ represents the force, per unit area, exerted across $\mathcal{S}$ *upon* the material on the negative side of $\mathcal{S}$ by the material on the positive side.[2]

Thus, with reference to Fig. 4.3,

$$\int_{\partial \mathcal{P}_t} \mathbf{t}(\mathbf{n})\, da \tag{4.2.1}$$

represents the *net* contact force exerted on the spatial region $\mathcal{P}_t$ at time t.

For points on the boundary of $\mathcal{B}_t$, $\mathbf{t}(\mathbf{n}, \mathbf{x}, t)$ — with $\mathbf{n}$ the outward unit normal to $\partial \mathcal{B}_t$ at $\mathbf{x}$ — gives the surface force, per unit area, exerted on the body at $\mathbf{x}$ by contact with the environment. The environment can also exert forces on interior points of $\mathcal{B}_t$, a classical example of such a

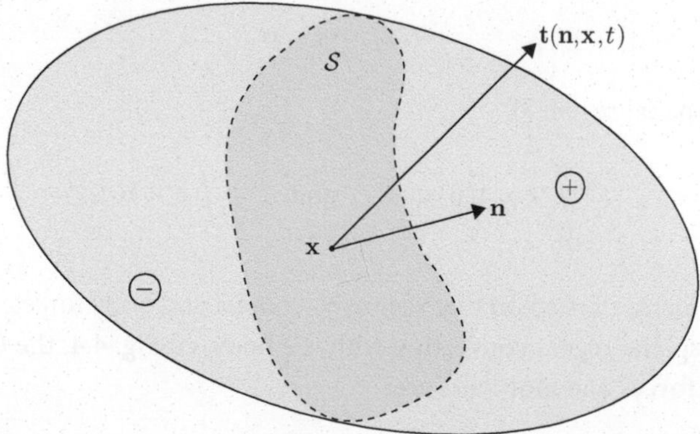

Fig. 4.2 Traction $\mathbf{t}$ at $\mathbf{x}$ exerted by material in the positive side on the material in the negative side of $\mathcal{S}$.

[2] By the positive side of $\mathcal{S}$ we mean the portion of $\mathcal{B}_t$ into which $\mathbf{n}$ points; similarly, the negative side of $\mathcal{S}$ is the portion of $\mathcal{B}_t$ out of which $\mathbf{n}$ points.

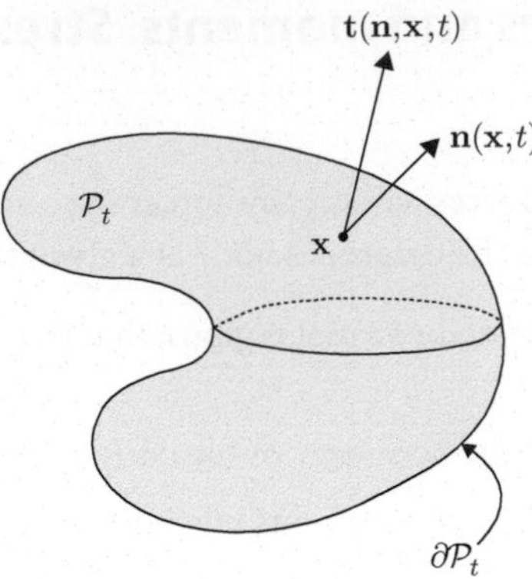

Fig. 4.3 Traction field $\mathbf{t}(\mathbf{n}, \mathbf{x}, t)$ on the boundary $\partial\mathcal{P}_t$ of a spatial region $\mathcal{P}_t$.

force being that due to gravity. Such forces are determined by a vector field $\mathbf{b}_0(\mathbf{x}, t)$, called a *conventional* body force, which gives the force, per unit volume, exerted by the environment on the material at $\mathbf{x}$.[3] Next, we introduce a **generalized body force b** defined by

$$\mathbf{b} = \mathbf{b}_0 + \boldsymbol{\imath}, \qquad \boldsymbol{\imath} = -\rho\dot{\mathbf{v}}, \tag{4.2.2}$$

in which $\boldsymbol{\imath}$ is referred to as the **inertial body force.**[4] Like $\mathbf{b}_0$, the body forces $\mathbf{b}$ and $\boldsymbol{\imath}$ are measured per unit volume in the deformed body. Thus, for any spatial region $\mathcal{P}_t$, the integral

$$\int\limits_{\mathcal{P}_t} \mathbf{b}\, dv$$

gives that part of the force on $\mathcal{P}_t$ not due to contact.

It is convenient to write $\mathbf{r}$ for the *position vector*

$$\mathbf{r}(\mathbf{x}) = \mathbf{x} - \mathbf{o}. \tag{4.2.3}$$

Then, given any spatial region $\mathcal{P}_t$,

$$\int\limits_{\partial\mathcal{P}_t} \mathbf{r} \times \mathbf{t}(\mathbf{n})\, da \qquad \text{and} \qquad \int\limits_{\mathcal{P}} \mathbf{r} \times \mathbf{b}\, dv \tag{4.2.4}$$

represent net moments exerted on that region by contact and body forces.

Thus for $\mathcal{P}_t$ a spatial region convecting with the body, cf. Fig. 4.4, the (generalized) global **balance laws for forces and moments** are:

[3] We use the adjective "conventional" to differentiate this body force from a body force that accounts also for the inertia (i.e., also for the d'Alembert force).

[4] Often also called the d'Alembert body force.

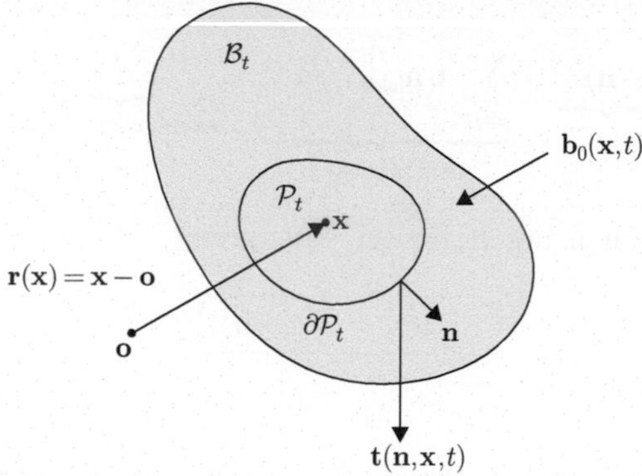

Fig. 4.4 Surface and body forces acting on $\mathcal{P}_t$.

$$\begin{aligned}
&\int_{\partial\mathcal{P}_t} \mathbf{t}(\mathbf{n})\,da + \int_{\mathcal{P}_t} \mathbf{b}\,dv = \mathbf{0}, \qquad \int_{\partial\mathcal{P}_t} t_i(\mathbf{n})\,da + \int_{\mathcal{P}_t} b_i\,dv = 0,\\
&\int_{\partial\mathcal{P}_t} \mathbf{r}\times\mathbf{t}(\mathbf{n})\,da + \int_{\mathcal{P}_t} \mathbf{r}\times\mathbf{b}\,dv = \mathbf{0}, \qquad \int_{\partial\mathcal{P}_t} e_{ijk}x_j t_k(\mathbf{n})\,da + \int_{\mathcal{P}_t} e_{ijk}x_j b_k\,dv = 0,
\end{aligned} \tag{4.2.5}$$

relations that are to hold for every convecting region $\mathcal{P}_t$.[5]

4.3 **Local balance of forces and moments**

The balance laws of forces and moments lead to a central result of continuum mechanics, embodied in Cauchy's theorem, which we discuss next.

4.3.1 **Pillbox construction**

First, consider a part $\mathcal{P}_t$ which is in the form of a infinitesimal rectangular parallelepiped, or pillbox, which is centered at $\mathbf{x}$, and has characteristic dimension δ, with $\mathcal{P}_t$ contained in the interior of $\mathcal{B}_t$ for all sufficiently small δ; cf. Fig. 4.5. Specifically, we take such a parallelepiped to have dimensions $\delta \times \delta \times \alpha\,\delta$, where $\alpha > 0$ is also a small number, and with $\mathbf{n}$ the outward normal to the face $\mathcal{S}_t$. The normal to the opposite face $\mathcal{S}'_t$ is $-\mathbf{n}$. Applying balance of forces $(4.2.5)_1$ to such an infinitesimal part gives[6]

[5] When $\mathbf{b}$ is expanded into its constituent parts, $\mathbf{b}_0$ and $\boldsymbol{\imath} = -\rho\dot{\mathbf{v}}$, these relations are referred to as the balance laws for linear and angular momentum.

[6] In (4.3.1) $\mathbf{t}(\mathbf{n})$ and $\mathbf{b}$ denote *average* tractions and body forces over each surface and the volume of the parallelepiped, respectively.

$$\underbrace{\mathbf{t}(\mathbf{n})\,\delta^2}_{\text{from } \mathcal{S}_t} + \underbrace{\mathbf{t}(-\mathbf{n})\,\delta^2}_{\text{from } \mathcal{S}'_t} + \underbrace{\sum_{i=1}^{4} \mathbf{t}(\mathbf{n}_{\text{lateral,i}})\,\alpha\delta^2}_{\text{from lateral faces}} + \underbrace{\mathbf{b}\,\alpha\delta^3}_{\text{volume contribution}} = \mathbf{0}. \tag{4.3.1}$$

Dividing (4.3.1) by δ^2 and taking the limit $\alpha \to 0$,[7] gives

$$\boxed{\mathbf{t}(\mathbf{n}) = -\mathbf{t}(-\mathbf{n}),} \tag{4.3.2}$$

for all unit vectors $\mathbf{n}$.

- *That is, the traction vectors* $\mathbf{t}(\mathbf{n})$ *and* $\mathbf{t}(-\mathbf{n})$ *acting on opposite sides of the same surface at a given point are equal in magnitude, but opposite in sign,* which is **Newton's law of action and reaction**; cf. Fig. 4.6. Note that in this context this law is derived from the balance of force law and is thus not a separate law.

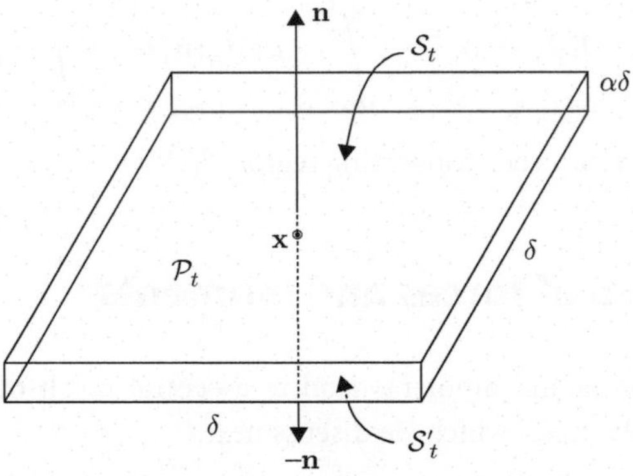

Fig. 4.5 Elemental rectangular parallelepiped of edge lengths δ, δ, and $\alpha\delta$.

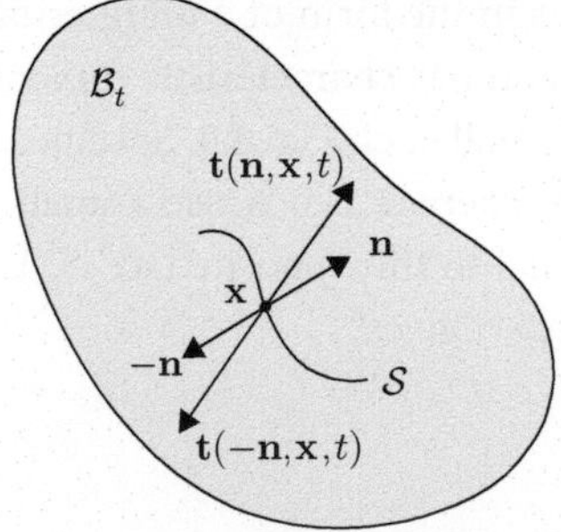

Fig. 4.6 Newton's law of action and reaction for traction vectors.

[7] We assume throughout that the body force $\mathbf{b}_0$ and the inertial body force $\boldsymbol{\imath} = -\rho\dot{\mathbf{v}}$ are bounded.

4.3.2 **Cauchy's tetrahedron construction**

Next, let $\mathbf{x}$ belong to the interior of $\mathcal{B}_t$. Given any orthonormal basis $\{\mathbf{e}_i\}$, and any unit vector $\mathbf{n}$ with

$$\mathbf{n} \cdot \mathbf{e}_i > 0, \quad i = 1, 2, 3, \tag{4.3.3}$$

we consider an elemental tetrahedron of a characteristic dimension δ as follows. Choose $\delta > 0$ and consider the tetrahedron $\mathcal{T}_\delta$ with the following properties: the faces of $\mathcal{T}_\delta$ are $\mathcal{S}_\delta$, $\mathcal{S}_{1\delta}$, $\mathcal{S}_{2\delta}$, and $\mathcal{S}_{3\delta}$, where $\mathbf{n}$ and $-\mathbf{e}_i$ are the outward unit normals on $\mathcal{S}_\delta$ and $\mathcal{S}_{i\delta}$, respectively; the vertex opposite to $\mathcal{S}_\delta$ is $\mathbf{x}$; the distance from $\mathbf{x}$ to $\mathcal{S}_\delta$ is δ; cf. Fig. 4.7. Then $\mathcal{T}_\delta$ is contained in the interior of $\mathcal{B}_t$ for all sufficiently small δ.

For this infinitesimal tetrahedron, let

$$A_\delta = \text{area}(\mathcal{S}_\delta);$$

then the areas of the faces perpendicular to the coordinate axes are

$$A_{i\,\delta} = \text{area}(\mathcal{S}_{i\,\delta}) = (\mathbf{n} \cdot \mathbf{e}_i) A_\delta,$$

and the volume of the tetrahedron is

$$\text{vol}(\mathcal{T}_\delta) = \frac{1}{3} A_\delta\, \delta.$$

Applying balance of forces $(4.2.5)_1$ to such an infinitesimal tetrahedron gives

$$\mathbf{t}(\mathbf{n}) A_\delta + \sum_{i=1}^{3} \mathbf{t}(-\mathbf{e}_i)\, (\mathbf{n} \cdot \mathbf{e}_i)\, A_\delta + \mathbf{b}\, \frac{1}{3} A_\delta\, \delta = \mathbf{0}, \tag{4.3.4}$$

where the tractions and body force are average values over their respective domains. Dividing (4.3.4) by A_δ and taking the limit $\delta \to 0$ gives

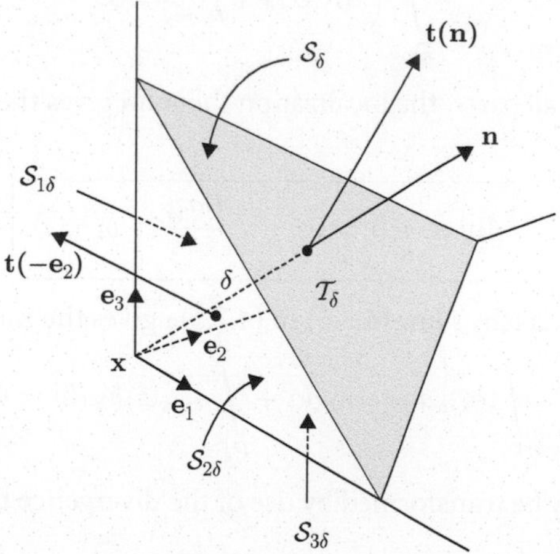

Fig. 4.7 Elemental tetrahedron $\mathcal{T}_\delta$.

$$\mathbf{t}(\mathbf{n},\mathbf{x}) = -\sum_{i=1}^{3}(\mathbf{n}\cdot\mathbf{e}_i)\,\mathbf{t}(-\mathbf{e}_i,\mathbf{x}). \tag{4.3.5}$$

Next, using (4.3.2) we have

$$\mathbf{t}(-\mathbf{e}_i) = -\mathbf{t}(\mathbf{e}_i),$$

and using this in (4.3.5) we obtain

$$\mathbf{t}(\mathbf{n},\mathbf{x}) = \sum_{i=1}^{3}(\mathbf{n}\cdot\mathbf{e}_i)\mathbf{t}(\mathbf{e}_i,\mathbf{x}) = \left(\sum_{i=1}^{3}\mathbf{t}(\mathbf{e}_i,\mathbf{x})\otimes\mathbf{e}_i\right)\mathbf{n},$$

or the first of Cauchy's results—Cauchy's law,

$$\boxed{\mathbf{t}(\mathbf{n},\mathbf{x}) = \boldsymbol{\sigma}(\mathbf{x})\mathbf{n}, \qquad t_i = \sigma_{ij}n_j,} \tag{4.3.6}$$

where $\boldsymbol{\sigma}(\mathbf{x})$ is **the Cauchy stress tensor** defined by

$$\boldsymbol{\sigma}(\mathbf{x}) \stackrel{\text{def}}{=} \sum_{i=1}^{3}\mathbf{t}(\mathbf{e}_i,\mathbf{x})\otimes\mathbf{e}_i. \tag{4.3.7}$$

4.3.3 Cauchy's local balance of forces and moments

Using Cauchy's law (4.3.6), balance of forces $(4.2.5)_1$ may be rewritten as

$$\int_{\partial\mathcal{P}_t}\boldsymbol{\sigma}\mathbf{n}\,da + \int_{\mathcal{P}_t}\mathbf{b}\,dv = \mathbf{0}, \tag{4.3.8}$$

which upon use of the divergence theorem gives

$$\int_{\mathcal{P}_t}\Big(\operatorname{div}\boldsymbol{\sigma} + \mathbf{b}\Big)\,dv = \mathbf{0}.$$

Since this must hold for all parts, the localization theorem gives the following **local statement of the balance of forces**

$$\boxed{\operatorname{div}\boldsymbol{\sigma} + \mathbf{b} = \mathbf{0}, \qquad \frac{\partial\sigma_{ij}}{\partial x_j} + b_i = 0.} \tag{4.3.9}$$

Next, substitution of Cauchy's law (4.3.6) in $(4.2.5)_2$ gives the moment balance as

$$\int_{\partial\mathcal{P}_t} e_{ijk}x_j\sigma_{kl}n_l\,da + \int_{\mathcal{P}_t} e_{ijk}x_jb_k\,dv = 0. \tag{4.3.10}$$

The surface integral may be transformed by use of the divergence theorem as follows:

$$\int_{\partial\mathcal{P}_t} e_{ijk}x_j\,\sigma_{kl}n_l\,da = \int_{\mathcal{P}_t}\left(e_{ijk}x_j\,\sigma_{kl}\right)_{,l}\,dv,$$

$$= \int_{\mathcal{P}_t} (e_{ijk}\delta_{jl}\,\sigma_{kl} + e_{ijk}x_j\,\sigma_{kl,l})\,dv,$$

$$= \int_{\mathcal{P}_t} (e_{ijk}\,\sigma_{kj} + e_{ijk}x_j\,\sigma_{kl,l})\,dv. \tag{4.3.11}$$

Using (4.3.11), the moment balance (4.3.10) may be written as

$$\int_{\mathcal{P}_t} e_{ijk}\,\sigma_{kj} dv + \int_{\mathcal{P}_t} e_{ijk}x_j\,(\sigma_{kl,l} + b_k)\,dv = 0,$$

for all parts $\mathcal{P}_t$. Since balance of forces (4.3.9) also holds for all parts, the last term vanishes, and we obtain from the localization theorem that

$$e_{ijk}\,\sigma_{kj} = 0. \tag{4.3.12}$$

Multiplying (4.3.12) with the alternating symbol e_{ipq} and using the epsilon–delta identity (1.1.12), we obtain

$$\begin{aligned} e_{ipq}e_{ijk}\,\sigma_{kj} &= 0, \\ (\delta_{pj}\delta_{qk} - \delta_{pk}\delta_{qj})\sigma_{kj} &= 0, \\ \sigma_{qp} - \sigma_{pq} &= 0, \end{aligned}$$

which implies that the Cauchy stress tensor $\boldsymbol{\sigma}$ must be *symmetric*

$$\boxed{\boldsymbol{\sigma} = \boldsymbol{\sigma}^{\top}, \qquad \sigma_{ij} = \sigma_{ji}.} \tag{4.3.13}$$

This is the **local statement of the balance of moments.**

4.3.4 **Cauchy's theorem**

The results obtained above are summarized in Cauchy's theorem, which is stated below.

CAUCHY'S THEOREM[8] *Let* $(\mathbf{t}(\mathbf{n}), \mathbf{b}_0)$ *be a system of forces for a body* $\mathcal{B}$ *during a motion. Then, satisfaction of global balance of forces and moments (4.2.5),*[9] *is equivalent to the existence of a spatial tensor field* $\boldsymbol{\sigma}$, *called the* **Cauchy stress**, *such that*

(i) *For each unit vector* $\mathbf{n}$,

$$\mathbf{t}(\mathbf{n}) = \boldsymbol{\sigma}\mathbf{n}, \qquad t_i(\mathbf{n}) = \sigma_{ij}n_j; \tag{4.3.14}$$

(ii) $\boldsymbol{\sigma}$ *is symmetric,*

$$\boldsymbol{\sigma} = \boldsymbol{\sigma}^{\top}, \qquad \sigma_{ij} = \sigma_{ji}, \tag{4.3.15}$$

and

(iii) $\boldsymbol{\sigma}$ *satisfies the local equation of motion*

$$\operatorname{div}\boldsymbol{\sigma} + \mathbf{b}_0 = \rho\dot{\mathbf{v}}, \qquad \sigma_{ij,j} + b_{0i} = \rho\dot{v}_i. \tag{4.3.16}$$

---

[8] This theorem is due to Cauchy (1789–1857) in 1827.

[9] More generally, satisfaction of balance of linear and angular momentum.

REMARKS 4.1

- In (4.3.16) we have used (4.2.2) to expand $\mathbf{b}$ in terms of the conventional body force and the d'Alembert force.

- The developments presented above prove this theorem in one direction, i.e. the global statements of force and moment balance imply Cauchy's law and the local statements of force and moment balance. The converse statement can be proved by reversing the steps of the proof.

- Henceforth, for simplicity, we will drop the subscript 0 on the conventional body force $\mathbf{b}_0$, and explicitly account for the inertial body force in the equation of motion (4.3.16). Further, we replace the acceleration $\dot{\mathbf{v}}$ with its equivalent $\ddot{\mathbf{u}}$ in terms of the displacement vector $\mathbf{u}$; hence we replace (4.3.16) by

$$\boxed{\operatorname{div}\boldsymbol{\sigma} + \mathbf{b} = \rho\ddot{\mathbf{u}}, \qquad \sigma_{ij,j} + b_i = \rho\ddot{u}_i.} \tag{4.3.17}$$

EXAMPLE 4.1 Signorini's Theorem

The volume average stress in a body $\mathcal{B}_t$ with volume vol $\mathcal{B}_t$ defined as

$$\bar{\boldsymbol{\sigma}} = \frac{1}{\operatorname{vol}\mathcal{B}_t}\int_{\mathcal{B}_t} \boldsymbol{\sigma}\, dv,$$

can be expressed in the quasi-static case ($\rho\ddot{\mathbf{u}} \approx \mathbf{0}$) as

$$\bar{\boldsymbol{\sigma}} = \frac{1}{\operatorname{vol}\mathcal{B}_t}\left[\int_{\partial\mathcal{B}_t} \mathbf{t}\otimes\mathbf{x}\, da + \int_{\mathcal{B}_t} \mathbf{b}\otimes\mathbf{x}\, dv\right]$$

This result is known as *Signorini's Theorem*. To show the result, start with the result and work backwards

$$\begin{aligned}
\frac{1}{\operatorname{vol}\mathcal{B}_t}\left[\int_{\partial\mathcal{B}_t} t_i x_k\, da + \int_{\mathcal{B}_t} b_i x_k\, dv\right] &= \frac{1}{\operatorname{vol}\mathcal{B}_t}\left[\int_{\partial\mathcal{B}_t} \sigma_{ij} n_j x_k\, da + \int_{\mathcal{B}_t} b_i x_k\, dv\right] \\
&= \frac{1}{\operatorname{vol}\mathcal{B}_t}\left[\int_{\mathcal{B}_t} \Big((\sigma_{ij} x_k)_{,j} + b_i x_k\Big)\, dv\right] \\
&= \frac{1}{\operatorname{vol}\mathcal{B}_t}\left[\int_{\mathcal{B}_t} \Big(\sigma_{ij,j} x_k + \sigma_{ij}\delta_{kj} + b_i x_k\Big)\, dv\right] \\
&= \frac{1}{\operatorname{vol}\mathcal{B}_t}\left[\int_{\mathcal{B}_t} \Big(-b_i x_k + \sigma_{ik} + b_i x_k\Big)\, dv\right] \\
&= \frac{1}{\operatorname{vol}\mathcal{B}_t}\left[\int_{\mathcal{B}_t} \sigma_{ik}\, dv\right] \\
&= \bar{\sigma}_{ik}.
\end{aligned}$$

4.4 **Physical interpretation of the components of stress**

Consider the state of stress in the interior of the body at a point $\mathbf{x}$ in the deformed body $\mathcal{B}_t$. Visualize an infinitesimal cubic element centered at point $\mathbf{x}$, and with sides parallel to the coordinate axes $\{\mathbf{e}_1, \mathbf{e}_2, \mathbf{e}_3\}$. Consider the traction vectors $\mathbf{t}(\mathbf{x}, \mathbf{e}_1)$, $\mathbf{t}(\mathbf{x}, \mathbf{e}_2)$ and $\mathbf{t}(\mathbf{x}, \mathbf{e}_3)$ on the faces of the cube with outward unit normals in the three positive coordinate directions; cf. Fig. 4.8. Suppressing the argument $\mathbf{x}$, the traction vectors $\mathbf{t}(\mathbf{e}_1)$, $\mathbf{t}(\mathbf{e}_2)$ and $\mathbf{t}(\mathbf{e}_3)$ have the following components with respect to the basis $\{\mathbf{e}_1, \mathbf{e}_2, \mathbf{e}_3\}$:

$$\left.\begin{aligned}
\mathbf{e}_1 \cdot \mathbf{t}(\mathbf{e}_1) &\equiv \sigma_{11}, & \mathbf{e}_2 \cdot \mathbf{t}(\mathbf{e}_1) &\equiv \sigma_{21}, & \mathbf{e}_3 \cdot \mathbf{t}(\mathbf{e}_1) &\equiv \sigma_{31}, \\
\mathbf{e}_1 \cdot \mathbf{t}(\mathbf{e}_2) &\equiv \sigma_{12}, & \mathbf{e}_2 \cdot \mathbf{t}(\mathbf{e}_2) &\equiv \sigma_{22}, & \mathbf{e}_3 \cdot \mathbf{t}(\mathbf{e}_2) &\equiv \sigma_{32}, \\
\mathbf{e}_1 \cdot \mathbf{t}(\mathbf{e}_3) &\equiv \sigma_{13}, & \mathbf{e}_2 \cdot \mathbf{t}(\mathbf{e}_3) &\equiv \sigma_{23}, & \mathbf{e}_3 \cdot \mathbf{t}(\mathbf{e}_3) &\equiv \sigma_{33}.
\end{aligned}\right\} \tag{4.4.1}$$

- The three quantities

$$\{\sigma_{11}, \sigma_{22}, \sigma_{33}\},$$

represent **normal stress components** on the coordinate planes, and the three quantities

$$\{\sigma_{12}(=\sigma_{21}), \sigma_{13}(=\sigma_{31}), \sigma_{23}(=\sigma_{23})\},$$

represent **shear stress components** on these same planes.

It is important to note that according to our construction and definitions[10]

- *the second suffix on the nine components σ_{ij} denotes the direction of the normal to the face, and the first suffix denotes the direction of the traction component.*

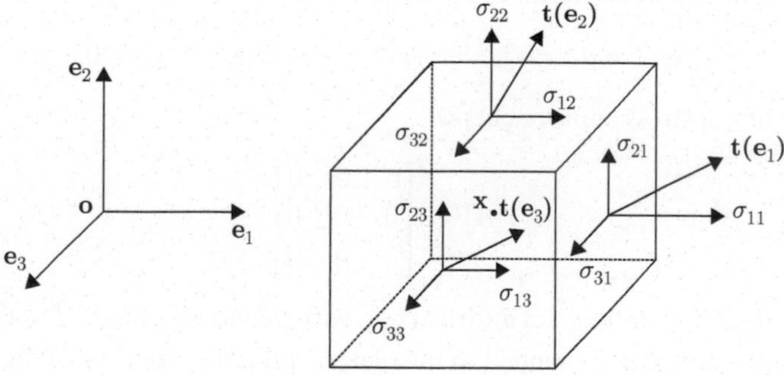

Fig. 4.8 Schematic of components of traction vectors on the faces of an infinitesimal cube surrounding a point $\mathbf{x}$ in the deformed body $\mathcal{B}_t$.

[10] Be careful! Some other texts reverse the order. In such texts the first subscript denotes the face normal, and the second subscript the direction of the component. The distinction does not really matter since the stress tensor is symmetric.

REMARK 4.2

The normal stress on an arbitrary plane with normal $\mathbf{n}$ is given by

$$\sigma_n = \mathbf{n} \cdot \boldsymbol{\sigma}\mathbf{n}.$$

This quantity is maximized by the eigenvector associated with the largest eigenvalue of $\boldsymbol{\sigma}$. The shear stress in the $\mathbf{m}$ direction on a plane with normal $\mathbf{n}$ is given by

$$\sigma_{mn} = \mathbf{m} \cdot \boldsymbol{\sigma}\mathbf{n},$$

wherein this quantity is maximized when $\mathbf{n} = (\boldsymbol{\mu}_1 + \boldsymbol{\mu}_3)/\sqrt{2}$ and $\mathbf{m} = (\boldsymbol{\mu}_1 - \boldsymbol{\mu}_3)/\sqrt{2}$, where $\boldsymbol{\mu}_1$ and $\boldsymbol{\mu}_3$ are, respectively, the eigenvectors associated to the algebraically largest and smallest eigenvalues of $\boldsymbol{\sigma}$.

4.5 Some simple states of stress

Next, we discuss some simple but important states of **homogeneous** stress, that is stress states $\boldsymbol{\sigma}$ which are *independent* of position $\mathbf{x}$ in the deformed configuration.

4.5.1 Pure tension or compression

In a homogeneous state of pure tension (or compression) with tensile stress σ in the $\mathbf{e}_1$-direction,

$$\boldsymbol{\sigma} = \sigma\, \mathbf{e}_1 \otimes \mathbf{e}_1.$$

The components of the stress tensor are then given by

$$\sigma_{11} = \sigma, \quad \sigma_{22} = \sigma_{33} = \sigma_{12} = \sigma_{13} = \sigma_{23} = 0,$$

and the matrix of the stress components is

$$[\boldsymbol{\sigma}] = \begin{bmatrix} \sigma & 0 & 0 \\ 0 & 0 & 0 \\ 0 & 0 & 0 \end{bmatrix}. \tag{4.5.1}$$

This gives the stress in a uniform cylindrical bar, with generators parallel to the $\mathbf{e}_1$-axis, under uniform forces per unit area σ, applied to the planar end cross-sections of the bar, Fig. 4.9. If σ is positive, then the bar is in tension, and if σ is negative, then the bar is in compression.

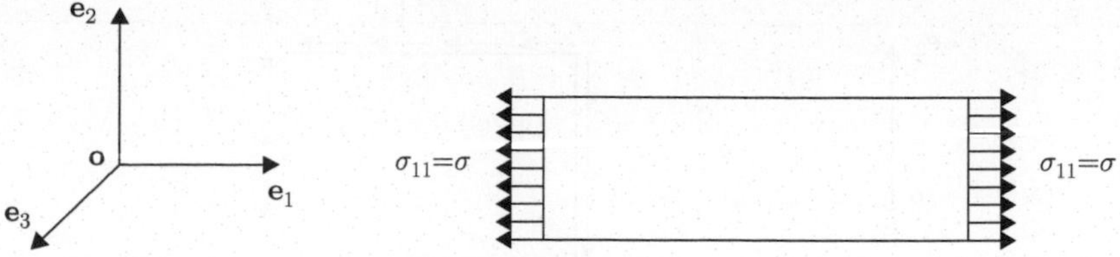

Fig. 4.9 Pure tension.

EXAMPLE 4.2 Traction components on a plane

For a bar in uniaxial tension, the traction vector on a plane which is at angle relative to the $\mathbf{e}_1$-direction, so that the normal to the plane is

$$\mathbf{n} = \cos\theta\mathbf{e}_1 + \sin\theta\mathbf{e}_2,$$

is given by

$$\mathbf{t} = \boldsymbol{\sigma}\mathbf{n} = \sigma(\cos\theta\mathbf{e}_1 + \sin\theta\mathbf{e}_2) = \sigma\,\cos\theta\mathbf{e}_1.$$

Thus on this surface there is a normal component

$$\sigma_n = \mathbf{t}\cdot\mathbf{n} = \mathbf{t}\cdot(\cos\theta\mathbf{e}_1 + \sin\theta\mathbf{e}_2) = \sigma\,\cos^2\theta,$$

and there is a shear component

$$\tau = \mathbf{t}\cdot(-\sin\theta\mathbf{e}_1 + \cos\theta\mathbf{e}_2) = -\sigma\,\sin\theta\cos\theta.$$

The shear component is maximum in absolute value for $\theta = \pm 45°$, and the normal component is maximum at $\theta = 0°$.

4.5.2 **Pure shear stress**

In a homogeneous state of pure shear stress τ in the $(\mathbf{e}_1, \mathbf{e}_2)$-plane, the stress tensor is given by

$$\boldsymbol{\sigma} = \tau\,(\mathbf{e}_1\otimes\mathbf{e}_2 + \mathbf{e}_2\otimes\mathbf{e}_1),$$

and the components of the stress tensor relative to an orthonormal basis $\{\mathbf{e}_i\}$ are,

$$\sigma_{12} = \sigma_{21} = \tau, \quad \sigma_{11} = \sigma_{22} = \sigma_{33} = \sigma_{13} = \sigma_{23} = 0,$$

or in matrix notation

$$[\sigma] = \begin{bmatrix} 0 & \tau & 0 \\ \tau & 0 & 0 \\ 0 & 0 & 0 \end{bmatrix}. \tag{4.5.2}$$

This stress state may occur, for example in laminar shear flow of a viscous fluid when the fluid flows in the $\mathbf{e}_1$-direction by shearing on planes x_2 = constant; cf. Fig. 4.10.

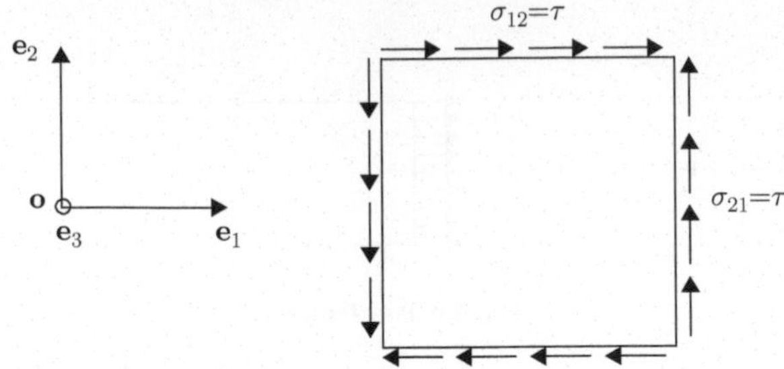

Fig. 4.10 Pure shear.

4.5.3 **Hydrostatic stress state**

In a hydrostatic stress state the stress tensor has the form

$$\boldsymbol{\sigma} = -p\mathbf{1},$$

and the components of the stress tensor relative to an orthonormal basis $\{\mathbf{e}_i\}$ are

$$\sigma_{11} = \sigma_{22} = \sigma_{33} = -p, \quad \text{and} \quad \sigma_{12} = \sigma_{13} = \sigma_{23} = 0.$$

In matrix notation

$$[\boldsymbol{\sigma}] = \begin{bmatrix} -p & 0 & 0 \\ 0 & -p & 0 \\ 0 & 0 & -p \end{bmatrix}. \tag{4.5.3}$$

This is the state of a stress in a fluid at rest. The scalar p is called the **pressure** of the fluid; cf. Fig. 4.11.

4.5.4 **Stress deviator**

Next we introduce the important concept of the **stress deviator** for an arbitrary state of stress at a point. First, the **mean normal stress** at a point is defined by

$$\frac{1}{3}\sigma_{kk}. \tag{4.5.4}$$

For example, (i) in a state of pure hydrostatic pressure the mean normal stress is $(1/3)\,\sigma_{kk} = -p$, (cf. (4.5.3)); (ii) in a state of pure tension $(1/3)\,\sigma_{kk} = (1/3)\sigma$ (cf. (4.5.1)); (iii) while in a state of pure shear the mean normal stress is $(1/3)\,\sigma_{kk} = 0$ (cf. (4.5.2)).

Corresponding to an arbitrary state of stress σ_{ij} we wish to define a state of stress in which the mean normal stress vanishes. Such a state of stress is called

- the **stress deviator**, and is defined by

$$\boldsymbol{\sigma}' \overset{\text{def}}{=} \boldsymbol{\sigma} - \frac{1}{3}(\operatorname{tr}\boldsymbol{\sigma})\mathbf{1}, \qquad \sigma'_{ij} \overset{\text{def}}{=} \sigma_{ij} - \frac{1}{3}\left(\sigma_{kk}\right)\delta_{ij}. \tag{4.5.5}$$

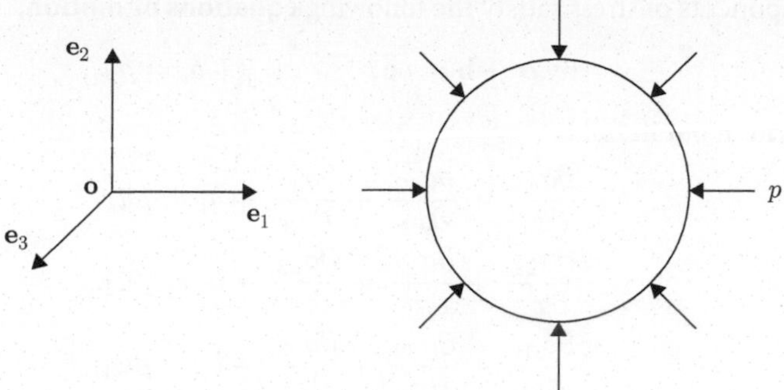

Fig. 4.11 Hydrostatic stress state.

The matrix of components of the stress deviator $\boldsymbol{\sigma}'$ is then given by

$$[\boldsymbol{\sigma}'] = \begin{bmatrix} \sigma'_{11} & \sigma'_{12} & \sigma'_{13} \\ \sigma'_{21} & \sigma'_{22} & \sigma'_{23} \\ \sigma'_{31} & \sigma'_{32} & \sigma'_{33} \end{bmatrix} = \begin{bmatrix} \sigma_{11} & \sigma_{12} & \sigma_{13} \\ \sigma_{21} & \sigma_{22} & \sigma_{23} \\ \sigma_{31} & \sigma_{32} & \sigma_{33} \end{bmatrix} - \frac{1}{3}\Big(\sigma_{11} + \sigma_{22} + \sigma_{33}\Big) \begin{bmatrix} 1 & 0 & 0 \\ 0 & 1 & 0 \\ 0 & 0 & 1 \end{bmatrix}. \tag{4.5.6}$$

Clearly,

$$\mathrm{tr}[\boldsymbol{\sigma}'] = \sigma'_{kk} = \Big(\sigma_{11} + \sigma_{22} + \sigma_{33}\Big) - \frac{1}{3}\Big(\sigma_{11} + \sigma_{22} + \sigma_{33}\Big) \times 3 = 0. \tag{4.5.7}$$

4.6 Summary of major concepts related to stress

Balance of linear and angular momentum requires that at each place $\mathbf{x}$ in the deformed body at a given time t there exists a **symmetric** stress tensor

$$\boldsymbol{\sigma} = \boldsymbol{\sigma}^{\top}, \qquad \sigma_{ij} = \sigma_{ji},$$

such that

1. For each oriented surface element with outward normal $\mathbf{n}$, the associated **surface traction vector** $\mathbf{t}(\mathbf{n})$ is given by Cauchy's law,

$$\mathbf{t}(\mathbf{n}) = \boldsymbol{\sigma}\mathbf{n}, \qquad t_i = \sigma_{ij} n_j, \tag{4.6.1}$$

which written out in full is

$$\begin{aligned} t_1 &= \sigma_{11}\, n_1 + \sigma_{12}\, n_2 + \sigma_{13}\, n_3, \\ t_2 &= \sigma_{21}\, n_1 + \sigma_{22}\, n_2 + \sigma_{23}\, n_3, \\ t_3 &= \sigma_{31}\, n_1 + \sigma_{32}\, n_2 + \sigma_{33}\, n_3. \end{aligned} \tag{4.6.2}$$

That is, if we know the six components of stress at a point, then we can determine the traction across any surface element passing through that point.

2. The components of stress satisfy the following **equations of motion,**

$$\operatorname{div}\boldsymbol{\sigma} + \mathbf{b} = \rho\ddot{\mathbf{u}}, \qquad \sigma_{ij,j} + b_i = \rho\ddot{u}_i, \tag{4.6.3}$$

which written out in full is

$$\begin{aligned}
\frac{\partial\sigma_{11}}{\partial x_1} + \frac{\partial\sigma_{12}}{\partial x_2} + \frac{\partial\sigma_{13}}{\partial x_3} + b_1 &= \rho\ddot{u}_1,\\
\frac{\partial\sigma_{21}}{\partial x_1} + \frac{\partial\sigma_{22}}{\partial x_2} + \frac{\partial\sigma_{23}}{\partial x_3} + b_2 &= \rho\ddot{u}_2,\\
\frac{\partial\sigma_{31}}{\partial x_1} + \frac{\partial\sigma_{32}}{\partial x_2} + \frac{\partial\sigma_{33}}{\partial x_3} + b_3 &= \rho\ddot{u}_3,
\end{aligned} \tag{4.6.4}$$

where b_i are the components of the body force field **b** per unit volume,[11] ρ is the mass density, and $\ddot{u}_i$ are the components of the acceleration $\ddot{\mathbf{u}}$, respectively, at position **x** and at time t.

3. Under **quasi-static** conditions inertial effects can be neglected, and under these conditions the equations of motion reduce to the following **equations of equilibrium**

$$\operatorname{div}\boldsymbol{\sigma} + \mathbf{b} = \mathbf{0}, \qquad \sigma_{ij,j} + b_i = 0, \tag{4.6.5}$$

or

$$\begin{aligned}
\frac{\partial\sigma_{11}}{\partial x_1} + \frac{\partial\sigma_{12}}{\partial x_2} + \frac{\partial\sigma_{13}}{\partial x_3} + b_1 &= 0,\\
\frac{\partial\sigma_{21}}{\partial x_1} + \frac{\partial\sigma_{22}}{\partial x_2} + \frac{\partial\sigma_{23}}{\partial x_3} + b_2 &= 0,\\
\frac{\partial\sigma_{31}}{\partial x_1} + \frac{\partial\sigma_{32}}{\partial x_2} + \frac{\partial\sigma_{33}}{\partial x_3} + b_3 &= 0.
\end{aligned} \tag{4.6.6}$$

Finally, the **stress deviator** corresponding to an arbitrary state of stress $\boldsymbol{\sigma}$ is defined by

$$\boldsymbol{\sigma}' \stackrel{\text{def}}{=} \boldsymbol{\sigma} - \frac{1}{3}(\operatorname{tr}\boldsymbol{\sigma})\mathbf{1}, \qquad \sigma'_{ij} \stackrel{\text{def}}{=} \sigma_{ij} - \frac{1}{3}(\sigma_{kk})\,\delta_{ij}. \tag{4.6.7}$$

[11] Recall that we have dropped the subscript 0 on the conventional body force.

5 Balance of energy and entropy imbalance

In addition to the laws of balance of mass and balance of forces and moments (linear and angular momentum), the response of a body is governed by

(i) the law of balance of energy, and

(ii) a law of imbalance of entropy.

These two laws are known as *the first and second laws of thermodynamics*. As in Chapter 4, we will first formulate these two laws globally for parts $\mathcal{P}_t$ that convect with the body, and then using the requirement that the underlying parts be arbitrary, we will derive local forms of the laws as partial differential equations.[1]

5.1 Balance of energy. First law of thermodynamics

The first law of thermodynamics represents a balance between the rate of change of the internal energy plus the rate of change of kinetic energy of $\mathcal{P}_t$, and the rate at which energy in the form of heat is transferred to $\mathcal{P}_t$ plus the mechanical power expended on $\mathcal{P}_t$.

Bearing Fig. 5.1 in mind, the first law is given a mathematical form as follows:

(i) The *net internal energy* $\mathcal{E}(\mathcal{P}_t)$ of $\mathcal{P}_t$ is taken to be given by

$$\mathcal{E}(\mathcal{P}_t) = \int_{\mathcal{P}_t} \rho \varepsilon_m \, dv, \tag{5.1.1}$$

where the scalar field $\varepsilon_m(\mathbf{x}, t)$ represents the internal energy measured per unit mass, or the *specific internal energy*.

(ii) The *kinetic energy* $\mathcal{K}(\mathcal{P}_t)$ of $\mathcal{P}_t$ is given by

$$\mathcal{K}(\mathcal{P}_t) \stackrel{\text{def}}{=} \int_{\mathcal{P}_t} \tfrac{1}{2}\rho |\dot{\mathbf{u}}|^2 \, dv, \tag{5.1.2}$$

with $\dot{\mathbf{u}}(\mathbf{x}, t)$ the velocity.

(iii) The *heat flow* $\mathcal{Q}(\mathcal{P}_t)$ into $\mathcal{P}_t$ is described by a vector *heat flux* $\mathbf{q}(\mathbf{x}, t)$ and a scalar *heat supply* $r(\mathbf{x}, t)$,

[1] A large fraction of this chapter is based on an exposition of these central topics in *The Mechanics and Thermodynamics of Continua* by Gurtin et al. (2010).

Continuum Mechanics of Solids. Lallit Anand and Sanjay Govindjee, Oxford University Press (2020).

DOI: 10.1093/oso/9780198864721.001.0001

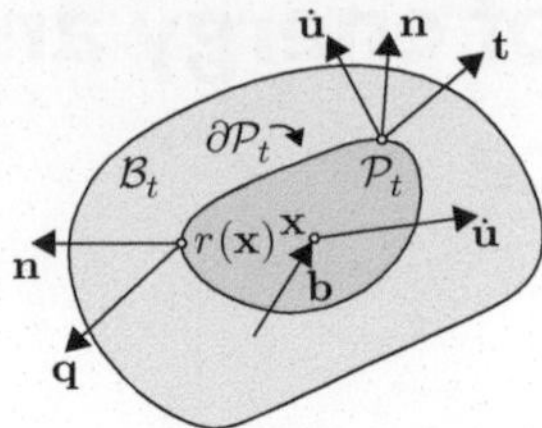

Fig. 5.1 Schematic figure for balance of energy for a part $\mathcal{P}_t$ with boundary $\partial\mathcal{P}_t$ and outward unit normal $\mathbf{n}$, energy or heat flux $\mathbf{q}$, volumetric heat supply r, subjected to surface tractions $\mathbf{t} = \boldsymbol{\sigma}\mathbf{n}$ and body forces $\mathbf{b}$, with a velocity field $\dot{\mathbf{u}}$.

$$\mathcal{Q}(\mathcal{P}_t) = -\int_{\partial\mathcal{P}_t} \mathbf{q}\cdot\mathbf{n}\, da + \int_{\mathcal{P}_t} r\, dv. \tag{5.1.3}$$

The term

$$-\int_{\partial\mathcal{P}_t} \mathbf{q}\cdot\mathbf{n}\, da$$

gives the rate at which heat is transferred *into* $\mathcal{P}_t$ across $\partial\mathcal{P}_t$; the minus sign is introduced because $\mathbf{n}$ is the *outward* unit normal to $\partial\mathcal{P}_t$. The term

$$\int_{\mathcal{P}_t} r\, dv$$

represents the rate at which heat is transferred to $\mathcal{P}_t$ by agencies *external* to $\mathcal{P}_t$, for example by radiation.

(iv) The *external power* $\mathcal{W}_{\rm ext}(\mathcal{P}_t)$ *expended* on $\mathcal{P}_t$ is

$$\mathcal{W}_{\rm ext}(\mathcal{P}_t) = \int_{\partial\mathcal{P}_t} \boldsymbol{\sigma}\mathbf{n}\cdot\dot{\mathbf{u}}\, da + \int_{\mathcal{P}_t} \mathbf{b}\cdot\dot{\mathbf{u}}\, dv, \tag{5.1.4}$$

with $\boldsymbol{\sigma}(\mathbf{x},t)\mathbf{n}(\mathbf{x},t)$ the traction exerted on $\partial\mathcal{P}_t$ and $\mathbf{b}(\mathbf{x},t)$ the non-inertial body force acting on $\mathcal{P}_t$.

Then, as stated earlier, the first law of thermodynamics represents a balance between the rate of change of the internal energy plus the rate of change of kinetic energy of $\mathcal{P}_t$, and the rate at which heat is transferred to $\mathcal{P}_t$ and power is expended on $\mathcal{P}_t$, which may be written symbolically as

$$\overline{\mathcal{E}(\mathcal{P}_t) + \mathcal{K}(\mathcal{P}_t)}^{\boldsymbol{\cdot}} = \mathcal{Q}(\mathcal{P}_t) + \mathcal{W}_{\rm ext}(\mathcal{P}_t). \tag{5.1.5}$$

Upon using (5.1.1), (5.1.2), (5.1.3), and (5.1.4) in (5.1.5) we obtain the **balance of energy** or the **first law of thermodynamics** in the form,

$$\boxed{\overline{\int_{\mathcal{P}_t} (\rho\varepsilon_m + \tfrac{1}{2}\rho\,|\dot{\mathbf{u}}|^2)\, dv}^{\boldsymbol{\cdot}} = -\int_{\partial\mathcal{P}_t} \mathbf{q}\cdot\mathbf{n}\, da + \int_{\mathcal{P}_t} r\, dv + \int_{\partial\mathcal{P}_t} \boldsymbol{\sigma}\mathbf{n}\cdot\dot{\mathbf{u}}\, da + \int_{\mathcal{P}_t} \mathbf{b}\cdot\dot{\mathbf{u}}\, dv.} \tag{5.1.6}$$

5.1.1 **Local form of balance of energy**

We next derive the local form of the first law of thermodynamics. Note that

$$\overline{\int_{\mathcal{P}_t} (\rho\varepsilon_m + \tfrac{1}{2}\,\rho\,|\dot{\mathbf{u}}|^2)\,dv}^{\;\centerdot} = \overline{\int_{\mathcal{P}} (\rho_{\mathrm{R}}\varepsilon_m + \tfrac{1}{2}\,\rho_{\mathrm{R}}\,|\dot{\mathbf{u}}|^2)\,dv_{\mathrm{R}}}^{\;\centerdot}$$

$$= \int_{\mathcal{P}} \big(\rho_{\mathrm{R}}\dot{\varepsilon}_m + \rho_{\mathrm{R}}\dot{\mathbf{u}}\cdot\ddot{\mathbf{u}}\big)\,dv_{\mathrm{R}}$$

$$= \int_{\mathcal{P}_t} \big(\rho\dot{\varepsilon}_m + \rho\dot{\mathbf{u}}\cdot\ddot{\mathbf{u}}\big)\,dv, \tag{5.1.7}$$

where we have exploited the fact that $J dv_{\mathrm{R}} = dv$, $J\rho = \rho_{\mathrm{R}}$, and ρ_{R} is independent of time. Use of the divergence theorem also gives

$$\int_{\partial\mathcal{P}_t} \mathbf{q}\cdot\mathbf{n}\,da = \int_{\mathcal{P}_t} \operatorname{div}\mathbf{q}\,dv. \tag{5.1.8}$$

Further, another use of the divergence theorem and the use of the symmetry of the Cauchy stress gives:

$$\int_{\partial\mathcal{P}_t} \boldsymbol{\sigma}\mathbf{n}\cdot\dot{\mathbf{u}}\,da = \int_{\partial\mathcal{P}_t} (\boldsymbol{\sigma}\dot{\mathbf{u}})\cdot\mathbf{n}\,da = \int_{\mathcal{P}_t} \operatorname{div}(\boldsymbol{\sigma}\dot{\mathbf{u}})\,dv$$

$$= \int_{\mathcal{P}_t} \operatorname{div}\boldsymbol{\sigma}\cdot\dot{\mathbf{u}}\,dv + \int_{\mathcal{P}_t} \boldsymbol{\sigma} : \operatorname{grad}\dot{\mathbf{u}}\,dv,$$

$$= \int_{\mathcal{P}_t} \operatorname{div}\boldsymbol{\sigma}\cdot\dot{\mathbf{u}}\,dv + \int_{\mathcal{P}_t} \boldsymbol{\sigma} : \operatorname{sym}[\operatorname{grad}\dot{\mathbf{u}}]\,dv. \tag{5.1.9}$$

Substituting (5.1.7)–(5.1.9) in (5.1.6) gives

$$\int_{\mathcal{P}_t} \Big(\rho\dot{\varepsilon}_m + \operatorname{div}\mathbf{q} - r - \boldsymbol{\sigma} : \operatorname{sym}[\operatorname{grad}\dot{\mathbf{u}}] - \big(\operatorname{div}\boldsymbol{\sigma} + \mathbf{b} - \rho\ddot{\mathbf{u}}\big)\cdot\dot{\mathbf{u}}\Big)\,dv = 0. \tag{5.1.10}$$

Using the localization theorem, cf. Sec. 2.4.1, and the balance of linear momentum, cf. (4.3.9), gives at every point in the body

$$\boxed{\rho\dot{\varepsilon}_m = \boldsymbol{\sigma} : \mathbf{D} - \operatorname{div}\mathbf{q} + r \qquad \rho\dot{\varepsilon}_m = \sigma_{ij}D_{ij} - \frac{\partial q_i}{\partial x_i} + r,} \tag{5.1.11}$$

where we have introduced the notation

$$\mathbf{D} \stackrel{\text{def}}{=} \operatorname{sym}[\operatorname{grad}\dot{\mathbf{u}}], \qquad D_{ij} = \tfrac{1}{2}(\dot{u}_{i,j} + \dot{u}_{j,i}), \tag{5.1.12}$$

for the *rate of deformation tensor.*

5.2 Entropy imbalance. Second law of thermodynamics

As with energy, parts $\mathcal{P}_t$ of a body possess *entropy*, and entropy is allowed to flow from one part to another, and into the part from the external world. But unlike energy, a part $\mathcal{P}_t$ of the body can *produce entropy*.

Bearing Fig. 5.2 in mind, we mathematize this idea as follows:

(i) The *net entropy* $\mathcal{S}(\mathcal{P}_t)$ of a part $\mathcal{P}_t$ is

$$\mathcal{S}(\mathcal{P}_t) = \int_{\mathcal{P}_t} \rho\eta_m \, dv, \tag{5.2.1}$$

where $\eta_m(\mathbf{x}, t)$ denotes the entropy per unit mass or the *specific entropy*.

(ii) The *entropy flow* $\mathcal{J}(\mathcal{P}_t)$ into $\mathcal{P}_t$ is,

$$\mathcal{J}(\mathcal{P}_t) = -\int_{\partial\mathcal{P}_t} \mathbf{j}\cdot\mathbf{n}\, da + \int_{\mathcal{P}_t} j\, dv, \tag{5.2.2}$$

where $\mathbf{j}$ is an *entropy flux* and j an *entropy supply*.

The basic premise that systems tend to increase their entropy manifests itself in the requirement that the rate at which the net entropy of $\mathcal{P}_t$ is changing is greater than or at a minimum equal to the entropy flow into $\mathcal{P}_t$,

$$\overline{\mathcal{S}(\mathcal{P}_t)}^{\,\cdot} \geq \mathcal{J}(\mathcal{P}_t). \tag{5.2.3}$$

Upon using (5.2.1) and (5.2.2) in the entropy imbalance (5.2.3) we obtain,

$$\overline{\int_{\mathcal{P}_t} \rho\eta_m \, dv}^{\,\cdot} \geq -\int_{\partial\mathcal{P}_t} \mathbf{j}\cdot\mathbf{n}\, da + \int_{\mathcal{P}_t} j\, dv. \tag{5.2.4}$$

A fundamental hypothesis of continuum thermodynamics relates entropy flow to heat flow and asserts that there is a scalar field

$$\vartheta > 0,$$

the (absolute) **temperature**, such that

$$\mathbf{j} = \frac{\mathbf{q}}{\vartheta} \quad \text{and} \quad j = \frac{r}{\vartheta}. \tag{5.2.5}$$

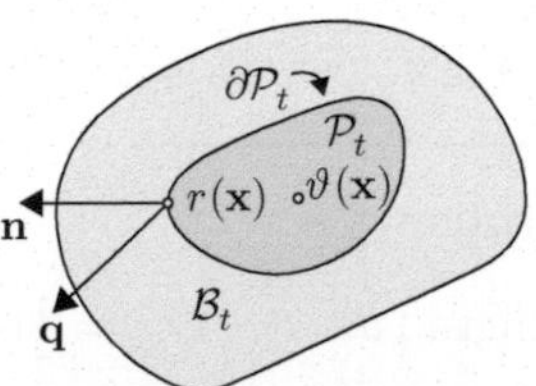

Fig. 5.2 Schematic figure for imbalance of entropy for a part $\mathcal{P}_t$ with boundary $\partial\mathcal{P}_t$ having a outward unit normal $\mathbf{n}$, subjected to an energy or heat flux $\mathbf{q}$, a volumetric heat supply r, and an absolute temperature field ϑ.

Using (5.2.5) we write the entropy imbalance (5.2.4) in the form

$$\boxed{\overline{\int_{\mathcal{P}_t} \rho\eta_m \, dv}^{\,\boldsymbol{\cdot}} \geq -\int_{\partial\mathcal{P}_t} \frac{\mathbf{q}}{\vartheta}\cdot\mathbf{n}\, da + \int_{\mathcal{P}_t} \frac{r}{\vartheta}\, dv,} \tag{5.2.6}$$

which is the **second law of thermodynamics**, and often referred to as the **Clausius–Duhem inequality**.

5.2.1 **Local form of entropy imbalance**

We next derive the local form of the second law of thermodynamics. First note that

$$\begin{aligned}\overline{\int_{\mathcal{P}_t} \rho\eta_m \, dv}^{\,\boldsymbol{\cdot}} &= \overline{\int_{\mathcal{P}} \rho_{\mathrm{R}}\eta_m \, dv_{\mathrm{R}}}^{\,\boldsymbol{\cdot}} \\ &= \int_{\mathcal{P}} \rho_{\mathrm{R}}\dot{\eta}_m \, dv_{\mathrm{R}} \\ &= \int_{\mathcal{P}_t} \rho\dot{\eta}_m \, dv.\end{aligned} \tag{5.2.7}$$

Employing the divergence theorem on the first term on the right-hand side of (5.2.6), (5.2.7) and the localization theorem then give the local form of the second law of thermodynamics,

$$\boxed{\rho\dot{\eta}_m \geq -\operatorname{div}\left(\frac{\mathbf{q}}{\vartheta}\right) + \frac{r}{\vartheta}, \qquad \rho\dot{\eta}_m \geq -\frac{\partial}{\partial x_i}\left(\frac{q_i}{\vartheta}\right) + \frac{r}{\vartheta}.} \tag{5.2.8}$$

5.3 **Free-energy imbalance. Dissipation**

By (5.1.11) and (5.2.8)

$$\begin{aligned}\rho(\vartheta\dot{\eta}_m - \dot{\varepsilon}_m) &\geq -\vartheta \operatorname{div}\left(\frac{\mathbf{q}}{\vartheta}\right) + \operatorname{div}\mathbf{q} - \boldsymbol{\sigma}:\mathbf{D} \\ &\geq \frac{1}{\vartheta}\mathbf{q}\cdot\operatorname{grad}\vartheta - \boldsymbol{\sigma}:\mathbf{D},\end{aligned}$$

or

$$\rho(\dot{\varepsilon}_m - \vartheta\dot{\eta}_m) - \boldsymbol{\sigma}:\mathbf{D} + \frac{1}{\vartheta}\mathbf{q}\cdot\operatorname{grad}\vartheta \leq 0.$$

Introducing now the (Helmholtz) *free energy*

$$\psi_m \stackrel{\text{def}}{=} \varepsilon_m - \vartheta\eta_m, \tag{5.3.1}$$

measured per unit mass, we are led to the local **free-energy imbalance**,

$$\boxed{\rho\dot{\psi}_m + \rho\eta_m\dot{\vartheta} - \boldsymbol{\sigma}:\mathbf{D} + \frac{1}{\vartheta}\mathbf{q}\cdot\operatorname{grad}\vartheta \leq 0,} \tag{5.3.2}$$

a result basic to much of what follows.

The free-energy imbalance may be alternatively written as

$$\boxed{\mathcal{D} \stackrel{\text{def}}{=} \boldsymbol{\sigma}:\mathbf{D} - \rho\eta_m\dot{\vartheta} - \frac{1}{\vartheta}\mathbf{q}\cdot\operatorname{grad}\vartheta - \rho\dot{\psi}_m \geq 0\,;} \tag{5.3.3}$$

equation (5.3.3) is known as the local **dissipation inequality**.

5.3.1 Free-energy imbalance and dissipation inequality for mechanical theories

For mechanical theories, that is for situations under which thermal influences are negligible and which are phrased under the assumption that $\vartheta(\mathbf{x}, t) = \text{constant}$, the free energy imbalance reduces to

$$\boxed{\rho\dot{\psi}_m - \boldsymbol{\sigma}:\mathbf{D} \leq 0,} \tag{5.3.4}$$

and correspondingly the dissipation inequality is

$$\boxed{\mathcal{D} \stackrel{\text{def}}{=} \boldsymbol{\sigma}:\mathbf{D} - \rho\dot{\psi}_m \geq 0.} \tag{5.3.5}$$

6 Balance laws for small deformations

The law for **balance of mass**, the **balance laws for forces and moments**, and the two laws of thermodynamics—the **balance of energy** and the **imbalance of entropy**, as formulated in the previous two chapters, hold for arbitrarily large deformations. However, for the major part of the book which follows, we will primarily be interested in the case of solids undergoing *small deformations*, and in this setting these basic laws take on simpler forms, which we discuss below.

6.1 Basic assumptions

Deformations are considered to be *small* when:

1. The displacement gradient is small everywhere in the body $\mathcal{B}$

$$|\mathbf{H}| = |\nabla\mathbf{u}| \ll 1, \tag{6.1.1}$$

   and that the strain-displacement relation

$$\boxed{\epsilon = \tfrac{1}{2}\big(\nabla\mathbf{u} + (\nabla\mathbf{u})^{\top}\big)} \tag{6.1.2}$$

   holds in $\mathcal{B}$.

2. Differentiation with respect to $\mathbf{x}$ is equivalent to differentiation with respect to $\mathbf{X}$, even though $\mathbf{x} = \boldsymbol{\chi}(\mathbf{X}, t)$. To see this, note that since

$$dx_i = F_{ij}dX_j = \underbrace{(\delta_{ij} + u_{i,j})}_{\approx\delta_{ij}}\,dX_j \approx dX_i$$

   for $|u_{i,j}| \ll 1$, it follows that

$$\frac{\partial}{\partial x_i} \approx \frac{\partial}{\partial X_i}.$$

   This allows us to consider all gradient and divergence operators that appear in the balance laws to be gradients and divergences with respect to the original (undeformed) locations of the material points.

3. The mass density in the deformed body is essentially the same as that in the reference body

$$\rho_{\mathrm{R}} = J\rho \approx \underbrace{(1 + \operatorname{div}\mathbf{u})}_{\approx 1}\,\rho \approx \rho, \tag{6.1.3}$$

   and is thus independent of time.

Continuum Mechanics of Solids. Lallit Anand and Sanjay Govindjee, Oxford University Press (2020).

DOI: 10.1093/oso/9780198864721.001.0001

To summarize, we consider a deformable medium occupying a (fixed) region $\mathcal{B}$, the body. For small deformations we will denote points in $\mathcal{B}$ by $\mathbf{x}$, and ∇ and div will denote the gradient and divergence with respect to $\mathbf{x}$. Further, for small deformations we use the following notation:

- $\rho(\mathbf{x}) > 0$ denotes the *constant* (time-independent) mass density of $\mathcal{B}$;
- $\mathbf{u}(\mathbf{x}, t)$ denotes the displacement field of $\mathcal{B}$;
- $\mathbf{H}(\mathbf{x}, t) = \nabla\mathbf{u}(\mathbf{x}, t)$ denotes the displacement gradient in $\mathcal{B}$, with $|\mathbf{H}| \ll 1$;
- $\boldsymbol{\epsilon} = \frac{1}{2}(\nabla\mathbf{u} + (\nabla\mathbf{u})^\top)$ denotes the symmetric small strain measure;
- $\boldsymbol{\sigma}(\mathbf{x}, t)$ denotes the (symmetric) stress;
- $\mathbf{b}(\mathbf{x}, t)$ denotes the (non-inertial) body force field on the body per unit volume in $\mathcal{B}$.
- $\varepsilon(\mathbf{x}, t) \stackrel{\text{def}}{=} \rho\varepsilon_m$ denotes the *internal energy* measured *per unit volume* in $\mathcal{B}$;
- $\eta(\mathbf{x}, t) \stackrel{\text{def}}{=} \rho\eta_m$ denotes the *entropy* measured *per unit volume* in $\mathcal{B}$;
- $\psi(\mathbf{x}, t) \stackrel{\text{def}}{=} \rho\psi_m$ denotes the *Helmholtz free energy* measured *per unit volume* in $\mathcal{B}$;
- $\vartheta(\mathbf{x}, t) > 0$ is the absolute temperature of $\mathcal{B}$;
- $\mathbf{q}(\mathbf{x}, t)$ denotes the *heat flux* measured *per unit area* in $\mathcal{B}$;
- $r(\mathbf{x}, t)$ denotes the *body heat supply* measured *per unit volume* in $\mathcal{B}$.

6.2 Local balance and imbalance laws for small deformations

Using this notation, for small deformations the balance laws for forces, moments, and energy, and the entropy imbalance, as well as the free-energy imbalance reduce to:

1. **Balance of forces and moments**: Balance of forces (linear momentum balance) becomes the requirement that

$$\boxed{\operatorname{div}\boldsymbol{\sigma} + \mathbf{b} = \rho\ddot{\mathbf{u}}, \qquad \frac{\partial\sigma_{ij}}{\partial x_j} + b_i = \rho\ddot{u}_i,} \tag{6.2.1}$$

hold in the reference body $\mathcal{B}$, and balance of moments (angular momentum balance) is the requirement that

$$\boldsymbol{\sigma}(\mathbf{x}, t) = \boldsymbol{\sigma}^\top(\mathbf{x}, t),$$

for all points $\mathbf{x} \in \mathcal{B}$.

2. **Energy balance**: The local form of the energy balance (5.1.11) reduces to

$$\boxed{\dot{\varepsilon} = \boldsymbol{\sigma} : \dot{\boldsymbol{\epsilon}} - \operatorname{div}\mathbf{q} + r, \qquad \dot{\varepsilon} = \sigma_{ij}\dot{\epsilon}_{ij} - \frac{\partial q_i}{\partial x_i} + r,} \tag{6.2.2}$$

where $\varepsilon \overset{\text{def}}{=} \rho\varepsilon_m$ is the internal energy per unit volume in the reference body, and we have exploited the fact that in this situation ρ is independent of time, and that for small deformations,

$$\mathbf{D} \equiv \dot{\boldsymbol{\epsilon}}.$$

3. **Entropy imbalance**: The second law of thermodynamics (5.2.8), **the entropy imbalance**, reduces to

$$\boxed{\dot{\eta} \geq -\operatorname{div}\left(\frac{\mathbf{q}}{\vartheta}\right) + \frac{r}{\vartheta}, \qquad \dot{\eta} \geq -\frac{\partial}{\partial x_i}\left(\frac{q_i}{\vartheta}\right) + \frac{r}{\vartheta},} \tag{6.2.3}$$

where $\eta \overset{\text{def}}{=} \rho\eta_m$ is the entropy density per unit reference volume. Note that in arriving at (6.2.3) we have used the fact that for small deformations ρ is independent of time.

4. **Free-energy imbalance. Dissipation inequality**

With $\psi \overset{\text{def}}{=} \rho\psi_m$ the free energy density per unit volume of $\mathcal{B}$, ρ independent of time, and $\mathbf{D} \equiv \dot{\boldsymbol{\epsilon}}$, the free-energy imbalance (5.3.2) takes on the form

$$\boxed{\dot{\psi} + \eta\dot{\vartheta} - \boldsymbol{\sigma} : \dot{\boldsymbol{\epsilon}} + \frac{1}{\vartheta}\mathbf{q} \cdot \nabla\vartheta \leq 0, \qquad \dot{\psi} + \eta\dot{\vartheta} - \sigma_{ij}\dot{\epsilon}_{ij} + \frac{1}{\vartheta}q_i\frac{\partial\vartheta}{\partial x_i} \leq 0,} \tag{6.2.4}$$

and correspondingly the dissipation inequality (5.3.3) becomes

$$\boxed{\mathcal{D} \overset{\text{def}}{=} \boldsymbol{\sigma} : \dot{\boldsymbol{\epsilon}} - \eta\dot{\vartheta} - \frac{1}{\vartheta}\mathbf{q} \cdot \nabla\vartheta - \dot{\psi} \geq 0, \qquad \mathcal{D} = \sigma_{ij}\dot{\epsilon}_{ij} - \eta\dot{\vartheta} - \frac{1}{\vartheta}q_i\frac{\partial\vartheta}{\partial x_i} - \dot{\psi} \geq 0.} \tag{6.2.5}$$

For mechanical theories, that is for situations where $\vartheta(\mathbf{x}, t) = \text{constant}$, the free-energy imbalance (6.2.4) reduces to

$$\boxed{\dot{\psi} - \boldsymbol{\sigma} : \dot{\boldsymbol{\epsilon}} \leq 0,} \tag{6.2.6}$$

and the dissipation inequality (6.2.5) becomes,

$$\boxed{\mathcal{D} \overset{\text{def}}{=} \boldsymbol{\sigma} : \dot{\boldsymbol{\epsilon}} - \dot{\psi} \geq 0.} \tag{6.2.7}$$

PART IV

Linear elasticity

7 Constitutive equations for linear elasticity

The concepts of kinematics and the balance and imbalance laws introduced apply to all solid bodies independent of their material constitution. *Also, the notions of stress and strain are applicable to all solid bodies.* However, physical experience tells us that when two bodies of the same size and shape, but made from different materials, are subjected to the same system of forces, then the shape change in general is different. For example, when two thin rods of the same length and diameter, one of steel and one of a polymer, are subjected to the same axial force, then the resulting elongations of the rods are different. We therefore introduce additional hypotheses,

- called **constitutive assumptions**, *which serve to distinguish different types of material behavior.*

The adjective "constitutive" is used because these are assumptions concerning the internal physical constitution of bodies. In this chapter we confine ourselves to a discussion of constitutive equations for materials which may be idealized to be **linearly elastic**.

7.1 Free-energy imbalance (second law): Elasticity

We restrict our attention to a purely mechanical theory based on a free-energy imbalance that represents the first two laws of thermodynamics under **isothermal** conditions. This imbalance requires that *the temporal increase in free energy of any part be less than or equal to the power expended on that part*. With ψ denoting the free energy per unit volume and $\boldsymbol{\sigma}:\dot{\boldsymbol{\epsilon}}$ representing the stress-power per unit volume, the local statement of this requirement is (cf. eq. (6.2.6))

$$\dot{\psi} \leq \boldsymbol{\sigma}:\dot{\boldsymbol{\epsilon}}. \tag{7.1.1}$$

Based on (7.1.1), the rate of dissipation per unit volume per unit time may be defined by

$$\mathcal{D} \stackrel{\text{def}}{=} \boldsymbol{\sigma}:\dot{\boldsymbol{\epsilon}} - \dot{\psi} \geq 0. \tag{7.1.2}$$

An *idealized* **elastic** *material* possesses the following characteristics:

- An elastic material is one in which the free energy ψ and stress $\boldsymbol{\sigma}$ depend only on the current value of the strain $\boldsymbol{\epsilon}$ in the body; they are independent of the past history of strain, as well as the rate at which the strain is changing with time. Accordingly, the constitutive equations for the free energy and stress are taken as

$$\left.\begin{aligned} \psi &= \hat{\psi}(\boldsymbol{\epsilon}), \\ \boldsymbol{\sigma} &= \hat{\boldsymbol{\sigma}}(\boldsymbol{\epsilon}). \end{aligned}\right\} \tag{7.1.3}$$

Continuum Mechanics of Solids. Lallit Anand and Sanjay Govindjee, Oxford University Press (2020).

DOI: 10.1093/oso/9780198864721.001.0001

- An elastic material **does not dissipate energy**. That is, for an elastic material

$$\mathcal{D} = \boldsymbol{\sigma} : \dot{\boldsymbol{\epsilon}} - \dot{\psi} = 0. \tag{7.1.4}$$

The requirement (7.1.4), when combined with the constitutive equations (7.1.3) is then equivalent to the requirement that

$$\left[\hat{\boldsymbol{\sigma}}(\boldsymbol{\epsilon}) - \frac{\partial \hat{\psi}(\boldsymbol{\epsilon})}{\partial \boldsymbol{\epsilon}} \right] : \dot{\boldsymbol{\epsilon}} = 0 \tag{7.1.5}$$

hold in all motions of the body. This requirement is satisfied provided the term within the brackets in (7.1.5) vanishes, or equivalently when the stress is given as the derivative of the free energy with respect to the strain:[1]

$$\boxed{\boldsymbol{\sigma} = \frac{\partial \hat{\psi}(\boldsymbol{\epsilon})}{\partial \boldsymbol{\epsilon}}, \qquad \sigma_{ij} = \frac{\partial \hat{\psi}(\boldsymbol{\epsilon})}{\partial \epsilon_{ij}}.} \tag{7.1.6}$$

An elastic material for which the stress $\boldsymbol{\sigma}$ is given as the derivative of a free energy with respect to the strain $\boldsymbol{\epsilon}$ is also known as a **hyperelastic** material.

For **linear elastic materials** we consider a Taylor series expansion of the free-energy function $\psi = \hat{\psi}(\boldsymbol{\epsilon})$ about $\boldsymbol{\epsilon} = \mathbf{0}$, up to terms which are *quadratic in strain*:

$$\psi = \hat{\psi}(\mathbf{0}) + \left.\frac{\partial \hat{\psi}(\boldsymbol{\epsilon})}{\partial \epsilon_{pq}}\right|_{\boldsymbol{\epsilon}=\mathbf{0}} \epsilon_{pq} + \frac{1}{2} \left.\frac{\partial^2 \hat{\psi}(\boldsymbol{\epsilon})}{\partial \epsilon_{pq} \partial \epsilon_{rs}}\right|_{\boldsymbol{\epsilon}=\mathbf{0}} \epsilon_{pq}\epsilon_{rs} + \cdots \tag{7.1.7}$$

We assume that when $\boldsymbol{\epsilon} = \mathbf{0}$, then so also is the free energy

$$\hat{\psi}(\mathbf{0}) = 0.$$

On account of $(7.1.3)_2$ and (7.1.6),

$$\left.\frac{\partial \hat{\psi}(\boldsymbol{\epsilon})}{\partial \epsilon_{pq}}\right|_{\boldsymbol{\epsilon}=\mathbf{0}} \equiv \hat{\boldsymbol{\sigma}}(\mathbf{0})$$

represents the *residual stress*, that is the stress when $\boldsymbol{\epsilon} = \mathbf{0}$. In what follows we assume that there is no residual stress. Under these assumptions the free-energy density ψ for a *linear elastic material* is quadratic in the strain $\boldsymbol{\epsilon}$, and given by

$$\boxed{\psi \equiv \frac{1}{2}\boldsymbol{\epsilon} : \mathbb{C}\boldsymbol{\epsilon} = \frac{1}{2} C_{pqrs}\epsilon_{pq}\epsilon_{rs},} \tag{7.1.8}$$

where $\mathbb{C}$ is the (fourth-order) **elasticity tensor** with components

$$C_{pqrs} \overset{\text{def}}{=} \left.\frac{\partial^2 \hat{\psi}(\boldsymbol{\epsilon})}{\partial \epsilon_{pq} \partial \epsilon_{rs}}\right|_{\boldsymbol{\epsilon}=\mathbf{0}}. \tag{7.1.9}$$

Note that since

$$\left.\frac{\partial^2 \hat{\psi}(\boldsymbol{\epsilon})}{\partial \epsilon_{pq} \partial \epsilon_{rs}}\right|_{\boldsymbol{\epsilon}=\mathbf{0}} = \left.\frac{\partial^2 \hat{\psi}(\boldsymbol{\epsilon})}{\partial \epsilon_{rs} \partial \epsilon_{pq}}\right|_{\boldsymbol{\epsilon}=\mathbf{0}}, \tag{7.1.10}$$

[1] The use of thermodynamics to restrict constitutive equations via this argument is generally referred to as the *Coleman–Noll procedure* (Coleman and Noll, 1963).

the elastic moduli C_{pqrs} possess the following *major symmetry*

$$C_{pqrs} = C_{rspq}. \tag{7.1.11}$$

On account of (7.1.8), and the physical requirement that the free energy is positive for all non-zero strains, we require that $\mathbb{C}$ is *positive definite*:

$$\mathbf{A}:\mathbb{C}\mathbf{A} > 0 \quad \text{for all symmetric tensors } \mathbf{A} \neq \mathbf{0}. \tag{7.1.12}$$

Next, on account of (7.1.6), the components of the stress $\boldsymbol{\sigma}$ are obtained as the derivative of the free-energy function (7.1.8) with respect to strain as follows:

$$\begin{aligned}\sigma_{ij} = \frac{\partial\hat{\psi}(\boldsymbol{\epsilon})}{\partial\epsilon_{ij}} &= \frac{1}{2}\Big(C_{pqrs}I^{\rm sym}_{pqij}\epsilon_{rs} + C_{pqrs}\epsilon_{pq}I^{\rm sym}_{rsij}\Big)\\ &= \frac{1}{4}\Big(C_{ijrs}\epsilon_{rs} + C_{jirs}\epsilon_{rs} + C_{pqij}\epsilon_{pq} + C_{pqji}\epsilon_{pq}\Big)\\ &= \frac{1}{2}\Big(C_{ijrs}\epsilon_{rs} + C_{ijpq}\epsilon_{pq}\Big) \qquad \text{[using (7.1.11) and (4.3.13)]}.\end{aligned}$$

Replacing the dummy indices pq and rs in the expression above by kl, we obtain the following classical relation for stress in terms of strain for a linear elastic material:

$$\boxed{\boldsymbol{\sigma} = \mathbb{C}\boldsymbol{\epsilon}, \qquad \sigma_{ij} = C_{ijkl}\epsilon_{kl}.} \tag{7.1.13}$$

The elastic constants C_{ijkl}, in addition to the major symmetries (cf., (7.1.11))

$$C_{ijkl} = C_{klij}, \tag{7.1.14}$$

also possess the *minor symmetries*

$$C_{ijkl} = C_{jikl} \qquad \text{and} \qquad C_{ijkl} = C_{ijlk} \tag{7.1.15}$$

on account of the symmetries of the stress ($\sigma_{ij} = \sigma_{ji}$) and the strain ($\epsilon_{kl} = \epsilon_{lk}$) tensors. Since the indices i, j, k, l each range from 1 to 3, there are $3^4 = 81$ elastic moduli C_{ijkl}. However, the major and minor symmetries, (7.1.14) and (7.1.15), respectively,

- *reduce the independent elastic constants for an arbitrarily anisotropic linear elastic material to 21.*

The elasticity tensor $\mathbb{C}$ is positive definite, and thus is invertible. Hence, the stress-strain relation (7.1.13) may be inverted to give

$$\boxed{\boldsymbol{\epsilon} = \mathbb{S}\boldsymbol{\sigma}, \qquad \epsilon_{ij} = S_{ijkl}\sigma_{kl}.} \tag{7.1.16}$$

The fourth-order tensor $\mathbb{S}$ is called the **compliance tensor.**

Next, recall that the free energy per unit volume for a linear elastic material is given by

$$\psi = \tfrac{1}{2}\boldsymbol{\epsilon}:\mathbb{C}\boldsymbol{\epsilon}. \tag{7.1.17}$$

Using the inverse relation (7.1.16), the free energy ψ may be expressed as a function of stress as follows:

$$
\begin{aligned}
\psi &= \tfrac{1}{2}\boldsymbol{\epsilon} : \mathbb{C}\boldsymbol{\epsilon}, \\
&= \tfrac{1}{2}(\mathbb{S}\boldsymbol{\sigma}) : \mathbb{C}(\mathbb{S}\boldsymbol{\sigma}), \\
&= \tfrac{1}{2}(\mathbb{S}\boldsymbol{\sigma}) : \underbrace{(\mathbb{C}\mathbb{S})}_{\equiv \mathbb{I}^{\rm sym}} \boldsymbol{\sigma},
\end{aligned}
$$

where $\mathbb{I}^{\rm sym}$ is the symmetric fourth-order identity tensor,

$$I^{\rm sym}_{ijkl} = \tfrac{1}{2}(\delta_{ik}\delta_{jl} + \delta_{il}\delta_{jk}). \tag{7.1.18}$$

Thus, in terms of the stress and the compliance tensor, the free energy is given by

$$\psi = \tfrac{1}{2}\boldsymbol{\sigma} : \mathbb{S}\boldsymbol{\sigma}, \tag{7.1.19}$$

from which, it is also clear that

$$\mathbb{S} = \frac{\partial^2 \psi}{\partial \boldsymbol{\sigma} \otimes \partial \boldsymbol{\sigma}}, \tag{7.1.20}$$

which implies that $\mathbb{S}$ has minor and major symmetries, and since on physical grounds the free energy is expected to be positive for all non-zero stresses, (7.1.19) implies that $\mathbb{S}$ is also positive definite. Thus, on account of (7.1.20), the components of the elastic compliance have the following symmetries:

$$S_{ijkl} = S_{klij}, \tag{7.1.21}$$

$$S_{ijkl} = S_{jikl}, \qquad S_{ijkl} = S_{ijlk}, \tag{7.1.22}$$

so that the number of independent elastic compliances for an arbitrary linear elastic material is also 21 in number.

7.1.1 Voigt single index notation

In what follows, for convenience, we write the component form of the stress-strain relation $\boldsymbol{\sigma} = \mathbb{C}\boldsymbol{\epsilon}$ with respect to an orthonormal basis $\{\mathbf{e}_i\}$ in the following matrix form:

$$
\begin{pmatrix} \sigma_{11} \\ \sigma_{22} \\ \sigma_{33} \\ \sigma_{23} \\ \sigma_{13} \\ \sigma_{12} \end{pmatrix}
=
\begin{pmatrix}
C_{1111} & C_{1122} & C_{1133} & C_{1123} & C_{1113} & C_{1112} \\
C_{2211} & C_{2222} & C_{2233} & C_{2223} & C_{2213} & C_{2212} \\
C_{3311} & C_{3322} & C_{3333} & C_{3323} & C_{3313} & C_{3312} \\
C_{2311} & C_{2322} & C_{2333} & C_{2323} & C_{2313} & C_{2312} \\
C_{1311} & C_{1322} & C_{1333} & C_{1323} & C_{1313} & C_{1312} \\
C_{1211} & C_{1222} & C_{1233} & C_{1223} & C_{1213} & C_{1212}
\end{pmatrix}
\begin{pmatrix} \epsilon_{11} \\ \epsilon_{22} \\ \epsilon_{33} \\ 2\epsilon_{23} \\ 2\epsilon_{13} \\ 2\epsilon_{12} \end{pmatrix};
\tag{7.1.23}
$$

also, the inverse relation $\boldsymbol{\epsilon} = \mathbb{S}\boldsymbol{\sigma}$ with respect to the same basis is written in matrix form as:

$$
\begin{pmatrix} \epsilon_{11} \\ \epsilon_{22} \\ \epsilon_{33} \\ 2\epsilon_{23} \\ 2\epsilon_{13} \\ 2\epsilon_{12} \end{pmatrix}
=
\begin{pmatrix}
S_{1111} & S_{1122} & S_{1133} & 2S_{1123} & 2S_{1113} & 2S_{1112} \\
S_{2211} & S_{2222} & S_{2233} & 2S_{2223} & 2S_{2213} & 2S_{2212} \\
S_{3311} & S_{3322} & S_{3333} & 2S_{3323} & 2S_{3313} & 2S_{3312} \\
2S_{2311} & 2S_{2322} & 2S_{2333} & 4S_{2323} & 4S_{2313} & 4S_{2312} \\
2S_{1311} & 2S_{1322} & 2S_{1333} & 4S_{1323} & 4S_{1313} & 4S_{1312} \\
2S_{1211} & 2S_{1222} & 2S_{1233} & 4S_{1223} & 4S_{1213} & 4S_{1212}
\end{pmatrix}
\begin{pmatrix} \sigma_{11} \\ \sigma_{22} \\ \sigma_{33} \\ \sigma_{23} \\ \sigma_{13} \\ \sigma_{12} \end{pmatrix}.
\tag{7.1.24}
$$

Expanding the summations in $\sigma_{ij} = C_{ijkl}\epsilon_{kl}$ and $\epsilon_{ij} = S_{ijkl}\sigma_{kl}$ and comparing to (7.1.23) and (7.1.24), respectively, shows the correctness of these matrix-vector relations.

Next, to make contact with the crystallography literature, we next introduce the Voigt "contracted notation" (Voigt, 1910) in which the components of stress and strain are written as

$$\begin{pmatrix}\sigma_1\\ \sigma_2\\ \sigma_3\\ \sigma_4\\ \sigma_5\\ \sigma_6\end{pmatrix} \stackrel{\text{def}}{=} \begin{pmatrix}\sigma_{11}\\ \sigma_{22}\\ \sigma_{33}\\ \sigma_{23}\\ \sigma_{13}\\ \sigma_{12}\end{pmatrix} \quad \text{and} \quad \begin{pmatrix}\epsilon_1\\ \epsilon_2\\ \epsilon_3\\ \epsilon_4\\ \epsilon_5\\ \epsilon_6\end{pmatrix} \stackrel{\text{def}}{=} \begin{pmatrix}\epsilon_{11}\\ \epsilon_{22}\\ \epsilon_{33}\\ 2\epsilon_{23}\\ 2\epsilon_{13}\\ 2\epsilon_{12}\end{pmatrix}, \tag{7.1.25}$$

respectively, the stiffness matrix in (7.1.23) is written as

$$\begin{pmatrix}\mathcal{C}_{11} & \mathcal{C}_{12} & \mathcal{C}_{13} & \mathcal{C}_{14} & \mathcal{C}_{15} & \mathcal{C}_{16}\\ \mathcal{C}_{21} & \mathcal{C}_{22} & \mathcal{C}_{23} & \mathcal{C}_{24} & \mathcal{C}_{25} & \mathcal{C}_{26}\\ \mathcal{C}_{31} & \mathcal{C}_{32} & \mathcal{C}_{33} & \mathcal{C}_{34} & \mathcal{C}_{35} & \mathcal{C}_{36}\\ \mathcal{C}_{41} & \mathcal{C}_{42} & \mathcal{C}_{43} & \mathcal{C}_{44} & \mathcal{C}_{45} & \mathcal{C}_{46}\\ \mathcal{C}_{51} & \mathcal{C}_{52} & \mathcal{C}_{53} & \mathcal{C}_{54} & \mathcal{C}_{55} & \mathcal{C}_{56}\\ \mathcal{C}_{61} & \mathcal{C}_{62} & \mathcal{C}_{63} & \mathcal{C}_{64} & \mathcal{C}_{65} & \mathcal{C}_{66}\end{pmatrix} \stackrel{\text{def}}{=} \begin{pmatrix}C_{1111} & C_{1122} & C_{1133} & C_{1123} & C_{1113} & C_{1112}\\ C_{2211} & C_{2222} & C_{2233} & C_{2223} & C_{2213} & C_{2212}\\ C_{3311} & C_{2233} & C_{3333} & C_{3323} & C_{3313} & C_{3312}\\ C_{2311} & C_{2322} & C_{2333} & C_{2323} & C_{2313} & C_{2312}\\ C_{1311} & C_{1322} & C_{1333} & C_{1323} & C_{1313} & C_{1312}\\ C_{1211} & C_{1222} & C_{1233} & C_{1223} & C_{1213} & C_{1212}\end{pmatrix}, \tag{7.1.26}$$

and the compliance matrix in (7.1.24) is written as

$$\begin{pmatrix}\mathcal{S}_{11} & \mathcal{S}_{12} & \mathcal{S}_{13} & \mathcal{S}_{14} & \mathcal{S}_{15} & \mathcal{S}_{16}\\ \mathcal{S}_{21} & \mathcal{S}_{22} & \mathcal{S}_{23} & \mathcal{S}_{24} & \mathcal{S}_{25} & \mathcal{S}_{26}\\ \mathcal{S}_{31} & \mathcal{S}_{32} & \mathcal{S}_{33} & \mathcal{S}_{34} & \mathcal{S}_{35} & \mathcal{S}_{36}\\ \mathcal{S}_{41} & \mathcal{S}_{42} & \mathcal{S}_{43} & \mathcal{S}_{44} & \mathcal{S}_{45} & \mathcal{S}_{46}\\ \mathcal{S}_{51} & \mathcal{S}_{52} & \mathcal{S}_{53} & \mathcal{S}_{54} & \mathcal{S}_{55} & \mathcal{S}_{56}\\ \mathcal{S}_{61} & \mathcal{S}_{62} & \mathcal{S}_{63} & \mathcal{S}_{64} & \mathcal{S}_{65} & \mathcal{S}_{66}\end{pmatrix} \stackrel{\text{def}}{=} \begin{pmatrix}S_{1111} & S_{1122} & S_{1133} & 2S_{1123} & 2S_{1113} & 2S_{1112}\\ S_{2211} & S_{2222} & S_{2233} & 2S_{2223} & 2S_{2213} & 2S_{2212}\\ S_{3311} & S_{2233} & S_{3333} & 2S_{3323} & 2S_{3313} & 2S_{3312}\\ 2S_{2311} & 2S_{2322} & 2S_{2333} & 4S_{2323} & 4S_{2313} & 4S_{2312}\\ 2S_{1311} & 2S_{1322} & 2S_{1333} & 4S_{1323} & 4S_{1313} & 4S_{1312}\\ 2S_{1211} & 2S_{1222} & 2S_{1233} & 4S_{1223} & 4S_{1213} & 4S_{1212}\end{pmatrix}. \tag{7.1.27}$$

Thus, in the Voigt contracted notation the stress-strain relation is

$$\begin{pmatrix}\sigma_1\\ \sigma_2\\ \sigma_3\\ \sigma_4\\ \sigma_5\\ \sigma_6\end{pmatrix} = \begin{pmatrix}\mathcal{C}_{11} & \mathcal{C}_{12} & \mathcal{C}_{13} & \mathcal{C}_{14} & \mathcal{C}_{15} & \mathcal{C}_{16}\\ \mathcal{C}_{21} & \mathcal{C}_{22} & \mathcal{C}_{23} & \mathcal{C}_{24} & \mathcal{C}_{25} & \mathcal{C}_{26}\\ \mathcal{C}_{31} & \mathcal{C}_{32} & \mathcal{C}_{33} & \mathcal{C}_{34} & \mathcal{C}_{35} & \mathcal{C}_{36}\\ \mathcal{C}_{41} & \mathcal{C}_{42} & \mathcal{C}_{43} & \mathcal{C}_{44} & \mathcal{C}_{45} & \mathcal{C}_{46}\\ \mathcal{C}_{51} & \mathcal{C}_{52} & \mathcal{C}_{53} & \mathcal{C}_{54} & \mathcal{C}_{55} & \mathcal{C}_{56}\\ \mathcal{C}_{61} & \mathcal{C}_{62} & \mathcal{C}_{63} & \mathcal{C}_{64} & \mathcal{C}_{65} & \mathcal{C}_{66}\end{pmatrix} \begin{pmatrix}\epsilon_1\\ \epsilon_2\\ \epsilon_3\\ \epsilon_4\\ \epsilon_5\\ \epsilon_6\end{pmatrix}, \tag{7.1.28}$$

or in index notation with summation convention as

$$\sigma_I = \mathcal{C}_{IJ}\epsilon_J, \qquad \mathcal{C}_{IJ} = \mathcal{C}_{JI}, \qquad I, J = 1, \ldots, 6, \tag{7.1.29}$$

and correspondingly, the inverse relation is

$$\begin{pmatrix}\epsilon_1\\ \epsilon_2\\ \epsilon_3\\ \epsilon_4\\ \epsilon_5\\ \epsilon_6\end{pmatrix} = \begin{pmatrix}\mathcal{S}_{11} & \mathcal{S}_{12} & \mathcal{S}_{13} & \mathcal{S}_{14} & \mathcal{S}_{15} & \mathcal{S}_{16}\\ \mathcal{S}_{21} & \mathcal{S}_{22} & \mathcal{S}_{23} & \mathcal{S}_{24} & \mathcal{S}_{25} & \mathcal{S}_{26}\\ \mathcal{S}_{31} & \mathcal{S}_{32} & \mathcal{S}_{33} & \mathcal{S}_{34} & \mathcal{S}_{35} & \mathcal{S}_{36}\\ \mathcal{S}_{41} & \mathcal{S}_{42} & \mathcal{S}_{43} & \mathcal{S}_{44} & \mathcal{S}_{45} & \mathcal{S}_{46}\\ \mathcal{S}_{51} & \mathcal{S}_{52} & \mathcal{S}_{53} & \mathcal{S}_{54} & \mathcal{S}_{55} & \mathcal{S}_{56}\\ \mathcal{S}_{61} & \mathcal{S}_{62} & \mathcal{S}_{63} & \mathcal{S}_{64} & \mathcal{S}_{65} & \mathcal{S}_{66}\end{pmatrix} \begin{pmatrix}\sigma_1\\ \sigma_2\\ \sigma_3\\ \sigma_4\\ \sigma_5\\ \sigma_6\end{pmatrix}, \tag{7.1.30}$$

or in index notation as

$$\epsilon_I = \mathcal{S}_{IJ}\sigma_J, \qquad \mathcal{S}_{IJ} = \mathcal{S}_{JI}, \qquad I, J = 1, \ldots, 6. \tag{7.1.31}$$

7.2 Material symmetry

Most solids exhibit symmetry properties with respect to certain rotations of the body, or reflection about one or more planes. These symmetries arise from the local microstructure of materials. For example, in a material with uni-directional reinforcing fibers there is a reflective plane of symmetry that is orthogonal to the fiber direction. The physical manifestation of this symmetry is that the stresses corresponding to a given state of strain will be the same whether or not one reflects the material about the symmetry plane. *Stated another way, this means that the components of the elasticity tensor do not change when expressed in two bases related by a material's symmetry transformation.* The effect of these (material) symmetries is to reduce the number of independent elastic constants from the number 21 for the most general anisotropic material. We discuss notions of material symmetry in this section on the basis of the invariance of the components of the elasticity tensor under changes of coordinate basis.

Let $\{\mathbf{e}_i\}$ denote an orthonormal basis, and consider a transformation of this basis to a new basis $\{\mathbf{e}_i^*\}$ given by

$$\mathbf{e}_i^* = \mathbf{Q}^\top \mathbf{e}_i \qquad i = 1, 2, 3, \tag{7.2.1}$$

where

$$\mathbf{Q} = \mathbf{e}_k \otimes \mathbf{e}_k^*. \tag{7.2.2}$$

The components of $\mathbf{Q}$ with respect to $\{\mathbf{e}_i\}$ are given by

$$Q_{ij} = \mathbf{e}_i \cdot (\mathbf{e}_k \otimes \mathbf{e}_k^*)\mathbf{e}_j = \delta_{ik}\mathbf{e}_k^* \cdot \mathbf{e}_j,$$

or

$$Q_{ij} = \mathbf{e}_i^* \cdot \mathbf{e}_j. \tag{7.2.3}$$

Hence,

$$\mathbf{e}_i^* = Q_{ij}\mathbf{e}_j, \tag{7.2.4}$$

and the matrix representation of the components of $\mathbf{Q}$ with respect to the basis $\{\mathbf{e}_i\}$ is

$$[\mathbf{Q}] = \begin{bmatrix} Q_{11} & Q_{12} & Q_{13} \\ Q_{21} & Q_{22} & Q_{23} \\ Q_{31} & Q_{32} & Q_{33} \end{bmatrix}.$$

If $[\mathbf{Q}]$ is proper orthogonal, then the two bases are related by a rotation, if it is improper orthogonal, then they are related by a reflection. As discussed in Sec. 1.2.19, under a change in basis the components of a fourth-order tensor transform as (cf. (1.2.63)),

$$C_{ijkl}^* = Q_{ip}Q_{jq}Q_{kr}Q_{ls}C_{pqrs}. \tag{7.2.5}$$

If after a specific transformation $\mathbf{Q}$ a value C_{ijkl}^* is identically equal to the value C_{ijkl} before the transformation, that is

$$C_{ijkl} = C_{ijkl}^*,$$

then $\mathbf{Q}$ represents a **symmetry transformation** for the material. Thus $\mathbf{Q}$ represents a symmetry transformation if

$$\boxed{C_{ijkl} = Q_{ip}Q_{jq}Q_{kr}Q_{ls}C_{pqrs}.} \tag{7.2.6}$$

A necessary outcome of material symmetry relation (7.2.6) is a reduction in the number of independent elastic constants.

A notational convention that can aid in the application of (7.2.6) is to observe that the components of the elasticity tensor can be expressed as

$$C_{ijkl} = \mathbb{C}(\mathbf{e}_i, \mathbf{e}_j, \mathbf{e}_k, \mathbf{e}_l) \stackrel{\text{def}}{=} (\mathbf{e}_i \otimes \mathbf{e}_j) : \mathbb{C}(\mathbf{e}_k \otimes \mathbf{e}_l),$$

where $\mathbb{C}$ depends linearly on each of its four arguments. Thus the condition (7.2.6) is equivalent to the requirement

$$\mathbb{C}(\mathbf{e}_i^*, \mathbf{e}_j^*, \mathbf{e}_k^*, \mathbf{e}_l^*) = \mathbb{C}(\mathbf{e}_i, \mathbf{e}_j, \mathbf{e}_k, \mathbf{e}_l). \tag{7.2.7}$$

EXAMPLE 7.1 n-Fold rotational symmetry transformation

For an n-fold rotational symmetry ($n = 2, 3, 4, \ldots$) about the $\mathbf{e}_3$-axis, the transformation matrix $[\mathbf{Q}]$ is given by

$$[\mathbf{Q}] = \begin{bmatrix} \cos\dfrac{2\pi}{n} & \sin\dfrac{2\pi}{n} & 0 \\ -\sin\dfrac{2\pi}{n} & \cos\dfrac{2\pi}{n} & 0 \\ 0 & 0 & 1 \end{bmatrix}.$$

EXAMPLE 7.2 Reflective symmetry transformation

For a reflection symmetry across the (x_1, x_2)-plane with normal $\mathbf{e}_3$, the transformation matrix $[\mathbf{Q}]$ is given by

$$[\mathbf{Q}] = \begin{bmatrix} 1 & 0 & 0 \\ 0 & 1 & 0 \\ 0 & 0 & -1 \end{bmatrix}.$$

7.3 **Forms of the elasticity and compliance tensors for some anisotropic linear elastic materials**

The most general anisotropic linear elastic materials are known as triclinic materials. Figure 7.1 shows the unit cell of a triclinic crystal. It is defined by the situation where all the lattice spacings and angles differ from one another: $a \neq b \neq c$ and $\alpha \neq \beta \neq \gamma$. In this situation there are no material symmetries and one necessarily has 21 independent elastic constants. Example materials that are triclinic include Ta_2O_5 and $CaSiO_3$ crystals, among many other compounds.

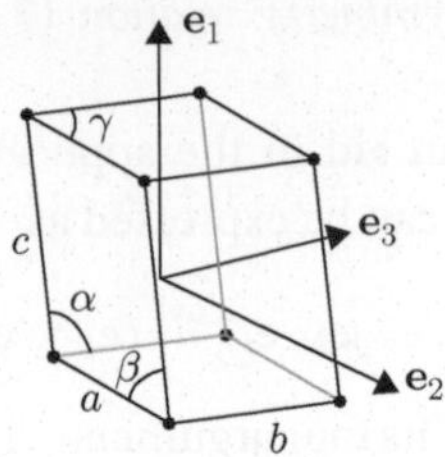

Fig. 7.1 A unit cell of a triclinic crystal. $a \neq b \neq c$, $\alpha \neq \beta \neq \gamma$.

7.3.1 **Monoclinic symmetry**

A material is said to possess **monoclinic** symmetry if it has one plane of reflection symmetry. Fig. 7.2 shows the unit cell of a monoclinic crystal. With respect to Fig. 7.2, consider an orthonormal basis $\{\mathbf{e}_1, \mathbf{e}_2, \mathbf{e}_3\}$ such that $\mathbf{e}_3$ is normal to the plane of reflection symmetry for the monoclinic crystal. Let $\{\mathbf{e}_1^*, \mathbf{e}_2^*, \mathbf{e}_3^*\}$ be a second basis such that

$$\mathbf{e}_1^* = \mathbf{e}_1, \quad \mathbf{e}_2^* = \mathbf{e}_2, \quad \mathbf{e}_3^* = -\mathbf{e}_3.$$

The corresponding orthogonal matrix $[\mathbf{Q}]$ of the components $Q_{ij} = \mathbf{e}_i^* \cdot \mathbf{e}_j$ is

$$[\mathbf{Q}] = \begin{bmatrix} 1 & 0 & 0 \\ 0 & 1 & 0 \\ 0 & 0 & -1 \end{bmatrix}. \tag{7.3.1}$$

Note that this is an improper orthogonal transformation, because $\det \mathbf{Q} = -1$. Recalling (7.2.7), one has that

$$\mathbb{C}(\cdot,\cdot,\cdot,\mathbf{e}_3^*) = \mathbb{C}(\cdot,\cdot,\cdot,\mathbf{e}_3) \Longrightarrow \mathbb{C}(\cdot,\cdot,\cdot,-\mathbf{e}_3) = \mathbb{C}(\cdot,\cdot,\cdot,\mathbf{e}_3).$$

This implies for example that $-C_{1123} = C_{1123} = 0$, since the only number equal to its negative is zero. From this reasoning we conclude that all components of C_{ijkl} which have the subscript 3 appearing an **odd number of times**, vanish. Thus for a **monoclinic** crystal/material we may write the stress-strain relation as

$$\begin{pmatrix} \sigma_{11} \\ \sigma_{22} \\ \sigma_{33} \\ \sigma_{23} \\ \sigma_{13} \\ \sigma_{12} \end{pmatrix} = \begin{pmatrix} C_{1111} & C_{1122} & C_{1133} & 0 & 0 & C_{1112} \\ C_{1122} & C_{2222} & C_{2233} & 0 & 0 & C_{2212} \\ C_{1133} & C_{2233} & C_{3333} & 0 & 0 & C_{3312} \\ 0 & 0 & 0 & C_{2323} & C_{2313} & 0 \\ 0 & 0 & 0 & C_{2313} & C_{1313} & 0 \\ C_{1112} & C_{2212} & C_{3312} & 0 & 0 & C_{1212} \end{pmatrix} \begin{pmatrix} \epsilon_{11} \\ \epsilon_{22} \\ \epsilon_{33} \\ 2\epsilon_{23} \\ 2\epsilon_{13} \\ 2\epsilon_{12} \end{pmatrix}, \tag{7.3.2}$$

when the $\mathbf{e}_3$-direction is orthogonal to the plane of reflective symmetry. Equivalently in the Voigt notation we have

$$\begin{pmatrix} \sigma_1 \\ \sigma_2 \\ \sigma_3 \\ \sigma_4 \\ \sigma_5 \\ \sigma_6 \end{pmatrix} = \begin{pmatrix} C_{11} & C_{12} & C_{13} & 0 & 0 & C_{16} \\ C_{12} & C_{22} & C_{23} & 0 & 0 & C_{26} \\ C_{13} & C_{23} & C_{33} & 0 & 0 & C_{36} \\ 0 & 0 & 0 & C_{44} & C_{45} & 0 \\ 0 & 0 & 0 & C_{45} & C_{55} & 0 \\ C_{16} & C_{26} & C_{36} & 0 & 0 & C_{66} \end{pmatrix} \begin{pmatrix} \epsilon_1 \\ \epsilon_2 \\ \epsilon_3 \\ \epsilon_4 \\ \epsilon_5 \\ \epsilon_6 \end{pmatrix}. \tag{7.3.3}$$

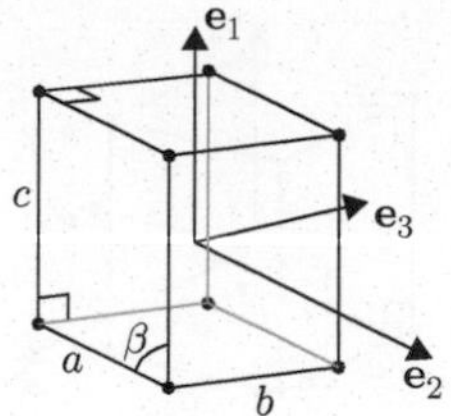

Fig. 7.2 A unit cell of a monoclinic crystal. $a \neq b \neq c$, $\beta \neq \pi/2 (\equiv 90°)$.

A monoclinic material has **13 independent elastic constants:**

$$\{C_{11}, C_{22}, C_{33}, C_{44}, C_{55}, C_{66}, C_{12}, C_{13}, C_{16}, C_{23}, C_{26}, C_{36}, C_{45}\}.$$

Materials that can display monoclinic symmetry include ZrO_2 crystals, as well as many shape memory alloys, e.g. the all-important Ni-Ti alloys.

7.3.2 Orthorhombic or orthotropic symmetry

A material is said to have **orthorhombic** or **orthotropic** symmetry if it has three mutually orthogonal planes of reflection symmetry. Figure 7.3 shows the unit cell of an orthorhombic crystal, where all the lattice angles are $\pi/2$ but the lattice spacings are all unequal, $a \neq b \neq c$. With reference to Fig. 7.3, consider a basis $\{\mathbf{e}_1, \mathbf{e}_2, \mathbf{e}_3\}$ such that the $\mathbf{e}_i$ are normal to the planes of symmetry. Then, reflection symmetry across planes normal to $\mathbf{e}_i$ requires that (7.2.6) holds for the three reflection transformations

$$[\mathbf{Q}] = \begin{bmatrix} 1 & 0 & 0 \\ 0 & 1 & 0 \\ 0 & 0 & -1 \end{bmatrix}, \qquad [\mathbf{Q}] = \begin{bmatrix} 1 & 0 & 0 \\ 0 & -1 & 0 \\ 0 & 0 & 1 \end{bmatrix}, \qquad [\mathbf{Q}] = \begin{bmatrix} -1 & 0 & 0 \\ 0 & 1 & 0 \\ 0 & 0 & 1 \end{bmatrix}. \tag{7.3.4}$$

Arguing as in the case for monoclinic symmetry, a direct consequence of this requirement is that

- all components of C_{ijkl} which have the subscript 1, 2, or 3 appearing an **odd number of times**, vanish.

Thus, for an **orthorhombic** or **orthotropic** material we may write the stress-strain relation as

$$\begin{pmatrix} \sigma_{11} \\ \sigma_{22} \\ \sigma_{33} \\ \sigma_{23} \\ \sigma_{13} \\ \sigma_{12} \end{pmatrix} = \begin{pmatrix} C_{1111} & C_{1122} & C_{1133} & 0 & 0 & 0 \\ C_{1122} & C_{2222} & C_{2233} & 0 & 0 & 0 \\ C_{1133} & C_{2233} & C_{3333} & 0 & 0 & 0 \\ 0 & 0 & 0 & C_{2323} & 0 & 0 \\ 0 & 0 & 0 & 0 & C_{1313} & 0 \\ 0 & 0 & 0 & 0 & 0 & C_{1212} \end{pmatrix} \begin{pmatrix} \epsilon_{11} \\ \epsilon_{22} \\ \epsilon_{33} \\ 2\epsilon_{23} \\ 2\epsilon_{13} \\ 2\epsilon_{12} \end{pmatrix}, \tag{7.3.5}$$

when expressed relative to a basis aligned with the orthotropic directions. Equivalently in the Voigt notation

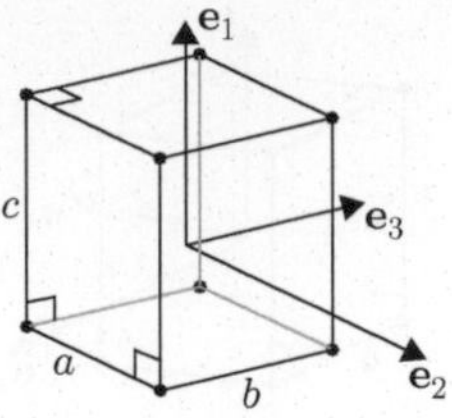

Fig. 7.3 A unit cell of an orthorhombic crystal, $a \neq b \neq c$.

$$\begin{pmatrix} \sigma_1 \\ \sigma_2 \\ \sigma_3 \\ \sigma_4 \\ \sigma_5 \\ \sigma_6 \end{pmatrix} = \begin{pmatrix} C_{11} & C_{12} & C_{13} & 0 & 0 & 0 \\ C_{12} & C_{22} & C_{23} & 0 & 0 & 0 \\ C_{13} & C_{23} & C_{33} & 0 & 0 & 0 \\ 0 & 0 & 0 & C_{44} & 0 & 0 \\ 0 & 0 & 0 & 0 & C_{55} & 0 \\ 0 & 0 & 0 & 0 & 0 & C_{66} \end{pmatrix} \begin{pmatrix} \epsilon_1 \\ \epsilon_2 \\ \epsilon_3 \\ \epsilon_4 \\ \epsilon_5 \\ \epsilon_6 \end{pmatrix}; \tag{7.3.6}$$

an orthotropic material has **9 independent elastic constants**

$$\{C_{11}, C_{22}, C_{33}, C_{44}, C_{55}, C_{66}, C_{12}, C_{13}, C_{23}\}.$$

Many composite materials display orthotropic symmetries as do Cu-Al-Ni alloys which display superelastic properties.

7.3.3 **Tetragonal symmetry**

A material is said to have **tetragonal symmetry** if it has three mutually orthogonal planes of reflection symmetry, plus an additional symmetry with respect to a 90° rotation about an axis. Figure 7.4 shows the unit cell of a tetragonal crystal, where all the lattice angles are $\pi/2$ and two of the lattice spacings are equal and dissimilar from the third. Thus, consider a basis $\{\mathbf{e}_1, \mathbf{e}_2, \mathbf{e}_3\}$ such that the $\mathbf{e}_i$ are normal to the planes of symmetry, with $\mathbf{e}_3$ being the axis about which there is a fourfold rotational symmetry. In this case we require that

$$C_{ijkl} = Q_{ip}Q_{jq}Q_{kr}Q_{ls}C_{pqrs} \tag{7.3.7}$$

hold for

$$[\mathbf{Q}] = \begin{bmatrix} 1 & 0 & 0 \\ 0 & 1 & 0 \\ 0 & 0 & -1 \end{bmatrix}, \qquad [\mathbf{Q}] = \begin{bmatrix} 1 & 0 & 0 \\ 0 & -1 & 0 \\ 0 & 0 & 1 \end{bmatrix}, \qquad [\mathbf{Q}] = \begin{bmatrix} -1 & 0 & 0 \\ 0 & 1 & 0 \\ 0 & 0 & 1 \end{bmatrix}, \tag{7.3.8}$$

and

$$[\mathbf{Q}] = \begin{bmatrix} 0 & 1 & 0 \\ -1 & 0 & 0 \\ 0 & 0 & 1 \end{bmatrix}, \tag{7.3.9}$$

which describes a 90° rotation about the $\mathbf{e}_3$-axis. Therefore, the elastic constants are subjected to the same restrictions as an orthotropic material, plus (7.3.9). Hence the restrictions on the C_{ijkl} from (7.3.8) lead to (7.3.5), viz.,

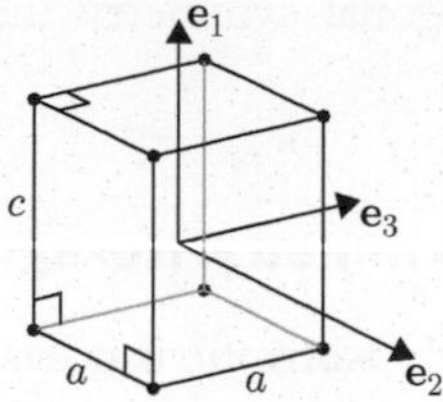

Fig. 7.4 A unit cell of a tetragonal crystal, $a \neq c$.

$$\begin{pmatrix}\sigma_{11}\\ \sigma_{22}\\ \sigma_{33}\\ \sigma_{23}\\ \sigma_{13}\\ \sigma_{12}\end{pmatrix} = \begin{pmatrix} C_{1111} & C_{1122} & C_{1133} & 0 & 0 & 0\\ C_{1122} & C_{2222} & C_{2233} & 0 & 0 & 0\\ C_{1133} & C_{2233} & C_{3333} & 0 & 0 & 0\\ 0 & 0 & 0 & C_{2323} & 0 & 0\\ 0 & 0 & 0 & 0 & C_{1313} & 0\\ 0 & 0 & 0 & 0 & 0 & C_{1212}\end{pmatrix}\begin{pmatrix}\epsilon_{11}\\ \epsilon_{22}\\ \epsilon_{33}\\ 2\epsilon_{23}\\ 2\epsilon_{13}\\ 2\epsilon_{12}\end{pmatrix},$$

and the additional requirements due to (7.3.9). In particular (7.3.9) indicates that $\mathbf{e}_1^* = \mathbf{e}_2$ and $\mathbf{e}_2^* = -\mathbf{e}_1$. Thus employing (7.2.7)

$$\begin{aligned} C_{1111} &= \mathbb{C}(\mathbf{e}_1, \mathbf{e}_1, \mathbf{e}_1, \mathbf{e}_1) = \mathbb{C}(\mathbf{e}_1^*, \mathbf{e}_1^*, \mathbf{e}_1^*, \mathbf{e}_1^*) = \mathbb{C}(\mathbf{e}_2, \mathbf{e}_2, \mathbf{e}_2, \mathbf{e}_2) = C_{2222},\\ C_{1133} &= \mathbb{C}(\mathbf{e}_1, \mathbf{e}_1, \mathbf{e}_3, \mathbf{e}_3) = \mathbb{C}(\mathbf{e}_1^*, \mathbf{e}_1^*, \mathbf{e}_3^*, \mathbf{e}_3^*) = \mathbb{C}(\mathbf{e}_2, \mathbf{e}_2, \mathbf{e}_3, \mathbf{e}_3) = C_{2233},\\ C_{2323} &= \mathbb{C}(\mathbf{e}_2, \mathbf{e}_3, \mathbf{e}_2, \mathbf{e}_3) = \mathbb{C}(\mathbf{e}_2^*, \mathbf{e}_3^*, \mathbf{e}_2^*, \mathbf{e}_3^*) = \mathbb{C}(-\mathbf{e}_1, \mathbf{e}_3, -\mathbf{e}_1, \mathbf{e}_3) = C_{1313}. \end{aligned}$$

So for a material with **tetragonal symmetry** the stress-strain relation may be written as

$$\begin{pmatrix}\sigma_{11}\\ \sigma_{22}\\ \sigma_{33}\\ \sigma_{23}\\ \sigma_{13}\\ \sigma_{12}\end{pmatrix} = \begin{pmatrix} C_{1111} & C_{1122} & C_{1133} & 0 & 0 & 0\\ C_{1122} & C_{1111} & C_{1133} & 0 & 0 & 0\\ C_{1133} & C_{1133} & C_{3333} & 0 & 0 & 0\\ 0 & 0 & 0 & C_{2323} & 0 & 0\\ 0 & 0 & 0 & 0 & C_{2323} & 0\\ 0 & 0 & 0 & 0 & 0 & C_{1212}\end{pmatrix}\begin{pmatrix}\epsilon_{11}\\ \epsilon_{22}\\ \epsilon_{33}\\ 2\epsilon_{23}\\ 2\epsilon_{13}\\ 2\epsilon_{12}\end{pmatrix}, \qquad (7.3.10)$$

when expressed in a basis aligned with the tetragonal directions and $\mathbf{e}_3$ being a fourfold rotational symmetry axis. Equivalently in the Voigt notation one has

$$\begin{pmatrix}\sigma_{1}\\ \sigma_{2}\\ \sigma_{3}\\ \sigma_{4}\\ \sigma_{5}\\ \sigma_{6}\end{pmatrix} = \begin{pmatrix} \mathcal{C}_{11} & \mathcal{C}_{12} & \mathcal{C}_{13} & 0 & 0 & 0\\ \mathcal{C}_{12} & \mathcal{C}_{11} & \mathcal{C}_{13} & 0 & 0 & 0\\ \mathcal{C}_{13} & \mathcal{C}_{13} & \mathcal{C}_{33} & 0 & 0 & 0\\ 0 & 0 & 0 & \mathcal{C}_{44} & 0 & 0\\ 0 & 0 & 0 & 0 & \mathcal{C}_{44} & 0\\ 0 & 0 & 0 & 0 & 0 & \mathcal{C}_{66}\end{pmatrix}\begin{pmatrix}\epsilon_{1}\\ \epsilon_{2}\\ \epsilon_{3}\\ \epsilon_{4}\\ \epsilon_{5}\\ \epsilon_{6}\end{pmatrix}; \qquad (7.3.11)$$

a material with tetragonal symmetry has **6 independent elastic constants**

$$\{\mathcal{C}_{11}, \mathcal{C}_{33}, \mathcal{C}_{44}, \mathcal{C}_{66}, \mathcal{C}_{12}, \mathcal{C}_{13}\}.$$

Some materials that display tetragonal symmetry include BC_4, Ni-Al alloys, and many minerals.

7.3.4 **Transversely isotropic symmetry. Hexagonal symmetry**

A material is said to be **transversely isotropic** if it has three mutually orthogonal planes of reflection symmetry, plus an additional symmetry with respect to all rotations about an axis perpendicular to one of the symmetry planes. Consider a basis $\{\mathbf{e}_1, \mathbf{e}_2, \mathbf{e}_3\}$ such that the $\mathbf{e}_i$ are normal to the planes of symmetry, with $\mathbf{e}_3$ being the axis about which there is rotational symmetry for any angle of rotation. In this case we require that

$$C_{ijkl} = Q_{ip}Q_{jq}Q_{kr}Q_{ls}C_{pqrs} \tag{7.3.12}$$

hold for

$$[\mathbf{Q}] = \begin{bmatrix} 1 & 0 & 0 \\ 0 & 1 & 0 \\ 0 & 0 & -1 \end{bmatrix}, \qquad [\mathbf{Q}] = \begin{bmatrix} 1 & 0 & 0 \\ 0 & -1 & 0 \\ 0 & 0 & 1 \end{bmatrix}, \qquad [\mathbf{Q}] = \begin{bmatrix} -1 & 0 & 0 \\ 0 & 1 & 0 \\ 0 & 0 & 1 \end{bmatrix}, \tag{7.3.13}$$

and

$$[\mathbf{Q}] = \begin{bmatrix} \cos\theta & \sin\theta & 0 \\ -\sin\theta & \cos\theta & 0 \\ 0 & 0 & 1 \end{bmatrix}, \quad 0 \le \theta < 2\pi. \tag{7.3.14}$$

Therefore, the elastic constants are subjected to the same restrictions as an orthotropic material, plus (7.3.14). Hence the restrictions on the C_{ijkl} from (7.3.13) lead to (7.3.5), viz.,

$$\begin{pmatrix} \sigma_{11} \\ \sigma_{22} \\ \sigma_{33} \\ \sigma_{23} \\ \sigma_{13} \\ \sigma_{12} \end{pmatrix} = \begin{pmatrix} C_{1111} & C_{1122} & C_{1133} & 0 & 0 & 0 \\ C_{1122} & C_{2222} & C_{2233} & 0 & 0 & 0 \\ C_{1133} & C_{2233} & C_{3333} & 0 & 0 & 0 \\ 0 & 0 & 0 & C_{2323} & 0 & 0 \\ 0 & 0 & 0 & 0 & C_{1313} & 0 \\ 0 & 0 & 0 & 0 & 0 & C_{1212} \end{pmatrix} \begin{pmatrix} \epsilon_{11} \\ \epsilon_{22} \\ \epsilon_{33} \\ 2\epsilon_{23} \\ 2\epsilon_{13} \\ 2\epsilon_{12} \end{pmatrix},$$

and the restrictions from (7.3.14) lead to (after a lengthy computation) the additional requirements,

$$C_{1111} = C_{2222}, \qquad C_{1133} = C_{2233}, \qquad C_{2323} = C_{1313}, \qquad C_{1212} = \tfrac{1}{2}\left(C_{1111} - C_{1122}\right), \tag{7.3.15}$$

which taken together require that for a **transversely isotropic** material the stress-strain relation may be written as

$$\begin{pmatrix} \sigma_{11} \\ \sigma_{22} \\ \sigma_{33} \\ \sigma_{23} \\ \sigma_{13} \\ \sigma_{12} \end{pmatrix} = \begin{pmatrix} C_{1111} & C_{1122} & C_{1133} & 0 & 0 & 0 \\ C_{1122} & C_{1111} & C_{1133} & 0 & 0 & 0 \\ C_{1133} & C_{1133} & C_{3333} & 0 & 0 & 0 \\ 0 & 0 & 0 & C_{2323} & 0 & 0 \\ 0 & 0 & 0 & 0 & C_{2323} & 0 \\ 0 & 0 & 0 & 0 & 0 & \frac{1}{2}\left(C_{1111} - C_{1122}\right) \end{pmatrix} \begin{pmatrix} \epsilon_{11} \\ \epsilon_{22} \\ \epsilon_{33} \\ 2\epsilon_{23} \\ 2\epsilon_{13} \\ 2\epsilon_{12} \end{pmatrix}, \tag{7.3.16}$$

when the $\mathbf{e}_3$-axis is the rotational symmetry axis. Equivalently in the Voigt notation one has

$$\begin{pmatrix}\sigma_1\\ \sigma_2\\ \sigma_3\\ \sigma_4\\ \sigma_5\\ \sigma_6\end{pmatrix} = \begin{pmatrix} C_{11} & C_{12} & C_{13} & 0 & 0 & 0\\ C_{12} & C_{11} & C_{13} & 0 & 0 & 0\\ C_{13} & C_{13} & C_{33} & 0 & 0 & 0\\ 0 & 0 & 0 & C_{44} & 0 & 0\\ 0 & 0 & 0 & 0 & C_{44} & 0\\ 0 & 0 & 0 & 0 & 0 & \frac{1}{2}(C_{11}-C_{12})\end{pmatrix}\begin{pmatrix}\epsilon_1\\ \epsilon_2\\ \epsilon_3\\ \epsilon_4\\ \epsilon_5\\ \epsilon_6\end{pmatrix}; \tag{7.3.17}$$

a transversely isotropic material has **5 independent elastic constants**

$$\{C_{11}, C_{33}, C_{44}, C_{12}, C_{13}\}.$$

It may be shown that:

- A material with **hexagonal lattice structure** has precisely the same symmetries as a **transversely isotropic** material, and hence also has five independent elastic constants. The axis of symmetry is perpendicular to the basal (0001) plane of the crystal.

- Representative values for elastic constants of some transversely isotropic hexagonal close-packed crystals are shown in Table 7.1.

Table 7.1 Representative values for elastic constants of transversely isotropic hexagonal close-packed crystals; data from Freund and Suresh (2004).

Crystal	C_{11} (GPa)	C_{33} (GPa)	C_{44} (GPa)	C_{12} (GPa)	C_{13} (GPa)
Be	292.3	336.4	162.5	26.7	14
C	1160	46.6	2.3	290	109
Cd	115.8	51.4	20.4	39.8	40.6
Co	307	358.1	78.3	165	103
Hf	181.1	196.9	55.7	77.2	66.1
Mg	59.7	61.7	16.4	26.2	21.7
Ti	162.4	180.7	46.7	92	69
Zn	161	61	38.3	34.2	50.1
Zr	143.4	164.8	32	72.8	65.3
ZnO	209.7	210.9	42.5	121.1	105.1

7.3.5 Cubic symmetry

A material is said to have **cubic** symmetry if it has three mutually orthogonal planes of reflection symmetry, plus additional symmetries with respect to rotations of 90° about the axes normal to these planes.

Consider a basis $\{\mathbf{e}_1, \mathbf{e}_2, \mathbf{e}_3\}$ such that the $\mathbf{e}_i$ are normal to the planes of reflection symmetry (Fig. 7.5). The $\mathbf{e}_i$ are also axes about which there is a 90° (or fourfold) rotational symmetry. In this case we require that

$$C_{ijkl} = Q_{ip}Q_{jq}Q_{kr}Q_{ls}C_{pqrs} \tag{7.3.18}$$

hold for the reflection symmetries with corresponding $[\mathbf{Q}]$ matrices,

$$[\mathbf{Q}] = \begin{bmatrix} 1 & 0 & 0 \\ 0 & 1 & 0 \\ 0 & 0 & -1 \end{bmatrix}, \quad [\mathbf{Q}] = \begin{bmatrix} 1 & 0 & 0 \\ 0 & -1 & 0 \\ 0 & 0 & 1 \end{bmatrix}, \quad [\mathbf{Q}] = \begin{bmatrix} -1 & 0 & 0 \\ 0 & 1 & 0 \\ 0 & 0 & 1 \end{bmatrix}, \tag{7.3.19}$$

and also for the rotation symmetries described by

$$[\mathbf{Q}] = \begin{bmatrix} 0 & 1 & 0 \\ -1 & 0 & 0 \\ 0 & 0 & 1 \end{bmatrix}, \quad [\mathbf{Q}] = \begin{bmatrix} 0 & 0 & 1 \\ 0 & 1 & 0 \\ -1 & 0 & 0 \end{bmatrix}, \quad [\mathbf{Q}] = \begin{bmatrix} 1 & 0 & 0 \\ 0 & 0 & 1 \\ 0 & -1 & 0 \end{bmatrix}. \tag{7.3.20}$$

By generalizing the analysis for tetragonal materials, one obtains the restrictions

$$\begin{aligned} C_{1111} &= C_{2222} = C_{3333}, \\ C_{1133} &= C_{2233} = C_{1122}, \\ C_{2323} &= C_{1313} = C_{1212}. \end{aligned}$$

Thus, for cubic materials we may write the stress-strain relation in matrix form as

$$\begin{pmatrix} \sigma_{11} \\ \sigma_{22} \\ \sigma_{33} \\ \sigma_{23} \\ \sigma_{13} \\ \sigma_{12} \end{pmatrix} = \begin{pmatrix} C_{1111} & C_{1122} & C_{1122} & 0 & 0 & 0 \\ C_{1122} & C_{1111} & C_{1122} & 0 & 0 & 0 \\ C_{1122} & C_{1122} & C_{1111} & 0 & 0 & 0 \\ 0 & 0 & 0 & C_{1212} & 0 & 0 \\ 0 & 0 & 0 & 0 & C_{1212} & 0 \\ 0 & 0 & 0 & 0 & 0 & C_{1212} \end{pmatrix} \begin{pmatrix} \epsilon_{11} \\ \epsilon_{22} \\ \epsilon_{33} \\ 2\epsilon_{23} \\ 2\epsilon_{13} \\ 2\epsilon_{12} \end{pmatrix}, \tag{7.3.21}$$

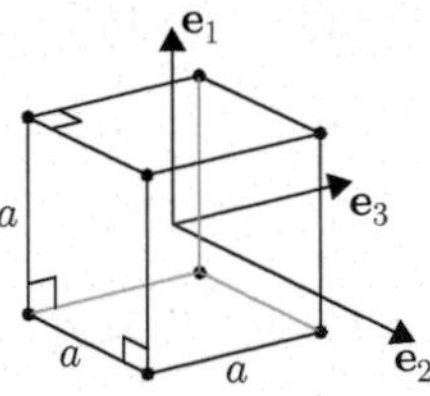

Fig. 7.5 A unit cell of cubic crystal.

when the basis is aligned with the cubic axes. Equivalently in the Voigt notation one has

$$\begin{pmatrix}\sigma_1\\ \sigma_2\\ \sigma_3\\ \sigma_4\\ \sigma_5\\ \sigma_6\end{pmatrix}=\begin{pmatrix}C_{11} & C_{12} & C_{12} & 0 & 0 & 0\\ C_{12} & C_{11} & C_{12} & 0 & 0 & 0\\ C_{12} & C_{12} & C_{11} & 0 & 0 & 0\\ 0 & 0 & 0 & C_{44} & 0 & 0\\ 0 & 0 & 0 & 0 & C_{44} & 0\\ 0 & 0 & 0 & 0 & 0 & C_{44}\end{pmatrix}\begin{pmatrix}\epsilon_1\\ \epsilon_2\\ \epsilon_3\\ \epsilon_4\\ \epsilon_5\\ \epsilon_6\end{pmatrix}; \tag{7.3.22}$$

a cubic material has **3 independent elastic constants**

$$\{C_{11}, C_{12}, C_{44}\}.$$

Values of the elastic stiffnesses $\{C_{11}, C_{12}, C_{44}\}$ and compliances $\{S_{11}, S_{12}, S_{44}\}$ for some cubic materials are listed in Table 7.2.

REMARK 7.1

We note that the elasticity tensor $\mathbb{C}$ for a cubic material can be expressed independent of coordinate basis as

$$\mathbb{C} = C_{12}\mathbf{1}\otimes\mathbf{1} + 2C_{44}\mathbb{I}^{\text{sym}} + (C_{11} - C_{12} - 2C_{44})\,(\mathbf{a}\otimes\mathbf{a}\otimes\mathbf{a}\otimes\mathbf{a} + \mathbf{b}\otimes\mathbf{b}\otimes\mathbf{b}\otimes\mathbf{b}$$
$$+\mathbf{c}\otimes\mathbf{c}\otimes\mathbf{c}\otimes\mathbf{c}),$$

where the vectors **a**, **b**, and **c** point in the three cubic directions.

Inter-relations between stiffnesses and compliances: Cubic materials

The inter-relations between the elastic stiffnesses $\{C_{11}, C_{12}, C_{44}\}$ and compliances $\{S_{11}, S_{12}, S_{44}\}$ for cubic materials may be derived by considering some simple states of stress, as follows:

1. Consider a uniaxial stress σ_1 applied in the $\mathbf{e}_1$-direction of the crystallographic axis. From

$$\begin{pmatrix}\epsilon_1\\ \epsilon_2\\ \epsilon_3\\ \epsilon_4\\ \epsilon_5\\ \epsilon_6\end{pmatrix}=\begin{pmatrix}S_{11} & S_{12} & S_{12} & 0 & 0 & 0\\ S_{12} & S_{11} & S_{12} & 0 & 0 & 0\\ S_{12} & S_{12} & S_{11} & 0 & 0 & 0\\ 0 & 0 & 0 & S_{44} & 0 & 0\\ 0 & 0 & 0 & 0 & S_{44} & 0\\ 0 & 0 & 0 & 0 & 0 & S_{44}\end{pmatrix}\begin{pmatrix}\sigma_1\\ 0\\ 0\\ 0\\ 0\\ 0\end{pmatrix},$$

$$\begin{pmatrix}\sigma_1\\ 0\\ 0\\ 0\\ 0\\ 0\end{pmatrix}=\begin{pmatrix}C_{11} & C_{12} & C_{12} & 0 & 0 & 0\\ C_{12} & C_{11} & C_{12} & 0 & 0 & 0\\ C_{12} & C_{12} & C_{11} & 0 & 0 & 0\\ 0 & 0 & 0 & C_{44} & 0 & 0\\ 0 & 0 & 0 & 0 & C_{44} & 0\\ 0 & 0 & 0 & 0 & 0 & C_{44}\end{pmatrix}\begin{pmatrix}\epsilon_1\\ \epsilon_2\\ \epsilon_3\\ \epsilon_4\\ \epsilon_5\\ \epsilon_6\end{pmatrix}, \tag{7.3.23}$$

Table 7.2 Representative elastic stiffness and compliance constants for some cubic crystals. Data from Simmons and Wang (1971).

Crystal	$\mathcal{C}_{11}$ (GPa)	$\mathcal{C}_{12}$ (GPa)	$\mathcal{C}_{44}$ (GPa)	$\mathcal{S}_{11}$ (TPa^{-1})	$\mathcal{S}_{12}$ (TPa^{-1})	$\mathcal{S}_{44}$ (TPa^{-1})
Cr	339.8	58.6	99.0	3.10	−0.46	10.10
Fe	231.4	134.7	116.4	7.56	−2.78	8.59
K	3.70	3.14	1.88	1223.9	56.19	53.19
Li	13.50	11.44	8.78	332.8	−152.7	113.9
Mo	441.6	172.7	121.9	2.90	−0.816	8.21
Na	6.15	4.96	5.92	581.0	−259.4	168.9
Nb	240.2	125.6	28.2	6.50	−2.23	35.44
Ta	260.2	154.5	82.6	6.89	−2.57	12.11
W	522.4	204.4	160.8	2.45	−0.69	6.22
Ag	122.2	90.7	45.4	22.26	−9.48	22.03
Al	107.3	60.9	28.3	15.82	−5.73	35.34
Au	192.9	163.8	41.5	23.55	−10.81	24.10
Cu	166.1	199.0	75.6	15.25	−6.39	13.23
Ni	248.1	154.9	124.2	7.75	−2.98	8.05
Pb	49.5	42.3	14.9	94.57	−43.56	67.11
Pd	227.1	176.0	71.7	13.63	5.95	13.94
Pt	346.7	250.7	76.5	7.34	−3.08	13.07
C	949.0	151.0	521.0	1.10	−1.51	1.92
Ge	128.4	48.2	66.7	9.80	−2.68	15.00
Si	166.2	64.4	79.7	7.67	−2.14	12.54
MgO	287.6	87.4	151.4	4.05	−0.94	6.60
MnO	223.0	120.0	79.0	7.19	−2.52	12.66
LiF	114.0	47.7	63.6	11.65	−3.43	15.71
KCl	39.5	4.9	6.3	26.00	−2.85	158.6
NaCl	49.0	12.6	12.7	22.80	−4.66	78.62
NaB	40.4	10.1	10.2	27.54	−5.53	98.52
NaI	30.1	9.12	7.33	38.72	−9.011	136.4

continued

Table 7.2 Continued

Crystal	$\mathcal{C}_{11}$ (GPa)	$\mathcal{C}_{12}$ (GPa)	$\mathcal{C}_{44}$ (GPa)	$\mathcal{S}_{11}$ (TPa^{-1})	$\mathcal{S}_{12}$ (TPa^{-1})	$\mathcal{S}_{44}$ (TPa^{-1})
NaF	97.0	23.8	28.2	11.41	−2.29	35.43
ZnS	103.2	64.8	46.2	18.77	−7.24	21.65
InP	102.2	57.6	46.0	16.48	−5.94	21.74
GaAs	118.8	53.7	59.4	11.72	−3.65	16.82

we have

$$\begin{aligned} \epsilon_1 = \mathcal{S}_{11}\sigma_1, \qquad \epsilon_2 = \mathcal{S}_{12}\sigma_1, \qquad \epsilon_3 = \mathcal{S}_{12}\sigma_1, \\ \sigma_1 = \mathcal{C}_{11}\epsilon_1 + \mathcal{C}_{12}\epsilon_2 + \mathcal{C}_{12}\epsilon_3. \end{aligned} \tag{7.3.24}$$

Substituting $(7.3.24)_1$ in $(7.3.24)_2$ we obtain

$$\sigma_1 = (\mathcal{C}_{11}\mathcal{S}_{11} + \mathcal{C}_{12}\mathcal{S}_{12} + \mathcal{C}_{12}\mathcal{S}_{12})\sigma_1$$

and since this must hold for arbitrary σ_1, the constitutive moduli $(\mathcal{C}_{11}, \mathcal{C}_{12})$ and $(\mathcal{S}_{11}, \mathcal{S}_{12})$ satisfy

$$\mathcal{C}_{11}\mathcal{S}_{11} + 2\mathcal{C}_{12}\mathcal{S}_{12} = 1. \tag{7.3.25}$$

2. Next consider the imposition of a hydrostatic stress

$$\sigma \equiv \sigma_1 = \sigma_2 = \sigma_3.$$

Then, from (7.3.23) we have

$$\begin{aligned} \epsilon(\equiv \epsilon_1 = \epsilon_2 = \epsilon_3) = (\mathcal{S}_{11} + 2\mathcal{S}_{12})\sigma, \\ \sigma(\equiv \sigma_1 = \sigma_2 = \sigma_3) = (\mathcal{C}_{11} + 2\mathcal{C}_{12})\epsilon. \end{aligned} \tag{7.3.26}$$

Substituting $(7.3.26)_1$ in $(7.3.26)_2$ we obtain

$$\sigma = (\mathcal{C}_{11} + 2\mathcal{C}_{12})(\mathcal{S}_{11} + 2\mathcal{S}_{12})\sigma$$

and since this must hold for arbitrary σ, the constitutive moduli $(\mathcal{C}_{11}, \mathcal{C}_{12})$ and $(\mathcal{S}_{11}, \mathcal{S}_{12})$ satisfy

$$(\mathcal{C}_{11} + 2\mathcal{C}_{12})(\mathcal{S}_{11} + 2\mathcal{S}_{12}) = 1. \tag{7.3.27}$$

Using (7.3.25) and (7.3.27) one can solve for $\mathcal{S}_{11}$ and $\mathcal{S}_{12}$ in terms of $\mathcal{C}_{11}$ and $\mathcal{C}_{12}$; straightforward algebra shows that

$$\boxed{\mathcal{S}_{11} = \frac{\mathcal{C}_{11} + \mathcal{C}_{12}}{(\mathcal{C}_{11} - \mathcal{C}_{12})(\mathcal{C}_{11} + 2\mathcal{C}_{12})}, \qquad \mathcal{S}_{12} = \frac{-\mathcal{C}_{12}}{(\mathcal{C}_{11} - \mathcal{C}_{12})(\mathcal{C}_{11} + 2\mathcal{C}_{12})}.} \tag{7.3.28}$$

3. Consider next a cubic single crystal subjected to pure shear stress $\sigma_4 \neq 0$. In this case constitutive equations (7.3.23) require that

$$\begin{aligned} \epsilon_4 = \mathcal{S}_{44}\sigma_4, \\ \sigma_4 = \mathcal{C}_{44}\epsilon_4. \end{aligned} \tag{7.3.29}$$

Substituting $(7.3.29)_1$ in $(7.3.29)_2$ we obtain

$$\sigma_4 = C_{44} S_{44} \sigma_4$$

and since this must hold for arbitrary σ_4, the constitutive moduli C_{44} and S_{44} are related as

$$\boxed{C_{44} = \frac{1}{S_{44}}.} \tag{7.3.30}$$

Together, (7.3.28) and (7.3.30) provide relationships between the three independent constants $\{S_{11}, S_{12}, S_{44}\}$ and $\{C_{11}, C_{12}, C_{44}\}$ for a cubic single crystal.

7.4 Directional elastic modulus

It is often of interest to determine an **effective elastic modulus**, $E_{\mathbf{d}}$, in a specific direction $\mathbf{d}$ (unit vector) of an anisotropic material: the effective elastic modulus is defined as the ratio of the magnitude of a uniaxial stress in the direction $\mathbf{d}$ to the resulting extensional (normal) strain in that direction. Thus, consider a uniaxial stress σ in the direction $\mathbf{d}$; in this case the corresponding stress tensor is

$$\boldsymbol{\sigma} = \sigma \mathbf{d} \otimes \mathbf{d},$$

and the components of the stress tensor with respect to a given basis $\{\mathbf{e}_1, \mathbf{e}_2, \mathbf{e}_3\}$ are

$$\sigma_{kl} = \sigma\, d_k d_l. \tag{7.4.1}$$

This stress will induce, via the compliance constants S_{ijkl}, a strain tensor with components

$$\epsilon_{ij} = S_{ijkl} \sigma_{kl} = \sigma S_{ijkl} d_k d_l,$$

and the normal strain in the direction $\mathbf{d}$ is given by (cf. (3.5.2))

$$\epsilon = \mathbf{d} \cdot \boldsymbol{\epsilon} \mathbf{d} = \epsilon_{ij} d_i d_j = \sigma d_i d_j S_{ijkl} d_k d_l. \tag{7.4.2}$$

Hence the **effective elastic modulus** in the direction $\mathbf{d}$, is defined by

$$E_{\mathbf{d}} \overset{\text{def}}{=} \frac{\sigma}{\epsilon} = \frac{1}{d_i d_j S_{ijkl} d_k d_l} = \frac{1}{\mathbf{d} \otimes \mathbf{d} : \mathbb{S}(\mathbf{d} \otimes \mathbf{d})}. \tag{7.4.3}$$

7.4.1 Directional elastic modulus: Cubic materials

In what follows, we calculate $E_{\mathbf{d}}$ specialized to the case of cubic materials. The components (7.4.1) of the uniaxial stress $\boldsymbol{\sigma} = \sigma \mathbf{d} \otimes \mathbf{d}$ may be written in the Voigt single index notations as

$$\begin{aligned}
\sigma_1 &= \sigma_{11} = \sigma d_1^2, \\
\sigma_2 &= \sigma_{22} = \sigma d_2^2, \\
\sigma_3 &= \sigma_{33} = \sigma d_3^2, \\
\sigma_4 &= \sigma_{23} = \sigma d_2 d_3, \\
\sigma_5 &= \sigma_{31} = \sigma d_3 d_1, \\
\sigma_6 &= \sigma_{12} = \sigma d_1 d_2.
\end{aligned} \tag{7.4.4}$$

Next, recall that for cubic materials the components of the strain in the Voigt notation are given by $(7.3.23)_1$; hence, using (7.4.4) the strain components are

$$\begin{aligned}
\epsilon_1 &= \sigma\left(S_{11}d_1^2 + S_{12}d_2^2 + S_{12}d_3^2\right),\\
\epsilon_2 &= \sigma\left(S_{12}d_1^2 + S_{11}d_2^2 + S_{12}d_3^2\right),\\
\epsilon_3 &= \sigma\left(S_{12}d_1^2 + S_{12}d_2^2 + S_{11}d_3^2\right),\\
\epsilon_4 &= \sigma\left(S_{44}d_2d_3\right),\\
\epsilon_5 &= \sigma\left(S_{44}d_3d_1\right),\\
\epsilon_6 &= \sigma\left(S_{44}d_1d_2\right).
\end{aligned} \tag{7.4.5}$$

Next, using (7.4.2), the strain ϵ in the $\mathbf{d}$ direction is given by

$$\begin{aligned}
\epsilon &= \epsilon_{11}d_1^2 + \epsilon_{22}d_2^2 + \epsilon_{33}d_3^2 + (2\epsilon_{23})d_2d_3 + (2\epsilon_{31})d_3d_1 + (2\epsilon_{12})d_1d_2\\
&= \epsilon_1 d_1^2 + \epsilon_2 d_2^2 + \epsilon_3 d_3^2 + \epsilon_4 d_2d_3 + \epsilon_5 d_3d_1 + \epsilon_6 d_1d_2.
\end{aligned} \tag{7.4.6}$$

Using equations (7.4.5) and (7.4.6) in (7.4.3) we obtain

$$\frac{1}{E_{\mathbf{d}}} = \frac{\epsilon}{\sigma} = S_{11}\left(d_1^4 + d_2^4 + d_3^4\right) + 2S_{12}\left(d_1^2d_2^2 + d_2^2d_3^2 + d_3^2d_1^2\right) + S_{44}\left(d_1^2d_2^2 + d_2^2d_3^2 + d_3^2d_1^2\right), \tag{7.4.7}$$

and since $\mathbf{d}$ is a unit vector,

$$d_1^2 + d_2^2 + d_3^2 = 1,$$

it is straightforward to rewrite (7.4.7) as

$$\begin{aligned}
\frac{1}{E_{\mathbf{d}}} &= S_{11}\left(1 - 2d_1^2d_2^2 - 2d_2^2d_3^2 - 2d_3^2d_1^2\right) + 2S_{12}\left(d_1^2d_2^2 + d_2^2d_3^2 + d_3^2d_1^2\right)\\
&\quad + S_{44}\left(d_1^2d_2^2 + d_2^2d_3^2 + d_3^2d_1^2\right)
\end{aligned} \tag{7.4.8}$$

or

$$\boxed{\frac{1}{E_{\mathbf{d}}} = S_{11} + (2S_{12} - 2S_{11} + S_{44})\left(d_1^2d_2^2 + d_2^2d_3^2 + d_3^2d_1^2\right).} \tag{7.4.9}$$

The direction $\mathbf{d}$ may be expressed in terms of a crystal's Miller's direction indices $[hkl]$ by

$$d_1 = h/\sqrt{h^2+k^2+l^2}, \qquad d_2 = k/\sqrt{h^2+k^2+l^2}, \qquad d_3 = l/\sqrt{h^2+k^2+l^2},$$

and therefore the effective compliance in the direction $[hkl]$ in a cubic crystal may be written as

$$\frac{1}{E_{[hkl]}} = S_{11} + (2S_{12} - 2S_{11} + S_{44})\frac{(h^2k^2 + k^2l^2 + l^2h^2)}{(h^2+k^2+l^2)^2}. \tag{7.4.10}$$

From amongst all possible directions $[hkl]$ in a cubic crystal, the maximum and minimum values of the effective elastic modulus are found to exist in either a $\langle 100\rangle$ direction or in a $\langle 111\rangle$ direction. The extreme values and their ratios are

$$E_{[100]} = \frac{1}{S_{11}}, \qquad E_{[111]} = \frac{3}{S_{11} + 2S_{12} + S_{44}}, \qquad \frac{E_{[111]}}{E_{[100]}} = \frac{3S_{11}}{S_{11} + 2S_{12} + S_{44}}. \tag{7.4.11}$$

Table 7.3 Directional modulus ratios for a few cubic materials.

Crystal	S_{11}, TPa^{-1}	S_{12}, TPa^{-1}	S_{44}, TPa^{-1}	$E_{[111]}/E_{[100]}$
Fe	7.56	−2.78	8.59	2.14
W	2.45	−0.69	6.22	1.01
Nb	6.50	−2.23	35.44	0.52

The ratio $E_{[111]}/E_{[100]}$ provides a measure of the degree of departure from isotropy in a given cubic crystal; the value of this ratio may be either greater than or less than unity, taking on the value of unity for isotropic materials. The ratio $E_{[111]}/E_{[100]}$ for a few cubic crystals are listed in Table 7.3.

Graphical representations of the variation in the magnitude of $E_{\rm d}$ with respect to the crystallographic axes for Fe, W, and Nb are plotted in Fig. 7.6. For Fe crystals, the effective modulus has the highest value along the $\langle 111 \rangle$ directions, and the lowest values along the $\langle 100 \rangle$ directions, while the reverse is true for Nb single crystals; on the other hand W single crystals with a ratio $E_{[111]}/E_{[100]} = 1.01$ are almost "isotropic" — the values of the directional modulus are almost uniform in space.

Anisotropy ratio for a cubic single crystal

If $E_{[hkl]}$ is independent of orientation, then the cubic crystal is "isotropic." From (7.4.10) we see that this condition holds if

$$\mathcal{S}_{44} = 2(\mathcal{S}_{11} - \mathcal{S}_{12}). \tag{7.4.12}$$

For a cubic material, using (7.3.28) and (7.3.30), the condition (7.4.12) for isotropy may be equivalently written as

$$\mathcal{C}_{44} = \tfrac{1}{2}(\mathcal{C}_{11} - \mathcal{C}_{12}). \tag{7.4.13}$$

The degree of departure from isotropy in the response of a cubic crystal is, thus, often also characterized by the **anisotropy ratio**, AR, defined by

$$AR \stackrel{\text{def}}{=} \frac{\mathcal{C}_{44}}{\frac{1}{2}(\mathcal{C}_{11} - \mathcal{C}_{12})} \equiv \frac{2(\mathcal{S}_{11} - \mathcal{S}_{12})}{\mathcal{S}_{44}}. \tag{7.4.14}$$

7.5 Constitutive equations for isotropic linear elastic materials

The physical idea of **isotropy** is that there are no *special* or *preferential* directions present in the material. The response of the material in any one direction is identical to any other direction.

- If (7.2.6) holds for *all* orthogonal tensors **Q**, then the linear tensor function $\mathbb{C}$ which maps symmetric tensors to symmetric tensors is an *isotropic linear tensor function*, and the linear elastic body is said to be **isotropic**.

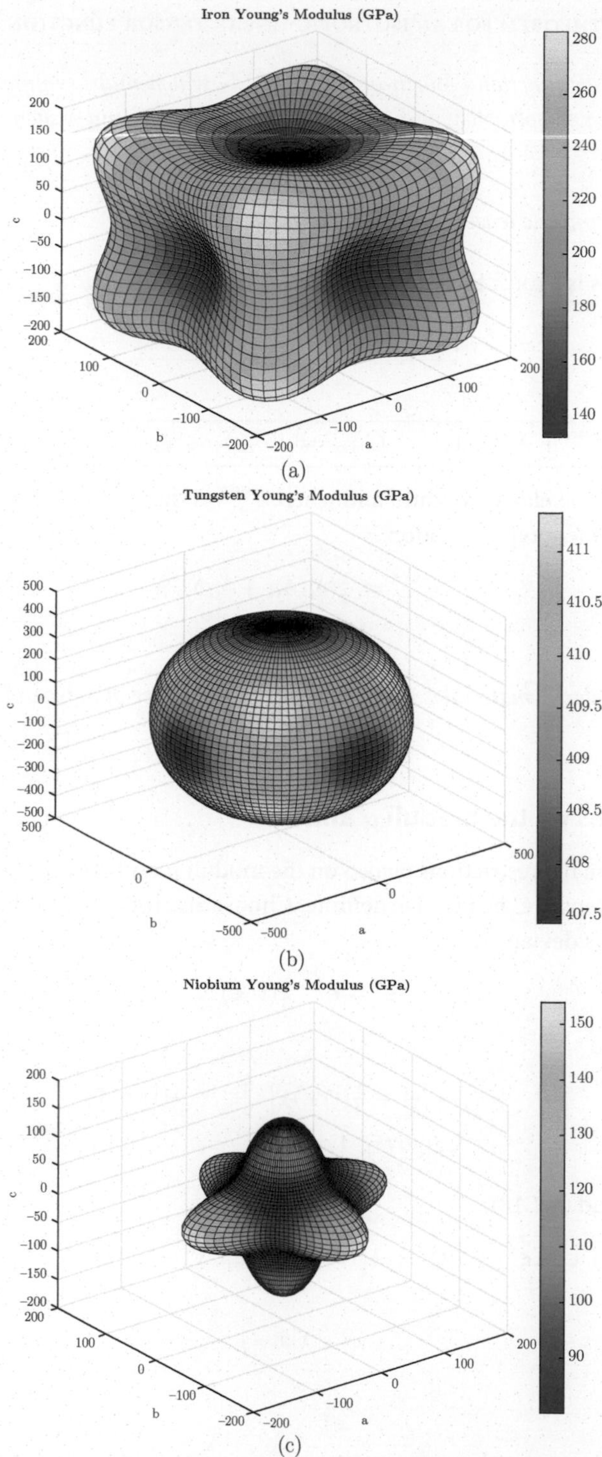

Fig. 7.6 Variation in the magnitudes of the effective modulus E_{d} with respect to the crystallographic axes of a cubic crystal for (a) Fe, (b) W, and (c) Nb crystals. The (X, Y, Z)-scales in the figure are in GPa.

REPRESENTATION THEOREM FOR AN ISOTROPIC LINEAR TENSOR FUNCTION[2]

- *A linear tensor function (fourth-order tensor) $\mathbb{G}$ which maps symmetric tensors to symmetric tensors is isotropic if and only if there are scalars μ and λ such that*

$$\mathbb{G}\mathbf{A} = 2\mu\mathbf{A} + \lambda(\operatorname{tr}\mathbf{A})\mathbf{1} \tag{7.5.1}$$

for every symmetric tensor $\mathbf{A}$.

Thus, if the body is isotropic then $\boldsymbol{\sigma} = \mathbb{C}\boldsymbol{\epsilon}$ has the specific form

$$\boxed{\boldsymbol{\sigma} = \mathbb{C}\boldsymbol{\epsilon} = 2\mu\boldsymbol{\epsilon} + \lambda(\operatorname{tr}\boldsymbol{\epsilon})\mathbf{1},} \tag{7.5.2}$$

with

$$\boxed{\mathbb{C} = 2\mu\mathbb{I}^{\text{sym}} + \lambda\,\mathbf{1}\otimes\mathbf{1}, \qquad C_{ijkl} = \mu(\delta_{ik}\delta_{jl} + \delta_{il}\delta_{jk}) + \lambda\delta_{ij}\delta_{kl},} \tag{7.5.3}$$

where μ and λ are the **elastic moduli**, and where $\mathbb{I}^{\text{sym}}$ is the fourth-order symmetric identity tensor with component representation

$$I^{\text{sym}}_{ijkl} = \tfrac{1}{2}(\delta_{ik}\delta_{jl} + \delta_{il}\delta_{jk}). \tag{7.5.4}$$

Hence,

- *if the linear elastic material is* **isotropic**, *then the number of independent elastic constants reduces to 2.*

7.5.1 Restrictions on the moduli μ and λ

Next, we determine the restrictions placed on the moduli μ and λ by the requirement (7.1.12) that the elasticity tensor $\mathbb{C}$ be positive definite. Choose an arbitrary symmetric strain tensor $\boldsymbol{\epsilon}$ and let $\boldsymbol{\epsilon}'$ denote its deviatoric part:

$$\boldsymbol{\epsilon}' = \boldsymbol{\epsilon} - \tfrac{1}{3}(\operatorname{tr}\boldsymbol{\epsilon})\mathbf{1}.$$

Then $\operatorname{tr}\boldsymbol{\epsilon}' = 0$ and

$$\begin{aligned}|\boldsymbol{\epsilon}|^2 &= \big[\boldsymbol{\epsilon}' + \tfrac{1}{3}(\operatorname{tr}\boldsymbol{\epsilon})\mathbf{1}\big] : \big[\boldsymbol{\epsilon}' + \tfrac{1}{3}(\operatorname{tr}\boldsymbol{\epsilon})\mathbf{1}\big]\\ &= |\boldsymbol{\epsilon}'|^2 + \tfrac{1}{3}(\operatorname{tr}\boldsymbol{\epsilon})^2.\end{aligned}$$

Thus, by (7.5.2) and (7.1.12),

$$\boldsymbol{\epsilon} : \mathbb{C}\boldsymbol{\epsilon} = 2\mu|\boldsymbol{\epsilon}|^2 + \lambda(\operatorname{tr}\boldsymbol{\epsilon})^2 = 2\mu|\boldsymbol{\epsilon}'|^2 + \kappa(\operatorname{tr}\boldsymbol{\epsilon})^2 > 0, \tag{7.5.5}$$

with

$$\kappa = \lambda + \tfrac{2}{3}\mu. \tag{7.5.6}$$

[2] Cf., e.g. Gurtin (1981) pp. 235–6.

Choosing $\boldsymbol{\epsilon}$ to be spherical (so that $\boldsymbol{\epsilon}' = \mathbf{0}$) yields $\kappa > 0$, and choosing $\boldsymbol{\epsilon}$ to be deviatoric (so that $\operatorname{tr}\boldsymbol{\epsilon} = \mathbf{0}$) yields $\mu > 0$. Thus the elastic moduli μ and λ satisfy

$$\boxed{\mu > 0, \qquad \kappa = \lambda + \tfrac{2}{3}\mu > 0.} \tag{7.5.7}$$

The scalars μ and λ are referred to as **Lamé moduli**.

In view of (7.5.6), the relation (7.5.2) may alternatively be written in terms of the scalars μ and κ as

$$\boxed{\boldsymbol{\sigma} = 2\mu\boldsymbol{\epsilon}' + \kappa(\operatorname{tr}\boldsymbol{\epsilon})\mathbf{1}, \qquad \sigma_{ij} = 2\mu\epsilon'_{ij} + \kappa\,(\epsilon_{kk})\delta_{ij}.} \tag{7.5.8}$$

Next we express (7.5.8) in the form $\boldsymbol{\sigma} = \mathbb{C}\boldsymbol{\epsilon}$, with the fourth-order elasticity tensor $\mathbb{C}$ expressed in terms of the elastic moduli μ and κ. To do so, we first introduce the fourth-order *volumetric identity tensor*

$$\mathbb{I}^{\text{vol}} \stackrel{\text{def}}{=} \tfrac{1}{3}\mathbf{1}\otimes\mathbf{1}, \tag{7.5.9}$$

which is the identity map for spherical tensors, i.e.

$$\mathbb{I}^{\text{vol}}\mathbf{A} = \mathbf{A}, \quad \text{for all tensors } \mathbf{A} \text{ with } \mathbf{A}' = 0.$$

Next we introduce the fourth-order *symmetric-deviatoric identity tensor*

$$\mathbb{I}^{\text{symdev}} \stackrel{\text{def}}{=} \mathbb{I}^{\text{sym}} - \mathbb{I}^{\text{vol}}, \tag{7.5.10}$$

which is the identity map for symmetric-deviatoric tensors, i.e.

$$\mathbb{I}^{\text{symdev}}\mathbf{B} = \mathbf{B}, \quad \text{for all symmetric tensors } \mathbf{B} \text{ with } \operatorname{tr}\mathbf{B} = 0.$$

Thus, for an arbitrary tensor $\mathbf{C}$,

$$\mathbb{I}^{\text{vol}}\mathbf{C} = \tfrac{1}{3}(\operatorname{tr}\mathbf{C})\mathbf{1}, \tag{7.5.11}$$

the spherical part of $\mathbf{C}$, and

$$\mathbb{I}^{\text{symdev}}\mathbf{C} = \operatorname{sym}\mathbf{C} - \tfrac{1}{3}(\operatorname{tr}\mathbf{C})\mathbf{1} \equiv (\operatorname{sym}\mathbf{C})' \tag{7.5.12}$$

the symmetric-deviatoric part of $\mathbf{C}$. Using this notation, we can write (7.5.8) as

$$\boldsymbol{\sigma} = \mathbb{C}\boldsymbol{\epsilon} = \underbrace{\left(2\mu\mathbb{I}^{\text{symdev}} + 3\kappa\mathbb{I}^{\text{vol}}\right)}_{\mathbb{C}}\boldsymbol{\epsilon}. \tag{7.5.13}$$

7.5.2 Inverted form of the stress-strain relation

Next, we determine the inverted form of the stress-strain relation (7.5.8). Since

$$\boldsymbol{\sigma}' = 2\mu\boldsymbol{\epsilon}', \qquad \operatorname{tr}\boldsymbol{\sigma} = 3\kappa(\operatorname{tr}\boldsymbol{\epsilon}), \tag{7.5.14}$$

we obtain

$$\boldsymbol{\epsilon}' = \frac{1}{2\mu}\boldsymbol{\sigma}', \qquad \operatorname{tr}\boldsymbol{\epsilon} = \frac{1}{3\kappa}\operatorname{tr}\boldsymbol{\sigma}, \tag{7.5.15}$$

and hence

$$\boxed{\epsilon = \frac{1}{2\mu}\boldsymbol{\sigma}' + \frac{1}{9\kappa}(\operatorname{tr}\boldsymbol{\sigma})\mathbf{1}, \qquad \epsilon_{ij} = \frac{1}{2\mu}\sigma'_{ij} + \frac{1}{9\kappa}(\sigma_{kk})\delta_{ij}.} \tag{7.5.16}$$

7.5.3 **Physical interpretation of the elastic constants in terms of local strain and stress states**

Consider a local strain state in the form of **simple shear**, in which the matrix of ϵ is

$$[\epsilon] = \begin{bmatrix} 0 & \frac{1}{2}\gamma & 0 \\ \frac{1}{2}\gamma & 0 & 0 \\ 0 & 0 & 0 \end{bmatrix}.$$

Substituting for this strain state in the stress-strain relations (7.5.2) gives the following local stress state

$$[\boldsymbol{\sigma}] = \begin{bmatrix} 0 & \tau & 0 \\ \tau & 0 & 0 \\ 0 & 0 & 0 \end{bmatrix},$$

with

$$\tau = \mu\gamma.$$

- Thus μ determines the response of the material in shear, and it is for this reason that it is called the **shear modulus.**

- Because an elastic material should respond to positive shearing strain by a positive shearing stress, we require that

$$\mu > 0,$$

  a requirement already deduced in $(7.5.7)_1$.

Next, consider a local strain state in the form of **uniform compaction**, in which the matrix representation of ϵ is

$$[\epsilon] = \begin{bmatrix} -\frac{1}{3}\Delta & 0 & 0 \\ 0 & -\frac{1}{3}\Delta & 0 \\ 0 & 0 & -\frac{1}{3}\Delta \end{bmatrix}.$$

Substituting for this strain state in the stress-strain relations (7.5.8) gives the following local stress state

$$[\boldsymbol{\sigma}] = \begin{bmatrix} -p & 0 & 0 \\ 0 & -p & 0 \\ 0 & 0 & -p \end{bmatrix},$$

with

$$p = \kappa\Delta.$$

- Thus κ determines the response of the body in compaction, and it is for this reason that it is called the **modulus of compaction** or the **bulk modulus**.

- Because an elastic body should require a positive pressure ($p > 0$) for a compaction ($\Delta > 0$), we require that

$$\kappa > 0,$$

a requirement already deduced in $(7.5.7)_2$.

For our third special case, we need the inverted form of the stress-strain relation (7.5.16). Consider a local stress state in the form of **pure tension**, in which the matrix representation of σ is

$$[\sigma] = \begin{bmatrix} \sigma & 0 & 0 \\ 0 & 0 & 0 \\ 0 & 0 & 0 \end{bmatrix}.$$

Substituting for this stress state in the strain-stress relations (7.5.16) gives the following local strain state

$$[\epsilon] = \frac{1}{2\mu}\begin{bmatrix} \sigma & 0 & 0 \\ 0 & 0 & 0 \\ 0 & 0 & 0 \end{bmatrix} - \left(\frac{3\kappa - 2\mu}{18\kappa\mu}\right)\sigma\begin{bmatrix} 1 & 0 & 0 \\ 0 & 1 & 0 \\ 0 & 0 & 1 \end{bmatrix},$$

or

$$[\epsilon] = \begin{bmatrix} \dfrac{1}{2\mu}\left(1 - \dfrac{3\kappa - 2\mu}{9\kappa}\right)\sigma & 0 & 0 \\ 0 & -\left(\dfrac{3\kappa - 2\mu}{18\kappa\mu}\right)\sigma & 0 \\ 0 & 0 & -\left(\dfrac{3\kappa - 2\mu}{18\kappa\mu}\right)\sigma \end{bmatrix}.$$

This matrix of strain components has the form

$$[\epsilon] = \begin{bmatrix} \epsilon & 0 & 0 \\ 0 & l & 0 \\ 0 & 0 & l \end{bmatrix}$$

with

$$\epsilon = \frac{\sigma}{E}, \qquad l = -\nu\epsilon,$$

and

$$\boxed{E \equiv \frac{9\kappa\mu}{3\kappa + \mu}, \qquad \nu \equiv \frac{1}{2}\left[\frac{3\kappa - 2\mu}{3\kappa + \mu}\right].} \tag{7.5.17}$$

Thus:

- The modulus E is obtained by dividing the tensile stress σ by the longitudinal strain ϵ produced by it. It is known as the **Young's modulus.**

- The symbol ν is the ratio of the lateral contraction to the longitudinal strain of a bar under pure tension. It is known as **Poisson's ratio.**

- Since an elastic solid should increase in length when subject to tensile stress,

$$E > 0.$$

The shear modulus μ and the bulk modulus κ may be expressed in terms of the Young's modulus and the Poisson's ratio as follows:

$$\mu = \frac{E}{2(1+\nu)}, \qquad \kappa = \frac{E}{3(1-2\nu)}. \tag{7.5.18}$$

Noting that no real materials should be infinitely stiff, we require that μ and κ be *bounded*: $\mu < \infty$ and $\kappa < \infty$. Thus, (7.5.18) requires that E and ν should obey the following set of inequalities

$$0 < E < \infty, \qquad \text{and} \qquad -1 < \nu < \tfrac{1}{2}. \tag{7.5.19}$$

For metallic materials it is commonly found that

$$\nu \approx \frac{1}{3};$$

in this case

$$\mu \approx \frac{3}{8}E, \qquad \kappa \approx E.$$

Limiting value of Poisson's ratio for incompressible materials

From $(7.5.18)_2$, the Poisson's ratio may be expressed in terms of (E, κ) as

$$\nu = \frac{1}{2}\left[1 - \frac{E}{3\kappa}\right].$$

In the limit of an "incompressible" elastic material, that is, in the limit

$$\kappa \to \infty,$$

the Poisson's ratio approaches

$$\nu \to \frac{1}{2}.$$

Note that no actual material is truly incompressible. However, when $\mu/\kappa \ll 1$, as for rubber-like materials, for mathematical convenience one can often model the material as incompressible.

7.5.4 Stress-strain relations in terms of E and ν

In terms of the Young's modulus, E, and the Poisson's ratio, ν, the constitutive relations (7.5.2) and (7.5.16) may be written in alternative useful forms as

$$\boxed{\boldsymbol{\sigma} = \frac{E}{(1+\nu)}\left[\boldsymbol{\epsilon} + \frac{\nu}{(1-2\nu)}\,(\operatorname{tr}\boldsymbol{\epsilon})\,\mathbf{1}\right], \qquad \sigma_{ij} = \frac{E}{(1+\nu)}\left[\epsilon_{ij} + \frac{\nu}{(1-2\nu)}\,(\epsilon_{kk})\,\delta_{ij}\right],} \tag{7.5.20}$$

and

$$\boxed{\boldsymbol{\epsilon} = \frac{1}{E}\left[(1+\nu)\boldsymbol{\sigma} - \nu\,(\operatorname{tr}\boldsymbol{\sigma})\,\mathbf{1}\right]. \qquad \epsilon_{ij} = \frac{1}{E}\left[(1+\nu)\sigma_{ij} - \nu\,(\sigma_{kk})\,\delta_{ij}\right].} \tag{7.5.21}$$

The expanded form of (7.5.21) is

$$\begin{aligned}
\epsilon_{11} &= \frac{1}{E}\Big[\sigma_{11} - \nu\Big(\sigma_{22} + \sigma_{33}\Big)\Big],\\
\epsilon_{22} &= \frac{1}{E}\Big[\sigma_{22} - \nu\Big(\sigma_{11} + \sigma_{33}\Big)\Big],\\
\epsilon_{33} &= \frac{1}{E}\Big[\sigma_{33} - \nu\Big(\sigma_{11} + \sigma_{22}\Big)\Big],\\
\epsilon_{12} &= \frac{1+\nu}{E}\sigma_{12},\\
\epsilon_{13} &= \frac{1+\nu}{E}\sigma_{13},\\
\epsilon_{23} &= \frac{1+\nu}{E}\sigma_{23}.
\end{aligned} \tag{7.5.22}$$

7.5.5 **Relations between various elastic moduli**

Table 7.4 gives the relations between the various isotropic elastic moduli. The important thing to remember is that **any two** isotropic linear elastic constants are sufficient to determine any other elastic constant.

Table 7.4 Relationships between $\mu, \kappa, E, \nu, \lambda$ for isotropic linear elastic materials. In this table, $R \stackrel{\text{def}}{=} \sqrt{E^2 + 9\lambda^2 + 2E\lambda}$.

	μ	κ	E	ν	λ
$\mu,\ E$		$\dfrac{\mu E}{3(3\mu - E)}$		$\dfrac{E-2\mu}{2\mu}$	$\dfrac{\mu(E-2\mu)}{3\mu - E}$
$\mu,\ \nu$		$\dfrac{2\mu(1+\nu)}{3(1-2\nu)}$	$2\mu(1+\nu)$		$\dfrac{2\mu\nu}{1-2\nu}$
$\mu,\ \kappa$			$\dfrac{9\kappa\mu}{3\kappa+\mu}$	$\dfrac{1}{2}\left[\dfrac{3\kappa-2\mu}{3\kappa+\mu}\right]$	$\kappa - \dfrac{2}{3}\mu$
$E,\ \nu$	$\dfrac{E}{2(1+\nu)}$	$\dfrac{E}{3(1-2\nu)}$			$\dfrac{E\nu}{(1+\nu)(1-2\nu)}$
$E,\ \kappa$	$\dfrac{3E\kappa}{9\kappa - E}$			$\dfrac{1}{2}\left[\dfrac{3\kappa - E}{3\kappa}\right]$	$\dfrac{3\kappa(3\kappa-E)}{9\kappa - E}$
$\nu,\ \kappa$	$\dfrac{3\kappa(1-2\nu)}{2(1+\nu)}$		$3\kappa(1-2\nu)$		$\dfrac{3\kappa\nu}{1+\nu}$
μ, λ		$\dfrac{3\lambda+2\mu}{3}$	$\dfrac{\mu(3\lambda+2\mu)}{\lambda+\mu}$	$\dfrac{\lambda}{2(\lambda+\mu)}$	
κ, λ	$\dfrac{3}{2}(\kappa-\lambda)$		$\dfrac{9\kappa(\kappa-\lambda)}{(3\kappa-\lambda)}$	$\dfrac{\lambda}{3\kappa-\lambda}$	
E, λ	$\dfrac{E-3\lambda+R}{4}$	$\dfrac{E+3\lambda+R}{6}$		$\dfrac{2\lambda}{E+\lambda+R}$	
ν, λ	$\dfrac{\lambda(1-2\nu)}{2\nu}$	$\dfrac{\lambda(1+\nu)}{3\nu}$	$\dfrac{\lambda(1+\nu)(1-2\nu)}{\nu}$		

REMARK 7.2

In later parts of the book when we consider some special applications of the theory of elasticity, in accordance with common convention for such applications, we will denote the shear modulus by the symbol G instead of μ, and the bulk modulus by the symbol K instead of κ.

7.6 Isotropic linear thermoelastic constitutive relations

We will more formally and more fully develop the theory of linear thermoelasticity in a subsequent chapter. Here, for brevity we record the stress-strain relations for isotropic linear thermoelastic materials.

Consider a stress-free reference configuration at a reference temperature ϑ_0. For small deformations under non-isothermal conditions the strain $\boldsymbol{\epsilon}$ caused by a stress change $(\boldsymbol{\sigma} - \mathbf{0})$ and a temperature change $(\vartheta - \vartheta_0)$ may be decomposed into mechanical and thermal components as

$$\boldsymbol{\epsilon} = \boldsymbol{\epsilon}^{\text{mechanical}} + \boldsymbol{\epsilon}^{\text{thermal}}, \tag{7.6.1}$$

where the mechanical strain is the strain caused by the change in stress, and the thermal strain is the strain caused by the change in temperature.

For linear thermoelastic materials the thermal strains are given by

$$\boldsymbol{\epsilon}^{\text{thermal}} = \mathbf{A}(\vartheta - \vartheta_0), \qquad \mathbf{A} = \mathbf{A}^{\top}, \tag{7.6.2}$$

where $\mathbf{A}$ is the *thermal-expansion tensor*. For isotropic (and cubic) materials

$$\mathbf{A} = \alpha \mathbf{1}, \tag{7.6.3}$$

where α is the *coefficient of thermal expansion*. It has dimensions of 1/[Temperature], and is usually stated in units of microstrain/K or 10^{-6}/K.

Thus, for isotropic materials

$$\boxed{\epsilon_{ij} = \frac{1}{E}\left[(1+\nu)\sigma_{ij} - \nu\,(\sigma_{kk})\,\delta_{ij}\right] + \alpha\,(\vartheta - \vartheta_0)\,\delta_{ij},} \tag{7.6.4}$$

which can be easily inverted to give

$$\boxed{\sigma_{ij} = \frac{E}{(1+\nu)}\left[\epsilon_{ij} + \frac{\nu}{(1-2\nu)}\,(\epsilon_{kk})\,\delta_{ij} - \frac{(1+\nu)}{(1-2\nu)}\alpha\,(\vartheta - \vartheta_0)\,\delta_{ij}\right].} \tag{7.6.5}$$

In writing these constitutive equations for isotropic thermoelasticity, the small dependence of the Young's modulus E and Poisson's ratio ν on temperature, for small temperature changes, is neglected. Representative values of the coefficient of thermal expansion α for some nominally isotropic materials are given in Tables 7.5–7.7 for some metallic, ceramic, and polymeric materials, respectively.

Table 7.5 Some typical values for E, ν, μ, and α for some nominally isotropic (polycrystalline) metallic materials.

Metals		E (GPa)	ν	μ (GPa)	α (10^{-6}/K)
Tungsten	W	397	0.284	153	4.3–4.7
Molybdenum	Mo	327	0.30	116	4.9
Chromium	Cr	243	0.209	117	6.2
Iron	Fe	210	0.279	82	10.6–12.8
Nickel	Ni	193	0.3	75	12.5
Copper	Cu	124	0.345	45	16.5
Titanium	Ti	106	0.345	39	8.6
Zinc	Zn	92	0.29	37	30.0
Silver	Ag	81	0.37	29	20.0
Gold	Au	78	0.425	28	13.0
Aluminum	Al	71	0.34	27	23.2
Tin	Sn	53	0.375	19	23
Magnesium	Mg	44	0.28	17	26.1
Lead	Pb	16	0.44	5.4	29.3

Table 7.6 Some typical values for E, ν, μ, and α for some nominally isotropic ceramic materials. All compositions are in weight percent.

Ceramics	E (GPa)	ν	μ (GPa)	α $(10^{-6}/\text{K})$
Diamond	1128	0.18	451	1.2
Metal-bonded Tungsten Carbide 94 WC, 6 Co	580	0.26	230	4.4
Self-bonded Silicon Carbide 90 SiC, 10 Si	410	0.24	165	4.3
Sintered Alumina 100 Al_2O_3	350	0.23	142	8.5
Hot-pressed Silicon Nitride 96 Si_3N_4, 4MgO	310	0.25	124	3.2
Low-expansion Glass Ceramic 2 (Ti, Zr) O_2, 4 Li_2O 20 Al_2O_3, 70 SiO_2	87	0.25	35	0.02
Soda-Lime Glass 13 Na_2O, 12(Ca, Mg)O, 72 SiO_2	73	0.21	30	8.5
Vitreous Silica 100 SiO_2	71	0.17	30	0.55
Low-expansion Borosilicate Glass 12 B_2O_3, 4 Na_2O, 2 Al_2O_3, 80 SiO_2	66	0.2	27.5	4.0
Machinable Glass Ceramic 65 Mica, 35 Glass	64	0.26	25	12.7
High-density Molded Graphite	9	0.11	4	2.5

Table 7.7 Some typical values for E, ν, μ, and α for some nominally isotropic polymeric materials.

Polymers		E (GPa)	ν	μ (GPa)	α (10^{-6}/K)
Polymethylmethacrylate					
(PMMA)	−125°C	6.3	0.26	2.5	
	25°C	3.7	0.33	1.39	54–72
Polystyrene					
(PS)	25°C	3.4	0.33	1.28	70–100
Polyethylene					
(low density)	25°C	2.4	0.38	0.87	160–190
Polycarbonate					
(PC)	25°C	2.3	0.2	0.96	65–70
Polyethylene terephthalate					
(PET)	25°C	2	0.35	0.74	20–80
Polyamide (nylon)					
(PA)	25°C	2.8	0.4	1.0	80–95
Vulcanized Natural Rubber					
(VNR)	25°C	0.0016	0.499	0.0005	600
Polyurethane Foam Rubber					
(EUFR)	25°C	0.0005	0.25	0.0002	600

8 Linear elastostatics

In this chapter we gather together the basic field equations for linear elastic problems in the static case, that is, neglecting inertial terms. Further, we discuss some typical boundary conditions which are necessary to formulate a complete boundary-value problem.

8.1 Basic equations of linear elasticity

The linear theory of elasticity is based on the **strain-displacement relation**

$$\boxed{\boldsymbol{\epsilon} = \tfrac{1}{2}(\nabla\mathbf{u} + \nabla\mathbf{u}^{\top}), \qquad \epsilon_{ij} = \frac{1}{2}(u_{i,j} + u_{j,i}),} \tag{8.1.1}$$

the **free energy**

$$\boxed{\psi = \tfrac{1}{2}\boldsymbol{\epsilon}:\mathbb{C}\boldsymbol{\epsilon}, \qquad \psi = \tfrac{1}{2}\epsilon_{ij}C_{ijkl}\epsilon_{kl},} \tag{8.1.2}$$

and the **stress-strain relation**

$$\boxed{\boldsymbol{\sigma} = \frac{\partial\psi(\boldsymbol{\epsilon})}{\partial\boldsymbol{\epsilon}} = \mathbb{C}\boldsymbol{\epsilon}, \qquad \sigma_{ij} = C_{ijkl}\epsilon_{kl}.} \tag{8.1.3}$$

The *elasticity tensor* $\mathbb{C}$ *is symmetric*, and the elastic moduli C_{ijkl} obey the major and minor symmetry requirements,

$$C_{ijkl} = C_{klij}, \qquad C_{ijkl} = C_{jikl}, \qquad C_{ijkl} = C_{ijlk}. \tag{8.1.4}$$

Further, in view of (8.1.2) and the physical requirement that the free energy be positive for non-zero strain and zero otherwise, we require that $\mathbb{C}$ be *positive definite*:

$$\mathbf{A}:\mathbb{C}\mathbf{A} > 0 \quad \text{for all symmetric tensors } \mathbf{A} \neq \mathbf{0}. \tag{8.1.5}$$

The *basic equations of the linear theory of elasticity* consist of (8.1.1), (8.1.3), and the **force equilibrium equation**:

$$\boxed{\operatorname{div}\boldsymbol{\sigma} + \mathbf{b} = \mathbf{0}, \qquad \sigma_{ij,j} + b_i = 0,} \tag{8.1.6}$$

along with the **moment equilibrium equation**

$$\boxed{\boldsymbol{\sigma} = \boldsymbol{\sigma}^{\top}, \qquad \sigma_{ij} = \sigma_{ji},} \tag{8.1.7}$$

which we will implicitly always assume to be satisfied in what follows.

Continuum Mechanics of Solids. Lallit Anand and Sanjay Govindjee, Oxford University Press (2020).

DOI: 10.1093/oso/9780198864721.001.0001

8.2 **Boundary conditions**

The basic problem in the linear theory of elasticity is to find fields $[\mathbf{u}, \boldsymbol{\epsilon}, \boldsymbol{\sigma}]$ that satisfy the field equations everywhere within the body $\mathcal{B}$, while meeting specified *boundary conditions* on the boundary $\partial\mathcal{B}$ of the body. In order for a boundary-value problem to be well-posed, suitable boundary conditions must be prescribed at *every* point on $\partial\mathcal{B}$. With $\hat{\mathbf{u}}$ and $\hat{\mathbf{t}}$ *prescribed* displacement and traction functions of $\mathbf{x}$ on $\partial\mathcal{B}$, the simplest set of boundary conditions consists of specifying

(i) displacements $\mathbf{u} = \hat{\mathbf{u}}$ everywhere on $\partial\mathcal{B}$; or

(ii) surface tractions $\boldsymbol{\sigma}\mathbf{n} = \hat{\mathbf{t}}$ everywhere on $\partial\mathcal{B}$; or

(iii) with $\mathcal{S}_1$ and $\mathcal{S}_2$ denoting *complementary subsurfaces* ($\mathcal{S}_1 \cup \mathcal{S}_2 = \partial\mathcal{B}$ and $\mathcal{S}_1 \cap \mathcal{S}_2 = \emptyset$) of the boundary $\partial\mathcal{B}$, the displacement is specified on $\mathcal{S}_1$ and the surface traction on $\mathcal{S}_2$:

$$\left.\begin{aligned} \mathbf{u} &= \hat{\mathbf{u}} \quad \text{on } \mathcal{S}_1, \\ \boldsymbol{\sigma}\mathbf{n} &= \hat{\mathbf{t}} \quad \text{on } \mathcal{S}_2. \end{aligned}\right\} \tag{8.2.1}$$

These three types of boundary conditions are called *displacement*, *traction*, and *mixed* respectively. Clearly, cases (i) and (ii) listed above, are special cases of case (iii) when $\mathcal{S}_2 = \emptyset$, or when $\mathcal{S}_1 = \emptyset$, respectively.

Another type of more general mixed-boundary condition often occurs. Choose a local Cartesian basis $\{\mathbf{e}_i^*\}$ on each point of the boundary, with one axis say $\mathbf{e}_3^*$ along the outward unit normal to the boundary. Then one can consider the situation that in each direction $\mathbf{e}_i^*$, one specifies either the displacement component or the traction component:

(a) u_1^* or t_1^*, but not both,

(b) u_2^* or t_2^*, but not both, and

(c) u_3^* or t_3^*, but not both.

This more general case includes the other cases (i), (ii), and (iii) listed above as special cases. However, for ease of presentation of the theory, we only consider the case of mixed boundary conditions (8.2.1).

REMARK 8.1

Traction boundary conditions are sometimes called "stress boundary conditions," but this is misleading. It is physically impossible to impose boundary conditions from outside on all the components of the stress tensor at a boundary point.

8.3 **Mixed, displacement, and traction problems of elastostatics**

The **mixed problem of elastostatics** may now be stated as follows:

Given: $\mathbb{C}$, **b**, and boundary conditions

$$\left.\begin{aligned} \mathbf{u} &= \hat{\mathbf{u}} \quad \text{on } \mathcal{S}_1, \\ \boldsymbol{\sigma}\mathbf{n} &= \hat{\mathbf{t}} \quad \text{on } \mathcal{S}_2, \end{aligned}\right\} \tag{8.3.1}$$

on *complementary subsurfaces* $\mathcal{S}_1$ and $\mathcal{S}_2$, respectively, of the boundary $\partial\mathcal{B}$ of the body $\mathcal{B}$,

Find: a displacement field **u**, a strain field ϵ, and a stress field $\boldsymbol{\sigma}$ that satisfy the field equations

$$\left.\begin{aligned} \boldsymbol{\epsilon} &= \tfrac{1}{2}(\nabla\mathbf{u} + (\nabla\mathbf{u})^{\top}), \\ \boldsymbol{\sigma} &= \mathbb{C}\boldsymbol{\epsilon}, \\ \operatorname{div}\boldsymbol{\sigma} + \mathbf{b} &= \mathbf{0}, \end{aligned}\right\} \quad \text{in } \mathcal{B} \tag{8.3.2}$$

and the boundary conditions (8.3.1).

- If $\mathcal{S}_1 = \emptyset$, so that $\boldsymbol{\sigma}\mathbf{n} = \hat{\mathbf{t}}$ on all of $\partial\mathcal{B}$, then the mixed problem reduces to a **traction problem of elastostatics.**
- On the other hand, if $\mathcal{S}_2 = \emptyset$, so that $\mathbf{u} = \hat{\mathbf{u}}$ on all of $\partial\mathcal{B}$, then the mixed problem reduces to a **displacement problem of elastostatics.**

8.4 **Uniqueness**

When a boundary-value problem is formulated, natural questions to ask are: Does a solution exist? And if so, is it unique? Although a discussion of existence of solutions is beyond the scope of this book, we consider the question of uniqueness in what follows.

In many applications of the theory, we come across situations in which solutions may be obtained by inverse methods. That is, we postulate certain forms of the displacement and stress fields, and then show that these fields satisfy the field equations (8.3.2) together with the prescribed boundary conditions (8.3.1). Having found a solution, it is then natural to question whether another solution of the field equations could be found satisfying the same boundary conditions. For materials with realistic material properties, we may answer in the negative—the solution is unique. The proof of this statement depends on the positive definiteness of the elasticity tensor $\mathbb{C}$.

UNIQUENESS THEOREM: *There is at most one solution* $[\boldsymbol{u}, \epsilon, \boldsymbol{\sigma}]$ *of equations* (8.3.2) *which satisfy the boundary conditions* (8.3.1), *except if* $\mathcal{S}_1 = \emptyset$, *that is if no displacement boundary conditions are specified; in this latter case any two solutions differ at most by a rigid displacement.*

Proof: Assume that two different solutions $[\mathbf{u}_1, \epsilon_1, \boldsymbol{\sigma}_1]$ and $[\mathbf{u}_2, \epsilon_2, \boldsymbol{\sigma}_2]$ exist to the same problem with *identical body forces and boundary conditions*. Next, define the *difference solution*

$$\mathbf{u} = \mathbf{u}_1 - \mathbf{u}_2, \qquad \epsilon = \epsilon_1 - \epsilon_2, \qquad \boldsymbol{\sigma} = \boldsymbol{\sigma}_1 - \boldsymbol{\sigma}_2. \tag{8.4.1}$$

Because the solutions $\boldsymbol{\sigma}_1$ and $\boldsymbol{\sigma}_2$ satisfy the equilibrium equation with the same body force, the difference solution must satisfy the equilibrium equation with null body force, $\mathbf{b} = \mathbf{0}$:

$$\operatorname{div}\boldsymbol{\sigma} = \mathbf{0}. \tag{8.4.2}$$

Likewise, the boundary conditions satisfied by the difference solution are:

$$\left.\begin{aligned} \mathbf{u} &= \mathbf{0} \quad \text{on } \mathcal{S}_1, \\ \boldsymbol{\sigma}\mathbf{n} &= \mathbf{0} \quad \text{on } \mathcal{S}_2. \end{aligned}\right\} \tag{8.4.3}$$

Then the difference $\mathbf{u} = \mathbf{u}_1 - \mathbf{u}_2$ represents a solution of a mixed problem in which the body force $\mathbf{b}$ vanishes, as do the prescribed surface displacement and surface traction. Hence,

$$\int_{\mathcal{B}} \mathbf{b}\cdot\mathbf{u}\,dv = 0,$$

and

$$\int_{\partial\mathcal{B}} \boldsymbol{\sigma}\mathbf{n}\cdot\mathbf{u}\,da = \int_{\mathcal{S}_1} \boldsymbol{\sigma}\mathbf{n}\cdot\mathbf{u}\,da + \int_{\mathcal{S}_2} \boldsymbol{\sigma}\mathbf{n}\cdot\mathbf{u}\,da = 0.$$

Thus,

$$0 = \int_{\partial\mathcal{B}} \boldsymbol{\sigma}\mathbf{n}\cdot\mathbf{u}\,da = \int_{\partial\mathcal{B}} (\boldsymbol{\sigma}\mathbf{u})\cdot\mathbf{n}\,da = \int_{\mathcal{B}} \operatorname{div}(\boldsymbol{\sigma}\mathbf{u})\,dv = \int_{\mathcal{B}} (\operatorname{div}\boldsymbol{\sigma})\cdot\mathbf{u}\,dv + \int_{\mathcal{B}} \boldsymbol{\sigma}:\nabla\mathbf{u}\,dv,$$

and using (8.4.2), the symmetry of $\boldsymbol{\sigma}$, and the stress-strain relation $\boldsymbol{\sigma} = \mathbb{C}\boldsymbol{\epsilon}$ gives

$$\int_{\mathcal{B}} \boldsymbol{\epsilon}:\mathbb{C}\boldsymbol{\epsilon}\,dv = 0. \tag{8.4.4}$$

Since the integrand is non-negative, we must have $\boldsymbol{\epsilon}:\mathbb{C}\boldsymbol{\epsilon} = 0$ at all points $\mathbf{x} \in \mathcal{B}$. However, since $\mathbb{C}$ is positive definite, we must have $\boldsymbol{\epsilon} = \mathbf{0}$, which implies $\boldsymbol{\epsilon}_1 = \boldsymbol{\epsilon}_2$, uniqueness of the strains. Further, $\boldsymbol{\sigma} = \mathbb{C}\boldsymbol{\epsilon} = \mathbf{0}$, which implies $\boldsymbol{\sigma}_1 = \boldsymbol{\sigma}_2$, uniqueness of the stresses. Note also, the vanishing of the difference strain $\boldsymbol{\epsilon}$ renders the difference displacement $\mathbf{u}$ to be rigid. However, if $\mathcal{S}_1 \neq \emptyset$, then the difference displacement field $\mathbf{u}$ must vanish everywhere. Thus,

$$\mathbf{u}_1 = \mathbf{u}_2, \qquad \boldsymbol{\epsilon}_1 = \boldsymbol{\epsilon}_2, \qquad \boldsymbol{\sigma}_1 = \boldsymbol{\sigma}_2, \tag{8.4.5}$$

and therefore any solution to the elastostatic boundary-value problem is unique. However, if tractions are prescribed over the entire boundary, then $\mathbf{u}_1$ and $\mathbf{u}_2$ may differ by an infinitesimal rigid displacement.

8.5 **Superposition**

A major consequence of the *linear* nature of the theory under consideration is that the superposition principle, stated below, holds.

SUPERPOSITION PRINCIPLE: If the triplet of fields $[\mathbf{u}_1, \boldsymbol{\epsilon}_1, \boldsymbol{\sigma}_1]$ is a solution to equations (8.3.2) with prescribed body forces $\mathbf{b}_1$, displacements $\hat{\mathbf{u}}_1$ on $\mathcal{S}_1$, and tractions $\hat{\mathbf{t}}_1$ on $\mathcal{S}_2$, and if $[\mathbf{u}_2, \boldsymbol{\epsilon}_2, \boldsymbol{\sigma}_2]$ is also a solution with prescribed body forces $\mathbf{b}_2$, displacements $\hat{\mathbf{u}}_2$ on $\mathcal{S}_1$, and tractions $\hat{\mathbf{t}}_2$ on $\mathcal{S}_2$, then the superposed fields

$$\mathbf{u} = \mathbf{u}_1 + \mathbf{u}_2,$$
$$\boldsymbol{\epsilon} = \boldsymbol{\epsilon}_1 + \boldsymbol{\epsilon}_2,$$
$$\boldsymbol{\sigma} = \boldsymbol{\sigma}_1 + \boldsymbol{\sigma}_2,$$

are also a solution[1] to (8.3.2) with body force $\mathbf{b} = \mathbf{b}_1 + \mathbf{b}_2$, and boundary conditions

$$\left.\begin{aligned} \mathbf{u} &= \hat{\mathbf{u}}_1 + \hat{\mathbf{u}}_2 \quad \text{on } \mathcal{S}_1, \\ \boldsymbol{\sigma}\mathbf{n} &= \hat{\mathbf{t}}_1 + \hat{\mathbf{t}}_2, \quad \text{on } \mathcal{S}_2. \end{aligned}\right\} \tag{8.5.1}$$

- Hence, for a given geometry of the body, the solutions to some simple problems can be combined to generate solutions to more complicated problems.

8.6 **Saint-Venant's principle**

In the solution of elastic boundary-value problems, it is often problematic to exactly satisfy the traction boundary conditions, or often the boundary conditions themselves are not fully specified in terms of pointwise traction distributions. Saint-Venant's principle helps in these situations to deliver meaningful solutions of practical utility.

Consider an elastic body $\mathcal{B}$ subjected to an arbitrary traction distribution $\mathbf{t}(\mathbf{n})$ on a portion $\mathcal{S}$ of its boundary $\partial\mathcal{B}$; cf. Fig. 8.1. Saint-Venant's Principle (Barré de Saint-Venant, 1855) embodies the physical expectation that at points of the body which are sufficiently far away from $\mathcal{S}$, the displacement, strain, and stress fields will depend on the *resultant* force $\mathbf{f}_{\text{resultant}}$ and moment $\mathbf{m}_{\text{resultant}}$ corresponding to $\mathbf{t}(\mathbf{n})$ rather than on the precise traction distribution $\mathbf{t}(\mathbf{n}, \mathbf{x})$ itself:

$$\mathbf{f}_{\text{resultant}} = \int_{\mathcal{S}} \mathbf{t}(\mathbf{n}, \mathbf{x})\, da$$
$$\mathbf{m}_{\text{resultant}} = \int_{\mathcal{S}} (\mathbf{x} - \mathbf{0}) \times \mathbf{t}(\mathbf{n}, \mathbf{x})\, da.$$

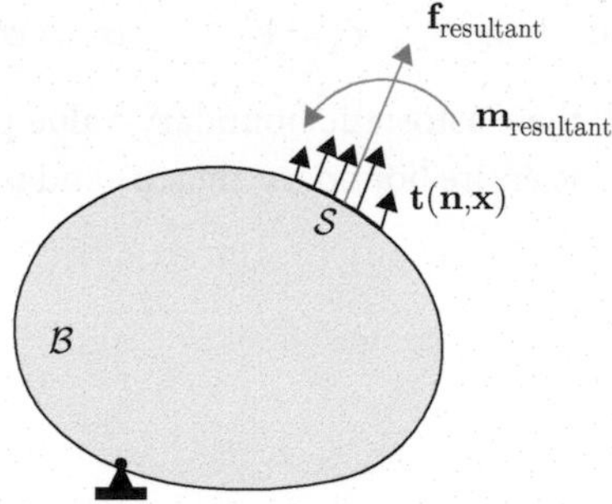

Fig. 8.1 Saint-Venant's principle. Resultant force and moment due to a distributed traction.

[1] Provided the prescribed tractions and body forces are independent of deformation, and provided the material remains linear elastic!

Thus, denoting two traction distributions to be **statically equivalent** if they have the same resultant force and moment, Saint-Venant's Principle may be stated as follows:

SAINT-VENANT'S PRINCIPLE: *The displacement, strain, and stress fields—caused by two different but statically equivalent force distributions on a portion of the boundary of a body—in parts of the body which are sufficiently far away from the loading points are approximately the same.*

REMARK 8.2

Note that this statement of Saint-Venant's principle is *qualitative* in nature since it includes terms such as "sufficiently far away" and "approximately the same". Saint-Venant introduced his "principle of elastic equivalence of statically equivalent systems of loads" only for the special case of perfect cylinders. A broader statement of the principle, was given by Love (1927),

> . . . the strains that are produced in a body by the application, to a small part of its surface, of a system of forces statically equivalent to zero force and zero couple, are of negligible magnitude at distances which are large compared with the linear dimensions of the part . . .

Since that time there has been a large body of literature concerning the *quantitative range of applicability* of the principle for bodies of more general shapes; for more recent discussions see Toupin (1965) and Horgan (1996).

8.7 Displacement formulation. The Navier equations

There are a number of different ways that one can approach an elastostatic boundary-value problem. In this section we develop a reduced set of field equations *only in terms of the displacement field* **u**. This is achieved by combining the strain-displacement relation (8.1.1), the stress-strain relation (8.1.3), and the equation of equilibrium (8.1.6) to form a single partial differential equation for the displacement field **u**—the **Navier displacement equation of equilibrium**. For simplicity, we limit our discussion here to a homogeneous body.

Starting from (8.1.6)

$$\sigma_{ij,j} + b_i = 0,$$

we can replace the stress using (8.1.3) to yield

$$C_{ijkl}\epsilon_{kl,j} + b_i = 0.$$

Next we can substitute for the strain in terms of the displacement using (8.1.1) to give

$$\boxed{C_{ijkl}u_{k,lj} + b_i = 0,} \tag{8.7.1}$$

which are known as **Navier's displacement equations of equilibrium**. In arriving at (8.7.1) we have taken advantage of the symmetry properties of the elasticity tensor. These equations represent the equilibrium equations of the body in terms of the displacement field, where the constitutive law and the strain-displacement relations are implicitly already accounted for. Note these are three partial differential equations for the three components of the displacement field.

When the body is *isotropic* the elasticity tensor has the simple form

$$C_{ijkl} = \mu(\delta_{ik}\delta_{jl} + \delta_{il}\delta_{jk}) + \lambda\delta_{ij}\delta_{kl},$$

and in this case the displacement equation of equilibrium takes the form

$$\boxed{\mu\Delta\mathbf{u} + (\lambda+\mu)\nabla\,\mathrm{div}\,\mathbf{u} + \mathbf{b} = 0, \qquad \mu\, u_{i,jj} + (\lambda+\mu)u_{j,ji} + b_i = 0,} \tag{8.7.2}$$

which must be satisfied in $\mathcal{B}$. The boundary conditions (8.3.1) for the displacement problem in the isotropic case may be written as

$$\boxed{\begin{aligned} \mathbf{u} &= \hat{\mathbf{u}} \quad \text{on } \mathcal{S}_1, \\ \Big(\mu(\nabla\mathbf{u} + (\nabla\mathbf{u})^{\top}) + \lambda(\mathrm{div}\,\mathbf{u})\mathbf{1}\Big)\mathbf{n} &= \hat{\mathbf{t}} \quad \text{on } \mathcal{S}_2. \end{aligned}} \tag{8.7.3}$$

Equation (8.7.2) may be written in an alternative form by using the identity

$$\mathrm{curl}\,\mathrm{curl}\,\mathbf{u} = \nabla\,\mathrm{div}\,\mathbf{u} - \Delta\mathbf{u}, \tag{8.7.4}$$

which is easily verified in components as follows:

$$\begin{aligned} (\mathrm{curl}(\mathrm{curl}\,\mathbf{u}))_i &= e_{ijk}(\mathrm{curl}\,\mathbf{u})_{k,j} \\ &= e_{ijk}e_{klm}u_{m,lj} = e_{kij}e_{klm}u_{m,lj} \\ &= (\delta_{il}\delta_{jm} - \delta_{im}\delta_{jl})u_{m,lj} \\ &= u_{j,ji} - u_{i,ll} \\ &= (\nabla\,\mathrm{div}\,\mathbf{u} - \Delta\mathbf{u})_i. \end{aligned}$$

Thus using (8.7.4) in (8.7.2) we obtain the alternate form of the equilibrium equations

$$\boxed{(\lambda+2\mu)\nabla\,\mathrm{div}\,\mathbf{u} - \mu\;\mathrm{curl}\,\mathrm{curl}\,\mathbf{u} + \mathbf{b} = \mathbf{0}, \qquad (\lambda+2\mu)u_{j,ji} - \mu\, e_{ijk}e_{klm}u_{m,lj} + b_i = 0,} \tag{8.7.5}$$

which must be satisfied in $\mathcal{B}$.

8.8 **Stress formulation. The Beltrami–Michell equations**

Often the boundary conditions are given only in terms of the tractions or components of the stress. In order to develop solution methodologies for such circumstances it is helpful to reformulate the general system of equations for elastostatics by eliminating the displacement and strain fields and cast the system solely in terms of the stress field.

The way we will do this is to start with a formulation that only involves stress and strain fields. In order to ensure that the determined strain field is compatible with a displacement field we will use the **compatibility relation** (cf., (3.B.6)) in the fundamental system of equations:

$$\boxed{\mathrm{curl}(\mathrm{curl}\,\boldsymbol{\epsilon}) = \mathbf{0}, \qquad e_{ipq}e_{jrs}\epsilon_{qs,rp} = 0.} \tag{8.8.1}$$

Lastly, we will eliminate the strain field using the constitutive relation. For simplicity we will restrict our attention to the isotropic case. Recall that for isotropic materials the inverted form of stress-strain relation may be written as

$$\boldsymbol{\epsilon} = \frac{1}{E}\left[(1+\nu)\boldsymbol{\sigma} - \nu\ (\operatorname{tr}\boldsymbol{\sigma})\ \mathbf{1}\right], \qquad \epsilon_{ij} = \frac{1}{E}\left[(1+\nu)\sigma_{ij} - \nu\ (\sigma_{kk})\ \delta_{ij}\right]. \tag{8.8.2}$$

By substituting the strain-stress relation (8.8.2) into the compatibility equation (8.8.1), and by using the equilibrium equation (8.1.6) one can deduce a set of compatibility equations in terms of the components of stress which are known as the **Beltrami–Michell compatibility equations**, which we derive below.

Substituting the the strain-stress relation (8.8.2) into the compatibility relation (8.8.1) gives

$$\begin{aligned}
0 &= e_{ipq}e_{jrs}\epsilon_{qs,rp}, \\
&= e_{ipq}e_{jrs}\Big((1+\nu)\sigma_{qs,rp} - \nu\,\sigma_{mm,rp}\delta_{qs}\Big), \\
&= (1+\nu)e_{ipq}e_{jrs}\sigma_{qs,rp} - \nu\, e_{ips}e_{jrs}\,\sigma_{mm,rp}, \\
&= (1+\nu)e_{ipq}e_{jrs}\sigma_{qs,rp} - \nu\,(\delta_{ij}\delta_{pr} - \delta_{ir}\delta_{pj})\,\sigma_{mm,rp}, \\
&= (1+\nu)e_{ipq}e_{jrs}\sigma_{qs,rp} - \nu\,(\delta_{ij}\sigma_{mm,pp} - \sigma_{mm,ij}).
\end{aligned} \tag{8.8.3}$$

Next, recalling (1.1.11), viz.

$$\begin{aligned}
e_{ipq}e_{jrs} &= \det\begin{bmatrix}\delta_{ij} & \delta_{ir} & \delta_{is}\\ \delta_{pj} & \delta_{pr} & \delta_{ps}\\ \delta_{qj} & \delta_{qr} & \delta_{qs}\end{bmatrix}, \\
&= \delta_{ij}(\delta_{pr}\delta_{qs} - \delta_{ps}\delta_{qr}) - \delta_{ir}(\delta_{pj}\delta_{qs} - \delta_{ps}\delta_{qj}) + \delta_{is}(\delta_{pj}\delta_{qr} - \delta_{pr}\delta_{qj}),
\end{aligned} \tag{8.8.4}$$

we find that

$$\begin{aligned}
e_{ipq}e_{jrs}\sigma_{qs,rp} &= \delta_{ij}(\delta_{pr}\delta_{qs} - \delta_{ps}\delta_{qr})\sigma_{qs,rp} - \delta_{ir}(\delta_{pj}\delta_{qs} - \delta_{ps}\delta_{qj})\sigma_{qs,rp} \\
&\quad + \delta_{is}(\delta_{pj}\delta_{qr} - \delta_{pr}\delta_{qj})\sigma_{qs,rp} \\
&= \delta_{ij}(\sigma_{qq,pp} - \sigma_{rs,rs}) - (\sigma_{ss,ij} - \sigma_{js,is}) + (\sigma_{ri,rj} - \sigma_{ji,rr}),
\end{aligned} \tag{8.8.5}$$

substitution of which in (8.8.3) simplifies that relation to

$$\delta_{ij}\sigma_{qq,pp} - \sigma_{ss,ij} - (1+\nu)(\delta_{ij}\sigma_{rs,sr} + \sigma_{ij,rr} - \sigma_{js,si} - \sigma_{ir,rj}) = 0. \tag{8.8.6}$$

Contracting the indices i and j in (8.8.6) gives

$$2\sigma_{qq,pp} - (1+\nu)(\sigma_{rs,rs} + \sigma_{ii,rr}) = 0, \tag{8.8.7}$$

which upon simplification gives

$$\sigma_{rr,ss} = \frac{(1+\nu)}{1-\nu}\sigma_{rs,rs}. \tag{8.8.8}$$

Substitution of (8.8.8) into (8.8.6) gives

$$\sigma_{ij,rr} + \frac{1}{1+\nu}\sigma_{ss,ij} = \frac{\nu}{1-\nu}\delta_{ij}\sigma_{rs,sr} + \sigma_{js,si} + \sigma_{ir,rj}. \tag{8.8.9}$$

Next, differentiating the equilibrium equation $\sigma_{ij,j} = -b_i$ with respect to x_k gives

$$\sigma_{ij,jk} = -b_{i,k},$$

use of which in (8.8.9) gives the following six compatibility equations in terms of the components of stress

$$\boxed{\sigma_{ij,kk} + \frac{1}{1+\nu}\sigma_{kk,ij} = -\frac{\nu}{1-\nu}\, b_{k,k}\delta_{ij} - b_{i,j} - b_{j,i},} \tag{8.8.10}$$

which are the **Beltrami–Michell equations.** It is also sometimes useful to observe that the trace of the stress satisfies the Poisson's equation

$$\Delta\sigma_{kk} = -\frac{1+\nu}{1-\nu}\, b_{k,k}. \tag{8.8.11}$$

When the body force is uniform, we have

$$\operatorname{div}\mathbf{b} = 0, \qquad \nabla\mathbf{b} = \mathbf{0};$$

under this circumstance or under the circumstance that the body forces vanish, we have that (8.8.10) and (8.8.11) reduce to

$$\boxed{\sigma_{ij,kk} + \frac{1}{1+\nu}\sigma_{kk,ij} = 0,} \tag{8.8.12}$$

and

$$\Delta\sigma_{kk} = 0, \tag{8.8.13}$$

respectively.

Note that actually only three of the Beltrami–Michell equations (8.8.10) *are independent equations.* To obtain the complete set of equations for the stress formulation we must add the three equations of equilibrium

$$\sigma_{ij,j} + b_i = 0 \tag{8.8.14}$$

to (8.8.10) to obtain a complete stress formulation for the six unknown components of the stress σ_{ij}.

REMARK 8.3

It is not convenient to solve these equations for the six components of stress, and very few solutions to the full three-dimensional problem based on the stress formulation have been attempted. A more convenient formulation in terms of **stress functions** is possible in certain special cases. The most widely used stress function, *the Airy stress function* of two-dimensional elasticity will be discussed later in Sec. 9.4.

8.9 Two-dimensional problems

Because of the complexity of the field equations of linear elasticity, it is often difficult to obtain analytical closed-form solutions to fully three-dimensional problems. Accordingly, in this section we study the formulation of two-dimensional problems. Because all real structures are three-dimensional in nature, the specialized two-dimensional theories discussed in this section represent approximate models, with the nature and accuracy of the approximation depending on the actual geometry of the body and the loading conditions. We shall focus on the two basic theories of *plane strain* and *plane stress* for isotropic materials.

We develop the two-dimensional theories in the $(\mathbf{e}_1, \mathbf{e}_2)$-plane, using Cartesian coordinates. In this case, when using indicial notation, we shall use Greek indices that range over the values $(1, 2)$. For example,

$$\sigma_{\alpha\beta,\beta} + b_\alpha = 0, \tag{8.9.1}$$

is equivalent to

$$\begin{aligned} \sigma_{11,1} + \sigma_{12,2} + b_1 &= 0, \\ \sigma_{21,1} + \sigma_{22,2} + b_2 &= 0. \end{aligned} \tag{8.9.2}$$

8.9.1 Plane strain

Consider a long cylindrical body with generators parallel to the $\mathbf{e}_3$-axis and cross-section in the $(\mathbf{e}_1, \mathbf{e}_2)$-plane. We denote the intersection of the cylindrical body with the $(\mathbf{e}_1, \mathbf{e}_2)$-plane as the region $\mathcal{R}$ with boundary curve $\mathcal{C}$; cf. Fig. 8.2. Such a body is said to be in a state of *plane strain* in the $(\mathbf{e}_1, \mathbf{e}_2)$-plane when the u_3 component of the displacement is identically zero everywhere, and when the u_1 and u_2 components of the displacement field are functions only of (x_1, x_2) and independent of x_3:

$$\boxed{u_\alpha = u_\alpha(x_1, x_2), \qquad u_3 = 0.} \tag{8.9.3}$$

Such a displacement field might arise in the cylindrical body only if the prescribed body forces, surface tractions, and surface displacements are all also independent of x_3.

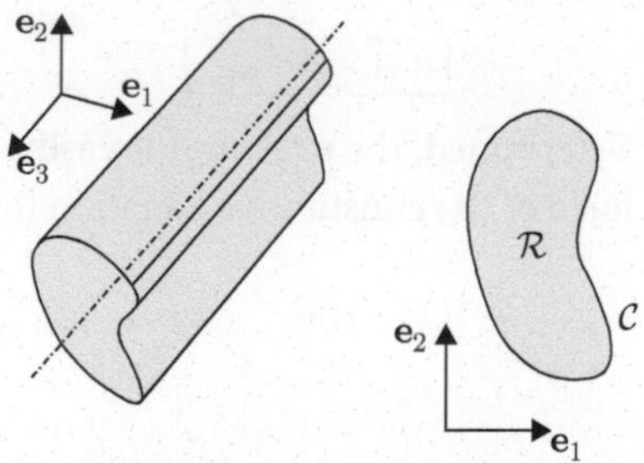

Fig. 8.2 A long cylindrical body with generators parallel to the $\mathbf{e}_3$-axis and cross-section in the $(\mathbf{e}_1, \mathbf{e}_2)$-plane.

Using the three-dimensional strain-displacement relation

$$\epsilon_{ij} = \tfrac{1}{2}(u_{i,j} + u_{j,i}), \tag{8.9.4}$$

we obtain that in plane strain

$$\boxed{\epsilon_{\alpha\beta} = \tfrac{1}{2}(u_{\alpha,\beta} + u_{\beta,\alpha}), \qquad \text{together with} \qquad \epsilon_{13} = \epsilon_{23} = \epsilon_{33} = 0.} \tag{8.9.5}$$

From the constitutive equation

$$\sigma_{ij} = \frac{E}{(1+\nu)}\left(\epsilon_{ij} + \frac{\nu}{(1-2\nu)}(\epsilon_{kk})\delta_{ij}\right), \tag{8.9.6}$$

equations (8.9.5), and noting that

$$\epsilon_{kk} = \epsilon_{11} + \epsilon_{22} = \epsilon_{\gamma\gamma},$$

we have

$$\sigma_{13} = \sigma_{23} = 0, \qquad \text{but} \qquad \sigma_{33} = \frac{E\nu}{(1+\nu)(1-2\nu)}\epsilon_{\gamma\gamma}, \tag{8.9.7}$$

and hence,

$$\boxed{\sigma_{\alpha\beta} = \frac{E}{(1+\nu)}\left(\epsilon_{\alpha\beta} + \frac{\nu}{(1-2\nu)}(\epsilon_{\gamma\gamma})\delta_{\alpha\beta}\right).} \tag{8.9.8}$$

Here

$$\delta_{\alpha\beta} = \begin{cases} 1 & \text{if } \alpha = \beta, \\ 0 & \text{if } \alpha \neq \beta, \end{cases} \tag{8.9.9}$$

is the two-dimensional Kronecker delta. Since

$$\delta_{\alpha\alpha} = 2,$$

from (8.9.8) we obtain

$$\sigma_{\alpha\alpha} = \frac{E}{(1+\nu)(1-2\nu)}(\epsilon_{\alpha\alpha}). \tag{8.9.10}$$

Thus, from (8.9.7) and (8.9.10) we obtain that the normal stress in the direction of the plane strain constraint is determined by the sum of the in-plane normal stresses,

$$\boxed{\sigma_{33} = \nu\sigma_{\alpha\alpha}.} \tag{8.9.11}$$

Hence, once σ_{11} and σ_{22} are determined, the stress σ_{33} is easily calculated using (8.9.11).

To obtain the plane strain form of the constitutive equation for strain in terms of stresses,

$$\epsilon_{ij} = \frac{1}{E}((1+\nu)\sigma_{ij} - \nu(\sigma_{kk})\delta_{ij}), \tag{8.9.12}$$

we note that

$$\sigma_{kk} = \sigma_{\gamma\gamma} + \sigma_{33} = (1+\nu)\sigma_{\gamma\gamma},$$

substitution of which in (8.9.12) gives

$$\boxed{\epsilon_{\alpha\beta} = \frac{1+\nu}{E}(\sigma_{\alpha\beta} - \nu(\sigma_{\gamma\gamma})\delta_{\alpha\beta}).} \tag{8.9.13}$$

The relations (8.9.8) and (8.9.13) are the *plane strain constitutive equations* of isotropic linear elasticity.

Using (8.9.8), the three-dimensional equilibrium equations

$$\sigma_{ij,j} + b_i = 0$$

reduce to the following *plane strain equations of equilibrium*,

$$\boxed{\sigma_{\alpha\beta,\beta} + b_\alpha = 0,} \tag{8.9.14}$$

provided the body force lies in the $(\mathbf{e}_1, \mathbf{e}_2)$-plane and is independent of the x_3-coordinate,

$$b_\alpha = b_\alpha(x_1, x_2), \qquad b_3 = 0.$$

Finally with $\mathcal{C}_1$ and $\mathcal{C}_2$ denoting *complementary subsurfaces* of the boundary $\mathcal{C}$ of the two-dimensional region $\mathcal{R}$ in the $(\mathbf{e}_1, \mathbf{e}_2)$-plane under consideration, typical boundary conditions are

$$\left.\begin{aligned} u_\alpha &= \hat{u}_\alpha \quad \text{on } \mathcal{C}_1, \\ \sigma_{\alpha\beta} n_\beta &= \hat{t}_\alpha \quad \text{on } \mathcal{C}_2, \end{aligned}\right\} \tag{8.9.15}$$

with $\hat{u}_\alpha = \hat{u}_\alpha(x_1, x_2)$ and $\hat{t}_\alpha = \hat{t}_\alpha(x_1, x_2)$ prescribed displacements and tractions.

The solution to the plane strain problem then involves the determination of the in-plane displacements, strains, and stresses, $[u_\alpha, \epsilon_{\alpha\beta}, \sigma_{\alpha\beta}]$, in $\mathcal{R}$, subject to the boundary condition (8.9.15). The out-of-plane stress σ_{33} is subsequently determined using (8.9.11).

Navier equation in plane strain

For plane strain the displacement equation of equilibrium (8.7.2) takes the form

$$\boxed{\left(\frac{E}{2(1+\nu)}\right) u_{\alpha,\beta\beta} + \left(\frac{E}{2(1+\nu)(1-2\nu)}\right) u_{\beta,\beta\alpha} + b_\alpha = 0.} \tag{8.9.16}$$

Equation of compatibility in plane strain

For plane strain, five of the six compatibility equations (3.B.7) are identically satisfied, and the remaining equation reduces to

$$\boxed{\epsilon_{11,22} + \epsilon_{22,11} - 2\epsilon_{12,12} = 0.} \tag{8.9.17}$$

Using the constitutive equation (8.9.13) this compatibility equation may be expressed in terms of the stress as follows:

$$(1-\nu)(\sigma_{11,22} + \sigma_{22,11}) - \nu(\sigma_{11,11} + \sigma_{22,22}) = 2\sigma_{12,12}. \tag{8.9.18}$$

However, differentiating the equilibrium equations (8.9.14) with respect to x_1 and x_2, respectively, gives

$$\begin{aligned} \sigma_{11,11} + \sigma_{12,12} + b_{1,1} &= 0, \\ \sigma_{12,12} + \sigma_{22,22} + b_{2,2} &= 0, \end{aligned}$$

which upon adding together give the following necessary condition for equilibrium:

$$2\sigma_{12,12} = -(\sigma_{11,11} + \sigma_{22,22}) - (b_{1,1} + b_{2,2}). \tag{8.9.19}$$

Using (8.9.19) in (8.9.18) gives

$$(\sigma_{11,11} + \sigma_{11,22} + \sigma_{22,11} + \sigma_{22,22}) = -\frac{1}{1-\nu}(b_{1,1} + b_{2,2}),$$
$$(\sigma_{11} + \sigma_{22})_{,11} + (\sigma_{11} + \sigma_{22})_{,22} = -\frac{1}{1-\nu}(b_{1,1} + b_{2,2}).$$

Hence the equation of compatibility when expressed in terms of stress and accounting for equilibrium reduces to

$$\boxed{\Delta(\sigma_{\alpha\alpha}) = -\frac{1}{1-\nu}\, b_{\alpha,\alpha},} \qquad (8.9.20)$$

where Δ is the two-dimensional Laplacian operator defined as

$$\Delta\phi \overset{\text{def}}{=} \phi_{,\beta\beta} = \phi_{,11} + \phi_{,22}.$$

8.9.2 **Plane stress**

For the second type of two-dimensional theory consider a very short cylindrical body with generators parallel to the $\mathbf{e}_3$-axis, and dimensions in the $(\mathbf{e}_1, \mathbf{e}_2)$-plane much larger than the thickness of the cylinder in the x_3-direction. The mid-plane of the cylindrical body is identified with the $(\mathbf{e}_1, \mathbf{e}_2)$-plane and denoted by $\mathcal{R}$; it is bounded by the curve $\mathcal{C}$, cf. Fig. 8.3.

Such a body is said to be in a state of of *plane stress* in the $(\mathbf{e}_1, \mathbf{e}_2)$-plane if

$$\boxed{\sigma_{\alpha\beta} = \sigma_{\alpha\beta}(x_1, x_2), \qquad \sigma_{33} = \sigma_{13} = \sigma_{23} = 0.} \qquad (8.9.21)$$

Such a stress field may arise in a thin sheet-like body, only if the prescribed body forces, surface tractions, and surface displacements are all independent of x_3.[2]

To obtain the plane stress form of the constitutive equation for strains in terms of stresses from the three-dimensional constitutive relation,

$$\epsilon_{ij} = \frac{(1+\nu)}{E}\left(\sigma_{ij} - \frac{\nu}{(1+\nu)}(\sigma_{kk})\delta_{ij}\right), \qquad (8.9.22)$$

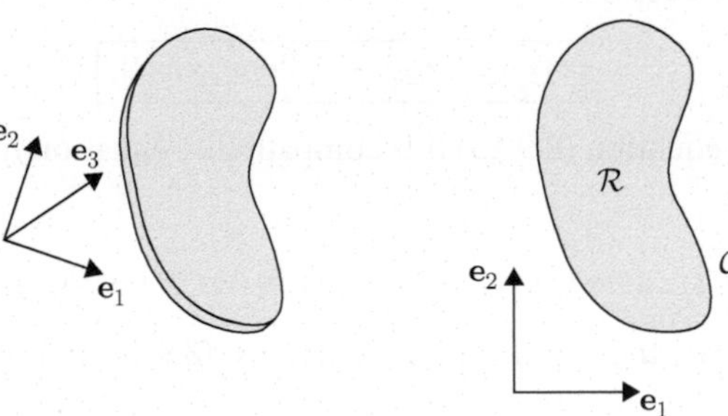

Fig. 8.3 A short cylindrical body with generators parallel to the $\mathbf{e}_3$-axis and cross-section in the $(\mathbf{e}_1, \mathbf{e}_2)$-plane.

[2] The components $\sigma_{\alpha\beta}$ usually vary somewhat with x_3. To get a two-dimensional problem we can take their average values through the thickness of the plate as *generalized plane stress components*.

we note that since

$$\sigma_{kk} = \sigma_{11} + \sigma_{22} = \sigma_{\gamma\gamma},$$

we have

$$\boxed{\epsilon_{\alpha\beta} = \frac{(1+\nu)}{E}\left(\sigma_{\alpha\beta} - \frac{\nu}{(1+\nu)}(\sigma_{\gamma\gamma})\delta_{\alpha\beta}\right),} \tag{8.9.23}$$

and

$$\epsilon_{13} = \epsilon_{23} = 0, \qquad \text{with} \qquad \epsilon_{33} = -\frac{\nu}{E}\sigma_{\gamma\gamma}, \tag{8.9.24}$$

where ϵ_{33} is a function only of (x_1, x_2). Next, from (8.9.23) and $(8.9.24)_2$

$$\epsilon_{\alpha\alpha} = \frac{1-\nu}{E}\sigma_{\alpha\alpha} = -\frac{1-\nu}{\nu}\epsilon_{33},$$

and hence

$$\boxed{\epsilon_{33} = -\frac{\nu}{1-\nu}\epsilon_{\alpha\alpha}.} \tag{8.9.25}$$

Thus the strain ϵ_{33} in the thickness direction is determined from a knowledge of the in-plane strains ϵ_{11} and ϵ_{22}.

The plane stress form of the equation for the stresses in terms of strains is obtained from the general equation

$$\sigma_{ij} = \frac{E}{(1+\nu)}\left(\epsilon_{ij} + \frac{\nu}{(1-2\nu)}(\epsilon_{kk})\delta_{ij}\right),$$

by using (8.9.25), and noting that

$$\epsilon_{kk} = \epsilon_{\gamma\gamma} + \epsilon_{33} = \epsilon_{\gamma\gamma} - \frac{\nu}{(1-\nu)}\epsilon_{\gamma\gamma} = \frac{(1-2\nu)}{(1-\nu)}\epsilon_{\gamma\gamma}, \tag{8.9.26}$$

and hence that

$$\boxed{\sigma_{\alpha\beta} = \frac{E}{(1+\nu)}\left(\epsilon_{\alpha\beta} + \frac{\nu}{(1-\nu)}(\epsilon_{\gamma\gamma})\delta_{\alpha\beta}\right).} \tag{8.9.27}$$

Equations (8.9.23) and (8.9.27) represent the *plane stress constitutive equations.*

Using (8.9.21) the three-dimensional equilibrium equations

$$\sigma_{ij,j} + b_i = 0$$

reduce to the following *plane stress equations of equilibrium,*

$$\boxed{\sigma_{\alpha\beta,\beta} + b_\alpha = 0,} \tag{8.9.28}$$

provided the body force lies in the $(\mathbf{e}_1, \mathbf{e}_2)$-plane and is independent of the x_3-coordinate:

$$b_\alpha = b_\alpha(x_1, x_2), \qquad b_3 = 0.$$

The equilibrium equations (8.9.28) for plane stress are identical in form to those for plane strain, equation (8.9.14).

Finally with $\mathcal{C}_1$ and $\mathcal{C}_2$ denoting *complementary subsurfaces* of the boundary $\mathcal{C}$ of the two-dimensional region $\mathcal{R}$ in the $(\mathbf{e}_1, \mathbf{e}_2)$-plane under consideration, typical boundary conditions are

$$\left.\begin{aligned} u_\alpha &= \hat{u}_\alpha \quad \text{on } \mathcal{C}_1, \\ \sigma_{\alpha\beta} n_\beta &= \hat{t}_\alpha \quad \text{on } \mathcal{C}_2, \end{aligned}\right\} \tag{8.9.29}$$

with $\hat{u}_\alpha = \hat{u}_\alpha(x_1, x_2)$ and $\hat{t}_\alpha = \hat{t}_\alpha(x_1, x_2)$ prescribed displacements and tractions.

The solution to the plane stress problem then involves the determination of the in-plane displacements, strains, and stresses $[u_\alpha, \epsilon_{\alpha\beta}, \sigma_{\alpha\beta}]$ in $\mathcal{R}$, subject to the boundary conditions (8.9.29). The out-of-plane strain ϵ_{33} is then determined by (8.9.25).

Navier equation in plane stress

In plane stress, using (8.9.27),

$$\sigma_{\alpha\beta,\beta} = \frac{E}{(1+\nu)} u_{\alpha,\beta\beta} + \frac{E}{2(1-\nu)} u_{\beta,\beta\alpha},$$

and hence the displacement equation of equilibrium takes the form

$$\boxed{\frac{E}{(1+\nu)} u_{\alpha,\beta\beta} + \frac{E}{2(1-\nu)} u_{\beta,\beta\alpha} + b_\alpha = 0.} \tag{8.9.30}$$

Compatibility equation in terms of stress

Next, on account of (8.9.24), for the case of plane stress the compatibility relations (3.B.7) reduce to

$$\boxed{\begin{aligned} \epsilon_{11,22} + \epsilon_{22,11} - 2\epsilon_{12,12} &= 0, \\ \epsilon_{33,22} &= 0, \\ \epsilon_{33,11} &= 0, \\ \epsilon_{33,12} &= 0. \end{aligned}} \tag{8.9.31}$$

- *In an engineering theory for plane stress, the last three compatibility equations involving ϵ_{33} are* neglected, *and this leads to an* approximate theory *which is acceptable for* thin bodies; *that is, bodies in which the thickness of the body in the x_3-direction is small relative to the in-plane dimensions. This is related to the fact that since the faces of such a body in the $\pm\mathbf{e}_3$-direction are traction-free, the stress σ_{33} is essentially zero for thin bodies, and may be neglected.*[3]

Thus, the remaining compatibility equation is $(8.9.31)_1$. Using the constitutive equation (8.9.23) this compatibility equation can be expressed in terms of the stress as follows:

$$(\sigma_{11,22} + \sigma_{22,11}) - \nu(\sigma_{11,11} + \sigma_{22,22}) = 2(1+\nu)\sigma_{12,12}. \tag{8.9.32}$$

[3] It is possible to develop a theory of "generalized" plane stress, in which such an assumption is not invoked; however, we do not go into such matters here; see e.g. Shames and Cozzarelli (1997, Sec. 13.A) for further details.

However, differentiating the equilibrium equations (8.9.28) with respect to x_1 and x_2, respectively, gives

$$\sigma_{11,11} + \sigma_{12,12} + b_{1,1} = 0,$$
$$\sigma_{12,12} + \sigma_{22,22} + b_{2,2} = 0,$$

which upon adding together give the following necessary condition for equilibrium:

$$2\sigma_{12,12} = -(\sigma_{11,11} + \sigma_{22,22}) - (b_{1,1} + b_{2,2}). \tag{8.9.33}$$

Using (8.9.33) in (8.9.32) gives

$$(\sigma_{11} + \sigma_{22})_{,11} + (\sigma_{11} + \sigma_{22})_{,22} = -(1+\nu)\,(b_{1,1} + b_{2,2}). \tag{8.9.34}$$

Hence the equation of compatibility for plane stress when expressed in terms of stress and accounting for equilibrium reduces to

$$\boxed{\Delta(\sigma_{\alpha\alpha}) = -(1+\nu)\,b_{\alpha,\alpha}.} \tag{8.9.35}$$

8.9.3 **Similarity of plane strain and plane stress equations**

Note that the set of equations for plane strain and the set of equations for plane stress are quite similar in nature, with the essential differences occurring in some coefficients involving the elastic constants:

- The plane strain constitutive equations (8.9.8) and (8.9.13),

$$\sigma_{\alpha\beta} = \frac{E}{(1+\nu)}\left(\epsilon_{\alpha\beta} + \frac{\nu}{(1-2\nu)}(\epsilon_{\gamma\gamma})\delta_{\alpha\beta}\right), \qquad \epsilon_{\alpha\beta} = \frac{1+\nu}{E}\Big(\sigma_{\alpha\beta} - \nu(\sigma_{\gamma\gamma})\delta_{\alpha\beta}\Big), \tag{8.9.36}$$

  compare with the following plane stress constitutive equations (8.9.27) and (8.9.23),

$$\sigma_{\alpha\beta} = \frac{E}{(1+\nu)}\left(\epsilon_{\alpha\beta} + \frac{\nu}{(1-\nu)}(\epsilon_{\gamma\gamma})\delta_{\alpha\beta}\right), \qquad \epsilon_{\alpha\beta} = \frac{1+\nu}{E}\left(\sigma_{\alpha\beta} - \frac{\nu}{(1+\nu)}(\sigma_{\gamma\gamma})\delta_{\alpha\beta}\right). \tag{8.9.37}$$

- The Navier displacement equation of equilibrium in plane strain

$$\left(\frac{E}{2(1+\nu)}\right)u_{\alpha,\beta\beta} + \left(\frac{E}{2(1+\nu)(1-2\nu)}\right)u_{\beta,\beta\alpha} + b_\alpha = 0, \tag{8.9.38}$$

  compares with the following equation in plane stress

$$\left(\frac{E}{2(1+\nu)}\right)u_{\alpha,\beta\beta} + \frac{E}{2(1-\nu)}u_{\beta,\beta\alpha} + b_\alpha = 0. \tag{8.9.39}$$

- The compatibility relation (8.9.20) for plane strain

$$\Delta(\sigma_{\alpha\alpha}) = -\frac{1}{(1-\nu)}\,b_{\alpha,\alpha}, \tag{8.9.40}$$

  compares with relation (8.9.35) for plane stress

$$\Delta(\sigma_{\alpha\alpha}) = -(1+\nu)\,b_{\alpha,\alpha}. \tag{8.9.41}$$

8.9.4 Succinct form of the governing equations for plane strain and plane stress

For later use, let

$$\boxed{s \stackrel{\text{def}}{=} \begin{cases} 1-\nu & \text{for plane strain,} \\ \dfrac{1}{1+\nu} & \text{for plane stress.} \end{cases}} \tag{8.9.42}$$

Using this parameter the plane strain and plane stress constitutive equations may be compactly written as

$$\boxed{\begin{aligned} \sigma_{\alpha\beta} &= \frac{E}{(1+\nu)}\left(\epsilon_{\alpha\beta} + \left(\frac{1-s}{2s-1}\right)(\epsilon_{\gamma\gamma})\delta_{\alpha\beta}\right), \\ \epsilon_{\alpha\beta} &= \frac{(1+\nu)}{E}\Big(\sigma_{\alpha\beta} - (1-s)(\sigma_{\gamma\gamma})\delta_{\alpha\beta}\Big). \end{aligned}} \tag{8.9.43}$$

The plane strain and plane stress equilibrium equations are identical

$$\boxed{\sigma_{\alpha\beta,\beta} + b_\alpha = 0,} \tag{8.9.44}$$

while the compatibility equation in terms of stresses may be written as

$$\boxed{\Delta(\sigma_{\alpha\alpha}) = -\frac{1}{s} b_{\alpha,\alpha}.} \tag{8.9.45}$$

As before the solution to the plane strain or plane stress problem involves the determination of the in-plane displacements, strains, and stresses $[u_\alpha, \epsilon_{\alpha\beta}, \sigma_{\alpha\beta}]$ in $\mathcal{R}$, subject to the boundary conditions of the type given in (8.9.29). Once a solution has been found in terms of the material parameter s, a distinction can be made between plane strain and plane stress by applying the definition (8.9.42), and recalling that

$$\boxed{\epsilon_{33} = \begin{cases} 0 & \text{for plane strain,} \\ -\dfrac{\nu}{1-\nu}\epsilon_{\alpha\alpha} & \text{for plane stress,} \end{cases}} \tag{8.9.46}$$

and that

$$\boxed{\sigma_{33} = \begin{cases} \nu\sigma_{\alpha\alpha} & \text{for plane strain,} \\ 0 & \text{for plane stress.} \end{cases}} \tag{8.9.47}$$

Of course in both cases $\epsilon_{13} = \epsilon_{23} = 0$ and $\sigma_{13} = \sigma_{23} = 0$.

9 Solutions to some classical problems in linear elastostatics

9.1 Spherical pressure vessel

Consider a thick-walled spherical pressure vessel of inner radius a, outer radius b, with no body forces, $\mathbf{b} = \mathbf{0}$, under internal and external pressures p_i and p_o, respectively, cf. Fig. 9.1. The pressure vessel is made from an *isotropic* linear elastic material. The problem is to find the displacement, strain, and stress fields which satisfy the boundary conditions. This is a problem in elastostatics. To solve this problem, we use a spherical coordinate system (r, θ, ϕ) (see Appendix A) with origin at the center of the spherical pressure vessel, so that the body occupies the region $a \le r \le b$.

For this problem there are only traction boundary conditions, so any solution to the displacement field that we shall determine will be modulo a rigid displacement. The prescribed tractions are

$$\begin{aligned} \hat{\mathbf{t}} &= \ \ p_i \mathbf{e}_r \qquad \text{on} \qquad r = a, \\ \hat{\mathbf{t}} &= -p_o \mathbf{e}_r \qquad \text{on} \qquad r = b. \end{aligned} \tag{9.1.1}$$

Since the *outward* unit normal at $r = a$ is $-\mathbf{e}_r$, and that on $r = b$ is $\mathbf{e}_r$, the traction boundary conditions are

$$\begin{aligned} \boldsymbol{\sigma}(-\mathbf{e}_r) &= \ \ p_i \mathbf{e}_r \qquad \text{on} \qquad r = a, \\ \boldsymbol{\sigma}(\mathbf{e}_r) &= -p_o \mathbf{e}_r \qquad \text{on} \qquad r = b, \end{aligned} \tag{9.1.2}$$

and hence

$$\begin{aligned} \sigma_{rr} &= -p_i \qquad \text{on} \qquad r = a, \\ \sigma_{rr} &= -p_o \qquad \text{on} \qquad r = b. \end{aligned} \tag{9.1.3}$$

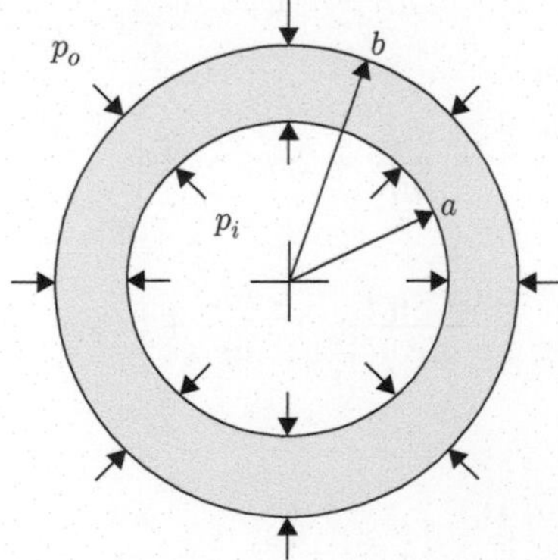

Fig. 9.1 Spherical pressure vessel.

Continuum Mechanics of Solids. Lallit Anand and Sanjay Govindjee, Oxford University Press (2020).

DOI: 10.1093/oso/9780198864721.001.0001

Because of the isotropy of the material, and because of the symmetry of the loading, we *assume* that the displacement field $\mathbf{u}$ is directed in the radial direction and depends only on r:

$$\mathbf{u} = u_r(r)\mathbf{e}_r \qquad \text{with} \qquad u_\theta = u_\phi = 0. \tag{9.1.4}$$

In the absence of body forces the displacement equation of equilibrium (8.7.5) becomes

$$(\lambda + 2\mu)\nabla \operatorname{div} \mathbf{u} - \mu \operatorname{curl} \operatorname{curl} \mathbf{u} = \mathbf{0}. \tag{9.1.5}$$

Considering expression (A.3.16) for the curl of a vector field $\mathbf{v}$ in spherical coordinates

$$\operatorname{curl} \mathbf{v} = \left(\frac{1}{r}\frac{\partial v_\phi}{\partial \theta} - \frac{1}{r\sin\theta}\frac{\partial v_\theta}{\partial \phi} + \cot\theta \frac{v_\phi}{r}\right)\mathbf{e}_r + \left(\frac{1}{r\sin\theta}\frac{\partial v_r}{\partial \phi} - \frac{\partial v_\phi}{\partial r} - \frac{v_\phi}{r}\right)\mathbf{e}_\theta$$
$$+ \left(\frac{\partial v_\theta}{\partial r} - \frac{1}{r}\frac{\partial v_r}{\partial \theta} + \frac{v_\theta}{r}\right)\mathbf{e}_\phi,$$

we see that for the presumed displacement field (9.1.4) the curl vanishes:

$$\operatorname{curl} \mathbf{u} = \mathbf{0}.$$

Using this in (9.1.5), and noting that $(\lambda + 2\mu) \neq 0$, we must have

$$\nabla \operatorname{div} \mathbf{u} = \mathbf{0}, \tag{9.1.6}$$

and hence that $\operatorname{div} \mathbf{u}$ is a *constant*. We write

$$\operatorname{div} \mathbf{u} = 3A, \tag{9.1.7}$$

where $3A$ is an integration constant, in which we have introduced the factor of 3 for later convenience. In a spherical coordinate system the divergence of a vector field $\mathbf{v}$ is given by (A.3.14):

$$\operatorname{div} \mathbf{v} = \frac{\partial v_r}{\partial r} + \frac{1}{r}\frac{\partial v_\theta}{\partial \theta} + \frac{1}{r\sin\theta}\frac{\partial v_\phi}{\partial \phi} + \frac{\cot\theta}{r}v_\theta + \frac{2v_r}{r}.$$

Hence, for the displacement field (9.1.4) we obtain

$$\operatorname{div} \mathbf{u} = \frac{du_r}{dr} + \frac{2u_r}{r}, \tag{9.1.8}$$

use of which in (9.1.7) gives

$$\frac{du_r}{dr} + \frac{2u_r}{r} = 3A. \tag{9.1.9}$$

Multiplying (9.1.9) by r^2 gives,

$$r^2\frac{du_r}{dr} + 2ru_r = 3Ar^2,$$

and noting that

$$\frac{d(r^2u_r)}{dr} = r^2\frac{du_r}{dr} + 2ru_r,$$

we obtain

$$\frac{d(r^2 u_r)}{dr} = 3Ar^2. \tag{9.1.10}$$

Integrating (9.1.10) we obtain

$$r^2 u_r = Ar^3 + B,$$

where B is another constant of integration, and hence

$$\boxed{u_r = Ar + \frac{B}{r^2}.} \tag{9.1.11}$$

It remains to find the constants A and B, which are determined by using the boundary conditions (9.1.3). To this end, from (A.3.19) the only non-zero strain components are

$$\epsilon_{rr} = \frac{du_r}{dr}, \qquad \epsilon_{\theta\theta} = \epsilon_{\phi\phi} = \frac{u_r}{r}. \tag{9.1.12}$$

Hence, using the constitutive equation for the stress (7.5.2) the radial stress component is

$$\begin{aligned}
\sigma_{rr} &= 2\mu\frac{du_r}{dr} + \lambda\left(\frac{du_r}{dr} + \frac{2u_r}{r}\right) \\
&= (2\mu+\lambda)\frac{du_r}{dr} + 2\lambda\left(\frac{u_r}{r}\right) \\
&= (2\mu+\lambda)\left(A - 2\frac{B}{r^3}\right) + 2\lambda\left(A + \frac{B}{r^3}\right) \qquad \text{(using (9.1.11))} \\
&= (2\mu+3\lambda)A - \frac{4\mu B}{r^3}.
\end{aligned} \tag{9.1.13}$$

Also, the hoop stress components are

$$\begin{aligned}
\sigma_{\theta\theta} = \sigma_{\phi\phi} &= 2\mu\frac{u_r}{r} + \lambda\left(\frac{du_r}{dr} + \frac{2u_r}{r}\right) \\
&= (2\mu+2\lambda)\frac{u_r}{r} + \lambda\left(\frac{du_r}{dr}\right) \\
&= (2\mu+2\lambda)\left(A + \frac{B}{r^3}\right) + \lambda\left(A - 2\frac{B}{r^3}\right) \qquad \text{(using (9.1.11))} \\
&= (2\mu+3\lambda)A + 2\mu\frac{B}{r^3},
\end{aligned} \tag{9.1.14}$$

while the shear stress components vanish:

$$\sigma_{r\theta} = \sigma_{\theta\phi} = \sigma_{\phi r} = 0. \tag{9.1.15}$$

Next, using (9.1.13) and the boundary conditions (9.1.3) we obtain

$$\begin{aligned}
(2\mu+3\lambda)A - \frac{4\mu B}{a^3} &= -p_i, \\
(2\mu+3\lambda)A - \frac{4\mu B}{b^3} &= -p_o
\end{aligned} \tag{9.1.16}$$

Solving (9.1.16) for the unknowns A and B, one obtains

$$A = \frac{1}{(2\mu + 3\lambda)}\left(\frac{a^3\,p_i - b^3\,p_o}{b^3 - a^3}\right), \qquad B = \frac{a^3b^3}{4\mu}\left(\frac{p_i - p_o}{b^3 - a^3}\right). \tag{9.1.17}$$

Finally, substituting for A and B in (9.1.11), (9.1.13), and (9.1.14), we obtain

$$\boxed{u_r = \frac{r}{(2\mu + 3\lambda)}\left(\frac{a^3\,p_i - b^3\,p_o}{b^3 - a^3}\right) + \frac{a^3b^3}{4\mu\,r^2}\left(\frac{p_i - p_o}{b^3 - a^3}\right), \qquad u_\theta = u_\phi = 0.} \tag{9.1.18}$$

$$\boxed{\begin{aligned} \sigma_{rr} &= \left(\frac{a^3\,p_i - b^3\,p_o}{b^3 - a^3}\right) - \frac{a^3b^3}{r^3}\left(\frac{p_i - p_o}{b^3 - a^3}\right), \\ \sigma_{\theta\theta} = \sigma_{\phi\phi} &= \left(\frac{a^3\,p_i - b^3\,p_o}{b^3 - a^3}\right) + \frac{a^3b^3}{2r^3}\left(\frac{p_i - p_o}{b^3 - a^3}\right), \\ \sigma_{r\theta} = \sigma_{\theta\phi} &= \sigma_{\phi r} = 0. \end{aligned}} \tag{9.1.19}$$

Since the shear stresses are all zero, $(9.1.19)_3$, the stresses σ_{rr} and $\sigma_{\theta\theta} = \sigma_{\phi\phi}$ are *principal stresses.*

REMARK 9.1

It is important to note that for this problem the solution for the stress is *independent of the elastic constants*! It is a so-called universal solution. That is, the same radial and hoop stress distribution is obtained for *all* isotropic elastic materials, regardless of the values of their elastic constants.

Zero external pressure: If there is no external pressure, $p_o = 0$, then

$$\boxed{\begin{aligned} \sigma_{rr} &= p_i\,\frac{a^3}{b^3 - a^3}\left(1 - \frac{b^3}{r^3}\right), \\ \sigma_{\theta\theta} = \sigma_{\phi\phi} &= p_i\,\frac{a^3}{b^3 - a^3}\left(1 + \frac{1}{2}\frac{b^3}{r^3}\right). \end{aligned}} \tag{9.1.20}$$

Zero internal pressure: On the other hand if the internal pressure is zero, $p_i = 0$, then

$$\boxed{\begin{aligned} \sigma_{rr} &= -p_o\,\frac{b^3}{b^3 - a^3}\left(1 - \frac{a^3}{r^3}\right), \\ \sigma_{\theta\theta} = \sigma_{\phi\phi} &= -p_o\,\frac{b^3}{b^3 - a^3}\left(1 + \frac{1}{2}\frac{a^3}{r^3}\right). \end{aligned}} \tag{9.1.21}$$

Unpressurized void under remote hydrostatic tension: A limiting case of interest is an unpressurized void in an infinite solid under remote *hydrostatic tension* $\Sigma > 0$:[1]

$$p_i = 0, \qquad p_o = -\Sigma, \qquad b \to \infty.$$

In this case (9.1.21) gives

$$\begin{aligned} \sigma_{rr} &= \Sigma\left(1 - \frac{a^3}{r^3}\right), \\ \sigma_{\theta\theta} = \sigma_{\phi\phi} &= \Sigma\left(1 + \frac{1}{2}\frac{a^3}{r^3}\right). \end{aligned} \tag{9.1.22}$$

At $r = a$,

$$\sigma_{rr} = 0, \qquad \sigma_{\theta\theta} = \sigma_{\phi\phi} = \frac{3}{2}\,\Sigma;$$

while far away from the void, $r \to \infty$:

$$\sigma_{rr} = \sigma_{\theta\theta} = \sigma_{\phi\phi} = \Sigma.$$

Observe that the hoop stresses $\sigma_{\theta\theta} = \sigma_{\phi\phi}$ have a value of $(3/2)\Sigma$ at the surface of the void, while far away from the void they have a value Σ. The ratio of the former over the latter represents a **stress concentration factor** of 1.5.

EXAMPLE 9.1 Pressurized void in a solid

To find the state of stress due to a pressurized void in a large body without additional load, we can consider the limit of (9.1.20) as $b \to \infty$. In that case, we find

$$\lim_{b\to\infty} \sigma_{rr} = -p_i \left(\frac{a}{r}\right)^3$$

$$\lim_{b\to\infty} \sigma_{\theta\theta} = \lim_{b\to\infty} \sigma_{\phi\phi} = \tfrac{1}{2} p_i \left(\frac{a}{r}\right)^3.$$

Observe that the stresses fall off rapidly, proportional to $1/r^3$, as one moves away from the void.

9.2 Cylindrical pressure vessel

Here we consider, cf. Fig. 9.2, an isotropic thick-walled cylinder with inner and outer radii a and b, respectively. The cylinder is subjected to an internal pressure, p_i, and external pressure, p_o. We wish to determine the displacement, strain, and stress fields in the cylinder. To that end we will employ a cylindrical coordinate system (Appendix A) centered upon the centerline of the cylinder.

[1] In this subsection, Σ is a scalar, and not the summation symbol.

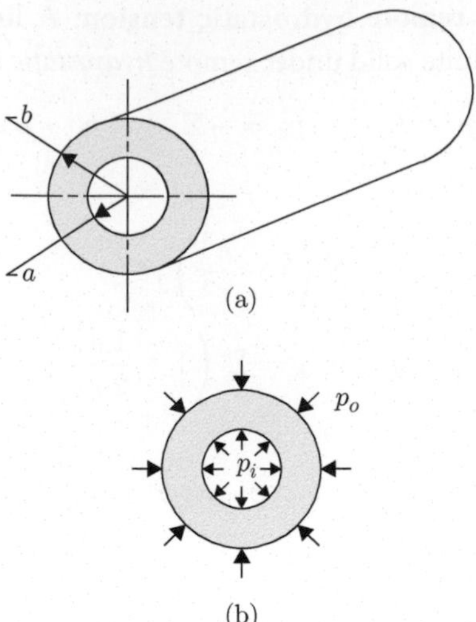

Fig. 9.2 (a) Thick-walled cylinder with inner radius a and outer radius b. (b) The cylinder is subjected to uniformly applied internal pressure p_i, and external pressure p_o.

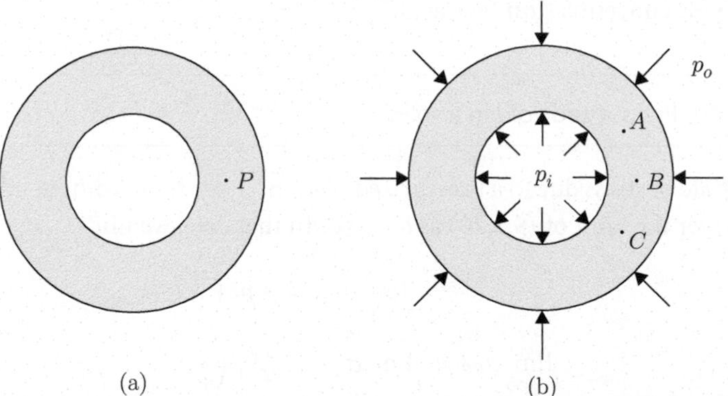

Fig. 9.3 (a) Before application of pressure. (b) After application of pressure.

Consider a typical point P which occupies the position indicated in Fig. 9.3(a) before any pressure is applied. After an internal pressure p_i and external pressure p_o is applied, it is clear that because the material is isotropic, and the geometry and loading are perfectly symmetric, the point P will occupy the position B on the same radial line, and not move to either the positions A or C which are off the radial line. Hence, the displacement field has the special form:

$$u_\theta = 0, \qquad u_r = \hat{u}_r(r), \qquad u_z = \hat{u}_z(z). \tag{9.2.1}$$

The only non-zero strain components corresponding to this displacement field are

$$\epsilon_{rr} = \frac{du_r}{dr}, \qquad \epsilon_{\theta\theta} = \frac{u_r}{r}, \qquad \epsilon_{zz} = \frac{du_z}{dz}, \tag{9.2.2}$$

and the corresponding stresses are

$$
\begin{aligned}
\sigma_{rr} &= 2\mu\frac{du_r}{dr} + \lambda\left(\frac{du_r}{dr} + \frac{u_r}{r} + \frac{du_z}{dz}\right),\\
\sigma_{\theta\theta} &= 2\mu\frac{u_r}{r} + \lambda\left(\frac{du_r}{dr} + \frac{u_r}{r} + \frac{du_z}{dz}\right),\\
\sigma_{zz} &= 2\mu\frac{du_z}{dz} + \lambda\left(\frac{du_r}{dr} + \frac{u_r}{r} + \frac{du_z}{dz}\right),\\
\sigma_{r\theta} &= \sigma_{\theta z} = \sigma_{z\theta} = 0.
\end{aligned} \tag{9.2.3}
$$

By following procedures essentially identical to those for the thick-walled sphere leading to (9.1.11), for the thick-walled cylinder we find that the equilibrium equations are satisfied provided

$$
\frac{du_z}{dz} = \epsilon_{zz} \equiv \epsilon_o \qquad \text{(a constant)}, \tag{9.2.4}
$$

which we assume, and that the displacement field $u_r = \hat{u}_r(r)$ is of the form:

$$
\boxed{u_r = Ar + Br^{-1},} \tag{9.2.5}
$$

where A and B are arbitrary constants. These constants are determined from the boundary conditions. Because pressures are specified, we will have to defer the evaluations of the constants until we obtain an expression for the radial stress σ_{rr}, which we obtain next.

We first calculate all the stress components in terms of A, B, and ϵ_o. From equations (9.2.3), (9.2.4), and (9.2.5)

$$
\begin{aligned}
\sigma_{rr} &= 2(\mu+\lambda)A - 2\mu\frac{B}{r^2} + \lambda\epsilon_o,\\
\sigma_{\theta\theta} &= 2(\mu+\lambda)A + 2\mu\frac{B}{r^2} + \lambda\epsilon_o,\\
\sigma_{zz} &= (2\mu+\lambda)\,\epsilon_o + 2\lambda A.
\end{aligned} \tag{9.2.6}
$$

Next, the traction boundary conditions concerning internal and external pressures lead to

$$
\sigma_{rr} = -p_i \quad \text{at} \quad r = a, \quad \text{and} \quad \sigma_{rr} = -p_o \quad \text{at} \quad r = b. \tag{9.2.7}
$$

With the use of the boundary conditions (9.2.7) in $(9.2.6)_1$, it is straightforward to determine that

$$
A = \frac{1}{2(\mu+\lambda)}\frac{(p_i a^2 - p_o b^2)}{(b^2-a^2)} - \frac{\lambda}{2(\mu+\lambda)}\epsilon_o, \qquad \text{and} \qquad B = \frac{1}{2\mu}\frac{a^2b^2}{(b^2-a^2)}(p_i - p_o). \tag{9.2.8}
$$

Substituting for A and B from (9.2.8) in (9.2.5) and (9.2.6) we obtain

$$
u_r = \left(\frac{1}{2(\mu+\lambda)}\frac{(p_i a^2 - p_o b^2)}{(b^2-a^2)}\right) r + \left(\frac{1}{2\mu}\frac{a^2b^2}{(b^2-a^2)}(p_i - p_o)\right)\frac{1}{r} - \frac{\lambda}{2(\mu+\lambda)} r\,\epsilon_o, \tag{9.2.9}
$$

and

$$\begin{aligned}\sigma_{rr} &= \left(\frac{p_i a^2 - p_o b^2}{b^2 - a^2}\right) - \left(\frac{a^2 b^2}{(b^2 - a^2)}(p_i - p_o)\right)\frac{1}{r^2},\\ \sigma_{\theta\theta} &= \left(\frac{p_i a^2 - p_o b^2}{b^2 - a^2}\right) + \left(\frac{a^2 b^2}{(b^2 - a^2)}(p_i - p_o)\right)\frac{1}{r^2},\\ \sigma_{zz} &= \frac{\lambda}{(\mu+\lambda)}\left(\frac{p_i a^2 - p_o b^2}{b^2 - a^2}\right) + \left(\frac{\mu(2\mu+3\lambda)}{(\mu+\lambda)}\right)\epsilon_o.\end{aligned} \tag{9.2.10}$$

Next, using the conversion-relations

$$2(\mu+\lambda) = \frac{E}{(1+\nu)(1-2\nu)}, \quad 2\mu = \frac{E}{(1+\nu)}, \quad \frac{\lambda}{(\mu+\lambda)} = 2\nu, \quad \frac{\mu(2\mu+3\lambda)}{(\mu+\lambda)} = E,$$

equations (9.2.9) and (9.2.10) above can be rewritten as

$$\boxed{u_r = \frac{(1+\nu)}{E}\frac{r}{\left(\left(\dfrac{b}{a}\right)^2 - 1\right)}\left[(1-2\nu)\left(p_i - p_o\left(\frac{b}{a}\right)^2\right) + (p_i - p_o)\left(\frac{b}{r}\right)^2\right] - \nu r \epsilon_o,} \tag{9.2.11}$$

and

$$\boxed{\begin{aligned}\sigma_{rr} &= \frac{1}{\left(\left(\dfrac{b}{a}\right)^2 - 1\right)}\left[\left(p_i - p_o\left(\frac{b}{a}\right)^2\right) - (p_i - p_o)\left(\frac{b}{r}\right)^2\right],\\ \sigma_{\theta\theta} &= \frac{1}{\left(\left(\dfrac{b}{a}\right)^2 - 1\right)}\left[\left(p_i - p_o\left(\frac{b}{a}\right)^2\right) + (p_i - p_o)\left(\frac{b}{r}\right)^2\right],\\ \sigma_{zz} &= \frac{2\nu}{\left(\left(\dfrac{b}{a}\right)^2 - 1\right)}\left[\left(p_i - p_o\left(\frac{b}{a}\right)^2\right)\right] + E\,\epsilon_o.\end{aligned}} \tag{9.2.12}$$

Thus, for a given ϵ_o the complete solutions for a thick-walled cylinder are given by equations (9.2.11) and (9.2.12) above.

- Note the very important result that *the solutions for σ_{rr} and $\sigma_{\theta\theta}$ are independent of the elastic moduli.* That is, the same radial and tangential stress distribution is obtained for *all* isotropic elastic materials.

- The radial displacement and the axial stress, however, do depend on the elastic moduli.

EXAMPLE 9.2 Thin-wall limit under internal pressure

To find the radial and hoop stresses for a *thin-walled* cylinder under internal pressure, consider (9.2.12) with $p_o = 0$, and substitute $b = a + t$ with $t/a \ll 1$. This implies

$$\sigma_{rr} = \frac{p_i}{\left(\left(\frac{b}{a}\right)^2 - 1\right)}\left[1 - \left(\frac{b}{r}\right)^2\right] \approx 0$$

and

$$\begin{aligned}\sigma_{\theta\theta} &= \frac{p_i}{\left(\left(\frac{b}{a}\right)^2 - 1\right)}\overbrace{\left[1 + \left(\frac{b}{r}\right)^2\right]}^{\approx 2} \\ &\approx \frac{2p_i}{\left(\left(\frac{b}{a}\right)^2 - 1\right)} = \frac{2p_i}{\left(\left(\frac{a+t}{a}\right)^2 - 1\right)} = \frac{2p_i}{\left(2\frac{t}{a} + \left(\frac{t}{a}\right)^2\right)} \\ &\approx \frac{p_i a}{t}.\end{aligned}$$

9.2.1 **Axial strain ϵ_0 for different end conditions**

The axial strain $\epsilon_{zz} = \epsilon_o$ must now be determined. There are three important cases to be considered:

1. *Plane strain*:

$$\epsilon_{zz} = \epsilon_o = 0.$$

   This condition is appropriate when the ends of the cylinder are prevented from moving by frictionless constraints.

2. *Unrestrained cylinder*: $\sigma_{zz} = 0$. If the internal pressure is contained by a piston (or pistons), or if expansion joints are provided in a piping system, then there is no net axial force carried by the walls of the cylinder and hence $\sigma_{zz} = 0$.

   Then, from $(9.2.12)_3$

$$0 = \frac{2\nu}{\left(\left(\frac{b}{a}\right)^2 - 1\right)}\left[\left(p_i - p_o\left(\frac{b}{a}\right)^2\right)\right] + E\,\epsilon_o$$

or

$$\boxed{\epsilon_0 = -\frac{2\nu}{E}\left(\frac{p_i a^2 - p_o b^2}{b^2 - a^2}\right).} \tag{9.2.13}$$

3. *Capped cylinder*: If the cylinder has sealed ends, then the total tensile force carried by the cross-section must balance the thrust of the pressure on the end-closures. Thus

$$\int_a^b \sigma_{zz}(2\pi r dr) = p_i \pi a^2 - p_o \pi b^2,$$

or, because σ_{zz} is independent of r (see equation (9.2.12)$_3$),

$$\sigma_{zz} 2\pi \left[\frac{r^2}{2}\right]_a^b = \pi(p_i a^2 - p_o b^2),$$

from which

$$\sigma_{zz} = (p_i a^2 - p_o b^2)/(b^2 - a^2).$$

In this case, using (9.2.12)$_3$, we obtain

$$\boxed{\epsilon_0 = \frac{(1-2\nu)}{E}\left(\frac{p_i a^2 - p_o b^2}{b^2 - a^2}\right).} \tag{9.2.14}$$

Once the axial strain ϵ_0 is determined, the displacement fields u_r and u_z, as well as the axial stress σ_{zz} are fully determined from (9.2.11) and (9.2.12)$_3$.

9.2.2 **Stress concentration in a thick-walled cylinder under internal pressure**

Let us consider the special case of a thick-walled cylinder under internal pressure p (no external pressure), under plane strain conditions. In this case equations (9.2.11) and (9.2.12) reduce to

$$\boxed{u_r = \frac{(1+\nu)}{E}\frac{pr}{\left(\left(\dfrac{b}{a}\right)^2 - 1\right)}\left[(1-2\nu) + \left(\frac{b}{r}\right)^2\right],} \tag{9.2.15}$$

and

$$\boxed{\begin{aligned} \sigma_{rr} &= \frac{p}{\left(\left(\dfrac{b}{a}\right)^2 - 1\right)}\left[1 - \left(\frac{b}{r}\right)^2\right], \\ \sigma_{\theta\theta} &= \frac{p}{\left(\left(\dfrac{b}{a}\right)^2 - 1\right)}\left[1 + \left(\frac{b}{r}\right)^2\right], \\ \sigma_{zz} &= 2\nu\frac{p}{\left(\left(\dfrac{b}{a}\right)^2 - 1\right)}. \end{aligned}} \tag{9.2.16}$$

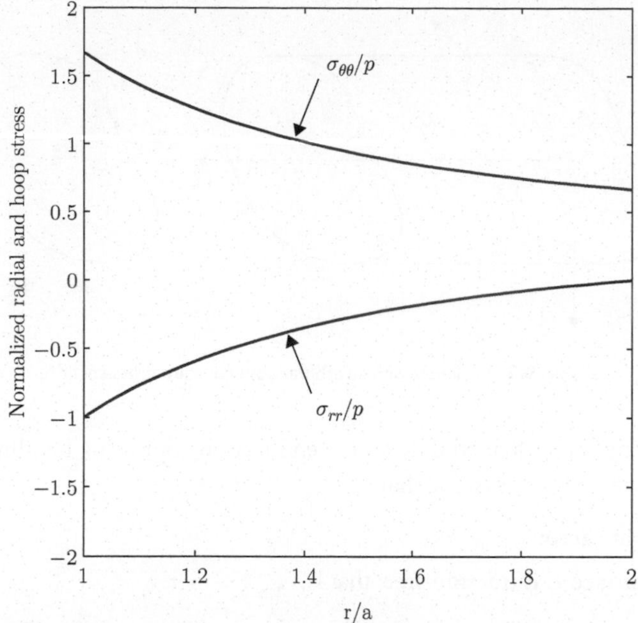

Fig. 9.4 Radial and hoop stress distribution in a thick-walled cylinder under internal pressure and plane strain conditions, for $(b/a) = 2$.

Figure 9.4 shows the normalized stress distribution σ_{rr}/p and $\sigma_{\theta\theta}/p$ for the special case of $(b/a) = 2$. Note the hoop stress component $\sigma_{\theta\theta}$ at $r = a$ has a value

$$\left.\frac{\sigma_{\theta\theta}}{p}\right|_{r=a} = \frac{5}{3}. \tag{9.2.17}$$

- Thus for an internal pressure p, the hoop stress $\sigma_{\theta\theta}$ is **tensile** and its value is $(5/3)p$, which is higher than the applied internal pressure p. This is a form of a **stress concentration.**

9.3 Bending and torsion

9.3.1 Bending of a bar

Consider a cylindrical bar of a homogeneous and linear elastic isotropic material with generators parallel to the x_3-axis, cf. Fig. 9.5 (left). Let the faces $\mathcal{S}_0$ and $\mathcal{S}_L$ be located at $x_3 = 0$ and $x_3 = L$, respectively, with origin at the centroid of $\mathcal{S}_0$, and the x_1- and x_2-axes being the *principal axes of inertia*, that is,

$$\int_{\mathcal{S}_0} x_1 \, da = \int_{\mathcal{S}_0} x_2 \, da = \int_{\mathcal{S}_0} x_1 x_2 \, da = 0. \tag{9.3.1}$$

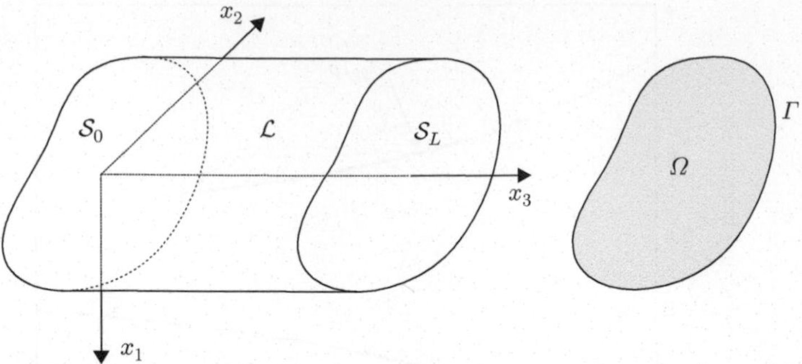

Fig. 9.5 A cylinder with an arbitrary but constant cross-section.

We assume that the bar is loaded only on the end faces by opposing bending moments about the x_2-axes. Specifically, we assume that

(a) body forces are zero;

(b) the lateral surface $\mathcal{L}$ is traction-free; that is,

$$\mathbf{t} = \boldsymbol{\sigma}\mathbf{n} = \mathbf{0} \quad \text{on} \quad \mathcal{L},$$

or equivalently, since $n_3 = 0$ on $\mathcal{L}$, and with Ω denoting a typical cross-section with boundary Γ, cf. Fig. 9.5 (right),

$$\left.\begin{aligned} t_1 &= \sigma_{11}n_1 + \sigma_{12}n_2 = 0 \\ t_2 &= \sigma_{21}n_1 + \sigma_{22}n_2 = 0 \\ t_3 &= \sigma_{31}n_1 + \sigma_{32}n_2 = 0 \end{aligned}\right\} \quad \text{on } \Gamma. \tag{9.3.2}$$

(c) The net force on $\mathcal{S}_0$ vanishes and the net moment on $\mathcal{S}_0$ is a moment of magnitude M about the negative x_2-axis; that is,

$$\begin{aligned} \int_{\mathcal{S}_0} \boldsymbol{\sigma}\mathbf{n}\, da &= \mathbf{0}, \\ \int_{\mathcal{S}_0} \mathbf{r} \times \boldsymbol{\sigma}\mathbf{n}\, da &= -M\mathbf{e}_2, \end{aligned} \tag{9.3.3}$$

or equivalently, since

$$\mathbf{n} = -\mathbf{e}_3, \qquad \mathbf{r} = \mathbf{x} - \mathbf{o} \quad \text{and} \quad x_3 = 0 \quad \text{on} \quad \mathcal{S}_0,$$

in components we have

$$\begin{aligned} \int_{\mathcal{S}_0} \sigma_{13}\, da &= 0, \\ \int_{\mathcal{S}_0} \sigma_{23}\, da &= 0, \\ \int_{\mathcal{S}_0} \sigma_{33}\, da &= 0, \end{aligned} \tag{9.3.4}$$

and

$$\int_{\mathcal{S}_0} x_2\,\sigma_{33}\,da = 0,$$
$$\int_{\mathcal{S}_0} (x_1\,\sigma_{23} - x_2\,\sigma_{13})\,da = 0, \tag{9.3.5}$$
$$\int_{\mathcal{S}_0} x_1\,\sigma_{33}\,da = -M.$$

We do not specify the loading on $\mathcal{S}_L$ because balance of forces and moments requires that

(d) the total force on $\mathcal{S}_L$ vanish and the total moment on $\mathcal{S}_L$ equal $M\mathbf{e}_2$; in fact (9.3.4) and (9.3.5) hold with $\mathcal{S}_0$ replaced by $\mathcal{S}_L$ with the appropriate change in sign in $(9.3.5)_3$.

There are many sets of boundary conditions consistent with (c) and (d). We take the simplest possible and assume that

$$\sigma_{13}(\mathbf{x}) = \sigma_{23}(\mathbf{x}) = 0, \qquad \sigma_{33}(\mathbf{x}) = \zeta_0 + \zeta_1 x_1 + \zeta_2 x_2 \qquad \text{at} \qquad x_3 = 0, L, \tag{9.3.6}$$

with ζ_0, ζ_1, and ζ_2 constants. Then $(9.3.4)_3$ requires that

$$\int_{\mathcal{S}_0} \sigma_{33}\,da = \zeta_0 \int_{\mathcal{S}_0} da + \zeta_1 \int_{\mathcal{S}_0} x_1\,da + \zeta_2 \int_{\mathcal{S}_0} x_2\,da = 0,$$

and on account of $(9.3.1)_{1,2}$ we have

$$\zeta_0 = 0.$$

Next, $(9.3.5)_1$ requires that

$$\int_{\mathcal{S}_0} x_2\,\sigma_{33}\,da = \zeta_1 \int_{\mathcal{S}_0} x_1 x_2 da + \zeta_2 \int_{\mathcal{S}_0} x_2^2\,da = 0,$$

and on account of $(9.3.1)_3$ we have

$$\zeta_2 = 0.$$

Finally, $(9.3.5)_3$ requires that

$$\int_{\mathcal{S}_0} x_1 \sigma_{33}\,da = \zeta_1 \int_{\mathcal{S}_0} x_1^2\,da = -M,$$

and hence that

$$\zeta_1 = \frac{-M}{I}$$

where

$$I \overset{\text{def}}{=} \int_{\mathcal{S}_0} x_1^2\,da \tag{9.3.7}$$

is the *moment of inertia* of the cross-section $\mathcal{S}_0$ of the beam about the x_2-axis. Our boundary conditions therefore consist of (9.3.2) and

$$\sigma_{13}(\mathbf{x}) = \sigma_{23}(\mathbf{x}) = 0, \qquad \sigma_{33} = -\frac{Mx_1}{I} \qquad \text{at} \qquad x_3 = 0, L. \tag{9.3.8}$$

A stress field compatible with the boundary conditions (9.3.2) and (9.3.8) is obtained by taking

$$\begin{aligned} \sigma_{11}(\mathbf{x}) &= \sigma_{22}(\mathbf{x}) = \sigma_{12}(\mathbf{x}) = \sigma_{13}(\mathbf{x}) = \sigma_{23}(\mathbf{x}) = 0, \\ \sigma_{33}(\mathbf{x}) &= -\frac{Mx_1}{I}, \end{aligned} \tag{9.3.9}$$

for all $\mathbf{x}$ in the body. This stress field clearly satisfies the equation of equilibrium

$$\operatorname{div} \boldsymbol{\sigma} = \mathbf{0}.$$

Thus to show that the stress field (9.3.9) is actually a solution of our problem, we need only construct the corresponding displacement field.

Using the stress-strain law in the form (7.5.21) we see that

$$\begin{aligned} \epsilon_{12}(\mathbf{x}) &= \epsilon_{13}(\mathbf{x}) = \epsilon_{23}(\mathbf{x}) = 0, \\ \epsilon_{11}(\mathbf{x}) &= \epsilon_{22}(\mathbf{x}) = \frac{\nu M x_1}{EI}, \\ \epsilon_{33}(\mathbf{x}) &= -\frac{Mx_1}{EI}. \end{aligned} \tag{9.3.10}$$

Note that in the three-dimensional theory of pure bending the strains $\epsilon_{11}(\mathbf{x})$ and $\epsilon_{22}(\mathbf{x})$ are not zero due to the Poisson effect. Integration of the strain-displacement equations (8.1.1), assuming (9.3.10), gives

$$\begin{aligned} u_1(\mathbf{x}) &= \frac{M}{2EI}\left(x_3^2 + \nu(x_1^2 - x_2^2)\right) + w_1(\mathbf{x}), \\ u_2(\mathbf{x}) &= \frac{M}{EI}\nu x_1 x_2 + w_2(\mathbf{x}), \\ u_3(\mathbf{x}) &= -\frac{M}{EI}\, x_1 x_3 + w_3(\mathbf{x}), \end{aligned} \tag{9.3.11}$$

with $\mathbf{w}(\mathbf{x})$ an arbitrary infinitesimal *rigid displacement*.[2]

To compare the result obtained here with the classical theory, *fix* $\mathcal{S}_0$ at $x_1 = x_2 = 0$ by requiring that

$$\mathbf{u}(\mathbf{o}) = \nabla \mathbf{u}(\mathbf{o}) = \mathbf{0}.$$

Then, $\mathbf{w}(\mathbf{x}) \equiv \mathbf{0}$, and the displacement of the *centroidal axis* (i.e., the x_3-axis) takes the form

$$\begin{aligned} u_1(0,0,x_3) &= \frac{Mx_3^2}{2EI}, \\ u_2(0,0,x_3) &= u_3(0,0,x_3) = 0. \end{aligned} \tag{9.3.14}$$

[2] Recalling (3.6.13), an (infinitesimal) rigid displacement is a displacement field $\mathbf{w}$ of the form

$$\mathbf{w}(\mathbf{x}) = \mathbf{w}_0 + \mathbf{W}_0(\mathbf{x} - \mathbf{x}_0), \tag{9.3.12}$$

with $\mathbf{x}_0$ a fixed point, $\mathbf{w}_0$ a constant vector, and $\mathbf{W}_0$ a constant *skew* tensor. In this case

$$\nabla \mathbf{w}(\mathbf{x}) = \mathbf{W}_0 = -\mathbf{W}_0^\top \quad \text{and} \quad \boldsymbol{\epsilon}(\mathbf{X}) = \mathbf{0}, \tag{9.3.13}$$

for every point $\mathbf{x}$ in the body. Thus the strain field corresponding to the rigid displacement field vanishes.

Note that (9.3.14) gives the classical relation

$$M = (EI)\underbrace{\frac{d^2u_1(0,0,x_3)}{dx_3^2}}_{\text{curvature}}$$

between the bending moment and the curvature of the centroidal axis in terms of the bending stiffness (EI).

In the solution the maximum stress occurs at points of $\mathcal{L}$ for which $|x_1|$ attains the largest value, say, c and this maximal stress is

$$\frac{Mc}{I}.$$

Further, the maximal deflection of the centroidal axis occurs at $x_3 = L$ and is given by

$$\frac{ML^2}{2EI}.$$

REMARK 9.2

If we had only been interested in the stresses, and had not integrated the strain-displacement relations, we would have needed to show that the stress field (9.3.9) satisfied the Beltrami–Michell compatibility equations (8.8.12), which is clearly true given the linearity of (9.3.9).

9.3.2 Torsion of a circular cylinder

Consider a circular cylinder of length L and radius R. As before, the axis of the cylinder coincides with the x_3-axis, while $\mathcal{S}_0$ and $\mathcal{S}_L$ correspond to $x_3 = 0$ and $x_3 = L$. We assume that the face $\mathcal{S}_0$ is held fixed, while the end face $\mathcal{S}_L$ is rigidly rotated about the x_3-axis through an angle β, cf. Fig. 9.6.

Under the applied twist, a generic point $\mathbf{x}$ in the (x_1, x_2)-plane displaces to $\mathbf{x}'$, and the vector $(\mathbf{x}-\mathbf{o})$ then rotates through a small angle β, and thus the arc length $|\mathbf{x}'-\mathbf{x}| \approx r\beta$. The in-plane displacements can thus be determined as

$$\begin{aligned} u_1 &= -r\beta\sin\theta = -\beta x_2, \\ u_2 &= r\beta\cos\theta = \beta x_1. \end{aligned} \tag{9.3.15}$$

Defining

$$\alpha \stackrel{\text{def}}{=} \frac{\beta}{L} \quad \text{as the } \textit{angle of twist} \text{ per unit length,} \tag{9.3.16}$$

we have the kinematic requirement that

$$\begin{aligned} &\mathbf{u}(\mathbf{x}) = \mathbf{0} \quad \text{at} \quad x_3 = 0, \\ &u_1(\mathbf{x}) = -\alpha L x_2, \qquad u_2(\mathbf{x}) = \alpha L x_1, \qquad u_3(\mathbf{x}) = 0 \quad \text{at} \quad x_3 = L. \end{aligned} \tag{9.3.17}$$

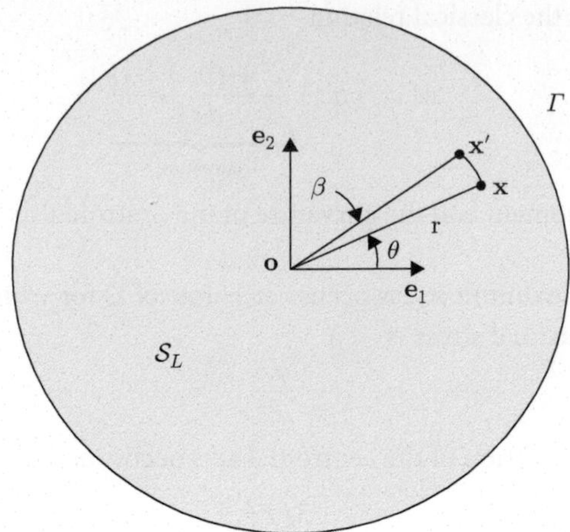

Fig. 9.6 End-face $\mathcal{S}_L$ of a cylinder with a circular cross-section. The material point at $\mathbf{x}$ rotates through an angle β to the point $\mathbf{x}'$.

Also, assuming that the lateral surface is traction-free, our traction boundary conditions consist of

$$\left.\begin{aligned} \sigma_{11}n_1 + \sigma_{12}n_2 &= 0, \\ \sigma_{21}n_1 + \sigma_{22}n_2 &= 0, \\ \sigma_{31}n_1 + \sigma_{32}n_2 &= 0, \end{aligned}\right\} \quad \text{on } \Gamma \tag{9.3.18}$$

where Γ is the boundary of a cross-section Ω of the cylinder.

It is reasonable to expect that under this type of loading the cylinder will twist uniformly along its length. Thus, we consider a displacement field of the form

$$\begin{aligned} u_1(\mathbf{x}) &= -\alpha x_2 x_3, \\ u_2(\mathbf{x}) &= \quad \alpha x_1 x_3, \\ u_3(\mathbf{x}) &= 0. \end{aligned} \tag{9.3.19}$$

The corresponding strain field is

$$\begin{aligned} \epsilon_{11}(\mathbf{x}) &= \epsilon_{22}(\mathbf{x}) = \epsilon_{33}(\mathbf{x}) = \epsilon_{12}(\mathbf{x}) = 0, \\ \epsilon_{23}(\mathbf{x}) &= \tfrac{1}{2}\alpha x_1, \\ \epsilon_{13}(\mathbf{x}) &= -\tfrac{1}{2}\alpha x_2, \end{aligned} \tag{9.3.20}$$

and the stress field is

$$\begin{aligned} \sigma_{11}(\mathbf{x}) &= \sigma_{22}(\mathbf{x}) = \sigma_{33}(\mathbf{x}) = \sigma_{12}(\mathbf{x}) = 0, \\ \sigma_{23}(\mathbf{x}) &= \mu\alpha x_1, \\ \sigma_{13}(\mathbf{x}) &= -\mu\alpha x_2. \end{aligned} \tag{9.3.21}$$

This stress field clearly satisfies

$$\operatorname{div}\boldsymbol{\sigma} = \mathbf{0}.$$

Further the stress field (9.3.21) trivially satisfies the boundary conditions $(9.3.18)_{1,2}$. Next, since the surface of the bar satisfies

$$\phi = x_1^2 + x_2^2 - R^2 = 0,$$

we have

$$\nabla\phi = 2x_1\mathbf{e}_1 + 2x_2\mathbf{e}_2,$$

and since the unit normal to the surface is

$$\mathbf{n} = \frac{\nabla\phi}{|\nabla\phi|},$$

we have

$$n_1 = \frac{x_1}{R}, \qquad n_2 = \frac{x_2}{R},$$

and hence it follows that

$$\sigma_{13}n_1 + \sigma_{23}n_2 = -\mu\alpha x_2\frac{x_1}{R} + \mu\alpha x_1\frac{x_2}{R} = 0 \qquad \text{on } \Gamma. \tag{9.3.22}$$

Thus $(9.3.18)_3$ is also satisfied. Taken together, (9.3.19)–(9.3.21) is seen to satisfy the governing equations for linear elastostatics along with the boundary conditions (9.3.17) and (9.3.18).

In order to understand the forces and moments that are required to impose the given boundary conditions, we can integrate the traction and moment of the traction over the ends of the cylinder. In particular, by (9.3.21) and (9.3.1),

$$\int_{\mathcal{S}_0} \sigma_{13} da = -\mu\alpha \int_{\mathcal{S}_0} x_2 da = 0,$$
$$\int_{\mathcal{S}_0} \sigma_{23} da = \mu\alpha \int_{\mathcal{S}_0} x_1 da = 0,$$
$$\int_{\mathcal{S}_0} \sigma_{33} da = 0$$

so that

$$\int_{\mathcal{S}_0} \boldsymbol{\sigma}\mathbf{n}\, da = \mathbf{0}; \tag{9.3.23}$$

thus the net force on the end vanishes. In addition,

$$\int_{\mathcal{S}_0} x_1\sigma_{33} da = \int_{\mathcal{S}_0} x_2\sigma_{33} = 0,$$
$$\int_{\mathcal{S}_0} (x_1\sigma_{23} - x_2\sigma_{13})\, da = \mu\alpha \int_{\mathcal{S}_0} (x_1^2 + x_2^2)\, da = \mu J\alpha,$$

where

$$J = \frac{\pi R^4}{2},$$

is the *polar moment of inertia* of the cross-section. Therefore the reaction torque on $\mathcal{S}_0$ is

$$\int_{\mathcal{S}_0} \mathbf{r} \times \boldsymbol{\sigma}\mathbf{n} da = -T\mathbf{e}_3 \tag{9.3.24}$$

with

$$T = \mu J \alpha. \tag{9.3.25}$$

Further, by balance of moments, the net torque on face $\mathcal{S}_L$ is

$$\int_{\mathcal{S}_L} \mathbf{r} \times \boldsymbol{\sigma}\mathbf{n}\, da = T\mathbf{e}_3. \tag{9.3.26}$$

Thus the twisting of the cylinder requires equal and opposite couples about the x_3-axis.

9.3.3 **Torsion of a cylinder of arbitrary cross-section**

Consider a cylindrical bar of a homogeneous and isotropic linear elastic material with generators parallel to the x_3-axis. Let the end faces of the bar $\mathcal{S}_0$ and $\mathcal{S}_L$ be located at $x_3 = 0$ and $x_3 = L$, respectively, cf. Fig. 9.5. A generic cross-section of the non-circular shaft is shown in Fig. 9.7, and, for convenience, in this figure the x_3-axis is along the *center of twist*, which is defined by the location where

$$u_1 = u_2 = 0;$$

the center of twist depends on the precise shape of the cross-section.

We assume that the face $\mathcal{S}_0$ is held such that,

$$u_1(\mathbf{x}) = 0 \quad \text{and} \quad u_2(\mathbf{x}) = 0 \quad \text{at} \quad x_3 = 0, \tag{9.3.27}$$

and the lateral surface is traction-free,

$$\left.\begin{aligned} \sigma_{11}n_1 + \sigma_{12}n_2 &= 0 \\ \sigma_{21}n_1 + \sigma_{22}n_2 &= 0 \\ \sigma_{31}n_1 + \sigma_{32}n_2 &= 0 \end{aligned}\right\} \quad \text{on } \Gamma, \tag{9.3.28}$$

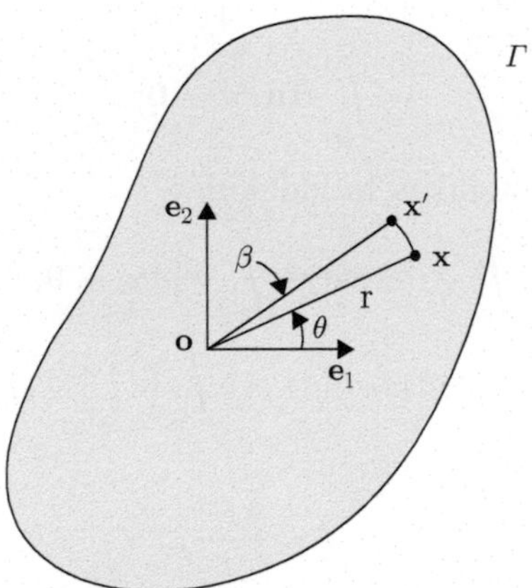

Fig. 9.7 In-plane displacements for the torsion problem.

while the face $\mathcal{S}_L$, and thereby $\mathcal{S}_0$, carries no net force

$$\int_{\mathcal{S}_L} \boldsymbol{\sigma}\mathbf{n}\, da = \mathbf{0}, \qquad \int_{\mathcal{S}_0} \boldsymbol{\sigma}\mathbf{n}\, da = \mathbf{0}, \tag{9.3.29}$$

but the face $\mathcal{S}_L$ is subjected to a twisting moment

$$\int_{\mathcal{S}_L} \mathbf{r} \times \boldsymbol{\sigma}\mathbf{n}\, da = T\mathbf{e}_3, \tag{9.3.30}$$

which of course must be balanced by a reaction moment on face $\mathcal{S}_0$,

$$\int_{\mathcal{S}_0} \mathbf{r} \times \boldsymbol{\sigma}\mathbf{n}\, da = -T\mathbf{e}_3, \tag{9.3.31}$$

with the origin $\mathbf{o}$ in $\mathbf{r} = \mathbf{x} - \mathbf{o}$ taken at the center of twist of each face.

Consider a generic cross-section of the shaft, Fig. 9.7. Under the applied twisting moment, a generic point $\mathbf{x}$ in the (x_1, x_2)-plane displaces to $\mathbf{x}'$, and the vector $(\mathbf{x}-\mathbf{o})$ then rotates through a small angle β, and thus the arc length $|\mathbf{x}' - \mathbf{x}| \approx r\beta$. The in-plane displacements can thus be guessed to be of the form

$$\begin{aligned} u_1 &= -r\beta \sin\theta = -\beta x_2, \\ u_2 &= r\beta \cos\theta = \beta x_1. \end{aligned} \tag{9.3.32}$$

Assuming that the section rotation is a linear function of the axial coordinate x_3, and since the cylinder is fixed in plane at $x_3 = 0$, we assume

$$\beta = \alpha x_3, \tag{9.3.33}$$

where

$$\alpha \quad \text{is the } \textit{angle of twist} \text{ per unit length.} \tag{9.3.34}$$

In torsion of cylinders of arbitrary plane cross sections, the cross sections are known *not to remain plane* after deformation, rather they undergo an out-of-plane *warping* displacement. The out-of-plane warping displacement is a function of the in-plane coordinates and vanishes when α vanishes. Collecting these observations, the assumed displacement field for the torsion problem may be written as

$$\begin{aligned} u_1(\mathbf{x}) &= -\alpha x_2 x_3, \\ u_2(\mathbf{x}) &= \alpha x_1 x_3, \\ u_3(\mathbf{x}) &= \alpha\varphi(x_1, x_2), \end{aligned} \tag{9.3.35}$$

where $\varphi(x_1, x_2)$ is called the *warping function* and is yet to be determined. We now proceed with the details of

- a *displacement*-based treatment, as well as a
- *stress*-based treatment

to solve the torsion problem for bars with arbitrary cross sections.

Displacement-based formulation

The strain-field corresponding to the displacement field (9.3.35) is

$$\begin{aligned} \epsilon_{11}(\mathbf{x}) &= \epsilon_{22}(\mathbf{x}) = \epsilon_{33}(\mathbf{x}) = \epsilon_{12}(\mathbf{x}) = 0, \\ \epsilon_{13}(\mathbf{x}) &= \frac{1}{2}\left(\frac{\partial \varphi}{\partial x_1} - x_2\right)\alpha, \\ \epsilon_{23}(\mathbf{x}) &= \frac{1}{2}\left(\frac{\partial \varphi}{\partial x_2} + x_1\right)\alpha, \end{aligned} \tag{9.3.36}$$

and using the constitutive equation for an isotropic linear elastic material the corresponding stresses are

$$\begin{aligned} \sigma_{11}(\mathbf{x}) &= \sigma_{22}(\mathbf{x}) = \sigma_{33}(\mathbf{x}) = \sigma_{12}(\mathbf{x}) = 0, \\ \sigma_{13}(\mathbf{x}) &= \mu\alpha\left(\frac{\partial \varphi}{\partial x_1} - x_2\right), \\ \sigma_{23}(\mathbf{x}) &= \mu\alpha\left(\frac{\partial \varphi}{\partial x_2} + x_1\right). \end{aligned} \tag{9.3.37}$$

Note that the non-zero shear stress components are functions only of x_1 and x_2. For this case, with zero body forces, the equilibrium equation $\operatorname{div} \boldsymbol{\sigma} = \mathbf{0}$ reduces to

$$\sigma_{13,1} + \sigma_{23,2} = 0 \quad \text{in } \Omega. \tag{9.3.38}$$

Using (9.3.37) in (9.3.38) gives

$$\Delta\varphi = 0 \quad \text{in } \Omega, \tag{9.3.39}$$

where

$$\Delta\varphi = \varphi_{,11} + \varphi_{,22} \tag{9.3.40}$$

is the two-dimensional Laplacian of φ. This is a form of Navier's equation for the torsion problem.

To complete the formulation we must now address the boundary conditions for the problem. As previously mentioned, the lateral surface is to be free of tractions. Since $\sigma_{11} = \sigma_{22} = \sigma_{12} = 0$, we are left with, cf. (9.3.28),

$$\sigma_{31}n_1 + \sigma_{32}n_2 = 0 \quad \text{on } \Gamma. \tag{9.3.41}$$

This boundary condition may be expressed using (9.3.37) in terms of the warping function as

$$\left(\frac{\partial \varphi}{\partial x_1} - x_2\right)n_1 + \left(\frac{\partial \varphi}{\partial x_2} + x_1\right)n_2 = 0 \qquad \text{on} \quad \Gamma \tag{9.3.42}$$

or

$$\underbrace{\frac{\partial \varphi}{\partial n} \equiv \frac{\partial \varphi}{\partial x_1}n_1 + \frac{\partial \varphi}{\partial x_2}n_2}_{\text{normal derivative of } \varphi \text{ on } \partial\Omega} = x_2 n_1 - x_1 n_2 \quad \text{on } \Gamma. \tag{9.3.43}$$

Thus the boundary-value problem that must be satisfied by the warping function φ is

$$\boxed{\Delta\varphi = 0 \quad \text{in } \Omega, \quad \text{with} \quad \frac{\partial \varphi}{\partial n} = x_2 n_1 - x_1 n_2 \quad \text{on } \Gamma.} \tag{9.3.44}$$

It is relatively difficult to determine analytical solutions to this Laplace equation with the complicated nature of the boundary condition. An alternate formulation that results in a Poisson's equation and a boundary condition which is much simpler in nature will be discussed shortly.

Notwithstanding the difficulty, once the warping function φ is determined it may be shown that boundary conditions (9.3.29) and (9.3.30) on $\mathcal{S}_L$, except for the torque in the $\mathbf{e}_3$ direction, are identically satisfied, and the remaining torque boundary condition requires that

$$T = \int_{\mathcal{S}_L} (x_1\sigma_{23} - x_2\sigma_{13})\, da. \tag{9.3.45}$$

Substituting (9.3.37) in the remaining boundary condition (9.3.45) for the resultant torque gives the requirement on the twist per unit length (the last remaining unknown) that

$$\alpha = \frac{T}{(\mu\bar{J})}, \tag{9.3.46}$$

where

$$\bar{J} \stackrel{\text{def}}{=} \int_{\Omega} \left(x_1^2 + x_2^2 + x_1\varphi_{,2} - x_2\varphi_{,1}\right) da, \tag{9.3.47}$$

is the *effective polar moment of inertia* for the non-circular cross-section (analog of the polar moment of inertia for a circular cross-section). The quantity $(\mu\bar{J})$ is the *torsional rigidity* of the cross-section.

Stress-based formulation. Prandtl stress function

First note that a necessary compatibility condition from $(9.3.36)_{2,3}$ is

$$\epsilon_{13,2} - \epsilon_{23,1} = -\alpha \quad \text{in } \Omega,$$

which implies, for linear elastic isotropic materials, that

$$\sigma_{13,2} - \sigma_{23,1} = -2\mu\alpha \quad \text{in } \Omega. \tag{9.3.48}$$

Recall now that the equilibrium equation for the problem under consideration reduces to

$$\sigma_{13,1} + \sigma_{23,2} = 0 \quad \text{in } \Omega. \tag{9.3.49}$$

The partial differential equations (9.3.48) and (9.3.49) represent the governing equations of the stress formulation for the torsion problem. The equilibrium equation (9.3.49) can be automatically satisfied by introducing an unknown scalar-valued function

$$\Psi(x_1, x_2) \quad \text{called the } \textit{Prandtl stress function},$$

with the properties that

$$\sigma_{13} = \frac{\partial\Psi}{\partial x_2}, \qquad \sigma_{23} = -\frac{\partial\Psi}{\partial x_1}. \tag{9.3.50}$$

While the compatibility condition (9.3.48) becomes

$$\Delta\Psi = -2\mu\alpha \quad \text{in } \Omega, \tag{9.3.51}$$

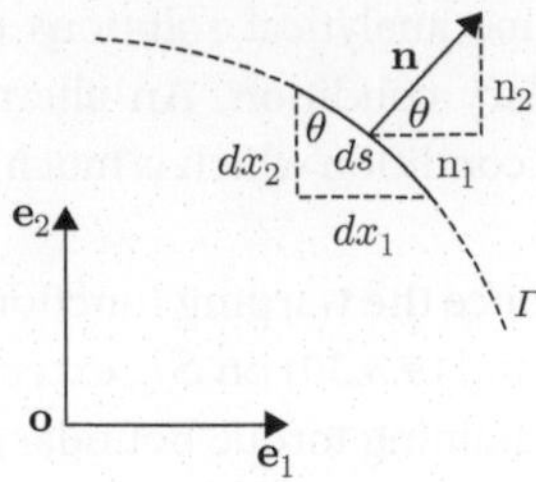

Fig. 9.8 An oriented surface element for the torsion problem.

where

$$\Delta\Psi = \Psi_{,11} + \Psi_{,22} \tag{9.3.52}$$

is the two-dimensional Laplacian of Ψ. Equation (9.3.51) is a *Poisson equation* that is amenable to several analytical solution techniques due to the simplicity of the applicable boundary conditions, which we discuss next.

As previously mentioned, the lateral surface is to be free of tractions, i.e. it should satisfy (9.3.28). The first two of these are identically satisfied since $\sigma_{11} = \sigma_{22} = \sigma_{12} = 0$, and we are left with

$$\sigma_{31}n_1 + \sigma_{32}n_2 = 0 \quad \text{on } \Gamma. \tag{9.3.53}$$

If we consider an arc length parameter s for the perimeter Γ of the cross-section, cf. Fig. 9.8, then

$$n_1 = \cos\theta = \frac{dx_2}{ds}, \qquad n_2 = \sin\theta = -\frac{dx_1}{ds}. \tag{9.3.54}$$

Substituting (9.3.50) and (9.3.54) into the boundary condition (9.3.53), we obtain

$$\frac{\partial\Psi}{\partial x_2}\frac{dx_2}{ds} + \frac{\partial\Psi}{\partial x_1}\frac{dx_1}{ds} \equiv \frac{d\Psi}{ds} = 0 \quad \text{on } \Gamma, \tag{9.3.55}$$

which implies that the stress function Ψ must be a *constant* on the boundary of the cross-section. For multiply connected regions the constants in general have different values on different boundaries. However, for *simply connected*[3] cross sections, we can arbitrarily set the constant to zero. Thus the boundary-value problem that must be satisfied by the stress function Ψ is

$$\boxed{\Delta\Psi = -2\mu\alpha \quad \text{in } \Omega, \qquad \text{subject to} \quad \Psi = 0 \quad \text{on } \Gamma.} \tag{9.3.56}$$

Consider next the boundary condition (9.3.29) on S_L. Since on this face $n_1 = n_2 = 0$ and $n_3 = 1$, the net force **P** in each of the three x_i-directions must vanish:

[3] A cross-section Ω is *simply connected* if any closed curve in Ω can be continuously shrunk to a point without ever leaving Ω.

$$P_1 = \int_{\mathcal{S}_L} \sigma_{13}\, da = 0,$$
$$P_2 = \int_{\mathcal{S}_L} \sigma_{23}\, da = 0, \tag{9.3.57}$$
$$P_3 = \int_{\mathcal{S}_L} \sigma_{33}\, da = 0.$$

Since $\sigma_{33} = 0$, $(9.3.57)_3$ is automatically satisfied. Consider next $(9.3.57)_{1,2}$ and apply Green's theorem (cf. (2.4.6)):

$$\int_{\mathcal{S}_L} \sigma_{13} da = \int_{\mathcal{S}_L} \frac{\partial \Psi}{\partial x_2}\, da \equiv \int_{\Gamma} \Psi n_2 ds, \quad \text{and}$$
$$\int_{\mathcal{S}_L} \sigma_{23} da = -\int_{\mathcal{S}_L} \frac{\partial \Psi}{\partial x_1}\, da \equiv -\int_{\Gamma} \Psi n_1 ds. \tag{9.3.58}$$

Because $\Psi = 0$ on the boundary Γ, the integrals in (9.3.58) also vanish, and the boundary condition (9.3.57) is satisfied.

Next consider the boundary condition (9.3.31) on $\mathcal{S}_L$. Since on this face $n_1 = n_2 = 0$, and $n_3 = 1$, the net moment $\mathbf{M}$ in each of the three x_i-directions must satisfy:

$$M_1 = \int_{\mathcal{S}_L} x_2 \sigma_{33} da = 0,$$
$$M_2 = \int_{\mathcal{S}_L} x_1 \sigma_{33} da = 0, \tag{9.3.59}$$
$$M_3 = \int_{\mathcal{S}_L} (x_1\sigma_{23} - x_2\sigma_{13})\, da = T,$$

Since $\sigma_{33} = 0$, $(9.3.59)_{1,2}$ are identically satisfied. Using (9.3.50) to substitute for σ_{13} and σ_{23} in terms of the stress function Ψ in $(9.3.59)_3$ we obtain

$$T = -\int_{\Omega} \left(x_1 \frac{\partial \Psi}{\partial x_1} + x_2 \frac{\partial \Psi}{\partial x_2} \right) da. \tag{9.3.60}$$

Next, since

$$\int_{\Omega} \frac{\partial}{\partial x_1} (x_1 \Psi)\, da = \int_{\Omega} \Psi\, da + \int_{\Omega} x_1 \frac{\partial \Psi}{\partial x_1}\, da,$$

we have, using Green's theorem and the fact that $\Psi = 0$ on Γ, that

$$\begin{aligned} \int_{\Omega} x_1 \frac{\partial \Psi}{\partial x_1}\, da &= \int_{\Omega} \frac{\partial}{\partial x_1} (x_1 \Psi)\, da - \int_{\Omega} \Psi\, da, \\ &= \int_{\Gamma} x_1 \Psi n_1\, ds - \int_{\Omega} \Psi\, da, \\ &= -\int_{\Omega} \Psi\, da, \end{aligned} \tag{9.3.61}$$

where in writing the last expression we have used the fact that $\Psi = 0$ on the boundary Γ. In an entirely analogous manner

$$
\begin{aligned}
\int_\Omega x_2 \frac{\partial \Psi}{\partial x_2}\, da &= \int_\Omega \frac{\partial}{\partial x_2}(x_2\Psi)\, da - \int_\Omega \Psi\, da, \\
&= \int_\Gamma x_2 \Psi n_2\, ds - \int_\Omega \Psi\, da, \\
&= -\int_\Omega \Psi\, da,
\end{aligned} \tag{9.3.62}
$$

Substituting (9.3.61) and (9.3.62) in (9.3.60) gives

$$
\boxed{T = 2\int_\Omega \Psi\, da.} \tag{9.3.63}
$$

- *In summary, what we have shown is that the assumed displacement field* (9.3.35) *produces a stress field that when represented by the Prandtl stress function* (9.3.50) *yields a governing Poisson's equation which must be satisfied in the cross-section and vanish on the boundary of the cross-section, cf.* (9.3.56). *All resultant boundary conditions on the lateral surfaces and on the ends of the cylinder are satisfied by the representation, and the overall torque is related to the stress function via* (9.3.63).

REMARK 9.3

Exact solutions of the torsion problem (9.3.56) are possible for a few special cases. For some cross sectional shapes classic methods of partial differential equations, like separation of variables or Fourier methods, are effective. In other cases it is sufficient to assume a solution of the form

$$
\Psi(x_1, x_2) = Af(x_1, x_2),
$$

where A is a constant, and f is a sufficiently differentiable function that is identically zero on the boundary. If Δf is a non-zero constant, say $-C$, so that AC can be equated to $2\mu\alpha$, then we can solve for A and obtain the complete solution. If Δf is not a constant, then an exact solution of this form is not possible, although an approximate solution may be obtained by using Galerkin's method.

EXAMPLE 9.3 Torsion of a shaft with elliptical cross-section

Consider a cylindrical shaft of elliptical cross-section Ω with boundary Γ, an ellipse with semi-axes a and b:

$$
\Gamma = \left\{ (x_1, x_2) \;\middle|\; \frac{x_1^2}{a^2} + \frac{x_2^2}{b^2} = 1 \right\}. \tag{9.3.64}
$$

To determine the Prandtl stress function, the shear stresses, the warping function and the displacement field, we select the Prandtl stress function to be of the form

$$\Psi(x_1, x_2) = A\left(\frac{x_1^2}{a^2} + \frac{x_2^2}{b^2} - 1\right), \tag{9.3.65}$$

where A is a constant to be determined such that (9.3.56) is satisfied. Since (9.3.65) automatically satisfies the boundary condition $\Psi = 0$ on Γ, we substitute (9.3.65) in $(9.3.56)_1$ to obtain

$$2A\left(\frac{1}{a^2} + \frac{1}{b^2}\right) = -2\mu\alpha \quad \Rightarrow \quad A = -\mu\alpha\frac{a^2b^2}{a^2+b^2}. \tag{9.3.66}$$

The Prandtl stress function is then given by

$$\Psi(x_1, x_2) = -\mu\alpha\frac{a^2b^2}{a^2+b^2}\left(\frac{x_1^2}{a^2} + \frac{x_2^2}{b^2} - 1\right). \tag{9.3.67}$$

The torque is then given by (9.3.63):

$$T = -2\mu\alpha\frac{a^2b^2}{a^2+b^2}\left(\frac{1}{a^2}\int_\Omega x_1^2 da + \frac{1}{b^2}\int_\Omega x_2^2 da - \int_\Omega da\right). \tag{9.3.68}$$

The integrals in this expression have the following simple meaning and evaluation:

$$\begin{aligned} \text{Area of section} &= \int_\Omega da = \pi ab, \\ \text{Moment of inertia about } x_1\text{-axis} &= \int_\Omega x_2^2 da = \frac{1}{4}\pi ab^3, \\ \text{Moment of inertia about } x_2\text{-axis} &= \int_\Omega x_1^2 da = \frac{1}{4}\pi ba^3, \end{aligned} \tag{9.3.69}$$

Substituting (9.3.69) back in (9.3.68) gives

$$\boxed{T = \mu\bar{J}\alpha, \qquad \text{with} \qquad \bar{J} \stackrel{\text{def}}{=} \frac{\pi a^3 b^3}{(a^2+b^2)}} \tag{9.3.70}$$

from which the twist-torque relation is

$$\boxed{\alpha = \frac{T}{\mu\bar{J}}.} \tag{9.3.71}$$

The resulting shear stresses are calculated from (9.3.50) and the stress function (9.3.67) as

$$\boxed{\begin{aligned} \sigma_{13} &= -\mu\alpha\frac{2a^2}{a^2+b^2}x_2 &&= -\frac{2Tx_2}{\pi ab^3}, \\ \sigma_{23} &= \mu\alpha\frac{2b^2}{a^2+b^2}x_1 &&= \frac{2Tx_1}{\pi ba^3}. \end{aligned}} \tag{9.3.72}$$

Using (9.3.72) we may calculate the *equivalent shear stress* defined by

$$\bar{\tau} \stackrel{\text{def}}{=} \sqrt{\sigma_{13}^2 + \sigma_{23}^2} = \frac{2T}{\pi ab}\sqrt{\frac{x_1^2}{a^4} + \frac{x_2^2}{b^4}}. \tag{9.3.73}$$

For $a > b$ the maximum value of the equivalent shear stress $\bar{\tau}$ occurs at $x_1 = 0$ and $x_2 = \pm b$ with a maximum value of[a]

$$\bar{\tau}\big|_{\max} = \frac{2T}{\pi a b^2}. \tag{9.3.74}$$

Next, recall the stress-relations $(9.3.37)_{2,3}$ in terms of the warping function φ:

$$\begin{aligned}\sigma_{13}(\mathbf{x}) &= \mu\alpha\left(\frac{\partial\varphi(x_1,x_2)}{\partial x_1} - x_2\right),\\ \sigma_{23}(\mathbf{x}) &= \mu\alpha\left(\frac{\partial\varphi(x_1,x_2)}{\partial x_2} + x_1\right).\end{aligned} \tag{9.3.75}$$

Substituting for the stresses from (9.3.72) in (9.3.75) we get

$$\begin{aligned}-\frac{2a^2}{a^2+b^2}x_2 &= \frac{\partial\varphi(x_1,x_2)}{\partial x_1} - x_2,\\ \frac{2b^2}{a^2+b^2}x_1 &= \frac{\partial\varphi(x_1,x_2)}{\partial x_2} + x_1.\end{aligned} \tag{9.3.76}$$

or

$$\begin{aligned}\frac{\partial\varphi(x_1,x_2)}{\partial x_1} &= \left(\frac{b^2-a^2}{a^2+b^2}\right)x_2,\\ \frac{\partial\varphi(x_1,x_2)}{\partial x_2} &= \left(\frac{b^2-a^2}{a^2+b^2}\right)x_1.\end{aligned} \tag{9.3.77}$$

which upon integration gives the warping function as

$$\boxed{\varphi(x_1,x_2) = \left(\frac{b^2-a^2}{a^2+b^2}\right)x_1x_2.} \tag{9.3.78}$$

With α and φ determined by (9.3.71) and (9.3.78) in terms of the geometry, the applied torque T and the shear modulus μ; the displacement field (9.3.35) is completely determined for the problem at hand.

---

[a] *This result is counterintuitive in the sense that based on ideas of elementary strength of materials one might have expected that for $a > b$ the maximum shear stress occurs at the boundary points most removed from the section's center; that is, at $x_1 = \pm a$ and $x_2 = 0$.*

EXAMPLE 9.4 Torsion with rectangular cross-section

Consider a torsion bar with a $c \times b$ rectangular cross-section, cf. Fig. 9.9.

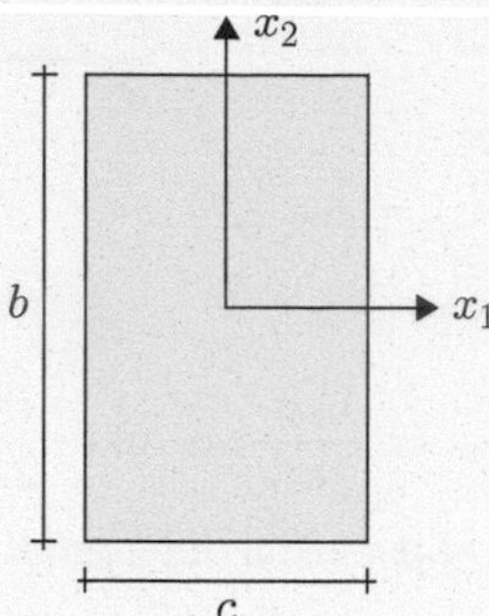

Fig. 9.9 Cross-section of rectangular torsion bar.

The solution to (9.3.56) can be approached using a Fourier series method with variable coefficients:

$$\Psi = \sum_{n \in \text{odd}} f_n(x_1) \cos\left(\frac{n\pi x_2}{b}\right), \tag{9.3.79}$$

where the coefficient *functions* $f_n(x_1)$ are unknown and need to be determined. The form of this guess is motivated by the following ideas:

1. Problem (9.3.56) for Ψ has the identical form as the problem that governs the deflection of a thin membrane subject to a transverse pressure (equal to the constant $-2\mu\alpha$). This observation, known as the *membrane analogy*, tells us that the shape of Ψ should be an even function of both x_1 and x_2 for this cross-section. Hence we have chosen an even Fourier series in x_2 which also satisfies the $\Psi = 0$ boundary condition at $x_2 = \pm b/2$.
2. We could have also chosen a similar guess for the functional form for x_1, i.e. a so-called double Fourier series, but instead we have chosen variable Fourier coefficients[a] as they are known to converge faster than the double series.

The essential thing to determine are the unknown functions $f_n(x_1)$. The procedure for doing this is to insert (9.3.79) into the governing partial differential equation $(9.3.56)_1$ to come up with equations that the $f_n(x_1)$ must satisfy. The utility of this procedure is that the resulting equations are straightforward to solve.

To begin, observe that we can also represent the term $-2\mu\alpha$ on the right-hand side of $(9.3.56)_1$ in the same form as the stress function:

$$-2\mu\alpha = \sum_{n \in \text{odd}} A_n \cos\left(\frac{n\pi x_2}{b}\right). \tag{9.3.80}$$

The coefficients A_n in this Fourier series are determined as follows. First we multiply both sides of (9.3.80) by $\cos(m\pi x_2/b)$ and integrate over the cross-section:

$$\int_{-b/2}^{b/2} -2\mu\alpha \cos\left(\frac{m\pi x_2}{b}\right) dx_2 = \sum_{n\in\text{odd}} A_n \underbrace{\int_{-b/2}^{b/2} \cos\left(\frac{n\pi x_2}{b}\right) \cos\left(\frac{m\pi x_2}{b}\right) dx_2}_{\frac{b}{2}\delta_{mn}}$$

$$-4\mu\alpha\frac{b}{m\pi}\sin\left(\frac{m\pi}{2}\right) = A_m\frac{b}{2},$$

which gives the Fourier coefficients as

$$A_m = -\frac{8}{m\pi}\mu\alpha\sin\left(\frac{m\pi}{2}\right). \tag{9.3.81}$$

Thus, using (9.3.80) and (9.3.81) the partial differential equation (9.3.36) becomes

$$\Delta\Psi = -\sum_{m\in\text{odd}} \frac{8\mu\alpha}{m\pi}\sin\left(\frac{m\pi}{2}\right)\cos\left(\frac{m\pi x_2}{b}\right). \tag{9.3.82}$$

Next, using (9.3.79) in (9.3.82), we find that

$$\sum_{m\in\text{odd}} \left(f_m'' - \left(\frac{m\pi}{b}\right)^2 f_m\right)\cos\left(\frac{m\pi x_2}{b}\right) = -\sum_{m\in\text{odd}} \frac{8\mu\alpha}{m\pi}\sin\left(\frac{m\pi}{2}\right)\cos\left(\frac{m\pi x_2}{b}\right). \tag{9.3.83}$$

We can now use the orthogonality of the cosine functions, viz.,

$$\int_{-b/2}^{b/2} \cos\left(\frac{n\pi x_2}{b}\right)\cos\left(\frac{m\pi x_2}{b}\right) dx_2 = \frac{b}{2}\delta_{nm}$$

to determine a governing equation for each function $f_m(x_1)$. Proceeding as we did to determine A_m, multiply both sides of (9.3.83) by $\cos(n\pi x_2/b)$ and integrate over the domain. We then find the governing problem for the coefficient functions in our assumed form (9.3.79) to be

$$f_n'' - \left(\frac{n\pi}{b}\right)^2 f_n + \frac{8\mu\alpha}{n\pi}\sin\left(\frac{n\pi}{2}\right) = 0, \tag{9.3.84}$$

and the condition $\Psi = 0$ on Γ gives us the boundary conditions

$$f_n\left(\pm\frac{c}{2}\right) = 0.$$

Equation (9.3.84) is an inhomogeneous ordinary differential equation whose solution is given by

$$f_n(x_1) = C_n\cosh\left(\frac{n\pi x_1}{b}\right) + D_n\sinh\left(\frac{n\pi x_1}{b}\right) + \left(\frac{b}{n\pi}\right)^2\frac{8\mu\alpha}{n\pi}\sin\left(\frac{n\pi}{2}\right),$$

where C_n and D_n are unknown coefficients. Note that since our solution needs to be an even function of x_1, D_n must equal zero. Using the boundary condition $f_n(\pm c/2) = 0$

then gives

$$C_n = -\frac{8\mu\alpha b^2 \sin(n\pi/2)}{n^3\pi^3 \cosh(n\pi c/2b)}.$$

Inserting into the expression for f_n and then into the Fourier series gives the final result of

$$\Psi = \sum_{n\in\text{odd}} \frac{8\mu\alpha b^2 \sin(n\pi/2)}{n^3\pi^3}\left[1 - \frac{\cosh(n\pi x_1/b)}{\cosh(n\pi c/2b)}\right]\cos\left(\frac{n\pi x_2}{b}\right),$$

which is a very quickly converging series.

Noting that $\Psi_{,2} = \mu\alpha(\varphi_{,1} - x_2)$ and $-\Psi_{,1} = \mu\alpha(\varphi_{,2} + x_1)$, one can integrate to solve for the warping function

$$\varphi = \sum_{n\in\text{odd}}\left[\frac{8b^2}{n^3\pi^3}\sin\left(\frac{n\pi}{2}\right)\sin\left(\frac{n\pi x_2}{b}\right)\frac{\sinh(n\pi x_1/b)}{\cosh(n\pi c/2b)}\right] - x_1x_2.$$

From these expressions for Ψ and φ one can easily determine the displacements, strains, and stresses in the bar.

[a] This method is sometimes known as Hencky's method, especially when applied to plate bending problems.

9.4 **Airy stress function and plane traction problems**

Several plane strain and plane stress traction problems may be solved by use of a scalar-valued function

$$\varphi(x_1, x_2),$$

called an **Airy stress function** (Airy, 1863). An Airy function is chosen such that

- it leads to a representation for the stress field which automatically satisfies the two-dimensional equations of equilibrium, and
- yields a single governing equation from the compatibility equation in terms of the stress function.

Using such a stress function, a two-dimensional (plane strain or plane stress) traction boundary-value problem is reduced to a single *biharmonic equation* in terms of φ. The resulting governing equation is then amenable to solution by several methods in applied mathematics.

For simplicity, we henceforth consider situations in which body forces may be neglected; in this case the equilibrium equations reduce to

$$\begin{aligned}\sigma_{11,1} + \sigma_{12,2} &= 0,\\ \sigma_{21,1} + \sigma_{22,2} &= 0.\end{aligned} \tag{9.4.1}$$

Let $\varphi(x_1, x_2)$ be a sufficiently smooth scalar field on the two-dimensional region $\mathcal{R}$ under consideration. We define the stress function to have the properties

$$\boxed{\sigma_{11} = \varphi_{,22}, \qquad \sigma_{22} = \varphi_{,11}, \qquad \sigma_{12} = -\varphi_{,12}.} \tag{9.4.2}$$

By direct substitution, it is easily verified that the stress components $\sigma_{\alpha\beta}$ so defined, satisfy the equilibrium equations (9.4.1).

- A scalar field φ with this property is called an *Airy stress function.*

For future use we note relations (9.4.2) may be written as

$$\sigma_{\alpha\beta} = e_{\alpha\lambda} e_{\beta\tau} \varphi_{,\lambda\tau}, \tag{9.4.3}$$

where $e_{\alpha\beta}$ is the two-dimensional alternator with values

$$e_{12} = 1, \qquad e_{21} = -1, \qquad e_{11} = e_{22} = 0. \tag{9.4.4}$$

Next recall the compatibility equation (8.9.34) for plane problems in the absence of body forces,

$$(\sigma_{11} + \sigma_{22})_{,11} + (\sigma_{11} + \sigma_{22})_{,22} = 0. \tag{9.4.5}$$

Using (9.4.2) to rewrite the compatibility equation (9.4.5) in terms of the Airy stress function φ gives

$$\varphi_{,1111} + 2\varphi_{,1122} + \varphi_{,2222} = 0, \tag{9.4.6}$$

which may be written as

$$\boxed{\Delta\Delta\varphi = 0.} \tag{9.4.7}$$

Equation (9.4.7) is known as the *biharmonic equation* of plane elasticity, and its solutions are known as *biharmonic functions.*

Thus, the plane traction problem (in the absence of body forces) for a simply connected region can be reduced to the problem of finding an Airy stress function $\varphi(x_1, x_2)$, that satisfies

$$\boxed{\begin{aligned} \Delta\Delta\varphi &= 0 \quad \text{on } \mathcal{R}, \\ (e_{\alpha\lambda} e_{\beta\tau} \varphi_{,\lambda\tau}) n_\beta &= \hat{t}_\alpha \quad \text{on } \mathcal{C}. \end{aligned}} \tag{9.4.8}$$

Stress fields derived from the Airy function that satisfies (9.4.8) will be in equilibrium and correspond to compatible strain fields, for which the displacements may be determined, as shown in the next subsection.

Displacements in terms of the Airy stress function

If the body force is negligible, then the displacement components can be expressed in the form

$$\boxed{\begin{aligned} u_1 &= \frac{(1+\nu)}{E}(-\varphi_{,1} + s\psi_{,2}) + w_1, \\ u_2 &= \frac{(1+\nu)}{E}(-\varphi_{,2} + s\psi_{,1}) + w_2, \end{aligned}} \tag{9.4.9}$$

where $\psi(x_1, x_2)$ is a potential function that satisfies

$$\boxed{\begin{aligned}\Delta\psi &= 0,\\ \psi_{,12} &= \Delta\varphi,\end{aligned}} \tag{9.4.10}$$

and **w** is a plane rigid displacement:

$$w_{\alpha,\beta} = -w_{\beta,\alpha} \qquad \Longrightarrow \qquad w_{1,1} = 0, \qquad w_{2,2} = 0, \qquad w_{1,2} + w_{2,1} = 0. \tag{9.4.11}$$

This assertion is verified below.

Verification: In terms of the material parameter s the plane constitutive equations for isotropic materials $(8.9.44)_1$ when written out in terms of displacements are

$$\begin{aligned}
\sigma_{11} &= \frac{E}{(1+\nu)}\left(u_{1,1} + \left(\frac{1-s}{2s-1}\right)(u_{1,1}+u_{2,2})\right),\\
\sigma_{22} &= \frac{E}{(1+\nu)}\left(u_{2,2} + \left(\frac{1-s}{2s-1}\right)(u_{1,1}+u_{2,2})\right),\\
\sigma_{12} &= \frac{E}{2(1+\nu)}(u_{1,2}+u_{2,1}).
\end{aligned} \tag{9.4.12}$$

Using (9.4.9), $(9.4.10)_2$, and $(9.4.11)_4$,

$$\frac{E}{(1+\nu)}(u_{1,1}+u_{2,2}) = -(\varphi_{,11}+\varphi_{,22}) + 2s\psi_{,12} = (2s-1)\Delta\varphi, \tag{9.4.13}$$

use of which in (9.4.12) gives

$$\begin{aligned}
\sigma_{11} &= -\varphi_{,11} + s\Delta\varphi + (1-s)\Delta\varphi &&= -\varphi_{,11} + \Delta\varphi &&= \varphi_{,22},\\
\sigma_{22} &= -\varphi_{,22} + s\Delta\varphi + (1-s)\Delta\varphi &&= -\varphi_{,22} + \Delta\varphi &&= \varphi_{,11},\\
\sigma_{12} &= \tfrac{1}{2}\left(-\varphi_{,12} + s\psi_{,22} - \varphi_{,21} + s\psi_{,11}\right) &&= -\varphi_{,12} + \tfrac{1}{2}s\Delta\psi &&= -\varphi_{,12}.
\end{aligned} \tag{9.4.14}$$

Thus the displacement field (9.4.9) is consistent with the definition of the Airy stress function.

9.4.1 Airy stress function and plane traction problems in polar coordinates

In what follows we shall use polar coordinates to obtain solutions to several important plane traction problems. Accordingly, in this section we develop the polar coordinate form of the equations discussed in the previous section.

In the absence of body forces the equations of equilibrium for plane problems reduce to (cf. (A.2.21))

$$\begin{aligned}
\frac{\partial\sigma_{rr}}{\partial r} + \frac{1}{r}\frac{\partial\sigma_{r\theta}}{\partial\theta} + \frac{1}{r}(\sigma_{rr}-\sigma_{\theta\theta}) &= 0,\\
\frac{\partial\sigma_{\theta r}}{\partial r} + \frac{1}{r}\frac{\partial\sigma_{\theta\theta}}{\partial\theta} + \frac{2}{r}\sigma_{\theta r} &= 0.
\end{aligned} \tag{9.4.15}$$

In polar coordinates an Airy function $\varphi(r,\theta)$ is defined such that

$$\boxed{\begin{aligned}\sigma_{rr} &= \frac{1}{r}\frac{\partial\varphi}{\partial r} + \frac{1}{r^2}\frac{\partial^2\varphi}{\partial\theta^2},\\ \sigma_{\theta\theta} &= \frac{\partial^2\varphi}{\partial r^2},\\ \sigma_{r\theta} &= -\frac{\partial}{\partial r}\left(\frac{1}{r}\frac{\partial\varphi}{\partial\theta}\right).\end{aligned}} \tag{9.4.16}$$

These relations can be derived from (9.4.2) via the change of coordinate relations $x_1 = r\cos\theta$ and $x_2 = r\sin\theta$. Substituting the defining equations for the Airy stress function into (9.4.15) shows that components of stress derived via (9.4.16) are guaranteed to be in equilibrium.

Note further, the tensor form of the compatibility condition is invariant, so that compatibility is satisfied if φ satisfies the biharmonic equation

$$\Delta\Delta\varphi = 0 \tag{9.4.17}$$

independent of the chosen coordinates. The expression for the Laplacian for cylindrical coordinates (A.2.18), when specialized to polar coordinates gives

$$\Delta\,\varphi = \frac{\partial^2\varphi}{\partial r^2} + \frac{1}{r^2}\frac{\partial^2\varphi}{\partial\theta^2} + \frac{1}{r}\frac{\partial\varphi}{\partial r}. \tag{9.4.18}$$

The biharmonic equation (9.4.17) in polar coordinates is therefore given by

$$\boxed{\left(\frac{\partial^2}{\partial r^2} + \frac{1}{r^2}\frac{\partial^2}{\partial\theta^2} + \frac{1}{r}\frac{\partial}{\partial r}\right)\left(\frac{\partial^2\varphi}{\partial r^2} + \frac{1}{r^2}\frac{\partial^2\varphi}{\partial\theta^2} + \frac{1}{r}\frac{\partial\varphi}{\partial r}\right) = 0.} \tag{9.4.19}$$

Displacements in terms of the Airy stress function in polar coordinates

If the body force is negligible, then the displacement components can be expressed in the form

$$\boxed{\begin{aligned}u_r &= \frac{(1+\nu)}{E}\left(-\varphi_{,r} + s\,r\,\psi_{,\theta}\right) + w_r,\\ u_\theta &= \frac{(1+\nu)}{E}\left(-\frac{1}{r}\varphi_{,\theta} + s\,r^2\,\psi_{,r}\right) + w_\theta,\end{aligned}} \tag{9.4.20}$$

where $\psi(r,\theta)$ is a potential function that satisfies

$$\boxed{\Delta\psi = 0, \qquad \text{and} \qquad (r\psi_{,\theta})_{,r} = \Delta\varphi,} \tag{9.4.21}$$

and $\mathbf{w}$ is a plane rigid displacement field $\nabla\mathbf{w} = -(\nabla\mathbf{w})^{\top}$.

9.4.2 Some biharmonic functions in polar coordinates

For plane problems in polar coordinates, some biharmonic functions $\varphi(r,\theta)$ that may be used as Airy stress functions are

$$\begin{aligned}\varphi &= C\theta,\\ \varphi &= Cr^2\theta,\\ \varphi &= Cr\theta\cos\theta,\\ \varphi &= Cr\theta\sin\theta,\end{aligned} \tag{9.4.22}$$

where C is a constant. Other useful forms of Airy stress functions are

$$\begin{aligned}\varphi &= f_n(r)\cos(n\theta),\\ \varphi &= f_n(r)\sin(n\theta),\end{aligned} \tag{9.4.23}$$

where in order for the functions (9.4.23) to be biharmonic, the functions $f_n(r)$ can be chosen from the form

$$\begin{aligned}f_0(r) &= a_0 r^2 + b_0 r^2 \ln r + c_0 + d_0 \ln r,\\ f_1(r) &= a_1 r^3 + b_1 r + c_1 r\ln r + d_1\, r^{-1},\\ f_n(r) &= a_n r^{n+2} + b_n r^n + c_n r^{-n+2} + d_n r^{-n}, \qquad \text{for} \qquad n > 1,\end{aligned} \tag{9.4.24}$$

where a_n, b_n, c_n, and d_n are constants to be determined from the boundary conditions.

Of course, sums of biharmonic functions are also biharmonic. A reasonably general biharmonic function $\varphi(r,\theta)$ has the form (Michell, 1899)

$$\begin{aligned}\varphi = {} & a_0 + a_1 \ln r + a_2 r^2 + a_3 r^2 \ln r,\\ & + (a_4 + a_5 \ln r + a_6 r^2 + a_7 r^2 \ln r)\theta\\ & + (a_{11} r + a_{12} r\ln r + a_{13} r^{-1} + a_{14} r^3 + a_{15} r\theta + a_{16} r\theta \ln r)\cos\theta\\ & + (b_{11} r + b_{12} r\ln r + b_{13} r^{-1} + b_{14} r^3 + b_{15} r\theta + b_{16} r\theta \ln r)\sin\theta\\ & + \sum_{n=2}^{\infty}(a_{n1} r^n + a_{n2} r^{2+n} + a_{n3} r^{-n} + a_{n4} r^{2-n})\cos(n\theta)\\ & + \sum_{n=2}^{\infty}(b_{n1} r^n + b_{n2} r^{2+n} + b_{n3} r^{-n} + b_{n4} r^{2-n})\sin(n\theta).\end{aligned} \tag{9.4.25}$$

9.4.3 Infinite medium with a hole under uniform far-field loading in tension

Consider a plate with a circular hole of radius a. The plate has infinite dimensions in the plane, and is subjected to a far-field uniaxial stress σ^{∞}, as shown in Fig. 9.10.

The traction boundary conditions in this problem are that the hole is traction-free. The outward unit normal to the body at the hole is $-\mathbf{e}_r$, and hence

$$\boldsymbol{\sigma}(-\mathbf{e}_r) = \mathbf{0} \qquad \text{at} \qquad r = a. \tag{9.4.26}$$

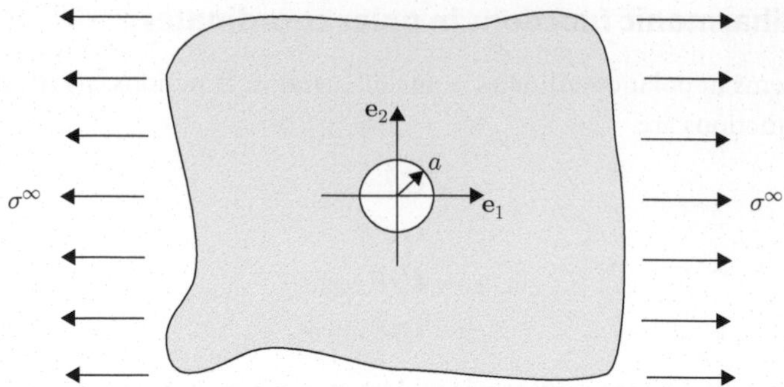

Fig. 9.10 A hole of radius a in an infinite plate subjected to a far-field uniaxial stress σ^∞.

so that

$$\begin{aligned}\sigma_{rr}(a,\theta) &= 0,\\ \sigma_{\theta r}(a,\theta) &= 0.\end{aligned} \tag{9.4.27}$$

While, far from the hole

$$\sigma_{11} = \sigma^\infty, \qquad \sigma_{22} = \sigma_{12} = 0, \qquad \text{at} \qquad r = \infty. \tag{9.4.28}$$

Using the transformation rule

$$\sigma^*_{\alpha\beta} = Q_{\alpha\gamma} Q_{\beta\delta} \sigma_{\gamma\delta} \tag{9.4.29}$$

under a change in basis, where

$$[Q] = \begin{bmatrix} \cos\theta & \sin\theta \\ -\sin\theta & \cos\theta \end{bmatrix} \tag{9.4.30}$$

is the transformation matrix from a rectangular to a polar coordinate system, for a uniaxial stress σ^∞ in the Cartesian system we have

$$\begin{aligned}\sigma_{rr}(\infty,\theta) &= \sigma^\infty \cos^2\theta &&= \tfrac{1}{2}\sigma^\infty(1+\cos 2\theta),\\ \sigma_{\theta\theta}(\infty,\theta) &= \sigma^\infty \sin^2\theta &&= \tfrac{1}{2}\sigma^\infty(1-\cos 2\theta),\\ \sigma_{r\theta}(\infty,\theta) &= -\sigma^\infty \sin\theta\cos\theta &&= -\tfrac{1}{2}\sigma^\infty \sin 2\theta.\end{aligned} \tag{9.4.31}$$

As a guess for the Airy stress function φ for this problem, note that since $\sigma_{11} = \varphi_{,22}$, we must have

$$\varphi_{,22} \to \sigma^\infty \qquad \text{at} \qquad r \to \infty,$$

and hence we should expect that the stress function looks something like

$$\varphi = \tfrac{1}{2}\sigma^\infty x_2^2 \qquad \text{as} \qquad r \to \infty,$$

and since

$$x_2 = r\sin\theta, \qquad x_2^2 = r^2\sin^2\theta = \tfrac{1}{2}r^2(1-\cos 2\theta),$$

that

$$\varphi = \frac{1}{4}\sigma^\infty r^2 - \frac{1}{4}\sigma^\infty r^2 \cos 2\theta \qquad \text{at} \qquad r \to \infty.$$

The presence of the hole acts to disturb the uniform far-field stress produced by such an Airy function. We expect that this disturbance is local in nature, and the disturbed field will decay to zero as we move away from the hole. Based on this, we choose a trial Airy function that includes the axisymmetric terms and the $\cos(2\theta)$ terms in the Michell biharmonic function (9.4.25):

$$\varphi(r,\theta) = \left(a_0 + a_1 \ln r + a_2 r^2 + a_3 r^2 \ln r\right) + \left(a_{21} r^2 + a_{22} r^4 + a_{23} r^{-2} + a_{24}\right)\cos(2\theta). \tag{9.4.32}$$

The stress components corresponding to this Airy stress function are

$$\begin{aligned} \sigma_{rr} &= a_3(1 + 2\ln r) + 2a_2 + a_1 r^{-2} - (2a_{21} + 6a_{23} r^{-4} + 4a_{24} r^{-2})\cos(2\theta), \\ \sigma_{\theta\theta} &= a_3(3 + 2\ln r) + 2a_2 - a_1 r^{-2} + (2a_{21} + 12a_{22} r^2 + 6a_{23} r^{-4})\cos(2\theta), \\ \sigma_{r\theta} &= (2a_{21} + 6a_{22} r^2 - 6a_{23} r^{-4} - 2a_{24} r^{-2})\sin(2\theta). \end{aligned} \tag{9.4.33}$$

For finite stresses at infinity, we must set

$$a_3 = a_{22} = 0,$$

so that (9.4.33) becomes

$$\begin{aligned} \sigma_{rr} &= 2a_2 + a_1 r^{-2} - (2a_{21} + 6a_{23} r^{-4} + 4a_{24} r^{-2})\cos(2\theta), \\ \sigma_{\theta\theta} &= 2a_2 - a_1 r^{-2} + (2a_{21} + 6a_{23} r^{-4})\cos(2\theta), \\ \sigma_{r\theta} &= (2a_{21} - 6a_{23} r^{-4} - 2a_{24} r^{-2})\sin(2\theta), \end{aligned} \tag{9.4.34}$$

and involves the five constants a_1, a_2, a_{21}, a_{23}, and a_{24}.

With this, applying the boundary conditions (9.4.27) at $r = a$ we require that

$$\begin{aligned} 2a_2 + a_1 a^{-2} &= 0, \\ 2a_{21} + 6a_{23} a^{-4} + 4a_{24} a^{-2} &= 0, \\ 2a_{21} - 6a_{23} a^{-4} - 2a_{24} a^{-2} &= 0, \end{aligned} \tag{9.4.35}$$

and applying the boundary conditions (9.4.31) at $r = \infty$ we require that

$$\begin{aligned} 2a_{21} &= -\tfrac{1}{2}\sigma^{\infty}, \\ 2a_2 &= \tfrac{1}{2}\sigma^{\infty}. \end{aligned} \tag{9.4.36}$$

The system of equations (9.4.35) and (9.4.36) may be solved for the constants, giving

$$a_1 = -\tfrac{1}{2}a^2\sigma^{\infty}, \quad a_2 = \frac{1}{4}\sigma^{\infty}, \quad a_{21} = -\frac{1}{4}\sigma^{\infty}, \quad a_{23} = -\frac{1}{4}a^4\sigma^{\infty}, \quad a_{24} = \tfrac{1}{2}a^2\sigma^{\infty}. \tag{9.4.37}$$

Thus the appropriate Airy stress function for the problem is[4]

$$\boxed{\varphi(r,\theta) = \frac{\sigma^{\infty}}{4}\left(\left(-2a^2\ln r + r^2\right) + \left(-r^2 - a^4 r^{-2} + 2a^2\right)\cos(2\theta)\right).} \tag{9.4.38}$$

[4] It should be observed that a plate with a hole is not a simply connected body. Thus a complete solution would involve the extra step of checking the additional conditions referred to in Appendix B of Chapter 3 for multiply connected bodies. Here we simply note that these additional conditions are satisfied by the Airy function (9.4.38), thus ensuring compatibility and the existence of a single valued displacement field.

Substituting the values of the coefficients (9.4.37) back into (9.4.34) gives the stress field

$$\boxed{\begin{aligned}\sigma_{rr} &= \frac{\sigma^\infty}{2}\left(1-\frac{a^2}{r^2}\right)+\frac{\sigma^\infty}{2}\left(1+\frac{3a^4}{r^4}-\frac{4a^2}{r^2}\right)\cos(2\theta),\\ \sigma_{\theta\theta} &= \frac{\sigma^\infty}{2}\left(1+\frac{a^2}{r^2}\right)-\frac{\sigma^\infty}{2}\left(1+\frac{3a^4}{r^4}\right)\cos(2\theta),\\ \sigma_{r\theta} &= -\frac{\sigma^\infty}{2}\left(1-\frac{3a^4}{r^4}+\frac{2a^2}{r^2}\right)\sin(2\theta).\end{aligned}} \tag{9.4.39}$$

EXAMPLE 9.5 Stress concentration around a hole in a plate subjected to uniform tension

From $(9.4.39)_2$, the hoop stress variation around the boundary of the hole is given by

$$\sigma_{\theta\theta}(a,\theta) = \sigma^\infty(1-2\cos 2\theta). \tag{9.4.40}$$

The normalized hoop stress distribution $\sigma_{\theta\theta}(a,\theta)/\sigma^\infty$ based on (9.4.40) is plotted in Fig. 9.11.

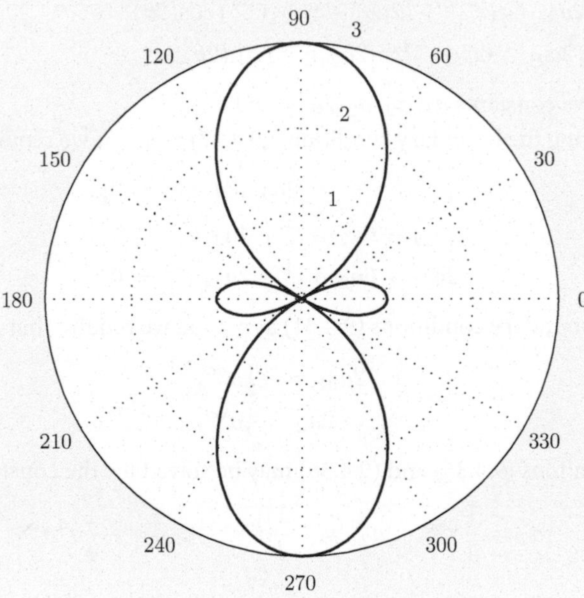

Fig. 9.11 Polar plot of the variation of the normalized hoop stress $\sigma_{\theta\theta}(a,\theta)/\sigma^\infty$ around the boundary of the hole, $r = a$.

This distribution indicates that the hoop stress actually vanishes at $\theta = 30°$, and has a maximum value at $\theta = \pm 90°$:

$$\boxed{\frac{\sigma_{\max}}{\sigma^{\infty}} = \frac{\sigma_{\theta\theta}(a, \pm\pi/2)}{\sigma^{\infty}} = 3 \qquad \text{stress concentration factor.}} \tag{9.4.41}$$

Also note from

$$\frac{\sigma_{\theta\theta}(r, \pi/2)}{\sigma^{\infty}} = \frac{1}{2}\left(2 + \frac{a^2}{r^2} + \frac{3a^4}{r^4}\right), \tag{9.4.42}$$

that the stress concentration decays rapidly from a factor of 3 at $r = a$ to ≈ 1.02 at $r = 5a$; cf. Fig. 9.12.

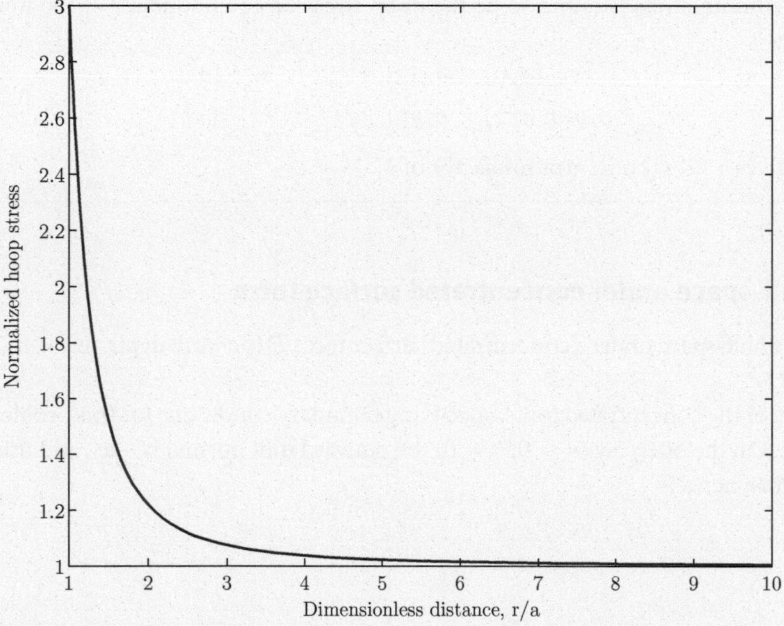

Fig. 9.12 Variation of the normalized hoop stress $\sigma_{\theta\theta}(r, \pi/2)/\sigma^{\infty}$ with the normalized radial distance r/a from the hole.

EXAMPLE 9.6 Stress concentration in a plate with a hole under biaxial/shear loading

Another interesting stress concentration problem corresponds to the biaxial loading case

$$\sigma_{11} = \sigma^{\infty} \qquad \sigma_{22} = -\sigma^{\infty} \qquad \text{as} \qquad r \to \infty. \tag{9.4.43}$$

This far-field loading is equivalent to a pure shear loading of σ^{∞} in a coordinate system that is rotated $45°$ with respect to the x_1, x_2 axes. The solution to the biaxial problem is easily found from the solution (9.4.39). This is done by superposing on the original stress field a stress field with the loading replaced by $-\sigma^\infty$ and having coordinate axes rotated by 90°. The final result is

$$\boxed{\begin{aligned}\sigma_{rr} &= \sigma^\infty\left(1+\frac{3a^4}{r^4}-\frac{4a^2}{r^2}\right)\cos(2\theta),\\ \sigma_{\theta\theta} &= -\sigma^\infty\left(1+\frac{3a^4}{r^4}\right)\cos(2\theta),\\ \sigma_{r\theta} &= -\sigma^\infty\left(1-\frac{3a^4}{r^4}+\frac{2a^2}{r^2}\right)\sin(2\theta).\end{aligned}} \tag{9.4.44}$$

The maximum stress is found to be the hoop stress on the boundary of the hole and is given by

$$\begin{aligned}\sigma_{\theta\theta}(a,0)=\sigma_{\theta\theta}(a,\pi) &= -4\sigma^\infty,\\ \sigma_{\theta\theta}(a,\pi/2)=\sigma_{\theta\theta}(a,3\pi/2) &= \;\;4\sigma^\infty,\end{aligned} \tag{9.4.45}$$

which gives a stress concentration factor of 4.

9.4.4 **Half-space under concentrated surface force**

Consider a half-space under a concentrated surface force P (per unit depth in the x_3 direction); cf. Fig. 9.13.[5]

Because of the concentrated force, specifying boundary conditions for this problem requires some care. On the open ray ($r > 0, \theta = 0$) the outward unit normal is $-\mathbf{e}_\theta$, and this surface is traction-free; hence,

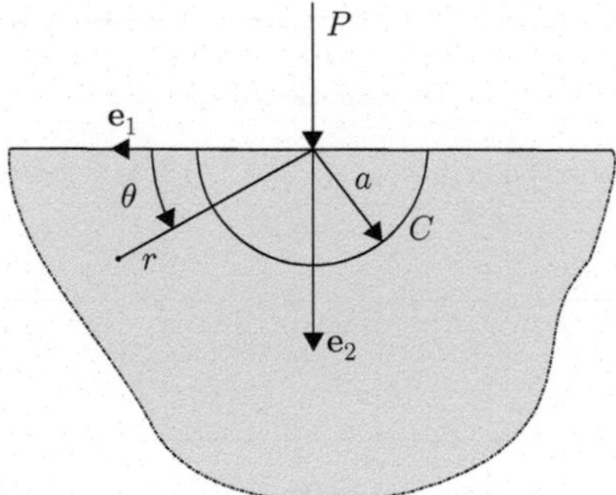

Fig. 9.13 A half-space under a concentrated load P.

[5] The solution to this classical problem closely follows that given in the book by Sadd (2014).

$$\boldsymbol{\sigma}(-\mathbf{e}_\theta) = \mathbf{0} \qquad \Longrightarrow \qquad \sigma_{\theta\theta} = \sigma_{r\theta} = 0 \qquad \text{on} \qquad (r > 0, \theta = 0). \tag{9.4.46}$$

Also the outward unit normal on the open ray $(r > 0, \theta = \pi)$ is $\mathbf{e}_\theta$, and on this surface we also have traction-free conditions:

$$\boldsymbol{\sigma}(\mathbf{e}_\theta) = \mathbf{0} \qquad \Longrightarrow \qquad \sigma_{\theta\theta} = \sigma_{r\theta} = 0 \qquad \text{on} \qquad (r > 0, \theta = \pi). \tag{9.4.47}$$

To account for the concentrated force at $r = 0$, we note that *the tractions on any semicircular arc C of radius a enclosing the origin must balance the applied load P*, Fig. 9.13. The outward unit normal to the curve C is $\mathbf{e}_r$, thus the tractions on this curve are $t_r = \sigma_{rr}$ and $t_\theta = \sigma_{\theta r}$. Therefore, in order for equilibrium to be satisfied, we must have

$$\begin{aligned}
\sum F_1 &= \int_0^\pi \mathbf{e}_1 \cdot \boldsymbol{\sigma}\mathbf{e}_r a\, d\theta = \int_0^\pi [\sigma_{rr}(a,\theta)\cos\theta - \sigma_{\theta r}(a,\theta)\sin\theta] a\, d\theta \underbrace{= 0}_{\text{force balance in the } \mathbf{e}_1\text{-direction}}, \\
\sum F_2 &= P + \int_0^\pi \mathbf{e}_2 \cdot \boldsymbol{\sigma}\mathbf{e}_r a\, d\theta = P + \int_0^\pi [\sigma_{rr}(a,\theta)\sin\theta + \sigma_{\theta r}(a,\theta)\cos\theta] a\, d\theta \underbrace{= 0}_{\text{force balance in the } \mathbf{e}_2\text{-direction}}, \\
\sum M_3 &= \int_0^\pi (\mathbf{x} - \mathbf{o}) \times \boldsymbol{\sigma}\mathbf{e}_r a\, d\theta = \int_0^\pi [a\sigma_{\theta r}(a,\theta)] a\, d\theta \underbrace{= 0}_{\text{moment balance about the } \mathbf{e}_3\text{-axis}}.
\end{aligned} \tag{9.4.48}$$

The boundary condition corresponding to the concentrated load is thus given *indirectly* by requiring that the solution satisfy (9.4.48) for all values of a.

Next, since the arc length of a curve C is proportional to r, the stresses at a distance r must be proportional to $1/r$ for equilibrium to hold. The appropriate terms in the Michell biharmonic function (9.4.25) which will give rise to a $1/r$ dependence for the stresses are

$$\varphi = (a_{12} r \ln r + a_{15} r\theta)\cos\theta + (b_{12} r \ln r + b_{15} r\theta)\sin\theta, \tag{9.4.49}$$

with corresponding stresses

$$\begin{aligned}
\sigma_{rr} &= \frac{1}{r}\left((a_{12} + 2b_{15})\cos\theta + (b_{12} - 2a_{15})\sin\theta\right), \\
\sigma_{\theta\theta} &= \frac{1}{r}\left(a_{12}\cos\theta + b_{12}\sin\theta\right), \\
\sigma_{r\theta} &= \frac{1}{r}\left(a_{12}\sin\theta - b_{12}\cos\theta\right).
\end{aligned} \tag{9.4.50}$$

To satisfy the boundary conditions $\sigma_{\theta\theta} = 0$ at $\theta = 0$ requires that

$$a_{12} = 0;$$

while satisfaction of $\sigma_{r\theta} = 0$ at $\theta = \pi$ requires that

$$b_{12} = 0.$$

Hence

$$\sigma_{\theta\theta} = \sigma_{r\theta} = 0 \qquad \text{everywhere,} \tag{9.4.51}$$

and the only non-zero stress is the radial stress, which is given by

$$\sigma_{rr} = \frac{1}{r}\left(2b_{15}\cos\theta - 2a_{15}\sin\theta\right). \tag{9.4.52}$$

Since the shear stress $\sigma_{r\theta}$ vanishes everywhere, the force and moment balance equations (9.4.48) reduce to

$$\begin{aligned} 0 &= \int_0^\pi [\sigma_{rr}(a,\theta)\cos\theta] a\, d\theta && \text{force balance in the } \mathbf{e}_1\text{-direction,} \\ P &= -\int_0^\pi [\sigma_{rr}(a,\theta)\sin\theta] a\, d\theta && \text{force balance in the } \mathbf{e}_2\text{-direction.} \end{aligned} \tag{9.4.53}$$

Substituting (9.4.52) in (9.4.53) we obtain

$$\begin{aligned} 0 &= \int_0^\pi \left(2b_{15}\cos^2\theta - 2a_{15}\sin\theta\cos\theta\right) d\theta \\ &= \left[b_{15}\left(\tfrac{1}{2}\sin\theta\cos\theta + \theta\right) - a_{15}\sin^2\theta\right]_0^\pi = \pi b_{15}, \\ P &= -\int_0^\pi \left(2b_{15}\sin\theta\cos\theta - 2a_{15}\sin^2\theta\right) d\theta \\ &= -\left[b_{15}\sin^2\theta - a_{15}(-\cos\theta\sin\theta + \theta)\right]_0^\pi = \pi a_{15}. \end{aligned} \tag{9.4.54}$$

Hence

$$b_{15} = 0, \qquad \text{and} \qquad a_{15} = \frac{P}{\pi}, \tag{9.4.55}$$

and therefore the Airy stress function for this problem is

$$\boxed{\varphi = \frac{P}{\pi} r\theta\cos\theta.} \tag{9.4.56}$$

and the stress field is fully characterized by

$$\boxed{\sigma_{rr} = -\frac{2P}{\pi r}\sin\theta, \qquad \sigma_{\theta\theta} = \sigma_{r\theta} = 0.} \tag{9.4.57}$$

As expected the stress field is singular at the origin directly under the point load.

Next, consider a circle of radius d centered at a distance $x_2 = d$ from the free surface. Then, geometry dictates that

$$\begin{aligned} d^2 &= (r\cos\theta)^2 + (d - r\sin\theta)^2 \\ &= r^2\cos^2\theta + d^2 - 2dr\sin\theta + r^2\sin^2\theta, \end{aligned}$$

which gives

$$\frac{\sin\theta}{r} = \frac{1}{2d}. \tag{9.4.58}$$

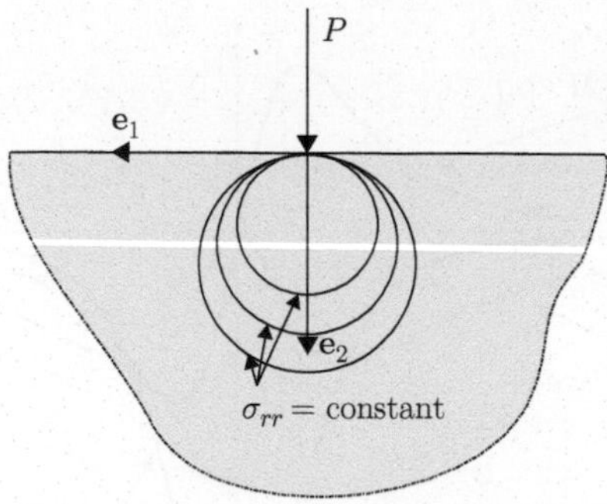

Fig. 9.14 Contours of σ_{rr} for a half-space under a concentrated load P.

Substitution of (9.4.58) in (9.4.57) gives

$$\sigma_{rr} = -\frac{P}{\pi d}. \tag{9.4.59}$$

Thus, from (9.4.59), at a given load P, the contour lines of constant radial stress σ_{rr} are circles which are tangent to the half-space surface at the loading point, Fig. 9.14, with σ_{rr} decreasing as d increases.

It is also useful to express the stress distribution in Cartesian coordinates. Note that in Cartesian coordinates, (9.4.56) is given by $\varphi = (P/\pi)x_1 \tan^{-1}(x_2/x_1)$. Thus,

$$\begin{aligned}
\sigma_{11} = \varphi_{,22} &= -\frac{2P}{\pi}\frac{x_1^2 x_2}{(x_1^2+x_2^2)^2},\\
\sigma_{22} = \varphi_{,11} &= -\frac{2P}{\pi}\frac{x_2^3}{(x_1^2+x_2^2)^2},\\
\sigma_{12} = -\varphi_{,12} &= -\frac{2P}{\pi}\frac{x_1 x_2^2}{(x_1^2+x_2^2)^2}.
\end{aligned} \tag{9.4.60}$$

At a distance $x_2 = a$ from the free surface, these Cartesian stress components have values

$$\begin{aligned}
\sigma_{11} &= -\frac{2P}{\pi a}\frac{(x_1/a)^2}{((x_1/a)^2+1)^2},\\
\sigma_{22} &= -\frac{2P}{\pi a}\frac{1}{((x_1/a)^2+1)^2},\\
\sigma_{12} &= -\frac{2P}{\pi a}\frac{(x_1/a)}{((x_1/a)^2+1)^2}.
\end{aligned} \tag{9.4.61}$$

These stresses, normalized by $(2P)/(\pi a)$, are plotted versus the normalized distance x_1/a in Fig. 9.15. The maximum normal stress is $|\sigma_{22}| = 2P/\pi a$, and occurs directly under the concentrated load. *Note that the effects of the concentrated load die away at horizontal distances larger than* $x_1/a = 5$.

We now wish to determine the displacement distribution for this problem. Assume *plane stress* conditions. In this case the constitutive equation $(8.9.37)_2$ gives

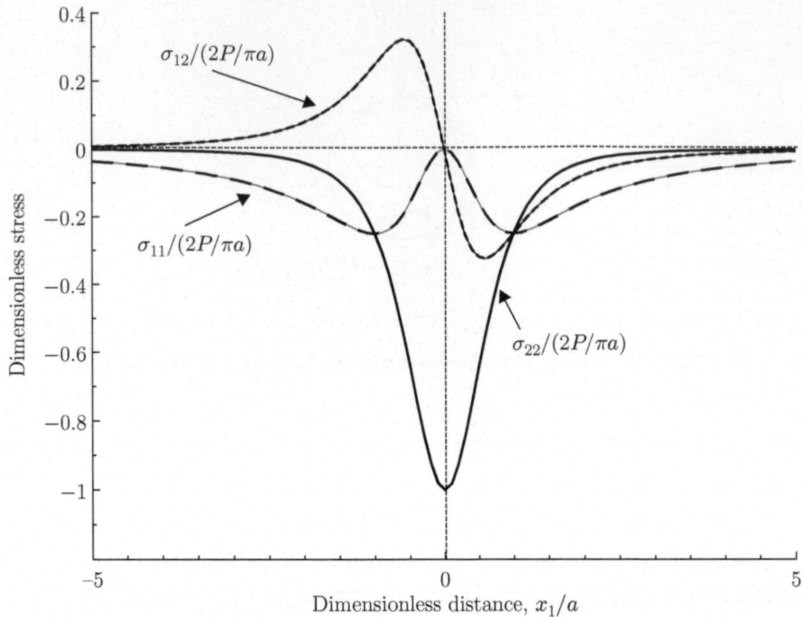

Fig. 9.15 Normalized stresses $\sigma_{\alpha\beta}/(2P)(\pi a)$ versus normalized distance x_1/a for a half-space subjected to a point force.

$$\begin{aligned}
\epsilon_{rr} &= \frac{1}{E}[\sigma_{rr} - \nu\sigma_{\theta\theta}], \\
\epsilon_{\theta\theta} &= \frac{1}{E}[\sigma_{\theta\theta} - \nu\sigma_{rr}], \\
\epsilon_{r\theta} &= \frac{1+\nu}{E}\sigma_{r\theta},
\end{aligned} \tag{9.4.62}$$

while the strain displacement relation (A.2.20) reduces to

$$\begin{aligned}
\epsilon_{rr} &= \frac{\partial u_r}{\partial r}, \\
\epsilon_{\theta\theta} &= \frac{1}{r}\frac{\partial u_\theta}{\partial \theta} + \frac{u_r}{r}, \\
\epsilon_{r\theta} &= \frac{1}{2}\left(\frac{1}{r}\frac{\partial u_r}{\partial \theta} + \frac{\partial u_\theta}{\partial r} - \frac{u_\theta}{r}\right).
\end{aligned} \tag{9.4.63}$$

Substituting the stress distribution (9.4.57) and the strain displacement relations (9.4.63) in the constitutive equations (9.4.62) gives

$$\begin{aligned}
\frac{\partial u_r}{\partial r} &= -\frac{2P}{\pi E r}\sin\theta, \\
\frac{\partial u_\theta}{\partial \theta} + u_r &= \frac{2\nu P}{\pi E}\sin\theta, \\
\frac{1}{r}\frac{\partial u_r}{\partial \theta} + \frac{\partial u_\theta}{\partial r} - \frac{u_\theta}{r} &= 0.
\end{aligned} \tag{9.4.64}$$

Integrating $(9.4.64)_1$ yields the radial displacement

$$u_r = -\frac{2P}{\pi E}\sin\theta \ln r + f(\theta), \tag{9.4.65}$$

where $f(\theta)$ is an arbitrary function of θ. Substituting (9.4.65) in $(9.4.64)_2$ gives

$$\frac{\partial u_\theta}{\partial \theta} = \frac{2\nu P}{\pi E}\sin\theta + \frac{2P}{\pi E}\sin\theta \ln r - f(\theta). \tag{9.4.66}$$

Integrating this last equation gives

$$u_\theta = -\frac{2\nu P}{\pi E}\cos\theta - \frac{2P}{\pi E}\cos\theta \ln r - \int f(\theta)d\theta + g(r), \tag{9.4.67}$$

where $g(r)$ is an arbitrary function of r. The arbitrary functions $f(\theta)$ and $g(r)$ are determined by substituting (9.4.66) and (9.4.67) in $(9.4.64)_3$. Since,

$$\begin{aligned}
\frac{1}{r}\frac{\partial u_r}{\partial \theta} &= -\frac{2P}{\pi E}\cos\theta\frac{\ln r}{r} + \frac{1}{r}f'(\theta),\\
\frac{\partial u_\theta}{\partial r} &= -\frac{2P}{\pi E r}\cos\theta + g'(r),\\
\frac{u_\theta}{r} &= -\frac{2\nu P}{\pi E r}\cos\theta - \frac{2P}{\pi E}\cos\theta\frac{\ln r}{r} - \frac{1}{r}\int f(\theta)d\theta + \frac{1}{r}g(r),
\end{aligned}$$

we have

$$\frac{1}{r}f'(\theta) - \frac{2(1-\nu)P}{\pi E r}\cos\theta + g'(r) + \frac{1}{r}\int f(\theta)d\theta - \frac{1}{r}g(r) = 0,$$

or multiplying through by r,

$$\underbrace{\left(-\frac{2(1-\nu)P}{\pi E}\cos\theta + f'(\theta) + \int f(\theta)d\theta\right)}_{\text{function of }\theta} - \underbrace{(g(r) - rg'(r))}_{\text{function of }r} = 0.$$

Thus we must have

$$\begin{aligned}
-\frac{2(1-\nu)P}{\pi E}\cos\theta + f'(\theta) + \int f(\theta)d\theta &= K,\\
g(r) - rg'(r) &= K
\end{aligned} \tag{9.4.68}$$

where K is an arbitrary constant. The two equations (9.4.68) admit the solution

$$\begin{aligned}
f(\theta) &= \frac{(1-\nu)P}{\pi E}\theta\cos\theta + A\sin\theta + B\cos\theta,\\
g(r) &= Cr + K,
\end{aligned} \tag{9.4.69}$$

where A, B, and C are constants of integration.

Collecting these results, the displacement components u_r and u_θ are

$$\begin{aligned}u_r &= -\frac{2P}{\pi E}\sin\theta\ln r + \frac{(1-\nu)P}{\pi E}\theta\cos\theta + A\sin\theta + B\cos\theta\\ u_\theta &= -\frac{2\nu P}{\pi E}\cos\theta - \frac{2P}{\pi E}\cos\theta\ln r\\ &\quad - \left(\frac{(1-\nu)P}{\pi E}(\theta\sin\theta + \cos\theta) - A\cos\theta + B\sin\theta\right) + Cr + K\end{aligned}$$

or

$$\begin{aligned}u_r &= -\frac{2P}{\pi E}\sin\theta\ln r + \frac{(1-\nu)P}{\pi E}\theta\cos\theta \underbrace{+A\sin\theta + B\cos\theta}_{\text{rigid body terms}}\\ u_\theta &= -\frac{(1+\nu)P}{\pi E}\cos\theta - \frac{2P}{\pi E}\cos\theta\ln r - \frac{(1-\nu)P}{\pi E}\theta\sin\theta \underbrace{-A\cos\theta + B\sin\theta + Cr + K}_{\text{rigid body terms}}.\end{aligned} \tag{9.4.70}$$

The terms involving the constants A, B, C, and K represent a *rigid displacement*. We use the following physical considerations to select suitable values for these constants. First, we enforce the symmetry condition

$$u_\theta(r, \pi/2) = 0;$$

this requires that

$$C = K = 0, \qquad \text{and} \qquad B = -\frac{(1-\nu)P}{2E}. \tag{9.4.71}$$

Next, to cancel out any vertical rigid displacement we set

$$u_r(r_0, \pi/2) = 0, \tag{9.4.72}$$

where r_0 is some arbitrarily chosen length deep below the concentrated load. This gives

$$A = \frac{2P}{\pi E}\ln r_0. \tag{9.4.73}$$

Thus using (9.4.71) and (9.4.73) in (9.4.70) we obtain

$$\begin{aligned}u_r &= -\frac{2P}{\pi E}\sin\theta\ln r + \frac{(1-\nu)P}{\pi E}\theta\cos\theta - \frac{(1-\nu)P}{2E}\cos\theta + \frac{2P}{\pi E}\sin\theta\ln r_0,\\ u_\theta &= -\frac{(1+\nu)P}{\pi E}\cos\theta - \frac{2P}{\pi E}\cos\theta\ln r - \frac{(1-\nu)P}{\pi E}\theta\sin\theta - \frac{2P}{\pi E}\cos\theta\ln r_0 - \frac{(1-\nu)P}{2E}\sin\theta,\end{aligned}$$

or[6]

$$\boxed{\begin{aligned}u_r &= \frac{P}{\pi E}\left((1-\nu)\left(\theta - \frac{\pi}{2}\right)\cos\theta - 2\sin\theta\ln\left(\frac{r}{r_0}\right)\right),\\ u_\theta &= \frac{P}{\pi E}\left(-(1-\nu)\left(\theta - \frac{\pi}{2}\right)\sin\theta - 2\cos\theta\ln\left(\frac{r}{r_0}\right) - (1+\nu)\cos\theta\right).\end{aligned}} \tag{9.4.74}$$

[6] Note that the displacements have a logarithmic singularity at $r = 0$ and $r = \infty$. While the singularity at $r = 0$ is to be expected due to the concentrated force, the unpleasant situation at $r = \infty$ is due to the two-dimensional nature of the model. However, if r_0 is taken to be suitably large, then this problem is somewhat mitigated.

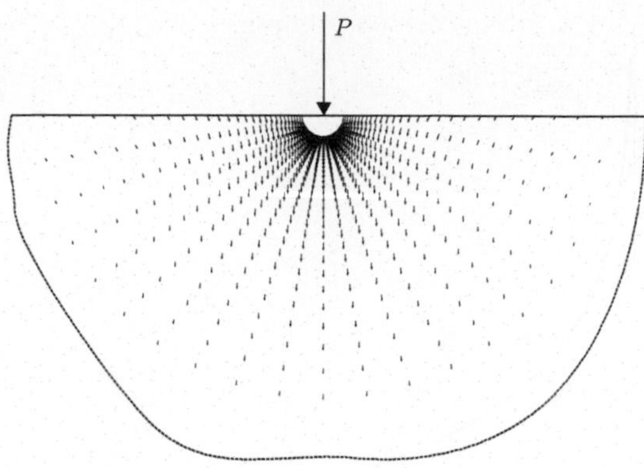

Fig. 9.16 Displacement field for a half-space subjected to a point force.

Fig. 9.16 shows the displacement vectors **u** with components (9.4.74) (in suitable units) for the near-field in the region $r \in [0, 0.5]$ for $P/E = 1$, $\nu = 0.3$, and cut-off radius $r_0 = 1$. The radial displacement along the free surface is given by

$$u_r(r, 0) = u_r(r, \pi) = -\frac{P}{2E}(1 - \nu). \tag{9.4.75}$$

Note that since $(1 - \nu) > 0$ (for most materials), we have the unexpected result that the horizontal displacement of all points on the free surface move an *equal amount* towards the loading point. The downward displacement component on the free surface is given by

$$u_\theta(r, 0) = -u_\theta(r, \pi) = -\frac{P}{\pi E}\left((1 + \nu) + 2\ln\left(\frac{r}{r_0}\right)\right), \tag{9.4.76}$$

which, as expected, is singular under the loading point.

Next, on a line $\theta = \pi/2$ directly under the concentrated load the horizontal displacement component vanishes $u_\theta = 0$, while

$$u_r(r, \pi/2) = -\frac{2P}{\pi E}\ln\left(\frac{r}{r_0}\right). \tag{9.4.77}$$

This displacement profile is sketched in Fig. 9.17. The logarithmic singularities at $r = 0$ and $r = \infty$ are accommodated by noting that, in reality: (i) there cannot be a concentrated load at $r = 0$ — the load must be distributed over some finite area; and (ii) the half-space must have some finite size, which may be taken as the distance r_0 introduced above.

EXAMPLE 9.7 Half-space under uniformly distributed surface load over a small region

Consider now the case of a uniformly distributed load p per unit length acting over a region $-a \le x_1 \le a$ of a half-space, Fig. 9.18. Distributed loadings on an elastic half-space are commonly used to simulate *contact mechanics* problems.[a]

This problem may be solved by using the previous example of a concentrated load, and the principle of superposition.[b]

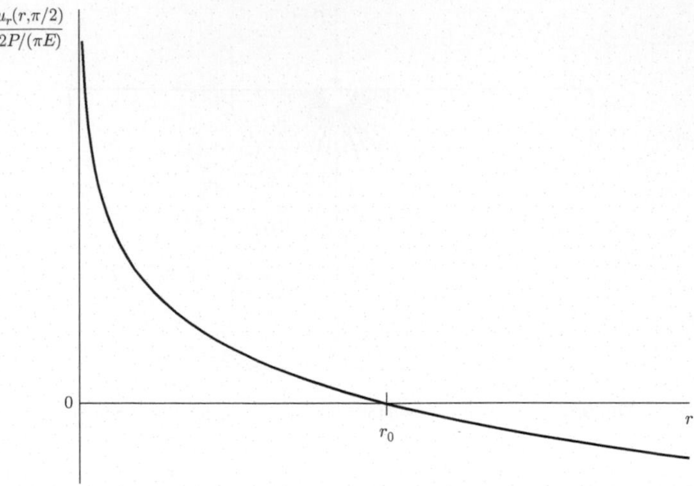

Fig. 9.17 Normalized displacement profile for the radial displacement u_r under the concentrated load.

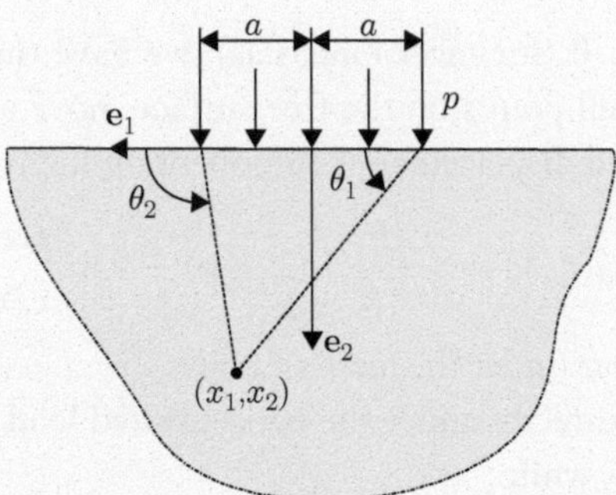

Fig. 9.18 Half-space under uniformly distributed load p per unit length acting over a region $-a \leq x_1 \leq a$.

Consider a material point at (r, θ). Recall that the stress at this point due to a single concentrated load of magnitude P is given in polar coordinates by (9.4.57):

$$\sigma_{rr}(r,\theta) = -\frac{2P}{\pi r}\sin\theta, \qquad \sigma_{\theta\theta}(r,\theta) = \sigma_{r\theta}(r,\theta) = 0. \tag{9.4.78}$$

Using standard transformation rules from polar to Cartesian coordinates, this stress state at the same material point with coordinates (x_1, x_2) is described by

$$\begin{aligned} \sigma_{11}(x_1,x_2) &= \sigma_{rr}(r,\theta)\cos^2\theta = -\frac{2P}{\pi r}\sin\theta\cos^2\theta \\ \sigma_{22}(x_1,x_2) &= \sigma_{rr}(r,\theta)\sin^2\theta = -\frac{2P}{\pi r}\sin^3\theta, \\ \sigma_{12}(x_1,x_2) &= \sigma_{rr}(r,\theta)\sin\theta\cos\theta = -\frac{2P}{\pi r}\sin^2\theta\cos\theta. \end{aligned} \tag{9.4.79}$$

For the case of distributed loading, consider a differential load

$$dP = p\,d\bar{x}_1 \tag{9.4.80}$$

acting over an infinitesimal element $d\bar{x}_1$ located at $\bar{x}_1$ on the free surface, cf. Fig. 9.19.

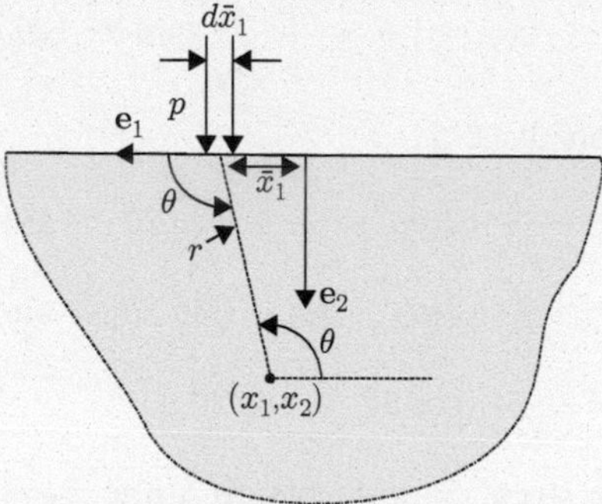

Fig. 9.19 Elemental geometry for a uniformly distributed load on a half-space.

Note that

$$\begin{aligned} x_2 &= r\cos(\theta - \pi/2) = r\sin\theta, \qquad \text{and} \\ \bar{x}_1 - x_1 &= r\sin(\theta - \pi/2) = -r\cos\theta. \end{aligned} \tag{9.4.81}$$

Thus, for a *fixed* point (x_1, x_2), $(9.4.81)_1$ gives

$$dr = -r\cot\theta d\theta,$$

and $(9.4.81)_2$ gives

$$d\bar{x}_1 = -dr\cos\theta + r\sin\theta d\theta = r\left(\frac{\cos^2\theta}{\sin\theta} + \sin\theta\right)d\theta = \frac{r}{\sin\theta}d\theta.$$

Combining with (9.4.80), we find

$$dP = p\,d\bar{x}_1 = \frac{pr}{\sin\theta}d\theta. \tag{9.4.82}$$

Using (9.4.82), the stress distribution in Cartesian components for a differential load $dP = pd\bar{x}_1$ becomes

$$\begin{aligned} d\sigma_{11}(x_1,x_2) &= -\frac{2p}{\pi}\cos^2\theta d\theta \\ d\sigma_{22}(x_1,x_2) &= -\frac{2p}{\pi}\sin^2\theta d\theta, \\ d\sigma_{12}(x_1,x_2) &= -\frac{2p}{\pi}\sin\theta\cos\theta d\theta. \end{aligned} \tag{9.4.83}$$

Then at a generic point (x_1, x_2) determined by the angles θ_1 and θ_2 defined in Fig. 9.18, the stress distribution is given by integrating (9.4.83) between the limits θ_1 and θ_2:

$$\begin{aligned}\sigma_{11}(x_1,x_2) &= -\frac{2p}{\pi}\int_{\theta_1}^{\theta_2}\cos^2\theta d\theta,\\ \sigma_{22}(x_1,x_2) &= -\frac{2p}{\pi}\int_{\theta_1}^{\theta_2}\sin^2\theta d\theta,\\ \sigma_{12}(x_1,x_2) &= -\frac{2p}{\pi}\int_{\theta_1}^{\theta_2}\sin\theta\cos\theta d\theta.\end{aligned} \tag{9.4.84}$$

Performing the integration gives

$$\begin{aligned}\sigma_{11}(x_1,x_2) &= -\frac{p}{2\pi}\left(2(\theta_2-\theta_1)+(\sin 2\theta_2-\sin 2\theta_1)\right),\\ \sigma_{22}(x_1,x_2) &= -\frac{p}{2\pi}\left(2(\theta_2-\theta_1)-(\sin 2\theta_2-\sin 2\theta_1)\right),\\ \sigma_{12}(x_1,x_2) &= -\frac{p}{2\pi}\left(\cos 2\theta_2-\cos 2\theta_1\right).\end{aligned} \tag{9.4.85}$$

The distribution of these stress components on a line $x_2 = a$ below the free surface is shown in Fig. 9.20. This distribution sufficiently far from the distributed load is similar to the stress distribution in Fig. 9.15 for the concentrated load—exemplifying an example of Saint-Venant's principle; see cf. Sec. 8.6.

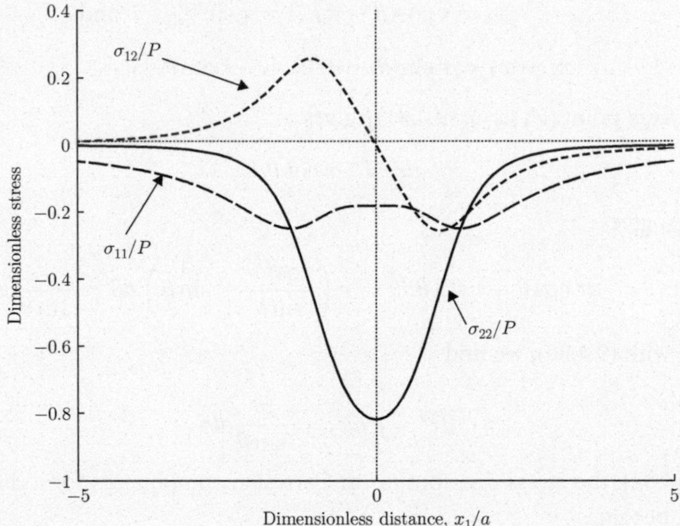

Fig. 9.20 Dimensionless stresses at $x_2 = a$ for a uniformly distributed load from $x_1 = -a$ to $x_1 = a$ on a half-space.

[a] Computations of this type are often best executed using symbolic mathematics programs, such as MATHEMATICA.

[b] The earliest studies of such contact problems were first conducted by Hertz (1882). For a current account see the book *Contact Mechanics* by Johnson (1985).

EXAMPLE 9.8 Shear load on a half-space

Consider a half-space subjected to a distributed shear load τ per unit length acting over a region $-a \leq x_1 \leq a$; cf. Fig. 9.21.

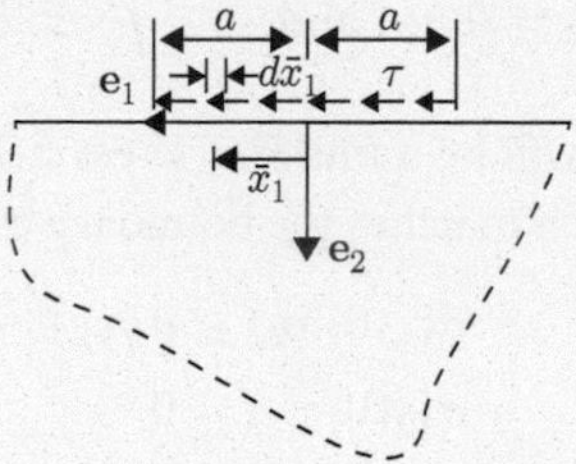

Fig. 9.21 Half-space subjected to a distributed shear traction.

The Airy stress function for this problem can be found by first observing that the Airy stress function for a point force of magnitude H acting tangential to the half-space at the origin is given by

$$\varphi = -\frac{H}{\pi} r\theta \sin\theta = -\frac{H}{\pi} x_2 \tan^{-1}\left(\frac{x_2}{x_1}\right);$$

see (9.4.49) and (9.4.54) under the consideration of the applied load now being in the x_1-direction. Focusing on the load in a small region $d\bar{x}_1$ located at $\bar{x}_1$, we have that the contribution to the Airy stress function (by superposition) is given by

$$d\varphi = -\frac{\tau d\bar{x}_1}{\pi} x_2 \tan^{-1}\left(\frac{x_2}{x_1 - \bar{x}_1}\right)$$

and the total stress function is given by[a]

$$\begin{aligned}
\varphi &= \int_{-a}^{a} -\frac{\tau}{\pi} x_2 \tan^{-1}\left(\frac{x_2}{x_1 - \bar{x}_1}\right) d\bar{x}_1 \\
&= \frac{\tau x_2}{\pi}\left[(x_1 - a)\tan^{-1}\left(\frac{x_2}{x_1 - a}\right) - (x_1 + a)\tan^{-1}\left(\frac{x_2}{x_1 + a}\right)\right. \\
&\quad \left. + \tfrac{1}{2}\log\left(\frac{(x_1 - a)^2 + {x_2}^2}{(x_1 + a)^2 + {x_2}^2}\right)\right].
\end{aligned}$$

From this expression one can compute any of the desired stresses.

[a] The solution to this problem closely follows that given in the book by Sadd (2014).

9.4.5 **A detailed example of Saint-Venant's principle**

In what follows we demonstrate Saint-Venant's principle (cf. Sec. 8.6) with a particular simple example. Consider a bar of width $2d$ and length L, cf. Fig. 9.22. Assume that the loading is a uniform end load, σ_o, plus a *self-equilibrated* contribution $\bar{\sigma}(x_2)$, that is,

$$\int_{-d}^{d} \bar{\sigma}(x_2)\,dx_2 = 0 \qquad \text{and} \qquad \int_{-d}^{d} x_2\bar{\sigma}(x_2)\,dx_2 = 0.$$

The solution to this problem will be a uniform stress field $\boldsymbol{\sigma}(x_1, x_2) = \sigma_o \mathbf{e}_1 \otimes \mathbf{e}_1$ plus a correction field, $\boldsymbol{\sigma}^c(x_1, x_2)$, which matches the boundary conditions:

$$\begin{aligned} \sigma^c_{11}(0, x_2) &= \bar{\sigma}(x_2) \\ \sigma^c_{12}(0, x_2) &= 0 \\ \sigma^c_{22}(x_1, \pm d) &= 0 \\ \sigma^c_{12}(x_1, \pm d) &= 0. \end{aligned} \tag{9.4.86}$$

We will assume $L \gg d$ and consequently will anticipate that $\boldsymbol{\sigma}^c \to \mathbf{0}$ as $x \to L$. There is of course a similar end-effect solution from the loading on the right end of the bar.

In keeping with the notion that the solution should drop off rapidly from the end of the bar, let us assume an Airy stress function φ^c of the form

$$\varphi^c = K \exp[-\gamma x_1/d] f(x_2) \tag{9.4.87}$$

for the correction stress field $\boldsymbol{\sigma}^c$. Here K and γ are unknown constants, and $f(x_2)$ is an unknown function. Our goal is to determine γ so that we can come to an understanding of how local the correction stresses really are. From (9.4.87) we have that:

$$\begin{aligned} \sigma^c_{11} &= \varphi^c_{,22} = K \exp[-\gamma x_1/d] f''(x_2), \\ \sigma^c_{22} &= \varphi^c_{,11} = K(\gamma/d)^2 \exp[-\gamma x_1/d] f(x_2), \\ \sigma^c_{12} &= -\varphi^c_{,12} = K(\gamma/d) \exp[-\gamma x_1/d] f'(x_2). \end{aligned} \tag{9.4.88}$$

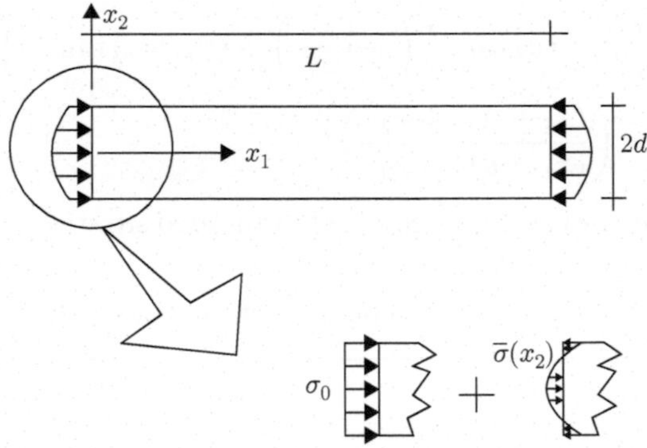

Fig. 9.22 End-loaded bar with a uniform end traction σ_0 plus a self-equilibrated loading pattern $\bar{\sigma}(x_2)$.

We know that φ^c must also satisfy the biharmonic equation. If we substitute our assumption (9.4.87) into $\Delta\Delta\varphi^c = 0$, we find that:

$$(\gamma/d)^4 f(x_2) + 2(\gamma/d)^2 f''(x_2) + f''''(x_2) = 0. \qquad (9.4.89)$$

This is a fourth-order ordinary differential equation with constant coefficients; its solution is known to be of the form

$$f(x_2) = A\exp[sx_2]. \qquad (9.4.90)$$

Inserting (9.4.90) into (9.4.89) yields the following polynomial for s:

$$(\gamma/d)^4 + 2(\gamma/d)^2 s^2 + s^4 = 0. \qquad (9.4.91)$$

This polynomial in s has only two unique roots $\pm i(\gamma/d)$, where $i = \sqrt{-1}$; thus, we need to use variation of parameters to generate two more linearly independent solutions (of the form $Ax_2\exp[sx_2]$). Combining these two additional solutions together with our first two solutions, we can write the functional form for our unknown function $f(x_2)$ in our assumed form (9.4.87) as:

$$f(x_2) = A\cosh(i\gamma x_2/d) + B\sinh(i\gamma x_2/d) + Cx_2\cosh(i\gamma x_2/d) + Dx_2\sinh(i\gamma x_2/d). \qquad (9.4.92)$$

In order to satisfy the lateral boundary conditions $(9.4.86)_{3,4}$ on σ^c_{22} and σ^c_{12} for all values of x_1, from (9.4.88) we must have that $f(\pm d) = 0$ and that $f'(\pm d) = 0$. These four conditions give us four equations for the four constants A, B, C, and D in (9.4.92):

$$\begin{bmatrix} \cosh(i\gamma) & \sinh(i\gamma) & d\cosh(i\gamma) & d\sinh(i\gamma) \\ \cosh(i\gamma) & -\sinh(i\gamma) & -d\cosh(i\gamma) & d\sinh(i\gamma) \\ i\frac{\gamma}{d}\sinh(i\gamma) & i\frac{\gamma}{d}\cosh(i\gamma) & \cosh(i\gamma)+i\gamma\sinh(i\gamma) & \sinh(i\gamma)+i\gamma\cosh(i\gamma) \\ -i\frac{\gamma}{d}\sinh(i\gamma) & i\frac{\gamma}{d}\cosh(i\gamma) & \cosh(i\gamma)+i\gamma\sinh(i\gamma) & -\sinh(i\gamma)-i\gamma\cosh(i\gamma) \end{bmatrix} \begin{pmatrix} A \\ B \\ C \\ D \end{pmatrix} = \begin{pmatrix} 0 \\ 0 \\ 0 \\ 0 \end{pmatrix}. \qquad (9.4.93)$$

If these equations are to have a non-trivial solution, then the determinant of the matrix in (9.4.93) must equal zero. The determinant of this matrix, after some lengthy but straightforward algebra, can be found to be:

$$\det[\cdot] = 2\gamma + \sin(2\gamma) = 0. \qquad (9.4.94)$$

Equation (9.4.94) gives a relation that γ must satisfy in order for the lateral boundary conditions $(9.4.86)_{3,4}$ to hold true. Equation (9.4.94) has multiple complex-valued roots, the first two of which are:

$$\gamma_{1,2} = 2.1061 \pm i1.1254. \qquad (9.4.95)$$

REMARKS 9.4

1. The first thing that we see, using (9.4.88), is that the correction stress components associated with these roots will decay very rapidly as one moves away from the end of the bar. In particular,

$$\sigma^c_{(\cdot\cdot)} \sim \exp[-2.1061 x_1/d]\,(\cos(1.1254 x_1/d) \pm i\sin(1.1254 x_1/d))\, g(x_2),$$

where $g(x_2)$ is either $f(x_2)$, $f'(x_2)$, or $f''(x_2)$ depending upon the stress component. In numerical terms, when $x_1/d = 1$ the influence of the correction stress is just 12% and for $x_1/d = 2$ the influence is only 1.5%.

2. Equation (9.4.95) has further roots all of which have $\mathrm{Re}[\gamma] > 2.1061$. Thus these additional roots give rise to local effects that decay even faster than those from (9.4.95). The next two roots, for example, are $\gamma_{3,4} = 5.3536 \pm i1.5516$.

3. The complete solution to the problem, if desired, can be found by using linear combinations of the solutions associated with all the roots. This provides the needed degrees of freedom to satisfy the remaining boundary conditions at $x_1 = 0$. This result can then be combined with a similar one from the other end of the bar to give a complete solution to the full boundary-value problem.

9.5 Asymptotic crack-tip stress fields. Stress intensity factors

An important application of the Airy stress function is the determination of the stress and deformation fields near the tips of sharp cracks. In this section we develop *asymptotic solutions* for these fields based on the theory of isotropic linear elasticity. Consider a sharp crack in a prismatic isotropic linear elastic body. There are three basic loading modes associated with relative crack-face displacements for a cracked body. These are described with respect to Fig. 9.23 as[7]

(a) the tensile opening mode, or Mode I;

(b) the in-plane sliding mode, or Mode II; and

(c) the anti-plane tearing mode, or Mode III.

We develop the asymptotic crack-tip solutions for these three basic modes in what follows.

[7] The symbols “I”, “II”, and “III” are Roman numerals, and should therefore be read as “one”, “two”, and “three”.

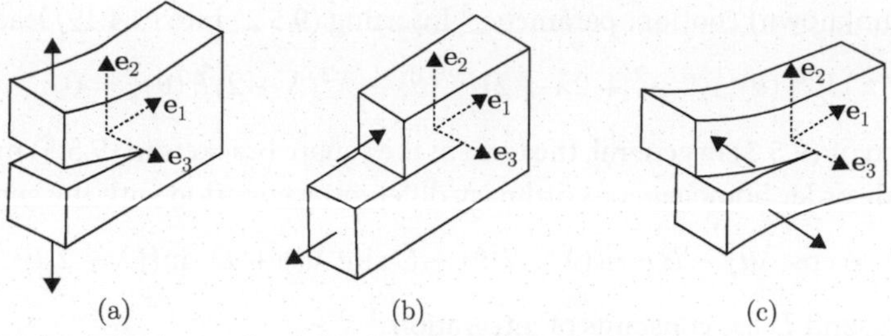

Fig. 9.23 Three basic loading modes for a cracked body: (a) Mode I, tensile opening mode. (b) Mode II, in-plane sliding mode. (c) Mode III, anti-plane tearing mode.

9.5.1 **Asymptotic stress fields in Mode I and Mode II**

The asymptotic crack problem for both Mode I and Mode II is shown in Fig. 9.24. Here, the boundary conditions are that the crack-faces at $\theta = \pm\pi$ are *traction-free*. This leads to

$$\begin{aligned} \sigma_{\theta\theta}(r, \pm\pi) &= 0, \\ \sigma_{r\theta}(r, \pm\pi) &= 0. \end{aligned} \tag{9.5.1}$$

Following Williams (1952, 1957) we consider a separable Airy stress function of the form

$$\varphi(r, \theta) = r^{\lambda+2} F(\theta), \tag{9.5.2}$$

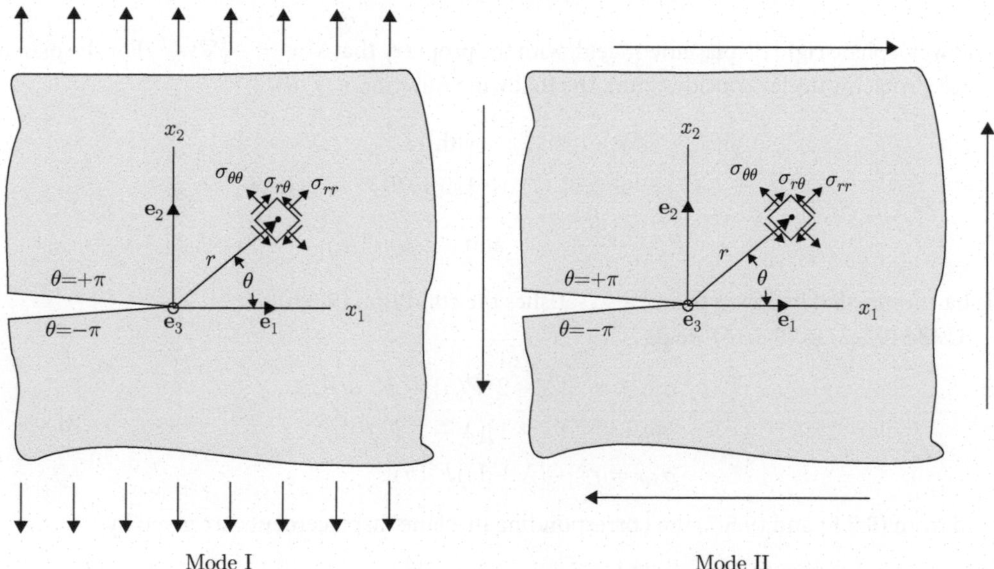

Fig. 9.24 Schematic of Mode I and Mode II loading. The stress components are with respect to a coordinate system with origin on the crack front. The faces of the sharp crack are at $\theta = \pm\pi$. However, in this schematic, for ease of visualization, the crack-faces are shown with finite opening.

where λ is a (unknown) constant parameter.[8] Inserting (9.5.2) into (9.4.19) leads to

$$r^{\lambda-2}\left[F''''(\theta)+2(\lambda^2+2\lambda+2)F''(\theta)+\lambda^2(\lambda+2)F(\theta)\right]=0. \tag{9.5.3}$$

For satisfaction of (9.5.3) in general, the term in the square brackets in (9.5.3) must equal zero. This is a fourth-order homogeneous ordinary differential equation with the general solution

$$F(\theta)=A\cos(\lambda\theta)+B\cos((\lambda+2)\theta)+C\sin(\lambda\theta)+D\sin((\lambda+2)\theta), \tag{9.5.4}$$

where A, B, C, and D are constants of integration.

Also, recall the general result (9.4.20) that the corresponding displacements may be determined from

$$\begin{aligned} u_r &= \frac{1}{2\mu}(-\varphi_{,r}+s\,r\,\psi_{,\theta})+w_r,\\ u_\theta &= \frac{1}{2\mu}\left(-\frac{1}{r}\varphi_{,\theta}+s\,r^2\,\psi_{,r}\right)+w_\theta, \end{aligned} \tag{9.5.5}$$

where

$$s \overset{\text{def}}{=} \begin{cases} 1-\nu & \text{for plane strain,}\\ \dfrac{1}{1+\nu} & \text{for plane stress,} \end{cases}$$

$\psi(r,\theta)$ is a potential function that satisfies

$$\begin{aligned} \Delta\psi &= 0,\\ (r\psi_{,\theta})_{,r} &= \Delta\varphi, \end{aligned} \tag{9.5.6}$$

and $\mathbf{w}$ is a plane rigid displacement field, with the property that $\nabla\mathbf{w}=-(\nabla\mathbf{w})^{\top}$. For the planar crack problems under consideration, the following function $\psi(r,\theta)$,

$$\begin{aligned} &\psi(r,\theta)\overset{\text{def}}{=}r^{\lambda}G(\theta), \qquad \text{with}\\ &G(\theta)=P\cos(\lambda\theta)+Q\sin(\lambda\theta), \qquad \text{where}\\ &P=-\frac{4C}{\lambda}, \qquad \text{and} \qquad Q=\frac{4A}{\lambda}, \end{aligned} \tag{9.5.7}$$

is harmonic, that is $\Delta\psi=0$, and also satisfies the condition $(9.5.6)_2$.

Using (9.5.2) in (9.4.16) yields

$$\begin{aligned} \sigma_{rr} &= r^{\lambda}[(\lambda+2)F(\theta)+F''(\theta)],\\ \sigma_{\theta\theta} &= r^{\lambda}[(\lambda+1)(\lambda+2)F(\theta)],\\ \sigma_{r\theta} &= r^{\lambda}[-(\lambda+1))F'(\theta)]. \end{aligned} \tag{9.5.8}$$

and from (9.5.5) and (9.5.7) the corresponding in-plane displacements are given by

$$\begin{aligned} u_r &= \frac{r^{\lambda+1}}{2\mu}[-(\lambda+2)F(\theta)+sG'(\theta)]+w_r,\\ u_\theta &= \frac{r^{\lambda+1}}{2\mu}[-F'(\theta)+s\,\lambda G(\theta)]+w_\theta. \end{aligned} \tag{9.5.9}$$

[8] Here λ is a scalar parameter, not the Lamé modulus.

With (9.5.8) and (9.5.9) in hand, we now separately investigate the asymptotic stress and displacement fields for Mode I and Mode II.

Asymptotic Mode I field

For Mode I, that is for a symmetric tension applied transverse to the length of the crack, we require that the $\sigma_{\theta\theta}$ component of the stress satisfy the symmetry condition

$$\sigma_{\theta\theta}(r,\theta) = \sigma_{\theta\theta}(r,-\theta). \tag{9.5.10}$$

From $(9.5.8)_2$, for the symmetry condition (9.5.10) to hold, we must have

$$F(\theta) = F(-\theta). \tag{9.5.11}$$

Since $\sin(\theta)$ is an odd function of θ, for (9.5.11) to hold, the constants C and D in $F(\theta)$, cf. (9.5.4), must vanish:

$$C = D = 0. \tag{9.5.12}$$

Thus for Mode I, the function $F(\theta)$ reduces to

$$F(\theta) = A\cos(\lambda\theta) + B\cos((\lambda+2)\theta)). \tag{9.5.13}$$

Hence,

$$\begin{aligned} F'(\theta) &= -\lambda A\sin(\lambda\theta) - (\lambda+2)B\sin((\lambda+2)\theta), \\ F''(\theta) &= -\lambda^2 A\,\cos(\lambda\theta) - (\lambda+2)^2 B\cos((\lambda+2)\theta), \end{aligned} \tag{9.5.14}$$

substitution of which in (9.5.8) gives the following expressions for the stress components:

$$\begin{aligned} \sigma_{rr} &= r^{\lambda}\left[\left((\lambda+2)-\lambda^2\right)A\cos(\lambda\theta) + \left((\lambda+2)-(\lambda+2)^2\right)B\cos((\lambda+2)\theta)\right], \\ \sigma_{\theta\theta} &= r^{\lambda}\left[(\lambda+1)(\lambda+2)\left(A\cos(\lambda\theta) + \ B\cos((\lambda+2)\theta)\right)\right], \\ \sigma_{r\theta} &= r^{\lambda}\left[(\lambda+1)\left(\lambda A\sin(\lambda\theta) + (\lambda+2)B\sin((\lambda+2)\theta)\right)\right]. \end{aligned} \tag{9.5.15}$$

Next, the boundary condition $(9.5.1)_1$ requires that

$$\sigma_{\theta\theta}(r,\pm\pi) = r^{\lambda}\left[(\lambda+1)(\lambda+2)(A+B)\cos(\lambda\pi)\right] = 0,$$

and hence that

$$(\lambda+1)(\lambda+2)(A+B)\cos(\lambda\pi) = 0. \tag{9.5.16}$$

Also, the boundary condition $(9.5.1)_2$ requires that

$$\sigma_{r\theta}(r,\pm\pi) = \pm r^{\lambda}(\lambda+1)\left(\lambda A + (\lambda+2)B\right)\sin(\lambda\pi) = 0,$$

or that

$$(\lambda+1)\left(\lambda A + (\lambda+2)B\right)\sin(\lambda\pi) = 0. \tag{9.5.17}$$

The conditions (9.5.16) and (9.5.17) must be satisfied *simultaneously*. We therefore have two possible solutions:

Solution (a):

$$\cos(\lambda\pi) = 0 \qquad \text{and} \qquad B = -\frac{\lambda}{\lambda+2}A, \tag{9.5.18}$$

which therefore requires that

$$\lambda = \ldots, -\frac{5}{2}, -\frac{3}{2}, -\frac{1}{2}, +\frac{1}{2}, +\frac{3}{2}, +\frac{5}{2}, \ldots \tag{9.5.19}$$

Solution (b):

$$\sin(\lambda\pi) = 0 \qquad \text{and} \qquad B = -A, \tag{9.5.20}$$

which requires that[9]

$$\lambda = \ldots, -2, -1, 0, +1, +2, \ldots \tag{9.5.21}$$

Negative values of λ *are interesting, because it is for such values that we will get a* stress singularity, *that is the stresses will go to infinity as* r *goes to zero!*

We choose an appropriate negative value of λ based on a physical argument that total free energy in any disc $\mathcal{D}_\delta$ of radius δ centered at the crack-tip, cf. Fig. 9.25, must be finite in value. The total free energy of such a region is

$$\Psi(\mathcal{D}_\delta) = \int_0^\delta \int_{-\pi}^{\pi} \psi\, r d\theta dr,$$

here ψ is the free-energy density per unit area (and per unit thickness).[10] Recall that for a linear elastic material, in three dimensions

$$\psi = \tfrac{1}{2}\boldsymbol{\epsilon} : \mathbb{C}\boldsymbol{\epsilon} = \tfrac{1}{2}\boldsymbol{\sigma} : \mathbb{C}^{-1}\boldsymbol{\sigma}.$$

Hence, for the two-dimensional case under consideration

$$\psi \propto \sigma_{\alpha\beta}^2, \qquad \text{but} \qquad \sigma_{\alpha\beta} \propto r^\lambda, \qquad \text{so} \qquad \psi \propto r^{2\lambda},$$

and therefore

$$\Psi(\mathcal{D}_\delta) = \int_0^\delta \int_{-\pi}^{\pi} (\cdots)\, r^{2\lambda+1} d\theta dr = \int_0^\delta (\cdots)\, r^{2\lambda+1} dr,$$

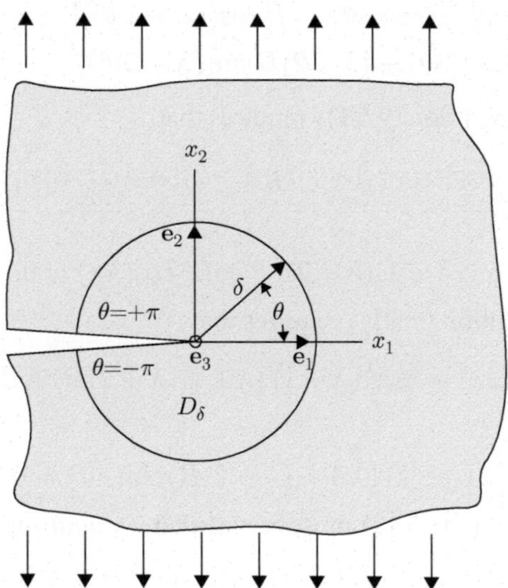

Fig. 9.25 A disc-like region D_δ of radius δ centered at the crack-tip.

[9] In the special cases of $\lambda = -1$ or $\lambda = -2$, A and B are arbitrary.

[10] There is some duplication of notation here. Note that the ψ in (9.5.7) is not the free-energy function.

or

$$\Psi(\mathcal{D}_\delta) = (\cdots)\left.\frac{r^{2\lambda+2}}{2\lambda+2}\right|_0^\delta.$$

For $\Psi(\mathcal{D}_\delta)$ to be finite, we therefore require that

$$(2\lambda + 2) > 0 \qquad \text{or} \qquad \lambda > -1.$$

From (9.5.19) and (9.5.21) we thus note that the only physically admissible value of λ which will give rise to a stress singularity is

$$\lambda = -\tfrac{1}{2}, \tag{9.5.22}$$

and for this case, from (9.5.18) we must also have

$$B = \frac{1}{3}A. \tag{9.5.23}$$

Substituting (9.5.22) and (9.5.23) into (9.5.15) we obtain that the asymptotic stress field at the tip of a crack in Mode I:

$$\begin{pmatrix} \sigma_{rr} \\ \\ \sigma_{\theta\theta} \\ \\ \sigma_{r\theta} \end{pmatrix} = \frac{A}{\sqrt{r}} \begin{pmatrix} \dfrac{5}{4}\cos\left(\dfrac{\theta}{2}\right) - \dfrac{1}{4}\cos\left(\dfrac{3\theta}{2}\right) \\ \\ \dfrac{3}{4}\cos\left(\dfrac{\theta}{2}\right) + \dfrac{1}{4}\cos\left(\dfrac{3\theta}{2}\right) \\ \\ \dfrac{1}{4}\sin\left(\dfrac{\theta}{2}\right) + \dfrac{1}{4}\sin\left(\dfrac{3\theta}{2}\right) \end{pmatrix} + \text{bounded terms} \qquad \text{as } r \to 0. \tag{9.5.24}$$

To determine the corresponding displacements, we note that using

$$C = D = 0, \qquad \lambda = -\tfrac{1}{2}, \qquad \text{and} \qquad B = \frac{1}{3}A,$$

in (9.5.4) and (9.5.7) we have that

$$\begin{aligned} F(\theta) &= A\left[\cos\left(\frac{\theta}{2}\right) + \frac{1}{3}\cos\left(\frac{3\theta}{2}\right)\right], \\ G(\theta) &= 8A\sin\left(\frac{\theta}{2}\right), \end{aligned} \tag{9.5.25}$$

use of which in (9.5.9) gives the displacement field for plane strain

$$\begin{pmatrix} u_r \\ \\ u_\theta \end{pmatrix} = \frac{A\sqrt{r}}{4\mu} \begin{pmatrix} (5-8\nu)\cos\left(\dfrac{\theta}{2}\right) - \cos\left(\dfrac{3\theta}{2}\right) \\ \\ (8\nu-7)\sin\left(\dfrac{\theta}{2}\right) + \sin\left(\dfrac{3\theta}{2}\right) \end{pmatrix} + \begin{pmatrix} w_r \\ w_\theta \end{pmatrix} \qquad \text{as } r \to 0. \tag{9.5.26}$$

- We define

$$K_{\mathrm{I}} \stackrel{\text{def}}{=} A\sqrt{2\pi}, \tag{9.5.27}$$

  and call it the *Mode I stress intensity factor.*

It will be proportional to the loading in a given problem and will depend on the crack geometry.

In terms of K_{I}, the asymptotic stress and displacement fields may be written as

$$\begin{aligned}
\sigma_{\alpha\beta} &= \frac{K_{\mathrm{I}}}{\sqrt{2\pi r}} f_{\alpha\beta}(\theta) + \text{bounded terms},\\
u_\alpha &= \frac{K_{\mathrm{I}}}{4\mu}\sqrt{\frac{r}{2\pi}}\, g_\alpha(\theta;\nu) + \ \ldots + \text{rigid body terms},
\end{aligned}$$

$$\begin{pmatrix} f_{rr} \\ f_{\theta\theta} \\ f_{r\theta} \end{pmatrix} = \begin{pmatrix} \frac{5}{4}\cos\left(\frac{\theta}{2}\right) - \frac{1}{4}\cos\left(\frac{3\theta}{2}\right) \\ \frac{3}{4}\cos\left(\frac{\theta}{2}\right) + \frac{1}{4}\cos\left(\frac{3\theta}{2}\right) \\ \frac{1}{4}\sin\left(\frac{\theta}{2}\right) + \frac{1}{4}\sin\left(\frac{3\theta}{2}\right) \end{pmatrix}, \tag{9.5.28}$$

$$\begin{pmatrix} g_r \\ g_\theta \end{pmatrix} = \begin{pmatrix} (5-8\nu)\cos\left(\frac{\theta}{2}\right) - \cos\left(\frac{3\theta}{2}\right) \\ (8\nu-7)\sin\left(\frac{\theta}{2}\right) + \sin\left(\frac{3\theta}{2}\right) \end{pmatrix},$$

as $r \to 0$. Also,

- for plane strain $\sigma_{33} = \nu(\sigma_{rr} + \sigma_{\theta\theta})$.
- The same singular stress field also holds for plane stress, but then $\sigma_{33} = 0$. Also, for plane stress ν is replaced by $\nu/(1+\nu)$ in the relation for displacements.
- Note that the units of the stress intensity factor are stress times square-root of length.

Asymptotic Mode II field

For Mode II, that is for an anti-symmetric field with respect to the crack-tip, we require that the $\sigma_{\theta\theta}$ component of the stress satisfy the anti-symmetry condition

$$\sigma_{\theta\theta}(r,\theta) = -\sigma_{\theta\theta}(r,-\theta). \tag{9.5.29}$$

From $(9.5.8)_2$, for the anti-symmetry condition (9.5.29) to hold, we must have

$$F(\theta) = -F(-\theta). \tag{9.5.30}$$

Since $\cos(\theta)$ is an even function of θ, for (9.5.30) to hold, the constants A and B in $F(\theta)$, cf. (9.5.4), must vanish:

$$A = B = 0. \tag{9.5.31}$$

Thus for Mode II, the function $F(\theta)$ reduces to

$$F(\theta) = C\sin(\lambda\theta) + D\sin((\lambda+2)\theta)). \tag{9.5.32}$$

Hence,

$$\begin{aligned} F'(\theta) &= \lambda C\cos(\lambda\theta) + (\lambda+2)D\cos((\lambda+2)\theta), \\ F''(\theta) &= -\lambda^2 C\sin(\lambda\theta) - (\lambda+2)^2 D\sin((\lambda+2)\theta), \end{aligned} \tag{9.5.33}$$

substitution of which in (9.5.8) gives the following expressions for the stress components:

$$\begin{aligned} \sigma_{rr} &= r^\lambda\left[\left((\lambda+2)-\lambda^2\right)C\sin(\lambda\theta) + \left((\lambda+2)-(\lambda+2)^2\right)D\sin((\lambda+2)\theta)\right], \\ \sigma_{\theta\theta} &= r^\lambda\left[(\lambda+1)(\lambda+2)\left(C\sin(\lambda\theta) + D\sin((\lambda+2)\theta)\right)\right], \\ \sigma_{r\theta} &= r^\lambda\left[-(\lambda+1)\left(\lambda C\cos(\lambda\theta) + (\lambda+2)D\cos((\lambda+2)\theta)\right)\right]. \end{aligned} \tag{9.5.34}$$

Next, the boundary condition $(9.5.1)_1$ requires that

$$\sigma_{\theta\theta}(r, \pm\pi) = \pm r^\lambda\left[(\lambda+1)(\lambda+2)(C+D)\sin(\lambda\pi)\right] = 0,$$

and hence that

$$(\lambda+1)(\lambda+2)(C+D)\sin(\lambda\pi) = 0. \tag{9.5.35}$$

Also, the boundary condition $(9.5.1)_2$ requires that

$$\sigma_{r\theta}(r, \pm\pi) = r^\lambda\left[-(\lambda+1)\left(\lambda C + (\lambda+2)D\right)\cos(\lambda\pi)\right] = 0,$$

or that

$$(\lambda+1)\left(\lambda C + (\lambda+2)D\right)\cos(\lambda\pi) = 0. \tag{9.5.36}$$

The conditions (9.5.35) and (9.5.36) must be satisfied *simultaneously*. We therefore have two possible solutions:

Solution (a):

$$\cos(\lambda\pi) = 0 \qquad \text{and} \qquad D = -C \tag{9.5.37}$$

which therefore requires that

$$\lambda = \ldots, -\frac{5}{2}, -\frac{3}{2}, -\frac{1}{2}, +\frac{1}{2}, +\frac{3}{2}, +\frac{5}{2}, \ldots \tag{9.5.38}$$

Solution (b):

$$\sin(\lambda\pi) = 0 \qquad \text{and} \qquad D = -\frac{\lambda}{\lambda+2}C, \tag{9.5.39}$$

which requires that[11]

$$\lambda = \ldots, -2, -1, 0, +1, +2, \ldots \tag{9.5.40}$$

---

[11] In the special case $\lambda = -1$, C and D are arbitrary. In the special case $\lambda = -2$, $C = 0$ and D is arbitrary.

As before negative values of λ give rise to a stress singularity, and for $\Psi(\mathcal{D}_\delta)$ to be finite in value, we require that

$$\lambda > -1.$$

Thus the first solution of interest is

$$\lambda = -\tfrac{1}{2}, \qquad D = -C. \tag{9.5.41}$$

Substituting (9.5.41) into (9.5.34) we obtain that the asymptotic stress field at the tip of a crack in Mode II is given by

$$\boxed{\begin{pmatrix}\sigma_{rr}\\ \\ \sigma_{\theta\theta}\\ \\ \sigma_{r\theta}\end{pmatrix} = \frac{C}{\sqrt{r}}\begin{pmatrix}-\dfrac{5}{4}\sin\left(\dfrac{\theta}{2}\right)+\dfrac{3}{4}\sin\left(\dfrac{3\theta}{2}\right)\\ \\ -\dfrac{3}{4}\sin\left(\dfrac{\theta}{2}\right)-\dfrac{3}{4}\sin\left(\dfrac{3\theta}{2}\right)\\ \\ \dfrac{1}{4}\cos\left(\dfrac{\theta}{2}\right)+\dfrac{3}{4}\cos\left(\dfrac{3\theta}{2}\right)\end{pmatrix} + \text{bounded terms} \qquad \text{as } r\to 0.} \tag{9.5.42}$$

To determine the corresponding displacements, we note that using

$$A = B = 0, \qquad \lambda = -\tfrac{1}{2}, \qquad \text{and} \qquad D = -C,$$

in (9.5.4) and (9.5.7) we have that

$$\begin{aligned} F(\theta) &= -C\left[\sin\left(\frac{\theta}{2}\right)+\sin\left(\frac{3\theta}{2}\right)\right],\\ G(\theta) &= 8C\cos\left(\frac{\theta}{2}\right), \end{aligned} \tag{9.5.43}$$

use of which in (9.5.9) gives the displacement field for plane strain

$$\boxed{\begin{pmatrix}u_r\\ \\ u_\theta\end{pmatrix} = \frac{C\sqrt{r}}{4\mu}\begin{pmatrix}(8\nu-5)\sin\left(\dfrac{\theta}{2}\right)+3\sin\left(\dfrac{3\theta}{2}\right)\\ \\ (8\nu-7)\cos\left(\dfrac{\theta}{2}\right)+3\cos\left(\dfrac{3\theta}{2}\right)\end{pmatrix} + \begin{pmatrix}w_r\\ w_\theta\end{pmatrix} \qquad \text{as } r\to 0.} \tag{9.5.44}$$

We define

$$K_{\mathrm{II}} \stackrel{\text{def}}{=} C\sqrt{2\pi}, \tag{9.5.45}$$

and call it the *Mode II stress intensity factor.*

In terms of $K_{\rm II}$ the asymptotic stress and displacement fields are

$$\boxed{\begin{aligned}
\sigma_{\alpha\beta} &= \frac{K_{\rm II}}{\sqrt{2\pi r}}\, f_{\alpha\beta}(\theta) + \text{bounded terms},\\
u_{\alpha} &= \frac{K_{\rm II}}{4\mu}\sqrt{\frac{r}{2\pi}}\, g_{\alpha}(\theta;\nu) + \ \ldots\ + \text{rigid body terms},\\
\begin{pmatrix} f_{rr}\\ f_{\theta\theta}\\ f_{r\theta}\end{pmatrix} &= \begin{pmatrix} -\dfrac{5}{4}\sin\left(\dfrac{\theta}{2}\right) + \dfrac{3}{4}\sin\left(\dfrac{3\theta}{2}\right)\\[2ex] -\dfrac{3}{4}\sin\left(\dfrac{\theta}{2}\right) - \dfrac{3}{4}\sin\left(\dfrac{3\theta}{2}\right)\\[2ex] \dfrac{1}{4}\cos\left(\dfrac{\theta}{2}\right) + \dfrac{3}{4}\cos\left(\dfrac{3\theta}{2}\right)\end{pmatrix},\\
\begin{pmatrix} g_{r}\\ g_{\theta}\end{pmatrix} &= \begin{pmatrix} (8\nu-5)\sin\left(\dfrac{\theta}{2}\right) + 3\sin\left(\dfrac{3\theta}{2}\right)\\[2ex] (8\nu-7)\cos\left(\dfrac{\theta}{2}\right) + 3\cos\left(\dfrac{3\theta}{2}\right)\end{pmatrix},
\end{aligned}} \tag{9.5.46}$$

as $r \to 0$.

- Also, for plane strain $\sigma_{33} = \nu(\sigma_{rr} + \sigma_{\theta\theta})$.
- The same singular stress field holds for plane stress, but then $\sigma_{33} = 0$. Also, for plane stress ν is replaced by $\nu/(1+\nu)$ in the relation for displacements.

9.5.2 **Asymptotic stress field in Mode III**

Consider a crack with **traction-free** faces in the $(\mathbf{e}_1, \mathbf{e}_3)$-plane at $x_2 = \pm 0$, and with tip along the $\mathbf{e}_3$-axis., which is loaded so that surrounding medium is loaded in *anti-plane strain*—Mode III, as schematically shown in Fig. 9.26. We will determine the stress components with respect to a cylindrical coordinate system with origin on the crack front. The faces of the sharp crack are at $\theta = \pm\pi$. However, in this schematic, for ease of visualization, the crack-faces are shown with finite opening.

In *anti-plane strain* the displacement field has the form

$$u_z = u_z(r,\theta), \quad u_r = u_\theta = 0. \tag{9.5.47}$$

For the displacement field (9.5.47), the non-zero strain components are

$$\epsilon_{z\theta} = \frac{1}{2}\left(\frac{1}{r}\frac{\partial u_z}{\partial \theta}\right), \quad \epsilon_{zr} = \frac{1}{2}\frac{\partial u_z}{\partial r}, \tag{9.5.48}$$

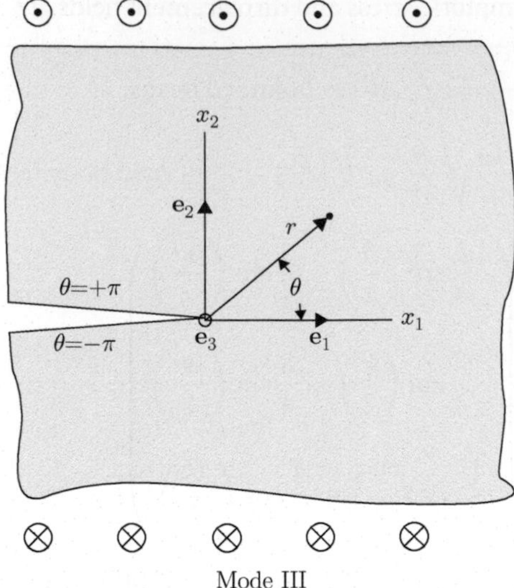

Fig. 9.26 Schematic of Mode III loading. The circles with dots represent arrows coming out of the plane of the paper, while the circles with crosses represent arrows pointing into the plane of the paper.

and the non-zero stress components are

$$\sigma_{z\theta} = 2\mu\,\epsilon_{z\theta} \quad \text{and} \quad \sigma_{zr} = 2\mu\,\epsilon_{zr}. \tag{9.5.49}$$

In the absence of body forces, the only pertinent equation of equilibrium is

$$\frac{\partial \sigma_{zr}}{\partial r} + \frac{1}{r}\frac{\partial \sigma_{z\theta}}{\partial \theta} + \frac{\sigma_{zr}}{r} = 0. \tag{9.5.50}$$

Using (9.5.48) and (9.5.49), the equilibrium equation (9.5.50) may be written as

$$\begin{aligned}
\frac{\partial \sigma_{zr}}{\partial r} + \frac{1}{r}\frac{\partial \sigma_{z\theta}}{\partial \theta} + \frac{\sigma_{zr}}{r} &= 0,\\
2\mu\left[\frac{\partial \epsilon_{zr}}{\partial r} + \frac{1}{r}\frac{\partial \epsilon_{z\theta}}{\partial \theta} + \frac{\epsilon_{zr}}{r}\right] &= 0,\\
2\mu\left[\frac{\partial}{\partial r}\left(\frac{1}{2}\frac{\partial u_z}{\partial r}\right) + \frac{1}{r}\frac{\partial}{\partial \theta}\left(\frac{1}{2}\frac{1}{r}\frac{\partial u_z}{\partial \theta}\right) + \frac{1}{r}\frac{1}{2}\frac{\partial u_z}{\partial r}\right] &= 0,\\
\mu\left[\frac{\partial^2 u_z}{\partial r^2} + \frac{1}{r^2}\frac{\partial^2 u_z}{\partial \theta^2} + \frac{1}{r}\frac{\partial u_z}{\partial r}\right] &= 0,
\end{aligned}$$

or, since $\mu > 0$,

$$\Delta u_z = 0, \tag{9.5.51}$$

where

$$\Delta u_z = \frac{\partial^2 u_z}{\partial r^2} + \frac{1}{r^2}\frac{\partial^2 u_z}{\partial \theta^2} + \frac{1}{r}\frac{\partial u_z}{\partial r},$$

is the two-dimensional Laplacian.

This partial differential equation for u_z is to be solved in the vicinity of the crack-tip. The three main conditions that are required from such a solution are

1. *Anti-symmetry* of Mode III requires that

$$u_z(r,\theta) = -u_z(r,-\theta). \tag{9.5.52}$$

2. The crack-faces are *traction-free*. Note that on the crack-face at $\theta = +\pi$, the outward unit normal to the cracked body is $\mathbf{n} = \mathbf{e}_\theta$, and the vanishing of the traction $\boldsymbol{\sigma}(r, \theta = +\pi)\mathbf{e}_\theta = \mathbf{0}$ on this crack-face implies that

$$\sigma_{z\theta}(r, \theta = +\pi) = 0. \tag{9.5.53}$$

Similarly, on the crack-face at $\theta = -\pi$, the outward unit normal to the cracked body is $\mathbf{n} = -\mathbf{e}_\theta$, and the vanishing of the traction $-\boldsymbol{\sigma}(r, \theta = -\pi)\mathbf{e}_\theta = \mathbf{0}$ on this crack-face implies that

$$\sigma_{z\theta}(r, \theta = -\pi) = 0. \tag{9.5.54}$$

3. The total free energy in any disc $\mathcal{D}_\delta$ of radius δ centered at the crack-tip must be finite in value. The total free energy of such a region is

$$\Psi(\mathcal{D}_\delta) = \int_0^\delta \int_{-\pi}^{\pi} \psi\, r d\theta dr, \tag{9.5.55}$$

where ψ is the free-energy density per unit area (and per unit thickness).

Solution for $u_z(r,\theta)$ in the vicinity of the crack

We now seek a solution for the displacement field $u_z(r,\theta)$ in the vicinity of the crack-tip which satisfies the three conditions listed above. First note that because of the nature of the partial differential equation (9.5.51), the displacement component u_z in anti-plane strain is *harmonic*. We can seek separable solutions of the form

$$u_z(r,\theta) = r^{\lambda+1} F(\theta)/\mu, \tag{9.5.56}$$

where μ is the shear modulus. Inserting this form into (9.5.51), yields the result that the displacement field is of the form

$$u_z = \frac{r^{\lambda+1}}{\mu}\left[A\cos\left((\lambda+1)\theta\right) + B\sin\left((\lambda+1)\theta\right)\right], \tag{9.5.57}$$

with A, B, and λ constants.[12]

Anti-symmetry requires that (9.5.52) be satisfied. This condition requires that the displacement field (9.5.57) satisfy

$$[A\cos((\lambda+1)\theta) + B\sin((\lambda+1)\theta)] = -\left[A\cos((\lambda+1)\theta) - B\sin((\lambda+1)\theta)\right],$$

[12] Note that here λ is not the Lamé modulus.

which implies that $A\cos((\lambda+1)\theta)=0$, and hence u_z has the form

$$u_z = \frac{B}{\mu} r^{\lambda+1} \sin((\lambda+1)\theta). \tag{9.5.58}$$

Using (9.5.48) and (9.5.58), the non-zero strain components are

$$\begin{aligned} \epsilon_{zr} &= \frac{1}{2}\frac{\partial u_z}{\partial r} &&= \frac{B}{2\mu}(\lambda+1)r^{\lambda}\sin((\lambda+1)\theta),\\ \epsilon_{z\theta} &= \frac{1}{2r}\frac{\partial u_z}{\partial \theta} &&= \frac{B}{2\mu}(\lambda+1)r^{\lambda}\cos((\lambda+1)\theta), \end{aligned} \tag{9.5.59}$$

and hence, using (9.5.49) the non-zero stress components are

$$\begin{aligned} \sigma_{zr} &= B(\lambda+1)r^{\lambda}\sin((\lambda+1)\theta),\\ \sigma_{z\theta} &= B(\lambda+1)r^{\lambda}\cos((\lambda+1)\theta). \end{aligned} \tag{9.5.60}$$

We now consider the restrictions placed on the solution u_z by the condition (9.5.55) of the boundedness of the free energy. Recall that for a linear elastic material, in three dimensions

$$\psi = \tfrac{1}{2}\boldsymbol{\epsilon}:\mathbb{C}\boldsymbol{\epsilon} = \tfrac{1}{2}\boldsymbol{\sigma}:\mathbb{C}^{-1}\boldsymbol{\sigma}.$$

Hence, for the two-dimensional case under consideration

$$\psi \propto \sigma_{ij}^2 \quad \text{but as} \quad \sigma_{ij}\propto r^{\lambda}, \quad \text{we obtain} \quad \psi \propto r^{2\lambda}.$$

Therefore, the total free energy Ψ in any disk of radius δ, centered at the crack-tip will be

$$\Psi(\mathcal{D}_\delta) = \int_0^\delta\int_{-\pi}^{\pi}\psi\, r d\theta\, dr = \int_0^\delta\int_{-\pi}^{\pi}(...)r^{2\lambda+1}d\theta\, dr = \int_0^\delta (...)r^{2\lambda+1}\, dr = (...)\left.\frac{r^{2\lambda+2}}{2\lambda+2}\right|_0^\delta.$$

For $\Psi(\mathcal{D}_\delta)$ to be finite, we require that

$$(2\lambda+2)>0 \quad \text{or} \quad \lambda > -1. \tag{9.5.61}$$

Next, we study the restriction imposed by the traction-free condition on the crack-faces, (9.5.53) and (9.5.54). Using $(9.5.60)_2$ requires that

$$B\,(\lambda+1)\,r^{\lambda}\,\cos((\lambda+1)\pi)=0, \tag{9.5.62}$$

which is satisfied if $\lambda=-1$ or if $\cos((\lambda+1)\pi)=0$, namely

$$\lambda+1 = \ldots, -\frac{3}{2}, -\frac{1}{2}, +\frac{1}{2}, +\frac{3}{2}, \ldots.$$

which implies

$$\lambda = \ldots, -\frac{3}{2}, -\frac{1}{2}, +\frac{1}{2}, +\frac{3}{2}, \ldots \tag{9.5.63}$$

Restrictions (9.5.61) and (9.5.63) imply that the smallest allowed value of λ is

$$\lambda = -\frac{1}{2}. \tag{9.5.64}$$

This value of λ corresponds to an asymptotic singular field at the crack-tip. Writing

$$K_{\text{III}} \overset{\text{def}}{=} B\sqrt{\frac{\pi}{2}} \tag{9.5.65}$$

in (9.5.58), the displacement field is

$$\boxed{u_z = 2\frac{K_{\text{III}}}{\mu}\sqrt{\frac{r}{2\pi}}\sin\left(\frac{\theta}{2}\right) + \ldots,} \tag{9.5.66}$$

and the corresponding singular stress field has the form

$$\boxed{\begin{aligned} \sigma_{\theta z} &= \frac{K_{\text{III}}}{\sqrt{2\pi r}}\cos\left(\frac{\theta}{2}\right) + \text{bounded terms}, \\ \sigma_{rz} &= \frac{K_{\text{III}}}{\sqrt{2\pi r}}\sin\left(\frac{\theta}{2}\right) + \text{bounded terms}, \end{aligned}} \tag{9.5.67}$$

where K_{III} is the *Mode III stress intensity factor.*

9.6 Cartesian component expressions at crack-tip

The results in the previous section for the asymptotic stress and displacement fields are expressed below in terms of components σ_{ij} and u_i of the stress and displacement fields with respect to a rectangular coordinate system with origin at a point along the crack front. However, for ease of presentation, we shall identify the position in terms of (r, θ)-coordinates of a cylindrical coordinate system (Fig. 9.27).

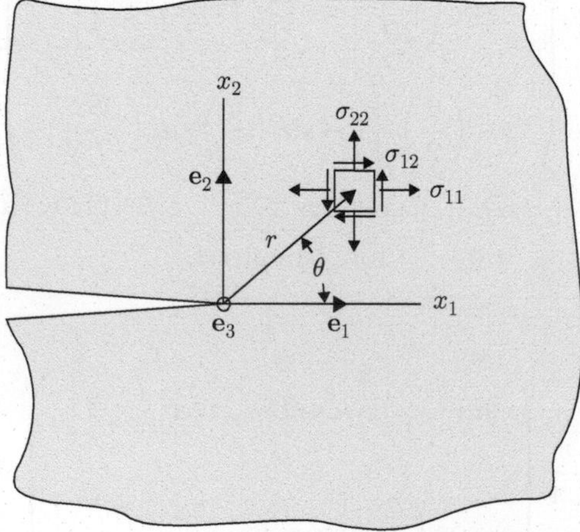

Fig. 9.27 Stress components with respect to a Cartesian coordinate system with origin on the crack front.

1. **Mode I, tensile opening mode:**

$$\left.\begin{aligned}
&\begin{pmatrix}\sigma_{11}\\ \sigma_{22}\\ \sigma_{12}\end{pmatrix} = \frac{K_{\rm I}}{\sqrt{2\pi r}}\cos\left(\frac{\theta}{2}\right)\begin{pmatrix}1-\sin\left(\frac{\theta}{2}\right)\sin\left(\frac{3\theta}{2}\right)\\ 1+\sin\left(\frac{\theta}{2}\right)\sin\left(\frac{3\theta}{2}\right)\\ \sin\left(\frac{\theta}{2}\right)\cos\left(\frac{3\theta}{2}\right)\end{pmatrix} + \text{bounded terms},\\
&\sigma_{33}=\nu\left(\sigma_{11}+\sigma_{22}\right),\quad \sigma_{13}=\sigma_{23}=0,\qquad \text{for plane strain},\\
&\sigma_{33}=\sigma_{13}=\sigma_{23}=0,\qquad \text{for plane stress.}
\end{aligned}\right\} \tag{9.6.1}$$

$$\begin{aligned}
&\begin{pmatrix}u_1\\ u_2\end{pmatrix} = \frac{K_{\rm I}}{2\mu}\sqrt{\frac{r}{2\pi}}\begin{pmatrix}\cos\left(\frac{\theta}{2}\right)\left[\varkappa-1+2\sin^2\left(\frac{\theta}{2}\right)\right]\\ \sin\left(\frac{\theta}{2}\right)\left[\varkappa+1-2\cos^2\left(\frac{\theta}{2}\right)\right]\end{pmatrix} + \text{rigid displacement}\\
&u_3=0,\quad \varkappa=3-4\nu\quad \text{for plane strain}\\
&\bar{u}_3=0,\qquad \bar{\epsilon}_{33}\neq 0,\quad \varkappa=\frac{3-\nu}{1+\nu}\qquad \text{for plane stress.}
\end{aligned} \tag{9.6.2}$$

2. **Mode II, in-plane sliding mode:**

$$\left.\begin{aligned}
&\begin{pmatrix}\sigma_{11}\\ \sigma_{22}\\ \sigma_{12}\end{pmatrix} = \frac{K_{\rm II}}{\sqrt{2\pi r}}\begin{pmatrix}-\sin\left(\frac{\theta}{2}\right)\left[2+\cos\left(\frac{\theta}{2}\right)\cos\left(\frac{3\theta}{2}\right)\right]\\ \sin\left(\frac{\theta}{2}\right)\left[\cos\left(\frac{\theta}{2}\right)\cos\left(\frac{3\theta}{2}\right)\right]\\ \cos\left(\frac{\theta}{2}\right)\left[1-\sin\left(\frac{\theta}{2}\right)\sin\left(\frac{3\theta}{2}\right)\right]\end{pmatrix} + \text{bounded terms},\\
&\sigma_{33}=\nu\left(\sigma_{11}+\sigma_{22}\right),\quad \sigma_{13}=\sigma_{23}=0,\qquad \text{for plane strain},\\
&\sigma_{33}=\sigma_{13}=\sigma_{23}=0,\qquad \text{for plane stress.}
\end{aligned}\right\} \tag{9.6.3}$$

$$\begin{aligned}
&\begin{pmatrix}u_1\\ u_2\end{pmatrix} = \frac{K_{\rm II}}{2\mu}\sqrt{\frac{r}{2\pi}}\begin{pmatrix}\sin\left(\frac{\theta}{2}\right)\left[\varkappa+1+2\cos^2\left(\frac{\theta}{2}\right)\right]\\ -\cos\left(\frac{\theta}{2}\right)\left[\varkappa-1-2\sin^2\left(\frac{\theta}{2}\right)\right]\end{pmatrix} + \text{rigid displacement}\\
&u_3=0,\quad \varkappa=3-4\nu\quad \text{for plane strain}\\
&\bar{u}_3=0,\qquad \bar{\epsilon}_{33}\neq 0,\quad \varkappa=\frac{3-\nu}{1+\nu}\qquad \text{for plane stress.}
\end{aligned} \tag{9.6.4}$$

3. **Mode III, anti-plane tearing mode**:

$$\left.\begin{aligned}\begin{pmatrix}\sigma_{13}\\ \\ \sigma_{23}\end{pmatrix} &= \frac{K_{\mathrm{III}}}{\sqrt{2\pi r}}\begin{pmatrix}-\sin\left(\dfrac{\theta}{2}\right)\\ \\ \cos\left(\dfrac{\theta}{2}\right)\end{pmatrix} + \text{bounded terms,}\\ \sigma_{11} &= \sigma_{22} = \sigma_{33} = \sigma_{12} = 0,\end{aligned}\right\} \tag{9.6.5}$$

$$\left(u_3\right) = \frac{K_{\mathrm{III}}}{2\mu}\sqrt{\frac{r}{2\pi}}\left(4\sin\left(\frac{\theta}{2}\right)\right) + \text{rigid displacement.} \tag{9.6.6}$$

The higher order terms, called "bounded terms", are omitted, since they are not of significant magnitude compared to the leading terms near the crack-tip. The equations above predict that the magnitudes of the stress components increase rapidly as one approaches the crack-tip. Indeed,

- since the non-zero stress components are proportional to $1/\sqrt{r}$, they approach infinity as $r \to 0$. That is, a **mathematical singularity** exists at the crack-tip!

The non-zero stress components in each of the three modes are proportional to the parameters K_{I}, K_{II}, and K_{III}, respectively[13], and the remaining terms only give the variation with r and θ. Thus, the **magnitude** of the stress components near the crack-tip can be characterized by giving the values of K_{I}, K_{II}, and K_{III}. It is for this reason that these quantities are called **the stress intensity factors** for Modes I, II, and III, respectively. Actual expressions for the stress intensity factors depend upon the constants of integration which are determined from the precise geometry of the body containing the crack and the nature of the loads applied to it.

REMARK 9.5

The qualifiers **plane strain** and **plane stress** in the equations above refer to the two-dimensional specializations of the general three-dimensional equations of linear elasticity, under which the asymptotic stress fields are derived.

9.6.1 **Crack-opening displacements**

For later use we note the following results for **crack-opening displacements** δ_2, δ_1, and δ_3 for the three modes based on the solutions (9.6.2), (9.6.4), and (9.6.6) for the displacement fields:

1. **Mode I**:

$$\delta_2 \stackrel{\text{def}}{=} u_2(r,+\pi) - u_2(r,-\pi) = \frac{4}{\mu}(1-\nu)K_{\mathrm{I}}\sqrt{\frac{r}{2\pi}}. \tag{9.6.7}$$

[13] The subscripts "I", "II", and "III" are Roman numerals, and should therefore be read as "one", "two", and "three".

2. **Mode II:**

$$\delta_1 \overset{\text{def}}{=} u_1(r, +\pi) - u_1(r, -\pi) = \frac{4}{\mu}(1-\nu)K_{\mathrm{II}}\sqrt{\frac{r}{2\pi}}. \tag{9.6.8}$$

3. **Mode III:**

$$\delta_3 \overset{\text{def}}{=} u_3(r, +\pi) - u_3(r, -\pi) = \frac{4}{\mu}K_{\mathrm{III}}\sqrt{\frac{r}{2\pi}}. \tag{9.6.9}$$

The results for the crack-opening displacements δ_2 and δ_1 for Modes I and II respectively are for plane strain. For plane stress replace ν by $\nu/(1+\nu)$.

PART V

Variational formulations

10 Variational formulation of boundary-value problems

As a prelude to our discussion of a *variational formulation*—or a *weak formulation*—of boundary-value problems, we discuss the important *principle of virtual power*. In Sec. 4.3 we derived Cauchy's equations of equilibrium $\operatorname{div}\boldsymbol{\sigma} + \mathbf{b} = \mathbf{0}$, moment balance $\boldsymbol{\sigma} = \boldsymbol{\sigma}^{\top}$, as well his traction relation $\mathbf{t}(\mathbf{n}) = \boldsymbol{\sigma}\mathbf{n}$, based on *balance of forces and moments* on an arbitrary part of a body. In this section we show that these relations are *derivable* starting from the *principle of virtual power* (Germain, 1973; Gurtin, 2002; Gurtin et al., 2010).

10.1 Principle of virtual power

We denote by $\mathcal{P}$ an arbitrary part of the reference body $\mathcal{B}$ with $\mathbf{n}$ the outward unit normal on the boundary $\partial\mathcal{P}$ of $\mathcal{P}$ (Fig. 10.1). Given a displacement $\mathbf{u}$ with displacement gradient $\mathbf{H} = \nabla\mathbf{u}$, let $(\dot{\mathbf{u}}, \dot{\mathbf{H}})$ denote the corresponding rates which are constrained by

$$\nabla\dot{\mathbf{u}} = \dot{\mathbf{H}}. \tag{10.1.1}$$

With each motion of the part we associate an external force system defined by,

- a traction $\mathbf{t}(\mathbf{n})$ (for each unit vector $\mathbf{n}$) that expends power over the velocity $\dot{\mathbf{u}}$, and
- an external body force $\mathbf{b}$, measured per unit volume in the reference body, that also expends power over $\dot{\mathbf{u}}$,[1]

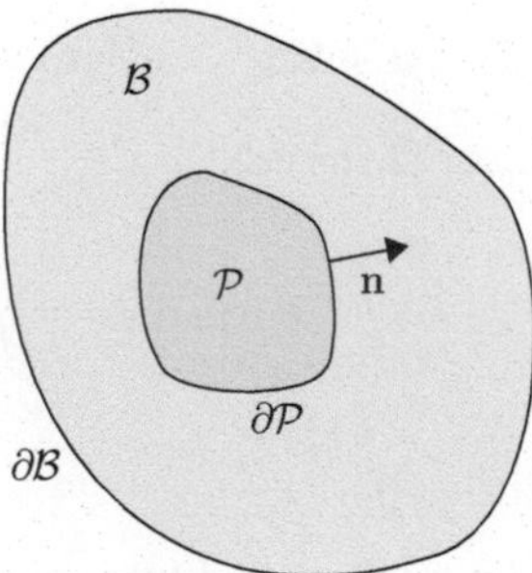

Fig. 10.1 An arbitrary part $\mathcal{P}$ of the reference body $\mathcal{B}$, with outward unit normal $\mathbf{n}$ on the boundary $\partial\mathcal{P}$ of $\mathcal{P}$.

[1] The body force may be assumed to account for inertia in the sense that $\mathbf{b} = \mathbf{b}_0 - \rho\ddot{\mathbf{u}}$.

Continuum Mechanics of Solids. Lallit Anand and Sanjay Govindjee, Oxford University Press (2020).

DOI: 10.1093/oso/9780198864721.001.0001

such that

$$\mathcal{W}_{\text{ext}}(\mathcal{P}) = \int_{\partial\mathcal{P}} \mathbf{t}(\mathbf{n}) \cdot \dot{\mathbf{u}}\, da + \int_{\mathcal{P}} \mathbf{b} \cdot \dot{\mathbf{u}}\, dv, \tag{10.1.2}$$

represents the *external power* expended on $\mathcal{P}$.

The external power is accompanied by power expended internally by

- a tensor field $\boldsymbol{\sigma}$ which is power-conjugate to $\dot{\mathbf{H}}$,

and we write the *internal power* as

$$\mathcal{W}_{\text{int}}(\mathcal{P}) = \int_{\mathcal{P}} \boldsymbol{\sigma} : \dot{\mathbf{H}}\, dv. \tag{10.1.3}$$

Note that at this stage the tensor field $\boldsymbol{\sigma}$ is not restricted to be symmetric. Also, power conjugacy means that the scalar product $\boldsymbol{\sigma} : \dot{\mathbf{H}}$ represents the rate of work expended (per unit volume) in the interior of the body. Shortly we will find that the tensor field $\boldsymbol{\sigma}$ satisfies all the requirements of the Cauchy stress tensor, and it is for this reason that we have chosen to represent it by the symbol $\boldsymbol{\sigma}$.[2]

10.1.1 Virtual power

Consider now an arbitrary velocity field $\tilde{\mathbf{u}}$, unrelated to the actual velocity field $\dot{\mathbf{u}}$, along with its associated velocity gradient field

$$\tilde{\mathbf{H}} = \nabla\tilde{\mathbf{u}}. \tag{10.1.4}$$

Since $\tilde{\mathbf{u}}$ and $\tilde{\mathbf{H}}$ are arbitrarily imagined, we will call them *virtual* velocities and *virtual* velocity gradients, respectively. Further, we define a generalized virtual velocity to be a list

$$\mathcal{V} = (\tilde{\mathbf{u}}, \tilde{\mathbf{H}}),$$

consistent with (10.1.4). We refer to $\mathcal{V}$ as *rigid* if, for any $\mathbf{x}$ in $\mathcal{B}$ (cf. (3.6.13))

$$\tilde{\mathbf{u}}(\mathbf{x}) = \mathbf{a} + \boldsymbol{\Omega}\mathbf{x}, \qquad \tilde{\mathbf{H}}(\mathbf{x}) = \boldsymbol{\Omega}, \tag{10.1.5}$$

with the vector $\mathbf{a}$ and the skew tensor $\boldsymbol{\Omega}$ *constant*.

Writing

$$\begin{aligned} \mathcal{W}_{\text{ext}}(\mathcal{P}, \mathcal{V}) &= \int_{\partial\mathcal{P}} \mathbf{t}(\mathbf{n}) \cdot \tilde{\mathbf{u}}\, da + \int_{\mathcal{P}} \mathbf{b} \cdot \tilde{\mathbf{u}}\, dv, \\ \mathcal{W}_{\text{int}}(\mathcal{P}, \mathcal{V}) &= \int_{\mathcal{P}} \boldsymbol{\sigma} : \tilde{\mathbf{H}}\, dv, \end{aligned} \tag{10.1.6}$$

respectively, for the external and internal expenditures of *virtual power*, we state the principle of virtual power below.

[2] In the virtual power method the existence of the stress tensor need not be proved, it is assumed to exist from the beginning as power conjugate to $\dot{\mathbf{H}}$.

THE PRINCIPLE OF VIRTUAL POWER *This principle consists of two basic requirements:*

(V1) *(Power balance) Given any part* $\mathcal{P} \subseteq \mathcal{B}$,[3]

$$\mathcal{W}_{\text{ext}}(\mathcal{P}, \mathcal{V}) = \mathcal{W}_{\text{int}}(\mathcal{P}, \mathcal{V}) \quad \textit{for all generalized virtual velocities } \mathcal{V}. \tag{10.1.7}$$

(V2) *(Rigid motion hypothesis) Given any part* $\mathcal{P} \subseteq \mathcal{B}$,

$$\mathcal{W}_{\text{int}}(\mathcal{P}, \mathcal{V}) = 0 \textit{ whenever } \mathcal{V} \textit{ is a rigid generalized virtual velocity.} \tag{10.1.8}$$

REMARKS 10.1

1. The *generalized virtual velocity* $\mathcal{V} = (\tilde{\mathbf{u}}, \tilde{\mathbf{H}})$ of a material point is a fictitious velocity, constrained only by (10.1.4). It is emphasized that the forces $\mathbf{t}(\mathbf{n})$ and $\mathbf{b}$, and the power-conjugate tensor field $\boldsymbol{\sigma}$ acting at a point remain unaffected by $\tilde{\mathbf{u}}$ and $\tilde{\mathbf{H}}$. The power expended by the *actual forces and the power-conjugate tensor field* during the *virtual velocities* is called the *virtual power*.
2. We may also interpret $\tilde{\mathbf{u}}$ as a *virtual displacement*, in which case the principle is called the *principle of virtual work*.

10.1.2 **Consequences of the principle of virtual power**

First, we deduce the consequences of requirement (V1) of the principle of virtual power, which recalling (10.1.6) is that

$$\int_{\partial\mathcal{P}} \mathbf{t}(\mathbf{n}) \cdot \tilde{\mathbf{u}}\, da + \int_{\mathcal{P}} \mathbf{b} \cdot \tilde{\mathbf{u}}\, dv = \int_{\mathcal{P}} \boldsymbol{\sigma} : \tilde{\mathbf{H}}\, dv. \tag{10.1.9}$$

Using (10.1.4) yields

$$\int_{\partial\mathcal{P}} \mathbf{t}(\mathbf{n}) \cdot \tilde{\mathbf{u}}\, da + \int_{\mathcal{P}} \mathbf{b} \cdot \tilde{\mathbf{u}}\, dv = \int_{\mathcal{P}} \boldsymbol{\sigma} : \nabla \tilde{\mathbf{u}}\, dv, \tag{10.1.10}$$

to be satisfied for all $\tilde{\mathbf{u}}$ and all $\mathcal{P}$. Further, using the divergence theorem in the form

$$\int_{\mathcal{P}} \boldsymbol{\sigma} : \nabla \tilde{\mathbf{u}}\, dv = \int_{\partial\mathcal{P}} \boldsymbol{\sigma}\mathbf{n} \cdot \tilde{\mathbf{u}}\, da - \int_{\mathcal{P}} \operatorname{div} \boldsymbol{\sigma} \cdot \tilde{\mathbf{u}}\, dv,$$

we may conclude that

$$\int_{\partial\mathcal{P}} \big(\mathbf{t}(\mathbf{n}) - \boldsymbol{\sigma}\mathbf{n}\big) \cdot \tilde{\mathbf{u}}\, da + \int_{\mathcal{P}} (\operatorname{div} \boldsymbol{\sigma} + \mathbf{b}) \cdot \tilde{\mathbf{u}}\, dv = 0.$$

---

[3] Note that we require the principle hold for all parts $\mathcal{P}$, and not just for $\mathcal{P} = \mathcal{B}$; this requirement is basic to what follows.

Since this relation must hold for all $\mathcal{P}$ and all $\tilde{\mathbf{u}}$, standard variational arguments[4] yield the condition

$$\boxed{\mathbf{t}(\mathbf{n}) = \boldsymbol{\sigma}\mathbf{n},} \tag{10.1.11}$$

and the requirement that

$$\boxed{\operatorname{div}\boldsymbol{\sigma} + \mathbf{b} = \mathbf{0},} \tag{10.1.12}$$

everywhere in $\mathcal{B}$.

Next, we deduce the consequences of requirement (V2) of the principle of virtual power. For a rigid $\mathcal{V}$, cf. (10.1.5), relation (10.1.8) requires that

$$\int_{\mathcal{P}} \boldsymbol{\sigma} : \boldsymbol{\Omega}\, dv = 0 \tag{10.1.13}$$

for all parts $\mathcal{P}$ and all skew tensors $\boldsymbol{\Omega}$. Since $\mathcal{P}$ is arbitrary the localization theorem requires

$$\boldsymbol{\sigma} : \boldsymbol{\Omega} = 0. \tag{10.1.14}$$

Since this must also hold for all skew tensors $\boldsymbol{\Omega}$, $\boldsymbol{\sigma}$ must be symmetric:

$$\boxed{\boldsymbol{\sigma} = \boldsymbol{\sigma}^{\top}.} \tag{10.1.15}$$

Looking at (10.1.11), (10.1.12), and (10.1.15), we see that the tensor field $\boldsymbol{\sigma}$—which we had introduced as a tensor power-conjugate to $\dot{\mathbf{H}}$—satisfies all the properties of the classical Cauchy stress, with (10.1.12) and (10.1.15) representing the local force and moment balances, and (10.1.11) Cauchy's traction relation.

Next, we denote the symmetric part of $\tilde{\mathbf{H}}$ by $\tilde{\boldsymbol{\epsilon}}$ and conclude from the symmetry of $\boldsymbol{\sigma}$ that

$$\boldsymbol{\sigma} : \tilde{\mathbf{H}} = \boldsymbol{\sigma} : \tilde{\boldsymbol{\epsilon}}.$$

Thus we may rewrite the internal virtual power $(10.1.6)_2$ as

$$\mathcal{W}_{\text{int}}(\mathcal{P}, \mathcal{V}) = \int_{\mathcal{P}} \boldsymbol{\sigma} : \tilde{\boldsymbol{\epsilon}}\, dv. \tag{10.1.16}$$

Further, the virtual-power relation (10.1.10) may now be written as

$$\boxed{\int_{\partial\mathcal{P}} \mathbf{t}(\mathbf{n}) \cdot \tilde{\mathbf{u}}\, da + \int_{\mathcal{P}} \mathbf{b} \cdot \tilde{\mathbf{u}}\, dv = \int_{\mathcal{P}} \boldsymbol{\sigma} : \tilde{\boldsymbol{\epsilon}}\, dv, \qquad \text{with} \qquad \tilde{\boldsymbol{\epsilon}} \stackrel{\text{def}}{=} \operatorname{sym}(\tilde{\mathbf{H}}).} \tag{10.1.17}$$

We can now summarize the consequences of the principle of virtual power:

Consequences of the principle of virtual power *Assume that for any part $\mathcal{P} \subseteq \mathcal{B}$ the virtual power balance requirements* (10.1.7) *and* (10.1.8) *are satisfied. Then, at all points of the body:*

[4] To show this result, one considers special cases for $\tilde{\mathbf{u}}$. For example, to show (10.1.11) one can choose $\tilde{\mathbf{u}} = \mathbf{0}$ in the interior of $\mathcal{P}$ and $\tilde{\mathbf{u}} = \mathbf{t}(\mathbf{n}) - \boldsymbol{\sigma}\mathbf{n}$ on $\partial\mathcal{P}$. This results in $\int_{\partial\mathcal{P}} |\mathbf{t}(\mathbf{n}) - \boldsymbol{\sigma}\mathbf{n}|^2\, da = 0$, from which it follows that $\mathbf{t}(\mathbf{n}) - \boldsymbol{\sigma}\mathbf{n} = \mathbf{0}$ on $\partial\mathcal{P}$ due to the positivity of the integrand—if the integral of a non-negative integrand is zero, then the integrand must be zero at all points in the domain. To show (10.1.12), one can choose $\tilde{\mathbf{u}} = \mathbf{0}$ on $\partial\mathcal{P}$ and $\tilde{\mathbf{u}} = \operatorname{div}\boldsymbol{\sigma} + \mathbf{b}$ in $\mathcal{P}$. This then leads to $\int_{\mathcal{P}} |\operatorname{div}\boldsymbol{\sigma} + \mathbf{b}|^2\, dv = 0$ and thus $\operatorname{div}\boldsymbol{\sigma} + \mathbf{b} = \mathbf{0}$ in $\mathcal{P}$. Since these results hold for all $\mathcal{P} \subseteq \mathcal{B}$, (10.1.11) and (10.1.12) must hold everywhere in $\mathcal{B}$; see Hughes (2000, §1.4) or Gurtin et al. (2010, §22.2.2) for additional technical details.

(i) *The traction* $\mathbf{t}(\mathbf{n})$ *and the power-conjugate tensor field* $\boldsymbol{\sigma}$, *the stress, are related through the relation*

$$\mathbf{t}(\mathbf{n}) = \boldsymbol{\sigma}\mathbf{n} \tag{10.1.18}$$

for every unit vector $\mathbf{n}$.

(ii) $\boldsymbol{\sigma}$ *and* $\mathbf{b}$ *satisfy the local force and moment balances*

$$\operatorname{div}\boldsymbol{\sigma} + \mathbf{b} = \mathbf{0} \qquad \text{and} \qquad \boldsymbol{\sigma} = \boldsymbol{\sigma}^{\top}. \tag{10.1.19}$$

The foregoing results demonstrate the all-encompassing nature of the principle of virtual power:

- *The principle of virtual power* **encapsulates** *the local force balance* (10.1.12), *the local moment balance* (10.1.15), *and Cauchy's law* (10.1.11).

It should further be observed that the steps leading to (10.1.11), (10.1.12), and (10.1.15) are all reversible. Thus,

- The force and moment equilibrium equations, along with Cauchy's traction relation, also imply the principle of virtual power.

EXAMPLE 10.1 Global force balance for a part $\mathcal{P}$

The principle of virtual power can also be shown to imply global force balance for any part $\mathcal{P}$ of a body. Consider an arbitrary constant vector $\mathbf{a}$, and set $\tilde{\mathbf{u}}(\mathbf{x}) = \mathbf{a}$. For this virtual velocity field $\tilde{\boldsymbol{\epsilon}} = \mathbf{0}$ and (10.1.17) reduces to

$$\int_{\partial\mathcal{P}} \mathbf{t}(\mathbf{n})\cdot\mathbf{a}\,da + \int_{\mathcal{P}} \mathbf{b}\cdot\mathbf{a}\,dv = \mathbf{a}\cdot\left(\int_{\partial\mathcal{P}} \mathbf{t}(\mathbf{n})\,da + \int_{\mathcal{P}} \mathbf{b}\,dv\right) = 0.$$

Since $\mathbf{a}$ is an arbitrary constant vector, this implies that

$$\int_{\partial\mathcal{P}} \mathbf{t}(\mathbf{n})\,da + \int_{\mathcal{P}} \mathbf{b}\,dv = \mathbf{0},$$

which represents a global force balance for the part $\mathcal{P}$.

10.1.3 Application of the principle of virtual power to boundary-value problems

Recall from Sec. 8.2 that in linear elastostatics, typically, displacement is prescribed on a portion $\mathcal{S}_1$ of the boundary $\partial\mathcal{B}$, while on another portion $\mathcal{S}_2$ traction is prescribed (Fig. 10.2). Both the displacement and traction cannot be prescribed on the same part of the boundary, so

$$\mathcal{S}_1 \cap \mathcal{S}_2 = \emptyset.$$

However, either the displacement or the traction must be prescribed at every boundary point, so

$$\mathcal{S}_1 \cup \mathcal{S}_2 = \partial\mathcal{B}.$$

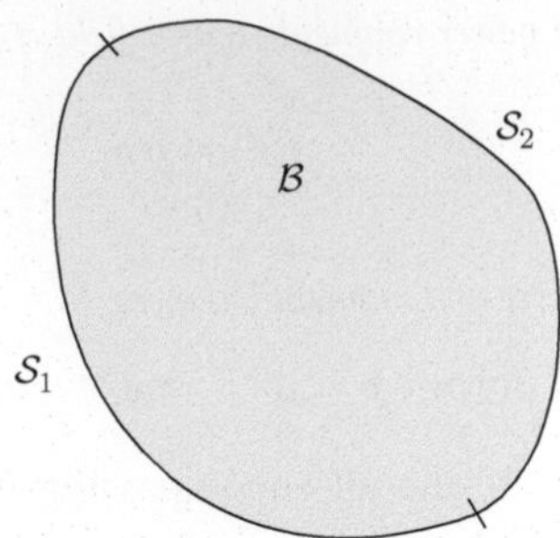

Fig. 10.2 The body $\mathcal{B}$ with boundary $\partial\mathcal{B} = \mathcal{S}_1 \cup \mathcal{S}_2$ and $\mathcal{S}_1 \cap \mathcal{S}_2 = \emptyset$.

Without proof, we state that these points hold in general, independent of the material, i.e. the material need not be elastic.

Since the principle of virtual power encapsulates both the equation of equilibrium and the Cauchy relation for tractions, an application of it can be used to formulate and solve mechanical boundary-value problems. First, we will consider the case where the arbitrary part $\mathcal{P}$ is in fact the *entire* body $\mathcal{B}$, and augment it by the traction condition

$$\boldsymbol{\sigma}\mathbf{n} = \hat{\mathbf{t}} \quad \text{on } \mathcal{S}_2, \tag{10.1.20}$$

in which $\hat{\mathbf{t}}$ is a prescribed function on $\mathcal{S}_2 \subset \partial\mathcal{B}$. Further, for ease of notation denote a virtual velocity $\tilde{\mathbf{u}}$ simply by $\mathbf{w}$, and limit attention to virtual fields that satisfy

$$\mathbf{w} = \mathbf{0} \quad \text{on } \mathcal{S}_1; \tag{10.1.21}$$

we refer to virtual velocities which satisfy (10.1.21) as *admissible*.

Granted (10.1.20) and (10.1.21), the virtual power balance (10.1.10) has the form

$$\int_{\mathcal{B}} \boldsymbol{\sigma} : \nabla\mathbf{w}\, dv = \int_{\mathcal{S}_2} \hat{\mathbf{t}} \cdot \mathbf{w}\, da + \int_{\mathcal{B}} \mathbf{b} \cdot \mathbf{w}\, dv. \tag{10.1.22}$$

Since the virtual power balance holds for all generalized virtual velocities, it must in particular hold for all admissible $\mathbf{w}$. This allows us to establish the following result:

WEAK FORM OF THE EQUATION OF EQUILIBRIUM AND THE TRACTION CONDITION: *The virtual power balance, that is*

$$\int_{\mathcal{B}} \boldsymbol{\sigma} : \nabla\mathbf{w}\, dv = \int_{\mathcal{S}_2} \hat{\mathbf{t}} \cdot \mathbf{w}\, da + \int_{\mathcal{B}} \mathbf{b} \cdot \mathbf{w}\, dv \qquad \forall \text{ virtual fields } \mathbf{w} \textit{ that vanish on } \mathcal{S}_1, \tag{10.1.23}$$

is satisfied if and only if

$$\operatorname{div}\boldsymbol{\sigma} + \mathbf{b} = \mathbf{0} \qquad \text{in } \mathcal{B}, \tag{10.1.24}$$

and

$$\boldsymbol{\sigma}\mathbf{n} = \hat{\mathbf{t}} \qquad \text{on } \mathcal{S}_2. \tag{10.1.25}$$

REMARK 10.2

Note that in our discussion of the principle of virtual power and its consequence which lead to the weak form (10.1.23) of the equation of equilibrium and traction condition, *we did not invoke any constitutive assumptions.* In the next section we apply these ideas to the displacement problem of linear elastostatics.

10.2 Strong and weak forms of the displacement problem of linear elastostatics

We return now to our discussion of the displacement problem of linear elastostatics, and show how it may be formulated variationally using the principle of virtual power. Recall from Sec. 8.7 that when phrased in terms of displacements the system of field equations for the static behavior of a linear elastic body consists of:

- **the strain-displacement relation**

$$\epsilon = \tfrac{1}{2}(\nabla \mathbf{u} + \nabla \mathbf{u}^{\top}), \tag{10.2.1}$$

- **the stress-strain relation**

$$\boldsymbol{\sigma} = \mathbb{C}\epsilon = \mathbb{C}(\operatorname{sym} \nabla \mathbf{u}), \tag{10.2.2}$$

- and **the equation of equilibrium**

$$\operatorname{div}(\mathbb{C}(\operatorname{sym} \nabla \mathbf{u})) + \mathbf{b} = \mathbf{0}. \tag{10.2.3}$$

With $\mathcal{S}_1$ and $\mathcal{S}_2$ *complementary subsurfaces* of the boundary $\partial\mathcal{B}$ of the body $\mathcal{B}$, typical mixed **boundary conditions** are

$$\left.\begin{aligned} \mathbf{u} &= \hat{\mathbf{u}} \quad \text{on } \mathcal{S}_1, \\ (\mathbb{C}(\operatorname{sym} \nabla \mathbf{u}))\mathbf{n} &= \hat{\mathbf{t}} \quad \text{on } \mathcal{S}_2, \end{aligned}\right. \tag{10.2.4}$$

with $\hat{\mathbf{u}}$ and $\hat{\mathbf{t}}$ *prescribed* functions of $\mathbf{x}$, and the **displacement problem** of elastostatics is stated as follows.

STRONG FORM OF THE DISPLACEMENT PROBLEM OF ELASTOSTATICS *Given* $\mathbb{C}$, $\mathbf{b}$, *and boundary data* $\hat{\mathbf{u}}$ *and* $\hat{\mathbf{t}}$, *find a displacement field* $\mathbf{u}$ *that satisfies*

$$\operatorname{div}(\mathbb{C}(\operatorname{sym} \nabla \mathbf{u})) + \mathbf{b} = \mathbf{0} \qquad \text{in } \mathcal{B} \tag{10.2.5}$$

and the boundary conditions

$$\left.\begin{aligned} \mathbf{u} &= \hat{\mathbf{u}} \quad \text{on } \mathcal{S}_1, \\ (\mathbb{C}(\operatorname{sym} \nabla \mathbf{u}))\mathbf{n} &= \hat{\mathbf{t}} \quad \text{on } \mathcal{S}_2. \end{aligned}\right. \tag{10.2.6}$$

Assuming that the constitutive equation

$$\boldsymbol{\sigma} = \mathbb{C}(\mathrm{sym}\,\nabla\mathbf{u})$$

is satisfied, then the balance

$$\mathrm{div}\,(\mathbb{C}(\mathrm{sym}\,\nabla\mathbf{u})) + \mathbf{b} = \mathbf{0} \qquad \text{in } \mathcal{B},$$

and the traction boundary condition

$$(\mathbb{C}(\mathrm{sym}\,\nabla\mathbf{u}))\mathbf{n} = \hat{\mathbf{t}} \qquad \text{on } \mathcal{S}_2$$

are together equivalent to the requirement that

$$\boxed{\int_{\mathcal{B}} (\mathrm{sym}\,\nabla\mathbf{w}) : \mathbb{C}(\mathrm{sym}\,\nabla\mathbf{u})\,dv = \int_{\mathcal{S}_2} \hat{\mathbf{t}}\cdot\mathbf{w}\,da + \int_{\mathcal{B}} \mathbf{b}\cdot\mathbf{w}\,dv \qquad \forall \text{ virtual fields } \mathbf{w} \textit{ that vanish on } \mathcal{S}_1.}$$

(10.2.7)

Finally, the weak form of the displacement problem of linear elastostatics may be stated as follows:

WEAK FORM OF THE DISPLACEMENT PROBLEM OF ELASTOSTATICS *Given* $\mathbb{C}$, $\mathbf{b}$, *and boundary data* $\hat{\mathbf{u}}$ *and* $\hat{\mathbf{t}}$, *find a displacement field* $\mathbf{u}$, *equal to* $\hat{\mathbf{u}}$ *on* $\mathcal{S}_1$, such that

$$\int_{\mathcal{B}} (\mathrm{sym}\,\nabla\mathbf{w}) : \mathbb{C}(\mathrm{sym}\,\nabla\mathbf{u})dv - \int_{\mathcal{S}_2} \hat{\mathbf{t}}\cdot\mathbf{w}da - \int_{\mathcal{B}} \mathbf{b}\cdot\mathbf{w}dv = 0 \qquad \forall \text{ admissible } \mathbf{w}. \quad (10.2.8)$$

- This is the basis of most displacement-based finite element procedures for the solution of boundary-value problems in linear elastostatics.

11 Introduction to the finite element method for linear elastostatics

In this section we briefly introduce the widely used finite element method for solving boundary-value problems in solid mechanics. We restrict our discussion to two-dimensional plane stress or plane strain problems, and skip numerous details. For a more complete discussion cf., e.g., Hughes (2000), Fish and Belytschko (2007), or Zienkiewicz and Taylor (2013).

Consider a two-dimensional domain Ω with boundary Γ discretized with non-overlapping two-dimensional elements Ω^e (triangles or quadrilaterals), geometric regions whose vertices are called nodes (Fig. 11.1), such that

$$\Omega = \bigcup_{e=1}^{n_{\rm el}} \Omega^e. \tag{11.1.1}$$

In what follows

$$\begin{aligned} n_{\rm np} &= \text{number of nodal points,} \\ n_{\rm el} &= \text{number of elements,} \\ n_{\rm en} &= \text{number of nodes in an element, and} \\ e &= \text{element number.} \end{aligned}$$

Also,

- *In this section boldface letters* $\mathbf{u}$, $\boldsymbol{\epsilon}$, $\boldsymbol{\sigma}$, *etc. denote* matrices *of the Cartesian components of the corresponding vectors and tensors, and not the vectors and tensors themselves.*
- For ease of notation, we employ Cartesian labels x and y in place of x_1 and x_2.
- Likewise we will drop the superposed tilde on the virtual fields.

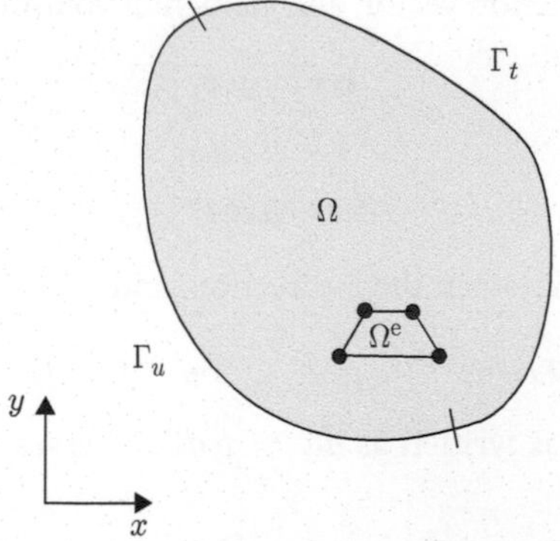

Fig. 11.1 A two-dimensional domain Ω showing a quadrilateral finite element Ω^e with 4 nodes.

Continuum Mechanics of Solids. Lallit Anand and Sanjay Govindjee, Oxford University Press (2020).

DOI: 10.1093/oso/9780198864721.001.0001

On a portion Γ_u of Γ the displacement is prescribed, while on another portion Γ_t traction is prescribed. As before, the displacement and traction cannot both be prescribed on the same part of the boundary, so

$$\Gamma_u \cap \Gamma_t = \emptyset.$$

However, on any portion of the boundary, either the displacement or the traction must be prescribed, so

$$\Gamma_u \cup \Gamma_t = \Gamma.$$

Let

$$\mathbf{u} = [u_x, u_y]^\top \tag{11.1.2}$$

denote the displacement matrix where u_x and u_y are the x and y components of the displacement. The stresses and strains in the Voigt matrix notation (in two dimensions) are then written as

$$\begin{aligned} \boldsymbol{\epsilon} &= [\epsilon_{xx}, \epsilon_{yy}, \gamma_{xy}]^\top, \\ \boldsymbol{\sigma} &= [\sigma_{xx}, \sigma_{yy}, \sigma_{xy}]^\top, \end{aligned} \tag{11.1.3}$$

where $\gamma_{xy} = 2\epsilon_{xy}$. Also, the relation between the strains and displacements may be written as

$$\boldsymbol{\epsilon} = \begin{bmatrix} \epsilon_{xx} \\ \epsilon_{yy} \\ \gamma_{xy} \end{bmatrix} = \begin{bmatrix} \dfrac{\partial}{\partial x} & 0 \\ 0 & \dfrac{\partial}{\partial y} \\ \dfrac{\partial}{\partial y} & \dfrac{\partial}{\partial x} \end{bmatrix} \begin{bmatrix} u_x \\ u_y \end{bmatrix} \equiv \nabla_s \begin{bmatrix} u_x \\ u_y \end{bmatrix} = \nabla_s \mathbf{u}, \tag{11.1.4}$$

where ∇_s is a symmetric gradient matrix operator defined by

$$\nabla_s \stackrel{\text{def}}{=} \begin{bmatrix} \dfrac{\partial}{\partial x} & 0 \\ 0 & \dfrac{\partial}{\partial y} \\ \dfrac{\partial}{\partial y} & \dfrac{\partial}{\partial x} \end{bmatrix}. \tag{11.1.5}$$

The body force vector, traction vector, and elemental volume are written as

$$\begin{aligned} \mathbf{b} &= [b_x, b_y]^\top, \\ \mathbf{t} &= [t_x, t_y]^\top, \\ dv &= h\, d\Omega, \end{aligned} \tag{11.1.6}$$

where h is the constant thickness in the z-direction, and

$$t_x = \sigma_{xx} n_x + \sigma_{xy} n_y, \qquad t_y = \sigma_{xy} n_x + \sigma_{yy} n_y. \tag{11.1.7}$$

The elemental surface area is written as $da = h\, d\Gamma$. Stresses and strains are then related in Voigt matrix notation by

$$\boldsymbol{\sigma} = \mathbf{D}\boldsymbol{\epsilon}, \tag{11.1.8}$$

where for an isotropic material the **D** matrix in plane stress and plane strain is, respectively, given by

$$\mathbf{D} = \frac{E}{1-\nu^2}\begin{bmatrix} 1 & \nu & 0 \\ \nu & 1 & 0 \\ 0 & 0 & (1-\nu)/2 \end{bmatrix} \qquad \text{(plane stress)}, \tag{11.1.9}$$

and

$$\mathbf{D} = \frac{E}{(1+\nu)(1-2\nu)}\begin{bmatrix} 1-\nu & \nu & 0 \\ \nu & 1-\nu & 0 \\ 0 & 0 & (1-2\nu)/2 \end{bmatrix} \qquad \text{(plane strain)}. \tag{11.1.10}$$

For two dimensions the weak form (10.2.8) in matrix notation becomes

$$\int_\Omega (\nabla_s \mathbf{w})^\top \mathbf{D}(\nabla_s \mathbf{u})\,d\Omega - \int_{\Gamma_t} \mathbf{w}^\top \hat{\mathbf{t}}\,d\Gamma - \int_\Omega \mathbf{w}^\top \mathbf{b}\,d\Omega = 0 \qquad \forall \text{ admissible } \mathbf{w}, \tag{11.1.11}$$

where the constant thickness h cancels out from all the terms after integration through the thickness.

The finite element solution is constructed by approximating the displacement field and the virtual velocity field on each element as

$$\begin{aligned} \mathbf{u}(x,y) &\approx \mathbf{u}^e(x,y) = \mathbf{N}^e(x,y)\mathbf{d}^e, \qquad (x,y) \in \Omega^e, \\ \mathbf{w}(x,y) &\approx \mathbf{w}^e(x,y) = \mathbf{N}^e(x,y)\mathbf{w}^e, \qquad (x,y) \in \Omega^e, \end{aligned} \tag{11.1.12}$$

where $\mathbf{N}^e(x,y)$ is a known element shape function matrix given by

$$\mathbf{N}^e(x,y) = \begin{bmatrix} N_1^e(x,y) & 0 & N_2^e(x,y) & 0 & \dots & N_{n_{\text{en}}}^e(x,y) & 0 \\ 0 & N_1^e(x,y) & 0 & N_2^e(x,y) & \dots & 0 & N_{n_{\text{en}}}^e(x,y) \end{bmatrix} \tag{11.1.13}$$

where n_{en} is the total number of nodes in the element (e.g., for a linear triangular element $n_{\text{en}} = 3$),

$$\mathbf{d}^e = \begin{bmatrix} u_{x1}^e & u_{y1}^e & u_{x2}^e & u_{y2}^e & \dots & u_{xn_{\text{en}}}^e & u_{yn_{\text{en}}}^e \end{bmatrix}^\top \tag{11.1.14}$$

are the *nodal displacements*, and

$$\mathbf{w}^e = \begin{bmatrix} w_{x1}^e & w_{y1}^e & w_{x2}^e & w_{y2}^e & \dots & w_{xn_{\text{en}}}^e & w_{yn_{\text{en}}}^e \end{bmatrix}^\top \tag{11.1.15}$$

are the nodal values of the virtual velocities. Note that in (11.1.12) we use the same symbol for the field $\mathbf{w}^e(x,y)$ as we do for the vector $\mathbf{w}^e$ in (11.1.15). We do so to minimize the amount of extra notation, and, in context, there should be no confusion.

With n_{np} denoting the total number of nodes in the finite element mesh *the global nodal displacement matrix* is

$$\mathbf{d} = \begin{bmatrix} u_{x1} & u_{y1} & u_{x2} & u_{y2} & \dots & u_{xn_{\text{np}}} & u_{yn_{\text{np}}} \end{bmatrix}^\top. \tag{11.1.16}$$

Similarly,

$$\mathbf{w} = \begin{bmatrix} w_{x1} & w_{y1} & w_{x2} & w_{y2} & \dots & w_{xn_{\text{np}}} & w_{yn_{\text{np}}} \end{bmatrix}^\top \tag{11.1.17}$$

denotes the global finite element virtual velocity matrix. The element and global matrices are related by a *gather matrix* $\mathbf{L}^e$ such that

$$\mathbf{d}^e = \mathbf{L}^e\mathbf{d} \qquad \text{and} \qquad \mathbf{w}^e = \mathbf{L}^e\mathbf{w}. \tag{11.1.18}$$

The gather matrix follows from the relation between local and global node numbers in the finite element approximation. The name *gather* originates from the fact that these matrices gather (or extract) the nodal displacements of each element from the global matrix.

The integration in (11.1.11) is computed over each element Ω^e and then summed to compute the integration over Ω:

$$\sum_{e=1}^{n_{\text{el}}}\left\{\int_{\Omega^e}(\nabla_s\mathbf{w}^e)^\top\mathbf{D}(\nabla_s\mathbf{u}^e)d\Omega - \int_{\Gamma_t^e}\mathbf{w}^{e\top}\hat{\mathbf{t}}d\Gamma - \int_{\Omega^e}\mathbf{w}^{e\top}\mathbf{b}d\Omega\right\} = 0. \tag{11.1.19}$$

Next we express the strains in terms of the element shape functions and the nodal displacements. Recall the strain-displacement relations (11.1.4) in terms of the symmetric gradient operator (11.1.5). Applying ∇_s to the matrix $\mathbf{N}^e$ gives

$$\boldsymbol{\epsilon} = \begin{bmatrix}\epsilon_{xx}\\ \epsilon_{yy}\\ \gamma_{xy}\end{bmatrix} \approx \boldsymbol{\epsilon}^e = \nabla_s\mathbf{u}^e(x,y) = (\nabla_s\mathbf{N}^e(x,y))\mathbf{d}^e \equiv \mathbf{B}^e(x,y)\mathbf{d}^e, \tag{11.1.20}$$

where the strain-displacement matrix $\mathbf{B}^e$ is defined as

$$\mathbf{B}^e \overset{\text{def}}{=} \nabla_s\mathbf{N}^e = \begin{bmatrix} \frac{\partial N_1^e}{\partial x} & 0 & \frac{\partial N_2^e}{\partial x} & 0 & \cdots & \frac{\partial N_{n_{\text{en}}}^e}{\partial x} & 0 \\ 0 & \frac{\partial N_1^e}{\partial y} & 0 & \frac{\partial N_2^e}{\partial y} & \cdots & 0 & \frac{\partial N_{n_{\text{en}}}^e}{\partial y} \\ \frac{\partial N_1^e}{\partial y} & \frac{\partial N_1^e}{\partial x} & \frac{\partial N_2^e}{\partial y} & \frac{\partial N_2^e}{\partial x} & \cdots & \frac{\partial N_{n_{\text{en}}}^e}{\partial y} & \frac{\partial N_{n_{\text{en}}}^e}{\partial x} \end{bmatrix}. \tag{11.1.21}$$

Similarly the derivatives of the virtual velocities are

$$(\nabla_s\mathbf{w}^e(x,y))^\top = (\mathbf{B}^e(x,y)\mathbf{w}^e)^\top = \mathbf{w}^{e\top}\mathbf{B}^{e\top}(x,y), \tag{11.1.22}$$

recalling that $(\mathbf{ab})^\top = \mathbf{b}^\top\mathbf{a}^\top$.

Substituting $(11.1.12)_2$, (11.1.20), and (11.1.22) in (11.1.19)

$$\sum_{e=1}^{n_{\text{el}}}\mathbf{w}^{e\top}\left(\int_{\Omega^e}\mathbf{B}^{e\top}\mathbf{D}\mathbf{B}^e d\Omega\right)\mathbf{d}^e - \sum_{e=1}^{n_{\text{el}}}\mathbf{w}^{e\top}\left\{\int_{\Gamma_t^e}\mathbf{N}^{e\top}\hat{\mathbf{t}}d\Gamma + \int_{\Omega^e}\mathbf{N}^{e\top}\mathbf{b}d\Omega\right\} = 0. \tag{11.1.23}$$

Defining the *element stiffness matrix* and the *element external force vector*, respectively, by

$$\mathbf{K}^e \overset{\text{def}}{=} \int_{\Omega^e}\mathbf{B}^{e\top}\mathbf{D}\mathbf{B}^e d\Omega \qquad \text{(element stiffness matrix)} \tag{11.1.24}$$

and

$$\mathbf{f}^e \overset{\text{def}}{=} \int_{\Gamma_t^e}\mathbf{N}^{e\top}\hat{\mathbf{t}}d\Gamma + \int_{\Omega^e}\mathbf{N}^{e\top}\mathbf{b}d\Omega \qquad \text{(element external force vector)}, \tag{11.1.25}$$

equation (11.1.23) may be written as

$$\sum_{e=1}^{n_{\rm el}} \mathbf{w}^{e\top}\mathbf{K}^e\mathbf{d}^e - \sum_{e=1}^{n_{\rm el}} \mathbf{w}^{e\top}\mathbf{f}^e = 0. \tag{11.1.26}$$

Next, using the gather matrix $\mathbf{L}^e$, as defined in (11.1.18), equation (11.1.26) becomes

$$\mathbf{w}^\top\left[\left(\sum_{e=1}^{n_{\rm el}} \mathbf{L}^{e\top}\mathbf{K}^e\mathbf{L}^e\right)\mathbf{d} - \left(\sum_{e=1}^{n_{\rm el}} \mathbf{L}^{e\top}\mathbf{f}^e\right)\right] = 0 \qquad \forall \mathbf{w}. \tag{11.1.27}$$

Finally, defining the *global stiffness matrix* and the *global external force vector*, respectively by

$$\mathbf{K} \stackrel{\rm def}{=} \sum_{e=1}^{n_{\rm el}} \mathbf{L}^{e\top}\mathbf{K}^e\mathbf{L}^e \qquad \text{(global stiffness matrix)}, \tag{11.1.28}$$

and

$$\mathbf{f} \stackrel{\rm def}{=} \sum_{e=1}^{n_{\rm el}} \mathbf{L}^{e\top}\mathbf{f}^e \qquad \text{(global external force vector)}, \tag{11.1.29}$$

equation (11.1.27) may be written as

$$\mathbf{w}^\top[\mathbf{K}\mathbf{d} - \mathbf{f}] = 0 \qquad \forall \mathbf{w}, \tag{11.1.30}$$

and since $\mathbf{w}$ is arbitrary, we are led to the linear system of equations

$$\boxed{\mathbf{K}\mathbf{d} = \mathbf{f},} \tag{11.1.31}$$

which may be solved numerically for the unknown global nodal displacement vector $\mathbf{d}$. This then leads to an approximation to the displacement field over each element as $\mathbf{u}^e(x,y) = \mathbf{N}^e(x,y)\mathbf{L}^e\mathbf{d}$, the strain field over each element as $\boldsymbol{\epsilon}^e(x,y) = \mathbf{B}^e(x,y)\mathbf{L}^e\mathbf{d}$, and the stress field over each element as $\boldsymbol{\sigma}^e(x,y) = \mathbf{D}\boldsymbol{\epsilon}^e(x,y)$.

While we have illustrated the main structure of the finite element method in this brief chapter using the equations of linear elasticity, the method can also be applied to a wide variety of other problems in science and engineering.

12 Principles of minimum potential energy and complementary energy

The principles of minimum potential energy and minimum complementary energy form the basis of certain *variational methods* in elasticity. These minimization principles may be used to derive the governing differential equations and boundary conditions for specialized classes of problems in elasticity. Another important application of energy methods is in finding approximate solutions to elasticity problems. Since the *calculus of variations* is an important tool in minimization problems, we begin with a brief review of this important mathematical topic.

12.1 A brief digression on calculus of variations

To begin, consider the prototypical problem of the calculus of variations in which one is asked to find a function $u(x)$ that minimizes the functional[1]

$$I\{u(x)\} = \int_{x_0}^{x_1} F(x, u(x), u'(x))\, dx, \tag{12.1.1}$$

over a given set of functions, $\mathcal{C}$, for example the set of all continuous functions on the interval $[x_0, x_1]$. In (12.1.1) the integrand $F(x, u(x), u'(x))$ is assumed to be a known function.

Let $w(x)$ be an arbitrary function, and let ζ be a real variable in the range $-\zeta_0 < \zeta < \zeta_0$, with ζ_0 a small parameter, such that

$$\|\zeta w(x)\| < h, \tag{12.1.2}$$

for some function norm[2] $\|\cdot\|$. Then a function $u(x)$ is said to (uniquely) minimize the functional (12.1.1) over the set $\mathcal{C}$ in a neighborhood of size h around $u(x)$, if

$$I\{u(x) + \zeta w(x)\} \geq I\{u(x)\}, \tag{12.1.3}$$

[1] As a notational convention, we will use braces around the arguments to functionals—that is functions of functions. And, as usual, we will use parenthesis around the arguments to regular functions of scalar, vector, or tensor variables.

[2] The choice of the precise function norm can affect the solution to the problem. For the problems we will treat, we do not have to worry about such issues; for further details, see Smith (1998) and Troutman (1983) for elementary presentations and Dacorogna (1989) for an advanced presentation.

Continuum Mechanics of Solids. Lallit Anand and Sanjay Govindjee, Oxford University Press (2020).

DOI: 10.1093/oso/9780198864721.001.0001

where the equality in (12.1.3) holds only when $\zeta w(x) = 0$ everywhere on the interval $[x_0, x_1]$. The function $\zeta w(x)$ is known as the **variation of** u. It is important to observe that $u(x)+\zeta w(x)$ must also be in the set C over which the minimization is being performed for (12.1.3) to make sense. Functions $w(x)$ that ensure this property are termed *admissible.*

From (12.1.3) we note that for small positive ζ we have

$$\frac{I\{u(x) + \zeta w(x)\} - I\{u(x)\}}{\zeta} \geq 0, \tag{12.1.4}$$

and for small negative ζ we have

$$\frac{I\{u(x) + \zeta w(x)\} - I\{u(x)\}}{\zeta} \leq 0. \tag{12.1.5}$$

Thus we may conclude that

$$\lim_{\zeta \to 0} \frac{I\{u(x) + \zeta w(x)\} - I\{u(x)\}}{\zeta} \equiv \frac{d}{d\zeta} I\{u(x) + \zeta w(x)\}\bigg|_{\zeta=0} = 0 \tag{12.1.6}$$

must hold for every such minimizer $u(x)$ (provided the limit exists). The quantity on the left-hand side of (12.1.6) is known as the *first variation* of I and is usually denoted as

$$\delta I\{u; w\} \stackrel{\text{def}}{=} \frac{d}{d\zeta} I\{u(x) + \zeta w(x)\}\bigg|_{\zeta=0}. \tag{12.1.7}$$

The condition

$$\delta I\{u; w\} = 0 \tag{12.1.8}$$

for all admissible w, is thus seen to be a *necessary* condition for $u(x)$ to be a minimizer of I.

REMARKS 12.1

1. Definition (12.1.7) for the first variation of I is seen to closely resemble the conventional directional derivative formula (2.1.1) for ordinary functions. In the context of (12.1.8), we then have the interpretation that at a minimizer of I, the value of I does not change for infinitesimally small perturbations in any direction w about u—the functional has "zero slope" at the minimizer.
2. Since (12.1.8) holds for all admissible w, it is often possible to eliminate w from the expression, leaving one with a necessary condition involving only the minimizer u. We discuss this point next.

12.1.1 Application of the necessary condition $\delta I = 0$

Computing the first variation of the functional (12.1.1) we obtain

$$\frac{d}{d\zeta}I\{u(x)+\zeta w(x)\} = \int_{x_0}^{x_1}\left[\frac{\partial F(x,u+\zeta w,u'+\zeta w')}{\partial(u+\zeta w)}\,w + \frac{\partial F(x,u+\zeta w,u'+\zeta w')}{\partial(u'+\zeta w')}\,w'\right]dx, \tag{12.1.9}$$

and hence

$$\delta I\{u;w\} = \frac{d}{d\zeta}I\{u(x)+\zeta w(x)\}\bigg|_{\zeta=0} = \int_{x_0}^{x_1}\left[\frac{\partial F(x,u,u')}{\partial u}\,w + \frac{\partial F(x,u,u')}{\partial u'}\,w'\right]dx. \tag{12.1.10}$$

Consider the second term in the integrand of (12.1.10), and integrate it by parts to obtain

$$\int_{x_0}^{x_1}\frac{\partial F}{\partial u'}\,\underbrace{w'\,dx}_{dw} = \frac{\partial F}{\partial u'}\,w\bigg|_{x_0}^{x_1} - \int_{x_0}^{x_1} w\frac{d}{dx}\left(\frac{\partial F}{\partial u'}\right)dx. \tag{12.1.11}$$

Using (12.1.11) in (12.1.10), we find that the necessary condition $\delta I = 0$, for u to minimize I, requires that

$$\frac{\partial F}{\partial u'}\,w\bigg|_{x_0}^{x_1} + \int_{x_0}^{x_1}\left[\frac{\partial F}{\partial u} - \frac{d}{dx}\left(\frac{\partial F}{\partial u'}\right)\right]w\,dx = 0 \tag{12.1.12}$$

for all admissible functions w. This is true if and only if u satisfies the differential equation

$$\boxed{\frac{\partial F}{\partial u} - \frac{d}{dx}\left(\frac{\partial F}{\partial u'}\right) = 0 \quad \forall x \in (x_0, x_1),} \tag{12.1.13}$$

and the boundary conditions

$$\boxed{\frac{\partial F}{\partial u'}\bigg|_{x=x_0} = 0 \quad \text{or} \quad w(x_0) = 0} \tag{12.1.14}$$

and

$$\boxed{\frac{\partial F}{\partial u'}\bigg|_{x=x_1} = 0 \quad \text{or} \quad w(x_1) = 0.} \tag{12.1.15}$$

Euler–Lagrange equation

The differential equation (12.1.13) is often referred to as the **Euler–Lagrange equation** for the functional I.

Essential and natural boundary conditions

Note that the boundary condition (12.1.14) may be satisfied in one of two ways. Suppose that the original minimization problem is performed over the set of functions $\mathcal{C}$ such that $u(x_0) = y_0$. Then, we must also require that

$$u(x_0) + \zeta w(x_0) = y_0,$$

or that we consider only those variations for which

$$\zeta w(x_0) = 0.$$

For such variations the boundary condition at x_0 is automatically satisfied; such a boundary condition is known as an **essential (or imposed) boundary condition**. Alternatively, if a boundary condition at x_0 is not imposed on $u(x)$, then the condition

$$\left.\frac{\partial F}{\partial u'}\right|_{x=x_0} = 0 \tag{12.1.16}$$

is a necessary condition for u to minimize I. This is known as a **natural boundary condition**. A similar situation exists for the boundary condition (12.1.15) at x_1.

REMARKS 12.2

1. The condition $\delta I = 0$ is only a *necessary* condition for minimizing the functional I, and as such, a function $u(x)$ that satisfies (12.1.13) to (12.1.15) is said to render the functional I *stationary*. In general, a function that satisfies $\delta I = 0$ need not minimize the functional I; it may, in fact, maximize the functional or correspond to a saddle point. Further, a solution $u(x)$ based on this condition may only represent a *local*, rather than a global minimum.
2. To mathematically determine whether such a function minimizes I requires, in general, the use of advanced methods in the calculus of variations; see Smith (1998, Chap. 8) or Dacorogna (1989, Chap. 3). However, in practice, when dealing with functionals derived from physical problems, minimization can be established if the functional can be shown to be convex. In our context, *convexity* is defined by the property

   $$I\{u+w\} - I\{u\} \geq \delta I\{u; w\}.$$

   Thus with convexity *and* $\delta I\{u; w\} = 0$ for all admissible w, one has the condition that u is a minimizer.

EXAMPLE 12.1 Minimization of a functional

Consider the functional

$$I\{u(x)\} = \int_0^1 \left(\frac{1}{2}u'(x)^2 + au(x)\right) dx, \tag{12.1.17}$$

where a is a fixed constant and $u(0) = u(1) = 0$. To find the function $u(x)$ that possibly minimizes (12.1.17), one can note that the integrand is expressible as $F(x, u(x), u'(x)) = au(x) + \frac{1}{2}u'(x)^2$; thus, $\partial F/\partial u = a$ and $\partial F/\partial u' = u'(x)$. By (12.1.13), the necessary condition for the minimizer is then

$$u''(x) = a,$$

subject to $u(0) = u(1) = 0$. Solving this ordinary differential equation and using the boundary conditions gives

$$u(x) = \frac{a}{2} x(x-1)$$

as a stationary function of (12.1.17) and a possible minimizer—which in this case it is, since $I\{u(x)\}$ is convex, viz. $I\{u+w\} - I\{u\} \geq \delta I\{u; w\}$.

12.1.2 **Treating δ as a mathematical operator. The process of taking "variations"**

In order to expedite the steps involved in deriving the Euler–Lagrange equation corresponding to a functional minimization problem, one usually exploits the fact that definition (12.1.7) allows one to treat δ as a type of variational operator.

Let $H\{u\}$ be any quantity that depends on u. Using definition (12.1.7), we have

$$\delta H\{u; w\} \stackrel{\text{def}}{=} \frac{d}{d\zeta} H\{u + \zeta w\}\bigg|_{\zeta=0}.$$

Consider now H to be the identity operator $H\{u\} = u$, in which case

$$\delta H \equiv \delta u \stackrel{\text{def}}{=} \frac{d}{d\zeta} H\{u + \zeta w\}\bigg|_{\zeta=0} = \frac{d}{d\zeta}(u + \zeta w)\bigg|_{\zeta=0} = w.$$

Since the variation of u, i.e. δu, is equal to w, in what follows it is convenient to use the alternate definition

$$\delta H\{u; \delta u\} \stackrel{\text{def}}{=} \frac{d}{d\zeta} H\{u + \zeta \delta u\}\bigg|_{\zeta=0}.$$

Next, if we consider H to be the first derivative operator $H\{u\} = u'$, then we see that

$$\delta H \equiv \delta(u') \stackrel{\text{def}}{=} \frac{d}{d\zeta} H\{u + \zeta \delta u\}\bigg|_{\zeta=0} = \frac{d}{d\zeta}(u' + \zeta(\delta u)')\bigg|_{\zeta=0} = (\delta u)';$$

hence the variational operator can be interchanged with the differential operator.

In this notational convention,

$$\delta F = \left(\frac{\partial F}{\partial u}\right) \delta u + \left(\frac{\partial F}{\partial u'}\right) \delta u' \qquad (12.1.18)$$

defines the first variation of $F(x, u, u')$.

The laws of variation of sums, products, ratios and powers, and so forth are completely analogous to the corresponding laws of differentiation. For example, if $F_1 = F_1(u)$ and $F_2 = F_2(u)$ are two functions of u, we have:

$$\begin{aligned}
\delta(F_1 \pm F_2) &= \delta F_1 \pm \delta F_2, \\
\delta(F_1 F_2) &= \delta F_1 F_2 + F_1 \delta F_2, \\
\delta\left(\frac{F_1}{F_2}\right) &= \frac{\delta F_1 F_2 - F_1 \delta F_2}{F_2^2}, \\
\delta\left(F_1^n\right) &= n\left(F_1^{(n-1)}\right)\delta F_1.
\end{aligned} \tag{12.1.19}$$

Also, if $G = G(u, v, w)$ is a function of several independent variables (and possibly their derivatives), the total variation is the sum of partial variations:

$$\delta G = \delta_u G + \delta_v G + \delta_w G, \tag{12.1.20}$$

where δ_u, δ_v, and δ_w denote the partial variations with respect to u, v, and w, respectively.

Next,

$$\begin{aligned}
\delta I = \delta \int_{x_0}^{x_1} F(x, u, u')\,dx &= \frac{d}{d\zeta}\left[\int_{x_0}^{x_1} F(x, u + \zeta\delta u, u' + \zeta\delta u')\,dx\right]_{\zeta=0} \\
&= \int_{x_0}^{x_1} \frac{d}{d\zeta} F(x, u + \zeta\delta u, u' + \zeta\delta u')\bigg|_{\zeta=0} dx \\
&= \int_{x_0}^{x_1} \delta F\,dx.
\end{aligned} \tag{12.1.21}$$

Hence, from (12.1.21)

$$\delta I = \delta \int_{x_0}^{x_1} F\,dx = \int_{x_0}^{x_1} \delta F\,dx; \tag{12.1.22}$$

that is, the variational operator δ can also be interchanged with the integral operator.

Finally, the necessary condition for a minimum of I may be written as

$$\delta I\{u; \delta u\} = 0 \qquad \text{for all admissible variations } \delta u. \tag{12.1.23}$$

The process of taking "variations"

Let us now repeat our previous derivation of the Euler–Lagrange equation and the boundary conditions using our new notational convention. Consider a functional

$$I\{u(x)\} = \int_{x_0}^{x_1} F(x, u, u')\,dx. \tag{12.1.24}$$

Using (12.1.22) and (12.1.18) the corresponding "variation" of the functional I is

$$\begin{aligned}
\delta I &\equiv \int_{x_0}^{x_1} \delta F(x, u, u')\,dx \\
&= \int_{x_0}^{x_1} \left(\frac{\partial F}{\partial u}\,\delta u + \frac{\partial F}{\partial u'}\,\delta u'\right) dx.
\end{aligned} \tag{12.1.25}$$

Integrating the second term in (12.1.25) by parts we get

$$\int_{x_0}^{x_1} \frac{\partial F}{\partial u'} \underbrace{\delta u'\,dx}_{d\delta u} = \frac{\partial F}{\partial u'}\,\delta u\bigg|_{x_0}^{x_1} - \int_{x_0}^{x_1} (\delta u)\,\frac{d}{dx}\left(\frac{\partial F}{\partial u'}\right) dx,$$

substitution of which back in (12.1.25) gives

$$\begin{aligned}\delta I &\equiv \int_{x_0}^{x_1} \delta F(x,u,u')\,dx \\ &= \int_{x_0}^{x_1} \left(\frac{\partial F}{\partial u}\,\delta u + \frac{\partial F}{\partial u'}\,\delta u'\right)\,dx, \\ &= \frac{\partial F}{\partial u'}\,\delta u\bigg|_{x_0}^{x_1} + \int_{x_0}^{x_1}\left[\frac{\partial F}{\partial u} - \frac{d}{dx}\left(\frac{\partial F}{\partial u'}\right)\right]\,\delta u\,dx.\end{aligned} \tag{12.1.26}$$

From here on we can proceed as before, setting $\delta I = 0$, and using the arbitrariness of an admissible variation $\delta u(x)$ we can derive the Euler–Lagrange equation, and determine the essential and natural boundary conditions exactly as we did before in Sec. 12.1.1.

EXAMPLE 12.2 Taking a variation

Consider the functional

$$I\{u(x)\} = \int_0^1 \left(u(x)^2 + u(x)^3\right)\,dx$$

and find the possible minimizers over the space of continuous functions on $[0,1]$. The first variation is given by

$$\delta I\{u;\delta u\} = \int_0^1 \left(2u(x)\delta u(x) + 3u(x)^2\delta u(x)\right)\,dx = \int_0^1 \left[2u(x) + 3u(x)^2\right]\delta u(x)\,dx. \tag{12.1.27}$$

In order for $\delta I\{u;\delta u\} = 0$ for all δu, one must have

$$u(x)\left(2 + 3u(x)\right) = 0.$$

Thus there are two possibilities $u(x) = 0$ or $u(x) = -\frac{2}{3}$.

EXAMPLE 12.3 Variations in three dimensions

The notion of a variation may be generalized to three dimensions as follows. Consider the functional

$$\Psi\{\rho(\mathbf{x})\} = \int_{\mathcal{B}} \psi(\rho(\mathbf{x}), \nabla\rho(\mathbf{x}))\,dv, \tag{12.1.28}$$

defined on a region $\mathcal{B} \subset \mathbb{R}^3$. The variation of this functional for fields $\rho(\mathbf{x})$ which, for example, are constant on the boundary $\partial\mathcal{B}$ is

$$\delta\Psi\{\rho;\delta\rho\} = \int_{\mathcal{B}} \left[\frac{\partial\psi}{\partial\rho} - \operatorname{div}\left(\frac{\partial\psi}{\partial\nabla\rho}\right)\right]\,\delta\rho\,dv. \tag{12.1.29}$$

For Ψ to be stationary at $\rho(\mathbf{x})$, viz. $\delta\Psi\{\rho;\delta\rho\}=0$ for all admissible $\delta\rho$, $\rho(\mathbf{x})$ must satisfy the Euler–Lagrange equation

$$\frac{\partial\psi}{\partial\rho} - \text{div}\left(\frac{\partial\psi}{\partial\nabla\rho}\right) = 0. \tag{12.1.30}$$

12.2 Principle of minimum potential energy

Here we introduce a stationary energy principle in which the displacement field $\mathbf{u}$ is taken to be the only fundamental unknown field.

Recall that in linear elastostatics one considers that the displacement is prescribed on a portion $\mathcal{S}_1$ of the boundary $\partial\mathcal{B}$, while on another portion $\mathcal{S}_2$ traction is prescribed:

$$\begin{aligned} \mathbf{u} &= \hat{\mathbf{u}} \quad \text{on } \mathcal{S}_1, \\ \boldsymbol{\sigma}\mathbf{n} &= \hat{\mathbf{t}} \quad \text{on } \mathcal{S}_2, \end{aligned} \tag{12.2.1}$$

with $\hat{\mathbf{u}}$ and $\hat{\mathbf{t}}$ *prescribed* functions of $\mathbf{x}$, and the body force field $\mathbf{b}$ is given on $\mathcal{B}$.

We now show that a solution of the mixed problem of linear elastostatics, if it exists, can be characterized as the minimum value of a certain functional of $\mathbf{u}$. Before stating the principle of stationary potential energy, we consider some definitions. First, by a **kinematically admissible displacement field** we mean an arbitrary field $\mathbf{u}$ on $\mathcal{B}$ that satisfies the *displacement boundary condition* $(12.2.1)_1$,

$$\mathbf{u} = \hat{\mathbf{u}} \quad \text{on } \mathcal{S}_1, \tag{12.2.2}$$

and that the strain field $\boldsymbol{\epsilon}$ is compatible with the displacement field; that is, the strain-displacement relation

$$\boldsymbol{\epsilon} = \tfrac{1}{2}(\nabla\mathbf{u} + \nabla\mathbf{u}^{\top}) \tag{12.2.3}$$

holds everywhere on $\mathcal{B}$. Next, let $\mathcal{F}$ denote the net free energy of the body

$$\mathcal{F}(\boldsymbol{\epsilon}) = \tfrac{1}{2}\int_{\mathcal{B}} \boldsymbol{\epsilon} : \mathbb{C}\boldsymbol{\epsilon}\, dv \tag{12.2.4}$$

corresponding to a strain field $\boldsymbol{\epsilon}$. Finally, we restrict our attention to circumstances for which the loads $(\hat{\mathbf{t}}, \mathbf{b})$ do not depend on the deformation of the body; such loads are referred to as **dead loads**.[3] Then, the *potential energy* functional Π is defined on the set of kinematically admissible displacement fields $\mathbf{u}$ by

$$\boxed{\Pi\{\mathbf{u}\} = \mathcal{F}(\boldsymbol{\epsilon}) - \int_{\mathcal{S}_2} \hat{\mathbf{t}}\cdot\mathbf{u}\, da - \int_{\mathcal{B}} \mathbf{b}\cdot\mathbf{u}\, dv.} \tag{12.2.5}$$

[3] In contrast, a **follower load** depends on the motion of the body. For example, a pressure load that always remains normal to a surface, even as it deforms.

The principle of minimum potential energy may now be stated as follows.

PRINCIPLE OF MINIMUM POTENTIAL ENERGY *Let* $\mathbf{u}$ *denote a solution of the mixed problem of elastostatics. Then*

$$\Pi\{\mathbf{u}\} \leq \Pi\{\tilde{\mathbf{u}}\} \tag{12.2.6}$$

for every kinematically admissible displacement field $\tilde{\mathbf{u}}$*, and equality holds only if* $\mathbf{u}$ *and* $\tilde{\mathbf{u}}$ *differ by a rigid displacement field.*

In words, the principle of minimum potential energy asserts that the difference between the free energy and the work done by the prescribed surface tractions and the body forces assumes a smaller value for the solution of the mixed problem than for any other kinematically admissible state.

To establish this principle choose a kinematically admissible displacement field $\tilde{\mathbf{u}}$ and let

$$\bar{\mathbf{w}} = \tilde{\mathbf{u}} - \mathbf{u},$$

$$\bar{\epsilon} = \tilde{\epsilon} - \epsilon.$$

Then, since $\mathbf{u}$ is a solution and $\tilde{\mathbf{u}}$ is kinematically admissible

$$\begin{aligned} \bar{\mathbf{w}} &\overset{\text{def}}{=} \mathbf{0} \quad \text{on } \mathcal{S}_1, \\ \bar{\epsilon} &\overset{\text{def}}{=} \tfrac{1}{2}\left(\nabla\bar{\mathbf{w}} + (\nabla\bar{\mathbf{w}})^{\top}\right). \end{aligned} \tag{12.2.7}$$

Further, since $\mathbb{C}$ is symmetric and $\sigma = \mathbb{C}\epsilon$,

$$\tilde{\epsilon} : \mathbb{C}\tilde{\epsilon} = (\epsilon + \bar{\epsilon}) : \mathbb{C}(\epsilon + \bar{\epsilon}) \tag{12.2.8}$$

$$= \epsilon : \mathbb{C}\epsilon + \bar{\epsilon} : \mathbb{C}\bar{\epsilon} + \epsilon : \mathbb{C}\bar{\epsilon} + \bar{\epsilon} : \mathbb{C}\epsilon$$

$$= \epsilon : \mathbb{C}\epsilon + \bar{\epsilon} : \mathbb{C}\bar{\epsilon} + 2\sigma : \bar{\epsilon}; \tag{12.2.9}$$

hence, by (12.2.4) and (12.2.9),

$$\begin{aligned} \mathcal{F}(\tilde{\epsilon}) &= \tfrac{1}{2}\int_{\mathcal{B}} \tilde{\epsilon} : \mathbb{C}\tilde{\epsilon}\, dv, \\ &= \tfrac{1}{2}\int_{\mathcal{B}} \epsilon : \mathbb{C}\epsilon\, dv + \tfrac{1}{2}\int_{\mathcal{B}} \bar{\epsilon} : \mathbb{C}\bar{\epsilon}\, dv + \int_{\mathcal{B}} \sigma : \bar{\epsilon}\, dv, \\ &= \mathcal{F}(\epsilon) + \mathcal{F}(\bar{\epsilon}) + \int_{\mathcal{B}} \sigma : \bar{\epsilon}\, dv, \end{aligned}$$

or

$$\mathcal{F}(\tilde{\epsilon}) - \mathcal{F}(\epsilon) = \mathcal{F}(\bar{\epsilon}) + \int_{\mathcal{B}} \sigma : \bar{\epsilon}\, dv. \tag{12.2.10}$$

On the other hand,

$$
\begin{aligned}
\int_{\mathcal{B}} \boldsymbol{\sigma} : \bar{\boldsymbol{\epsilon}}\, dv &= \int_{\mathcal{B}} \boldsymbol{\sigma} : \nabla \bar{\mathbf{w}}\, dv && \text{(using } (12.2.7)_2 \text{ and the symmetry of } \boldsymbol{\sigma}\text{)}\\
&= \int_{\partial\mathcal{B}} \boldsymbol{\sigma}\mathbf{n} \cdot \bar{\mathbf{w}}\, da - \int_{\mathcal{B}} \operatorname{div} \boldsymbol{\sigma} \cdot \bar{\mathbf{w}}\, dv && \text{(using the divergence theorem)}\\
&= \int_{\partial\mathcal{B}} \boldsymbol{\sigma}\mathbf{n} \cdot \bar{\mathbf{w}}\, da + \int_{\mathcal{B}} \mathbf{b} \cdot \bar{\mathbf{w}}\, dv && \text{(using } \operatorname{div}\boldsymbol{\sigma} = -\mathbf{b}\text{)}\\
&= \int_{\mathcal{S}_2} \hat{\mathbf{t}} \cdot \bar{\mathbf{w}}\, da + \int_{\mathcal{B}} \mathbf{b} \cdot \bar{\mathbf{w}}\, dv. && \text{(using } (12.2.1)_2 \text{ and } (12.2.7)_1\text{)}
\end{aligned}
\tag{12.2.11}
$$

In view of (12.2.5) and (12.2.11),

$$
\begin{aligned}
\Pi(\tilde{\mathbf{u}}) - \Pi(\mathbf{u}) &= \left[\mathcal{F}(\tilde{\boldsymbol{\epsilon}}) - \int_{\mathcal{S}_2} \hat{\mathbf{t}} \cdot \tilde{\mathbf{u}}\, da - \int_{\mathcal{B}} \mathbf{b} \cdot \tilde{\mathbf{u}}\, dv \right] - \left[\mathcal{F}(\boldsymbol{\epsilon}) - \int_{\mathcal{S}_2} \hat{\mathbf{t}} \cdot \mathbf{u}\, da - \int_{\mathcal{B}} \mathbf{b} \cdot \mathbf{u}\, dv \right],\\
&= \mathcal{F}(\tilde{\boldsymbol{\epsilon}}) - \mathcal{F}(\boldsymbol{\epsilon}) - \int_{\mathcal{S}_2} \hat{\mathbf{t}} \cdot \bar{\mathbf{w}}\, da - \int_{\mathcal{B}} \mathbf{b} \cdot \bar{\mathbf{w}}\, dv,\\
&= \mathcal{F}(\tilde{\boldsymbol{\epsilon}}) - \mathcal{F}(\boldsymbol{\epsilon}) - \int_{\mathcal{B}} \boldsymbol{\sigma} : \bar{\boldsymbol{\epsilon}}\, dv,
\end{aligned}
$$

and using (12.2.10),

$$
\Pi(\tilde{\mathbf{u}}) - \Pi(\mathbf{u}) = \mathcal{F}(\bar{\boldsymbol{\epsilon}}).
$$

Thus, since $\mathbb{C}$ is positive definite, $\mathcal{F}(\bar{\boldsymbol{\epsilon}}) \geq 0$, and hence

$$
\Pi(\mathbf{u}) \leq \Pi(\tilde{\mathbf{u}}),
$$

and

$$
\Pi(\mathbf{u}) = \Pi(\tilde{\mathbf{u}})
$$

only when $\bar{\boldsymbol{\epsilon}} = \mathbf{0}$, that is only when $\bar{\mathbf{w}} = \tilde{\mathbf{u}} - \mathbf{u}$ is a *rigid displacement*. This completes the proof of the principle of minimum potential energy.

REMARK 12.3

The principle itself provides an alternate means of solving mixed boundary-value problems in elastostatics. In particular, instead of solving the governing Navier form of the equilibrium equation, one can search for functions $\mathbf{u}$ such that $\delta\Pi\{\mathbf{u}; \delta\mathbf{u}\} = 0$ for all admissible variations $\delta\mathbf{u}$. It is also important to note that in linear elastostatics the functional Π is convex, and thus satisfying the necessary condition is also sufficient to provide the minimizer. The validity of these assertions is examined in further detail in the next section.

12.3 Principle of minimum potential energy and the weak form of the displacement problem of elastostatics

The potential energy functional for the actual displacement field (since it is kinematically admissible) is

$$\Pi\{\mathbf{u}\} = \mathcal{F}(\boldsymbol{\epsilon}) - \int_{\mathcal{S}_2} \hat{\mathbf{t}} \cdot \mathbf{u}\, da - \int_{\mathcal{B}} \mathbf{b} \cdot \mathbf{u}\, dv. \tag{12.3.1}$$

Taking its variation we obtain

$$\delta\Pi = \delta\mathcal{F}(\boldsymbol{\epsilon}) - \int_{\mathcal{S}_2} \hat{\mathbf{t}} \cdot \delta\mathbf{u}\, da - \int_{\mathcal{B}} \mathbf{b} \cdot \delta\mathbf{u}\, dv. \tag{12.3.2}$$

Next, using the symmetry of $\mathbb{C}$,

$$\delta\mathcal{F}(\boldsymbol{\epsilon}) = \frac{1}{2}\int_{\mathcal{B}} \Big(\delta\boldsymbol{\epsilon} : \mathbb{C}\boldsymbol{\epsilon} + \boldsymbol{\epsilon} : \mathbb{C}\delta\boldsymbol{\epsilon}\Big)\, dv = \int_{\mathcal{B}} \delta\boldsymbol{\epsilon} : \mathbb{C}\boldsymbol{\epsilon}\, dv, \tag{12.3.3}$$

where

$$\delta\boldsymbol{\epsilon} = \tfrac{1}{2}(\nabla\delta\mathbf{u} + (\nabla\delta\mathbf{u})^{\top}). \tag{12.3.4}$$

Thus,

$$\delta\mathcal{F}(\boldsymbol{\epsilon}) = \int_{\mathcal{B}} (\mathrm{sym}\,\nabla\delta\mathbf{u}) : \mathbb{C}(\mathrm{sym}\,\nabla\mathbf{u})\, dv, \tag{12.3.5}$$

use of which in (12.3.2) gives

$$\delta\Pi = \int_{\mathcal{B}} (\mathrm{sym}\,\nabla\delta\mathbf{u}) : \mathbb{C}(\mathrm{sym}\,\nabla\mathbf{u})\, dv - \int_{\mathcal{S}_2} \hat{\mathbf{t}} \cdot \delta\mathbf{u}\, da - \int_{\mathcal{B}} \mathbf{b} \cdot \delta\mathbf{u}\, dv. \tag{12.3.6}$$

Therefore, the necessary condition for a minimizer $\delta\Pi = 0$ of the functional Π gives

$$\int_{\mathcal{B}} (\mathrm{sym}\,\nabla\delta\mathbf{u}) : \mathbb{C}(\mathrm{sym}\,\nabla\mathbf{u})\, dv - \int_{\mathcal{S}_2} \hat{\mathbf{t}} \cdot \delta\mathbf{u}\, da - \int_{\mathcal{B}} \mathbf{b} \cdot \delta\mathbf{u}\, dv = 0, \tag{12.3.7}$$

which is identical to the weak form (10.2.8) *of the mixed problem of elastostatics, provided the variation* $\delta\mathbf{u}$ *is replaced by the virtual velocity (or weight function)* $\mathbf{w}$.

Next, bearing the symmetry properties of the moduli $\mathbb{C}$ in mind and using the divergence theorem in the form

$$\int_{\mathcal{B}} (\mathbb{C}(\nabla\mathbf{u})) : \nabla\delta\mathbf{u}\, dv = \int_{\partial\mathcal{B}} (\mathbb{C}(\nabla\mathbf{u}))\mathbf{n} \cdot \delta\mathbf{u}\, da - \int_{\mathcal{B}} \mathrm{div}(\mathbb{C}(\nabla\mathbf{u})) \cdot \delta\mathbf{u}\, dv \tag{12.3.8}$$

in (12.3.7), and using the fact that $\delta\mathbf{u} = \mathbf{0}$ on $\mathcal{S}_1$ gives

$$\int_{\mathcal{S}_2} (\mathbb{C}(\nabla\mathbf{u})\mathbf{n} - \hat{\mathbf{t}}) \cdot \delta\mathbf{u}\, da - \int_{\mathcal{B}} (\mathrm{div}\,\mathbb{C}(\nabla\mathbf{u}) + \mathbf{b}) \cdot \delta\mathbf{u}\, dv = 0. \tag{12.3.9}$$

Finally since $\delta\mathbf{u}$ is arbitrary, we have

$$\begin{aligned}\mathbb{C}(\nabla\mathbf{u})\mathbf{n} &= \hat{\mathbf{t}} \quad \text{on} \quad \mathcal{S}_2,\\ \operatorname{div}\mathbb{C}(\nabla\mathbf{u}) + \mathbf{b} &= \mathbf{0} \quad \text{in} \quad \mathcal{B},\end{aligned} \tag{12.3.10}$$

and the solution to the necessary conditions $\delta\Pi = 0$ also implies the strong form (12.3.10).

REMARK 12.4

The principle of minimum potential energy makes it possible to concentrate in a single functional Π all the intrinsic features of the problem of elastostatics. Further, it provides a natural means for seeking approximate solutions.

12.4 **Principle of minimum complementary energy**

Recall that the elasticity tensor $\mathbb{C}$ is symmetric, that is,

$$\mathbf{A} : \mathbb{C}\mathbf{B} = \mathbf{B} : \mathbb{C}\mathbf{A} \tag{12.4.1}$$

for every pair of symmetric tensors $\mathbf{A}$ and $\mathbf{B}$, and it is positive definite, that is,

$$\mathbf{A} : \mathbb{C}\mathbf{A} > 0 \tag{12.4.2}$$

for every non-zero symmetric tensor $\mathbf{A}$. Thus, $\mathbb{C}$ restricted to operate on symmetric tensors is invertible. Let $\mathbb{S}$ denote the corresponding inverse, so that the stress-strain relation $\boldsymbol{\sigma} = \mathbb{C}\boldsymbol{\epsilon}$ may be inverted to give

$$\boldsymbol{\epsilon} = \mathbb{S}\boldsymbol{\sigma}, \tag{12.4.3}$$

with $\mathbb{S}$ the fourth-order *compliance tensor.*

Next, recall that the free energy per unit volume for a linear elastic material is given by

$$\psi = \tfrac{1}{2}\boldsymbol{\epsilon} : \mathbb{C}\boldsymbol{\epsilon}. \tag{12.4.4}$$

Using the inverse relation (12.4.3), the free energy ψ may be expressed as a function of stress as follows:

$$\begin{aligned}\psi &= \tfrac{1}{2}\boldsymbol{\epsilon} : \mathbb{C}\boldsymbol{\epsilon},\\ &= \tfrac{1}{2}(\mathbb{S}\boldsymbol{\sigma}) : \mathbb{C}(\mathbb{S}\boldsymbol{\sigma}),\\ &= \tfrac{1}{2}(\mathbb{S}\boldsymbol{\sigma}) : \underbrace{(\mathbb{C}\mathbb{S})}_{\equiv\mathbb{I}}\boldsymbol{\sigma},\end{aligned}$$

where $\mathbb{I}$ is the fourth-order identity tensor which maps symmetric tensors into themselves. Thus, in terms of the stress and the compliance tensor, the free energy is given by

$$\psi = \tfrac{1}{2}\boldsymbol{\sigma} : \mathbb{S}\boldsymbol{\sigma}, \tag{12.4.5}$$

from which it is also clear that

$$\mathbb{S} = \frac{\partial^2 \psi}{\partial \boldsymbol{\sigma} \otimes \partial \boldsymbol{\sigma}}, \tag{12.4.6}$$

which implies that $\mathbb{S}$ is symmetric, and since on physical grounds the free energy is expected to be positive for all non-zero stress, (12.4.5) implies that $\mathbb{S}$ is also positive definite.

Using (12.4.5), let

$$\mathcal{F}(\boldsymbol{\sigma}) = \tfrac{1}{2} \int_{\mathcal{B}} \boldsymbol{\sigma} : \mathbb{S}\boldsymbol{\sigma} \, dv \tag{12.4.7}$$

denote the net free energy of the body, when expressed in terms of the compliance tensor $\mathbb{S}$ and the stress $\boldsymbol{\sigma}$. Further, we call *any* smooth symmetric tensor field $\boldsymbol{\sigma}$ that satisfies the equilibrium equation

$$\operatorname{div} \boldsymbol{\sigma} + \mathbf{b} = \mathbf{0} \tag{12.4.8}$$

in $\mathcal{B}$, and the boundary condition

$$\boldsymbol{\sigma}\mathbf{n} = \hat{\mathbf{t}} \qquad \text{on} \quad \mathcal{S}_2, \tag{12.4.9}$$

a **statically admissible stress field.**

Next, we define the *complementary potential energy* functional Π^c over the set of statically admissible stress fields by

$$\boxed{\Pi^c\{\boldsymbol{\sigma}\} = \mathcal{F}(\boldsymbol{\sigma}) - \int_{\mathcal{S}_1} \boldsymbol{\sigma}\mathbf{n} \cdot \hat{\mathbf{u}} \, da.} \tag{12.4.10}$$

Using the definitions above, we state the *principle of minimum complementary energy* below.

PRINCIPLE OF MINIMUM COMPLEMENTARY ENERGY *Let $\boldsymbol{\sigma}$ denote the stress field in the solution of a mixed problem of elastostatics. Then*

$$\Pi^c\{\boldsymbol{\sigma}\} \le \Pi^c\{\tilde{\boldsymbol{\sigma}}\} \tag{12.4.11}$$

for every statically admissible stress field $\tilde{\boldsymbol{\sigma}}$, and equality holds only if $\boldsymbol{\sigma} = \tilde{\boldsymbol{\sigma}}$.

To establish this principle choose a statically admissible stress field $\tilde{\boldsymbol{\sigma}}$ and let

$$\bar{\boldsymbol{\sigma}} = \tilde{\boldsymbol{\sigma}} - \boldsymbol{\sigma}.$$

Then, since $\boldsymbol{\sigma}$ is a solution and $\tilde{\boldsymbol{\sigma}}$ is statically admissible, $\bar{\boldsymbol{\sigma}}$ satisfies

$$\bar{\boldsymbol{\sigma}}\mathbf{n} = \mathbf{0} \quad \text{on} \quad \mathcal{S}_2, \tag{12.4.12}$$

and also satisfies the equilibrium equation with zero body forces,

$$\operatorname{div} \bar{\boldsymbol{\sigma}} = \mathbf{0} \quad \text{in} \quad \mathcal{B}. \tag{12.4.13}$$

Further, since $\mathbb{S}$ is symmetric and $\boldsymbol{\epsilon} = \mathbb{S}\boldsymbol{\sigma}$,

$$\begin{aligned}
\tilde{\boldsymbol{\sigma}} : \mathbb{S}\tilde{\boldsymbol{\sigma}} &= (\boldsymbol{\sigma} + \bar{\boldsymbol{\sigma}}) : \mathbb{S}(\boldsymbol{\sigma} + \bar{\boldsymbol{\sigma}}) \\
&= \boldsymbol{\sigma} : \mathbb{S}\boldsymbol{\sigma} + \bar{\boldsymbol{\sigma}} : \mathbb{S}\bar{\boldsymbol{\sigma}} + \boldsymbol{\sigma} : \mathbb{S}\bar{\boldsymbol{\sigma}} + \bar{\boldsymbol{\sigma}} : \mathbb{S}\boldsymbol{\sigma} \\
&= \boldsymbol{\sigma} : \mathbb{S}\boldsymbol{\sigma} + \bar{\boldsymbol{\sigma}} : \mathbb{S}\bar{\boldsymbol{\sigma}} + 2\bar{\boldsymbol{\sigma}} : \boldsymbol{\epsilon};
\end{aligned} \tag{12.4.14}$$

hence, by (12.4.7) and (12.4.14),

$$\begin{aligned}\mathcal{F}(\tilde{\boldsymbol{\sigma}}) &= \tfrac{1}{2}\int_{\mathcal{B}} \tilde{\boldsymbol{\sigma}} : \mathbb{S}\tilde{\boldsymbol{\sigma}}\, dv,\\ &= \tfrac{1}{2}\int_{\mathcal{B}} \boldsymbol{\sigma} : \mathbb{S}\boldsymbol{\sigma}\, dv + \tfrac{1}{2}\int_{\mathcal{B}} \bar{\boldsymbol{\sigma}} : \mathbb{S}\bar{\boldsymbol{\sigma}}\, dv + \int_{\mathcal{B}} \bar{\boldsymbol{\sigma}} : \boldsymbol{\epsilon}\, dv,\\ &= \mathcal{F}(\boldsymbol{\sigma}) + \mathcal{F}(\bar{\boldsymbol{\sigma}}) + \int_{\mathcal{B}} \bar{\boldsymbol{\sigma}} : \boldsymbol{\epsilon}\, dv,\end{aligned}$$

or

$$\mathcal{F}(\tilde{\boldsymbol{\sigma}}) - \mathcal{F}(\boldsymbol{\sigma}) = \mathcal{F}(\bar{\boldsymbol{\sigma}}) + \int_{\mathcal{B}} \bar{\boldsymbol{\sigma}} : \boldsymbol{\epsilon}\, dv. \tag{12.4.15}$$

On the other hand, by the symmetry of $\bar{\boldsymbol{\sigma}}$ and the divergence theorem,

$$\begin{aligned}\int_{\mathcal{B}} \bar{\boldsymbol{\sigma}} : \boldsymbol{\epsilon}\, dv &= \int_{\mathcal{B}} \bar{\boldsymbol{\sigma}} : \nabla\mathbf{u}\, dv\\ &= \int_{\partial\mathcal{B}} \bar{\boldsymbol{\sigma}}\mathbf{n}\cdot\mathbf{u}\, da - \int_{\mathcal{B}} \operatorname{div}\bar{\boldsymbol{\sigma}}\cdot\mathbf{u}\, dv && \text{(using the divergence theorem)}\\ &= \int_{\partial\mathcal{B}} \bar{\boldsymbol{\sigma}}\mathbf{n}\cdot\mathbf{u}\, da && \text{(using (12.4.13))}\\ &= \int_{\mathcal{S}_1} \bar{\boldsymbol{\sigma}}\mathbf{n}\cdot\hat{\mathbf{u}}\, da. && \text{(using (12.4.12))}\end{aligned} \tag{12.4.16}$$

In view of (12.4.10) and (12.4.16),

$$\begin{aligned}\Pi^c\{\tilde{\boldsymbol{\sigma}}\} - \Pi^c\{\boldsymbol{\sigma}\} &= \left[\mathcal{F}(\tilde{\boldsymbol{\sigma}}) - \int_{\mathcal{S}_1} \tilde{\boldsymbol{\sigma}}\mathbf{n}\cdot\hat{\mathbf{u}}\, da\right] - \left[\mathcal{F}(\boldsymbol{\sigma}) - \int_{\mathcal{S}_1} \boldsymbol{\sigma}\mathbf{n}\cdot\hat{\mathbf{u}}\, da\right],\\ &= \mathcal{F}(\tilde{\boldsymbol{\sigma}}) - \mathcal{F}(\boldsymbol{\sigma}) - \int_{\mathcal{S}_1} \bar{\boldsymbol{\sigma}}\mathbf{n}\cdot\hat{\mathbf{u}}\, da,\\ &= \mathcal{F}(\tilde{\boldsymbol{\sigma}}) - \mathcal{F}(\boldsymbol{\sigma}) - \int_{\mathcal{B}} \bar{\boldsymbol{\sigma}} : \boldsymbol{\epsilon}\, dv,\end{aligned}$$

and using (12.4.15)

$$\Pi^c\{\tilde{\boldsymbol{\sigma}}\} - \Pi^c\{\boldsymbol{\sigma}\} = \mathcal{F}(\bar{\boldsymbol{\sigma}}). \tag{12.4.17}$$

Thus, since $\mathbb{S}$ is positive definite, $\mathcal{F}(\bar{\boldsymbol{\sigma}}) \geq 0$,

$$\Pi^c\{\boldsymbol{\sigma}\} \leq \Pi^c\{\tilde{\boldsymbol{\sigma}}\}, \tag{12.4.18}$$

and

$$\Pi^c\{\boldsymbol{\sigma}\} = \Pi^c\{\tilde{\boldsymbol{\sigma}}\} \tag{12.4.19}$$

only when $\bar{\boldsymbol{\sigma}} = \mathbf{0}$, that is only when $\tilde{\boldsymbol{\sigma}} = \boldsymbol{\sigma}$. This completes the proof of the principle of minimum complementary energy.

REMARK 12.5

As with the principle of minimum potential energy, the principle of minimum complementary energy can be used to solve aspects of boundary-value problems by determining the stress field that satisfies the stationary conditions $\delta\Pi^c\{\boldsymbol{\sigma};\delta\boldsymbol{\sigma}\} = 0$. Since the functional Π^c is convex, such a solution furnishes the true minimizing solution.

EXAMPLE 12.4 Torsion of a thin rectangular bar

The complementary energy for a torsion bar with a thin rectangular cross-section,

$$\Omega = [-\frac{b}{2}, \frac{b}{2}] \times [-\frac{c}{2}, \frac{c}{2}] \qquad \text{with} \qquad b \gg c,$$

and length L is given by

$$\Pi^c = L \int_\Omega \frac{1}{2\mu}\left(\sigma_{13}{}^2 + \sigma_{23}{}^2\right) da - \int_{S_L} \boldsymbol{\sigma}\mathbf{n} \cdot \hat{\mathbf{u}}\, da\,; \tag{12.4.20}$$

cf. Sec. 9.3.3. From (9.3.35), the displacement at $x_3 = L$ is given by

$$\mathbf{u}(\mathbf{x}) = \alpha L(-x_2\mathbf{e}_1 + x_1\mathbf{e}_2) + \alpha\varphi\mathbf{e}_3.$$

As shown in (9.3.63), this allows one to express the last integral in (12.4.20) as

$$2\alpha L \int_\Omega \Psi da, \tag{12.4.21}$$

where Ψ is the Prandtl stress function. Using (9.3.50) we can also express the first integral in (12.4.20) in terms of the Prandtl stress function:

$$L \int_\Omega \frac{1}{2\mu}\left(\Psi_{,1}{}^2 + \Psi_{,2}{}^2\right) da. \tag{12.4.22}$$

Inserting (12.4.21) and (12.4.22) in (12.4.20), we come to an expression for the complementary energy solely in terms of Ψ:

$$\Pi^c\{\Psi\} = \int_\Omega \frac{1}{2}\left(\Psi_{,1}{}^2 + \Psi_{,2}{}^2\right) - \int_\Omega 2\alpha\mu\Psi da, \tag{12.4.23}$$

where without loss of generality we have scaled Π^c by μ/L. The complementary energy (12.4.23) needs to be minimized with respect to Ψ over the set of functions $\mathcal{C} = \{\Psi \mid \Psi = 0 \text{ on } \partial\Omega\}$, cf. (9.3.55). Note that by introducing the Prandtl stress function the problem is automatically formulated in terms of statically admissible stress fields.

To complete the solution to the problem we now approximate the minimization problem by selecting a specific form for Ψ,

$$\Psi \approx A\left(x_2 - \frac{c}{2}\right)\left(x_2 + \frac{c}{2}\right), \tag{12.4.24}$$

in terms of one unknown scalar parameter A. In selecting this form we have used the *membrane analogy* (cf. Example 9.4) and have taken the liberty of ignoring the boundary conditions on the short edges $x_1 = \pm\frac{b}{2}$. This then gives

$$\begin{aligned}\Pi^c(A) &= \int_\Omega \frac{1}{2}\left(0 + 4A^2{x_2}^2\right) da - 2\alpha\mu A \int_\Omega \left(x_2 - \frac{c}{2}\right)\left(x_2 + \frac{c}{2}\right) da \\ &\Downarrow \\ \frac{d\Pi^c}{dA} &= \int_\Omega 4A{x_2}^2 da - 2\alpha\mu \int_\Omega \left(x_2 - \frac{c}{2}\right)\left(x_2 + \frac{c}{2}\right) da \\ &= \frac{1}{3}c^3 bA + \frac{1}{3}\alpha\mu c^3 b \end{aligned} \tag{12.4.25}$$

Using (12.4.25) in the stationarity condition $d\Pi^c/dA = 0$, determines the unknown parameter A as

$$A = -\alpha\mu, \tag{12.4.26}$$

so that

$$\Psi \approx -\alpha\mu\left(x_2 - \frac{c}{2}\right)\left(x_2 + \frac{c}{2}\right). \tag{12.4.27}$$

This gives the following approximate solution for the stresses,

$$\sigma_{13} \approx -2\alpha\mu x_2 \tag{12.4.28}$$

$$\sigma_{23} \approx 0, \tag{12.4.29}$$

and the effective polar moment of inertia is given by,

$$\bar{J} = \frac{T}{\alpha\mu} = \frac{2}{\alpha\mu}\int_\Omega \Psi \approx \frac{1}{3}c^3 b. \tag{12.4.30}$$

Results (12.4.28)–(12.4.30) are quite reasonable for $b/c > 10$.

12.5 **Application of the principles of minimum potential energy and complementary energy to structural problems**

The principles of minimum potential energy and minimum complementary energy can be applied directly to complex problems, or they can be applied to derive specialized principles which are particularly well suited to solving structural problems. We develop some of these results in what follows.

First, consider a concentrated force $\mathbf{P} = P\,\mathbf{e}$, with magnitude P acting in the direction of the unit vector $\mathbf{e}$, applied at the point $\mathbf{x}_a$ on the boundary $\partial\mathcal{B}$ of a body $\mathcal{B}$. In terms of a traction distribution this can be modeled as $\mathbf{t} = P\delta(\mathbf{x} - \mathbf{x}_a)\mathbf{e}$, where $\delta(\mathbf{x} - \mathbf{x}_a)$ is the (surface) Dirac delta function centered at the point of application of the force, $\mathbf{x}_a$. In this case,

$$\int_{\partial\mathcal{B}} \mathbf{t}\cdot\mathbf{u}da = \int_{\partial\mathcal{B}} P\delta(\mathbf{x} - \mathbf{x}_a)\mathbf{e}\cdot\mathbf{u}da = P\Delta, \tag{12.5.1}$$

where $\Delta = \mathbf{u}(\mathbf{x}_a)\cdot\mathbf{e}$ is the component of the displacement vector $\mathbf{u}$ at the point of application of the concentrated load $\mathbf{P}$ in the direction $\mathbf{e}$. Now consider N such instances of concentrated forces located at points $\mathbf{x}^{(k)}$ in directions $\mathbf{e}^{(k)}$, then

$$\int_{\partial\mathcal{B}} \mathbf{t}\cdot\mathbf{u}da = \sum_{k=1}^{N} P^{(k)}\Delta^{(k)}. \tag{12.5.2}$$

Indeed, we can consider $P^{(k)}$ and $\Delta^{(k)}$ as *generalized forces* and their corresponding *generalized displacements*. For example, two very closely spaced equal and opposite forces form a couple, and by an appropriate limiting process, they can be made to represent a local *concentrated moment* M. The equal and opposite displacements which correspond to the forces in the couple, in the limit, represent a local *rotation* θ. Thus, one of the $P^{(k)}$ may represent a moment M, and the corresponding $\Delta^{(k)}$ is the rotation θ in the direction of M.

Restricting ourselves to the case with no body forces, $\mathbf{b} = \mathbf{0}$, we can write the potential energy as

$$\Pi\{\mathbf{u}\} = \mathcal{F}\{\epsilon(\mathbf{u})\} - \sum_{k=1}^{N} P^{(k)}\Delta^{(k)}, \tag{12.5.3}$$

where $\epsilon(\mathbf{u}) = \frac{1}{2}(\nabla\mathbf{u} + \nabla\mathbf{u}^\top)$ and $\Delta^{(k)} = \mathbf{u}(\mathbf{x}^{(k)})\cdot\mathbf{e}^{(k)}$. Correspondingly, the complementary energy of the body is

$$\Pi^c\{\boldsymbol{\sigma}\} = \mathcal{F}\{\boldsymbol{\sigma}\} - \sum_{k=1}^{N} \Delta^{(k)}P^{(k)}. \tag{12.5.4}$$

The ready exploitation of (12.5.3) and (12.5.4) in structural mechanics is possible in the two special cases of kinematically determinate and statically determinate problems, respectively, which are defined as follows:

- In a **kinematically determinate** problem, it is possible to explicitly express the strain field as $\epsilon(\Delta^{(k)})$, i.e. solely in terms of the motion at the points of application of the generalized loads.
- In a **statically determinate** problem it is possible to explicitly express the stress field as $\boldsymbol{\sigma}(P^{(k)})$, i.e. solely in terms of the generalized point forces.

These two special cases lead to the celebrated theorems of *Castigliano*.

12.5.1 **Castigliano's first theorem**

In the kinematically determinate case

$$\Pi(\Delta^{(k)}) = \mathcal{F}(\Delta^{(k)}) - \sum_{k=1}^{N} P^{(k)} \Delta^{(k)}. \tag{12.5.5}$$

Taking the first variation of (12.5.5) and setting it to zero gives

$$\begin{aligned}
\delta\Pi(\Delta^{(k)}) &= \delta\Big(\mathcal{F}(\Delta^{(k)}) - \sum_{k=1}^{N} P^{(k)} \Delta^{(k)}\Big) = 0, \\
&= \sum_{k=1}^{N} \left[\frac{\partial \mathcal{F}}{\partial \Delta^{(k)}}\, \delta\Delta^{(k)} - P^{(k)}\, \delta\Delta^{(k)}\right] = 0, \\
&= \sum_{k=1}^{N} \left[\frac{\partial \mathcal{F}}{\partial \Delta^{(k)}} - P^{(k)}\right] \delta\Delta^{(k)} = 0.
\end{aligned}$$

Since this must hold for all variations $\delta\Delta^{(k)}$, we must have

$$\boxed{P^{(k)} = \frac{\partial \mathcal{F}}{\partial \Delta^{(k)}} \qquad \text{for all } k.} \tag{12.5.6}$$

This result is encapsulated in:

CASTIGLIANO'S FIRST THEOREM *Given an expression for the net free energy $\mathcal{F}$ of the body in terms of the generalized displacements $\Delta^{(k)}$, $k = 1, \ldots, N$, the corresponding generalized forces $P^{(k)}$ are given by* (12.5.6).

12.5.2 **Castigliano's second theorem**

In the statically determinate case

$$\Pi^c(P^{(k)}) = \mathcal{F}(P^{(k)}) - \sum_{k=1}^{N} \Delta^{(k)} P^{(k)}. \tag{12.5.7}$$

Taking the first variation of (12.5.7) and setting it to zero gives

$$\begin{aligned}
\delta\Pi^c(P^{(k)}) &= \delta\left(\mathcal{F}(P^{(k)}) - \sum_{k=1}^{N} \Delta^{(k)} P^{(k)}\right) = 0, \\
&= \sum_{k=1}^{N} \left[\frac{\partial \mathcal{F}}{\partial P^{(k)}}\, \delta P^{(k)} - \Delta^{(k)}\, \delta P^{(k)}\right] = 0, \\
&= \sum_{k=1}^{N} \left[\frac{\partial \mathcal{F}}{\partial P^{(k)}} - \Delta^{(k)}\right] \delta P^{(k)} = 0.
\end{aligned}$$

Since this must hold for all variations $\delta P^{(k)}$, we must have

$$\boxed{\Delta^{(k)} = \frac{\partial \mathcal{F}}{\partial P^{(k)}} \qquad \text{for all } k.} \tag{12.5.8}$$

This result is encapsulated in:

CASTIGLIANO'S SECOND THEOREM *Given an expression for the net free energy $\mathcal{F}$ of the body in terms of the generalized forces $P^{(k)}$, $k = 1, \ldots, N$, the corresponding generalized displacements $\Delta^{(k)}$ are given by* (12.5.8).

Castigliano's theorems are very powerful tools for determining forces and deflections in structures.

EXAMPLE 12.5 Castigliano's first theorem

Shown in Fig. 12.1 is a slender[a] elastic bar with cross sectional area A and Young's modulus E. We can determine the relation between the applied forces $P^{(1)}$ and $P^{(2)}$ and the resulting displacements $\Delta^{(1)}$ and $\Delta^{(2)}$, respectively, using Castigliano's first theorem, since the problem is kinematically determinate.[b]

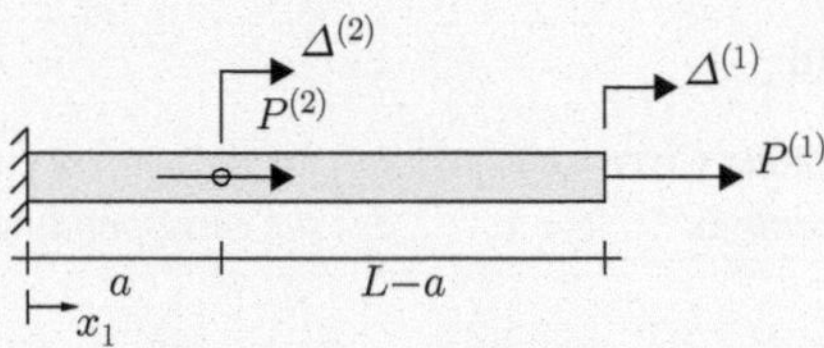

Fig. 12.1 Bar with two axial forces. Young's modulus E and cross sectional area A.

The free energy for the bar is to a good approximation given by:

$$\mathcal{F}\{\epsilon\} = \int_0^L \frac{1}{2} AE(\epsilon_{11})^2 \, dx_1. \tag{12.5.9}$$

Additionally, the strains are expressible in terms of the displacements $\Delta^{(1)}$ and $\Delta^{(2)}$:

$$\epsilon_{11} = \begin{cases} \dfrac{\Delta^{(2)}}{a} & 0 < x_1 < a, \\[2ex] \dfrac{\Delta^{(1)} - \Delta^{(2)}}{L-a} & a < x_1 < L. \end{cases} \tag{12.5.10}$$

Thus,

$$\mathcal{F}(\Delta^{(1)}, \Delta^{(2)}) = \int_0^a \frac{1}{2} AE \left(\frac{\Delta^{(2)}}{a} \right)^2 dx_1 + \int_a^L \frac{1}{2} AE \left(\frac{\Delta^{(1)} - \Delta^{(2)}}{L-a} \right)^2 dx_1. \tag{12.5.11}$$

Using Castigliano's first theorem then results in the desired relations:

$$\frac{AE(\Delta^{(1)} - \Delta^{(2)})}{L - a} = P^{(1)} \tag{12.5.12}$$

$$\frac{AE\Delta^{(2)}}{a} - \frac{AE(\Delta^{(1)} - \Delta^{(2)})}{L - a} = P^{(2)}. \tag{12.5.13}$$

The solution as given provides the required loads for known displacements. One can of course invert the relations (two equations in two unknowns) to find the displacements for given forces.

[a] The cross sectional dimensions are all much less than the length L.

[b] This problem is also statically determinate and thus we could have also used Castigliano's second theorem to solve it.

EXAMPLE 12.6 Castigliano's second theorem: Angle frame

Consider an elastic frame shown in Fig. 12.2. As a structural mechanics problem this system is statically determinate, and we can use Castigliano's second theorem to find the tip-deflection of the frame in the direction of the applied load.

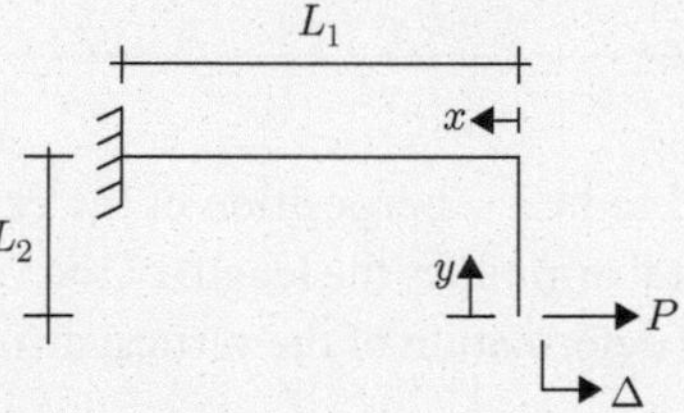

Fig. 12.2 Elastic angle frame with Young's modulus E, shear modulus μ, area A, and area moment of inertia I.

The frame is composed of two segments. From a free body diagram as shown in Fig. 12.3 we see that the horizontal segment is in a state of axial load and bending. From a free body diagram of the vertical segment, Fig. 12.3, we see that it is in a state of bending and direct shear.

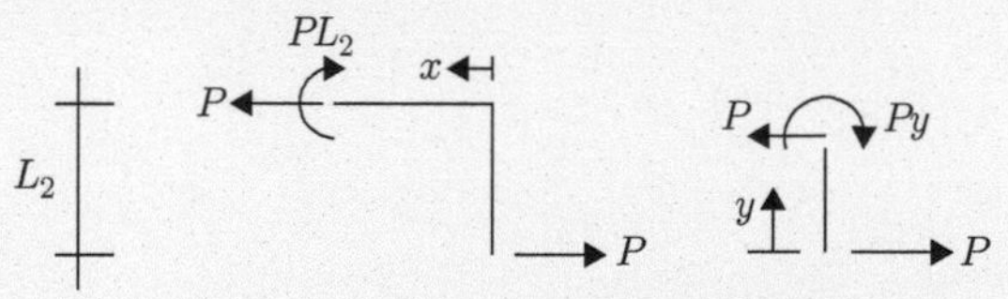

Fig. 12.3 Free body diagram of the horizontal and vertical segments of the angle frame.

To a good approximation the total free energy is then given by,

$$\mathcal{F}\{\sigma\} = \int_0^{L_1} \frac{M_1^2}{2EI}\,dx + \int_0^{L_1} \frac{R_1^2}{2AE}\,dx \tag{12.5.14}$$

$$+ \int_0^{L_2} \frac{M_2^2}{2EI}\,dy + \alpha \int_0^{L_2} \frac{V_2^2}{2\mu A}\,dy, \tag{12.5.15}$$

where M_1 and M_2 are the internal bending moments in the horizontal and vertical members, respectively, R_1 is the internal force in the horizontal member, V_2 is the internal shear force in the vertical member, and α is a geometric parameter dependent on the cross sectional shape.

The desired deflection is given by

$$\Delta = \frac{\partial \mathcal{F}}{\partial P} = \int_0^{L_1} \frac{M_1}{EI}\frac{\partial M_1}{\partial P}\,dx + \int_0^{L_1} \frac{R_1}{AE}\frac{\partial R_1}{\partial P}\,dx \tag{12.5.16}$$

$$+ \int_0^{L_2} \frac{M_2}{EI}\frac{\partial M_2}{\partial P}\,dy + \alpha \int_0^{L_2} \frac{V_2}{\mu A}\frac{\partial V_2}{\partial P}\,dy. \tag{12.5.17}$$

From equilibrium applied to the free body diagrams one has that $M_1(x) = PL_2$, $R_1(x) = P$, $M_2(y) = Py$, and $V_2(y) = P$. Inserting these expressions and their derivatives, one finds:

$$\Delta = \frac{PL_2^2 L_1}{EI} + \frac{PL_1}{AE} + \frac{PL_2^3}{3EI} + \alpha \frac{PL_2}{\mu A}. \tag{12.5.18}$$

REMARK 12.6

The final result is observed to be a superposition of the extension of the horizontal arm, the rotation of the horizontal arm times the length of the vertical arm, the bending of the vertical arm, and the shear deformation of the vertical arm.

PART VI

Elastodynamics. Sinusoidal progressive waves

13 Elastodynamics. Sinusoidal progressive waves

In this chapter we give a very brief discussion of problems in which inertial effects cannot be neglected. We only consider sinusoidal progressive waves in isotropic elastic materials, which form an important class of solutions to the equations of linear elastodynamics.

13.1 Mixed problem of elastodynamics

As before we consider **boundary conditions** in which the displacement is specified on $\mathcal{S}_1$ and the traction on $\mathcal{S}_2$:

$$\begin{aligned} \mathbf{u} &= \hat{\mathbf{u}} \quad \text{on } \mathcal{S}_2 \text{ for all times } (t \geq 0), \\ \boldsymbol{\sigma}\mathbf{n} &= \hat{\mathbf{t}} \quad \text{on } \mathcal{S}_1 \text{ for all times } (t \geq 0), \end{aligned} \tag{13.1.1}$$

with $\hat{\mathbf{u}}$ and $\hat{\mathbf{t}}$ *prescribed* functions of position and time. In dynamics these conditions are supplemented by **initial conditions** in which the displacement $\mathbf{u}$ and the velocity $\dot{\mathbf{u}}$ are specified initially:

$$\mathbf{u}(\mathbf{x}, 0) = \mathbf{u}_0(\mathbf{x}) \quad \text{and} \quad \dot{\mathbf{u}}(\mathbf{x}, 0) = \mathbf{v}_0(\mathbf{x}) \quad \text{for all } \mathbf{x} \text{ in } \mathcal{B}, \tag{13.1.2}$$

with $\mathbf{u}_0$ and $\mathbf{v}_0$ prescribed functions. Then the **mixed problem of elastodynamics** may be stated as follows: *given*[1] boundary data $\hat{\mathbf{u}}$ and $\hat{\mathbf{t}}$ and initial data $\mathbf{u}_0$ and $\mathbf{v}_0$, *find* a displacement field $\mathbf{u}$, a strain field $\boldsymbol{\epsilon}$, and a stress field $\boldsymbol{\sigma}$ that satisfy the field equations

$$\left.\begin{aligned} \boldsymbol{\epsilon} &= \tfrac{1}{2}(\nabla\mathbf{u} + (\nabla\mathbf{u})^{\top}), \\ \boldsymbol{\sigma} &= \mathbb{C}\boldsymbol{\epsilon}, \\ \operatorname{div}\boldsymbol{\sigma} + \mathbf{b} &= \rho\ddot{\mathbf{u}}, \end{aligned}\right\} \quad \text{in } \mathcal{B} \text{ for all times } (t \geq 0), \tag{13.1.3}$$

the boundary conditions (13.1.1), and the initial conditions (13.1.2).

Without proof we note the following result:

UNIQUENESS THEOREM *The mixed problem of elastodynamics has at most one solution.*

[1] In addition to $\mathbb{C}$, ρ, and $\mathbf{b}$.

Continuum Mechanics of Solids. Lallit Anand and Sanjay Govindjee, Oxford University Press (2020).

DOI: 10.1093/oso/9780198864721.001.0001

13.2 **Sinusoidal progressive waves**

Sinusoidal progressive waves form an important class of solutions to the equations of linear elastodynamics. We consider these waves for isotropic media and assume that the conventional body force **b** vanishes. The underlying field equation is then the displacement equation of motion

$$\mu\Delta\mathbf{u} + (\lambda + \mu)\nabla\,\mathrm{div}\,\mathbf{u} = \rho\ddot{\mathbf{u}}; \tag{13.2.1}$$

cf. (8.7.2). A displacement field **u** of the form

$$\mathbf{u}(\mathbf{x}, t) = \mathbf{a}\sin\left(k(\mathbf{r}\cdot\mathbf{m} - ct)\right), \qquad |\mathbf{m}| = 1, \qquad \mathbf{r} = \mathbf{x} - \mathbf{o}, \tag{13.2.2}$$

is called a sinusoidal progressive wave with amplitude **a**, direction **m**, wavenumber k (wavelength $\lambda = 2\pi/k$), and velocity c. Such a wave is *longitudinal* if **a** and **m** are parallel, and *transverse* if **a** and **m** are perpendicular.

We now determine conditions necessary and sufficient for (13.2.2) to satisfy (13.2.1). Applying the chain rule to (13.2.2) and writing $\varphi(\mathbf{x}, t) = k(\mathbf{r}\cdot\mathbf{m} - ct)$, we find that

$$\begin{aligned}\nabla\mathbf{u} &= \mathbf{a}\otimes k\mathbf{m}\cos\varphi,\\ \Delta\mathbf{u} &= -\mathbf{a}k^2\sin\varphi,\\ \nabla\,\mathrm{div}\,\mathbf{u} &= -(\mathbf{a}\cdot k\mathbf{m})k\mathbf{m}\sin\varphi,\\ \ddot{\mathbf{u}} &= -k^2c^2\mathbf{a}\sin\varphi.\end{aligned} \tag{13.2.3}$$

By $(13.2.3)_1$,

$$\begin{aligned}\mathrm{div}\,\mathbf{u} &= \mathbf{a}\cdot k\mathbf{m}\cos\varphi,\\ \mathrm{curl}\,\mathbf{u} &= k\mathbf{m}\times\mathbf{a}\cos\varphi,\end{aligned}$$

and it follows that the wave is longitudinal if and only if $\mathrm{curl}\,\mathbf{u} = \mathbf{0}$ and transverse if and only if $\mathrm{div}\,\mathbf{u} = 0$.

Next, by $(13.2.3)_{2\text{-}4}$, **u** as defined in (13.2.2) satisfies (13.2.1) if and only if

$$\rho c^2\mathbf{a} = \mu\mathbf{a} + (\lambda + \mu)(\mathbf{a}\cdot\mathbf{m})\mathbf{m}. \tag{13.2.4}$$

Defining the *acoustic tensor* $\mathbf{A}(\mathbf{m})$ via

$$\mathbf{A}(\mathbf{m}) = \mu\mathbf{1} + (\lambda + \mu)\mathbf{m}\otimes\mathbf{m},$$

allows us to rewrite (13.2.4) in the form of a propagation condition

$$\mathbf{A}(\mathbf{m})\mathbf{a} = \rho c^2\mathbf{a}. \tag{13.2.5}$$

Thus a condition necessary and sufficient for **u** to satisfy (13.2.1) is that **a** be an eigenvector and that ρc^2 be a corresponding eigenvalue of the acoustic tensor $\mathbf{A}(\mathbf{m})$.

A direct calculation shows that

$$\mathbf{A}(\mathbf{m}) = \mu(\mathbf{1} - \mathbf{m}\otimes\mathbf{m}) + (\lambda + 2\mu)\mathbf{m}\otimes\mathbf{m}.$$

But this is simply the spectral decomposition of $\mathbf{A}(\mathbf{m})$ (cf. (1.2.51)) and we may conclude from the Spectral Theorem that μ and $(\lambda + 2\mu)$ are the eigenvalues of $\mathbf{A}(\mathbf{m})$. Thus, a sinusoidal progressive wave with amplitude $\mathbf{a}$, direction $\mathbf{m}$, and velocity c will be a solution of the displacement equation of motion (13.2.1) if and only if either

- $c = \sqrt{\dfrac{\lambda + 2\mu}{\rho}}$ and the wave is longitudinal, or
- $c = \sqrt{\dfrac{\mu}{\rho}}$ and the wave is transverse.

This result shows that for an isotropic medium, only two types of sinusoidal progressive waves are possible: longitudinal and transverse. The corresponding wave speeds $\sqrt{(\lambda + 2\mu)/\rho}$ and $\sqrt{\mu/\rho}$ are called, respectively, the longitudinal and transverse sound speeds of the medium. Importantly, these speeds are real when the elasticity tensor is positive semi-definite.

For an anisotropic medium the situation is more complicated. A propagation condition of the form (13.2.5) can be obtained. However, the waves determined by that condition are generally neither longitudinal nor transverse and those waves may propagate with different speeds in different directions.

EXAMPLE 13.1 Energy transport by waves

Elastic sinusoidal progressive waves transport energy. The power transmitted across a surface with normal $\mathbf{n}$ is given by

$$P = (\boldsymbol{\sigma}\mathbf{n}) \cdot \mathbf{v}.$$

For a sinusoidal progressive wave the strain field is given by

$$\begin{aligned} \epsilon &= \frac{1}{2}\left(\nabla\mathbf{u} + \nabla\mathbf{u}^{\top}\right) \\ &= \frac{1}{2}k(\mathbf{a}\otimes\mathbf{m} + \mathbf{m}\otimes\mathbf{a})\cos(k(\mathbf{r}\cdot\mathbf{m} - ct)). \end{aligned}$$

The stress field in this case is

$$\boldsymbol{\sigma} = \mathbb{C}\epsilon = k\left[\mu(\mathbf{a}\otimes\mathbf{m} + \mathbf{m}\otimes\mathbf{a}) + \lambda(\mathbf{a}\cdot\mathbf{m})\mathbf{1}\right]\cos(k(\mathbf{r}\cdot\mathbf{m} - ct)).$$

The velocity field is given by

$$\mathbf{v} = \dot{\mathbf{u}} = -kc\,\mathbf{a}\cos(k(\mathbf{r}\cdot\mathbf{m} - ct)).$$

Combining, one finds the power transmitted across a surface with normal $\mathbf{n}$ to be

$$P(\mathbf{r}, t) = -k^2 c\left[(\lambda + \mu)(\mathbf{a}\cdot\mathbf{m})(\mathbf{a}\cdot\mathbf{n}) + \mu(\mathbf{m}\cdot\mathbf{n})(\mathbf{a}\cdot\mathbf{a})\right]\cos^2(k(\mathbf{r}\cdot\mathbf{m} - ct)).$$

The transmitted power oscillates in time with period $T = 2\pi/kc$; note $kc = \omega$, where ω is the circular frequency of the wave. The mean power transmitted over a single period is given by

$$\bar{P} = \frac{1}{T}\int_0^T P(\mathbf{r},t) = -\frac{1}{2}k^2c\left[(\lambda+\mu)(\mathbf{a}\cdot\mathbf{m})(\mathbf{a}\cdot\mathbf{n}) + \mu(\mathbf{m}\cdot\mathbf{n})(\mathbf{a}\cdot\mathbf{a})\right].$$

Note that the power computed here is the power transmitted by the material on the positive side of the surface with normal $\mathbf{n}$ to the material on the negative side of the surface. Thus, for example, if the wave is longitudinal and $\mathbf{n} = \mathbf{m}$, then the algebraic sign of $\bar{P}$ will be negative, indicating that power is flowing in the positive $\mathbf{m}(= \mathbf{n})$ direction.

PART VII

Coupled theories

14 Linear thermoelasticity

14.1 Introduction

In this chapter we provide a framework for the coupled thermal and mechanical response of solids. We restrict our attention to situations in which the deformations are small and elastic. A framework known as the theory of linear thermoelasticity.[1]

We consider a *deformable, heat conducting* body occupying a (fixed) region $\mathcal{B}$. As in the linear theory of elasticity, $\mathbf{x}$ denotes points in $\mathcal{B}$, ∇ and div denote the gradient and divergence with respect to $\mathbf{x}$. We use the following notation:

- $\rho(\mathbf{x}) > 0$ denotes the *constant* mass density of $\mathcal{B}$;
- $\varepsilon(\mathbf{x}, t)$ the *internal energy* measured *per unit volume* in $\mathcal{B}$;
- $\eta(\mathbf{x}, t)$ the *entropy* measured *per unit volume* in $\mathcal{B}$;
- $\vartheta(\mathbf{x}, t) > 0$ the absolute temperature of $\mathcal{B}$;
- $\psi = \varepsilon - \vartheta\eta$ the Helmholtz free-energy measured *per unit volume* in $\mathcal{B}$;
- $\nabla\vartheta(\mathbf{x}, t)$ the temperature gradient field in $\mathcal{B}$;
- $\mathbf{q}(\mathbf{x}, t)$ the *heat flux* measured *per unit area* in $\mathcal{B}$;
- $r(\mathbf{x}, t)$ the *body heat supply* measured *per unit volume* in $\mathcal{B}$.
- $\mathbf{u}(\mathbf{x}, t)$ the displacement field of $\mathcal{B}$;
- $\boldsymbol{\epsilon}(\mathbf{x}, t) = \frac{1}{2}\left(\nabla\mathbf{u}(\mathbf{x}, t) + \nabla\mathbf{u}(\mathbf{x}, t)^{\top}\right)$ the small strain field in $\mathcal{B}$;
- $\boldsymbol{\sigma}(\mathbf{x}, t)$ the stress field in $\mathcal{B}$;
- $\mathbf{b}(\mathbf{x}, t)$ (non-inertial) body force field on $\mathcal{B}$.

We allow for the body to deform and conduct heat. As a first step we formulate a *non-linear theory* in terms of the small strain tensor $\boldsymbol{\epsilon}$, and then specialize the theory to the fully *linear* case in which the following hypotheses are valid:

[1] The discussion of linear thermoelasticity in this chapter is a simplified version of Part XI on *Thermoelasticity* in Gurtin et al. (2010). The reader is referred to that book for a more rigorous treatment of the subject under both large and small elastic deformations. Also see the classic book on the subject by Boley and Weiner (1960).

Continuum Mechanics of Solids. Lallit Anand and Sanjay Govindjee, Oxford University Press (2020).

DOI: 10.1093/oso/9780198864721.001.0001

- the reference body $\mathcal{B}$ is *stress-free* at a (constant) reference temperature ϑ_0;
- the temperature ϑ is everywhere close to ϑ_0;
- and the magnitude of the non-dimensional temperature gradient $\ell\,\nabla\vartheta/\vartheta_0$ is everywhere *small*. Here, ℓ is a characteristic length scale associated with the reference body $\mathcal{B}$.

14.2 Basic equations

We take as the starting point the kinematical assumptions of the linear theory of elasticity. Many of the equations derived there are independent of constitutive relations and hence applicable to a wide class of materials under the assumption of small deformations. Of particular importance to our discussion of linear thermoelasticity are the following relations which still hold true:

(i) the *strain-displacement relation*

$$\boxed{\boldsymbol{\epsilon} = \tfrac{1}{2}\,(\nabla\mathbf{u} + (\nabla\mathbf{u})^{\top}),} \tag{14.2.1}$$

(ii) the *momentum balances*

$$\boxed{\operatorname{div}\boldsymbol{\sigma} + \mathbf{b} = \rho\,\ddot{\mathbf{u}}, \qquad \boldsymbol{\sigma} = \boldsymbol{\sigma}^{\top},} \tag{14.2.2}$$

(iii) the *energy balance*[2]

$$\boxed{\dot{\varepsilon} = \boldsymbol{\sigma} : \dot{\boldsymbol{\epsilon}} - \operatorname{div}\mathbf{q} + r,} \tag{14.2.3}$$

(iv) and the *free-energy imbalance*[3]

$$\boxed{\dot{\psi} + \eta\dot{\vartheta} - \boldsymbol{\sigma} : \dot{\boldsymbol{\epsilon}} + \frac{1}{\vartheta}\mathbf{q}\cdot\nabla\vartheta \le 0,} \tag{14.2.4}$$

where the superposed dot indicates differentiation with respect to time.

14.3 Constitutive equations

Assume for the moment that the strain and temperature fields are known. Then, guided by the free-energy imbalance (14.2.4), we introduce the following constitutive response functions for the free energy ψ, the stress $\boldsymbol{\sigma}$, and the entropy η in terms of $\boldsymbol{\epsilon}$ and ϑ,

$$\begin{aligned} \psi &= \hat{\psi}(\boldsymbol{\epsilon}, \vartheta), \\ \boldsymbol{\sigma} &= \hat{\boldsymbol{\sigma}}(\boldsymbol{\epsilon}, \vartheta), \\ \eta &= \hat{\eta}(\boldsymbol{\epsilon}, \vartheta), \end{aligned} \tag{14.3.1}$$

---

[2] Cf. (6.2.2).

[3] Cf. (6.2.4).

to which we append a further constitutive assumption of a simple Fourier's law for the heat flux,

$$\mathbf{q} = -\mathbf{K}\nabla\vartheta, \tag{14.3.2}$$

with

$$\mathbf{K} = \hat{\mathbf{K}}(\boldsymbol{\epsilon}, \vartheta) \tag{14.3.3}$$

the second-order **thermal conductivity tensor**, which is a material property. The relations $(14.3.1)_{1,3}$ yield an auxiliary constitutive relation

$$\varepsilon = \hat{\varepsilon}(\boldsymbol{\epsilon}, \vartheta) = \hat{\psi}(\boldsymbol{\epsilon}, \vartheta) + \vartheta\,\hat{\eta}(\boldsymbol{\epsilon}, \vartheta) \tag{14.3.4}$$

for the internal energy ε.

REMARK 14.1

Although our final goal is the formulation of a *linear* theory, we find it convenient to first develop the theory with constitutive equations (14.3.1) and (14.3.2) for the free energy, stress, entropy, and the heat flux—constitutive equations which are in general *non-linear* in the small strain tensor $\boldsymbol{\epsilon}$ and the temperature ϑ. We will discuss the linear theory in Sec. 14.4.

The free energy imbalance restricts the possible constitutive response of the thermoelastic system, and in particular it restricts the possible constitutive equations for the stress, entropy, and the heat flux. By $(14.3.1)_1$,

$$\dot{\psi} = \frac{\partial\hat{\psi}(\boldsymbol{\epsilon}, \vartheta)}{\partial\boldsymbol{\epsilon}} : \dot{\boldsymbol{\epsilon}} + \frac{\partial\hat{\psi}(\vartheta)}{\partial\vartheta}\,\dot{\vartheta}. \tag{14.3.5}$$

Hence the free-energy imbalance (14.2.4) is equivalent to the requirement that

$$\left(\frac{\partial\hat{\psi}(\boldsymbol{\epsilon}, \vartheta)}{\partial\boldsymbol{\epsilon}} - \hat{\boldsymbol{\sigma}}(\boldsymbol{\epsilon}, \vartheta)\right) : \dot{\boldsymbol{\epsilon}} + \left(\frac{\partial\hat{\psi}(\boldsymbol{\epsilon}, \vartheta)}{\partial\vartheta} + \hat{\eta}(\boldsymbol{\epsilon}, \vartheta)\right)\dot{\vartheta} - \frac{1}{\vartheta}\nabla\vartheta\cdot\mathbf{K}\nabla\vartheta \leq 0 \tag{14.3.6}$$

for all strain and temperature fields. Given any point $\mathbf{x}_0$ in $\mathcal{B}$ and any time t_0, it is possible to find a strain and temperature field such that

$$\boldsymbol{\epsilon},\ \dot{\boldsymbol{\epsilon}},\ \vartheta,\ \nabla\vartheta,\ \text{and}\ \dot{\vartheta} \tag{14.3.7}$$

have arbitrarily prescribed values at $(\mathbf{x}_0, t_0)$. Granted this, the coefficients of $\dot{\boldsymbol{\epsilon}}$ and $\dot{\vartheta}$ must vanish, for otherwise these rates may be chosen to violate inequality (14.3.6). We therefore have the **thermodynamic restrictions**[4]:

[4] As in Sec. 7.1, where we discussed elasticity, this deductive process is known as the Coleman and Noll (1963) procedure.

(i) *The free energy determines the stress through the* **stress relation**

$$\hat{\boldsymbol{\sigma}}(\boldsymbol{\epsilon},\vartheta) = \frac{\partial\hat{\psi}(\boldsymbol{\epsilon},\vartheta)}{\partial\boldsymbol{\epsilon}}. \tag{14.3.8}$$

(ii) *The free energy determines the entropy through the* **entropy relation**

$$\hat{\eta}(\boldsymbol{\epsilon},\vartheta) = -\frac{\partial\hat{\psi}(\boldsymbol{\epsilon},\vartheta)}{\partial\vartheta}. \tag{14.3.9}$$

(iii) *The thermal conductivity tensor* **K** *satisfies the heat-conduction inequality*

$$\nabla\vartheta\cdot\mathbf{K}\nabla\vartheta \geq 0 \qquad \text{for all } \nabla\vartheta. \tag{14.3.10}$$

We assume that the heat-conduction inequality as **strict**, that is

$$\nabla\vartheta\cdot\mathbf{K}\nabla\vartheta > 0 \qquad \text{if } \nabla\vartheta\neq\mathbf{0}; \tag{14.3.11}$$

this assumption is the manifestation of the observation that there is always non-zero heat flow whenever there is a temperature gradient. Assumption (14.3.11) implies that the thermal conductivity tensor **K** is *positive definite.*

14.3.1 Maxwell and Gibbs relations

The state relations (14.3.8) and (14.3.9) yield the *Maxwell relation*

$$\frac{\partial\hat{\boldsymbol{\sigma}}(\boldsymbol{\epsilon},\vartheta)}{\partial\vartheta} = -\frac{\partial\hat{\eta}(\boldsymbol{\epsilon},\vartheta)}{\partial\boldsymbol{\epsilon}}. \tag{14.3.12}$$

Further, by (14.3.5), (14.3.8), and (14.3.9), we have the *first Gibbs relation*

$$\dot{\psi} = \boldsymbol{\sigma}:\dot{\boldsymbol{\epsilon}} - \eta\dot{\vartheta}. \tag{14.3.13}$$

Next, since $\psi = \varepsilon - \vartheta\eta$,

$$\dot{\psi} + \eta\dot{\vartheta} = \dot{\varepsilon} - \vartheta\dot{\eta},$$

and use of (14.3.13) yields a *second Gibbs relation*

$$\dot{\varepsilon} = \boldsymbol{\sigma}:\dot{\boldsymbol{\epsilon}} + \vartheta\dot{\eta}. \tag{14.3.14}$$

Hence, (14.3.14) and the energy balance (14.2.3) imply that

$$\begin{aligned}\vartheta\dot{\eta} &= \dot{\varepsilon} - \boldsymbol{\sigma}:\dot{\boldsymbol{\epsilon}}\\ &= (\boldsymbol{\sigma}:\dot{\boldsymbol{\epsilon}} - \operatorname{div}\mathbf{q} + r) - \boldsymbol{\sigma}:\dot{\boldsymbol{\epsilon}}\\ &= -\operatorname{div}\mathbf{q} + r.\end{aligned}$$

Thus, an important consequence of the Gibbs relations is that the energy balance may be written as

$$\dot{\eta} = -\frac{1}{\vartheta}\operatorname{div}\mathbf{q} + \frac{r}{\vartheta}. \tag{14.3.15}$$

That is, granted the thermodynamically restricted constitutive relations (14.3.8) and (14.3.9), the entropy relation (14.3.15) is equivalent to balance of energy. This relation shows that if $\mathbf{q} = \mathbf{0}$ and $r = 0$, that is, if the process is *adiabatic*, then it is *isentropic*, $\dot{\eta} = 0$.

14.3.2 **Specific heat**

Next, recall the constitutive relation (14.3.4) for the internal energy, viz.

$$\varepsilon = \hat{\varepsilon}(\epsilon, \vartheta) = \hat{\psi}(\epsilon, \vartheta) + \vartheta\hat{\eta}(\epsilon, \vartheta). \tag{14.3.16}$$

An important consequence of this relation and (14.3.9) is that

$$\frac{\partial \hat{\varepsilon}(\epsilon, \vartheta)}{\partial \vartheta} = \vartheta \frac{\partial \hat{\eta}(\epsilon, \vartheta)}{\partial \vartheta}. \tag{14.3.17}$$

The *specific heat* (at fixed strain) is defined by[5]

$$c = c(\epsilon, \vartheta) \stackrel{\text{def}}{=} \frac{\partial \hat{\varepsilon}(\epsilon, \vartheta)}{\partial \vartheta}. \tag{14.3.18}$$

Using (14.3.17) and (14.3.9) the specific heat is alternatively given by

$$\begin{aligned} c(\epsilon, \vartheta) &= \vartheta \, \frac{\partial \hat{\eta}(\epsilon, \vartheta)}{\partial \vartheta} \\ &= -\vartheta \, \frac{\partial^2 \hat{\psi}(\epsilon, \vartheta)}{\partial \vartheta^2}. \end{aligned} \tag{14.3.19}$$

We assume that the specific heat is strictly positive,

$$c(\epsilon, \vartheta) > 0 \tag{14.3.20}$$

This assumption, together with $\vartheta > 0$, is equivalent to assuming that entropy $\hat{\eta}(\epsilon, \vartheta)$ is a strictly increasing function of ϑ, as well as $\partial^2 \hat{\psi}(\epsilon, \vartheta)/\partial \vartheta^2 < 0$ for all ϑ, so that $\hat{\psi}(\epsilon, \vartheta)$ is concave in ϑ.

14.3.3 **Elasticity tensor. Stress-temperature modulus. Evolution equation for temperature**

Let $\epsilon(t)$ be a time-dependent strain tensor, let $\vartheta(t)$ be a time-dependent temperature, and write

$$\boldsymbol{\sigma}(t) = \hat{\boldsymbol{\sigma}}(\epsilon(t), \vartheta(t)).$$

The chain-rule then yields the relation,

$$\dot{\boldsymbol{\sigma}} = \frac{\partial \hat{\boldsymbol{\sigma}}(\epsilon, \vartheta)}{\partial \epsilon} \dot{\epsilon} + \frac{\partial \hat{\boldsymbol{\sigma}}(\epsilon, \vartheta)}{\partial \vartheta} \dot{\vartheta},$$

which suggests the introduction of two constitutive moduli:

---

[5] The specific heat is sometimes known as the heat capacity.

- The fourth-order *elasticity tensor* $\mathbb{C}(\epsilon, \vartheta)$ defined by

$$\mathbb{C}(\epsilon, \vartheta) \stackrel{\text{def}}{=} \frac{\partial \hat{\sigma}(\epsilon, \vartheta)}{\partial \epsilon} = \frac{\partial^2 \hat{\psi}(\epsilon, \vartheta)}{\partial \epsilon^2} \quad \text{(at fixed } \vartheta\text{)}. \tag{14.3.21}$$

The elasticity tensor $\mathbb{C}$ represents elastic moduli under conditions in which ϑ is held *constant*, and accordingly we call $\mathbb{C}$ the *isothermal elasticity tensor.*

- The second-order *stress-temperature tensor* $\mathbf{M}(\epsilon, \vartheta)$ defined by

$$\mathbf{M}(\epsilon, \vartheta) \stackrel{\text{def}}{=} \frac{\partial \hat{\sigma}(\epsilon, \vartheta)}{\partial \vartheta} = \frac{\partial^2 \hat{\psi}(\epsilon, \vartheta)}{\partial \epsilon \partial \vartheta}. \tag{14.3.22}$$

The Maxwell relation (14.3.12) implies that

$$\mathbf{M}(\epsilon, \vartheta) = -\frac{\partial \hat{\eta}(\epsilon, \vartheta)}{\partial \epsilon}. \tag{14.3.23}$$

For each (ϵ, ϑ) the elasticity tensor $\mathbb{C}(\epsilon, \vartheta)$ — a linear transformation that maps symmetric tensors to symmetric tensors — has the minor symmetry properties

$$C_{ijkl} = C_{jikl} = C_{ijlk}, \tag{14.3.24}$$

as well as the major symmetry property

$$C_{ijkl} = C_{klij}. \tag{14.3.25}$$

Further, since the stress $\boldsymbol{\sigma}$ is symmetric, we may conclude from (14.3.22) that the stress-temperature modulus $\mathbf{M}(\epsilon, \vartheta)$ is a symmetric tensor; this tensor measures the change in stress due to a change in temperature holding the strain fixed.

Proceeding as above, let

$$\eta(t) = \hat{\eta}(\epsilon(t), \vartheta(t)).$$

Then, by the chain rule and (14.3.23),

$$\begin{aligned} \dot{\eta} &= \frac{\partial \hat{\eta}(\epsilon, \vartheta)}{\partial \epsilon} : \dot{\epsilon} + \frac{\partial \hat{\eta}(\epsilon, \vartheta)}{\partial \vartheta} \dot{\vartheta}, \\ &= -\mathbf{M}(\epsilon, \vartheta) : \dot{\epsilon} + \frac{\partial \hat{\eta}(\epsilon, \vartheta)}{\partial \vartheta} \dot{\vartheta}. \end{aligned}$$

Thus, using $(14.3.19)_1$

$$\vartheta \dot{\eta} = -\vartheta \mathbf{M}(\epsilon, \vartheta) : \dot{\epsilon} + c(\epsilon, \vartheta) \dot{\vartheta}. \tag{14.3.26}$$

The identity (14.3.26), the constitutive equation (14.3.2) for the heat flux, and the entropy balance (14.3.15) yield the following important evolution equation for temperature,

$$c(\epsilon, \vartheta) \dot{\vartheta} = \operatorname{div}(\mathbf{K}(\epsilon, \vartheta) \nabla \vartheta) + \vartheta \mathbf{M}(\epsilon, \vartheta) : \dot{\epsilon} + r. \tag{14.3.27}$$

Classical simple models of heat conduction assume that c and $\mathbf{K}$ are *constant.* In this case, (14.3.27) takes the form

$$c \dot{\vartheta} = \mathbf{K} : \nabla\nabla\vartheta + \vartheta \mathbf{M}(\epsilon, \vartheta) : \dot{\epsilon} + r, \tag{14.3.28}$$

which is the classical anisotropic heat equation augmented by the term $\vartheta \mathbf{M}(\epsilon, \vartheta) : \dot{\epsilon}$ which represents a local stress-power expenditure.

14.3.4 **Entropy as independent variable**

For problems involving heat conduction, the strain ϵ and the temperature ϑ are the natural choice of independent constitutive variables. However, for processes that occur over short time-scales so that heat conduction is negligible, it is often preferable to replace the constitutive dependence upon ϑ by a constitutive dependence upon the entropy η.

Recall that we have assumed that the specific heat is strictly positive $c(\epsilon, \vartheta) > 0$ for all (ϵ, ϑ). Since $\vartheta > 0$, it follows from $(14.3.19)_1$ that

$$\frac{\partial \hat{\eta}(\epsilon, \vartheta)}{\partial \vartheta} = \frac{c(\epsilon, \vartheta)}{\vartheta} > 0. \tag{14.3.29}$$

Assuming that $c(\epsilon, \vartheta)$ is continuous in ϑ allows us to conclude that, for each fixed ϵ, the relation

$$\eta = \hat{\eta}(\epsilon, \vartheta) \tag{14.3.30}$$

is invertible in ϑ, so that

$$\vartheta = \breve{\vartheta}(\epsilon, \eta). \tag{14.3.31}$$

Note that by (14.3.29),

$$\frac{\partial \breve{\vartheta}(\epsilon, \eta)}{\partial \eta} = \left(\frac{\partial \hat{\eta}(\epsilon, \vartheta)}{\partial \vartheta}\right)^{-1} = \frac{\vartheta}{c(\epsilon, \vartheta)}, \tag{14.3.32}$$

with $\vartheta = \breve{\vartheta}(\epsilon, \eta)$. Then, using (14.3.4) the internal energy may be expressed as function of strain and entropy,

$$\begin{aligned} \varepsilon &= \breve{\varepsilon}(\epsilon, \eta) \\ &= \hat{\psi}(\epsilon, \breve{\vartheta}(\epsilon, \eta)) + \breve{\vartheta}(\epsilon, \eta)\eta, \end{aligned} \tag{14.3.33}$$

while (14.3.8) yields

$$\boldsymbol{\sigma} = \breve{\boldsymbol{\sigma}}(\epsilon, \eta) \tag{14.3.34}$$

$$= \hat{\boldsymbol{\sigma}}(\epsilon, \breve{\vartheta}(\epsilon, \eta)). \tag{14.3.35}$$

Thus, bearing in mind that a "breve" denotes a function of (ϵ, η) while a "hat" denotes a function of (ϵ, ϑ), we find, using (14.3.9), that

$$\frac{\partial \breve{\varepsilon}}{\partial \epsilon} = \frac{\partial \hat{\psi}}{\partial \epsilon} + \underbrace{\left(\frac{\partial \hat{\psi}}{\partial \vartheta} + \eta\right)}_{=0} \frac{\partial \breve{\vartheta}}{\partial \epsilon} = \breve{\boldsymbol{\sigma}},$$

and

$$\frac{\partial \breve{\varepsilon}}{\partial \eta} = \underbrace{\left(\frac{\partial \hat{\psi}}{\partial \vartheta} + \eta\right)}_{=0} \frac{\partial \breve{\vartheta}}{\partial \eta} + \breve{\vartheta} = \breve{\vartheta}.$$

The stress and temperature are therefore determined by the response function for the internal energy $\varepsilon = \breve{\varepsilon}(\epsilon, \eta)$ via the relations

$$\begin{aligned}\boldsymbol{\sigma} = \breve{\boldsymbol{\sigma}}(\boldsymbol{\epsilon}, \eta) &= \frac{\partial \breve{\varepsilon}(\boldsymbol{\epsilon}, \eta)}{\partial \boldsymbol{\epsilon}}, \\ \vartheta = \breve{\vartheta}(\boldsymbol{\epsilon}, \eta) &= \frac{\partial \breve{\varepsilon}(\boldsymbol{\epsilon}, \eta)}{\partial \eta}.\end{aligned} \tag{14.3.36}$$

An immediate consequence of (14.3.33) and (14.3.36) is the *Gibbs relation*

$$\dot{\varepsilon} = \boldsymbol{\sigma} : \dot{\boldsymbol{\epsilon}} + \vartheta \dot{\eta}, \tag{14.3.37}$$

and the *Maxwell relation*

$$\frac{\partial \breve{\boldsymbol{\sigma}}(\boldsymbol{\epsilon}, \eta)}{\partial \eta} = \frac{\partial \breve{\vartheta}(\boldsymbol{\epsilon}, \eta)}{\partial \boldsymbol{\epsilon}}. \tag{14.3.38}$$

The isothermal elasticity tensor $\mathbb{C}(\boldsymbol{\epsilon}, \vartheta)$ and stress-temperature tensor $\mathbf{M}(\boldsymbol{\epsilon}, \vartheta)$ have natural counterparts in the theory with entropy as independent variable; they are the *elasticity tensor*

$$\mathbb{C}^{\text{ent}}(\boldsymbol{\epsilon}, \eta) = \frac{\partial \breve{\boldsymbol{\sigma}}(\boldsymbol{\epsilon}, \eta)}{\partial \boldsymbol{\epsilon}} \quad \text{(at fixed } \eta), \tag{14.3.39}$$

and the *stress-entropy modulus*

$$\mathbf{M}^{\text{ent}}(\boldsymbol{\epsilon}, \eta) = \frac{\partial \breve{\boldsymbol{\sigma}}(\boldsymbol{\epsilon}, \eta)}{\partial \eta} \quad \text{(at fixed } \boldsymbol{\epsilon}). \tag{14.3.40}$$

The elasticity tensor $\mathbb{C}^{\text{ent}}$ represents the elastic moduli under conditions in which the entropy is held constant, and there is no heat flow in or out of the body during deformation.

By $(14.3.36)_1$ and (14.3.39),

$$\mathbb{C}^{\text{ent}}(\boldsymbol{\epsilon}, \eta) = \frac{\partial^2 \breve{\varepsilon}(\boldsymbol{\epsilon}, \eta)}{\partial \boldsymbol{\epsilon}^2}, \tag{14.3.41}$$

while the Maxwell relation (14.3.38) implies that

$$\mathbf{M}^{\text{ent}}(\boldsymbol{\epsilon}, \eta) = \frac{\partial \breve{\vartheta}(\boldsymbol{\epsilon}, \eta)}{\partial \boldsymbol{\epsilon}}. \tag{14.3.42}$$

14.3.5 **Isentropic and isothermal inter-relations**

We now determine relations between the isothermal and isentropic material functions. Toward this end, we note, by (14.3.31) and (14.3.32), that we may relate the alternative descriptions of the stress in terms of $(\boldsymbol{\epsilon}, \vartheta)$ and $(\boldsymbol{\epsilon}, \eta)$ via

$$\breve{\boldsymbol{\sigma}}(\boldsymbol{\epsilon}, \eta) = \hat{\boldsymbol{\sigma}}(\boldsymbol{\epsilon}, \breve{\vartheta}(\boldsymbol{\epsilon}, \eta)).$$

Thus, by (14.3.40)

$$\begin{aligned}\mathbf{M}^{\text{ent}} &= \frac{\partial \breve{\boldsymbol{\sigma}}}{\partial \eta} \\ &= \frac{\partial \hat{\boldsymbol{\sigma}}}{\partial \vartheta} \frac{\partial \breve{\vartheta}}{\partial \eta},\end{aligned}$$

and, using (14.3.22) and (14.3.32), we conclude that the stress-entropy and stress-temperature moduli are related via

$$\mathbf{M}^{\text{ent}}(\boldsymbol{\epsilon},\eta)=\frac{\vartheta}{c(\boldsymbol{\epsilon},\vartheta)}\mathbf{M}(\boldsymbol{\epsilon},\vartheta), \tag{14.3.43}$$

for $\vartheta=\breve{\vartheta}(\boldsymbol{\epsilon},\eta)$.

The relation between the elasticity tensor $\mathbb{C}^{\text{ent}}$ at fixed entropy to the elasticity tensor $\mathbb{C}$ at fixed temperature is based on computing the partial derivative

$$\begin{aligned}\mathbb{C}^{\text{ent}}(\boldsymbol{\epsilon},\eta)&=\frac{\partial\breve{\boldsymbol{\sigma}}(\boldsymbol{\epsilon},\eta)}{\partial\boldsymbol{\epsilon}}\\&=\frac{\partial}{\partial\boldsymbol{\epsilon}}\left(\hat{\boldsymbol{\sigma}}(\boldsymbol{\epsilon},\breve{\vartheta}(\boldsymbol{\epsilon},\eta))\right)\end{aligned}$$

with respect to $\boldsymbol{\epsilon}$ holding η fixed. Suppressing arguments and using components, this derivative of $\hat{\boldsymbol{\sigma}}(\boldsymbol{\epsilon},\breve{\vartheta}(\boldsymbol{\epsilon},\eta))$ is given by

$$\frac{\partial(\hat{\boldsymbol{\sigma}})_{ij}}{\partial\epsilon_{kl}}+\frac{\partial(\hat{\boldsymbol{\sigma}})_{ij}}{\partial\vartheta}\frac{\partial\breve{\vartheta}}{\partial\epsilon_{kl}}.$$

Thus, since the term

$$\frac{\partial(\hat{\boldsymbol{\sigma}})_{ij}}{\partial\vartheta}\frac{\partial\breve{\vartheta}}{\partial\epsilon_{kl}}\qquad\text{is the component form of}\qquad\frac{\partial\hat{\boldsymbol{\sigma}}}{\partial\vartheta}\otimes\frac{\partial\breve{\vartheta}}{\partial\boldsymbol{\epsilon}}$$

we find, with the aid of (14.3.21), (14.3.22), (14.3.42), and (14.3.43), that

$$\mathbb{C}^{\text{ent}}(\boldsymbol{\epsilon},\eta)=\mathbb{C}(\boldsymbol{\epsilon},\vartheta)+\frac{\vartheta}{c(\boldsymbol{\epsilon},\vartheta)}\mathbf{M}(\boldsymbol{\epsilon},\vartheta)\otimes\mathbf{M}(\boldsymbol{\epsilon},\vartheta), \tag{14.3.44}$$

for $\vartheta=\breve{\vartheta}(\boldsymbol{\epsilon},\eta)$; equivalently, in components, suppressing arguments,

$$C^{\text{ent}}_{ijkl}=C_{ijkl}+\frac{\vartheta}{c}M_{ij}M_{kl}. \tag{14.3.45}$$

The identity (14.3.44) has two important consequence. First, given symmetric tensors $\mathbf{A}$ and $\mathbf{G}$

$$\begin{aligned}\mathbf{A}:(\mathbf{M}\otimes\mathbf{M})\mathbf{G}&=(\mathbf{A}:\mathbf{M})(\mathbf{M}:\mathbf{G}) &(14.3.46)\\&=\mathbf{G}:(\mathbf{M}\otimes\mathbf{M})\mathbf{A}. &(14.3.47)\end{aligned}$$

Given this property, noting further that $\mathbf{M}$ is symmetric and $\mathbb{C}$ possesses minor and major symmetries, we see that the tensor $\mathbb{C}^{\text{ent}}$ also *possesses major and minor symmetries*. Further, by (14.3.44) and (14.3.46),

$$\mathbf{A}:\mathbb{C}^{\text{ent}}(\boldsymbol{\epsilon},\eta)\mathbf{A}-\mathbf{A}:\mathbb{C}(\boldsymbol{\epsilon},\vartheta)\mathbf{A}=\frac{\vartheta}{c}(\mathbf{A}:\mathbf{M})^2,$$

for any tensor $\mathbf{A}$, and, since ϑ and c are both positive,

$$\mathbf{A}:\mathbb{C}^{\text{ent}}(\boldsymbol{\epsilon},\eta)\mathbf{A}>\mathbf{A}:\mathbb{C}(\boldsymbol{\epsilon},\vartheta)\mathbf{A}\quad\text{for all }\mathbf{A}\neq\mathbf{0}. \tag{14.3.48}$$

Thus $\mathbb{C}^{\text{ent}}$ is positive definite whenever $\mathbb{C}$ is positive definite. *We assume henceforth that both $\mathbb{C}^{\text{ent}}$ and $\mathbb{C}$ are positive definite.* Positive definiteness is a condition sufficient to ensure that these elasticity tensors are invertible.

REMARK 14.2

The quantities C_{ijkl} are often known as the "isothermal elastic constants" and C^{ent}_{ijkl} as the "adiabatic elastic constants." This terminology is unfortunate because it gives the impression that the former can only be used when isothermal conditions prevail and the latter only when adiabatic conditions prevail. As can be seen from their derivation, the isothermal constants can be used for the analysis of a general thermoelastic process when (ϵ, ϑ) are used as independent variables, and the adiabatic constants can be used for the same process with the use of (ϵ, η) as independent variables. Of course, if it is known *a priori* that the particular elastic process to be studied is isothermal to a good degree of approximation, then it is convenient to use the isothermal constants and treat the constant temperature as a parameter. In this way one field equation can be eliminated from the problem formulation. A corresponding simplification can be made when it is known that a particular process is adiabatic.

14.4 **Linear thermoelasticity**

Guided by the discussion in the previous section we next consider a *linear* theory. For **linear thermoelastic materials** we consider a Taylor series expansion of the free energy function $\psi = \hat{\psi}(\epsilon, \vartheta)$ about $\epsilon = \mathbf{0}$ and $\vartheta = \vartheta_0$, up to terms which are *quadratic in strain* ϵ *and the temperature difference* $(\vartheta - \vartheta_0)$:

$$
\begin{aligned}
\psi = \hat{\psi}(\mathbf{0}, \vartheta_0) &+ \left(\frac{\partial \hat{\psi}(\epsilon, \vartheta)}{\partial \epsilon_{pq}} \bigg|_{\epsilon=\mathbf{0}, \vartheta=\vartheta_0} \right) \epsilon_{pq} + \left(\frac{\partial \hat{\psi}(\epsilon, \vartheta)}{\partial \vartheta} \bigg|_{\epsilon=\mathbf{0}, \vartheta=\vartheta_0} \right) (\vartheta - \vartheta_0) \\
&+ \frac{1}{2} \left(\frac{\partial^2 \hat{\psi}(\epsilon, \vartheta)}{\partial \epsilon_{pq} \partial \epsilon_{rs}} \bigg|_{\epsilon=\mathbf{0}, \vartheta=\vartheta_0} \right) \epsilon_{pq} \epsilon_{rs} + \left(\frac{\partial^2 \hat{\psi}(\epsilon, \vartheta)}{\partial \epsilon_{pq} \partial \vartheta} \bigg|_{\epsilon=\mathbf{0}, \vartheta=\vartheta_0} \right) \epsilon_{pq} (\vartheta - \vartheta_0) \\
&+ \frac{1}{2} \left(\frac{\partial^2 \hat{\psi}(\epsilon, \vartheta)}{\partial \vartheta^2} \bigg|_{\epsilon=\mathbf{0}, \vartheta=\vartheta_0} \right) (\vartheta - \vartheta_0)^2 + \ldots
\end{aligned}
\tag{14.4.1}
$$

We assume, as a datum, that when $\epsilon = \mathbf{0}$ and $\vartheta = \vartheta_0$, then the free energy is zero:

$$
\hat{\psi}(\mathbf{0}, \vartheta_0) \equiv \psi_0 = 0.
$$

On account of (14.3.8),

$$
\frac{\partial \hat{\psi}(\epsilon, \vartheta)}{\partial \epsilon_{pq}} \bigg|_{\epsilon=\mathbf{0}, \vartheta=\vartheta_0} \equiv \hat{\sigma}_{pq}(\mathbf{0}, \vartheta_0) \equiv (\boldsymbol{\sigma}_0)_{pq}
$$

represents the *residual stress*, that is the stress when $\epsilon = \mathbf{0}$ and the body is at its reference temperature ϑ_0; we assume that there is no residual stress, $\boldsymbol{\sigma}_0 = \mathbf{0}$. Also, on account of (14.3.9),

$$-\left.\frac{\partial\hat{\psi}(\boldsymbol{\epsilon},\vartheta)}{\partial\vartheta}\right|_{\boldsymbol{\epsilon}=\mathbf{0},\vartheta=\vartheta_0} \equiv \hat{\eta}(\mathbf{0},\vartheta_0) \equiv \eta_0,$$

represents the *entropy* when $\boldsymbol{\epsilon} = \mathbf{0}$ and the body is at its reference temperature ϑ_0; again, as a datum, we set this entropy to zero, $\eta_0 = 0$. Under these assumptions the free-energy density ψ for a *linear thermoelastic material* is quadratic in the strain $\boldsymbol{\epsilon}$ and the temperature difference $(\vartheta - \vartheta_0)$, and is given by

$$\boxed{\psi \equiv \frac{1}{2}\boldsymbol{\epsilon}:\mathbb{C}\boldsymbol{\epsilon} + (\vartheta - \vartheta_0)\mathbf{M}:\boldsymbol{\epsilon} - \frac{1}{2}\frac{c}{\vartheta_0}(\vartheta - \vartheta_0)^2.} \tag{14.4.2}$$

In (14.4.2) we have introduced the following quantities:

(a) A fourth-order (isothermal) **elasticity tensor** $\mathbb{C}$ with components

$$C_{pqrs} \stackrel{\text{def}}{=} \left.\frac{\partial^2\hat{\psi}(\boldsymbol{\epsilon},\vartheta)}{\partial\epsilon_{pq}\partial\epsilon_{rs}}\right|_{\boldsymbol{\epsilon}=\mathbf{0},\vartheta=\vartheta_0}. \tag{14.4.3}$$

These elastic moduli C_{pqrs} possess the following major symmetry

$$C_{pqrs} = C_{rspq}, \tag{14.4.4}$$

as well as the minor symmetries

$$C_{pqrs} = C_{qprs}, \qquad C_{pqrs} = C_{pqsr}. \tag{14.4.5}$$

Also, on account of the physical requirement that the free energy is positive for all non-zero strains, we require that $\mathbb{C}$ be *positive definite*:

$$\mathbf{A}:\mathbb{C}\mathbf{A} > 0 \quad \text{for all symmetric tensors } \mathbf{A} \neq \mathbf{0}. \tag{14.4.6}$$

(b) A *symmetric* second-order **stress-temperature** tensor $\mathbf{M}$, with components

$$M_{pq} \stackrel{\text{def}}{=} \left.\frac{\partial^2\hat{\psi}(\boldsymbol{\epsilon},\vartheta)}{\partial\epsilon_{pq}\partial\vartheta}\right|_{\boldsymbol{\epsilon}=\mathbf{0},\vartheta=\vartheta_0}. \tag{14.4.7}$$

(c) A coefficient c, the **specific heat**, defined by

$$c \stackrel{\text{def}}{=} \left.\frac{\partial\hat{\varepsilon}(\boldsymbol{\epsilon},\vartheta)}{\partial\vartheta}\right|_{\boldsymbol{\epsilon}=\mathbf{0},\vartheta=\vartheta_0}, \tag{14.4.8}$$

Using (14.3.4) and using the entropy relation (14.3.9), it follows that an equivalent definition of c is given by,

$$c \stackrel{\text{def}}{=} -\vartheta_0 \left.\frac{\partial^2\hat{\psi}(\boldsymbol{\epsilon},\vartheta)}{\partial\vartheta^2}\right|_{\boldsymbol{\epsilon}=\mathbf{0},\vartheta=\vartheta_0}. \tag{14.4.9}$$

We require that c be positive,

$$c > 0, \tag{14.4.10}$$

so that a positive change in temperature is required to raise the internal energy of a material point.

The free energy (14.4.2), the stress relation (14.3.8), and the entropy relation (14.3.9) then yield the following constitutive equations for the stress and the entropy:

$$\boxed{\boldsymbol{\sigma} = \mathbb{C}\boldsymbol{\epsilon} + \mathbf{M}(\vartheta - \vartheta_0),} \tag{14.4.11}$$

and

$$\boxed{\eta = -\mathbf{M}:\boldsymbol{\epsilon} + \frac{c}{\vartheta_0}(\vartheta - \vartheta_0).} \tag{14.4.12}$$

14.5 **Basic field equations of linear thermoelasticity**

The linear theory of thermoelasticity is based on the **strain-displacement relation**

$$\boldsymbol{\epsilon} = \tfrac{1}{2}(\nabla\mathbf{u} + (\nabla\mathbf{u})^{\top}), \tag{14.5.1}$$

and the **constitutive equations**

$$\boxed{\begin{aligned}
&\psi = \frac{1}{2}\boldsymbol{\epsilon}:\mathbb{C}\boldsymbol{\epsilon} + (\vartheta - \vartheta_0)\mathbf{M}:\boldsymbol{\epsilon} - \frac{c}{2\vartheta_0}(\vartheta - \vartheta_0)^2\\
&\boldsymbol{\sigma} = \mathbb{C}\boldsymbol{\epsilon} + \mathbf{M}(\vartheta - \vartheta_0),\\
&\eta = -\mathbf{M}:\boldsymbol{\epsilon} + \frac{c}{\vartheta_0}(\vartheta - \vartheta_0),\\
&\mathbf{q} = -\mathbf{K}\nabla\vartheta,
\end{aligned}} \tag{14.5.2}$$

where $\mathbb{C}$, $\mathbf{M}$, c, and $\mathbf{K}$ are, respectively, the elasticity tensor, the stress-temperature modulus, the specific heat, and the thermal conductivity tensor at the reference temperature ϑ_0. The elasticity tensor $\mathbb{C}$ is symmetric and *positive definite*,

$$\mathbf{A}:\mathbb{C}\mathbf{A} > 0 \quad \text{for all symmetric tensors } \mathbf{A} \neq \mathbf{0}. \tag{14.5.3}$$

The stress-temperature tensor is *symmetric*,

$$\mathbf{M} = \mathbf{M}^{\top}. \tag{14.5.4}$$

The specific heat c is *positive*,

$$c > 0. \tag{14.5.5}$$

Finally, the thermal conductivity tensor $\mathbf{K}$ is symmetric and positive definite,

$$\mathbf{a}\cdot\mathbf{K}\mathbf{a} > 0 \quad \text{for all vectors } \mathbf{a} \neq \mathbf{0}. \tag{14.5.6}$$

Substituting for the stress from $(14.5.2)_2$ into the balance of linear momentum (14.2.2) gives the **local momentum balance**:

$$\boxed{\rho\ddot{\mathbf{u}} = \operatorname{div}(\mathbb{C}\boldsymbol{\epsilon} + \mathbf{M}(\vartheta - \vartheta_0)) + \mathbf{b}.} \tag{14.5.7}$$

Next, from the constitutive equation (14.3.4) for the internal energy

$$\dot{\varepsilon} = \frac{\partial\hat{\psi}}{\partial\boldsymbol{\epsilon}}:\dot{\boldsymbol{\epsilon}} + \frac{\partial\hat{\psi}}{\partial\vartheta}\dot{\vartheta} + \eta\dot{\vartheta} + \vartheta\frac{\partial\hat{\eta}}{\partial\boldsymbol{\epsilon}}:\dot{\boldsymbol{\epsilon}} + \vartheta\frac{\partial\hat{\eta}}{\partial\vartheta}\dot{\vartheta}, \tag{14.5.8}$$

and using (14.3.8) and (14.3.9) this reduces to

$$\dot{\varepsilon} = \boldsymbol{\sigma} : \dot{\boldsymbol{\epsilon}} + \vartheta \frac{\partial \hat{\eta}}{\partial \boldsymbol{\epsilon}} : \dot{\boldsymbol{\epsilon}} - \vartheta \frac{\partial^2 \hat{\psi}}{\partial \vartheta^2} \dot{\vartheta}. \tag{14.5.9}$$

For *linear thermoelastic* materials (with ϑ close to ϑ_0), use of (14.4.9) and (14.4.12) reduces this to

$$\dot{\varepsilon} = \boldsymbol{\sigma} : \dot{\boldsymbol{\epsilon}} - \vartheta_0 \mathbf{M} : \dot{\boldsymbol{\epsilon}} + c\dot{\vartheta}. \tag{14.5.10}$$

Finally substituting (14.5.10) in (14.2.3) gives the **local energy balance** as an evolution equation for the temperature,

$$\boxed{c\dot{\vartheta} = \mathbf{K} : \nabla\nabla\vartheta + \vartheta_0 \mathbf{M} : \dot{\boldsymbol{\epsilon}} + r.} \tag{14.5.11}$$

The *basic equations of the linear theory of anisotropic thermoelasticity* consist of (14.5.1), (14.5.2), the local momentum balance (14.5.7), and the local energy balance (14.5.11).

14.6 **Isotropic linear thermoelasticity**

If the body is isotropic then $\mathbb{C}$, $\mathbf{M}$, and $\mathbf{K}$ have the specific forms

$$\left.\begin{aligned} \mathbb{C} &= 2\mu\mathbb{I}^{\text{sym}} + \lambda \mathbf{1} \otimes \mathbf{1}, \\ \mathbf{M} &= \beta \mathbf{1}, \\ \mathbf{K} &= k\mathbf{1}, \end{aligned}\right\} \tag{14.6.1}$$

with μ and λ **elastic Lamé moduli**,[6] β the **stress-temperature modulus**, and k the (scalar) **thermal conductivity**.

Next, as in linear elasticity, the positive definiteness requirement (14.5.3) dictates that the Lamé moduli satisfy,

$$\mu > 0, \qquad \lambda + \frac{2}{3}\mu > 0. \tag{14.6.2}$$

Relation $(14.6.1)_1$ may alternatively be written in terms of the scalars μ and κ as[7]

$$\mathbb{C} = 2\mu\mathbb{I}^{\text{symdev}} + 3\kappa\mathbb{I}^{\text{vol}}, \tag{14.6.3}$$

where

$$\mathbb{I}^{\text{symdev}} = \mathbb{I}^{\text{sym}} - \mathbb{I}^{\text{vol}} \qquad \text{with} \qquad \mathbb{I}^{\text{vol}} = \tfrac{1}{3}\mathbf{1} \otimes \mathbf{1}.$$

In view of (14.6.3), the Lamé modulus μ is also known as the **isothermal shear modulus**, while

$$\kappa = \lambda + \frac{2}{3}\mu, \tag{14.6.4}$$

is the **isothermal bulk modulus**.

---

[6] Cf. (7.5.3).

[7] Cf. Sec. 7.5.1.

Further, since by (14.5.6) $\mathbf{K}$ is positive definite, the thermal conductivity k must be positive

$$k > 0. \tag{14.6.5}$$

By (14.6.1), the defining constitutive equations for an isotropic, linear thermoelastic solid are,

$$\boxed{\begin{aligned} \psi &= \mu|\epsilon|^2 + \frac{\lambda}{2}(\operatorname{tr}\epsilon)^2 + \beta(\vartheta - \vartheta_0)\operatorname{tr}\epsilon - \frac{c}{2\vartheta_0}(\vartheta - \vartheta_0)^2, \\ \boldsymbol{\sigma} &= 2\mu\epsilon + \lambda(\operatorname{tr}\epsilon)\mathbf{1} + \beta(\vartheta - \vartheta_0)\,\mathbf{1}, \\ \eta &= -\beta\operatorname{tr}\epsilon + \frac{c}{\vartheta_0}(\vartheta - \vartheta_0), \\ \mathbf{q} &= -k\nabla\vartheta. \end{aligned}} \tag{14.6.6}$$

Granted (14.6.2), the stress-strain relation $(14.6.6)_2$ may be inverted to give

$$\boxed{\epsilon = \frac{1}{2\mu}\left[\boldsymbol{\sigma} - \frac{\lambda}{2\mu + 3\lambda}(\operatorname{tr}\boldsymbol{\sigma})\mathbf{1}\right] + \alpha(\vartheta - \vartheta_0)\mathbf{1},} \tag{14.6.7}$$

where

$$\boxed{\alpha \stackrel{\text{def}}{=} -\frac{\beta}{2\mu + 3\lambda},} \tag{14.6.8}$$

and equivalently,

$$\boxed{\beta = -3\kappa\alpha.} \tag{14.6.9}$$

The coefficient α is called the *coefficient of thermal expansion.*

In applications it is often useful to express the stress-strain relation in terms of (E, ν, α) rather that (μ, λ, β). Upon using appropriate conversion relations (see Table 7.4) we find that

$$\boxed{\epsilon = \frac{1}{E}\left[(1 + \nu)\boldsymbol{\sigma} - \nu(\operatorname{tr}\boldsymbol{\sigma})\mathbf{1}\right] + \alpha(\vartheta - \vartheta_0)\mathbf{1},} \tag{14.6.10}$$

which can be inverted to give

$$\boxed{\boldsymbol{\sigma} = \frac{E}{(1 + \nu)}\left[\epsilon + \frac{\nu}{(1 - 2\nu)}(\operatorname{tr}\epsilon)\,\mathbf{1} - \frac{(1 + \nu)}{(1 - 2\nu)}\alpha\,(\vartheta - \vartheta_0)\,\mathbf{1}\right].} \tag{14.6.11}$$

14.6.1 **Homogeneous and isotropic linear thermoelasticity**

Next, when $\mathcal{B}$ is homogeneous and isotropic, then ρ, μ, λ, β, and k are *constants*, independent of position in $\mathcal{B}$. In this case, since

$$\begin{aligned} 2\operatorname{div}\epsilon &= \operatorname{div}(\nabla\mathbf{u} + (\nabla\mathbf{u})^{\top}) \\ &= \Delta\mathbf{u} + \nabla\operatorname{div}\mathbf{u}, \end{aligned}$$

and

$$\begin{aligned}\operatorname{div}((\operatorname{tr}\boldsymbol{\epsilon})\mathbf{1}) &= \operatorname{div}((\operatorname{div}\mathbf{u})\mathbf{1})\\ &= \nabla \operatorname{div}\mathbf{u},\end{aligned}$$

the momentum balance (14.5.7) yields

$$\boxed{\mu\Delta\mathbf{u} + (\lambda+\mu)\nabla\operatorname{div}\mathbf{u} - (3\lambda+2\mu)\alpha\nabla\tilde{\vartheta} + \mathbf{b} = \rho\,\ddot{\mathbf{u}},} \tag{14.6.12}$$

where

$$\tilde{\vartheta} \stackrel{\text{def}}{=} \vartheta - \vartheta_0. \tag{14.6.13}$$

This equation may be written in an alternative form by using the identity

$$\operatorname{curl}\operatorname{curl}\mathbf{u} = \nabla\operatorname{div}\mathbf{u} - \Delta\mathbf{u}, \tag{14.6.14}$$

as

$$\boxed{(\lambda+2\mu)\nabla\operatorname{div}\mathbf{u} - \mu\operatorname{curl}\operatorname{curl}\mathbf{u} - (3\lambda+2\mu)\alpha\nabla\tilde{\vartheta} + \mathbf{b} = \rho\,\ddot{\mathbf{u}}.} \tag{14.6.15}$$

Further, the balance of energy (14.5.11), in the homogeneous and isotropic case becomes,

$$\boxed{c\dot{\vartheta} = k\Delta\vartheta + \beta\vartheta_0\operatorname{div}\dot{\mathbf{u}} + r.} \tag{14.6.16}$$

In engineering practice, the coupling term

$$\beta\vartheta_0\operatorname{div}\dot{\mathbf{u}},$$

in the partial differential equation (14.6.16) for ϑ is often neglected due to its small magnitude for many materials. Under such an approximation, the resulting theory is referred to as the *uncoupled theory* (or sometimes as the *weakly coupled or one-way coupled* theory) of isotropic linear thermoelasticity—the temperature affects the mechanical response, but the deformation does not effect the thermal response. The complete set of governing equations for this important practical case are summarized in the next section.

14.6.2 **Uncoupled theory of isotropic linear thermoelasticity**

The general uncoupled problem of (homogeneous) isotropic thermoelasticity consists of

(i) The strain-displacement relation

$$\boldsymbol{\epsilon} = \frac{1}{2}\left(\nabla\mathbf{u} + (\nabla\mathbf{u})^{\top}\right), \qquad \epsilon_{ij} = \frac{1}{2}(u_{i,j} + u_{j,i}). \tag{14.6.17}$$

(ii) The stress-strain-temperature relation

$$\begin{aligned}\boldsymbol{\sigma} &= 2\mu\boldsymbol{\epsilon} + \lambda(\operatorname{tr}\boldsymbol{\epsilon})\mathbf{1} - (3\lambda+2\mu)\alpha\tilde{\vartheta}\mathbf{1},\\ \sigma_{ij} &= 2\mu\epsilon_{ij} + \lambda\epsilon_{kk}\delta_{ij} - (3\lambda+2\mu)\alpha\tilde{\vartheta}\delta_{ij},\end{aligned} \tag{14.6.18}$$

where

$$\tilde{\vartheta} \stackrel{\text{def}}{=} \vartheta - \vartheta_0. \tag{14.6.19}$$

The inverted form of (14.6.18) is

$$\begin{aligned}\boldsymbol{\epsilon} &= \frac{(1+\nu)}{E}\boldsymbol{\sigma} - \frac{\nu}{E}\sigma_{kk}\mathbf{1} + \alpha\tilde{\vartheta}\mathbf{1},\\ \epsilon_{ij} &= \frac{(1+\nu)}{E}\sigma_{ij} - \frac{\nu}{E}\sigma_{kk}\delta_{ij} + \alpha\tilde{\vartheta}\delta_{ij}.\end{aligned} \tag{14.6.20}$$

(iii) The equation of motion

$$\operatorname{div}\boldsymbol{\sigma} + \mathbf{b} = \rho\,\ddot{\mathbf{u}}, \qquad \sigma_{ij,j} + b_i = \rho\,\ddot{u}_i. \tag{14.6.21}$$

Using the strain-displacement relation and the stress-strain-temperature relation this may be written as

$$\mu\Delta\mathbf{u} + (\lambda+\mu)\nabla\operatorname{div}\mathbf{u} - (3\lambda+2\mu)\alpha\nabla\tilde{\vartheta} + \mathbf{b} = \rho\,\ddot{\mathbf{u}} \tag{14.6.22}$$

or equivalently as

$$(\lambda+2\mu)\nabla\operatorname{div}\mathbf{u} - \mu\operatorname{curl}\operatorname{curl}\mathbf{u} - (3\lambda+2\mu)\alpha\nabla\tilde{\vartheta} + \mathbf{b} = \rho\,\ddot{\mathbf{u}}. \tag{14.6.23}$$

(iv) The energy equation

$$\rho c\dot{\vartheta} = k\Delta\vartheta + r, \qquad \rho c\dot{\vartheta} = k\vartheta_{,ii} + r. \tag{14.6.24}$$

REMARK 14.3

In (14.6.24) we have reverted to tradition and introduced the mass density ρ, since the specific heat c is usually specified in units of J/(kg K), rather than J/(m^3K) as it appears in (14.6.16).

14.7 Boundary and initial conditions

Let $\mathcal{S}_\mathbf{u}$ and $\mathcal{S}_\mathbf{t}$ denote complementary subsurfaces of the boundary $\partial\mathcal{B}$ of the body $\mathcal{B}$. Then, as is standard, for a time interval $t \in [0, T]$ we consider a pair of simple boundary conditions in which the displacement is specified on $\mathcal{S}_\mathbf{u}$ and the surface traction on $\mathcal{S}_\mathbf{t}$:

$$\begin{aligned}\mathbf{u} &= \bar{\mathbf{u}} \quad \text{on } \mathcal{S}_\mathbf{u} \times [0, T],\\ \boldsymbol{\sigma}\mathbf{n} &= \bar{\mathbf{t}} \quad \text{on } \mathcal{S}_\mathbf{t} \times [0, T].\end{aligned} \tag{14.7.1}$$

The initial conditions for the mechanical problem are as before, displacements and velocities throughout the body.

The thermal boundary conditions take the form of either prescribed temperatures or normal heat-flux $\mathbf{q}\cdot\mathbf{n}$ on the boundary surface, as discussed below. For the thermal problem one needs to specify the initial temperature distribution throughout the body, which is most often constant. As boundary conditions, temperature is prescribed on a portion $\mathcal{S}_\vartheta$ of the boundary

$\partial\mathcal{B}$, while on another portion $\mathcal{S}_q$ the normal component of the heat flux is prescribed. The temperature and heat flux cannot both be prescribed on a given portion of the boundary, so

$$\mathcal{S}_\vartheta \cap \mathcal{S}_q = \emptyset.$$

However, on every portion of the boundary either the temperature or the heat flux must be prescribed, so

$$\mathcal{S}_\vartheta \cup \mathcal{S}_q = \partial\mathcal{B}.$$

Thus, with $\mathcal{S}_\vartheta$ and $\mathcal{S}_q$ *complementary subsurfaces* of the boundary $\partial\mathcal{B}$ of the body $\mathcal{B}$, typical **boundary conditions** are

$$\begin{aligned} \vartheta(\mathbf{x},t) &= \bar{\vartheta}(\mathbf{x},t) \quad \text{on } \mathcal{S}_\vartheta \times [0,T], \\ q_n(\equiv \mathbf{q}\cdot\mathbf{n}) &= \bar{q}(\mathbf{x},t) \quad \text{on } \mathcal{S}_q \times [0,T], \end{aligned} \tag{14.7.2}$$

with $\bar{\vartheta}(\mathbf{x},t)$ and $\bar{q}(\mathbf{x},t)$ *prescribed* functions of $\mathbf{x}$ and t.

Some common (idealized) boundary conditions for the normal heat flux are as follows:

1. **Perfectly insulated surface**:
$$\bar{q}(\mathbf{x},t) = 0. \tag{14.7.3}$$

2. **Convection boundary condition**:
At a solid-fluid interface the heat flux is related linearly to the difference between the temperature of the surface $\vartheta(\mathbf{x},t)$ and the far field ambient fluid (sink) temperature ϑ^∞:
$$\bar{q}(\mathbf{x},t) = h(\mathbf{x},t)\,(\vartheta(\mathbf{x},t) - \vartheta^\infty), \tag{14.7.4}$$
where $h(\mathbf{x},t)$ is called the *film coefficient*. This relation is known as *Newton's law of cooling*.

3. **Radiation boundary condition**:
If the surface of the body is exposed to a high-temperature source, it will receive heat by radiation in accordance with a law of the form
$$\bar{q}(\mathbf{x},t) = A(\mathbf{x},t)\,\big((\vartheta(\mathbf{x},t))^4 - (\vartheta^\infty)^4\big), \tag{14.7.5}$$
where $\vartheta(\mathbf{x},t)$ is the temperature of the surface, ϑ^∞ is the temperature of the source, and $A(\mathbf{x},t)$ is the *effective* radiation parameter (that is, the emissivity of the source, including the view factor, times the Stefan–Boltzmann constant).

REMARK 14.4

We present these common boundary conditions in reasonable generality but they often also appear in more complex forms; see e.g. Lienhard IV and Lienhard V (2007) or Annaratone (2010) for further details.

14.8 **Example problems**

EXAMPLE 14.1 Thermally induced stresses in a thick-walled sphere

Consider a thick-walled sphere made from a homogeneous, isotropic, linear thermoelastic material. The inner surface of the sphere at $r = r_1$ is held at a constant temperature ϑ_1, and the outer surface $r = r_2$ is held at a constant temperature $\vartheta_2 < \vartheta_1$, and assume that the temperature field in the sphere is steady, $\vartheta = \hat{\vartheta}(\mathbf{x})$ independent of time. Determine the stress distribution in the sphere (neglect inertial and body forces).

To solve this problem, we use a spherical coordinate system (r, θ, ϕ) with origin at the center of the sphere, so that the body occupies the region $r_1 \leq r \leq r_2$. For this problem, there are only null traction boundary conditions, so any solution to the displacement field that we shall determine will be modulo a rigid displacement. Since there are no prescribed tractions at $r = r_1$ and $r = r_2$, we have

$$\begin{aligned} \sigma_{rr} = \sigma_{\theta r} = \sigma_{\phi r} = 0 \quad &\text{on} \quad r = r_1, \\ \sigma_{rr} = \sigma_{\theta r} = \sigma_{\phi r} = 0 \quad &\text{on} \quad r = r_2. \end{aligned} \tag{14.8.1}$$

The prescribed temperatures are

$$\begin{aligned} \vartheta = \vartheta_1 \quad &\text{on} \quad r = r_1, \\ \vartheta = \vartheta_2 \quad &\text{on} \quad r = r_2. \end{aligned} \tag{14.8.2}$$

Because of the isotropy of the material, and because of the symmetry of the loading, we *assume* that the displacement field $\mathbf{u}$ is directed in the radial direction and depends only on r:

$$\mathbf{u} = u_r(r)\mathbf{e}_r \qquad \text{with} \qquad u_\theta = u_\phi = 0. \tag{14.8.3}$$

In the absence of body forces, and neglect of inertia, (14.6.22) reduces to

$$(\lambda + 2\mu)\nabla \operatorname{div} \mathbf{u} - \mu \operatorname{curl} \operatorname{curl} \mathbf{u} - (3\lambda + 2\mu)\alpha \nabla \tilde{\vartheta} = \mathbf{0}, \tag{14.8.4}$$

where we recall the notation

$$\tilde{\vartheta} \stackrel{\text{def}}{=} \vartheta - \vartheta_0, \tag{14.8.5}$$

with ϑ_0 a constant reference temperature. Also, recalling the expression (A.3.16) for the curl of a vector field $\mathbf{v}$ in spherical coordinates

$$\begin{aligned} \operatorname{curl} \mathbf{v} = &\left\{ \frac{1}{r}\frac{\partial v_\phi}{\partial \theta} - \frac{1}{r \sin\theta}\frac{\partial v_\theta}{\partial \phi} + \cot\theta \frac{v_\phi}{r} \right\} \mathbf{e}_r + \left\{ \frac{1}{r \sin\theta}\frac{\partial v_r}{\partial \phi} - \frac{\partial v_\phi}{\partial r} - \frac{v_\phi}{r} \right\} \mathbf{e}_\theta \\ &+ \left\{ \frac{\partial v_\theta}{\partial r} - \frac{1}{r}\frac{\partial v_r}{\partial \theta} + \frac{v_\theta}{r} \right\} \mathbf{e}_\phi, \end{aligned}$$

we see that for the presumed displacement field (14.8.3) the curl vanishes:

$$\operatorname{curl} \mathbf{u} = \mathbf{0}.$$

Using this in (14.8.4), and noting that $(\lambda + 2\mu) \neq 0$, we must have

$$\nabla \operatorname{div} \mathbf{u} = \frac{(3\lambda + 2\mu)}{(\lambda + 2\mu)} \alpha \nabla \tilde{\vartheta}. \tag{14.8.6}$$

In a spherical coordinate system the divergence of a vector field $\mathbf{v}$ is given by (A.3.14):

$$\operatorname{div} \mathbf{v} = \frac{\partial v_r}{\partial r} + \frac{1}{r}\frac{\partial v_\theta}{\partial \theta} + \frac{1}{r \sin\theta}\frac{\partial v_\phi}{\partial \phi} + \frac{\cot\theta}{r} v_\theta + \frac{2 v_r}{r}.$$

Hence, for the displacement field (14.8.3) we obtain

$$\operatorname{div} \mathbf{u} = \frac{d u_r}{dr} + \frac{2 u_r}{r}. \tag{14.8.7}$$

Further, in a spherical coordinate system the gradient of a scalar field ψ is

$$\nabla \psi = \frac{\partial \psi}{\partial r}\mathbf{e}_r + \frac{1}{r}\frac{\partial \psi}{\partial \theta}\mathbf{e}_\theta + \frac{1}{r\sin\theta}\frac{\partial \psi}{\partial \phi}\mathbf{e}_\phi. \tag{14.8.8}$$

Since u_r is only a function of r we have

$$\nabla \operatorname{div} \mathbf{u} = \frac{d}{dr}\left(\frac{d u_r}{dr} + \frac{2 u_r}{r}\right)\mathbf{e}_r. \tag{14.8.9}$$

Further, because of isotropy and the symmetry of the loading, the temperature field in the sphere also depends only on r

$$\vartheta(r).$$

Hence

$$\nabla \tilde{\vartheta} = \frac{d\tilde{\vartheta}}{dr}\mathbf{e}_r. \tag{14.8.10}$$

Use of (14.8.9) and (14.8.10) gives

$$\frac{d}{dr}\left(\frac{d u_r}{dr} + \frac{2 u_r}{r}\right) = \frac{(3\lambda + 2\mu)}{(\lambda + 2\mu)} \alpha \frac{d\tilde{\vartheta}}{dr} \tag{14.8.11}$$

or equivalently

$$\frac{d u_r}{dr} + \frac{2 u_r}{r} = 3A + \frac{(3\lambda + 2\mu)}{(\lambda + 2\mu)} \alpha\, \tilde{\vartheta}, \tag{14.8.12}$$

where we have selected the constant of integration to be $3A$ for later convenience. Multiplying (14.8.12) by r^2 gives,

$$r^2 \frac{d u_r}{dr} + 2 r u_r = 3 A r^2 + \frac{(3\lambda + 2\mu)}{(\lambda + 2\mu)} \alpha\, r^2 \tilde{\vartheta}.$$

Noting that

$$\frac{d(r^2 u_r)}{dr} = r^2 \frac{d u_r}{dr} + 2 r u_r,$$

we obtain

$$\frac{d(r^2 u_r)}{dr} = 3 A r^2 + \frac{(3\lambda + 2\mu)}{(\lambda + 2\mu)} \alpha\, r^2 \tilde{\vartheta}. \tag{14.8.13}$$

Integrating (14.8.13) we obtain

$$r^2 u_r(r) = Ar^3 + B + \frac{(3\lambda + 2\mu)}{(\lambda + 2\mu)} \alpha \int_{r_1}^{r} x^2 \tilde{\vartheta}(x)\, dx,$$

where B is another constant of integration, and hence

$$u_r(r) = Ar + \frac{B}{r^2} + \frac{(3\lambda + 2\mu)}{(\lambda + 2\mu)} \frac{\alpha}{r^2} \underbrace{\int_{r_1}^{r} x^2 \tilde{\vartheta}(x)\, dx}_{\equiv Y(r)} . \tag{14.8.14}$$

It remains to find the constants A and B, as well as $Y(r)$.

To this end, from (A.3.19) the only non-zero strain components are

$$\begin{aligned} \epsilon_{rr}(r) &= \frac{du_r}{dr} = A - 2\frac{B}{r^3} - \frac{2(3\lambda + 2\mu)}{(\lambda + 2\mu)} \frac{\alpha}{r^3} Y(r) + \frac{(3\lambda + 2\mu)}{(\lambda + 2\mu)} \alpha \tilde{\vartheta}(r), \\ \epsilon_{\theta\theta}(r) = \epsilon_{\phi\phi}(r) &= \frac{u_r}{r} = A + \frac{B}{r^3} + \frac{(3\lambda + 2\mu)}{(\lambda + 2\mu)} \frac{\alpha}{r^3} Y(r). \end{aligned} \tag{14.8.15}$$

Hence, using the constitutive equation for the stress (14.6.18) the radial stress component is

$$\begin{aligned} \sigma_{rr}(r) &= 2\mu \frac{du_r}{dr} + \lambda \left(\frac{du_r}{dr} + \frac{2u_r}{r} \right) - (3\lambda + 2\mu)\alpha\tilde{\vartheta}(r) \\ &= (2\mu + \lambda)\frac{du_r}{dr} + 2\lambda \left(\frac{u_r}{r} \right) - (3\lambda + 2\mu)\alpha\tilde{\vartheta}(r) \\ &= (2\mu + \lambda) \left(A - 2\frac{B}{r^3} \right) - 2(3\lambda + 2\mu)\frac{\alpha}{r^3} Y(r) + (3\lambda + 2\mu)\alpha\tilde{\vartheta}(r) \\ &\quad + 2\lambda \left(A + \frac{B}{r^3} \right) + 2\lambda \frac{(3\lambda + 2\mu)}{(\lambda + 2\mu)} \frac{\alpha}{r^3} Y(r), \\ &\quad - (3\lambda + 2\mu)\alpha\tilde{\vartheta}(r) \\ &= (2\mu + \lambda) \left(A - 2\frac{B}{r^3} \right) + 2\lambda \left(A + \frac{B}{r^3} \right) - 4\mu \frac{(3\lambda + 2\mu)}{(\lambda + 2\mu)} \frac{\alpha}{r^3} Y(r) \end{aligned}$$

or

$$\sigma_{rr}(r) = (2\mu + 3\lambda)A - \frac{4\mu B}{r^3} - 4\mu \frac{(3\lambda + 2\mu)}{(\lambda + 2\mu)} \frac{\alpha}{r^3} Y(r). \tag{14.8.16}$$

Also, the hoop stress components are

$$\begin{aligned} \sigma_{\theta\theta}(r) = \sigma_{\phi\phi}(r) &= 2\mu \frac{u_r}{r} + \lambda \left(\frac{du_r}{dr} + \frac{2u_r}{r} \right) - (3\lambda + 2\mu)\alpha\tilde{\vartheta}(r) \\ &= (2\mu + 2\lambda)\frac{u_r}{r} + \lambda \left(\frac{du_r}{dr} \right) - (3\lambda + 2\mu)\alpha\tilde{\vartheta}(r) \\ &= (2\mu + 2\lambda) \left(A + \frac{B}{r^3} \right) + (2\mu + 2\lambda)\frac{(3\lambda + 2\mu)}{(\lambda + 2\mu)} \frac{\alpha}{r^3} Y(r) \end{aligned}$$

$$+\lambda\left(A-2\frac{B}{r^3}\right)-\lambda\frac{2(3\lambda+2\mu)}{(\lambda+2\mu)}\frac{\alpha}{r^3}Y(r)+\lambda\frac{(3\lambda+2\mu)}{(\lambda+2\mu)}\alpha\tilde{\vartheta}(r)$$
$$-(3\lambda+2\mu)\alpha\tilde{\vartheta}(r)$$

or

$$\boxed{\sigma_{\theta\theta}(r)=\sigma_{\phi\phi}(r)=(2\mu+3\lambda)A+2\mu\frac{B}{r^3}+2\mu\frac{2(3\lambda+2\mu)}{(\lambda+2\mu)}\frac{\alpha}{r^3}Y(r)-2\mu\frac{2(3\lambda+2\mu)}{(\lambda+2\mu)}\alpha\tilde{\vartheta}(r),} \tag{14.8.17}$$

while the shear stress components vanish:

$$\sigma_{r\theta}=\sigma_{\theta\phi}=\sigma_{\phi r}=0. \tag{14.8.18}$$

It is useful to write the equations for u_r, and the non-zero stresses in terms of (E,ν) rather than (μ,λ). Using the relations

$$\mu=\frac{E}{2(1+\nu)},\qquad \lambda=\frac{E\nu}{(1+\nu)(1-2\nu)}, \tag{14.8.19}$$

a little algebra gives

$$\boxed{u_r(r)=Ar+\frac{B}{r^2}+\frac{(1+\nu)}{(1-\nu)}\frac{\alpha}{r^2}\underbrace{\int_{r_1}^{r}x^2\tilde{\vartheta}(x)\,dx}_{\equiv Y(r)},} \tag{14.8.20}$$

$$\boxed{\sigma_{rr}(r)=\frac{E}{(1-2\nu)}A-\frac{2E}{(1+\nu)}\frac{B}{r^3}-\frac{2E}{(1-\nu)}\frac{\alpha}{r^3}Y(r),} \tag{14.8.21}$$

and

$$\boxed{\sigma_{\theta\theta}(r)=\sigma_{\phi\phi}(r)=\frac{E}{(1-2\nu)}A+\frac{E}{(1+\nu)}\frac{B}{r^3}+\frac{E}{(1-\nu)}\frac{\alpha}{r^3}Y(r)-\frac{E}{(1-\nu)}\alpha\tilde{\vartheta}(r).} \tag{14.8.22}$$

Next, using (14.8.21) and the boundary conditions (14.8.1) we obtain

$$\begin{aligned}
&\frac{E}{(1-2\nu)}A-\frac{2E}{(1+\nu)}\frac{B}{r_1^3}=0,\\
&\frac{E}{(1-2\nu)}A-\frac{2E}{(1+\nu)}\frac{B}{r_2^3}=\frac{2E}{(1-\nu)}\frac{\alpha}{r^3}\underbrace{\int_{r_1}^{r_2}x^2\tilde{\vartheta}(x)\,dx}_{\equiv Z}.
\end{aligned} \tag{14.8.23}$$

Solving (14.8.23) for the unknowns A and B, one obtains

$$\begin{aligned}
A&=\frac{2\alpha(1-2\nu)}{(1-\nu)}\frac{1}{r_2^3-r_1^3}Z,\\
B&=\frac{\alpha(1+\nu)}{(1-\nu)}\frac{r_1^3}{r_2^3-r_1^3}Z.
\end{aligned} \tag{14.8.24}$$

To complete the solution, we need to determine the temperature distribution $\tilde{\vartheta}(r)$, $Y(r)$, and Z; once this is done the displacement and stress fields are determined from the equations above. For *steady* heat flow in the absence of a source term, the temperature field satisfies Laplace's equation

$$\Delta\vartheta = 0,$$

which in spherical coordinates, from (A.3.17) for a field which is only a function of r, is

$$\frac{d^2\vartheta}{dr^2} + \frac{2}{r}\frac{d\vartheta}{dr} = 0. \tag{14.8.25}$$

Since the heat flux in the sphere is

$$q = -k\frac{d\vartheta}{dr}, \tag{14.8.26}$$

we can rewrite (14.8.25)

$$\frac{dq}{dr} + \frac{2}{r}q = 0. \tag{14.8.27}$$

Thus, separating variables and integrating

$$\int \frac{dq}{q} = -2\int \frac{dr}{r}, \tag{14.8.28}$$

which gives

$$\ln q = -2\ln r + \ln C = \ln\frac{C}{r^2} \qquad \Longrightarrow \qquad q = \frac{C}{r^2}. \tag{14.8.29}$$

Hence

$$-k\frac{d\vartheta}{dr} = \frac{C}{r^2} \qquad \Longrightarrow \qquad -k\int d\vartheta = C\int r^{-2}dr \qquad \Longrightarrow \qquad -k\vartheta(r) = -\frac{C}{r} - D, \tag{14.8.30}$$

or absorbing the constant k,

$$\boxed{\vartheta(r) = \frac{C}{r} + D.} \tag{14.8.31}$$

Thus, using the boundary conditions (14.8.2)

$$\boxed{\vartheta(r) = \left(\frac{r_1 r_2(\vartheta_1 - \vartheta_2)}{(r_2 - r_1)}\right)\frac{1}{r} + \left(\frac{r_2\vartheta_2 - r_1\vartheta_1}{r_2 - r_1}\right).} \tag{14.8.32}$$

From which

$$\boxed{\tilde{\vartheta}(r) \equiv \vartheta(r) - \vartheta_0 = \left(\frac{r_1 r_2(\vartheta_1 - \vartheta_2)}{(r_2 - r_1)}\right)\frac{1}{r} + \left(\frac{r_2\vartheta_2 - r_1\vartheta_1}{r_2 - r_1} - \vartheta_0\right).} \tag{14.8.33}$$

Hence

$$\boxed{Y(r) \equiv \int_{r_1}^{r} x^2\tilde{\vartheta}(x)dx = \frac{1}{2}\left(\frac{r_1 r_2(\vartheta_1 - \vartheta_2)}{(r_2 - r_1)}\right)(r^2 - r_1^2) + \frac{1}{3}\left(\frac{r_2\vartheta_2 - r_1\vartheta_1}{r_2 - r_1} - \vartheta_0\right)(r^3 - r_1^3),} \tag{14.8.34}$$

and

$$Z = Y(r_2) \equiv \int_{r_1}^{r_2} x^2 \tilde{\vartheta}(x)dx = \frac{1}{2}\left(\frac{r_1 r_2(\vartheta_1 - \vartheta_2)}{(r_2 - r_1)}\right)(r_2^2 - r_1^2) + \frac{1}{3}\left(\frac{r_2\vartheta_2 - r_1\vartheta_1}{r_2 - r_1} - \vartheta_0\right)(r_2^3 - r_1^3). \tag{14.8.35}$$

This specifies the integration constants A and B above, and hence the required stress distribution in the sphere.

EXAMPLE 14.2 Thermoelastic spherical inclusion in an infinite solid

Consider an isotropic thermoelastic spherical inclusion which is embedded in an isotropic thermoelastic material with different material properties as shown in Fig. 14.1.

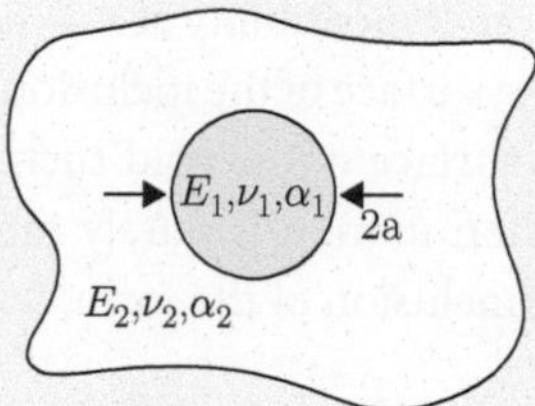

Fig. 14.1 Thermoelastic spherical inclusion of radius a in an infinite solid.

Assume that at a given reference temperature ϑ_0 that the inclusion has a radius a and that the composite system is stress free. Let us determine the state of deformation, strain, and stress, if the temperature were to uniformly change by an amount

$$\tilde{\vartheta} \stackrel{\text{def}}{=} \vartheta - \vartheta_0. \tag{14.8.36}$$

We can solve this problem by first considering what would happen to each material if the other were not present, the unconstrained or free expansion case. First assume a spherical coordinate system centered at the center of the inclusion. In the free expansion case, one can verify that in material 1 (an isolated sphere) the strain field

$$\boldsymbol{\epsilon}^{(f,1)} = \alpha_1 \tilde{\vartheta} \mathbf{1},$$

with corresponding stress field

$$\boldsymbol{\sigma}^{(f,1)} = \mathbf{0},$$

and displacement field

$$\mathbf{u}^{(f,1)} = \alpha_1 \tilde{\vartheta}\, r\mathbf{e}_r,$$

satisfies the strain-displacement, equilibrium, and constitutive relations assuming that the surface of the inclusion is traction free. The energy balance equation is trivially satisfied since the temperature is assumed constant throughout the body. Similarly, one can verify that in the free expansion case for material 2 (an infinite solid with a spherical void) the strain field

$$\boldsymbol{\epsilon}^{(f,2)} = \alpha_2 \tilde{\vartheta} \mathbf{1},$$

with corresponding stress field

$$\boldsymbol{\sigma}^{(f,2)} = \mathbf{0},$$

and displacement field

$$\mathbf{u}^{(f,2)} = \alpha_2 \tilde{\vartheta}\, r \mathbf{e}_r,$$

satisfies the strain-displacement, equilibrium, and constitutive relations assuming that the surface of the void were traction free.

We seemingly have a solution to our problem of interest, except for the fact that when $\alpha_1 \neq \alpha_2$, the displacement of material 1 at $r = a$ does not match that of material 2,

$$\mathbf{u}^{(f,1)}(a) \neq \mathbf{u}^{(f,2)}(a).$$

To correct for the displacement incompatibility at $r = a$, we search for a set of correction tractions that we can apply to the surface of the inclusion along with an equal and opposite set of tractions to apply to the surface of the void such that the two materials exactly fit together. Since the free expansion motion is purely radial, and uniform, it is natural to consider a traction field on the inclusion of the form

$$\hat{\mathbf{t}}^{(c,1)} = -p \mathbf{e}_r$$

and on the surface of the void

$$\hat{\mathbf{t}}^{(c,2)} = p \mathbf{e}_r.$$

The expectation is that the inclusion and the infinite solid will respond to these tractions with purely radial motions that are parameterized by the unknown correction parameter p, viz.

$$\mathbf{u}^{(c,1)} = u_r^{(c,1)}(p) \mathbf{e}_r \qquad \text{and} \qquad \mathbf{u}^{(c,2)} = u_r^{(c,2)}(p) \mathbf{e}_r.$$

The value of the correction parameter will then be set by the compatibility requirement that

$$\underbrace{u_r^{(1)}(a)}_{u_r^{(f,1)}(a)+u_r^{(c,1)}(a)} = \underbrace{u_r^{(2)}(a)}_{u_r^{(f,2)}(a)+u_r^{(c,2)}(a)}. \tag{14.8.37}$$

Note that the parameter p will then represent the interface pressure (normal traction) between the inclusion and the infinite solid (in the actual full problem), and that we are exploiting the fact that superposition holds in linear thermoelasticity.

To proceed, first observe that for the inclusion under the action of the boundary conditions

$$\hat{\mathbf{t}}^{(c,1)} = -p \mathbf{e}_r \quad \text{at} \quad r = a,$$

a consistent stress field is

$$\boldsymbol{\sigma}^{(c,1)} = -p \mathbf{1}$$

with corresponding strains

$$\boldsymbol{\epsilon}^{(c,1)} = \frac{1+\nu_1}{E_1}\boldsymbol{\sigma}^{(c,1)} - \frac{\nu_1}{E_1}(\operatorname{tr}\boldsymbol{\sigma}^{(c,1)})\mathbf{1} = -p\,\frac{1-2\nu_1}{E_1}\mathbf{1},$$

and displacements

$$\mathbf{u}^{(c,1)} = -p\,\frac{1-2\nu_1}{E_1}r\mathbf{e}_r.$$

The response of the infinite solid can be found from the solution to the thick-walled sphere with zero external pressure, with the outer radius going to infinity. In particular from (9.1.18), with $b \to \infty$, $p_0 = 0$, and $p_i = p$, we have

$$\mathbf{u}^{(c,2)} = p\frac{1+\nu_2}{2E_2}\frac{a^3}{r^2},$$

where use has been made of Table 7.4 to convert the Lamé moduli to Young's modulus, E, and Poisson's ratio, ν.

Using (14.8.37) we find

$$\alpha_1\tilde{\vartheta}a - p\frac{1-2\nu_1}{E_1}a = \alpha_2\tilde{\vartheta}a + p\frac{1+\nu_2}{2E_2}a.$$

Solving for p gives,

$$p = \frac{2E_1E_2(\alpha_1-\alpha_2)\tilde{\vartheta}}{(1+\nu_2)E_1 + 2E_2(1-2\nu_1)}.$$

The complete solution fields in the inclusion ($r \leq a$) are

$$\mathbf{u}^{(1)} = \alpha_1\tilde{\vartheta}r - p\frac{1-2\nu_1}{E_1}r,$$

$$\boldsymbol{\epsilon}^{(1)} = \alpha_1\tilde{\vartheta}\mathbf{1} - p\frac{1-2\nu_1}{E_1}\mathbf{1},$$

$$\boldsymbol{\sigma}^{(1)} = -p\mathbf{1},$$

and in the infinite solid ($r \geq a$)

$$\mathbf{u}^{(2)} = \alpha_2\tilde{\vartheta}r + p\frac{1+\nu_2}{E_2}\frac{a^3}{r^2},$$

$$\boldsymbol{\epsilon}^{(2)} = \alpha_2\tilde{\vartheta}\mathbf{1} + p\frac{1+\nu_2}{E_2}\frac{a^3}{r^3}\left[-2\mathbf{e}_r\otimes\mathbf{e}_r + \mathbf{e}_\theta\otimes\mathbf{e}_\theta + \mathbf{e}_\phi\otimes\mathbf{e}_\phi\right],$$

$$\boldsymbol{\sigma}^{(2)} = p\frac{a^3}{r^3}\left[-2\mathbf{e}_r\otimes\mathbf{e}_r + \mathbf{e}_\theta\otimes\mathbf{e}_\theta + \mathbf{e}_\phi\otimes\mathbf{e}_\phi\right].$$

EXAMPLE 14.3 Analysis of thin films with thermal mismatch

Consider a cylindrical substrate of radius R and thickness t_s, with $R \gg t_s$. A thin film of thickness t_f, with $t_f \ll t_s$, is perfectly bonded on the upper surface of the substrate as shown in Fig. 14.2. The film and the substrate typically have different mechanical and thermal properties. One important consequence of this is that upon a change in temperature of the system from a stress-free state at reference temperature ϑ_0, the composite will bend — inducing a curvature in the system. Of interest is the derivation of formulae that describe the resulting stress state in the film and the substrate, as well as formulae describing the induced curvature.

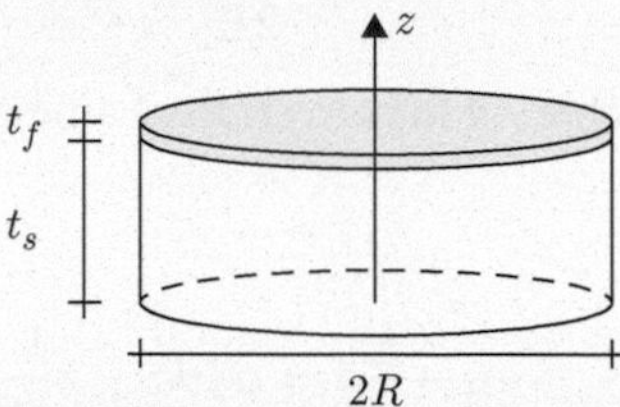

Fig. 14.2 A circular thin substrate with a thin film attached on its upper surface (not to scale).

To proceed, we will follow the logical developments of Sec. 9.3.1 by first considering the traction boundary conditions on the composite system, whose satisfaction we will assume only on average. This will lead to a hypothesis on the stress state in the body and subsequently to the strains and displacements. To effect the analysis we will employ a cylindrical coordinate system with the reference $z = 0$ plane on the bottom face of the substrate as shown in Fig. 14.2.

First note that at $r = R$ the composite body is traction free. Thus on the lateral surface $(r = R)$

$$\boldsymbol{\sigma}\mathbf{e}_r = \mathbf{0} \qquad \Rightarrow \qquad \sigma_{rr} = \sigma_{\theta r} = \sigma_{zr} = 0. \tag{14.8.38}$$

In particular this means that the net force and the net moment (per unit perimeter length) must be zero on the lateral surface:

$$\mathbf{F} = \int_0^{t_s+t_f} \boldsymbol{\sigma}\mathbf{e}_r \, dx_3 = \mathbf{0} \tag{14.8.39}$$

$$\mathbf{M} = \int_0^{t_s+t_f} z\mathbf{e}_z \times \boldsymbol{\sigma}\mathbf{e}_r \, dx_3 = \mathbf{0}. \tag{14.8.40}$$

Instead of enforcing the pointwise boundary conditions (14.8.38), we will only enforce (14.8.39) and (14.8.40). There are a wide number of traction distributions that will satisfy (14.8.39) and (14.8.40) and following Sec. 9.3.1 we will assume a linear distribution for the radial normal stress on the edge. However we need to acknowledge that the mechanical properties of the thin film differ from the substrate. In particular, we should allow for a jump in the radial normal stress as one crosses the interface between the two materials.

Also, since the film is very much thinner than the substrate we will assume the radial normal stress in the film to be a constant, σ_f (which is as yet unknown), as shown in Fig. 14.3.

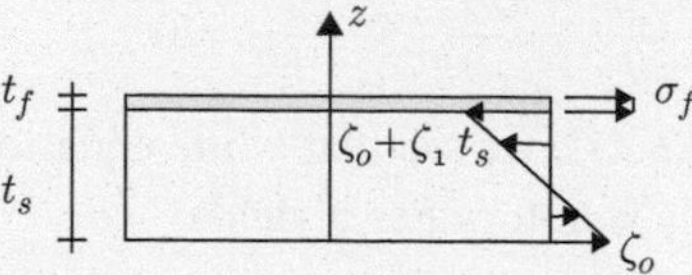

Fig. 14.3 Assumed stress distribution in the composite body (not to scale).

Thus we assume that at $r = R$,

$$\sigma_{\theta r} = \sigma_{zr} = 0,$$
$$\sigma_{rr} = \begin{cases} \sigma_f & z > t_s, \\ \zeta_o + \zeta_1 z & 0 \le z < t_s. \end{cases} \tag{14.8.41}$$

The constants ζ_o and ζ_1 can be determined from (14.8.39) and (14.8.40). Substituting (14.8.41) into (14.8.39) and (14.8.40) gives two equations for ζ_o and ζ_1:

$$\begin{aligned} \zeta_o t_s + \frac{1}{2}\zeta_1 t_s^2 + \sigma_f t_f &= 0, \\ \frac{1}{2}\zeta_o t_s^2 + \frac{1}{3}\zeta_1 t_s^3 + \sigma_f t_f t_s &= 0. \end{aligned} \tag{14.8.42}$$

The pair of linear equations (14.8.42) may be solved to yield:

$$\boxed{\zeta_o = 2\sigma_f \frac{t_f}{t_s} \qquad \text{and} \qquad \zeta_1 = -6\sigma_f \frac{t_f}{t_s^2}.} \tag{14.8.43}$$

With these values for ζ_o and ζ_1, the assumed stress distribution satisfies the average boundary conditions (14.8.39) and (14.8.40).

REMARK 14.5

Note that ζ_o is proportional to t_f/t_s which is a very small number. Thus the stresses are dominated by the film stress. The stresses in the substrate are much smaller.

Proceeding again as in Sec. 9.3.1 we will assume that the stress distribution (14.8.41) applies throughout the entire composite body, with the remaining components assumed to be zero ($\sigma_{\theta\theta} = \sigma_{zz} = \sigma_{z\theta} = 0$). To see if this is permissible, we substitute the assumed stress field into the equilibrium equations (cf. (A.2.21) with zero body force and neglect of inertia):

$$\begin{aligned}
&\frac{\partial \sigma_{rr}}{\partial r} + \frac{1}{r}\frac{\partial \sigma_{r\theta}}{\partial \theta} + \frac{\partial \sigma_{rz}}{\partial z} + \frac{1}{r}(\sigma_{rr} - \sigma_{\theta\theta}) = 0,\\
&\frac{\partial \sigma_{\theta r}}{\partial r} + \frac{1}{r}\frac{\partial \sigma_{\theta\theta}}{\partial \theta} + \frac{\partial \sigma_{\theta z}}{\partial z} + \frac{2}{r}\sigma_{\theta r} = 0,\\
&\frac{\partial \sigma_{zr}}{\partial r} + \frac{1}{r}\frac{\partial \sigma_{z\theta}}{\partial \theta} + \frac{\partial \sigma_{zz}}{\partial z} + \frac{\sigma_{zr}}{r} = 0.
\end{aligned} \tag{14.8.44}$$

With the assumed field (14.8.41), we see that while equations $(14.8.44)_{2,3}$ are identically satisfied, equation $(14.8.44)_1$ is not; rather it yields

$$\frac{1}{r}(\sigma_{rr} - \sigma_{\theta\theta}) = 0.$$

Thus we need to modify our assumption (14.8.41) of the assumed stress field to read:

$$\boxed{\begin{aligned}
&\sigma_{\theta r} = \sigma_{zr} = \sigma_{z\theta} = \sigma_{zz} = 0\\
&\sigma_{rr} = \begin{cases} \sigma_f & z > t_s \\ \zeta_o + \zeta_1 z & 0 \le z < t_s \end{cases}\\
&\sigma_{\theta\theta} = \sigma_{rr}.
\end{aligned}} \tag{14.8.45}$$

This stress field, which is biaxial, satisfies the equilibrium equations (14.8.44) and the average boundary conditions (14.8.39) and (14.8.40).

The corresponding strain field is given by:

$$\boldsymbol{\epsilon} = \frac{1}{E}\left((1+\nu)\,\boldsymbol{\sigma} - \nu\,(\operatorname{tr}\boldsymbol{\sigma})\,\mathbf{1}\right) + \alpha(\vartheta - \vartheta_0)\mathbf{1} \tag{14.8.46}$$

or in components

$$\begin{aligned}
\epsilon_{rr} &= \frac{1-\nu}{E}\sigma_{rr} + \alpha(\vartheta - \vartheta_0)\\
\epsilon_{\theta\theta} &= \frac{1-\nu}{E}\sigma_{rr} + \alpha(\vartheta - \vartheta_0)\\
\epsilon_{zz} &= -\frac{2\nu}{E}\sigma_{rr} + \alpha(\vartheta - \vartheta_0)\\
\epsilon_{r\theta} &= \epsilon_{rz} = \epsilon_{\theta z} = 0.
\end{aligned} \tag{14.8.47}$$

It is important to note that at the interface between the film and the substrate, the radial and the hoop strains must match when evaluated on either side of the interface, otherwise we will not have compatibility of the material at the interface. This implies that at $z = t_s$

$$\frac{1-\nu_f}{E_f}\sigma_{rr}(t_s^+) + \alpha_f(\vartheta - \vartheta_0) = \frac{1-\nu_s}{E_s}\sigma_{rr}(t_s^-) + \alpha_s(\vartheta - \vartheta_0), \tag{14.8.48}$$

or by using (14.8.45) and (14.8.43)

$$\frac{1-\nu_f}{E_f}\sigma_f + \alpha_f(\vartheta - \vartheta_0) = -4\frac{1-\nu_s}{E_s}\sigma_f\frac{t_f}{t_s} + \alpha_s(\vartheta - \vartheta_0). \tag{14.8.49}$$

Equation (14.8.49) can be solved to find an expression for the assumed constant stress in the film in terms of the material properties and the temperature change:

$$\boxed{\sigma_f = \frac{(\alpha_s - \alpha_f)(\vartheta - \vartheta_0)}{\frac{1-\nu_f}{E_f} + 4\frac{1-\nu_s}{E_s}\frac{t_f}{t_s}}.} \tag{14.8.50}$$

To find the deformation of the substrate, one needs to integrate the strain-displacement relations (cf. (A.2.20)):

$$\begin{aligned}
\epsilon_{rr} &= \frac{\partial u_r}{\partial r}, \qquad \epsilon_{\theta\theta} = \frac{1}{r}\frac{\partial u_\theta}{\partial \theta} + \frac{u_r}{r}, \qquad \epsilon_{zz} = \frac{\partial u_z}{\partial z},\\
\epsilon_{r\theta} &= \frac{1}{2}\left(\frac{1}{r}\frac{\partial u_r}{\partial \theta} + \frac{\partial u_\theta}{\partial r} - \frac{u_\theta}{r}\right) = \epsilon_{\theta r},\\
\epsilon_{\theta z} &= \frac{1}{2}\left(\frac{1}{r}\frac{\partial u_z}{\partial \theta} + \frac{\partial u_\theta}{\partial z}\right) = \epsilon_{z\theta},\\
\epsilon_{zr} &= \frac{1}{2}\left(\frac{\partial u_z}{\partial r} + \frac{\partial u_r}{\partial z}\right) = \epsilon_{rz}.
\end{aligned} \tag{14.8.51}$$

By assuming all the fields are independent of θ (since the problem is axi-symmetric), one finds:

$$u_r = \frac{1-\nu_s}{E_s}(\zeta_o + \zeta_1 z r) + \alpha_s(\vartheta - \vartheta_0)r, \tag{14.8.52}$$

$$u_\theta = 0, \tag{14.8.53}$$

$$u_z = -\frac{2\nu_s}{E_s}(\zeta_o z + \frac{1}{2}\zeta_1 z^2) + \alpha_s(\vartheta - \vartheta_0)z - \frac{1-\nu_s}{2E_s}\zeta_1 r^2, \tag{14.8.54}$$

plus any arbitrary rigid displacement. In particular this yields the famous Stoney formula which relates the substrate's curvature, $\kappa \approx d^2u_z/dr^2$, to the film stress (Stoney, 1909):

$$\boxed{\kappa = 6\frac{1-\nu_s}{E_s}\sigma_f\frac{t_f}{t_s^2}.} \tag{14.8.55}$$

15 Small deformation theory of species diffusion coupled to elasticity

15.1 Introduction

This chapter presents a coupled theory for transport of a chemical species through a solid that deforms elastically. We limit our considerations to a *single* species under isothermal considerations, and restrict attention to small deformations. The species in question may be atomic or molecular. *Underlying our approach is the notion that the solid can deform elastically but it retains its connectivity and does not itself diffuse* (Gibbs, 1878; Larche and Cahn, 1985). Two common examples of such phenomena include interstitial diffusion of alloying elements in crystalline solids and solvents and plasticizers diffusing through the interstices of a solid polymer network. The phenomenon of species diffusion-induced stress generation has been widely studied since the early 1960s, both from a fundamental chemomechanics point of view, as well as in the context of a variety of applications (cf., e.g., Prussin, 1961; Alfrey Jr. et al., 1966; Li, 1978; Larche and Cahn, 1985; Govindjee and Simo, 1993).

The most unfamiliar and conceptually challenging feature of the theory is associated with the need to account for the energy flow due to species transport and the introduction of the notion of *chemical potential*. Even though in our treatment we shall neglect any chemical reactions of the diffusing species with the host material, for lack of better terminology we call the theory discussed here "chemoelasticity".[1]

15.2 Kinematics and force and moment balances

We consider a deformable body occupying a (fixed) region $\mathcal{B}$. We restrict attention to *isothermal* situations and *small deformations*, and since timescales associated with species diffusion are usually considerably longer than those associated with wave propagation, we *neglect all inertial effects*.

We take as the starting point the kinematical assumptions of the linear theory of elasticity. As in the theory of linear elasticity, $\mathbf{x}$ denotes a material point in $\mathcal{B}$, ∇ and div denote the gradient and divergence with respect to $\mathbf{x}$. We use the following notation:

[1] This chapter is based on a paper by Anand (2015). The linear chemoelasticity theory presented here is itself a specialization of the large deformation theory presented in Gurtin et al. (2010). The reader is referred to this latter book for a more rigorous treatment of the subject under both large and small deformations.

Continuum Mechanics of Solids. Lallit Anand and Sanjay Govindjee, Oxford University Press (2020).

DOI: 10.1093/oso/9780198864721.001.0001

- $\mathbf{u}(\mathbf{x}, t)$ displacement field of $\mathcal{B}$;
- $\nabla\mathbf{u}(\mathbf{x}, t)$ displacement gradient in $\mathcal{B}$;
- $\boldsymbol{\epsilon}(\mathbf{x}, t) = (1/2)\big(\nabla\mathbf{u}(\mathbf{x}, t) + (\nabla\mathbf{u}(\mathbf{x}, t))^{\top}\big)$, strain in $\mathcal{B}$;
- $\boldsymbol{\sigma}(\mathbf{x}, t)$ stress;
- $\mathbf{b}(\mathbf{x}, t)$ (non-inertial) body force field in $\mathcal{B}$,

and recall the following standard relations for force and moment balances from the theory of linear elasticity, which are also valid for the theory considered here,

$$\operatorname{div}\boldsymbol{\sigma} + \mathbf{b} = \mathbf{0}, \qquad \boldsymbol{\sigma} = \boldsymbol{\sigma}^{\top}. \tag{15.2.1}$$

15.3 **Balance law for the diffusing species**

Throughout, we denote by $\mathcal{P}$ an arbitrary part of the reference body $\mathcal{B}$ with $\mathbf{n}$ the outward unit normal on $\partial\mathcal{P}$. Let

$$c(\mathbf{x}, t) \tag{15.3.1}$$

denote the *total number of moles of the diffusing species per unit reference volume*. Changes in c in a part $\mathcal{P}$ are brought about by diffusion across the boundary $\partial\mathcal{P}$, which is characterized by a flux $\mathbf{j}(\mathbf{x}, t)$, the number of moles of diffusing species measured per unit area per unit time, so that

$$-\int_{\partial\mathcal{P}} \mathbf{j}\cdot\mathbf{n}\, da$$

represents the number of moles of the diffusing species entering $\mathcal{P}$ across $\partial\mathcal{P}$ per unit time. The balance law for the diffusing species therefore takes the form

$$\overline{\int_{\mathcal{P}} c\, dv}^{\boldsymbol{\cdot}} = -\int_{\partial\mathcal{P}} \mathbf{j}\cdot\mathbf{n}\, da, \tag{15.3.2}$$

for every part $\mathcal{P}$. Bringing the time derivative in (15.3.2) inside the integral and using the divergence theorem on the integral over $\partial\mathcal{P}$, we find that

$$\int_{\mathcal{P}} (\dot{c} + \operatorname{div}\mathbf{j})\, dv = 0. \tag{15.3.3}$$

Since $\mathcal{P}$ is arbitrary, by the localization theorem, this leads to the following local species balance:

$$\dot{c} = -\operatorname{div}\mathbf{j}. \tag{15.3.4}$$

15.4 **Free-energy imbalance**

We develop the theory within a framework that accounts for the first two laws of thermodynamics. For isothermal processes the first two laws collapse into a single dissipation inequality which asserts that temporal changes in free energy of a part $\mathcal{P}$ be not greater than the power expended on $\mathcal{P}$, plus the flux of energy carried into $\mathcal{P}$.

Thus, let $\psi(\mathbf{x}, t)$ denote the free energy density per unit reference volume. Then, neglecting inertial effects, the free-energy imbalance under isothermal conditions requires that for each part $\mathcal{P} \subseteq \mathcal{B}$,

$$\overline{\int_{\mathcal{P}} \psi \, dv}^{\,\boldsymbol{\cdot}} \leq \int_{\partial\mathcal{P}} \boldsymbol{\sigma}\mathbf{n} \cdot \dot{\mathbf{u}} \, da + \int_{\mathcal{P}} \mathbf{b} \cdot \dot{\mathbf{u}} \, dv - \int_{\partial\mathcal{P}} \mu \mathbf{j} \cdot \mathbf{n} \, da, \tag{15.4.1}$$

where the left-hand side of (15.4.1) represents the rate of change of the free energy of the part $\mathcal{P}$, and the right-hand side represents the power expended on $\mathcal{P}$ by the surface tractions and body force, plus the flux of energy carried into $\mathcal{P}$ by the flux $\mathbf{j}$ of diffusing species, where to characterize the energy carried into part $\mathcal{P}$ by species transport, we have introduced a *chemical potential field* $\mu(\mathbf{x}, t)$. It is a *primitive quantity* associated with the diffusing species and solid combination being modeled.

Bringing the time derivative in (15.4.1) inside the integral and using the divergence theorem on the integrals over $\partial\mathcal{P}$ gives

$$\int_{\mathcal{P}} \Big(\dot{\psi} - \boldsymbol{\sigma} : \dot{\boldsymbol{\epsilon}} - \big(\operatorname{div} \boldsymbol{\sigma} + \mathbf{b}\big) \cdot \dot{\mathbf{u}} + \mu \operatorname{div} \mathbf{j} + \mathbf{j} \cdot \nabla\mu\Big) \, dv \leq 0. \tag{15.4.2}$$

Thus, (15.4.2), balance of forces $(15.2.1)_1$, species balance (15.3.4), and the arbitrary nature of $\mathcal{P}$, together yield the following local form of the *free-energy imbalance*,

$$\dot{\psi} - \boldsymbol{\sigma} : \dot{\boldsymbol{\epsilon}} - \mu \dot{c} + \mathbf{j} \cdot \nabla\mu \leq 0. \tag{15.4.3}$$

15.5 **Constitutive equations**

Guided by the free-energy imbalance (15.4.3) we introduce the following constitutive response functions for the free energy ψ, the stress $\boldsymbol{\sigma}$, and the chemical potential μ, when the strain $\boldsymbol{\epsilon}$ and the species concentration c are known:

$$\begin{aligned} \psi &= \hat{\psi}(\boldsymbol{\epsilon}, c), \\ \boldsymbol{\sigma} &= \hat{\boldsymbol{\sigma}}(\boldsymbol{\epsilon}, c), \\ \mu &= \hat{\mu}(\boldsymbol{\epsilon}, c). \end{aligned} \tag{15.5.1}$$

To these constitutive equations we append a simple Fick's law for the species flux (Fick, 1855),

$$\mathbf{j} = -\mathbf{M}\nabla\mu, \qquad \text{with} \qquad \mathbf{M} = \hat{\mathbf{M}}(\boldsymbol{\epsilon}, c) \tag{15.5.2}$$

the *species mobility tensor*.

15.5.1 **Thermodynamic restrictions**

The free energy imbalance (15.4.3) restricts the constitutive equations. By $(15.5.1)_1$,

$$\dot{\psi} = \frac{\partial\hat{\psi}(\boldsymbol{\epsilon}, c)}{\partial\boldsymbol{\epsilon}} : \dot{\boldsymbol{\epsilon}} + \frac{\partial\hat{\psi}(\boldsymbol{\epsilon}, c)}{\partial c}\dot{c}. \tag{15.5.3}$$

Hence, the free-energy imbalance (15.4.3) is equivalent to the requirement that

$$\left(\frac{\partial\hat{\psi}(\boldsymbol{\epsilon}, c)}{\partial\boldsymbol{\epsilon}} - \hat{\boldsymbol{\sigma}}(\boldsymbol{\epsilon}, c)\right) : \dot{\boldsymbol{\epsilon}} + \left(\frac{\partial\hat{\psi}(\boldsymbol{\epsilon}, c)}{\partial c} - \hat{\mu}(\boldsymbol{\epsilon}, c)\right)\dot{c} - \nabla\mu \cdot \hat{\mathbf{M}}(\boldsymbol{\epsilon}, c)\nabla\mu \leq 0 \tag{15.5.4}$$

for all stress, strain, species concentration, and chemical potential fields. Given any point $\mathbf{x}_0$ in $\mathcal{B}$ and any time t_0, it is possible to find a strain and a species concentration field such that

$$\boldsymbol{\epsilon},\ \dot{\boldsymbol{\epsilon}},\ c,\ \text{and}\ \dot{c} \tag{15.5.5}$$

have arbitrarily prescribed values at $(\mathbf{x}_0, t_0)$. Granted this, the coefficients of $\dot{\boldsymbol{\epsilon}}$ and $\dot{c}$ must vanish, otherwise these rates may be chosen to violate inequality (15.5.4). We therefore have the thermodynamic restrictions:

(i) The free energy determines the stress through the *stress relation*,

$$\boldsymbol{\sigma} = \hat{\boldsymbol{\sigma}}(\boldsymbol{\epsilon}, c) = \frac{\partial\hat{\psi}(\boldsymbol{\epsilon}, c)}{\partial\boldsymbol{\epsilon}}. \tag{15.5.6}$$

(ii) The free energy determines the chemical potential through the *chemical potential relation*,

$$\mu = \hat{\mu}(\boldsymbol{\epsilon}, c) = \frac{\partial\hat{\psi}(\boldsymbol{\epsilon}, c)}{\partial c}. \tag{15.5.7}$$

(iii) The species flux satisfies the *species-transport inequality*,

$$\nabla\mu \cdot \hat{\mathbf{M}}(\boldsymbol{\epsilon}, c)\nabla\mu \geq 0 \qquad \text{for all } (\boldsymbol{\epsilon}, c, \nabla\mu); \tag{15.5.8}$$

thus the mobility tensor $\mathbf{M}$ is positive semi-definite.

15.5.2 **Consequences of the thermodynamic restrictions**

By (15.5.3), (15.5.6), and (15.5.7), we have the *Gibbs relation*

$$\dot{\psi} = \boldsymbol{\sigma} : \dot{\boldsymbol{\epsilon}} + \mu\dot{c}, \tag{15.5.9}$$

while (15.5.6) and (15.5.7) yield the *Maxwell relation*

$$\frac{\partial\hat{\boldsymbol{\sigma}}(\boldsymbol{\epsilon}, c)}{\partial c} = \frac{\partial\hat{\mu}(\boldsymbol{\epsilon}, c)}{\partial\boldsymbol{\epsilon}}. \tag{15.5.10}$$

Let $\boldsymbol{\epsilon}(t)$ be a time-dependent strain tensor, let $c(t)$ be a time-dependent species concentration, and write

$$\boldsymbol{\sigma}(t) = \hat{\boldsymbol{\sigma}}(\boldsymbol{\epsilon}(t), c(t)).$$

The chain rule then yields the relation

$$\dot{\boldsymbol{\sigma}} = \frac{\partial \hat{\boldsymbol{\sigma}}(\boldsymbol{\epsilon}, c)}{\partial \boldsymbol{\epsilon}} \dot{\boldsymbol{\epsilon}} + \frac{\partial \hat{\boldsymbol{\sigma}}(\boldsymbol{\epsilon}, c)}{\partial c} \dot{c},$$

which suggests the introduction of two constitutive moduli:

- The fourth-order *elasticity tensor* $\mathbb{C}^{\mathrm{u}}(\boldsymbol{\epsilon}, c)$ defined by

$$\mathbb{C}^{\mathrm{u}}(\boldsymbol{\epsilon}, c) = \frac{\partial \hat{\boldsymbol{\sigma}}(\boldsymbol{\epsilon}, c)}{\partial \boldsymbol{\epsilon}} = \frac{\partial^2 \hat{\psi}(\boldsymbol{\epsilon}, c)}{\partial \boldsymbol{\epsilon}^2} \quad \text{(at fixed } c\text{)}. \tag{15.5.11}$$

  The elasticity tensor $\mathbb{C}^{\mathrm{u}}$ represents elastic moduli under conditions in which c is held *constant*, that is conditions under which the species is constrained from flowing into or out of the body during deformation. In the study of diffusion of a fluid through porous materials, conditions under which c is constant are called *undrained* (Rice and Cleary, 1976), and accordingly we call $\mathbb{C}^{\mathrm{u}}$ the *undrained elasticity tensor*. In practice, undrained conditions are approached for deformations which are rapid relative to the timescale for species diffusion (though inertially quasi-static).

- The second-order *stress-chemistry tensor* $\mathbf{S}^{c}(\boldsymbol{\epsilon}, c)$ defined by

$$\mathbf{S}^{c}(\boldsymbol{\epsilon}, c) = \frac{\partial \hat{\boldsymbol{\sigma}}(\boldsymbol{\epsilon}, c)}{\partial c} = \frac{\partial^2 \hat{\psi}(\boldsymbol{\epsilon}, c)}{\partial \boldsymbol{\epsilon} \partial c}. \tag{15.5.12}$$

For each $(\boldsymbol{\epsilon}, c)$, the elasticity tensor $\mathbb{C}^{\mathrm{u}}(\boldsymbol{\epsilon}, c)$—a linear transformation that maps symmetric tensors to symmetric tensors—has the minor symmetry properties

$$C^{\mathrm{u}}_{ijkl} = C^{\mathrm{u}}_{jikl} = C^{\mathrm{u}}_{ijlk}, \tag{15.5.13}$$

as well as the major symmetry property

$$C^{\mathrm{u}}_{ijkl} = C^{\mathrm{u}}_{klij}. \tag{15.5.14}$$

Further, since the stress $\boldsymbol{\sigma}$ is symmetric, we may conclude from (15.5.12) that the stress-chemistry modulus $\mathbf{S}^{c}(\boldsymbol{\epsilon}, c)$ is a symmetric tensor; this tensor measures the change in stress due to a change in species concentration holding the strain fixed.

Proceeding as above, let

$$\mu(t) = \hat{\mu}(\boldsymbol{\epsilon}(t), c(t)).$$

Then, by the chain rule, (15.5.10) and (15.5.12),

$$\begin{aligned}\dot{\mu} &= \frac{\partial \hat{\mu}(\boldsymbol{\epsilon}, c)}{\partial \boldsymbol{\epsilon}} : \dot{\boldsymbol{\epsilon}} + \frac{\partial \hat{\mu}(\boldsymbol{\epsilon}, c)}{\partial c} \dot{c}, \\ &= \mathbf{S}^{c}(\boldsymbol{\epsilon}, c) : \dot{\boldsymbol{\epsilon}} + \frac{\partial \hat{\mu}(\boldsymbol{\epsilon}, c)}{\partial c} \dot{c}.\end{aligned}$$

This suggests the introduction of:

- a scalar *chemistry modulus* $\Lambda(\epsilon, c)$ defined by

$$\Lambda(\epsilon, c) = \frac{\partial \hat{\mu}(\epsilon, c)}{\partial c}. \tag{15.5.15}$$

Unless specified otherwise, we assume that the chemistry modulus is strictly positive for all (ϵ, c),

$$\Lambda(\epsilon, c) > 0. \tag{15.5.16}$$

15.5.3 **Chemical potential as independent variable**

For problems involving little or no species diffusion, the strain ϵ and the concentration c are the natural choice of independent constitutive variables. However, for processes that occur over long timescales so that species diffusion is important, it is preferable to replace constitutive dependence upon c by constitutive dependence upon the chemical potential μ.

Note from (15.5.15) and (15.5.16) that we have assumed that

$$\frac{\partial \hat{\mu}(\epsilon, c)}{\partial c} > 0. \tag{15.5.17}$$

Further assuming continuity of this derivative allows us to conclude that, for each fixed ϵ, the relation

$$\mu = \hat{\mu}(\epsilon, c) \tag{15.5.18}$$

is invertible in c, so that we may write c as a function of (ϵ, μ):

$$c = \breve{c}(\epsilon, \mu). \tag{15.5.19}$$

Let

$$\omega \stackrel{\text{def}}{=} \psi - c\mu \tag{15.5.20}$$

define a *grand-canonical energy* per unit reference volume (Gurtin et al., 2010, Chap. 62). Then,

$$\begin{aligned} \omega &= \breve{\omega}(\epsilon, \mu) \\ &\stackrel{\text{def}}{=} \hat{\psi}(\epsilon, \breve{c}(\epsilon, \mu)) - \breve{c}(\epsilon, \mu)\mu, \end{aligned} \tag{15.5.21}$$

while (15.5.6) yields

$$\boldsymbol{\sigma} = \breve{\boldsymbol{\sigma}}(\epsilon, c) \tag{15.5.22}$$

$$\stackrel{\text{def}}{=} \hat{\boldsymbol{\sigma}}(\epsilon, \breve{c}(\epsilon, \mu)). \tag{15.5.23}$$

Thus, bearing in mind that a "breve" denotes a function of (ϵ, μ) while a "hat" denotes a function of (ϵ, c), we find, using (15.5.7), that

$$\frac{\partial \breve{\omega}}{\partial \epsilon} = \frac{\partial \hat{\psi}}{\partial \epsilon} + \underbrace{\left(\frac{\partial \hat{\psi}}{\partial c} - \mu\right)}_{=0} \frac{\partial \breve{c}}{\partial \epsilon} = \breve{\boldsymbol{\sigma}},$$

and

$$\frac{\partial \breve{\omega}}{\partial \mu} = \underbrace{\left(\frac{\partial \hat{\psi}}{\partial c} - \mu\right)}_{=0} \frac{\partial \breve{c}}{\partial \mu} - \breve{c} = -\breve{c}.$$

The stress and the species concentration are therefore determined by the response function $\omega = \breve{\omega}(\boldsymbol{\epsilon}, \mu)$ for the grand-canonical energy via the relations

$$\begin{aligned} \boldsymbol{\sigma} = \breve{\boldsymbol{\sigma}}(\boldsymbol{\epsilon}, \mu) &= \frac{\partial \breve{\omega}(\boldsymbol{\epsilon}, \mu)}{\partial \boldsymbol{\epsilon}}, \\ c = \breve{c}(\boldsymbol{\epsilon}, \mu) &= -\frac{\partial \breve{\omega}(\boldsymbol{\epsilon}, \mu)}{\partial \mu}. \end{aligned} \tag{15.5.24}$$

An immediate consequence of (15.5.24) is the *Gibbs relation*

$$\dot{\omega} = \boldsymbol{\sigma} : \dot{\boldsymbol{\epsilon}} - c\dot{\mu}, \tag{15.5.25}$$

and the *Maxwell relation*

$$\frac{\partial \breve{\boldsymbol{\sigma}}(\boldsymbol{\epsilon}, \mu)}{\partial \mu} = -\frac{\partial \breve{c}(\boldsymbol{\epsilon}, \mu)}{\partial \boldsymbol{\epsilon}}. \tag{15.5.26}$$

The undrained elasticity and stress-chemistry tensors $\mathbb{C}^{\mathrm{u}}(\boldsymbol{\epsilon}, c)$ and $\mathbf{S}^{c}(\boldsymbol{\epsilon}, c)$ have natural counterparts in the theory with chemical potential as independent variable; they are the *elasticity tensor*

$$\mathbb{C}^{\mathrm{d}}(\boldsymbol{\epsilon}, \mu) = \frac{\partial \breve{\boldsymbol{\sigma}}(\boldsymbol{\epsilon}, \mu)}{\partial \boldsymbol{\epsilon}} \quad \text{(at fixed } \mu\text{)}, \tag{15.5.27}$$

and the *stress-chemical potential modulus*

$$\mathbf{S}^{\mu}(\boldsymbol{\epsilon}, \mu) = \frac{\partial \breve{\boldsymbol{\sigma}}(\boldsymbol{\epsilon}, \mu)}{\partial \mu} \quad \text{(at fixed } \boldsymbol{\epsilon}\text{)}. \tag{15.5.28}$$

The elasticity tensor $\mathbb{C}^{\mathrm{d}}$ represents the elastic moduli under conditions in which the chemical potential is held constant, and the species is *not constrained* from flowing in or out of the body during deformation. In the poroelasticity literature such conditions are called *drained* (cf., e.g. Rice and Cleary, 1976), and accordingly we call $\mathbb{C}^{\mathrm{d}}$ the *drained elasticity tensor*. A "drained" response is attained for deformations that are slow relative to the timescale of species diffusion.

By $(15.5.24)_1$ and (15.5.27),

$$\mathbb{C}^{\mathrm{d}}(\boldsymbol{\epsilon}, \mu) = \frac{\partial^2 \breve{\omega}(\boldsymbol{\epsilon}, \mu)}{\partial \boldsymbol{\epsilon}^2}, \tag{15.5.29}$$

while the Maxwell relation (15.5.26) implies that

$$\mathbf{S}^{\mu}(\boldsymbol{\epsilon}, \mu) = -\frac{\partial \breve{c}(\boldsymbol{\epsilon}, \mu)}{\partial \boldsymbol{\epsilon}}. \tag{15.5.30}$$

15.5.4 **Drained and undrained inter-relations**

We now determine relations between these various material functions. Toward this end, we note, by (15.5.19), that we may relate the alternative descriptions of the stress in terms of (ϵ, c) and (ϵ, μ) via

$$\breve{\sigma}(\epsilon,\mu) = \hat{\sigma}(\epsilon, \breve{c}(\epsilon,\mu)).$$

Thus, by (15.5.28)

$$\begin{aligned}\mathbf{S}^{\mu} &= \frac{\partial \breve{\sigma}}{\partial \mu}\\ &= \frac{\partial \hat{\sigma}}{\partial c}\frac{\partial \breve{c}}{\partial \mu},\end{aligned}$$

and, using (15.5.12) and (15.5.15), we conclude that the stress-chemical potential and stress-chemistry moduli are related via

$$\mathbf{S}^{\mu}(\epsilon,\mu) = \frac{1}{\Lambda(\epsilon,c)}\mathbf{S}^{c}(\epsilon,c), \tag{15.5.31}$$

where $c = \breve{c}(\epsilon,\mu)$.

The relation between the elasticity tensor $\mathbb{C}^{\mathrm{d}}$ at fixed chemical potential to the elasticity tensor $\mathbb{C}^{\mathrm{u}}$ at fixed concentration is based on computing the partial derivative

$$\begin{aligned}\mathbb{C}^{\mathrm{d}}(\epsilon,\mu) &= \frac{\partial \breve{\sigma}(\epsilon,\mu)}{\partial \epsilon}\\ &= \frac{\partial}{\partial \epsilon}\left(\hat{\sigma}(\epsilon, \breve{c}(\epsilon,\mu))\right)\end{aligned}$$

while holding μ fixed. Suppressing arguments and using components, this derivative of $\hat{\sigma}(\epsilon, \breve{c}(\epsilon,\mu))$ is given by

$$\frac{\partial(\hat{\sigma})_{ij}}{\partial \epsilon_{kl}} + \frac{\partial(\hat{\sigma})_{ij}}{\partial c}\frac{\partial \breve{c}}{\partial \epsilon_{kl}}.$$

Thus, since the term

$$\frac{\partial(\hat{\sigma})_{ij}}{\partial c}\frac{\partial \breve{c}}{\partial \epsilon_{kl}} \quad \text{is the component form of} \quad \frac{\partial \hat{\sigma}}{\partial c} \otimes \frac{\partial \breve{c}}{\partial \epsilon}$$

we find, with the aid of (15.5.11), (15.5.12), (15.5.30), and (15.5.31), that

$$\mathbb{C}^{\mathrm{d}}(\epsilon,\mu) = \mathbb{C}^{\mathrm{u}}(\epsilon,c) - \frac{1}{\Lambda(\epsilon,c)}\mathbf{S}^{c}(\epsilon,c)\otimes\mathbf{S}^{c}(\epsilon,c), \tag{15.5.32}$$

where $c = \breve{c}(\epsilon,\mu)$; equivalently, in components, suppressing arguments,

$$C^{\mathrm{d}}_{ijkl} = C^{\mathrm{u}}_{ijkl} - \frac{1}{\Lambda}S^{\mathrm{c}}_{ij}S^{\mathrm{c}}_{kl}. \tag{15.5.33}$$

The identity (15.5.32) has two important consequence. First, given symmetric tensors **A** and **G**

$$\mathbf{A}:(\mathbf{S}^c \otimes \mathbf{S}^c)\mathbf{G} = (\mathbf{A}:\mathbf{S}^c)(\mathbf{S}^c:\mathbf{G}) \tag{15.5.34}$$

$$= \mathbf{G}:(\mathbf{S}^c \otimes \mathbf{S}^c)\mathbf{A}. \tag{15.5.35}$$

Thus, since $\mathbb{C}^{\mathrm{u}}$ has major and minor symmetries and since the fourth-order tensor $\mathbf{S}^c \otimes \mathbf{S}^c$ has major and minor symmetries, the tensor $\mathbb{C}^{\mathrm{d}}$ *has major and minor symmetries*. Further, by (15.5.32) and (15.5.34),

$$\mathbf{A}:\mathbb{C}^{\mathrm{u}}(\epsilon, c)\mathbf{A} - \mathbf{A}:\mathbb{C}^{\mathrm{d}}(\epsilon, \mu)\mathbf{A} = \frac{1}{\Lambda}(\mathbf{A}:\mathbf{S}^c)^2,$$

for any tensor $\mathbf{A}$, and, since Λ is positive,

$$\mathbf{A}:\mathbb{C}^{\mathrm{u}}(\epsilon, c)\mathbf{A} > \mathbf{A}:\mathbb{C}^{\mathrm{d}}(\epsilon, \mu)\mathbf{A} \quad \text{for all } \mathbf{A} \neq \mathbf{0}. \tag{15.5.36}$$

Thus $\mathbb{C}^{\mathrm{u}}$ is positive definite whenever $\mathbb{C}^{\mathrm{d}}$ is positive definite. *We assume henceforth that both $\mathbb{C}^{\mathrm{d}}$ and $\mathbb{C}^{\mathrm{u}}$ are positive definite.* Positive definiteness is a condition sufficient to ensure that these elasticity tensors are invertible.

REMARK 15.1

The drained elasticity tensor $\mathbb{C}^{\mathrm{d}}(\epsilon, \mu)$ and the undrained elasticity tensor $\mathbb{C}^{\mathrm{u}}(\epsilon, c)$ are directly analogous to $\mathbb{C}(\epsilon, \vartheta)$ the isothermal elasticity tensor and $\mathbb{C}^{\mathrm{ent}}(\epsilon, \eta)$ the elasticity tensor at fixed entropy, respectively, in the theory of thermoelasticity; cf. Chapter 14.

15.6 **Linear chemoelasticity**

Guided by the discussion in the previous section, we next consider a *linear* theory. Using the strain ϵ and the concentration c as the governing variables, we take the constitutive equations in the linear theory to be given by

$$\begin{aligned}
\psi &= \mu_0\,(c - c_0) + \frac{1}{2}\epsilon:\mathbb{C}^{\mathrm{u}}\epsilon + (c - c_0)\mathbf{S}^c:\epsilon + \tfrac{1}{2}\Lambda(c - c_0)^2,\\
\boldsymbol{\sigma} &= \mathbb{C}^{\mathrm{u}}\epsilon + \mathbf{S}^c(c - c_0),\\
\mu &= \mu_0 + \mathbf{S}^c:\epsilon + \Lambda(c - c_0),\\
\mathbf{j} &= -\mathbf{M}\nabla\mu,
\end{aligned} \tag{15.6.1}$$

where $\mathbb{C}^{\mathrm{u}}$, $\mathbf{S}^c$, Λ, and $\mathbf{M}$ are, respectively, the undrained elasticity tensor, the stress-chemistry modulus, the chemistry modulus, and the species mobility tensor at the reference concentration c_0. Also, μ_0 is a reference chemical potential, and the reference configuration is taken to be stress-free. By (15.5.13), (15.5.14), and (15.5.36), the undrained elasticity tensor $\mathbb{C}^{\mathrm{u}}$ is symmetric and positive definite, and by (15.5.16) Λ is positive. Finally, by (15.5.8), the mobility tensor $\mathbf{M}$ is positive semi-definite.

The basic equations of the *linear theory of anisotropic chemoelasticity* consist of the constitutive equations (15.6.1), the local force balance

$$\operatorname{div}[\mathbb{C}^{\mathrm{u}}\boldsymbol{\epsilon} + \mathbf{S}^{c}(c - c_0)] + \mathbf{b} = \mathbf{0}, \tag{15.6.2}$$

and the local species balance

$$\dot{c} = \operatorname{div}(\mathbf{M}\nabla\mu). \tag{15.6.3}$$

15.6.1 **Isotropic linear chemoelasticity**

If the body is isotropic, then $\mathbb{C}^{\mathrm{u}}$, $\mathbf{S}^{c}$, and $\mathbf{M}$ have the specific forms

$$\begin{aligned}
\mathbb{C}^{\mathrm{u}} &= 2G^{\mathrm{u}}(\mathbb{I}^{\mathrm{sym}} - \tfrac{1}{3}\mathbf{1}\otimes\mathbf{1}) + K^{\mathrm{u}}\mathbf{1}\otimes\mathbf{1},\\
\mathbf{S}^{c} &= \varkappa\mathbf{1},\\
\mathbf{M} &= m\mathbf{1},
\end{aligned} \tag{15.6.4}$$

with G^{u} and K^{u} *undrained elastic shear and bulk moduli*, respectively, $\varkappa$ the *stress-chemistry modulus*, and m the *species mobility*.[2] Positive definiteness of the elasticity tensor $\mathbb{C}^{\mathrm{u}}$ requires that

$$G^{\mathrm{u}} > 0, \qquad K^{\mathrm{u}} > 0, \tag{15.6.5}$$

Further, since $\mathbf{M}$ is positive semi-definite, the species mobility m must be non-negative,

$$m \geq 0. \tag{15.6.6}$$

REMARK 15.2

Since in this chapter we have denoted the chemical potential by the symbol μ, the shear modulus is denoted here by the symbol G. Also the bulk modulus, which was denoted by κ in the theory of linear elasticity, is denoted here by K.

By (15.6.1) and (15.6.4), the defining constitutive equations for an isotropic linear chemoelastic solid with $\boldsymbol{\epsilon}$ and c as independent variables are,

$$\boxed{\begin{aligned}
&\psi = \mu_0\,(c - c_0) + G^{\mathrm{u}}|\boldsymbol{\epsilon}|^2 + \frac{1}{2}\left(K^{\mathrm{u}} - \frac{2}{3}G^{\mathrm{u}}\right)(\operatorname{tr}\boldsymbol{\epsilon})^2 + \varkappa(c - c_0)(\operatorname{tr}\boldsymbol{\epsilon}) + \tfrac{1}{2}\Lambda(c - c_0)^2,\\
&\boldsymbol{\sigma} = 2G^{\mathrm{u}}\boldsymbol{\epsilon} + \left(K^{\mathrm{u}} - \frac{2}{3}G^{\mathrm{u}}\right)(\operatorname{tr}\boldsymbol{\epsilon})\mathbf{1} + \varkappa(c - c_0)\,\mathbf{1},\\
&\mu = \mu_0 + \varkappa(\operatorname{tr}\boldsymbol{\epsilon}) + \Lambda(c - c_0),\\
&\mathbf{j} = -m\nabla\mu.
\end{aligned}} \tag{15.6.7}$$

[2] These results follow from standard representation theorems for isotropic functions; see e.g. Gurtin (1981).

Granted (15.6.5), the stress-strain relation $(15.6.7)_2$ may be inverted to give

$$\boxed{\epsilon = \frac{1}{2G^{\mathrm{u}}}\boldsymbol{\sigma}' + \frac{1}{9K^{\mathrm{u}}}(\operatorname{tr}\boldsymbol{\sigma})\mathbf{1} + \beta(c - c_0)\mathbf{1},} \tag{15.6.8}$$

where

$$\boxed{\beta \stackrel{\text{def}}{=} -\frac{\varkappa}{3K^{\mathrm{u}}}} \tag{15.6.9}$$

is the *coefficient of chemical expansion*. Thus the stress-chemistry modulus $\varkappa$ is related to the undrained bulk modulus K^{u} and the coefficient of chemical expansion β by

$$\boxed{\varkappa = -3K^{\mathrm{u}}\beta.} \tag{15.6.10}$$

Recall from (15.5.32) that the drained elasticity tensor $\mathbb{C}^{\mathrm{d}}$ is related to the undrained elastic tensor $\mathbb{C}^{\mathrm{u}}$ by

$$\mathbb{C}^{\mathrm{d}}(\epsilon, \mu) = \mathbb{C}^{\mathrm{u}}(\epsilon, c) - \frac{1}{\Lambda(\epsilon, c)}\mathbf{S}^{c}(\epsilon, c) \otimes \mathbf{S}^{c}(\epsilon, c). \tag{15.6.11}$$

Thus, using $(15.6.4)_{1,2}$ and (15.6.10),

$$\mathbb{C}^{\mathrm{d}} = 2G^{\mathrm{u}}\left(\mathbb{I}^{\mathrm{sym}} - \frac{1}{3}\mathbf{1}\otimes\mathbf{1}\right) + K^{\mathrm{d}}\mathbf{1}\otimes\mathbf{1}, \tag{15.6.12}$$

where

$$\boxed{K^{\mathrm{d}} = K^{\mathrm{u}} - \frac{\varkappa^2}{\Lambda}.} \tag{15.6.13}$$

- *Therefore the undrained and drained shear moduli are identical, while the drained bulk modulus K^{d} is related to the undrained bulk modulus K^{u} through* (15.6.13).

Since the undrained and drained shear moduli are identical we henceforth write the shear modulus for either case simply as

$$\boxed{G^{\mathrm{d}} \equiv G^{\mathrm{u}} \equiv G.} \tag{15.6.14}$$

Also, from $(15.6.4)_2$ and (15.5.31),

$$\mathbf{S}^{\mu} = \frac{\varkappa}{\Lambda}\mathbf{1}, \tag{15.6.15}$$

so that

$$\begin{aligned} \boldsymbol{\sigma} &= \mathbb{C}^{\mathrm{d}}\epsilon + \mathbf{S}^{\mu}(\mu - \mu_0), \\ &= 2G\epsilon + \left(K^{\mathrm{d}} - \frac{2}{3}G\right)(\operatorname{tr}\epsilon)\mathbf{1} + \frac{\varkappa}{\Lambda}(\mu - \mu_0)\mathbf{1}. \end{aligned} \tag{15.6.16}$$

Also, from $(15.6.7)_3$

$$c - c_0 = -\frac{\varkappa}{\Lambda}(\operatorname{tr}\epsilon) + \frac{1}{\Lambda}(\mu - \mu_0). \tag{15.6.17}$$

Thus the constitutive equations for isotropic linear chemoelasticity, with ϵ and μ as independent variables, are given in terms of the grand-canonical energy, $\omega = \psi - c\mu$, by

$$\boxed{\begin{aligned}
\omega &= -c_0(\mu-\mu_0) + G|\boldsymbol{\epsilon}|^2 + \frac{1}{2}\left(K^{\mathrm{d}} - \frac{2}{3}G\right)(\operatorname{tr}\boldsymbol{\epsilon})^2 + \frac{\varkappa}{\Lambda}(\mu-\mu_0)\operatorname{tr}\boldsymbol{\epsilon} - \frac{1}{2}\frac{1}{\Lambda}(\mu-\mu_0)^2,\\
\boldsymbol{\sigma} &= 2G\boldsymbol{\epsilon} + \left(K^{\mathrm{d}} - \frac{2}{3}G\right)(\operatorname{tr}\boldsymbol{\epsilon})\mathbf{1} + \frac{\varkappa}{\Lambda}(\mu-\mu_0)\mathbf{1},\\
c &= c_0 - \frac{\varkappa}{\Lambda}(\operatorname{tr}\boldsymbol{\epsilon}) + \frac{1}{\Lambda}(\mu-\mu_0),\\
\mathbf{j} &= -m\nabla\mu.
\end{aligned}} \tag{15.6.18}$$

15.6.2 **Governing partial differential equations**

Next, when $\mathcal{B}$ is homogeneous and isotropic, then G, K^{d}, $\varkappa$, Λ, and m are constants. In this case, since

$$2\operatorname{div}\boldsymbol{\epsilon} = \operatorname{div}(\nabla\mathbf{u} + (\nabla\mathbf{u})^{\top}) = \Delta\mathbf{u} + \nabla\operatorname{div}\mathbf{u},$$

and

$$\operatorname{div}[(\operatorname{tr}\boldsymbol{\epsilon})\mathbf{1}] = \operatorname{div}[(\operatorname{div}\mathbf{u})\mathbf{1}] = \nabla\operatorname{div}\mathbf{u},$$

the force balance (15.2.1) and (15.6.18)$_2$ yield

$$\boxed{G\Delta\mathbf{u} + (\lambda^{\mathrm{d}} + G)\nabla\operatorname{div}\mathbf{u} + \frac{\varkappa}{\Lambda}\nabla\mu + \mathbf{b} = \mathbf{0},} \tag{15.6.19}$$

with

$$\lambda^{\mathrm{d}} \stackrel{\text{def}}{=} K^{\mathrm{d}} - \frac{2}{3}G, \tag{15.6.20}$$

a drained Lamé modulus.

Further, the species balance (15.3.4) and (15.6.18)$_3$ yield

$$\boxed{\frac{1}{\Lambda}\dot{\mu} = m\Delta\mu + \frac{\varkappa}{\Lambda}\operatorname{div}\dot{\mathbf{u}}.} \tag{15.6.21}$$

REMARK 15.3

The system of coupled partial differential equations (15.6.19) and (15.6.21) for isotropic linear chemoelasticity are formally equivalent to (14.6.12) and (14.6.16) in the classical coupled theory of isotropic linear *thermoelasticity* (cf. Sec. 14.6.1). In that context, μ denotes the *temperature*, $(1/\Lambda) > 0$ is the *specific heat*, and $m > 0$ is the *thermal conductivity*. There, the term $-(3\lambda + 2\mu)\alpha\nabla\tilde{\vartheta}$ analogous to $(\varkappa/\Lambda)\nabla\mu$ arises from the *thermal stress*, and the term $\beta\vartheta_0\operatorname{div}\dot{\mathbf{u}}$ analogous to $(\varkappa/\Lambda)\operatorname{div}\dot{\mathbf{u}}$ corresponds to the *internal heating* due to the *dilatation rate*.

Boundary and initial conditions

Let $\mathcal{S}_\mathrm{u}$ and $\mathcal{S}_\mathrm{t}$ denote complementary subsurfaces of the boundary $\partial\mathcal{B}$ of the body $\mathcal{B}$. Then, as is standard, for a time interval $t \in [0, T]$ we consider a pair of simple boundary conditions in which the displacement is specified on $\mathcal{S}_\mathrm{u}$ and the surface traction on $\mathcal{S}_\mathrm{t}$:

$$\begin{aligned} \mathbf{u} &= \bar{\mathbf{u}} \quad \text{on } \mathcal{S}_\mathrm{u} \times [0, T], \\ \boldsymbol{\sigma}\mathbf{n} &= \bar{\mathbf{t}} \quad \text{on } \mathcal{S}_\mathrm{t} \times [0, T], \end{aligned} \tag{15.6.22}$$

where superposed bars denote prescribed boundary quantities.

Next, the flux of the diffusing species from the material in contact with the body changes the energy of the body according to (cf. (15.4.1))

$$-\int_{\partial\mathcal{B}} \mu\,\mathbf{j}\cdot\mathbf{n}\,da.$$

Thus, letting $\mathcal{S}_\mu$ and $\mathcal{S}_\mathrm{j}$ denote complementary subsurfaces of the boundary, we consider a pair of simple boundary conditions in which the species chemical potential is specified on $\mathcal{S}_\mu$ and the species flux on $\mathcal{S}_\mathrm{j}$:

$$\begin{aligned} \mu &= \bar{\mu} \quad \text{on } \mathcal{S}_\mu \times [0, T], \\ \mathbf{j}\cdot\mathbf{n} &= \bar{\jmath} \quad \text{on } \mathcal{S}_\mathrm{j} \times [0, T]. \end{aligned} \tag{15.6.23}$$

The initial data is taken as

$$\mathbf{u}(\mathbf{x}, 0) = \mathbf{0}, \quad \mu(\mathbf{x}, 0) = \mu_0 \quad \text{in} \quad \mathcal{B}. \tag{15.6.24}$$

The coupled set of partial differential equations (15.6.19) and (15.6.21), together with the boundary conditions (15.6.22) and (15.6.23) and the initial conditions (15.6.24), yield an initial/boundary-value problem for the displacement $\mathbf{u}(\mathbf{x}, t)$ and the chemical potential $\mu(\mathbf{x}, t)$.

REMARK 15.4

In the preceding development of the theory we have allowed for the chemical potential to have a value μ_0 in the reference state in which $\boldsymbol{\epsilon} = \mathbf{0}$ and $c = c_0$. However, the theory is unaltered if we choose to set $\mu_0 = 0$.

---

16 Linear poroelasticity

16.1 Introduction

A theory for addressing coupled problems concerning fluid-saturated deformable porous solids with diffusive fluid transport is of importance for a variety of engineering applications, such as geomechanics, petroleum engineering, and hydrogeology. The classical linear isotropic poroelasticity theory which accounts for the influence of pore fluid diffusion on the quasi-static deformation of porous media is due to (Biot, 1935, 1941; Biot and Willis, 1957). Biot's theory is intended to represent the coupled deformation-diffusion response of a material which at a microscopic scale consists of a porous solid skeleton and a freely moving fluid in a fully connected pore space. Over the last forty years the theory has been widely applied to geotechnical problems (Rice and Cleary (1976); Detournay and Cheng (1993); Wang (2000); Rudnicki (2001); Guéguen et al. (2004) among others), and it has also been applied to bone by many authors (see e.g., Cowin, 1999).

In this chapter we show that Biot's theory of linear poroelasticity *may be deduced as a special case of the theory of linear chemoelasticity* discussed in Chapter 15.[1] Following Biot, the fluid-solid mixture is treated as a single homogenized continuum body which allows for a mass flux of the fluid, and not as a multi-component mixture as in the "theory of mixtures" (Bowen, 1982; Coussy, 1995). We consider the general case when neither the solid nor the fluid phase is incompressible; however, we restrict our discussion to isotropic materials.

For a porous body a "material point" $\mathbf{x} \in \mathcal{B}$ is an infinitesimal element of volume which is assumed to be large enough to be representative of the material (solid skeleton together with pores and fluid), but small enough so that the strains and stresses in this infinitesimal volume are adequately approximated to be homogeneous.

[1] This chapter is based on a paper by Anand (2015).

Continuum Mechanics of Solids. Lallit Anand and Sanjay Govindjee, Oxford University Press (2020).

DOI: 10.1093/oso/9780198864721.001.0001

16.2 **Biot's theory**

In this section we show that Biot's classical theory of linear isotropic poroelasticity (Biot, 1935, 1941; Biot and Willis, 1957), may be obtained as a special case of the linear isotropic chemoelasticity theory detailed in Sec. 15.6.1.[2]

We denote the molar mass of the fluid, a *constant*, by $\mathcal{M}^f$ (kg/mol), a *constant* referential mass density of the fluid by ρ^f_{R} (kg/m^3), and a *constant* molar volume by

$$\Omega \stackrel{\text{def}}{=} \frac{\mathcal{M}^f}{\rho^f_{\mathrm{R}}} \quad (\mathrm{m}^3/\mathrm{mol}). \tag{16.2.1}$$

Then with c denoting the *number of moles of the fluid per unit reference volume* and c_0 its initial value, the quantity

$$J^s = 1 + \Omega(c - c_0) \tag{16.2.2}$$

is assumed to represent a volumetric swelling ratio due to the insertion of fluid into the pore space of the porous material. The dimensionless quantity $(J^s - 1)$, denoted by

$$\boxed{\zeta \stackrel{\text{def}}{=} \Omega(c - c_0),} \tag{16.2.3}$$

then represents *a strain-like measure of the change in fluid volume per unit reference volume of the porous solid.* Consistent with Biot's terminology (Biot, 1941), we call ζ the **variation in fluid content.**

Thus, consider a free-energy function in terms of the strain ϵ and the concentration c, but with the dependence on c expressed in terms of a dependence on the variation of fluid content ζ, to quadratic order,

$$\psi = \hat{\psi}(\epsilon, \zeta) = \frac{\mu_0}{\Omega}\zeta + G|\epsilon|^2 + \tfrac{1}{2}\left(K^{\mathrm{u}} - (2/3)G\right)(\mathrm{tr}\,\epsilon)^2 - \alpha\, M\,(\mathrm{tr}\,\epsilon)\zeta + \tfrac{1}{2}M\zeta^2, \tag{16.2.4}$$

where

- M is a Biot modulus, and
- α is a Biot effective stress coefficient.

The free energy (16.2.4) gives the stress and chemical potential as

$$\begin{aligned} \boldsymbol{\sigma} &= \frac{\partial \hat{\psi}}{\partial \epsilon} = 2G\epsilon + (K^{\mathrm{u}} - (2/3)G)(\mathrm{tr}\,\epsilon)\mathbf{1} - \alpha M\zeta\,\mathbf{1}, \\ \mu &= \frac{\partial \hat{\psi}}{\partial c} = \Omega\frac{\partial \hat{\psi}}{\partial \zeta} = \mu_0 + (-\alpha M\Omega)\,(\mathrm{tr}\,\epsilon) + M\Omega\,\zeta. \end{aligned} \tag{16.2.5}$$

Comparing (16.2.4) and (16.2.5) to (15.6.7) the stress-chemistry modulus $\varkappa$ and the chemistry modulus Λ are given by,

$$\boxed{\varkappa = -\alpha M\,\Omega \qquad \text{and} \qquad \Lambda = M\Omega^2,} \tag{16.2.6}$$

[2] The notation employed for poroelasticity theory in this chapter has been selected to coincide as far as possible with the notation in Detournay and Cheng (1993), Rice and Cleary (1976), and Biot (1941).

respectively. As before, positive definiteness of the undrained elasticity tensor $\mathbb{C}^{\mathrm{u}}$ requires that

$$G > 0 \quad \text{and} \quad K^{\mathrm{u}} > 0, \tag{16.2.7}$$

and positivity of the chemistry modulus Λ requires that the Biot modulus be positive,

$$M > 0. \tag{16.2.8}$$

Next, recalling (15.6.13), the drained bulk modulus is given by

$$\boxed{K^{\mathrm{d}} \stackrel{\text{def}}{=} K^{\mathrm{u}} - \alpha^2 M,} \tag{16.2.9}$$

and positive definiteness of $\mathbb{C}^{\mathrm{d}}$ requires that

$$K^{\mathrm{d}} > 0. \tag{16.2.10}$$

Hence, using $(\boldsymbol{\epsilon}, \mu)$ as the independent variables, (15.6.18) and (16.2.6) give the constitutive equations for $\boldsymbol{\sigma}$ and ζ as

$$\begin{aligned} \boldsymbol{\sigma} &= 2G\boldsymbol{\epsilon} + (K^{\mathrm{d}} - (2/3)G)(\operatorname{tr}\boldsymbol{\epsilon})\mathbf{1} - \alpha\left(\frac{\mu - \mu_0}{\Omega}\right)\mathbf{1}, \\ \zeta &= \alpha\,(\operatorname{tr}\boldsymbol{\epsilon}) + \frac{1}{M}\left(\frac{\mu - \mu_0}{\Omega}\right). \end{aligned} \tag{16.2.11}$$

Since the chemical potential μ has units of J/mol, and since the molar volume of the fluid, Ω, has units of m^3/mol, the quantity

$$\boxed{p \stackrel{\text{def}}{=} \frac{\mu - \mu_0}{\Omega} \qquad \text{which has units of N/m}^2,} \tag{16.2.12}$$

may be identified as a **pore pressure**. Using (16.2.12) the constitutive equations (16.2.5) and (16.2.11) may be written as

$$\boxed{\begin{aligned} \boldsymbol{\sigma} &= 2G\boldsymbol{\epsilon} + (K^{\mathrm{u}} - (2/3)G)(\operatorname{tr}\boldsymbol{\epsilon})\mathbf{1} - \alpha M\zeta\,\mathbf{1}, \\ p &= -\alpha M\,(\operatorname{tr}\boldsymbol{\epsilon}) + M\zeta, \end{aligned}} \tag{16.2.13}$$

and

$$\boxed{\begin{aligned} \boldsymbol{\sigma} &= 2G\boldsymbol{\epsilon} + (K^{\mathrm{d}} - (2/3)G)(\operatorname{tr}\boldsymbol{\epsilon})\mathbf{1} - \alpha p\,\mathbf{1}, \\ \zeta &= \alpha\,(\operatorname{tr}\boldsymbol{\epsilon}) + \frac{1}{M}p. \end{aligned}} \tag{16.2.14}$$

These constitutive equations are **identical** to the classical constitutive equations of Biot's linear theory of isotropic poroelasticity.[3]

From $(16.2.14)_2$ we see that,

$$\alpha \equiv \left.\frac{\partial\zeta}{\partial(\operatorname{tr}\boldsymbol{\epsilon})}\right|_{p=\text{constant}}. \tag{16.2.15}$$

[3] In Sec. 16.4 we discuss an alternative form of the constitutive equations (16.2.13) in terms of a change in the porosity ϕ_{R}, rather than in terms of the variation of fluid content ζ.

Thus the Biot coefficient α determines the ratio of the change in the fluid content to the change in the macroscopic volumetric strain when the pore pressure p is held constant. We assume that (cf. (16.4.31))

$$0 < \alpha \leq 1. \tag{16.2.16}$$

The stress-strain relation $(16.2.14)_1$ may be equivalently written as

$$\boldsymbol{\sigma}_{\text{eff}} = 2G\boldsymbol{\epsilon} + (K^{\text{d}} - (2/3)G)(\operatorname{tr}\boldsymbol{\epsilon})\mathbf{1}, \tag{16.2.17}$$

where

$$\boldsymbol{\sigma}_{\text{eff}} \stackrel{\text{def}}{=} \boldsymbol{\sigma} + \alpha\, p\mathbf{1}, \tag{16.2.18}$$

defines an *effective stress*. The concept of effective stress is due to Terzaghi (cf. e.g., Terzaghi and Froölicj, 1936), and is of great importance in poroelasticity.[4]

16.2.1 **Fluid flux**

Recall from $(15.6.7)_4$, that for isotropic linear materials the fluid flux is given by,

$$\mathbf{j} = -m\nabla\mu \qquad \text{with} \qquad m \geq 0. \tag{16.2.19}$$

Also, from (16.2.12) we have that $\mu - \mu_0 = \Omega\, p$, use of which in (16.2.19) gives the *molar fluid flux* as

$$\mathbf{j} = -(m\,\Omega)\,\nabla p \qquad (\text{mol}/(m^2\text{s})). \tag{16.2.20}$$

The *volumetric fluid flux* is given by

$$\mathbf{q} \stackrel{\text{def}}{=} \Omega\,\mathbf{j} \qquad (m^3/(m^2\text{s})). \tag{16.2.21}$$

Using (16.2.20) in (16.2.21) gives Darcy's law (Darcy, 1856),

$$\boxed{\mathbf{q} = -k\,\nabla p,} \tag{16.2.22}$$

where

$$k \stackrel{\text{def}}{=} m\Omega^2 \tag{16.2.23}$$

is the *permeability* of the porous solid. Because the permeability of the fluid through a porous solid depends on the fluid viscosity and the geometry of the fluid pathways, in the poroelasticity literature k is often expressed as

$$k \equiv \frac{k_{\text{da}}}{\eta^f}, \tag{16.2.24}$$

where

- η^f is the *viscosity* of the pore fluid (Pa-s), and
- k_{da} is a *permeability coefficient* with dimensions of length squared. Values of k_{da} are often given in *darcies* (1 darcy $= 9.869233 \times 10^{-13}\,\text{m}^2 \approx 10^{-12}\,\text{m}^2$).

---

[4] In Terzaghi's original work, the effective stress was defined only for $\alpha = 1$.

Finally the balance law for species concentration (15.3.4) may be written as

$$\boxed{\dot{\zeta} = -\operatorname{div}\mathbf{q}.} \tag{16.2.25}$$

16.2.2 Material properties for poroelastic materials

The Biot theory discussed above introduces the shear modulus G, the drained bulk modulus K^{d}, and two new parameters α and K^{u}, of which the latter is the undrained bulk modulus, to describe the elastic response of fluid-infiltrated materials.[5]

Other parameters are also used sometimes. For example, a measurable material property is the *Skempton coefficient* which represents the pore pressure which is induced when stresses are applied under undrained conditions. Thus, note that the inverted form of (16.2.14) is

$$\boxed{\begin{aligned} \boldsymbol{\epsilon} &= \frac{1}{2G}\boldsymbol{\sigma} + \left(\frac{1}{9K^{\mathrm{d}}} - \frac{1}{6G}\right)(\operatorname{tr}\boldsymbol{\sigma})\mathbf{1} + \frac{\alpha}{3K^{\mathrm{d}}}\,p\mathbf{1}, \\ \zeta &= \frac{\alpha}{K^{\mathrm{d}}}\left(\frac{1}{B}p + \left(\frac{1}{3}\operatorname{tr}\boldsymbol{\sigma}\right)\right), \end{aligned}} \tag{16.2.26}$$

where

$$\boxed{B \stackrel{\text{def}}{=} \frac{1}{\alpha}\left(1 - \frac{K^{\mathrm{d}}}{K^{\mathrm{u}}}\right).} \tag{16.2.27}$$

From $(16.2.26)_2$ note that

$$B = \left.\frac{\partial p}{\partial(-\frac{1}{3}\operatorname{tr}\boldsymbol{\sigma})}\right|_{\zeta\,\equiv\,\text{constant}}. \tag{16.2.28}$$

Thus, the coefficient B represents the change in the pore pressure to the change in the external pressure under undrained conditions, $\zeta \equiv$ constant. The parameter B is called the *Skempton compressibility coefficient*, which typically lies in the range

$$0 \le B \le 1.$$

Also sometimes it is convenient to treat the shear modulus G and the Poisson ratio ν as the two primary elastic variables. Using the standard relation

$$\nu = \tfrac{1}{2}\left(\frac{3K - 2G}{3K + G}\right) \tag{16.2.29}$$

in the theory of isotropic linear elasticity, we may define *drained and undrained Poisson ratios* by the relations

$$\nu^{\mathrm{d}} \stackrel{\text{def}}{=} \tfrac{1}{2}\left(\frac{3K^{\mathrm{d}} - 2G}{3K^{\mathrm{d}} + G}\right), \qquad \nu^{\mathrm{u}} \stackrel{\text{def}}{=} \tfrac{1}{2}\left(\frac{3K^{\mathrm{u}} - 2G}{3K^{\mathrm{u}} + G}\right). \tag{16.2.30}$$

---

[5] While G and K^{d} are usually experimentally measured, the Biot coefficient α and the undrained bulk modulus K^{u} are typically not directly measured, but calculated using the estimates (16.4.30) and (16.4.33) and measured values of the bulk modulus of the solid constituent K^{s} and that of the fluid constituent K^{f}, as well as the initial porosity ϕ_0.

16.3 Summary of the theory of linear isotropic poroelasticity

The theory of isotropic linear poroelasticity relates the following basic fields:

$\mathbf{u}$	displacement field;
$\boldsymbol{\epsilon} = \frac{1}{2}(\nabla\mathbf{u} + (\nabla\mathbf{u})^\top)$	strain-displacement relation;
$\boldsymbol{\sigma}$	Cauchy stress;
ζ	variation in fluid content;
p	pore pressure;
$\mathbf{q}$	fluid volumetric flux.

16.3.1 Constitutive equations

The set of constitutive equations are:

1. **Equations for the stress and pore pressure:**

$$\begin{aligned} \boldsymbol{\sigma} &= 2G\boldsymbol{\epsilon} + (K^{\mathrm{u}} - (2/3)G)(\operatorname{tr}\boldsymbol{\epsilon})\mathbf{1} - \alpha M\zeta\,\mathbf{1}, \\ p &= -\alpha M\,(\operatorname{tr}\boldsymbol{\epsilon}) + M\zeta. \end{aligned} \tag{16.3.1}$$

2. **Fluid flux:**
   The fluid volumetric flux is given by

$$\mathbf{q} = -k\,\nabla p. \tag{16.3.2}$$

To complete the constitutive model for a particular material the material parameters[6] that need to be specified are

$$(G, K^{\mathrm{u}}, \alpha, M, k).$$

The Biot modulus M is related to the undrained bulk modulus K^{u}, the drained bulk modulus K^{d}, and the Biot coefficient α by

$$M = \frac{K^{\mathrm{u}} - K^{\mathrm{d}}}{\alpha^2}. \tag{16.3.3}$$

Also, the scalar permeability k is related to the viscosity of the pore-fluid η^f and the permeability coefficient k_{da} by

$$k = \frac{k_{\mathrm{da}}}{\eta^f}. \tag{16.3.4}$$

[6] The notation for the material property symbols we have used is standard in much of the poroelasticity literature; however, it is also common to find other conventions that report properties in terms of $R = MK^{\mathrm{d}}/K^{\mathrm{u}}$, $H = K^{\mathrm{d}}/\alpha$, and $Q = M$ (see e.g. Biot, 1941). In the hydrogeology literature one also sees the hydraulic conductivity K which in our notation is equal to $k\rho g$, where ρ is the mass density of the fluid and g is the gravitational constant (see e.g. Domenico and Schwartz, 1990).

16.3.2 **Governing partial differential equations**

The governing partial differential equations consist of:

1. **The force balance**:

$$\operatorname{div} \boldsymbol{\sigma} + \mathbf{b} = \mathbf{0}, \tag{16.3.5}$$

where the stress $\boldsymbol{\sigma}$ is given by $(16.3.1)_1$, and $\mathbf{b}$ is the non-inertial body force.

2. **The fluid balance**:

$$\dot{\zeta} = -\operatorname{div} \mathbf{q}, \tag{16.3.6}$$

with the fluid flux $\mathbf{q}$ given by (16.3.2), and the pore pressure p by $(16.3.1)_2$.

Alternate forms of the governing partial differential equations

When $\mathcal{B}$ is homogeneous and isotropic, then G, K^{d}, α, M, and k are constants, and in this case the governing equations may be written as:

1. **The force balance**:
Using $(16.2.14)_1$ in (16.3.5) we obtain the following differential equation for the displacement field,

$$\boxed{G\Delta\mathbf{u} + (\lambda^{\mathrm{d}} + G)\nabla \operatorname{div} \mathbf{u} - \alpha\nabla p + \mathbf{b} = \mathbf{0},} \tag{16.3.7}$$

where

$$\lambda^{\mathrm{d}} \stackrel{\text{def}}{=} K^{\mathrm{d}} - \frac{2}{3}G, \tag{16.3.8}$$

is a drained Lamé modulus.

2. **The fluid balance**:
From (15.6.21), (16.2.6), and (16.2.12) we obtain the following evolution equation for the pore pressure,

$$\boxed{\frac{1}{M}\dot{p} = k\Delta p - \alpha \operatorname{div} \dot{\mathbf{u}}.} \tag{16.3.9}$$

REMARK 16.1

The system of coupled partial differential equations (16.3.7) and (16.3.9) for isotropic linear poroelasticity are formally equivalent to (14.6.12) and (14.6.16) in the classical coupled theory of isotropic linear *thermoelasticity* (cf. Sec. 14.6.1). In that context, p denotes the *temperature*, $(1/M) > 0$ is the *specific heat*, and $k > 0$ is the *thermal conductivity*. There, the term $(3\lambda + 2\mu)\alpha\nabla\tilde{\vartheta}$ analogous to $\alpha\nabla p$ arises from the *thermal stress*, and the term $\beta\vartheta_0 \operatorname{div} \dot{\mathbf{u}}$ analogous to $(-\alpha \operatorname{div} \dot{\mathbf{u}})$ corresponds to the *internal heating* due to the *dilatation rate*.

16.3.3 **Boundary conditions**

With $\mathcal{S}_u$ and $\mathcal{S}_t$ denoting complementary subsurfaces of the boundary $\partial\mathcal{B}$ of the deformed body $\mathcal{B}$, as boundary conditions we consider a pair of simple boundary conditions in which the displacement $\mathbf{u}$ is specified on $\mathcal{S}_u$ and the surface traction on $\mathcal{S}_t$:

$$\begin{aligned} \mathbf{u} &= \bar{\mathbf{u}} \quad &\text{on} \quad \mathcal{S}_u \times [0, T], \\ \boldsymbol{\sigma}\mathbf{n} &= \bar{\mathbf{t}} \quad &\text{on} \quad \mathcal{S}_t \times [0, T], \end{aligned} \tag{16.3.10}$$

where quantities with an overbar are given boundary data. With $\mathcal{S}_p$ and $\mathcal{S}_q$ another pair of complementary subsurfaces of the boundary $\partial\mathcal{B}$, we also consider boundary conditions in which the pore pressure is specified on $\mathcal{S}_p$ and the normal fluid flux on $\mathcal{S}_q$

$$\begin{aligned} p &= \bar{p} \quad &\text{on} \quad \mathcal{S}_p \times [0, T], \\ \mathbf{q} \cdot \mathbf{n} &= \bar{q} \quad &\text{on} \quad \mathcal{S}_q \times [0, T]. \end{aligned} \tag{16.3.11}$$

The initial data is taken as

$$\mathbf{u}(\mathbf{x}, 0) = \mathbf{0} \quad \text{and} \quad p(\mathbf{x}, 0) = p_0 \quad \text{in} \quad \mathcal{B}. \tag{16.3.12}$$

The coupled set of equations (16.3.5) and (16.3.6), together with (16.3.10), (16.3.11), and (16.3.12), yield an initial/boundary-value problem for the displacement $\mathbf{u}(\mathbf{x}, t)$ and the pore pressure $p(\mathbf{x}, t)$.

REMARK 16.2

The books by Wang (2000) and Cheng (2016) are recommended to see the numerous applications of the linear theory of poroelasticity presented here. These books also list representative values of the poroelastic moduli for different rock types.

EXAMPLE 16.1 One-dimensional consolidation

As an example application of the theory just developed, let us consider the one-dimensional consolidation problem. This problem is often known as the Terzaghi consolidation problem as it was first elucidated in Terzaghi (1923).

In this problem we consider a one-dimensional column of poroelastic material of length L; see Fig. 16.1. By one-dimensional we mean that the displacement field is $\mathbf{u}(\mathbf{x}, t) = u(z, t)\mathbf{e}_z$. The column of material is constrained from motion in the lateral directions, and at $z = L$ we have $u(L, t) = 0$ for all times t. Similarly, the lateral sides and the bottom are impermeable so that the flux vanishes there, $\mathbf{j} \cdot \mathbf{n} = 0$, where $\mathbf{n}$ is either the unit normal to the lateral sides or the bottom of the column. At time $t = 0$, the top of the column is subjected to a sudden load such that the traction

$$\boldsymbol{\sigma}(-\mathbf{e}_z) = P\mathbf{e}_z.$$

The load is held constant for all times $t > 0$. The pore pressure at the top of the column is taken to be zero for all positive times, $p(0,t) = 0$ for all $t > 0$. We are interested in determining the pore-pressure field $p(z,t)$ and the displacement field $u(z,t)$ in the z-direction for all $t \geq 0$.

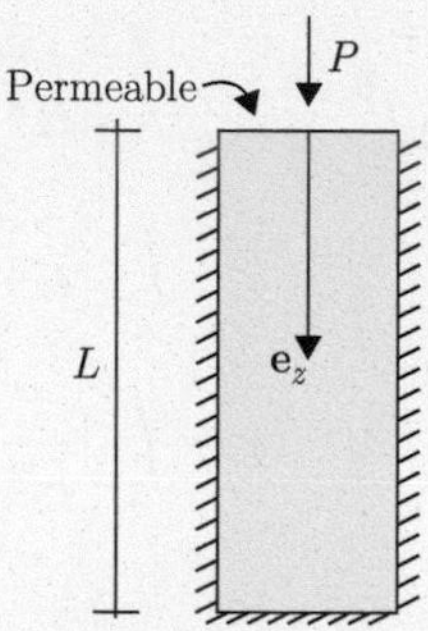

Fig. 16.1 Schematic of the Terzaghi one-dimensional consolidation problem.

In one dimension, the constitutive relations (16.2.13) and (16.2.20) reduce to

$$\sigma = \left(K^{\mathrm{u}} + \frac{4}{3}G\right)\epsilon - \alpha M\zeta, \tag{16.3.13}$$

$$p = -\alpha M\epsilon + M\zeta, \tag{16.3.14}$$

$$j = -m\Omega p_{,z}, \tag{16.3.15}$$

where the flux j is now a scalar and the strain is $\epsilon = u_{,z}$. The balance of forces (16.3.5) (in the absence of body forces) and the balance of mass (16.3.6) in this one-dimensional setting reduce to

$$\sigma_{,z} = 0, \tag{16.3.16}$$

$$\dot{\zeta} = -\Omega j_{,z}. \tag{16.3.17}$$

The problem as stated is not yet complete, since we have not yet completely specified the initial conditions. We will defer their specification until later when they are needed (and we know more about the solution).

We begin by recasting (16.3.13) in terms of strain and pore pressure by using (16.3.14) and (16.3.3) to eliminate the variation in fluid content, ζ:

$$\sigma = \left(K^{\mathrm{d}} + \frac{4}{3}G\right)\epsilon - \alpha p. \tag{16.3.18}$$

If we now use (16.3.16), we see that σ can at most be a function of time: $\sigma = \hat{\sigma}(t)$. However, we know that $\sigma(0,t) = -P$ from the boundary condition and thus

$$\sigma(z,t) = -P \tag{16.3.19}$$

for all z and all time $t \geq 0$. Noting that $\epsilon = u_{,z}$ allows us to conclude from (16.3.18) that

$$u_{,z} = \frac{\alpha p - P}{K^{\mathrm{d}} + \frac{4}{3}G}. \tag{16.3.20}$$

Integrating (16.3.20) will give us an expression for the displacement in terms of the pore pressure, the load, and the material properties:

$$\int_z^L u_{,z}\, dz = \int_z^L \frac{\alpha p - P}{K^{\mathrm{d}} + \frac{4}{3}G}\, dz,$$

$$\underbrace{u(L,t)}_{=0} - u(z,t) = \int_z^L \frac{\alpha p - P}{K^{\mathrm{d}} + \frac{4}{3}G}\, dz,$$

$$u(z,t) = \int_z^L \frac{P - \alpha p}{K^{\mathrm{d}} + \frac{4}{3}G}\, dz,$$

or

$$u(z,t) = \frac{P(L-z)}{K^{\mathrm{d}} + \frac{4}{3}G} - \int_z^L \frac{\alpha p}{K^{\mathrm{d}} + \frac{4}{3}G}\, dz. \tag{16.3.21}$$

Let us now turn our attention to the mass balance (16.3.17) and begin by again eliminating the variation in fluid content. From (16.3.14), $\dot{\zeta} = \dot{p}/M + \alpha\dot{\epsilon}$. Thus using (16.3.15) we can re-write the mass balance relation as

$$\dot{p} = Mkp_{,zz} - \alpha M\dot{\epsilon}, \tag{16.3.22}$$

where $k = m\Omega^2$. Observe now from (16.3.21) that

$$\dot{\epsilon} = \dot{u}_{,z} = \frac{\alpha\dot{p}}{K^{\mathrm{d}} + \frac{4}{3}G}, \tag{16.3.23}$$

which allows us to eliminate the strain-rate from (16.3.22), giving the following differential equation for the pore pressure,

$$\dot{p} = c_v p_{,zz}, \tag{16.3.24}$$

with

$$c_v \overset{\text{def}}{=} \frac{Mk}{1 + \frac{\alpha^2 M}{K^{\mathrm{d}} + \frac{4}{3}G}}, \tag{16.3.25}$$

the *hydraulic diffusivity*. The boundary conditions for the differential equation (16.3.24) are

$$p(0,t) = 0 \quad \text{and} \quad j(L,t) = -m\Omega p_{,z}(L,t) = 0.$$

The classical method to solve (16.3.24) is by separation of variables (see e.g. Hildebrand, 1976). One begins by assuming that $p(z,t)$ is a separable function of space and time:

$$p(z,t) = Z(z)T(t). \tag{16.3.26}$$

Substituting (16.3.26) into (16.3.24) we obtain

$$Z\dot{T} = c_v Z_{,zz} T,$$

or

$$\frac{\dot{T}}{c_v T} = \frac{Z_{,zz}}{Z}. \tag{16.3.27}$$

Notice that the left side of the equality (16.3.27) is a function of t whereas the right side is a function of z. The only way for this to be possible is for both sides to be equal to a constant. With no loss of generality we call this constant $-\beta^2$. This then implies

$$\dot{T} + c_v \beta^2 T = 0 \qquad \text{and} \qquad Z_{,zz} + \beta^2 Z = 0. \tag{16.3.28}$$

These two equations are trivially solved to give,

$$T(t) = Ae^{-c_v \beta^2 t}, \tag{16.3.29}$$

$$Z(z) = B\cos(\beta z) + C\sin(\beta z), \tag{16.3.30}$$

where A, B, C are unknown constants. Using the boundary condition at $z = 0$ we see

$$0 = p(0,t) = T(t)Z(0) = Ae^{-c_v \beta^2 t}(B \cdot 1 + C \cdot 0). \tag{16.3.31}$$

To satisfy this relation (in a non-trivial way) we require $B = 0$. The boundary condition at $z = L$ requires

$$0 = p_{,z}(L,t) = T(t)Z_{,z}(L) = Ae^{-c_v \beta^2 t}\beta C\cos(\beta L). \tag{16.3.32}$$

For a non-trivial solution we find that the *separation* constant β must be of the form

$$\beta_n = \frac{\pi(\frac{1}{2} + n)}{L}, \tag{16.3.33}$$

where $n \geq 0$ is an integer; there are an infinity of solutions of the form that we have assumed. Since any or all can be a solution, we find the general form of the solution for the pore pressure to be

$$p(z,t) = \sum_{n=0}^{\infty} P_n e^{-c_v \beta_n^2 t}\sin(\beta_n z), \tag{16.3.34}$$

where we have combined the constants A and C together into a constant P_n for each value of n. The values of the P_n are at this moment still unknown. To find them we need to consider the initial conditions for the problem.

The initial conditions for the column assume that the variation in fluid content, ζ, is zero. Since ζ is governed by a diffusion equation it cannot change instantly and it follows that right after the load is applied $\zeta(z, 0^+) = 0$. Knowing that the stress in the system is $-P$, cf. (16.3.19), we can exploit (16.3.13) to see that

$$-P = \left(K^{\mathrm{u}} + \frac{4}{3}G\right)\epsilon(z, 0^+) \tag{16.3.35}$$

and from (16.3.14) that

$$p(z, 0^+) = -\alpha M \epsilon(z, 0^+). \tag{16.3.36}$$

Eliminating the initial strain we see that we have the initial condition

$$p(z,0^+) = \gamma P, \tag{16.3.37}$$

where

$$\gamma \overset{\text{def}}{=} \frac{\alpha M}{K^{\mathrm{u}} + \frac{4}{3}G}, \tag{16.3.38}$$

is a *loading coefficient*.

From (16.3.34) and (16.3.37), we find that at $t = 0$,

$$\gamma P = \sum_{n=0}^{\infty} P_n \sin(\beta_n z). \tag{16.3.39}$$

This equation can be used to solve for the as yet unknown constants P_n. To do so, multiply both sides of (16.3.39) by $\sin(\beta_m z)$ and integrate with respect to z over the domain,

$$\gamma P \int_0^L \sin(\beta_m z)\, dz = \sum_{n=0}^{\infty} P_n \int_0^L \sin(\beta_m z) \sin(\beta_n z)\, dz. \tag{16.3.40}$$

Due to the orthogonality of the trigonometric functions

$$\int_0^L \sin(\beta_m z) \sin(\beta_n z)\, dz = \frac{L}{2}\delta_{mn},$$

and since

$$\int_0^L \sin(\beta_m z)\, dz = -\frac{1}{\beta_m}\left[\cos(\pi(\tfrac{1}{2} + m)) - \cos(0)\right] = \frac{1}{\beta_m},$$

we find that (16.3.40) reduces to

$$\frac{\gamma P}{\beta_m} = \frac{L}{2} P_m, \tag{16.3.41}$$

which upon replacing the dummy index m by n gives

$$P_n = \frac{2\gamma P}{\beta_n L}. \tag{16.3.42}$$

The complete solution for the pore-pressure field is thus given by

$$\boxed{p(z,t) = \sum_{n=0}^{\infty} P_n e^{-c_v \beta_n^2 t} \sin(\beta_n z),} \tag{16.3.43}$$

with

$$\boxed{P_n = \frac{2\gamma P}{\beta_n L}, \qquad \gamma = \frac{\alpha M}{K^{\mathrm{u}} + \frac{4}{3}G}, \qquad c_v = \frac{Mk}{1 + \frac{\alpha^2 M}{K^{\mathrm{d}} + \frac{4}{3}G}}, \qquad \beta_n = \frac{\pi(\frac{1}{2} + n)}{L}.} \tag{16.3.44}$$

Finally, using (16.3.43) in (16.3.21) and integrating, we obtain the displacement field as,

$$u(z,t) = \frac{P(L-z)}{K^{\mathrm{d}} + \frac{4}{3}G} - \frac{\alpha}{K^{\mathrm{d}} + \frac{4}{3}G}\sum_{n=0}^{\infty} \frac{P_n}{\beta_n} e^{-c_v \beta_n^2 t}\cos(\beta_n z). \tag{16.3.45}$$

16.4 Microstructural considerations

In the poroelasticity literature the *fine-scale pore space* of the material is often *explicitly accounted for*, and it is typically assumed that that porosity is fully saturated with the fluid. We introduce below some definitions used in microstructural discussions in the theory of poroelasticity.

Porosity: Consider an infinitesimal volume element dv_{R} of the reference microporous body $\mathcal{B}$, let dv_0^{pore} denote the initial volume of the connected pores in dv_{R},[7] and let

$$\phi_0 \stackrel{\mathrm{def}}{=} \frac{dv_0^{\mathrm{pore}}}{dv_{\mathrm{R}}} \qquad \text{in } \mathcal{B} \tag{16.4.1}$$

denote the *initial porosity*. The volume element dv_{R} is mapped to a volume

$$dv = J\,dv_{\mathrm{R}} \qquad \text{in } \mathcal{B}_t, \tag{16.4.2}$$

where, as is standard for small deformations,

$$J \approx 1 + \operatorname{tr}\boldsymbol{\epsilon}. \tag{16.4.3}$$

Let dv^{pore} denote the volume of the connected pores in dv, and let

$$\phi \stackrel{\mathrm{def}}{=} \frac{dv^{\mathrm{pore}}}{dv} \qquad \text{in } \mathcal{B}_t \tag{16.4.4}$$

denote the *porosity* in the deformed body. Using (16.4.2) we define a *referential measure of the porosity* by

$$\phi_{\mathrm{R}} \stackrel{\mathrm{def}}{=} \frac{dv^{\mathrm{pore}}}{dv_{\mathrm{R}}} \equiv J\phi; \tag{16.4.5}$$

the initial value of ϕ_{R} is

$$\phi_{\mathrm{R}0} \equiv \phi_0. \tag{16.4.6}$$

Molar mass, mass density, molar volume of fluid: We denote (i) the molar mass of the fluid, a *constant*, by $\mathcal{M}^f$ (kg/mol); (ii) the *variable* mass density of the fluid by ρ^f (kg/m^3); and (iii) the *variable* molar volume by

[7] Excluding occluded pores. In fact, for simplicity, we assume henceforth that the body does not contain any occluded pores.

$$\tilde{\Omega} \stackrel{\text{def}}{=} \frac{\mathcal{M}^f}{\rho^f}. \tag{16.4.7}$$

Recall in contrast, the *constant* referential molar volume is denoted by

$$\Omega \stackrel{\text{def}}{=} \frac{\mathcal{M}^f}{\rho^f_{\text{R}}}, \tag{16.4.8}$$

in terms of the molar mass $\mathcal{M}^f$ but divided by the *constant* referential mass density ρ^f_{R}.

Concentration of fluid: The concentration c of the fluid, measured in terms of the number of moles of the fluid per unit volume of the deformed body, is taken to be given by

$$\tilde{c} \stackrel{\text{def}}{=} \frac{\phi}{\tilde{\Omega}} \equiv \frac{\phi\rho^f}{\mathcal{M}^f} \qquad \text{mol/m}^3 \qquad \text{in } \mathcal{B}_t. \tag{16.4.9}$$

Physically, this definition of $\tilde{c}$ implies that when a given mass of fluid is present at a macroscopic material point, then all of the microscale pore space at that material point is presumed to be saturated with the fluid of density ρ^f, and that there is no microscale pore space which is devoid of the fluid.

Then,

$$c \stackrel{\text{def}}{=} J\,\tilde{c} = \frac{\phi_{\text{R}}}{\tilde{\Omega}} \qquad \text{in } \mathcal{B}, \tag{16.4.10}$$

represents the *number of moles of the fluid per unit reference volume*. The initial value of c is

$$c_0 = \frac{\phi_0}{\Omega}. \tag{16.4.11}$$

Referential mass content of fluid: The *referential fluid mass content* is defined by

$$m^f_{\text{R}} \stackrel{\text{def}}{=} \mathcal{M}^f\,c = \rho^f\tilde{\Omega}c = \rho^f\,\phi_{\text{R}} = J\rho^f\,\phi, \tag{16.4.12}$$

and its initial value is

$$m^f_{\text{R}0} \stackrel{\text{def}}{=} \mathcal{M}^f\,c_0 = \rho^f_{\text{R}}\Omega c_0 = \rho^f_{\text{R}}\phi_0. \tag{16.4.13}$$

Variation in fluid content ζ: The dimensionless quantity

$$\boxed{\zeta \stackrel{\text{def}}{=} \left(\frac{m^f_{\text{R}} - m^f_{\text{R}0}}{\rho^f_{\text{R}}}\right) \equiv \frac{\mathcal{M}^f}{\rho^f_{\text{R}}}(c - c_0) \equiv \Omega(c - c_0),} \tag{16.4.14}$$

represents a *normalized measure of the variation of the fluid mass content*. As already introduced in (16.2.3), we call ζ the **variation in fluid content**; *it is a measure of the change in fluid volume per unit reference volume.*

Next we consider an alternative form of the constitutive equations of poroelasticity in terms of the change in porosity. We also discuss how, under certain circumstances, the material constants α, K^{u}, and M may be estimated in terms of the bulk moduli of the fluid and solid constituents.

16.4.1 **Constitutive equations of poroelasticity in terms of the change in porosity**

The variable ζ defined in (16.4.14) may alternatively be expressed as

$$\begin{aligned}\zeta = \Omega(c - c_0) &= \frac{\mathcal{M}^f}{\rho^f_{\mathrm{R}}}(c - c_0) = \frac{m^f_{\mathrm{R}} - m^f_{\mathrm{R}0}}{\rho^f_{\mathrm{R}}}\\ &= \left(\frac{\rho^f}{\rho^f_{\mathrm{R}}}\right)(\phi_{\mathrm{R}} - \phi_0),\end{aligned} \tag{16.4.15}$$

where in writing the last of (16.4.15) we have used the fact that $\phi_{\mathrm{R}0} \equiv \phi_0$. Hence, note that

- *ζ accounts for a change in the porosity, as well as a change in the density of the fluid.*

From (16.4.15)

$$\begin{aligned}\zeta &= \left(\frac{\rho^f}{\rho^f_{\mathrm{R}}}\right)(\phi_{\mathrm{R}} - \phi_0),\\ &= (\phi_{\mathrm{R}} - \phi_0) + \left(\frac{\rho^f - \rho^f_{\mathrm{R}}}{\rho^f_{\mathrm{R}}}\right)\phi_{\mathrm{R}},\\ &= (\phi_{\mathrm{R}} - \phi_0) + \left(\frac{\rho^f - \rho^f_{\mathrm{R}}}{\rho^f_{\mathrm{R}}}\right)(\phi_0 + (\phi_{\mathrm{R}} - \phi_0)).\end{aligned} \tag{16.4.16}$$

A "linearized" version of (16.4.16), which only retains terms to first order in the changes $(\rho^f - \rho^f_{\mathrm{R}})$ and $(\phi_{\mathrm{R}} - \phi_0)$ gives

$$\zeta = (\phi_{\mathrm{R}} - \phi_0) + \left(\frac{\rho^f - \rho^f_{\mathrm{R}}}{\rho^f_{\mathrm{R}}}\right)\phi_0. \tag{16.4.17}$$

For small changes in fluid density we adopt a simple constitutive equation of the form

$$\frac{\rho^f - \rho^f_{\mathrm{R}}}{\rho^f_{\mathrm{R}}} = \frac{p}{K^f}, \tag{16.4.18}$$

with p the pore pressure and $K^f > 0$ a *bulk modulus for the fluid.* Use of (16.4.18) in (16.4.17) gives

$$\zeta = (\phi_{\mathrm{R}} - \phi_0) + \frac{\phi_0}{K^f}p. \tag{16.4.19}$$

Substituting (16.4.19) in $(16.2.14)_2$, viz.

$$\zeta = \alpha(\operatorname{tr}\epsilon) + \frac{p}{M},$$

gives

$$(\phi_{\mathrm{R}} - \phi_0) = \alpha(\operatorname{tr}\epsilon) + \frac{p}{M_b}, \tag{16.4.20}$$

where we have introduced another Biot modulus M_b defined as

$$\frac{1}{M_b} \overset{\text{def}}{=} \frac{1}{M} - \frac{\phi_0}{K^f}. \tag{16.4.21}$$

For later use, from (16.4.21) and (16.2.9), note that

$$\frac{1}{M_b} \overset{\text{def}}{=} \frac{\alpha^2}{K^{\mathrm{u}} - K^{\mathrm{d}}} - \frac{\phi_0}{K^f}. \tag{16.4.22}$$

Thus using (16.4.20), the constitutive equations (16.2.14) may be rewritten as

$$\boxed{\begin{aligned} \boldsymbol{\sigma}' &= 2G\boldsymbol{\epsilon}', \\ \tfrac{1}{3}\operatorname{tr}\boldsymbol{\sigma} &= K^{\mathrm{d}}(\operatorname{tr}\boldsymbol{\epsilon}) - \alpha p, \\ p &= -\alpha M_b(\operatorname{tr}\boldsymbol{\epsilon}) + M_b(\phi_{\mathrm{R}} - \phi_0). \end{aligned}} \tag{16.4.23}$$

16.4.2 Estimates for the material parameters α, K^{u}, and M

The discussion in this section closely follows Rice and Cleary (1976) and Rice (1998). There is a simple but often applicable situation in which the constants α, K^{u}, and M may be estimated in terms of the bulk moduli of the fluid and solid constituents. Suppose that all the pore space is fluid infiltrated, and all the solid phase consists of material elements which respond isotropically to pure pressure states, with the same bulk modulus K^s. Suppose we simultaneously apply a pore pressure of known magnitude $p^\dagger$,

$$p = p^\dagger, \tag{16.4.24}$$

as well as a macroscopic stress amounting to

$$\boldsymbol{\sigma} = -p^\dagger \mathbf{1}$$

at each point in the solid phase, so that each point in the solid part of the porous material is subjected to the same isotropic pressure.

$$-\tfrac{1}{3}\operatorname{tr}\boldsymbol{\sigma} = p^\dagger. \tag{16.4.25}$$

Because of the homogeneous state of pressure in the saturated porous material, the fluid phase could be replaced by the solid phase without any modification of the stress state. The medium behaves exactly as if it was composed of a single phase of bulk modulus K^s. This means that all linear dimensions of the material—*including those characterizing the void size*—reduce by the fractional amount $p^\dagger/3K^s$, causing the macroscopic volumetric strain

$$\operatorname{tr}\boldsymbol{\epsilon} = -\frac{p^\dagger}{K^s}, \tag{16.4.26}$$

and change in porosity

$$\frac{\phi_{\mathrm{R}} - \phi_0}{\phi_0} = -\frac{p^\dagger}{K^s}, \tag{16.4.27}$$

or

$$\phi_{\mathrm{R}} - \phi_0 = \phi_0\left(-\frac{p^\dagger}{K^s}\right). \tag{16.4.28}$$

The stress-strain-pressure relation $(16.4.23)_2$, viz.

$$\tfrac{1}{3}\operatorname{tr}\boldsymbol{\sigma} = K^{\mathrm{d}}(\operatorname{tr}\boldsymbol{\epsilon}) - \alpha p,$$

must be consistent with the special state just discussed, and by substituting (16.4.24), (16.4.25), and (16.4.26) in it we obtain

$$-p^{\dagger} = K^{\rm d}\left(\frac{-p^{\dagger}}{K^s}\right) - \alpha p^{\dagger}, \tag{16.4.29}$$

which upon rearranging yields the following estimate for the Biot coefficient,

$$\boxed{\alpha = 1 - \frac{K^{\rm d}}{K^s}.} \tag{16.4.30}$$

Note that α is independent of the properties of the fluid. Since $K^{\rm d} \leq K^s$, (16.4.30) implies that

$$0 < \alpha \leq 1. \tag{16.4.31}$$

Next, the constitutive equation $(16.4.23)_3$, viz.

$$(\phi_{\rm R} - \phi_0) = \alpha(\operatorname{tr}\epsilon) + \frac{p}{M_b},$$

must also be consistent with the special state just discussed, and by substituting (16.4.24), (16.4.26), (16.4.27), and (16.4.22) into it we obtain

$$\phi_0\left(\frac{-p^{\dagger}}{K^s}\right) = \alpha\left(\frac{-p^{\dagger}}{K^s}\right) + \left(\frac{\alpha^2}{K^{\rm u} - K^{\rm d}} - \frac{\phi_0}{K^f}\right)p^{\dagger}, \tag{16.4.32}$$

from which we obtain the following estimate for the undrained bulk modulus,

$$\boxed{K^{\rm u} = K^{\rm d} + \frac{\alpha^2 K^s K^f}{K^s\phi_0 + K^f(\alpha - \phi_0)}.} \tag{16.4.33}$$

This equation relates $K^{\rm u}$ to $K^{\rm d}$, the porosity ϕ_0, and the bulk moduli of the solid and fluid phases, K^s and K^f, respectively. As expected intuitively, when the fluid cannot flow out of the porous material, the porous material is stiffer so that

$$K^{\rm u} > K^{\rm d}.$$

Finally, from (16.4.33) and (16.3.3) we obtain the following estimate for M,

$$\boxed{\frac{1}{M} = \frac{\phi_0}{K^f} + \frac{\alpha - \phi_0}{K^s}.} \tag{16.4.34}$$

16.5 **Linear poroelasticity of polymer gels: Incompressible materials**

The presentation so far has considered the case of materials with finite compressibility. We now treat the technologically important case of materials with *incompressible* fluid and solid phases. Our motivating example is an elastomeric gel, a network of covalently cross-linked polymer chains, swollen with a solvent. Gels are used in many diverse applications, including drug delivery, tissue engineering, and oil-field management. In a gel, the chains in the network can change conformation and enable large and reversible entropic-elastic deformations, while the solvent can migrate through the network and enable mass transport. The coupled deformation-diffusion process typically involves large deformations. Describing such large deformation requires a non-linear theory which has been developed in recent years (Hong et al., 2008; Duda et al., 2010; Chester and Anand, 2010, 2011; Chester et al., 2015). *Here we consider a specialized theory for small deformations superposed on a previously homogeneously swollen gel.* The coupled deformation-diffusion theory discussed below is often employed in the small deformation analysis of previously swollen polymer gels.

The starting points for this theory are the Biot constitutive equations (cf. (16.2.14)):

$$\begin{aligned} \boldsymbol{\sigma} &= 2G\boldsymbol{\epsilon} + (K^{\mathrm{d}} - (2/3)G)(\operatorname{tr}\boldsymbol{\epsilon})\mathbf{1} - \alpha p\,\mathbf{1}, \\ \zeta &= \alpha\,(\operatorname{tr}\boldsymbol{\epsilon}) + \frac{1}{M}p. \end{aligned} \tag{16.5.1}$$

The previously swollen state is taken as the reference state in which the strain $\boldsymbol{\epsilon}$ is set to zero, the pore pressure is taken to be zero, and hence the stress $\boldsymbol{\sigma}$ and variation in fluid content ζ are also zero. Next, recall the estimates (16.4.30) and (16.4.34) for the Biot moduli α and M,

$$\alpha = 1 - \frac{K^{\mathrm{d}}}{K^s}, \qquad \frac{1}{M} = \frac{\phi_0}{K^f} + \frac{\alpha - \phi_0}{K^s}. \tag{16.5.2}$$

Recall also the relation (16.3.3)

$$K^{\mathrm{u}} = K^{\mathrm{d}} + M\alpha^2. \tag{16.5.3}$$

The swollen polymer gel is often considered to be made from an incompressible solid and an incompressible fluid,

$$K^s \to \infty \quad \text{and} \quad K^f \to \infty. \tag{16.5.4}$$

In this limit, from (16.5.2) and (16.5.3) we have that

$$\alpha = 1, \qquad M \to \infty, \qquad \text{and} \qquad K^{\mathrm{u}} \to \infty, \tag{16.5.5}$$

and the constitutive equations (16.5.1) reduce to the single constitutive relation

$$\boldsymbol{\sigma} = 2G\boldsymbol{\epsilon} + (K^{\mathrm{d}} - (2/3)G)(\operatorname{tr}\boldsymbol{\epsilon})\mathbf{1} - p\,\mathbf{1} \tag{16.5.6}$$

and the *constraint*

$$(\operatorname{tr}\boldsymbol{\epsilon}) = \zeta. \tag{16.5.7}$$

Note that in this limit the pore pressure p is no longer constitutively defined. The constraint (16.5.7) and the fluid balance serve to determine it. Using

$$K^{\mathrm{d}} = \frac{2G(1+\nu^{\mathrm{d}})}{3(1-2\nu^{\mathrm{d}})}, \tag{16.5.8}$$

we may rewrite the constitutive equation (16.5.6) in terms of the shear modulus G and the drained Poisson's ratio ν^{d} as,

$$\boxed{\boldsymbol{\sigma} = 2G\left[\boldsymbol{\epsilon} + \frac{\nu^{\mathrm{d}}}{1-2\nu^{\mathrm{d}}}(\operatorname{tr}\boldsymbol{\epsilon})\mathbf{1}\right] - p\,\mathbf{1},} \tag{16.5.9}$$

which must *still* be supplemented by the constitutive constraint

$$\boxed{\zeta = (\operatorname{tr} \epsilon).} \tag{16.5.10}$$

Note from (16.5.10) that there is a strong constraint in the theory in that the change in volumetric strain of the gel is equal to volumetric strain caused by the absorbed solvent. Under this constraint the concentration of the solvent is no longer an independent field, but is determined by the displacement field by way of the identity $\operatorname{tr} \epsilon = \operatorname{div} \mathbf{u}$.

16.5.1 Governing partial differential equations

The governing partial differential equations are the balance of forces,[8]

$$\operatorname{div} \boldsymbol{\sigma} = \mathbf{0}, \tag{16.5.11}$$

and the fluid balance law

$$\dot{\zeta} = -\operatorname{div} \mathbf{q}, \tag{16.5.12}$$

with the volumetric fluid flux $\mathbf{q}$ given by the constitutive relation,

$$\mathbf{q} = -k \nabla p, \tag{16.5.13}$$

where $k > 0$ is a constant scalar permeability, related to the mobility by the identity $k = m\Omega^2$.

Boundary and initial conditions

Let $\mathcal{S}_{\mathbf{u}}$ and $\mathcal{S}_{\mathbf{t}}$ denote complementary subsurfaces of the boundary $\partial\mathcal{B}$ of the body $\mathcal{B}$. Then, as is standard, for a time interval $t \in [0, T]$ we consider a pair of simple boundary conditions in which the displacement is specified on $\mathcal{S}_{\mathbf{u}}$ and the surface traction on $\mathcal{S}_{\mathbf{t}}$:

$$\begin{aligned} \mathbf{u} &= \bar{\mathbf{u}} \quad \text{on } \mathcal{S}_{\mathbf{u}} \times [0, T], \\ \boldsymbol{\sigma}\mathbf{n} &= \bar{\mathbf{t}} \quad \text{on } \mathcal{S}_{\mathbf{t}} \times [0, T], \end{aligned} \tag{16.5.14}$$

where superposed bars indicate given quantities.

Letting $\mathcal{S}_p$ and $\mathcal{S}_{\mathbf{q}}$ denote complementary subsurfaces of the boundary, we consider a pair of simple boundary conditions in which the pore pressure is specified on $\mathcal{S}_p$ and the normal volumetric flux on $\mathcal{S}_{\mathbf{q}}$:

$$\begin{aligned} p &= \bar{p} \quad \text{on } \mathcal{S}_p \times [0, T], \\ \mathbf{q} \cdot \mathbf{n} &= \bar{q} \quad \text{on } \mathcal{S}_{\mathbf{q}} \times [0, T]. \end{aligned} \tag{16.5.15}$$

Note that since $p = (\mu - \mu_0)/\Omega$, specifying pore pressure is equivalent to specifying chemical potential. Likewise, since as $\mathbf{q} = \Omega \mathbf{j}$, specifying normal volumetric flux $\mathbf{q} \cdot \mathbf{n}$ is equivalent to specifying normal fluid flux $\mathbf{j} \cdot \mathbf{n}$.

The initial data is taken as

$$\mathbf{u}(\mathbf{x}, 0) = \mathbf{0}, \quad p(\mathbf{x}, 0) = 0 \quad \text{in} \quad \mathcal{B}. \tag{16.5.16}$$

[8] Neglecting body and inertial forces.

The coupled set of partial differential equations (16.5.11) and (16.5.12), together with the boundary conditions (16.5.14) and (16.5.15) and the initial conditions (16.5.16), yield an initial/boundary-value problem for the displacement $\mathbf{u}(\mathbf{x}, t)$ and the pore pressure $p(\mathbf{x}, t)$.

16.5.2 **Alternate forms for the governing partial differential equations**

Since

$$2\,\mathrm{div}\,\boldsymbol{\epsilon} = \mathrm{div}(\nabla\mathbf{u} + (\nabla\mathbf{u})^{\top}) = \Delta\mathbf{u} + \nabla\,\mathrm{div}\,\mathbf{u},$$

$$\mathrm{div}[(\mathrm{tr}\,\boldsymbol{\epsilon})\mathbf{1}] = \mathrm{div}[(\mathrm{div}\,\mathbf{u})\mathbf{1}] = \nabla\,\mathrm{div}\,\mathbf{u}, \quad \text{and} \quad \mathrm{div}\,(p\,\mathbf{1}) = \nabla p,$$

(16.5.11) yields

$$\mathrm{div}\,\boldsymbol{\sigma} = G\Delta\mathbf{u} + G\left(\frac{1}{1-2\nu^{\mathrm{d}}}\right)\nabla\,\mathrm{div}\,\mathbf{u} - \nabla p,$$

and hence the force balance (16.5.11) may be written as

$$\boxed{\Delta\mathbf{u} + \left(\frac{1}{1-2\nu^{\mathrm{d}}}\right)\nabla\,\mathrm{div}\,\mathbf{u} = \frac{1}{G}\nabla p.} \tag{16.5.17}$$

Next, using (16.5.17) in (16.5.13) gives

$$\mathbf{q} = -\,(kG)\left(\Delta\mathbf{u} + \left(\frac{1}{1-2\nu^{\mathrm{d}}}\right)\nabla\,\mathrm{div}\,\mathbf{u}\right),$$

which in turn gives

$$\mathrm{div}\,\mathbf{q} = -\,(kG)\left(\mathrm{div}(\Delta\mathbf{u}) + \left(\frac{1}{1-2\nu^{\mathrm{d}}}\right)\mathrm{div}(\nabla\,\mathrm{div}\,\mathbf{u})\right),$$

or in components

$$\frac{\partial q_k}{\partial x_k} = -\,(kG)\left(\frac{2(1-\nu^{\mathrm{d}})}{1-2\nu^{\mathrm{d}}}\right)\frac{\partial^2}{\partial x_k \partial x_k}\left(\frac{\partial u_i}{\partial x_i}\right),$$

and hence

$$\mathrm{div}\,\mathbf{q} = -\,(kG)\left(\frac{2(1-\nu^{\mathrm{d}})}{1-2\nu^{\mathrm{d}}}\right)\Delta(\mathrm{tr}\,\boldsymbol{\epsilon}). \tag{16.5.18}$$

Further substituting for $(\mathrm{tr}\,\boldsymbol{\epsilon})$ from (16.5.10) in (16.5.18) gives

$$\mathrm{div}\,\mathbf{q} = -\,(kG)\left(\frac{2(1-\nu^{\mathrm{d}})}{1-2\nu^{\mathrm{d}}}\right)\Delta\zeta. \tag{16.5.19}$$

A final substitution of (16.5.19) in the balance (16.5.12) and recalling that $\zeta = \Omega(c - c_0)$ gives the species balance in the classical form of a diffusion equation for the concentration field:

$$\boxed{\dot{c} = D^{*}\Delta c,} \tag{16.5.20}$$

with

$$\boxed{D^* \overset{\text{def}}{=} 2G\left(\frac{1-\nu^{\text{d}}}{1-2\nu^{\text{d}}}\right) k} \tag{16.5.21}$$

as the *effective diffusivity* for the gel.

REMARKS 16.3

1. The coupled partial differential equations of the theory take the forms (16.5.17) and (16.5.20). Equation (16.5.20) appears to be uncoupled from (16.5.17), but it is not. The diffusion equation (16.5.20) cannot be solved by itself because the boundary conditions involve the pore pressure, equivalently the chemical potential, and the displacement.
2. It should also be noted that the naive boundary conditions for (16.5.20), viz. imposed concentration and concentration gradient, are in general not possible to control in a physical experiment. Thus, while (16.5.20) is mathematically appealing, using the form in terms of pore pressure is to be preferred. In particular using (16.5.10), (16.5.12), and (16.5.13) yields

$$\operatorname{tr}\dot{\boldsymbol{\epsilon}} = k\nabla^2 p \tag{16.5.22}$$

for which the physically realizable boundary conditions (16.5.15) apply.
3. With $e \overset{\text{def}}{=} \operatorname{tr}\boldsymbol{\epsilon}$ denoting the volumetric strain and noting that $\zeta = \Omega(c - c_0) = e$, we see that (16.5.20) may also be written in the form of a *consolidation equation*,

$$\boxed{\dot{e} = D^*\Delta e \qquad \text{with} \qquad D^* \overset{\text{def}}{=} 2G\left(\frac{1-\nu^{\text{d}}}{1-2\nu^{\text{d}}}\right) k,} \tag{16.5.23}$$

where D^* now represents the Terzaghi *consolidation coefficient* (cf. e.g., Terzaghi and Froölicj, 1936).[9] The governing equations in this form have long been used in the theory of consolidation in soil mechanics, in which the constitutive equations for a soil are defined as the limiting case of the Biot constitutive equations for a fluid-infiltrated porous solid with an incompressible solid and an incompressible fluid (cf. e.g., McNamee and Gibson (1960)).

---

[9] In the soil mechanics literature the consolidation coefficient is denoted by "c", but we do not use that notation here because we have used "c" to denote the concentration of the fluid in this book.

17 A small deformation chemoelasticity theory for energy storage materials

17.1 Introduction

In this chapter we develop thermodynamically consistent constitutive equations for chemoelasticity which while limited to small deformations, *allow for possibly large changes in species concentration*. In this sense the "quasi-linear" theory formulated below is different from the *linear chemoelasticity* theory discussed in Sec. 15.6 as well as Biot's *linear poroelasticity* theory discussed in Sec. 16.2.

A technologically important area of application of the theory to be developed in this chapter is in the chemo-mechanical analysis of the evolution of large stresses which develop because of the volume changes associated with the diffusion of lithium ions in the active storage particles in the electrodes of lithium-ion batteries during charge-discharge cycles. The insertion and extraction of lithium into storage particles results in significant volume changes, which, over time, are thought to lead to the comminution of the storage particles and degradation of the performance of a lithium-ion battery (cf., e.g., Christensen and Newman, 2005; Zhang et al., 2007; Purkayastha and McMeeking, 2012; Bohn et al., 2013).

17.2 Basic equations

We take as the starting point the small-deformation non-linear theory of chemoelasticity discussed in Chapter 15, and begin with:

(i) the *strain-displacement relation*:

$$\epsilon = \tfrac{1}{2}\left(\nabla \mathbf{u} + (\nabla \mathbf{u})^{\top}\right); \tag{17.2.1}$$

(ii) the *balance of forces and moments*:[1]

$$\operatorname{div} \boldsymbol{\sigma} + \mathbf{b} = \mathbf{0}, \qquad \boldsymbol{\sigma} = \boldsymbol{\sigma}^{\top}, \tag{17.2.2}$$

with $\boldsymbol{\sigma}$ the *stress* and $\mathbf{b}$ the *body force*;

[1] As before, we neglect inertial body forces.

Continuum Mechanics of Solids. Lallit Anand and Sanjay Govindjee, Oxford University Press (2020).

DOI: 10.1093/oso/9780198864721.001.0001

(iii) the *species balance*:

$$\dot{c} = -\operatorname{div}\mathbf{j}, \tag{17.2.3}$$

with c the *number of moles of species, say, lithium, per unit volume*, and $\mathbf{j}$ the species flux;

(iv) the *free-energy imbalance*:

$$\dot{\psi} - \boldsymbol{\sigma}:\dot{\boldsymbol{\epsilon}} - \mu\,\dot{c} + \mathbf{j}\cdot\nabla\mu \le 0, \tag{17.2.4}$$

with ψ the *free energy* per unit volume, and μ the chemical potential;

(v) the *state relations*:

$$\boldsymbol{\sigma} = \frac{\partial\psi(\boldsymbol{\epsilon}, c)}{\partial\boldsymbol{\epsilon}} \quad \text{and} \quad \mu = \frac{\partial\psi(\boldsymbol{\epsilon}, c)}{\partial c}; \tag{17.2.5}$$

(vi) and *Fick's law* for the species flux:

$$\mathbf{j} = -\mathbf{M}\nabla\mu, \qquad \text{with} \qquad \mathbf{M} = \hat{\mathbf{M}}(\boldsymbol{\epsilon}, c) \tag{17.2.6}$$

the *species mobility tensor* which is positive semi-definite.

17.3 Specialization of the constitutive equations

17.3.1 Strain decomposition

Underlying our discussion of species diffusion coupled to elasticity is a physical picture that associates with the solid a *microscopic structure*, such as a crystal lattice, that may be deformed, together with an atomic or molecular chemical species capable of migrating through that structure. Accordingly, we introduce a kinematical constitutive assumption that the small strain tensor admits the decomposition,

$$\boldsymbol{\epsilon} = \boldsymbol{\epsilon}^{\mathrm{e}} + \boldsymbol{\epsilon}^{\mathrm{c}}, \tag{17.3.1}$$

in which:

(i) $\boldsymbol{\epsilon}^{\mathrm{c}}$, the **chemical strain**, represents the local strain of the material due to the presence and migration of the chemical species through that structure, and

(ii) $\boldsymbol{\epsilon}^{\mathrm{e}}$, the **elastic strain**, represents local strain of the underlying microscopic structure due to elastic deformation.

We assume that the chemical strain depends on the concentration of the migrating species,

$$\boldsymbol{\epsilon}^{\mathrm{c}} = \hat{\boldsymbol{\epsilon}}^{\mathrm{c}}(c) \quad \text{with} \quad \hat{\boldsymbol{\epsilon}}^{\mathrm{c}}(c_0) = \mathbf{0}, \tag{17.3.2}$$

where c_0 represents an initial value of c at which $\boldsymbol{\epsilon}^{\mathrm{c}}$ is taken to vanish. Thus the rate of change of $\boldsymbol{\epsilon}^{\mathrm{c}}$ is

$$\dot{\boldsymbol{\epsilon}}^{\mathrm{c}} = \mathbf{S}\,\dot{c}, \tag{17.3.3}$$

where the symmetric tensor

$$\mathbf{S} = \mathbf{S}(c) \stackrel{\text{def}}{=} \frac{d\hat{\boldsymbol{\epsilon}}^{\text{c}}(c)}{dc} \tag{17.3.4}$$

represents a *chemical expansion tensor* at a given concentration c.

17.3.2 **Free energy**

Next, we consider a free energy of the form

$$\psi(\boldsymbol{\epsilon}, c) = \psi^{\text{c}}(c) + \psi^{\text{e}}(\boldsymbol{\epsilon}, c). \tag{17.3.5}$$

Here,

(i) As a simple continuum approximation for the chemical free energy ψ^{c}, we take it to be given by:

$$\psi^{\text{c}} = \mu_0(c - c_0) + R\vartheta\, c_{\text{max}}\Big(\bar{c}\ln\bar{c} + (1-\bar{c})\ln(1-\bar{c})\Big). \tag{17.3.6}$$

In (17.3.6):

- μ_0 is a reference value of the chemical potential of the diffusing species;
- $\bar{c}$ is a *normalized species concentration,*

$$\bar{c} \stackrel{\text{def}}{=} \frac{c}{c_{\text{max}}}, \qquad 0 \le \bar{c} \le 1, \tag{17.3.7}$$

  where c_{max} is the concentration of the species in moles per unit volume when all possible accommodating sites in the host material are filled; further
- the term in (17.3.6), involving $R\vartheta$, with R the universal gas constant and ϑ the (constant) temperature, represents the entropy of mixing for an ideal solid solution; see e.g. DeHoff (2006).

(ii) ψ^{e} is the contribution to the change in the free energy due to the elastic deformation of the host material, and is taken to be given by

$$\psi^{\text{e}} = \tfrac{1}{2}\boldsymbol{\epsilon}^{\text{e}} : \mathbb{C}(c)\boldsymbol{\epsilon}^{\text{e}} \qquad \text{with} \qquad \boldsymbol{\epsilon}^{\text{e}} = \boldsymbol{\epsilon} - \hat{\boldsymbol{\epsilon}}^{\text{c}}(c), \tag{17.3.8}$$

where $\mathbb{C}(c)$ is a concentration-dependent positive definite elasticity tensor with the usual symmetries.

Thus, using (17.3.6) and (17.3.8) in (17.3.5) a simple form of the free energy function which accounts for the combined effects of mixing, chemical expansion, and straining is

$$\begin{aligned}\psi = \mu_0\,(c - c_0) + R\vartheta\, c_{\text{max}}\Big(\bar{c}\ln\bar{c} + (1-\bar{c})\ln(1-\bar{c})\Big) + \tfrac{1}{2}\boldsymbol{\epsilon}^{\text{e}} : \mathbb{C}(c)\boldsymbol{\epsilon}^{\text{e}}\\ \text{with} \qquad \boldsymbol{\epsilon}^{\text{e}} = \boldsymbol{\epsilon} - \hat{\boldsymbol{\epsilon}}^{\text{c}}(c).\end{aligned} \tag{17.3.9}$$

17.3.3 Stress. Chemical potential

Using (17.3.9) and $(17.2.5)_1$ we find that the stress is given by

$$\boxed{\boldsymbol{\sigma} = \mathbb{C}(c)\boldsymbol{\epsilon}^{\mathrm{e}} \qquad \text{with} \qquad \boldsymbol{\epsilon}^{\mathrm{e}} = \boldsymbol{\epsilon} - \hat{\boldsymbol{\epsilon}}^{\mathrm{c}}(c).} \tag{17.3.10}$$

Also, using (17.3.9), $(17.2.5)_2$, (17.3.10), and (17.3.4) the chemical potential μ is given by

$$\boxed{\begin{aligned}\mu &= \mu_0 + R\vartheta \ln\left(\frac{\bar{c}}{1-\bar{c}}\right) - \frac{d\hat{\boldsymbol{\epsilon}}^{\mathrm{c}}(c)}{dc} : \mathbb{C}\boldsymbol{\epsilon}^{\mathrm{e}} + \tfrac{1}{2}\boldsymbol{\epsilon}^{\mathrm{e}} : \left(\frac{d\mathbb{C}(c)}{dc}\right)\boldsymbol{\epsilon}^{\mathrm{e}}, \\ &= \mu_0 + R\vartheta \ln\left(\frac{\bar{c}}{1-\bar{c}}\right) - \frac{d\hat{\boldsymbol{\epsilon}}^{\mathrm{c}}(c)}{dc} : \boldsymbol{\sigma} + \tfrac{1}{2}\boldsymbol{\epsilon}^{\mathrm{e}} : \left(\frac{d\mathbb{C}(c)}{dc}\right)\boldsymbol{\epsilon}^{\mathrm{e}}, \\ &= \mu_0 + R\vartheta \ln\left(\frac{\bar{c}}{1-\bar{c}}\right) - \mathbf{S}(c) : \boldsymbol{\sigma} + \tfrac{1}{2}\boldsymbol{\epsilon}^{\mathrm{e}} : \left(\frac{d\mathbb{C}(c)}{dc}\right)\boldsymbol{\epsilon}^{\mathrm{e}}.\end{aligned}} \tag{17.3.11}$$

REMARKS 17.1

1. In most of the existing literature on chemoelasticity for Li-ion battery materials (cf., e.g., Zhang et al., 2007; Purkayastha and McMeeking, 2012; Bohn et al., 2013), the species concentration dependence of the elasticity tensor $\mathbb{C}(c)$ and the chemical expansion tensor $\mathbf{S}(c)$ are neglected without comment—primarily because such dependencies have not yet been adequately characterized, either experimentally or theoretically.
2. Many solid solutions of interest for battery materials are highly non-ideal, in the sense that their chemical potentials are not well represented by the regular solution model. In this case the term $R\vartheta \ln(\bar{c}/(1-\bar{c}))$ in (17.3.11) is modified to read as

$$R\vartheta \ln\left(\gamma\frac{\bar{c}}{1-\bar{c}}\right),$$

where γ is an "activity coefficient". The correction term $(R\vartheta \ln \gamma)$ is often expressed in the form of a polynomial expression in $\bar{c}$, with the polynomial coefficients determined by fitting to suitable experimental data. We do not go into the details of such modifications here.

The constitutive equations (17.3.10) and (17.3.11) for the stress and the chemical potential, and (17.2.6) for the species flux characterize an *anisotropic* chemoelastic material. In the next section we further specialize these constitutive equations to represent an *isotropic* chemoelastic material. In the specialization considered below we will also *neglect* the concentration dependence of the elasticity tensor $\mathbb{C}$ and the chemical expansion tensor $\mathbf{S}$.

17.4 Isotropic idealization

17.4.1 Chemical strain

For the constitutive equation (17.3.2) for $\boldsymbol{\epsilon}^{\mathrm{c}}$, we assume that for isotropic materials it is given by,

$$\boldsymbol{\epsilon}^{\mathrm{c}} = \frac{1}{3}\Omega(c - c_0)\mathbf{1}, \tag{17.4.1}$$

with $\Omega > 0$ a *constant*. In this case the chemical expansion tensor $\mathbf{S}$ defined in (17.3.4) is independent of the species concentration c and given by

$$\mathbf{S} = \frac{1}{3}\Omega\mathbf{1}. \tag{17.4.2}$$

Also, from (17.4.1) we may identify $(1/3)\Omega$ as a *chemical expansion coefficient*. Thus,

$$\operatorname{tr}\boldsymbol{\epsilon}^{\mathrm{c}} = \Omega(c - c_0), \tag{17.4.3}$$

and hence

$$\Omega \equiv \frac{d\operatorname{tr}\boldsymbol{\epsilon}^{\mathrm{c}}}{dc} > 0 \tag{17.4.4}$$

represents the *molar volume* of the migrating species when embedded in the host solid.

17.4.2 Elasticity tensor

For an isotropic material the positive definite elasticity tensor $\mathbb{C}$ has the representation

$$\mathbb{C} \stackrel{\text{def}}{=} 2G\left(\mathbb{I}^{\text{sym}} - \frac{1}{3}\mathbf{1}\otimes\mathbf{1}\right) + K\mathbf{1}\otimes\mathbf{1}, \tag{17.4.5}$$

with $\mathbb{I}^{\text{sym}}$ and $\mathbf{1}$ the symmetric fourth- and second-order identity tensors, and the parameters

$$G > 0, \qquad K > 0, \tag{17.4.6}$$

are the concentration-independent shear modulus and bulk modulus, respectively.

17.4.3 Species flux

Neglecting a dependence of the species mobility on the strain $\boldsymbol{\epsilon}$, for isotropic materials the mobility tensor (17.2.6) at a fixed temperature ϑ has the representation

$$\mathbf{M}(c,\vartheta) = m(c,\vartheta)\mathbf{1}, \tag{17.4.7}$$

where $m(c,\vartheta) > 0$ is a scalar *species mobility*. We take the scalar mobility as the following function of the species concentration and temperature,

$$m(c,\vartheta) = \frac{D}{R\vartheta}\,c\,(1-\bar{c})\,, \tag{17.4.8}$$

where $D > 0$ is a concentration independent *diffusivity*. Expression (17.4.8) represents the physical requirement that at limiting values of the species concentrations, $c = 0$ and $\bar{c} = 1$, the mobility vanishes.

17.4.4 **Stress. Chemical potential. Species flux**

Using (17.4.1) and (17.4.5) in (17.3.10) gives the stress as

$$\boxed{\boldsymbol{\sigma} = 2G\boldsymbol{\epsilon}' + K\,(\mathrm{tr}\,\boldsymbol{\epsilon} - \Omega(c - c_0))\,\mathbf{1}.} \tag{17.4.9}$$

Next, using $(17.3.11)_3$ and (17.4.2) gives the chemical potential as

$$\boxed{\mu = \mu_0 + R\vartheta\,\ln\left(\frac{\bar{c}}{1-\bar{c}}\right) - \Omega\left(\frac{1}{3}\,\mathrm{tr}\,\boldsymbol{\sigma}\right).} \tag{17.4.10}$$

Finally, using (17.4.7) and (17.4.8) in (17.2.6) for the species flux we obtain,

$$\boxed{\mathbf{j} = -\left(\frac{D}{R\vartheta}\,c\,(1-\bar{c})\right)\,\nabla\mu.} \tag{17.4.11}$$

REMARK 17.2

In the *absence* of mechanical coupling the gradient of the chemical potential (17.3.11) is given by

$$\nabla\mu = \left(\frac{R\vartheta}{\bar{c}(1-\bar{c})}\right)\nabla\bar{c}. \tag{17.4.12}$$

Using (17.4.12) and (17.4.8) in the species balance (17.2.3) gives the classical diffusion equation

$$\dot{c} = D\,\Delta c, \tag{17.4.13}$$

where D denotes the *diffusivity* of the chemical species.

17.5 **Governing partial differential equations for the isotropic theory**

Partial differential equations

The governing partial differential equations consist of:

1. The force balance $(17.2.2)_1$,

$$\mathrm{div}\,\boldsymbol{\sigma} + \mathbf{b} = \mathbf{0}, \tag{17.5.1}$$

   where $\mathbf{b}$ is the non-inertial body force and $\boldsymbol{\sigma}$ given by (17.4.9).

2. The balance equation for the species concentration (17.2.3) and the constitutive equations for the species flux (17.4.11) gives

$$\dot{c} = \mathrm{div}\,(m\nabla\mu) \quad \text{with} \quad m = \frac{D}{R\vartheta}\,c\,(1-\bar{c}), \tag{17.5.2}$$

   and with the chemical potential given by (17.4.10).

Boundary and initial conditions

Let $\mathcal{S}_{\mathrm{u}}$ and $\mathcal{S}_{\mathrm{t}}$ denote complementary subsurfaces of the boundary $\partial\mathcal{B}$ of the body $\mathcal{B}$. Then, as is standard, for a time interval $t \in [0, T]$ we consider a pair of simple boundary conditions in which the displacement is specified on $\mathcal{S}_{\mathrm{u}}$ and the surface traction on $\mathcal{S}_{\mathrm{t}}$:

$$\begin{aligned} \mathbf{u} &= \bar{\mathbf{u}} \quad \text{on} \quad \mathcal{S}_{\mathrm{u}} \times [0, T], \\ \boldsymbol{\sigma}\mathbf{n} &= \bar{\mathbf{t}} \quad \text{on} \quad \mathcal{S}_{\mathrm{t}} \times [0, T], \end{aligned} \tag{17.5.3}$$

where quantities with a superposed bar are given boundary data.

Further, for the species balance, let $\mathcal{S}_{\mu}$ and $\mathcal{S}_{\mathrm{j}}$ denote complementary subsurfaces of the boundary; we then consider a pair of simple boundary conditions in which the species chemical potential is specified on $\mathcal{S}_{\mu}$ and the normal species flux on $\mathcal{S}_{\mathrm{j}}$:

$$\begin{aligned} \mu &= \bar{\mu} \quad \text{on} \quad \mathcal{S}_{\mu} \times [0, T], \\ \mathbf{j} \cdot \mathbf{n} &= \bar{j} \quad \text{on} \quad \mathcal{S}_{\mathrm{j}} \times [0, T], \end{aligned} \tag{17.5.4}$$

where quantities with a superposed bar are given boundary data.

The initial data is taken as

$$\mathbf{u}(\mathbf{x}, 0) = \mathbf{0}, \quad \mu(\mathbf{x}, 0) = \mu_0 \quad \text{in} \quad \mathcal{B}. \tag{17.5.5}$$

The coupled set of partial differential equations (17.5.1) and (17.5.2) together with the boundary conditions (17.5.3) and (17.5.4), and the initial conditions (17.5.5) yield an initial/boundary-value problem for the displacement $\mathbf{u}(\mathbf{x}, t)$ and the chemical potential $\mu(\mathbf{x}, t)$.

18 Linear piezoelectricity

18.1 Introduction

In this chapter we consider the coupling of an electric field to mechanical deformation. In particular we consider the modeling of piezoelectric materials. When a piezoelectric material is subjected to a mechanical deformation, there is a change in the polarization of the material. Conversely, when such a material is subjected to an electric field, it deforms. The phenomenon of piezoelectricity was first observed by the brothers Pierre and Jacques Curie in 1880 (Curie and Curie, 1880). The piezoelectricity effect occurs in several anisotropic crystals such as quartz, tourmaline, and Rochelle salt. The unit cells of single crystals of such materials do not possess a center of symmetry with regard to the locations of the centers of net positive and negative charges of their constituent ions; that is, the unit cells of such materials exhibit an *intrinsic polarization*. Figure 18.1 shows a schematic of a unit cell of an undeformed crystal of a material with a polar unit cell. The arrow indicates the intrinsic polarization, which is present even though the crystal lattice possesses no net electric charge.

When a crystal with a polar unit cell is subjected to a mechanical strain, the relative positions of the centers of the charges in its unit cell will change, and this displacement will cause a change in the polarization of the crystal. The change in polarization induced by a mechanical strain is known as the *direct piezoelectric effect*. This phenomenon is widely employed in applications of piezoelectric materials as *sensors*.

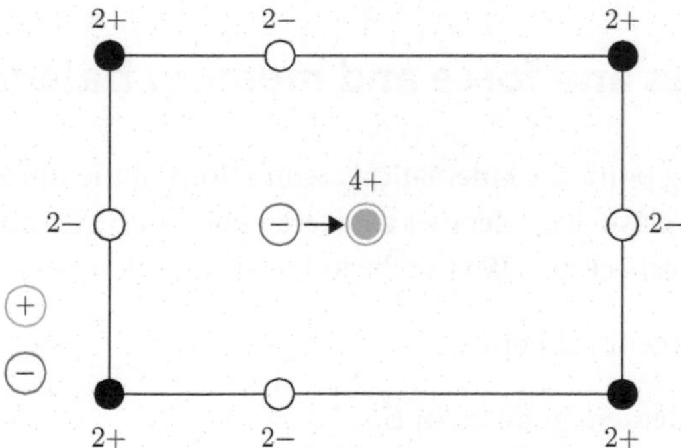

Fig. 18.1 Schematic of a unit cell of a polar crystal with different locations of the centers of positive and negative charges. The arrow indicates the intrinsic polarization, which is present even though the crystal lattice possesses no net electric charge (adapted from Govorukha et al., 2017).

Continuum Mechanics of Solids. Lallit Anand and Sanjay Govindjee, Oxford University Press (2020).

DOI: 10.1093/oso/9780198864721.001.0001

Conversely if an electric field is applied to a mechanically unloaded specimen of a crystal with a polar unit cell, then the crystal deforms. The electrically induced distortion depends on the direction of the applied electric field relative to the intrinsic polarization of the crystal. An electric field with the same direction as the spontaneous polarization will move the charge centers further apart from each other, while an electric field with the opposite direction will bring them closer together. The shifting of the charge centers is accompanied by a corresponding mechanical strain in the crystal. The production of a strain induced by an electric field is known as the *inverse piezoelectric effect*. This phenomenon is widely employed in applications of piezoelectric materials as *actuators*.

Natural occurring piezoelectric single crystals—such as quartz, tourmaline, and Rochelle salt—were widely used as sensors and actuators until about the mid-1940s. Since that time several synthetic polycrystalline ceramic materials which exhibit a strong piezoelectric effect have been developed. These materials are called *piezoceramics*. Barium-titanate, $BaTiO_3$, and lead-zirconate-titanate, PZT, are the most prominent materials in this class. The powder processing technique used to make polycrystalline piezoceramics initially produces materials with a microstructure in which the ceramic grains (and the electric polarization domains within grains) have random orientations. This material is subsequently subjected to a special processing step called *poling*, in which the polycrystalline aggregate with randomly oriented intrinsic polarization is subjected to a suitably high electric field which causes the polarization of the unit cells in the individual grains of the polycrystalline aggregate to (approximately) align in the direction of an applied electric field to produce a *transversely isotropic polycrystalline piezoelectric material*; see Jaffe et al. (1971) for a detailed discussion regarding the processing of modern piezoceramics. "Poled"-piezoceramics, such as PZT, are currently the preferred piezoelectric materials for engineering applications because of their ability to be processed into complex shapes, as well as the enhanced piezoelectric effect that these materials possess relative to the natural single crystal materials.[1]

In this chapter we give a derivation of the basic equations of the classical coupled theory of linear electroelasticity which forms the basis for design and analysis of piezoelectric devices.

18.2 Kinematics and force and moment balances

We take as the starting point the kinematical assumptions of the linear theory of elasticity. As in the theory of linear elasticity, $\mathbf{x}$ denotes a material point in $\mathcal{B}$, ∇ and div denote the gradient and divergence with respect to $\mathbf{x}$. We use the following notation:

- $\mathbf{u}(\mathbf{x}, t)$ displacement field of $\mathcal{B}$;
- $\nabla\mathbf{u}(\mathbf{x}, t)$ displacement gradient in $\mathcal{B}$;
- $\epsilon(\mathbf{x}, t) = (1/2)\big(\nabla\mathbf{u}(\mathbf{x}, t) + (\nabla\mathbf{u}(\mathbf{x}, t))^{\top}\big)$, strain in $\mathcal{B}$;

[1] In addition to piezoceramics, piezoelectric polymers can also be fabricated. One such popular material is polyvinylidene fluoride (see e.g. Murayama et al., 1976).

- $\boldsymbol{\sigma}(\mathbf{x}, t)$ stress;
- $\mathbf{b}(\mathbf{x}, t)$ body force field on $\mathcal{B}$;
- $\rho(\mathbf{x}, t)$ mass density of $\mathcal{B}$;

and recall the following standard relations for force and moment balances from the theory of linear elasticity, which are also valid for the theory considered here,

$$\operatorname{div}\boldsymbol{\sigma} + \mathbf{b} = \mathbf{0}, \qquad \boldsymbol{\sigma} = \boldsymbol{\sigma}^{\top}. \tag{18.2.1}$$

In (18.2.1) the body force $\mathbf{b}$ is presumed to account for inertia, that is,

$$\boxed{\mathbf{b} = \mathbf{b}_0 - \rho\ddot{\mathbf{u}},} \tag{18.2.2}$$

where $\mathbf{b}_0$ represents the conventional body force per unit volume, and $(-\rho\ddot{\mathbf{u}})$ represents the inertial body force, with $\ddot{\mathbf{u}}$ the acceleration.

18.3 Basic equations of electrostatics

The *electric field* in $\mathcal{B}$ is denoted by $\mathbf{e}(\mathbf{x}, t)$. In a polarizable material there is a vector field in $\mathcal{B}$ denoted by $\mathbf{d}(\mathbf{x}, t)$ called the *electric displacement*. In the theory formulated in this chapter we only consider the *electrostatic* (quasi-static) limit as far as the electrical variables are concerned, so that time t only plays the role of a history parameter. However, time t will have real meaning as far as inertial effects are concerned (if they are not neglected). See Appendix 18.A for the precise meaning of the quasi-static limit for the electrical fields in the context of this chapter.

Under electrostatic conditions there are two governing Maxwell equations that must be satisfied. The first of these is *Faraday's law*, which in the electrostatic limit reads,

$$\boxed{\operatorname{curl}\mathbf{e} = \mathbf{0}.} \tag{18.3.1}$$

Satisfaction of (18.3.1) occurs automatically if we allow the electric field $\mathbf{e}(\mathbf{x}, t)$ to be represented as the gradient of an *electrostatic potential* $\phi(\mathbf{x}, t)$,[2]

$$\boxed{\mathbf{e}(\mathbf{x}, t) \overset{\text{def}}{=} -\nabla\phi(\mathbf{x}, t).} \tag{18.3.2}$$

The negative sign in (18.3.1) follows the convention that the electric field points from a region of high potential to low potential, whereas the vector $\nabla\phi$ by definition points in the direction of increasing ϕ.

The second equation of electrostatics is given by *Gauss's law*. Let the *free charge* per unit volume of the deformed body be denoted by $q(\mathbf{x}, t)$. Gauss's law states that the divergence of the electric displacement $\mathbf{d}$ must be equal to the density of the free charge per unit volume,

$$\boxed{\operatorname{div}\mathbf{d} = q.} \tag{18.3.3}$$

---

[2] Recall that for any scalar field ϕ, $\operatorname{curl}(\nabla\phi) = \mathbf{0}$.

Equations (18.3.1) and (18.3.3) constitute the governing equations for electrostatics in matter. The system of equations is closed by specifying a constitutive relation between **d** and **e**. In the case of piezoelectricity, to which we will adapt these relations, the constitutive law will connect the electric displacement **d** to the electric field **e** *and* the mechanical strain $\boldsymbol{\epsilon}$.

18.4 **External and internal expenditures of power. Power balance**

We denote by $\mathcal{P}$ an arbitrary *part* of the body $\mathcal{B}$ with $\mathbf{n}$ the outward unit normal on the boundary $\partial\mathcal{P}$ of $\mathcal{P}$. At each fixed time we associate a mechanical and an electrical force system acting on $\mathcal{P}$. The mechanical system is characterized by:

(a) The traction $\boldsymbol{\sigma}\mathbf{n}$ at each point on the boundary $\partial\mathcal{P}$ which expends power over the velocity $\dot{\mathbf{u}}$; and

(b) the external body force $\mathbf{b}$ which also expends power over $\dot{\mathbf{u}}$.[3]

In the coupled piezoelectric case, we also need to account for the electric power that is expended on $\mathcal{P}$ by charges external to $\mathcal{P}$. In electromagnetic field theory this is characterized by,[4]

(c) the negative normal component of an electromagnetic energy flux $-\mathbf{s}\cdot\mathbf{n}$, where the vector $\mathbf{s}$ has dimensions of energy per unit area per unit time, and is known as the *Poynting vector*. As discussed in Appendix 18.A, in the electrostatic limit we can write

$$-\mathbf{s}\cdot\mathbf{n} \approx -\phi(\dot{\mathbf{d}}+\mathbf{j})\cdot\mathbf{n},$$

where $\mathbf{j}$ represents the free charge current (charge per unit area per unit time).[5]

Combining contributions (a)–(c) we obtain the following expression for the power expended on $\mathcal{P}$ by agents external to $\mathcal{P}$:

$$\boxed{\mathcal{W}_{\text{ext}}(\mathcal{P}) = \int_{\partial\mathcal{P}} (\boldsymbol{\sigma}\mathbf{n})\cdot\dot{\mathbf{u}}\,da + \int_{\mathcal{P}} \mathbf{b}\cdot\dot{\mathbf{u}}\,dv - \int_{\partial\mathcal{P}} \phi(\dot{\mathbf{d}}+\mathbf{j})\cdot\mathbf{n}\,da.} \tag{18.4.1}$$

Using the divergence theorem on the terms involving the integral over $\partial\mathcal{P}$, the symmetry of $\boldsymbol{\sigma}$, $\dot{\boldsymbol{\epsilon}} = (1/2)\big(\nabla\dot{\mathbf{u}} + (\nabla\dot{\mathbf{u}})^{\top}\big)$, and (18.3.2) we obtain

[3] The external body force **b** is presumed to account for inertia, cf. (18.2.2).

[4] Cf. e.g. Kovetz (2000, Chap. 15), Jackson (1999, Sec. 6.7), or Auld (1990, Chap. 5).

[5] Cf. (18.A.17).

$$\begin{aligned}\int_{\partial\mathcal{P}} (\boldsymbol{\sigma}\mathbf{n})\cdot\dot{\mathbf{u}}\,da &= \int_{\mathcal{P}} (\operatorname{div}\boldsymbol{\sigma})\cdot\dot{\mathbf{u}}\,dv + \int_{\mathcal{P}} \boldsymbol{\sigma}:\nabla\dot{\mathbf{u}}\,dv \\ &= \int_{\mathcal{P}} (\operatorname{div}\boldsymbol{\sigma})\cdot\dot{\mathbf{u}}\,dv + \int_{\mathcal{P}} \boldsymbol{\sigma}:\dot{\boldsymbol{\epsilon}}\,dv,\end{aligned} \tag{18.4.2}$$

$$\begin{aligned}-\int_{\partial\mathcal{P}} \phi(\dot{\mathbf{d}}+\mathbf{j})\cdot\mathbf{n}\,da &= -\int_{\mathcal{P}} \phi\,\operatorname{div}(\dot{\mathbf{d}}+\mathbf{j})\,dv - \int_{\mathcal{P}} (\nabla\phi)\cdot(\dot{\mathbf{d}}+\mathbf{j})\,dv \\ &= -\int_{\mathcal{P}} \phi(\operatorname{div}\dot{\mathbf{d}}+\operatorname{div}\mathbf{j})\,dv + \int_{\mathcal{P}} (\mathbf{e}\cdot\dot{\mathbf{d}}+\mathbf{e}\cdot\mathbf{j})\,dv.\end{aligned}$$

Now, balance of charge requires that $\operatorname{div}\mathbf{j}+\dot{q}=0$.[6] Using this balance, and substituting (18.4.2) in (18.4.1) we obtain

$$\mathcal{W}_{\text{ext}}(\mathcal{P}) = \int_{\mathcal{P}} \boldsymbol{\sigma}:\dot{\boldsymbol{\epsilon}}\,dv + \int_{\mathcal{P}} (\operatorname{div}\boldsymbol{\sigma}+\mathbf{b})\cdot\dot{\mathbf{u}}\,dv + \int_{\mathcal{P}} \mathbf{e}\cdot\dot{\mathbf{d}}\,dv - \int_{\mathcal{P}} \phi(\operatorname{div}\dot{\mathbf{d}}-\dot{q})\,dv + \int_{\mathcal{P}} \mathbf{e}\cdot\mathbf{j}\,dv, \tag{18.4.3}$$

and upon using the balance (18.2.1) and the time rate form of (18.3.3), we obtain the power balance

$$\mathcal{W}_{\text{ext}}(\mathcal{P}) = \mathcal{W}_{\text{int}}(\mathcal{P}) \tag{18.4.4}$$

where

$$\boxed{\mathcal{W}_{\text{int}}(\mathcal{P}) \stackrel{\text{def}}{=} \int_{\mathcal{P}} \Big(\boldsymbol{\sigma}:\dot{\boldsymbol{\epsilon}}+\mathbf{e}\cdot\dot{\mathbf{d}}+\mathbf{e}\cdot\mathbf{j}\Big)\,dv,} \tag{18.4.5}$$

represents the internal expenditure of *mechanical and electrical* power.

18.5 Dielectric materials

Dielectric materials do not possess free charges and are non-conducting, so that

$$q=0, \quad \text{and} \quad \mathbf{j}=\mathbf{0}. \tag{18.5.1}$$

Thus, for dielectric materials the relation (18.4.5) for $\mathcal{W}_{\text{int}}(\mathcal{P})$ reduces to,

$$\boxed{\mathcal{W}_{\text{int}}(\mathcal{P}) \stackrel{\text{def}}{=} \int_{\mathcal{P}} \Big(\boldsymbol{\sigma}:\dot{\boldsymbol{\epsilon}}+\mathbf{e}\cdot\dot{\mathbf{d}}\Big)\,dv.} \tag{18.5.2}$$

Note that piezoelectric materials are a special class of dielectric materials.

[6] Cf. (18.A.6).

18.6 Free-energy imbalance for dielectric materials

Under isothermal conditions the two laws of thermodynamics reduce to the statement that *the rate of increase in free energy of any part $\mathcal{P}$ is less than or equal to the power expended on $\mathcal{P}$.* Precisely, letting ψ denote the electromechanical *free energy* per unit volume, this requirement takes the form of a free-energy imbalance

$$\overline{\int_{\mathcal{P}} \psi\, dv}^{\,\boldsymbol{\cdot}} \leq \mathcal{W}_{\mathrm{ext}}(\mathcal{P}) = \mathcal{W}_{\mathrm{int}}(\mathcal{P}). \tag{18.6.1}$$

Since $\overline{\int_{\mathcal{P}} \psi dv}^{\,\boldsymbol{\cdot}} = \int_{\mathcal{P}} \dot{\psi}\, dv$, we may use (18.5.2) to localize (18.6.1); the result is the electromechanical *free-energy imbalance*

$$\dot{\psi} - \boldsymbol{\sigma} : \dot{\boldsymbol{\epsilon}} - \mathbf{e} \cdot \dot{\mathbf{d}} \leq 0. \tag{18.6.2}$$

18.7 Constitutive theory

Guided by the free-energy imbalance (18.6.2), we begin by assuming constitutive equations for the free energy ψ, the stress $\boldsymbol{\sigma}$, and the electric field $\mathbf{e}$ of the form

$$\psi = \hat{\psi}(\boldsymbol{\epsilon}, \mathbf{d}), \qquad \boldsymbol{\sigma} = \hat{\boldsymbol{\sigma}}(\boldsymbol{\epsilon}, \mathbf{d}), \qquad \mathbf{e} = \hat{\mathbf{e}}(\boldsymbol{\epsilon}, \mathbf{d}). \tag{18.7.1}$$

Then,

$$\dot{\psi} = \frac{\partial \hat{\psi}(\boldsymbol{\epsilon}, \mathbf{d})}{\partial \boldsymbol{\epsilon}} : \dot{\boldsymbol{\epsilon}} + \frac{\partial \hat{\psi}(\boldsymbol{\epsilon}, \mathbf{d})}{\partial \mathbf{d}} \cdot \dot{\mathbf{d}}. \tag{18.7.2}$$

Using (18.7.2) and substituting the constitutive equations (18.7.1) into the free-energy imbalance (18.6.2), we find that it may be written as

$$\left[\frac{\partial \hat{\psi}(\boldsymbol{\epsilon}, \mathbf{d})}{\partial \boldsymbol{\epsilon}} - \hat{\boldsymbol{\sigma}}(\boldsymbol{\epsilon}, \mathbf{d})\right] : \dot{\boldsymbol{\epsilon}} + \left[\frac{\partial \hat{\psi}(\boldsymbol{\epsilon}, \mathbf{d})}{\partial \mathbf{d}} - \hat{\mathbf{e}}(\boldsymbol{\epsilon}, \mathbf{d})\right] \cdot \dot{\mathbf{d}} \leq 0. \tag{18.7.3}$$

We may now use the Coleman–Noll argument to exploit this inequality. Given a point of the body and any time t, it is possible to find a motion and an electrical displacement field such that $\boldsymbol{\epsilon}$, $\mathbf{d}$ and their time derivatives $\dot{\boldsymbol{\epsilon}}$ and $\dot{\mathbf{d}}$ have arbitrarily prescribed values at that point and time. Granted this assertion, the coefficients of $\dot{\boldsymbol{\epsilon}}$ and $\dot{\mathbf{d}}$ must vanish, for otherwise the rates may be chosen to violate inequality (18.7.3). We therefore have the *thermodynamic restrictions* that the free energy delivers the stress $\boldsymbol{\sigma}$ and the electric field $\mathbf{e}$ through the state relations:

$$\boxed{\begin{aligned} \boldsymbol{\sigma} = \hat{\boldsymbol{\sigma}}(\boldsymbol{\epsilon}, \mathbf{d}) &= \frac{\partial \hat{\psi}(\boldsymbol{\epsilon}, \mathbf{d})}{\partial \boldsymbol{\epsilon}}, \\ \mathbf{e} = \hat{\mathbf{e}}(\boldsymbol{\epsilon}, \mathbf{d}) &= \frac{\partial \hat{\psi}(\boldsymbol{\epsilon}, \mathbf{d})}{\partial \mathbf{d}}, \end{aligned}} \tag{18.7.4}$$

and that the free-energy imbalance is satisfied with the material exhibiting no dissipation.

18.7.1 Electric field as independent variable

In applications, it is preferable to replace constitutive dependence upon the electric displacement $\mathbf{d}$ by constitutive dependence upon the electric field $\mathbf{e}$.

To effect this change, we assume that for each fixed ϵ, the relation

$$\mathbf{e} = \hat{\mathbf{e}}(\epsilon, \mathbf{d}) \tag{18.7.5}$$

is smoothly invertible in $\mathbf{d}$, so that[7]

$$\mathbf{d} = \breve{\mathbf{d}}(\epsilon, \mathbf{e}). \tag{18.7.6}$$

Let

$$\omega \stackrel{\text{def}}{=} \psi - \mathbf{d} \cdot \mathbf{e} \tag{18.7.7}$$

define an *electric enthalpy*. Then,

$$\begin{aligned} \omega &= \breve{\omega}(\epsilon, \mathbf{e}) \\ &= \hat{\psi}(\epsilon, \breve{\mathbf{d}}(\epsilon, \mathbf{e})) - \breve{\mathbf{d}}(\epsilon, \mathbf{e}) \cdot \mathbf{e}, \end{aligned} \tag{18.7.8}$$

while $(18.7.4)_1$ yields

$$\sigma = \hat{\sigma}(\epsilon, \mathbf{d}) \tag{18.7.9}$$

$$= \breve{\sigma}(\epsilon, \breve{\mathbf{d}}(\epsilon, \mathbf{e})). \tag{18.7.10}$$

Thus, bearing in mind that a "breve" denotes a function of $(\epsilon, \mathbf{e})$ while a "hat" denotes a function of $(\epsilon, \mathbf{d})$, we find, using (18.7.4) that

$$\frac{\partial \breve{\omega}}{\partial \epsilon} = \frac{\partial \hat{\psi}}{\partial \epsilon} + \underbrace{\left(\frac{\partial \hat{\psi}}{\partial \mathbf{d}} - \mathbf{e} \right)}_{=0} \frac{\partial \breve{\mathbf{d}}}{\partial \epsilon} = \breve{\sigma},$$

and

$$\frac{\partial \breve{\omega}}{\partial \mathbf{e}} = \underbrace{\left(\frac{\partial \hat{\psi}}{\partial \mathbf{d}} - \mathbf{e} \right)}_{=0} \frac{\partial \breve{\mathbf{d}}}{\partial \mathbf{e}} - \breve{\mathbf{d}} = -\breve{\mathbf{d}}.$$

The stress and the electric displacement are therefore determined by the response function $\omega = \breve{\omega}(\epsilon, \mathbf{e})$ via the relations

$$\begin{aligned} \sigma &= \breve{\sigma}(\epsilon, \mathbf{e}) = \frac{\partial \breve{\omega}(\epsilon, \mathbf{e})}{\partial \epsilon}, \\ \mathbf{d} &= \breve{\mathbf{d}}(\epsilon, \mathbf{e}) = -\frac{\partial \breve{\omega}(\epsilon, \mathbf{e})}{\partial \mathbf{e}}. \end{aligned} \tag{18.7.11}$$

---

[7] Quantities with a superposed "breve" indicate functions of strain ϵ and electric field $\mathbf{e}$.

An immediate consequence of (18.7.11) is the *Gibbs relation*

$$\dot{\omega} = \boldsymbol{\sigma} : \dot{\boldsymbol{\epsilon}} - \mathbf{d} \cdot \dot{\mathbf{e}}, \tag{18.7.12}$$

and the *Maxwell relation*

$$\frac{\partial \breve{\boldsymbol{\sigma}}(\boldsymbol{\epsilon}, \mathbf{e})}{\partial \mathbf{e}} = -\frac{\partial \breve{\mathbf{d}}(\boldsymbol{\epsilon}, \mathbf{e})}{\partial \boldsymbol{\epsilon}}. \tag{18.7.13}$$

18.8 Piezoelectricity tensor, permittivity tensor, elasticity tensor

Let $\boldsymbol{\epsilon}(t)$ be a time-dependent strain tensor, let $\mathbf{e}(t)$ be a time-dependent electric field, and let

$$\mathbf{d}(t) = \breve{\mathbf{d}}(\boldsymbol{\epsilon}(t), \mathbf{e}(t)).$$

Then, by the chain rule,

$$\dot{\mathbf{d}} = \frac{\partial \breve{\mathbf{d}}(\boldsymbol{\epsilon}, \mathbf{e})}{\partial \boldsymbol{\epsilon}} : \dot{\boldsymbol{\epsilon}} + \frac{\partial \breve{\mathbf{d}}(\boldsymbol{\epsilon}, \mathbf{e})}{\partial \mathbf{e}} \dot{\mathbf{e}}, \qquad \dot{d}_i = \frac{\partial \breve{d}_i}{\partial \epsilon_{jk}} \dot{\epsilon}_{jk} + \frac{\partial \breve{d}_i}{\partial e_j} \dot{e}_j.$$

This suggests the introduction of two constitutive moduli tensors:

- A third-order *piezoelectricity tensor* $\boldsymbol{\mathcal{E}}(\boldsymbol{\epsilon}, \mathbf{e})$ defined by[8]

$$\boldsymbol{\mathcal{E}}(\boldsymbol{\epsilon}, \mathbf{e}) = \frac{\partial \breve{\mathbf{d}}(\boldsymbol{\epsilon}, \mathbf{e})}{\partial \boldsymbol{\epsilon}} = -\frac{\partial^2 \breve{\omega}(\boldsymbol{\epsilon}, \mathbf{e})}{\partial \boldsymbol{\epsilon} \partial \mathbf{e}}, \qquad \mathcal{E}_{ijk} = \frac{\partial \breve{d}_i}{\partial \epsilon_{jk}} = -\frac{\partial^2 \breve{\omega}(\boldsymbol{\epsilon}, \mathbf{e})}{\partial e_i \partial \epsilon_{jk}}, \tag{18.8.1}$$

where in the component form we have taken advantage of the fact that the order of differentiation with respect to scalars is order independent. Further, since the strain $\boldsymbol{\epsilon}$ is symmetric, one may conclude from (18.8.1) that the third-order piezoelectricity tensor $\boldsymbol{\mathcal{E}}(\boldsymbol{\epsilon}, \mathbf{e})$ is symmetric in its last two indices:

$$\mathcal{E}_{ijk} = \mathcal{E}_{ikj}. \tag{18.8.2}$$

Thus in the most general anisotropic case one has 18 independent piezoelectric constants.

The *transpose* of the piezoelectricity tensor is

$$\boldsymbol{\mathcal{E}}^{\top}(\boldsymbol{\epsilon}, \mathbf{e}) = -\frac{\partial^2 \breve{\omega}(\boldsymbol{\epsilon}, \mathbf{e})}{\partial \mathbf{e} \partial \boldsymbol{\epsilon}} = -\frac{\partial \breve{\boldsymbol{\sigma}}(\boldsymbol{\epsilon}, \mathbf{e})}{\partial \mathbf{e}}, \tag{18.8.3}$$

or in component form

$$(\boldsymbol{\mathcal{E}}^{\top})_{ijk} = -\frac{\partial^2 \breve{\omega}(\boldsymbol{\epsilon}, \mathbf{e})}{\partial \epsilon_{ij} \partial e_k} = -\frac{\partial \breve{\sigma}_{ij}}{\partial e_k} = \mathcal{E}_{kij}, \tag{18.8.4}$$

again taking advantage of the interchangeability of order of differentiation with respect to scalars.

[8] For a brief discussion of the definition of a third-order tensor and its properties see Appendix 18.B.

- A second-order *permittivity tensor* $\boldsymbol{\kappa}(\boldsymbol{\epsilon}, \mathbf{e})$ (at constant strain) defined by[9]

$$\boldsymbol{\kappa}(\boldsymbol{\epsilon}, \mathbf{e}) = \frac{\partial \breve{\mathbf{d}}(\boldsymbol{\epsilon}, \mathbf{e})}{\partial \mathbf{e}} = -\frac{\partial^2 \breve{\omega}(\boldsymbol{\epsilon}, \mathbf{e})}{\partial \mathbf{e} \partial \mathbf{e}}, \qquad \kappa_{ij} = \frac{\partial \breve{d}_i}{\partial e_j} \quad \text{(at fixed } \boldsymbol{\epsilon}\text{)}. \tag{18.8.5}$$

The permittivity tensor $\boldsymbol{\kappa}$ is symmetric,

$$\kappa_{ij} = \kappa_{ji}$$

and has 6 independent components.

Combining together we have

$$\boxed{\dot{\mathbf{d}} = \boldsymbol{\mathcal{E}}(\boldsymbol{\epsilon}, \mathbf{e}) : \dot{\boldsymbol{\epsilon}} + \boldsymbol{\kappa}(\boldsymbol{\epsilon}, \mathbf{e}) \dot{\mathbf{e}} \qquad \dot{d}_i = \mathcal{E}_{ijk} \dot{\epsilon}_{jk} + \kappa_{ij} \dot{e}_j.} \tag{18.8.6}$$

Next consider,

$$\boldsymbol{\sigma}(t) = \breve{\boldsymbol{\sigma}}(\boldsymbol{\epsilon}(t), \mathbf{e}(t)).$$

The chain-rule and (18.8.3) then yield the relation

$$\dot{\boldsymbol{\sigma}} = \frac{\partial \breve{\boldsymbol{\sigma}}(\boldsymbol{\epsilon}, \mathbf{e})}{\partial \boldsymbol{\epsilon}} : \dot{\boldsymbol{\epsilon}} + \underbrace{\frac{\partial \breve{\boldsymbol{\sigma}}(\boldsymbol{\epsilon}, \mathbf{e})}{\partial \mathbf{e}}}_{-\boldsymbol{\mathcal{E}}^{\top}(\boldsymbol{\epsilon}, \mathbf{e})} \dot{\mathbf{e}},$$

which suggests the introduction of:

- The fourth-order *elasticity tensor* $\mathbb{C}(\boldsymbol{\epsilon}, \mathbf{e})$ (at fixed $\mathbf{e}$) defined by

$$\mathbb{C}(\boldsymbol{\epsilon}, \mathbf{e}) = \frac{\partial \breve{\boldsymbol{\sigma}}(\boldsymbol{\epsilon}, \mathbf{e})}{\partial \boldsymbol{\epsilon}} = \frac{\partial^2 \breve{\omega}(\boldsymbol{\epsilon}, \mathbf{e})}{\partial \boldsymbol{\epsilon}^2} \quad \text{(at fixed } \mathbf{e}\text{)}. \tag{18.8.7}$$

Thus we have,

$$\boxed{\dot{\boldsymbol{\sigma}} = \mathbb{C}(\boldsymbol{\epsilon}, \mathbf{e}) \dot{\boldsymbol{\epsilon}} - \boldsymbol{\mathcal{E}}^{\top}(\boldsymbol{\epsilon}, \mathbf{e}) \dot{\mathbf{e}} \qquad \dot{\sigma}_{ij} = C_{ijkl} \dot{\epsilon}_{kl} - \mathcal{E}_{kij} \dot{e}_k.} \tag{18.8.8}$$

For each $(\boldsymbol{\epsilon}, \mathbf{e})$, the elasticity tensor $\mathbb{C}(\boldsymbol{\epsilon}, \mathbf{e})$—a linear transformation that maps symmetric tensors to symmetric tensors—has the major symmetry property,

$$C_{ijkl} = C_{klij}, \tag{18.8.9}$$

as well as the minor symmetry properties,

$$C_{ijkl} = C_{jikl} \quad \text{and} \quad C_{ijkl} = C_{ijlk}. \tag{18.8.10}$$

Thus as in standard elasticity there are 21 independent elastic components C_{ijkl} in the most general anisotropic case.

---

9 The permittivity tensor is often denoted by $\boldsymbol{\epsilon}$ rather than $\boldsymbol{\kappa}$ in the literature. But in accordance with wide usage in the mechanics literature we have used the symbol $\boldsymbol{\epsilon}$ to denote the strain tensor, so here we use $\boldsymbol{\kappa}$ for the permittivity tensor.

18.9 Linear piezoelectricity

Guided by the discussion in the previous section, we next consider a *linear* theory. With the strain-displacement relation given by

$$\boldsymbol{\epsilon} = \tfrac{1}{2}(\nabla \mathbf{u} + (\nabla \mathbf{u})^{\top}), \tag{18.9.1}$$

and using the strain $\boldsymbol{\epsilon}$ and the electric field $\mathbf{e}$ as the governing variables, we take the constitutive equations in the linear theory to be given by

$$\left.\begin{aligned}
\omega &= \tfrac{1}{2}\boldsymbol{\epsilon}:\mathbb{C}\boldsymbol{\epsilon} - \underbrace{\mathbf{e}\cdot\boldsymbol{\mathcal{E}}:\boldsymbol{\epsilon}}_{\boldsymbol{\epsilon}:\boldsymbol{\mathcal{E}}^{\top}\mathbf{e}} - \tfrac{1}{2}\mathbf{e}\cdot\boldsymbol{\kappa}\mathbf{e} \qquad & \omega &= \frac{1}{2}C_{ijkl}\epsilon_{ij}\epsilon_{kl} - \underbrace{e_i\mathcal{E}_{ijk}\epsilon_{jk}}_{\epsilon_{ij}\mathcal{E}_{kij}e_k} - \tfrac{1}{2}\kappa_{ij}e_ie_j,\\
\boldsymbol{\sigma} &= \mathbb{C}\boldsymbol{\epsilon} - \boldsymbol{\mathcal{E}}^{\top}\mathbf{e} & \sigma_{ij} &= C_{ijkl}\epsilon_{kl} - \mathcal{E}_{kij}e_k,\\
\mathbf{d} &= \boldsymbol{\mathcal{E}}:\boldsymbol{\epsilon} + \boldsymbol{\kappa}\mathbf{e} & d_i &= \mathcal{E}_{ijk}\epsilon_{jk} + \kappa_{ij}e_j,
\end{aligned}\right\} \tag{18.9.2}$$

where $\mathbb{C}$ is the *elasticity tensor* (at constant $\mathbf{e}$), $\boldsymbol{\mathcal{E}}$ is the third-order *piezoelectricity tensor*, and $\boldsymbol{\kappa}$ is the second-order *permittivity tensor* (at constant $\boldsymbol{\epsilon}$)—all are *constants*. In total there are 21 independent elastic constants, 18 independent piezoelectric constants, and six independent dielectric constants in the most general case (triclinic crystal without center of symmetry). We assume henceforth that $\mathbb{C}$ and $\boldsymbol{\kappa}$ are positive definite.

REMARK 18.1

The constitutive equations (18.9.2) are in a form identical to that which are widely used in the piezoelectricity literature (cf., e.g., Tiersten (1969) or Auld (1990)) and the IEEE Standard on Piezoelectricity (Meitzler, 1988). However, in these references, please observe the following notational conventions:

- the components of the stress tensor σ_{ij} are denoted by T_{ij};
- the components of the strain tensor ϵ_{ij} are denoted by S_{ij};
- the components of piezoelectricity tensor $\mathcal{E}_{ijk}$ are denoted by e_{ijk}; and
- the components of the permittivity tensor κ_{ij} are denoted by ϵ^{s}_{ij}.

18.10 **Governing partial differential equations**

The basic equations of the linear theory of piezoelectricity consist of the strain-displacement relation (18.9.1), the constitutive equations (18.9.2), and the local force balance (18.2.1) (in the absence of body forces), which give

$$\operatorname{div}[\mathbb{C}\epsilon - \mathcal{E}^{\top}\mathbf{e}] = \rho\ddot{\mathbf{u}}, \tag{18.10.1}$$

and Gauss's law for a dielectric (18.3.3), which gives

$$\operatorname{div}(\mathcal{E}:\epsilon + \kappa\mathbf{e}) = 0. \tag{18.10.2}$$

Using the strain-displacement relation (18.9.1), the symmetries of the constitutive moduli C_{ijkl}, $\mathcal{E}_{ijk}$, and κ_{ij}, and (18.3.2), viz. $e_i = -\phi_{,i}$, the governing equations (18.10.1) and (18.10.2) may be written for homogeneous materials as

$$\begin{aligned} C_{ijkl}u_{k,lj} + \mathcal{E}_{kij}\phi_{,kj} &= \rho\ddot{u}_i, \\ \mathcal{E}_{ijk}u_{j,ki} - \kappa_{ij}\phi_{,ji} &= 0, \end{aligned} \tag{18.10.3}$$

in terms of the three components of the displacement u_i and the electric potential ϕ.

Boundary conditions

Let $\mathcal{S}_{\mathbf{u}}$ and $\mathcal{S}_{\mathbf{t}}$ be complementary subsurfaces of the boundary $\partial\mathcal{B}$ of the body $\mathcal{B}$. Then for a time interval $t \in [0, T]$ we consider a pair of boundary conditions:

$$\begin{aligned} \mathbf{u} &= \bar{\mathbf{u}} \quad \text{on} \quad \mathcal{S}_{\mathbf{u}} \times [0, T], \\ \boldsymbol{\sigma}\mathbf{n} &= \bar{\mathbf{t}} \quad \text{on} \quad \mathcal{S}_{\mathbf{t}} \times [0, T]. \end{aligned} \tag{18.10.4}$$

In the boundary conditions above $\bar{\mathbf{u}}$ and $\bar{\mathbf{t}}$ are functions (of $\mathbf{x}$ and t) describing the prescribed displacement and prescribed tractions, respectively.

Next, let $\mathcal{S}_{\phi}$ and $\mathcal{S}_{\mathbf{d}}$ be complementary subsurfaces of the boundary $\partial\mathcal{B}$ of the body $\mathcal{B}$. Then for a time interval $t \in [0, T]$ we consider a number of possible boundary conditions:

1. the prescription of a known voltage:

$$\phi = \bar{\phi} \qquad \text{on} \qquad \mathcal{S}_{\phi} \times [0, T]\,; \tag{18.10.5}$$

2. the case where the piezoelectric body is in contact with a conductor at $\mathcal{S}_{\mathbf{d}}$ *and* the surface charge of the conductor, $\bar{\beta}$, is a given function:[10]

$$-\,\mathbf{d}\cdot\mathbf{n} = \bar{\beta} \qquad \text{on} \qquad \mathcal{S}_{\mathbf{d}} \times [0, T]\,; \tag{18.10.6}$$

3. the case where the piezoelectric body is in contact with air at $\mathcal{S}_{\mathbf{d}}$ *and* the electric field, $\bar{\mathbf{e}}_{\text{air}}$, is known in the air at the boundary:

$$\mathbf{d}\cdot\mathbf{n} = \kappa_{\text{air}}\bar{\mathbf{e}}_{\text{air}}\cdot\mathbf{n} \qquad \text{on} \qquad \mathcal{S}_{\mathbf{d}} \times [0, T]. \tag{18.10.7}$$

In the boundary conditions above symbols with "bars" are given functions (of $\mathbf{x}$ and t).

---

[10] In the literature the usual symbol for the surface charge is σ; we avoid this since in mechanics this symbol is widely used for stress.

18.11 Alternate form of the constitutive equations for linear piezoelectricity with σ and e as independent variables

The elasticity tensor $\mathbb{C}$ is positive definite, and thus is invertible. Let,

$$\boxed{\mathbb{S} = \mathbb{C}^{-1}, \qquad S_{ijkl} = C^{-1}_{ijkl}.} \tag{18.11.1}$$

The fourth-order tensor $\mathbb{S}$ is called the **compliance tensor**. The components of the compliance tensor $\mathbb{S}$ have symmetries similar to those of the elasticity tensor:

$$S_{ijkl} = S_{klij}, \tag{18.11.2}$$

$$S_{ijkl} = S_{jikl}, \qquad S_{ijkl} = S_{ijlk}. \tag{18.11.3}$$

From $(18.9.2)_2$

$$\boldsymbol{\epsilon} = \mathbb{S}\boldsymbol{\sigma} + \mathbb{S}(\boldsymbol{\mathcal{E}}^{\top}\mathbf{e}), \tag{18.11.4}$$

use of which in $(18.9.2)_3$ gives

$$\begin{aligned} \mathbf{d} &= \boldsymbol{\mathcal{E}} : \big(\mathbb{S}\boldsymbol{\sigma} + \mathbb{S}(\boldsymbol{\mathcal{E}}^{\top}\mathbf{e})\big) + \boldsymbol{\kappa}\mathbf{e} \\ &= \boldsymbol{\mathcal{E}} : (\mathbb{S}\boldsymbol{\sigma}) + \boldsymbol{\mathcal{E}} : \mathbb{S}(\boldsymbol{\mathcal{E}}^{\top}\mathbf{e}) + \boldsymbol{\kappa}\mathbf{e}. \end{aligned} \tag{18.11.5}$$

Thus in terms of $\boldsymbol{\sigma}$ and $\mathbf{e}$ the constitutive equations for linear piezoelectricity may be written as

$$\begin{aligned} \boldsymbol{\epsilon} &= \mathbb{S}\boldsymbol{\sigma} + \mathbb{S}(\boldsymbol{\mathcal{E}}^{\top}\mathbf{e}), \\ \mathbf{d} &= \boldsymbol{\mathcal{E}} : (\mathbb{S}\boldsymbol{\sigma}) + \boldsymbol{\mathcal{E}} : \mathbb{S}(\boldsymbol{\mathcal{E}}^{\top}\mathbf{e}) + \boldsymbol{\kappa}\mathbf{e}. \end{aligned} \tag{18.11.6}$$

The component forms of these constitutive equations are

$$\begin{aligned} \epsilon_{ij} &= S_{ijkl}\sigma_{kl} + S_{ijkl}\mathcal{E}_{mkl}e_m = S_{ijkl}\sigma_{kl} + \mathcal{E}_{mkl}S_{klij}e_m, \\ d_i &= \mathcal{E}_{ipq}S_{pqjm}\sigma_{jm} + \mathcal{E}_{ipq}S_{pqrl}\mathcal{E}_{mrl}e_m + \kappa_{im}e_m. \end{aligned} \tag{18.11.7}$$

Let

$$\begin{aligned} (\boldsymbol{\mathcal{D}})_{ijm} &= \mathcal{D}_{ijm} \stackrel{\text{def}}{=} \mathcal{E}_{ipq}S_{pqjm}, \\ (\boldsymbol{\mathcal{D}}^{\top})_{ijm} &= \mathcal{D}_{mij} \stackrel{\text{def}}{=} \mathcal{E}_{mkl}S_{klij}, \\ (\boldsymbol{\kappa}^{\sigma})_{im} &= \kappa^{\sigma}_{im} \stackrel{\text{def}}{=} \kappa_{im} + \mathcal{E}_{ipq}S_{pqrl}\mathcal{E}_{mrl}. \end{aligned} \tag{18.11.8}$$

Then, in terms of the stress $\boldsymbol{\sigma}$ and the electric field $\mathbf{e}$, the strain ϵ and the electric displacement $\mathbf{d}$ are given by

$$\boxed{\begin{aligned} \boldsymbol{\epsilon} &= \mathbb{S}\boldsymbol{\sigma} + \boldsymbol{\mathcal{D}}^{\top}\mathbf{e}, & \epsilon_{ij} &= S_{ijkl}\sigma_{kl} + \mathcal{D}_{mij}e_m, \\ \mathbf{d} &= \boldsymbol{\mathcal{D}} : \boldsymbol{\sigma} + \boldsymbol{\kappa}^{\sigma}\mathbf{e}, & d_i &= \mathcal{D}_{ijm}\sigma_{jm} + \kappa^{\sigma}_{im}e_m. \end{aligned}} \tag{18.11.9}$$

18.12 Piezocrystals and poled piezoceramics

18.12.1 Voigt notation

Next, to make contact with the crystallography literature, and also from a practical computational point of view, we re-introduce the Voigt "contracted notation" (Voigt, 1910) now accounting for electric displacement and electric field (cf. Sec. 7.1.1), where

$$[\sigma] = \begin{bmatrix} \sigma_1 \\ \sigma_2 \\ \sigma_3 \\ \sigma_4 \\ \sigma_5 \\ \sigma_6 \end{bmatrix} \stackrel{\text{def}}{=} \begin{bmatrix} \sigma_{11} \\ \sigma_{22} \\ \sigma_{33} \\ \sigma_{23} \\ \sigma_{13} \\ \sigma_{12} \end{bmatrix} \qquad [\epsilon] = \begin{bmatrix} \epsilon_1 \\ \epsilon_2 \\ \epsilon_3 \\ \epsilon_4 \\ \epsilon_5 \\ \epsilon_6 \end{bmatrix} \stackrel{\text{def}}{=} \begin{bmatrix} \epsilon_{11} \\ \epsilon_{22} \\ \epsilon_{33} \\ 2\epsilon_{23} \\ 2\epsilon_{13} \\ 2\epsilon_{12} \end{bmatrix}, \qquad [d] \stackrel{\text{def}}{=} \begin{bmatrix} d_1 \\ d_2 \\ d_3 \end{bmatrix} \quad \text{and} \quad [e] \stackrel{\text{def}}{=} \begin{bmatrix} e_1 \\ e_2 \\ e_3 \end{bmatrix}, \tag{18.12.1}$$

respectively. A (6×6) matrix of elastic stiffnesses $[\mathcal{C}]$ is defined in terms of elastic moduli C_{ijkl} as

$$[\mathcal{C}] = \begin{bmatrix} \mathcal{C}_{11} & \mathcal{C}_{12} & \mathcal{C}_{13} & \mathcal{C}_{14} & \mathcal{C}_{15} & \mathcal{C}_{16} \\ \mathcal{C}_{21} & \mathcal{C}_{22} & \mathcal{C}_{23} & \mathcal{C}_{24} & \mathcal{C}_{25} & \mathcal{C}_{26} \\ \mathcal{C}_{31} & \mathcal{C}_{32} & \mathcal{C}_{33} & \mathcal{C}_{34} & \mathcal{C}_{35} & \mathcal{C}_{36} \\ \mathcal{C}_{41} & \mathcal{C}_{42} & \mathcal{C}_{43} & \mathcal{C}_{44} & \mathcal{C}_{45} & \mathcal{C}_{46} \\ \mathcal{C}_{51} & \mathcal{C}_{52} & \mathcal{C}_{53} & \mathcal{C}_{54} & \mathcal{C}_{55} & \mathcal{C}_{56} \\ \mathcal{C}_{61} & \mathcal{C}_{62} & \mathcal{C}_{63} & \mathcal{C}_{64} & \mathcal{C}_{65} & \mathcal{C}_{66} \end{bmatrix} \stackrel{\text{def}}{=} \begin{bmatrix} C_{1111} & C_{1122} & C_{1133} & C_{1123} & C_{1113} & C_{1112} \\ C_{2211} & C_{2222} & C_{2233} & C_{2223} & C_{2213} & C_{2212} \\ C_{3311} & C_{2233} & C_{3333} & C_{3323} & C_{3313} & C_{3312} \\ C_{2311} & C_{2322} & C_{2333} & C_{2323} & C_{2313} & C_{2312} \\ C_{1311} & C_{1322} & C_{1333} & C_{1323} & C_{1313} & C_{1312} \\ C_{1211} & C_{1222} & C_{1233} & C_{1223} & C_{1213} & C_{1212} \end{bmatrix}, \tag{18.12.2}$$

and a (6×6) matrix of elastic compliances $[\mathcal{S}]$ is defined in terms of elastic moduli S_{ijkl} as

$$[\mathcal{S}] = \begin{bmatrix} \mathcal{S}_{11} & \mathcal{S}_{12} & \mathcal{S}_{13} & \mathcal{S}_{14} & \mathcal{S}_{15} & \mathcal{S}_{16} \\ \mathcal{S}_{21} & \mathcal{S}_{22} & \mathcal{S}_{23} & \mathcal{S}_{24} & \mathcal{S}_{25} & \mathcal{S}_{26} \\ \mathcal{S}_{31} & \mathcal{S}_{32} & \mathcal{S}_{33} & \mathcal{S}_{34} & \mathcal{S}_{35} & \mathcal{S}_{36} \\ \mathcal{S}_{41} & \mathcal{S}_{42} & \mathcal{S}_{43} & \mathcal{S}_{44} & \mathcal{S}_{45} & \mathcal{S}_{46} \\ \mathcal{S}_{51} & \mathcal{S}_{52} & \mathcal{S}_{53} & \mathcal{S}_{54} & \mathcal{S}_{55} & \mathcal{S}_{56} \\ \mathcal{S}_{61} & \mathcal{S}_{62} & \mathcal{S}_{63} & \mathcal{S}_{64} & \mathcal{S}_{65} & \mathcal{S}_{66} \end{bmatrix} \stackrel{\text{def}}{=} \begin{bmatrix} S_{1111} & S_{1122} & S_{1133} & 2S_{1123} & 2S_{1113} & 2S_{1112} \\ S_{2211} & S_{2222} & S_{2233} & 2S_{2223} & 2S_{2213} & 2S_{2212} \\ S_{3311} & S_{2233} & S_{3333} & 2S_{3323} & 2S_{3313} & 2S_{3312} \\ 2S_{2311} & 2S_{2322} & 2S_{2333} & 4S_{2323} & 4S_{2313} & 4S_{2312} \\ 2S_{1311} & 2S_{1322} & 2S_{1333} & 4S_{1323} & 4S_{1313} & 4S_{1312} \\ 2S_{1211} & 2S_{1222} & 2S_{1233} & 4S_{1223} & 4S_{1213} & 4S_{1212} \end{bmatrix}. \tag{18.12.3}$$

A (3×6) matrix of piezoelectricity moduli $[E]$ is defined in terms of moduli $\mathcal{E}_{ijk}$ as

$$[E] = \begin{bmatrix} E_{11} & E_{12} & E_{13} & E_{14} & E_{15} & E_{16} \\ E_{21} & E_{22} & E_{23} & E_{24} & E_{25} & E_{26} \\ E_{31} & E_{32} & E_{33} & E_{34} & E_{35} & E_{36} \end{bmatrix} \stackrel{\text{def}}{=} \begin{bmatrix} \mathcal{E}_{111} & \mathcal{E}_{122} & \mathcal{E}_{133} & \mathcal{E}_{123} & \mathcal{E}_{113} & \mathcal{E}_{112} \\ \mathcal{E}_{211} & \mathcal{E}_{222} & \mathcal{E}_{233} & \mathcal{E}_{223} & \mathcal{E}_{213} & \mathcal{E}_{212} \\ \mathcal{E}_{311} & \mathcal{E}_{322} & \mathcal{E}_{333} & \mathcal{E}_{323} & \mathcal{E}_{313} & \mathcal{E}_{312} \end{bmatrix}, \tag{18.12.4}$$

and a (3×6) matrix of piezoelectricity moduli $[D]$ is defined in terms of the moduli $\mathcal{D}_{ijk}$ as

$$\begin{aligned} [D] &= \begin{bmatrix} D_{11} & D_{12} & D_{13} & D_{14} & D_{15} & D_{16} \\ D_{21} & D_{22} & D_{23} & D_{24} & D_{25} & D_{26} \\ D_{31} & D_{32} & D_{33} & D_{34} & D_{35} & D_{36} \end{bmatrix} \\ &\stackrel{\text{def}}{=} \begin{bmatrix} \mathcal{D}_{111} & \mathcal{D}_{122} & \mathcal{D}_{133} & \mathcal{D}_{123} & \mathcal{D}_{113} & \mathcal{D}_{112} \\ \mathcal{D}_{211} & \mathcal{D}_{222} & \mathcal{D}_{233} & \mathcal{D}_{223} & \mathcal{D}_{213} & \mathcal{D}_{212} \\ \mathcal{D}_{311} & \mathcal{D}_{322} & \mathcal{D}_{333} & \mathcal{D}_{323} & \mathcal{D}_{313} & \mathcal{D}_{312} \end{bmatrix}. \end{aligned} \tag{18.12.5}$$

Also, the matrices of the permittivity moduli κ_{ij} and κ^{σ}_{ij} are

$$[\kappa] = \begin{bmatrix} \kappa_{11} & \kappa_{12} & \kappa_{13} \\ \kappa_{21} & \kappa_{22} & \kappa_{23} \\ \kappa_{31} & \kappa_{32} & \kappa_{33} \end{bmatrix} \quad \text{and} \quad [\kappa^{\sigma}] = \begin{bmatrix} \kappa^{\sigma}_{11} & \kappa^{\sigma}_{12} & \kappa^{\sigma}_{13} \\ \kappa^{\sigma}_{21} & \kappa^{\sigma}_{22} & \kappa^{\sigma}_{23} \\ \kappa^{\sigma}_{31} & \kappa^{\sigma}_{32} & \kappa^{\sigma}_{33} \end{bmatrix}. \tag{18.12.6}$$

An arbitrarily anisotropic piezoelectric material has **45 independent elastic constants.**

Using the Voigt notation the piezoelectricity constitutive equation (18.9.2) may be written in matrix form as

$$\boxed{\begin{aligned} [\sigma] &= [\mathcal{C}][\epsilon] - [E]^{\top}[e], \\ [d] &= [E][\epsilon] + [\kappa][e]. \end{aligned}} \tag{18.12.7}$$

Correspondingly, the piezoelectricity constitutive equation (18.11.9) may be written in matrix form as

$$\boxed{\begin{aligned} [\epsilon] &= [\mathcal{S}][\sigma] + [D]^{\top}[e], \\ [d] &= [D][\sigma] + [\kappa^{\sigma}][e]. \end{aligned}} \tag{18.12.8}$$

18.12.2 **Properties of piezoelectric crystals**

For a discussion of the piezoelectric properties of quartz and other piezoelectric crystals see the IEEE Standard on Piezoelectricity (Meitzler, 1988).

18.12.3 **Properties of poled piezoceramics**

Poled piezoceramics typically exhibit a transversely isotropic response with the poling direction parallel to the axis of rotational symmetry. Assuming the $\mathbf{e}_3$-axis is the rotational symmetry axis, it may be shown that the matrices of the material properties $[\mathcal{C}]$, $[E]$, and $[\kappa]$ reduce to

$$\begin{aligned} [\mathcal{C}] &= \begin{bmatrix} \mathcal{C}_{11} & \mathcal{C}_{12} & \mathcal{C}_{13} & 0 & 0 & 0 \\ \mathcal{C}_{12} & \mathcal{C}_{11} & \mathcal{C}_{13} & 0 & 0 & 0 \\ \mathcal{C}_{13} & \mathcal{C}_{13} & \mathcal{C}_{33} & 0 & 0 & 0 \\ 0 & 0 & 0 & \mathcal{C}_{44} & 0 & 0 \\ 0 & 0 & 0 & 0 & \mathcal{C}_{44} & 0 \\ 0 & 0 & 0 & 0 & 0 & \frac{1}{2}(\mathcal{C}_{11} - \mathcal{C}_{12}) \end{bmatrix}, \\ [E] &= \begin{bmatrix} 0 & 0 & 0 & 0 & E_{15} & 0 \\ 0 & 0 & 0 & E_{15} & 0 & 0 \\ E_{31} & E_{31} & E_{33} & 0 & 0 & 0 \end{bmatrix}, \\ [\kappa] &= \begin{bmatrix} \kappa_{11} & 0 & 0 \\ 0 & \kappa_{11} & 0 \\ 0 & 0 & \kappa_{33} \end{bmatrix}. \end{aligned} \tag{18.12.9}$$

A transversely isotropic piezoelectric material has **10 independent elastic constants**

$$\{\mathcal{C}_{11}, \mathcal{C}_{33}, \mathcal{C}_{44}, \mathcal{C}_{12}, \mathcal{C}_{13}\}, \qquad \{E_{15}, E_{31}, E_{33}\}, \qquad \{\kappa_{11}, \kappa_{33}\}$$

Material parameters for some commercially available piezoelectric ceramics are listed in Table 18.1.

Table 18.1 Material parameters for some commercially available piezoelectric ceramics. (Adapted from Govorukha et al., 2017).

Symbol	Units	Piezoceramics				
		PZT-PIC 151	PZT-4	PZT-5H	PZT-5	$BaTiO_3$
$\mathcal{C}_{11}$	10^{10} N/m^2	10.8	13.9	12.6	12.1	15.0
$\mathcal{C}_{12}$		6.3	7.78	5.5	7.54	6.6
$\mathcal{C}_{13}$		6.4	7.43	5.3	7.52	6.6
$\mathcal{C}_{33}$		10.1	11.3	11.7	11.1	14.6
$\mathcal{C}_{44}$		2.0	2.56	3.53	2.11	4.4
E_{31}	C/m^2	−9.6	−6.98	−6.5	−5.4	−4.35
E_{33}		15.1	13.84	23.3	15.8	17.5
E_{15}		12.0	13.44	17.0	12.3	11.4
κ_{11}	10^{-10} C/(Vm)	98.2	60.0	151	81.1	98.7
κ_{33}		75.4	54.7	130	73.5	112

EXAMPLE 18.1 Motion of a piezoelectric button

Shown schematically in Fig. 18.2 is a cylindrical barium titanate button of height L and radius R. The bottom of the button is coated with a metal layer and grounded. The top of the button is coated with a metal layer and subjected to a voltage ϕ_o. Mechanically, the top and sides are free of loads and the bottom of the button is restrained from vertical motion in the z-direction and from rotational motion. It is desired to determine the displacement field in the button when the poling is in the z-direction.

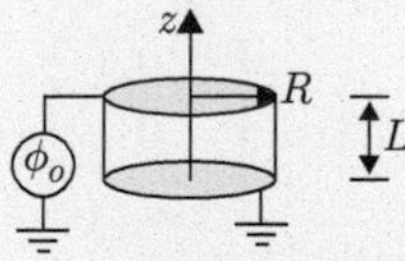

Fig. 18.2 Piezoelectric button.

The boundary conditions are given (assuming a polar coordinate frame) by:

$$z = 0 \qquad \phi = 0 \qquad u_z = u_\theta = 0 \tag{18.12.10}$$

$$z = L \qquad \phi = \phi_o \qquad \boldsymbol{\sigma}\mathbf{e}_z = \mathbf{0} \tag{18.12.11}$$

$$r = R \qquad \boldsymbol{\sigma}\mathbf{e}_r = 0 \qquad \mathbf{d}\cdot\mathbf{e}_r = \kappa_{\text{air}}\mathbf{e}_{\text{air}}\cdot\mathbf{e}_r \approx 0. \tag{18.12.12}$$

The electric displacement boundary condition $\mathbf{d}\cdot\mathbf{e}_r \approx 0$ is commonly used to avoid having to explicitly determine the electric field in the air around a piezoelectric material. In the current setting, this assumption is indeed justified due to the large value of the permittivity in the barium titanate in comparison to the air and the fact that the electric field in the air has a very small radial component near the sides of the button; its primary component is in the z-direction.

We will approach this problem using a semi-inverse method. In particular we will guess the basic form of the solution and then determine its unknown aspects requiring the guess to satisfy the governing balance equations (18.10.1) and (18.10.2) as well as the boundary conditions (18.12.10)–(18.12.12). As our guess we will assume that the potential field is linear in z and independent of r and θ. We will assume that the vertical displacement is linear in z, that the radial displacement is only a function of r, and that there is no tangential displacement:

$$\phi = \hat{\phi}(z) = A_1 z, \tag{18.12.13}$$

$$u_r = \hat{u}_r(r), \tag{18.12.14}$$

$$u_\theta = 0, \tag{18.12.15}$$

$$u_z = \hat{u}_z(z) = A_2 z\,, \tag{18.12.16}$$

where A_1, A_2 are constants to be determined and $\hat{u}_r(r)$ is also unknown. The first constant is easily seen to be $A_1 = \phi_o/L$ since $\phi(L) = \phi_o$ by the boundary conditions.

Using the assumed form for the displacement and potential fields, we have that the strain and the electric fields have the forms,

$$[\epsilon] = \begin{bmatrix} u_{r,r} \\ u_r/r \\ A_2 \\ 0 \\ 0 \\ 0 \end{bmatrix}, \tag{18.12.17}$$

and

$$[e] = \begin{bmatrix} 0 \\ 0 \\ -\phi_o/L \end{bmatrix}. \tag{18.12.18}$$

Using the material properties for barium titanate, see Table 18.1 and (18.12.9), in constitutive relations (18.12.7), we find the stresses to be given by

$$[\sigma] = \begin{bmatrix} \sigma_{rr} \\ \sigma_{\theta\theta} \\ \sigma_{zz} \\ \sigma_{\theta z} \\ \sigma_{zr} \\ \sigma_{r\theta} \end{bmatrix} = \begin{bmatrix} C_{11}u_{r,r} + C_{12}u_r/r + C_{13}A_2 + E_{31}\phi_o/L \\ C_{12}u_{r,r} + C_{11}u_r/r + C_{13}A_2 + E_{31}\phi_o/L \\ C_{13}u_{r,r} + C_{13}u_r/r + C_{33}A_2 + E_{33}\phi_o/L \\ 0 \\ 0 \\ 0 \end{bmatrix}, \tag{18.12.19}$$

and the electric displacement to be given by

$$[d] = \begin{bmatrix} d_r \\ d_\theta \\ d_z \end{bmatrix} = \begin{bmatrix} 0 \\ 0 \\ E_{31}u_{r,r} + E_{31}u_r/r + E_{33}A_2 - \kappa_{33}\phi_o/L \end{bmatrix}. \tag{18.12.20}$$

Let us now consider the balance equation $\operatorname{div}\mathbf{d} = 0$. For the present assumptions we find

$$\operatorname{div}\mathbf{d} = d_{z,z} = 0. \tag{18.12.21}$$

Thus Gauss's law for the dielectric is identically satisfied and the electrical components of the boundary conditions are also seen to be satisfied.

Considering now the balance of forces $\operatorname{div}\boldsymbol{\sigma} = \mathbf{0}$, we find that the only non-trivial equilibrium equation is the radial one:

$$C_{11}u_{r,rr} + C_{12}(u_r/r)_{,r} + \frac{(C_{11} - C_{12})u_{r,r} - (C_{11} - C_{12})u_r/r}{r} = 0. \tag{18.12.22}$$

Expanding the derivatives and canceling terms in (18.12.22) finally gives

$$r^2 u_{r,rr} + r u_{r,r} - u_r = 0. \tag{18.12.23}$$

The general solution for ordinary differential equations of the form

$$\sum_{k=0}^{n} a_k r^k \frac{d^k f}{dr^k} = 0, \quad \text{is} \quad f = Dr^m,$$

where D and m are constants (to be determined). Substituting the general form into (18.12.23) gives

$$m(m-1) + m - 1 = 0. \tag{18.12.24}$$

Thus $m = \pm 1$ and the radial displacement is given by

$$u_r = D_1 r + D_2 r^{-1}. \tag{18.12.25}$$

In order for the solution to be finite at $r = 0$ we must have $D_2 = 0$, and hence

$$u_r = D_1 r.$$

We can now use the traction-free boundary conditions on the top and the sides to determine D_1 and A_2. At the top we find

$$\mathcal{C}_{13}D_1 + \mathcal{C}_{13}D_1 + \mathcal{C}_{33}A_2 + E_{33}\phi_o/L = 0, \tag{18.12.26}$$

and on the sides we find,

$$\mathcal{C}_{11}D_1 + \mathcal{C}_{12}D_1 + \mathcal{C}_{13}A_2 + E_{31}\phi_o/L = 0. \tag{18.12.27}$$

Solving these two equations, we find

$$\begin{aligned} D_1 &= \frac{\mathcal{C}_{33}E_{31} - \mathcal{C}_{13}E_{33}}{2(\mathcal{C}_{13})^2 - \mathcal{C}_{33}(\mathcal{C}_{11} + \mathcal{C}_{12})}\frac{\phi_o}{L} = 78.4 \times 10^{-12}\,(\mathrm{m/V})\frac{\phi_o}{L} \\ A_2 &= \frac{(\mathcal{C}_{11} + \mathcal{C}_{12})E_{33} - 2\mathcal{C}_{13}E_{31}}{2(\mathcal{C}_{13})^2 - \mathcal{C}_{33}(\mathcal{C}_{11} + \mathcal{C}_{12})}\frac{\phi_o}{L} = -191 \times 10^{-12}\,(\mathrm{m/V})\frac{\phi_o}{L}. \end{aligned} \tag{18.12.28}$$

Thus,

$$u_z = A_2 z, \qquad \text{and} \qquad u_r = D_1 r,$$

with A_2 and D_1 given by (18.12.28). Hence, positive voltages are seen to compress the button and cause it to expand, whereas negative voltages do the opposite. It should also be observed that the motions are quite small even in the presence of substantial electric fields. The breakdown electric field, that is the electric field at which barium titanate will start to electrically short, is approximately 10^6 V/m. At such high electric fields, the non-dimensional factors D_1 and A_1 are seen to be maximally of the order of 10^{-4}.

EXAMPLE 18.2 Piezoelectric actuator

Shown schematically in Fig. 18.3 is an actuator composed of a PZT-4 cylinder with inner and outer radii R_i and R_o, respectively, and an inner rigid conducting core. The outer radius of the PZT-4 cylinder is grounded and the inner conducting core is connected to a voltage source of strength ϕ_o. The outer edge of the piezoelectric material is fixed from moving and the assemblage rests on a rigid insulating table, while being subjected to a downward force W due to a mass placed on top of the core. It is desired to determine the voltage required to lift the weight, and to determine the displacement of the weight as a function of voltage. Observe that initially the weight will attempt to move the core downward but that the rigid table will resist this motion. The weight will not be lifted until the piezoelectric forces are sufficient to fully overcome the weight of the mass. Further, the height L of the actuator can be assumed to be large relative to its radius, $L \gg R_0$. The poling axis of the piezoelectric material is in the vertical direction.

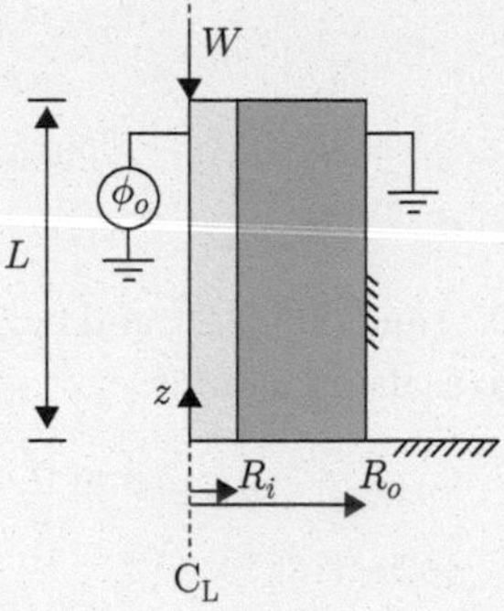

Fig. 18.3 Cylindrical piezoelectric actuator.

In solving this problem we will employ polar coordinates, where the z-direction is coincident with the poling 3-direction. The Voigt ordering is taken as

$$(11, 22, 33, 23, 31, 12) \equiv (rr, \theta\theta, zz, \theta z, zr, r\theta),$$

cf. Sec. 18.12.1. Since the geometry, material, and loads are rotationally symmetric about the centerline, it is reasonable to assume that both the displacement and potential fields are independent of the angle θ. Further, given that $L \gg R_o$, it is permissible to assume that the motion corresponds to anti-plane strain (cf. Sec. 9.5.2), which in the present context, implies that

$$\begin{aligned} u &\stackrel{\text{def}}{=} u_z = \hat{u}_z(r), \\ \phi &= \hat{\phi}(r). \end{aligned} \tag{18.12.29}$$

The boundary conditions are:

$$u(R_o) = 0, \tag{18.12.30}$$

$$-\boldsymbol{\sigma}(R_i)\mathbf{e}_r = -\frac{W}{2\pi R_i L}\mathbf{e}_z, \tag{18.12.31}$$

$$\phi(R_i) = \phi_o, \tag{18.12.32}$$

$$\phi(R_o) = 0. \tag{18.12.33}$$

Observe we ignore the conditions at the top and bottom faces of the cylinder. This is permissible due to our assumption of anti-plane strain. We also ignore the reaction force on the core from the rigid table; thus we will consider our solution only valid for values of ϕ_o such that $u(R_i) > 0$.

From assumption (18.12.29), one finds that there is only one non-zero component of strain and one non-zero component of electric field:

$$\epsilon_{zr} = \tfrac{1}{2}u_{,r}, \tag{18.12.34}$$

$$e_r = -\phi_{,r}. \tag{18.12.35}$$

Using the material properties for PZT-4, see Table 18.1 and (18.12.9), in the constitutive relations (18.12.7) gives that the only non-zero components for the stress and electric displacement are:

$$\sigma_{zr} = 2\mathcal{C}_{44}\epsilon_{zr} - E_{15}e_r, \tag{18.12.36}$$

$$d_r = 2E_{15}\epsilon_{zr} + \kappa_{11}e_r\,. \tag{18.12.37}$$

Substituting these into (18.10.1) and (18.10.2), and neglecting inertia, one finds the two balance principles reduce to two scalar equations:

$$\mathcal{C}_{44}\Delta u + E_{15}\Delta\phi = 0, \tag{18.12.38}$$

$$E_{15}\Delta u - \kappa_{11}\Delta\phi = 0, \tag{18.12.39}$$

where in the present setting the Laplacian of a scalar function $f(r)$ is given by

$$\Delta f = (1/r)(rf_{,r})_{,r}.$$

Assuming that $\mathcal{C}_{44}\kappa_{11} + (E_{15})^2 \neq 0$, the only way to satisfy (18.12.38) and (18.12.39) is for

$$\Delta u = 0, \tag{18.12.40}$$

$$\Delta\phi = 0. \tag{18.12.41}$$

Integrating (18.12.40) and (18.12.41) gives

$$u(r) = A_u + B_u \ln r, \tag{18.12.42}$$

$$\phi(r) = A_\phi + B_\phi \ln r, \tag{18.12.43}$$

where A_u, B_u, A_ϕ, B_ϕ are constants of integration which can be determined from the boundary conditions. In particular, applying (18.12.30), (18.12.32), and (18.12.33) shows that

$$u(r) = u(R_i)\frac{\ln(r/R_o)}{\ln(R_i/R_o)}, \tag{18.12.44}$$

$$\phi(r) = \phi_o\frac{\ln(r/R_o)}{\ln(R_i/R_o)}. \tag{18.12.45}$$

We can now apply the boundary condition (18.12.31) at $r = R_i$ in (18.12.36) to obtain,

$$\frac{W}{2\pi R_i L} = \mathcal{C}_{44}u_{,r}(R_i) + E_{15}\phi_{,r}(R_i), \tag{18.12.46}$$

$$= \mathcal{C}_{44}u(R_i)\frac{1}{R_i\ln(R_i/R_o)} + E_{15}\phi_o\frac{1}{R_i\ln(R_i/R_o)}, \tag{18.12.47}$$

which gives

$$u(R_i) = \left(\frac{W}{2\pi L E_{15}}\ln(R_i/R_o) - \phi_o\right)\frac{E_{15}}{\mathcal{C}_{44}}. \tag{18.12.48}$$

As noted earlier our solution is only valid for $u(R_i) > 0$ since we have ignored the reaction force of the rigid table. Since $\ln(R_i/R_o) < 0$, positive values of $u(R_i)$ will require a negative voltage ϕ_o. The voltage at which the mass will begin to rise will occur when $u(R_i) = 0$. We call this voltage the *lift off voltage*:

$$\phi_{\text{lift off}} = \frac{W}{2\pi L E_{15}} \ln(R_i/R_o). \tag{18.12.49}$$

The rate at which the mass rises per volt of actuation is given by

$$\frac{du(R_i)}{d\phi_o} = -\frac{E_{15}}{\mathcal{C}_{44}} = -\frac{13.44\,\text{C/m}^2}{2.56 \times 10^{10}\,\text{N/m}^2} = -0.525\,\text{nm/V}. \tag{18.12.50}$$

Note that the magnitude of the lift off voltage can be decreased by making the actuator taller or by bringing the inner and outer radii closer together. This second modification has the effect of increasing the electric field strength, thus inducing larger shear stresses to provide more lift.

EXAMPLE 18.3 $\mathbf{e}_1$-direction poling

Poled piezoceramics are not always used with the poling direction coinciding with the $\mathbf{e}_3$-direction, even though the properties as reported in the literature almost always make this assumption. In such situations it is necessary to rearrange the entries in the matrices of (18.12.9) to match the coordinate system being employed.

As an example, let us consider how (18.12.9) changes when the poling direction of the piezoceramic lines up with the $\mathbf{e}_1'$-direction as shown in Fig. 18.4 (right). The conventional case is shown in Fig. 18.4 (left). In the conventional case, the numerical values from tables such as Table 18.1 can be used directly in (18.12.9).[a]

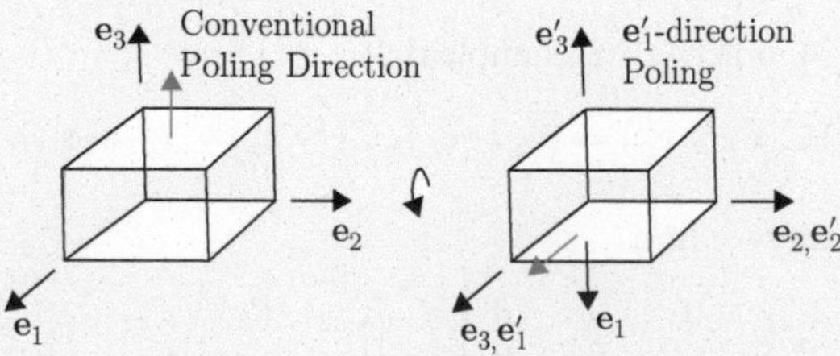

Fig. 18.4 Two poling orientations with respect to the coordinates axes in which a piezoelectric analysis is to be performed. Grey arrows indicate the poling direction.

One way to think of the case where the poling axis lines up with the $\mathbf{e}_1'$-direction is to consider it to occur due to a coordinate change generated by a 90-degree rotation about the $\mathbf{e}_2$-axis.

Starting with the second-order electric permittivity tensor, $\boldsymbol{\kappa}$, from (1.2.5) we have that

$$\kappa_{ij} = \mathbf{e}_i \cdot \boldsymbol{\kappa}\mathbf{e}_j. \tag{18.12.51}$$

Our interest is in the components of $\boldsymbol{\kappa}$ in the primed frame

$$\kappa'_{ij} = \mathbf{e}'_i \cdot \boldsymbol{\kappa}\mathbf{e}'_j. \tag{18.12.52}$$

From Fig. 18.4 (right), we observe that there is a correspondence between primed and unprimed coordinate directions, viz.

$$\begin{aligned} \mathbf{e}_1 &= -\mathbf{e}'_3 \\ \mathbf{e}_2 &= \ \ \mathbf{e}'_2 \\ \mathbf{e}_3 &= \ \ \mathbf{e}'_1. \end{aligned} \tag{18.12.53}$$

This correspondence can be used to find the correct components in the $\mathbf{e}'_1$-direction poling case. For example,

$$\kappa'_{11} = \mathbf{e}'_1 \cdot \boldsymbol{\kappa}\mathbf{e}'_1 = \mathbf{e}_3 \cdot \boldsymbol{\kappa}\mathbf{e}_3 = \kappa_{33}. \tag{18.12.54}$$

Continuing in this way, we find

$$[\kappa'] = \begin{bmatrix} \kappa'_{11} & 0 & 0 \\ 0 & \kappa'_{22} & 0 \\ 0 & 0 & \kappa'_{33} \end{bmatrix} = \begin{bmatrix} \kappa_{33} & 0 & 0 \\ 0 & \kappa_{22} & 0 \\ 0 & 0 & \kappa_{11} \end{bmatrix} = \begin{bmatrix} \kappa_{33} & 0 & 0 \\ 0 & \kappa_{11} & 0 \\ 0 & 0 & \kappa_{11} \end{bmatrix}, \tag{18.12.55}$$

where the values one finds in material property tables associated with $\mathbf{e}_3$-direction poling correspond to the unprimed values. Note that the order is permuted in comparison to $(18.12.9)_3$.

The same logic can now be applied to the elastic stiffness matrix and the piezoelectricity matrix. For the stiffness matrix we recall from (1.2.60) that

$$C_{ijkl} = (\mathbf{e}_i \otimes \mathbf{e}_j) : \mathbb{C}(\mathbf{e}_k \otimes \mathbf{e}_l) \tag{18.12.56}$$

and

$$C'_{ijkl} = (\mathbf{e}'_i \otimes \mathbf{e}'_j) : \mathbb{C}(\mathbf{e}'_k \otimes \mathbf{e}'_l). \tag{18.12.57}$$

Observing (18.12.53), one has for example that

$$C'_{1133} = (\mathbf{e}'_1 \otimes \mathbf{e}'_1) : \mathbb{C}(\mathbf{e}'_3 \otimes \mathbf{e}'_3) = (\mathbf{e}_3 \otimes \mathbf{e}_3) : \mathbb{C}((-\mathbf{e}_1) \otimes (-\mathbf{e}_1)) = C_{3311} = C_{31} = C_{13}. \tag{18.12.58}$$

Continuing in this way we find

$$[C'] = \begin{bmatrix} C'_{11} & C'_{12} & C'_{13} & 0 & 0 & 0 \\ C'_{12} & C'_{22} & C'_{23} & 0 & 0 & 0 \\ C'_{13} & C'_{23} & C'_{33} & 0 & 0 & 0 \\ 0 & 0 & 0 & C'_{44} & 0 & 0 \\ 0 & 0 & 0 & 0 & C'_{55} & 0 \\ 0 & 0 & 0 & 0 & 0 & C'_{66} \end{bmatrix} = \begin{bmatrix} C_{33} & C_{32} & C_{31} & 0 & 0 & 0 \\ C_{32} & C_{22} & C_{21} & 0 & 0 & 0 \\ C_{32} & C_{21} & C_{11} & 0 & 0 & 0 \\ 0 & 0 & 0 & C_{66} & 0 & 0 \\ 0 & 0 & 0 & 0 & C_{55} & 0 \\ 0 & 0 & 0 & 0 & 0 & C_{44} \end{bmatrix}$$

$$= \begin{bmatrix} C_{33} & C_{13} & C_{13} & 0 & 0 & 0 \\ C_{13} & C_{11} & C_{12} & 0 & 0 & 0 \\ C_{13} & C_{12} & C_{11} & 0 & 0 & 0 \\ 0 & 0 & 0 & \frac{1}{2}(C_{11} - C_{12}) & 0 & 0 \\ 0 & 0 & 0 & 0 & C_{44} & 0 \\ 0 & 0 & 0 & 0 & 0 & C_{44} \end{bmatrix}, \tag{18.12.59}$$

where the values one finds in material property tables associated with $\mathbf{e}_3$-direction poling correspond to the unprimed values. Note that the order is permuted in comparison to $(18.12.9)_1$.

We can now consider the piezoelectricity tensor, where according to (18.B.16)

$$\mathcal{E}_{ijk} = \mathbf{e}_i \cdot (\mathcal{E}\mathbf{e}_k)\mathbf{e}_j \tag{18.12.60}$$

and

$$\mathcal{E}'_{ijk} = \mathbf{e}'_i \cdot (\mathcal{E}\mathbf{e}'_k)\mathbf{e}'_j. \tag{18.12.61}$$

Now for example

$$\mathcal{E}'_{311} = \mathbf{e}'_3 \cdot (\mathcal{E}\mathbf{e}'_1)\mathbf{e}'_1 = (-\mathbf{e}_1) \cdot (\mathcal{E}\mathbf{e}_3)\mathbf{e}_3 = -\mathcal{E}_{133} = -E_{13}. \tag{18.12.62}$$

Following in this fashion, one finds

$$\begin{aligned}
[E'] &= \begin{bmatrix} \mathcal{E}'_{111} & \mathcal{E}'_{122} & \mathcal{E}'_{133} & \mathcal{E}'_{123} & \mathcal{E}'_{113} & \mathcal{E}'_{112} \\ \mathcal{E}'_{211} & \mathcal{E}'_{222} & \mathcal{E}'_{233} & \mathcal{E}'_{223} & \mathcal{E}'_{213} & \mathcal{E}'_{212} \\ \mathcal{E}'_{311} & \mathcal{E}'_{322} & \mathcal{E}'_{333} & \mathcal{E}'_{323} & \mathcal{E}'_{313} & \mathcal{E}'_{312} \end{bmatrix} \\
&= \begin{bmatrix} \mathcal{E}_{333} & \mathcal{E}_{322} & \mathcal{E}_{311} & -\mathcal{E}_{321} & -\mathcal{E}_{331} & \mathcal{E}_{332} \\ \mathcal{E}_{233} & \mathcal{E}_{222} & \mathcal{E}_{211} & -\mathcal{E}_{221} & -\mathcal{E}_{231} & \mathcal{E}_{232} \\ -\mathcal{E}_{133} & -\mathcal{E}_{122} & -\mathcal{E}_{111} & \mathcal{E}_{121} & \mathcal{E}_{131} & -\mathcal{E}_{132} \end{bmatrix} \\
&= \begin{bmatrix} E_{33} & E_{32} & E_{31} & -E_{36} & -E_{35} & E_{34} \\ E_{23} & E_{22} & E_{21} & -E_{26} & -E_{25} & E_{24} \\ -E_{13} & -E_{12} & -E_{11} & E_{16} & E_{15} & -E_{11} \end{bmatrix} \\
&= \begin{bmatrix} E_{33} & E_{31} & E_{31} & 0 & 0 & 0 \\ 0 & 0 & 0 & 0 & 0 & E_{15} \\ 0 & 0 & 0 & 0 & E_{15} & 0 \end{bmatrix},
\end{aligned} \tag{18.12.63}$$

where the values one finds in material property tables associated with $\mathbf{e}_3$-direction poling correspond to the unprimed values. Note that the order is permuted in comparison to $(18.12.9)_2$.

[a] Note that in the conventional case in Fig. 18.4 (left) we have denoted the coordinate directions without a prime to distinguish them from the case which we would like to consider, Fig. 18.4 (right), where we have denoted the coordinate directions with a prime.

Appendices

18.A Electromagnetics

The set of equations that govern the behavior of electromagnetic fields in matter (e.g. a region of space containing solid material) is given by Maxwell's equations (cf., e.g., Jackson, 1999):

- *Gauss's Law*

$$\operatorname{div} \mathbf{d} = q, \tag{18.A.1}$$

- *Maxwell-Ampère's Law*

$$\operatorname{curl} \mathbf{h} = \mathbf{j} + \dot{\mathbf{d}}, \tag{18.A.2}$$

- *Faraday's Law*

$$\operatorname{curl} \mathbf{e} = -\dot{\mathbf{b}}, \tag{18.A.3}$$

- *No Magnetic Monopole Law*

$$\operatorname{div} \mathbf{b} = 0. \tag{18.A.4}$$

In these relations $\mathbf{d}$ is the electric displacement, q is the free charge density,[11] $\mathbf{h}$ is the magnetic field, $\mathbf{j}$ is the free charge current, $\mathbf{e}$ is the electric field, and $\mathbf{b}$ is the magnetic induction. To close this system of equations, one requires constitutive laws for $\mathbf{d}$, $\mathbf{h}$, and $\mathbf{j}$ in terms of $\mathbf{e}$ and $\mathbf{b}$, i.e. we require knowledge of functions $\hat{\mathbf{d}}(\mathbf{e}, \mathbf{b})$, $\hat{\mathbf{h}}(\mathbf{e}, \mathbf{b})$, and $\hat{\mathbf{j}}(\mathbf{e}, \mathbf{b})$.[12]

18.A.1 Charge conservation

The governing equation of (free) charge conservation is contained within Maxwell's equations. First observe that the divergence of the curl of a vector field is always zero ($\operatorname{div} \operatorname{curl} \mathbf{v} \rightarrow e_{ijk} v_{k,ji} = 0$). If one takes the divergence on both sides of (18.A.2), then

$$\underbrace{\operatorname{div} \operatorname{curl} \mathbf{h}}_{=0} = \operatorname{div} \mathbf{j} + \operatorname{div} \dot{\mathbf{d}}. \tag{18.A.5}$$

Now observe that we can use the time derivative of (18.A.1) to replace the last term and arrive at

$$\dot{q} + \operatorname{div} \mathbf{j} = 0, \tag{18.A.6}$$

which is the classical law of free charge conservation.

---

[11] Free charge refers to charges that are not bound to atoms.

[12] Note that the literature on electromagnetics has a rich and sometimes conflicting variety of naming conventions for the governing equations as well as for the names of the vector fields. Our choice is a common one among several alternatives.

18.A.2 **Quasi-static limit**

The definition of the *quasi-static limit* of Maxwell's equations that we employ here, is one in which it is assumed that

$$\dot{\mathbf{b}} = 0. \tag{18.A.7}$$

With this assumption, Faraday's law becomes $\operatorname{curl}\mathbf{e} = \mathbf{0}$. Since the curl of any gradient field is zero ($\operatorname{curl}\nabla f \to e_{ijk}f_{,kj} = 0$), in the quasi-static limit we assure that (18.A.3) will always be satisfied by assuming that **e** is the gradient of some scalar field. By convention we write $\mathbf{e} = -\nabla\phi$, where ϕ is known as the potential (or voltage) field.

18.A.3 **Energy transport in the quasi-static limit**

The flux of electromagnetic energy is described by the *Poynting vector* (see, e.g., Jackson (1999, Sec. 6.7) or Kovetz (2000, Sec. 45 and Sec. 54)):

$$\mathbf{s} \stackrel{\text{def}}{=} \mathbf{e} \times \mathbf{h}. \tag{18.A.8}$$

This vector describes the flux of electromagnetic energy (energy per unit area per unit time). Thus the flux of electromagnetic energy *into* a region of material $\mathcal{P}$ is given by,

$$-\int_{\partial\mathcal{P}} \mathbf{s}\cdot\mathbf{n}\,da\,, \tag{18.A.9}$$

where **n** is the outward unit normal to its boundary $\partial\mathcal{P}$. Using the divergence theorem,

$$-\int_{\partial\mathcal{P}} \mathbf{s}\cdot\mathbf{n}\,da = -\int_{\mathcal{P}} \operatorname{div}\mathbf{s}\,dv = -\int_{\mathcal{P}} \operatorname{div}(\mathbf{e}\times\mathbf{h})\,dv. \tag{18.A.10}$$

Next, use of the vector identity

$$\operatorname{div}(\mathbf{e}\times\mathbf{h}) = \mathbf{h}\cdot(\operatorname{curl}\mathbf{e}) - \mathbf{e}\cdot(\operatorname{curl}\mathbf{h}), \tag{18.A.11}$$

allows us to write (18.A.10) as

$$-\int_{\partial\mathcal{P}} \mathbf{s}\cdot\mathbf{n}\,da = -\int_{\mathcal{P}} \big(\mathbf{h}\cdot(\operatorname{curl}\mathbf{e}) - \mathbf{e}\cdot(\operatorname{curl}\mathbf{h})\big)\,dv. \tag{18.A.12}$$

In the quasi-static limit $\operatorname{curl}\mathbf{e} = 0$, and since $\operatorname{curl}\mathbf{h} = \dot{\mathbf{d}} + \mathbf{j}$ (cf. (18.A.2)), the *inward* electromagnetic energy flux is given by,

$$\begin{aligned}
-\int_{\partial\mathcal{P}} \mathbf{s}\cdot\mathbf{n}\,da &= \int_{\mathcal{P}} \mathbf{e}\cdot(\dot{\mathbf{d}}+\mathbf{j})\,dv && (18.A.13)\\
&= -\int_{\mathcal{P}} \nabla\phi\cdot(\dot{\mathbf{d}}+\mathbf{j})\,dv && (18.A.14)\\
&= -\left[\int_{\mathcal{P}} \operatorname{div}(\phi(\dot{\mathbf{d}}+\mathbf{j}))\,dv - \int_{\mathcal{P}} \phi\operatorname{div}(\dot{\mathbf{d}}+\mathbf{j})\,dv\right] && (18.A.15)\\
&= -\left[\int_{\mathcal{P}} \operatorname{div}(\phi(\dot{\mathbf{d}}+\mathbf{j}))\,dv - \int_{\mathcal{P}} \phi\underbrace{\operatorname{div}(\operatorname{curl}\mathbf{h})}_{\equiv 0}\,dv\right], && (18.A.16)
\end{aligned}$$

or,

$$\boxed{-\int_{\partial\mathcal{P}} \mathbf{s}\cdot\mathbf{n}\,da = -\int_{\partial\mathcal{P}} \phi\,(\dot{\mathbf{d}}+\mathbf{j})\cdot\mathbf{n}\,da.} \tag{18.A.17}$$

18.B Third-order tensors

A **third-order tensor** $\boldsymbol{\mathcal{S}}$ is defined as a *linear* mapping of vectors to second-order tensors. That is, given a vector $\mathbf{a}$,

$$\mathbf{A} = \boldsymbol{\mathcal{S}}\mathbf{a} \tag{18.B.1}$$

is a second-order tensor. The **linearity** of a tensor $\boldsymbol{\mathcal{S}}$ is embodied by the requirements:

$$\boldsymbol{\mathcal{S}}(\mathbf{a}+\mathbf{b}) = \boldsymbol{\mathcal{S}}\mathbf{a} + \boldsymbol{\mathcal{S}}\mathbf{b} \qquad \text{for all vectors } \mathbf{a} \text{ and } \mathbf{b},$$

$$\boldsymbol{\mathcal{S}}(\alpha\mathbf{a}) = \alpha\boldsymbol{\mathcal{S}}\mathbf{a} \qquad \text{for all vectors } \mathbf{a} \text{ and scalars } \alpha.$$

An example of a third-order tensor is the tensor product $(\mathbf{a}\otimes\mathbf{b}\otimes\mathbf{c})$ of three vectors $\mathbf{a}$, $\mathbf{b}$, and $\mathbf{c}$, defined by

$$(\mathbf{a}\otimes\mathbf{b}\otimes\mathbf{c})\mathbf{u} = (\mathbf{c}\cdot\mathbf{u})(\mathbf{a}\otimes\mathbf{b}) \tag{18.B.2}$$

for all $\mathbf{u}$. By (18.B.2), the third-order tensor $(\mathbf{a}\otimes\mathbf{b}\otimes\mathbf{c})$ maps any vector $\mathbf{u}$ onto a scalar multiple of the second-order tensor $(\mathbf{a}\otimes\mathbf{b})$.

18.B.1 Components of a third-order tensor

Given a third-order tensor $\boldsymbol{\mathcal{S}}$, choose an arbitrary vector $\mathbf{a}$ and let

$$\mathbf{A} = \boldsymbol{\mathcal{S}}\mathbf{a}.$$

Then, with respect to an orthonormal basis $\{\mathbf{e}_i\}$,

$$\begin{aligned}
\mathbf{A} &= \boldsymbol{\mathcal{S}}(a_k\mathbf{e}_k),\\
\mathbf{e}_i\cdot\mathbf{A}\mathbf{e}_j &= a_k\mathbf{e}_i\cdot(\boldsymbol{\mathcal{S}}\mathbf{e}_k)\mathbf{e}_j,\\
A_{ij} &= \big(\mathbf{e}_i\cdot(\boldsymbol{\mathcal{S}}\mathbf{e}_k)\mathbf{e}_j\big)a_k.
\end{aligned}$$

Thus, defining the **components** $S_{ijk} = (\boldsymbol{\mathcal{S}})_{ijk}$ of $\boldsymbol{\mathcal{S}}$ with respect to the basis $\{\mathbf{e}_i\}$ by

$$\boxed{S_{ijk} \stackrel{\text{def}}{=} \mathbf{e}_i\cdot(\boldsymbol{\mathcal{S}}\mathbf{e}_k)\mathbf{e}_j,} \tag{18.B.3}$$

we see that the component form of the relation $\mathbf{A} = \boldsymbol{\mathcal{S}}\mathbf{a}$ is

$$\boxed{A_{ij} = S_{ijk}a_k.} \tag{18.B.4}$$

Further, this relation implies that

$$\begin{aligned}
\mathbf{A} &= A_{ij}\mathbf{e}_i\otimes\mathbf{e}_j\\
&= \big(S_{ijk}a_k\big)\mathbf{e}_i\otimes\mathbf{e}_j\\
&= \big(S_{ijk}\mathbf{a}\cdot\mathbf{e}_k\big)\mathbf{e}_i\otimes\mathbf{e}_j\\
&= \big(S_{ijk}\mathbf{e}_i\otimes\mathbf{e}_j\otimes\mathbf{e}_k\big)\mathbf{a}
\end{aligned}$$

and since $\mathbf{A} = \boldsymbol{\mathcal{S}}\mathbf{a}$ we must have

$$\boldsymbol{\mathcal{S}}\mathbf{a} = \big(S_{ijk}\mathbf{e}_i \otimes \mathbf{e}_j \otimes \mathbf{e}_k\big)\mathbf{a}.$$

Thus, since the vector $\mathbf{a}$ was arbitrarily chosen, we may conclude that $\boldsymbol{\mathcal{S}}$ has the representation

$$\boxed{\boldsymbol{\mathcal{S}} = S_{ijk}\mathbf{e}_i \otimes \mathbf{e}_j \otimes \mathbf{e}_k,} \tag{18.B.5}$$

with respect to an orthonormal basis $\{\mathbf{e}_i\}$.

18.B.2 Action of a third-order tensor on a second-order tensor

A third-order tensor $\boldsymbol{\mathcal{S}}$ also *linearly* maps second-order tensors to vectors. That is, given a second-order tensor $\mathbf{A}$,

$$\boxed{\mathbf{a} = \boldsymbol{\mathcal{S}}:\mathbf{A}} \tag{18.B.6}$$

is a vector, where the inner product ":" of the third-order tensor $\boldsymbol{\mathcal{S}}$ with the second-order tensor $\mathbf{A}$ is defined by

$$\boldsymbol{\mathcal{S}}:\mathbf{A} \overset{\text{def}}{=} \operatorname{tr}\big(\boldsymbol{\mathcal{S}}\mathbf{A}^{\top}\big). \tag{18.B.7}$$

The linearity of a tensor $\boldsymbol{\mathcal{S}}$ is embodied by the requirements:

$$\boldsymbol{\mathcal{S}}:(\mathbf{A}+\mathbf{B}) = \boldsymbol{\mathcal{S}}:\mathbf{A} + \boldsymbol{\mathcal{S}}:\mathbf{B} \quad \text{for all second-order tensors } \mathbf{A} \text{ and } \mathbf{B},$$

$$\boldsymbol{\mathcal{S}}:(\alpha\mathbf{A}) = \alpha\boldsymbol{\mathcal{S}}:\mathbf{A} \qquad \text{for all second-order tensors } \mathbf{A} \text{ and scalars } \alpha.$$

Thus, from the definition (18.B.7), the action of a third-order tensor $(\mathbf{a}\otimes\mathbf{b}\otimes\mathbf{c})$ on a second-order tensor $(\mathbf{u}\otimes\mathbf{v})$ gives

$$(\mathbf{a}\otimes\mathbf{b}\otimes\mathbf{c}):(\mathbf{u}\otimes\mathbf{v}) = \operatorname{tr}\Big((\mathbf{a}\otimes\mathbf{b}\otimes\mathbf{c})(\mathbf{v}\otimes\mathbf{u})\Big) = (\mathbf{c}\cdot\mathbf{v})(\mathbf{b}\cdot\mathbf{u})\mathbf{a}, \tag{18.B.8}$$

for all $\mathbf{u}\otimes\mathbf{v}$. Hence, the tensor $\mathbf{a}\otimes\mathbf{b}\otimes\mathbf{c}$ maps any second-order tensor $\mathbf{u}\otimes\mathbf{v}$ onto a scalar multiple of the vector $\mathbf{a}$.

The component form of (18.B.6) is obtained as follows:

$$\begin{aligned}
a_i\mathbf{e}_i &= \boldsymbol{\mathcal{S}}:\mathbf{A}\\
&= \operatorname{tr}\big(\boldsymbol{\mathcal{S}}\mathbf{A}^{\top}\big)\\
&= \operatorname{tr}\big((S_{ijk}\mathbf{e}_i\otimes\mathbf{e}_j\otimes\mathbf{e}_k)(A_{lm}\mathbf{e}_l\otimes\mathbf{e}_m)^{\top}\big)\\
&= S_{ijk}A_{lm}\operatorname{tr}\big((\mathbf{e}_i\otimes\mathbf{e}_j\otimes\mathbf{e}_k)(\mathbf{e}_m\otimes\mathbf{e}_l)\big)\\
&= S_{ijk}A_{lm}\delta_{km}\delta_{jl}\,\mathbf{e}_i\\
&= S_{ijk}A_{jk}\mathbf{e}_i;
\end{aligned} \tag{18.B.9}$$

and hence the component form of the relation $\mathbf{a} = \boldsymbol{\mathcal{S}}:\mathbf{A}$ is

$$\boxed{a_i = S_{ijk}A_{jk}.} \tag{18.B.10}$$

18.B.3 Transpose of a third-order tensor

The **transpose** $\boldsymbol{\mathcal{S}}^{\top}$ of a third-order tensor $\boldsymbol{\mathcal{S}}$ can be defined as the unique third-order tensor with the property that

$$\mathbf{a} \cdot \boldsymbol{\mathcal{S}} : \mathbf{A} = (\boldsymbol{\mathcal{S}}^{\top}\mathbf{a}) : \mathbf{A} \tag{18.B.11}$$

for all vectors **a** and second-order tensors **A**.

Equation (18.B.11) has the component representation

$$\begin{aligned} a_k S_{kij} A_{ij} &= (\boldsymbol{\mathcal{S}}^{\top}\mathbf{a})_{ij} A_{ij} \\ &= (\boldsymbol{\mathcal{S}}^{\top})_{ijk} a_k A_{ij} \end{aligned} \tag{18.B.12}$$

and since **a** and **A** are arbitrary, the components of $\boldsymbol{\mathcal{S}}^{\top}$ are

$$\boxed{(\boldsymbol{\mathcal{S}}^{\top})_{ijk} = S_{kij},} \tag{18.B.13}$$

and hence $\boldsymbol{\mathcal{S}}^{\top}$ has the representation

$$\boxed{\boldsymbol{\mathcal{S}}^{\top} = S_{kij} \mathbf{e}_i \otimes \mathbf{e}_j \otimes \mathbf{e}_k.} \tag{18.B.14}$$

Also, the component form of the second-order tensor $\boldsymbol{\mathcal{S}}^{\top}\mathbf{a}$ is

$$\boxed{(\boldsymbol{\mathcal{S}}^{\top}\mathbf{a})_{ij} = S_{kij} a_k.} \tag{18.B.15}$$

18.B.4 Summary of a third-order tensor and its action on vectors and tensors

Summarizing, for a third-order tensor $\boldsymbol{\mathcal{S}}$,

$$\boxed{\begin{aligned} S_{ijk} = \mathbf{e}_i \cdot (\boldsymbol{\mathcal{S}}\mathbf{e}_k)\mathbf{e}_j, \quad (\boldsymbol{\mathcal{S}}^{\top})_{ijk} = S_{kij}, \\ (\boldsymbol{\mathcal{S}}\mathbf{a})_{ij} = S_{ijk} a_k, \quad (\boldsymbol{\mathcal{S}}^{\top}\mathbf{a})_{ij} = S_{kij} a_k, \quad (\boldsymbol{\mathcal{S}} : \mathbf{A})_i = S_{ijk} A_{jk}. \end{aligned}} \tag{18.B.16}$$

PART VIII

Limits to elastic response. Yielding and plasticity

19 Limits to elastic response. Yielding and failure

19.1 Introduction

In addition to the small displacement gradient restrictions in the theory of linear elasticity, we need to also explicitly introduce criteria for yielding or fracture of materials—criteria which bound the levels of stresses beyond which the theory of linear elasticity is no longer valid.

Figure 19.1(a) schematically represents an idealized response of a "brittle" material, such as an engineering ceramic, in a simple tension or compression test, while Fig. 19.1(b) schematically represents a corresponding idealized response of a "ductile" material, such as an engineering metal. In a one-dimensional situation for brittle materials, as indicated in Fig. 19.1(a), the value of the stress T at which the material fails by fracture in tension, with essentially no permanent plastic strain, that is in a "brittle" fashion, is much smaller (often 15 times smaller) than the magnitude C of the failure stress in compression. After initiation of failure in compression the magnitude of the stress decreases gradually until the material eventually loses all stress-carrying capacity. The fracture surface in tension is usually oriented perpendicular to the direction of the tensile stress, while failure in compression occurs by splitting-type micro-fractures parallel, or sometimes angled acutely, to the axis of compression. The response in

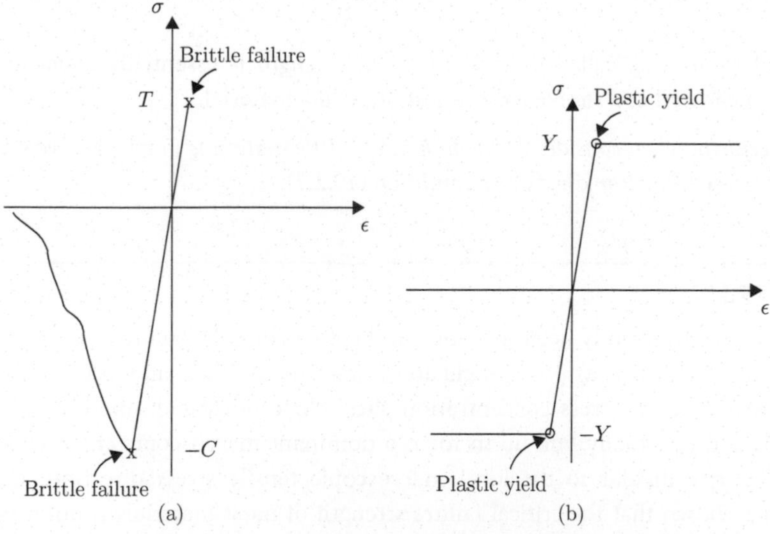

Fig. 19.1 (a) Failure of brittle materials. (b) Yield of ductile materials.

Continuum Mechanics of Solids. Lallit Anand and Sanjay Govindjee, Oxford University Press (2020).

DOI: 10.1093/oso/9780198864721.001.0001

tension $\sigma > 0$ for a brittle material is vastly more detrimental and is reached at a much lower magnitude of stress; for such materials a simple *failure criterion* is

$$\sigma \le T, \qquad (T \ll C). \tag{19.1.1}$$

In contrast, a ductile metal typically starts to deform plastically at essentially the same magnitude of the stress in tension or compression, and for such materials a suitable *yield criterion*, which sets the limit to elastic response is:

$$|\sigma| \le Y. \tag{19.1.2}$$

Next, we discuss generalization of these simple one-dimensional ideas to arbitrary three-dimensional situations. *Note that our discussion here does not explicitly account for pre-existing cracks; we postpone our discussion on the effects of cracks to Chapter 26, where we discuss fracture mechanics.*

19.2 **Failure criterion for brittle materials in tension**

Recall, that the principal values of the stress $\boldsymbol{\sigma}$ at a point in the body are

$$\{\sigma_1, \sigma_2, \sigma_3\} \qquad \text{with} \qquad \sigma_1 \ge \sigma_2 \ge \sigma_3. \tag{19.2.1}$$

As before, we take the principal stresses to be strictly ordered as in $(19.2.1)_2$. The simplest failure criterion for brittle materials—usually credited to William Rankine (Rankine, 1857)—is that,

- *failure in a brittle material will initiate when the maximum principal stress σ_1 at a point in the body reaches a critical value,*

$$\boxed{\sigma_1 \le \sigma_{1\text{cr}}.} \tag{19.2.2}$$

The critical value $\sigma_{1\text{cr}}$, called the *critical failure strength* is essentially equal to the failure strength T measured in a simple tension test on such a material.

- *Henceforth, when considering the failure of brittle materials in tension, we shall limit our discussion to the Rankine failure condition* (19.2.2).

REMARK 19.1

Every material is inherently heterogeneous. In brittle materials, failure normally nucleates at microscopic defects such as pores, inclusions, grain boundaries, and grain-boundary triple junctions which act as stress concentration sites. Once nucleated, micro-cracks propagate and branch and eventually link-up to form a dominant macroscopic crack which typically propagates perpendicular to the local macroscopic tensile stress direction. Experimental evidence has shown that the critical failure strength of most materials is not a unique well-defined number, but follows a distribution called the Weibull distribution (Weibull, 1939).

19.3 Yield criterion for ductile isotropic materials

In a three-dimensional situation, for ductile isotropic materials, a simple statement of a yield condition is

$$f(\boldsymbol{\sigma}, \text{internal state of the material}) \leq 0, \tag{19.3.1}$$

where f is a scalar-valued function of the applied stress $\boldsymbol{\sigma}$, and some *scalar measure* (yet to be specified) of the resistance offered to plastic deformation by the internal state of the material.

Isotropy requires that the dependence on $\boldsymbol{\sigma}$ in the function $f(\boldsymbol{\sigma})$ can only appear in terms of

- either the **principal values** of $\boldsymbol{\sigma}$, or
- the **invariants** of $\boldsymbol{\sigma}$,

since both the principal values and the invariants are by definition *independent of the choice of the basis* $\{\mathbf{e}_i\}$ with respect to which the components $\sigma_{ij} = \mathbf{e}_i \cdot \boldsymbol{\sigma}\mathbf{e}_j$ of the stress $\boldsymbol{\sigma}$ may be expressed. As in our discussion of the Rankine condition, we will assume that the principal stresses are strictly ordered:

$$\{\sigma_1, \sigma_2, \sigma_3\} \qquad \text{with} \qquad \sigma_1 \geq \sigma_2 \geq \sigma_3. \tag{19.3.2}$$

Numerous different stress invariants may be defined. The invariants that are most widely used to formulate yield conditions are as follows:

1. The invariant

$$\boxed{\bar{p} = -\frac{1}{3}\operatorname{tr}\boldsymbol{\sigma} = -\frac{1}{3}(\sigma_{kk})} \tag{19.3.3}$$

is called the **mean normal pressure** or the **equivalent pressure stress.**

2. The invariant

$$\boxed{\bar{\tau} = \frac{1}{\sqrt{2}}|\boldsymbol{\sigma}'| = \sqrt{\frac{1}{2}\operatorname{tr}(\boldsymbol{\sigma}'^2)} = \sqrt{\frac{1}{2}\sigma'_{ij}\sigma'_{ij}},} \tag{19.3.4}$$

is called the **equivalent shear stress.** Another invariant, which differs from $\bar{\tau}$ by a factor of $\sqrt{3}$ is the invariant

$$\boxed{\bar{\sigma} = \sqrt{\frac{3}{2}}|\boldsymbol{\sigma}'| = \sqrt{\frac{3}{2}\operatorname{tr}(\boldsymbol{\sigma}'^2)} = \sqrt{\frac{3}{2}\sigma'_{ij}\sigma'_{ij}},} \tag{19.3.5}$$

which is called the **equivalent tensile stress** or the **Mises equivalent stress.** Note that

$$\bar{\sigma} = \sqrt{3}\bar{\tau}. \tag{19.3.6}$$

3. The invariant

$$\boxed{\bar{r} = \left(\frac{9}{2}\operatorname{tr}(\boldsymbol{\sigma}'^3)\right)^{\frac{1}{3}} = \left(\frac{9}{2}\sigma'_{ik}\sigma'_{kj}\sigma'_{ji}\right)^{\frac{1}{3}},} \tag{19.3.7}$$

is simply called the **third stress invariant.**

The invariants $\bar{p}$, $\bar{\tau}$, and $\bar{\sigma}$, when written out in full in terms of the components σ_{ij} of the stress tensor $\boldsymbol{\sigma}$, take the forms:

1. **Mean normal pressure:**

$$\boxed{\bar{p} = -\frac{1}{3}\left(\sigma_{11} + \sigma_{22} + \sigma_{33}\right).} \tag{19.3.8}$$

2. **Equivalent shear stress:**

$$\boxed{\bar{\tau} = \left|\left[\frac{1}{6}\left((\sigma_{11} - \sigma_{22})^2 + (\sigma_{22} - \sigma_{33})^2 + (\sigma_{33} - \sigma_{11})^2\right) + \left(\sigma_{12}^2 + \sigma_{23}^2 + \sigma_{31}^2\right)\right]^{1/2}\right|.} \tag{19.3.9}$$

3. **Equivalent tensile stress:**

$$\boxed{\bar{\sigma} = \sqrt{3}\bar{\tau} = \left|\left[\frac{1}{2}\left((\sigma_{11} - \sigma_{22})^2 + (\sigma_{22} - \sigma_{33})^2 + (\sigma_{33} - \sigma_{11})^2\right) + 3\left(\sigma_{12}^2 + \sigma_{23}^2 + \sigma_{31}^2\right)\right]^{1/2}\right|.} \tag{19.3.10}$$

The terminology for these invariants becomes clear when we note that[1]

- in the case of a state of hydrostatic pressure, $\sigma_{11} = \sigma_{22} = \sigma_{33} = -p$ and all other $\sigma_{ij} = 0$, the mean normal pressure $\bar{p} = p$;
- in the case of pure shear, say $\sigma_{12} \neq 0$, all other $\sigma_{ij} = 0$, the equivalent shear stress $\bar{\tau} = |\sigma_{12}|$.
- and in the case of pure tension, say $\sigma_{11} \neq 0$, all other $\sigma_{ij} = 0$, the equivalent tensile stress $\bar{\sigma} = |\sigma_{11}|$.

In terms of the principal values of stress $(\sigma_1, \sigma_2, \sigma_3)$, the list $(\bar{p}, \bar{\tau}, \bar{\sigma}, \bar{r})$ is completely characterized by

$$\boxed{\begin{aligned}
\bar{p} &= -\frac{1}{3}\Big(\sigma_1 + \sigma_2 + \sigma_3\Big), \\
\bar{\tau} &= \sqrt{\frac{1}{2}\left(\sigma_1'^2 + \sigma_2'^2 + \sigma_3'^2\right)} = \sqrt{\frac{1}{6}\left((\sigma_1 - \sigma_2)^2 + (\sigma_2 - \sigma_3)^2 + (\sigma_3 - \sigma_1)^2\right)} \\
\bar{\sigma} &= \sqrt{\frac{3}{2}\left(\sigma_1'^2 + \sigma_2'^2 + \sigma_3'^2\right)} = \sqrt{\frac{1}{2}\left((\sigma_1 - \sigma_2)^2 + (\sigma_2 - \sigma_3)^2 + (\sigma_3 - \sigma_1)^2\right)} \\
\bar{r} &= \left(\frac{9}{2}\left(\sigma_1'^3 + \sigma_2'^3 + \sigma_3'^3\right)\right)^{\frac{1}{3}},
\end{aligned}} \tag{19.3.11}$$

where

$$\sigma_1' = \sigma_1 - \frac{1}{3}(\sigma_1 + \sigma_2 + \sigma_3), \qquad \sigma_2' = \sigma_2 - \frac{1}{3}(\sigma_1 + \sigma_2 + \sigma_3), \qquad \sigma_3' = \sigma_3 - \frac{1}{3}(\sigma_1 + \sigma_2 + \sigma_3). \tag{19.3.12}$$

[1] The third invariant $\bar{r}$ does not have a straightforward interpretation in terms of simple stress states.

19.3.1 **Mises yield condition**

Yield condition (19.3.1) for isotropic materials may be expressed in terms of the invariants $(\bar{p}, \bar{\tau}, \bar{r})$,

$$f(\bar{p}, \bar{\tau}, \bar{r}, \text{internal state of the material}) \leq 0.$$

For ductile metallic polycrystalline materials,

- *failure of the elastic response occurs when line defects called dislocations move large distances through the crystals of a material to produce significant permanent deformation.*

It has been found experimentally that for this mechanism of yield for polycrystalline metals,

- *the function*

$$f(\bar{p}, \bar{\tau}, \bar{r}, \text{internal state of the material}) \leq 0,$$

can, to a very good approximation, be taken to be **independent** *of the mean normal pressure $\bar{p}$ and the third invariant $\bar{r}$.*

The simplest such function takes the form

$$\boxed{\bar{\tau} \leq \tau_{y,\text{Mises}},} \tag{19.3.13}$$

where $\tau_{y,\text{Mises}}$ is a **shear yield strength** of the material. This yield condition was first proposed by Richard von Mises (von Mises, 1913) and is known as the **Mises yield condition**. A similar yield condition had previously been proposed by Huber (1904).[2]

The material constant $\tau_{y,\text{Mises}}$ may be found by considering a simple tension test in which $\sigma_{11} \neq 0$ and all other $\sigma_{ij} = 0$. In this case, (19.3.9) gives

$$\bar{\tau} = \frac{1}{\sqrt{3}}|\sigma_{11}|,$$

substitution of which in (19.3.13) gives

$$\frac{1}{\sqrt{3}}|\sigma_{11}| \leq \tau_{y,\text{Mises}}.$$

Denoting the value of the tensile stress at yield by

$$|\sigma_{11}| = Y,$$

with Y the **tensile yield strength** of the material, we obtain

$$\boxed{\tau_{y,\text{Mises}} \stackrel{\text{def}}{=} \frac{1}{\sqrt{3}}Y.} \tag{19.3.14}$$

Then, recalling that $\bar{\sigma} = \sqrt{3}\,\bar{\tau}$, the yield condition (19.3.13) may be alternatively be written as

$$\boxed{\bar{\sigma} \leq Y,} \tag{19.3.15}$$

and it is in this form that the Mises yield condition is *most widely expressed.*

[2] As noted by Hill (1950): "von Mises' criterion was anticipated, to some extent, by Huber (1904) in a paper in Polish which did not attract general attention until nearly twenty years later." Huber's contribution was also cited by Hencky (1924), who employs an argument quite similar to Huber's. According to Hill, this criterion was also anticipated ". . . by Clerk Maxwell . . . [in a] letter to W. Thomson, 18 Dec. 1856."

- The Mises yield condition stands for the physical notion that, as long as the applied equivalent tensile stress $\bar{\sigma}$ is less than the **material property** Y, dislocations would not have moved large enough distances through the crystals of a polycrystalline material to have produced significant permanent deformation.
- The **strength** Y is typically identified with the 0.2% offset **yield strength** in a tension test and is defined as the stress level from which unloading to zero stress would result in a permanent axial strain of 0.2%.

EXAMPLE 19.1 Yield due to Mises condition

Consider a one-parameter state of stress at a point in a solid under given boundary conditions,

$$[\sigma(\zeta)] = \begin{bmatrix} a\zeta & 0 & b\zeta \\ 0 & 0 & 0 \\ b\zeta & 0 & 0 \end{bmatrix}, \tag{19.3.16}$$

where a and b are given constants, and ζ is a parameter characterizing the magnitude of the boundary load. To determine the elastic limit of the load parameter we can apply the Mises condition.

The Mises equivalent stress is given by (19.3.10), which for the stress state (19.3.16) results in

$$\bar{\sigma} = \left[\tfrac{1}{2}\{a^2\zeta^2 + a^2\zeta^2\} + 3b^2\zeta^2\right]^{1/2} = \sqrt{a^2+3b^2}\,|\zeta|.$$

Thus we find according to the Mises condition (19.3.15) that the load parameter is limited to the following range:

$$|\zeta| \leq \frac{Y}{\sqrt{a^2+3b^2}}. \tag{19.3.17}$$

19.3.2 Tresca yield condition

This yield condition is phrased in terms of the principal values of the stress $\{\sigma_1, \sigma_2, \sigma_3\}$. In terms of the principal stresses, the **maximum shear stress** in a material at a given point is given by

$$\boxed{\tau_{\max} \overset{\text{def}}{=} \frac{1}{2}(\sigma_1 - \sigma_3) \geq 0.} \tag{19.3.18}$$

Henri Édouard Tresca (Tresca, 1864), based on his experimental observations, proposed that

- *yield in a ductile material will initiate when the maximum shear stress at a point in the body reaches a critical value,*[3]

[3] Tresca's yield condition does not depend on the intermediate principal stress σ_2.

$$\boxed{\frac{1}{2}\Big(\sigma_1 - \sigma_3\Big) \leq \tau_{y,\text{Tresca}},} \tag{19.3.19}$$

where $\tau_{y,\text{Tresca}}$ is a parameter called the **shear yield strength** of the material.

The parameter $\tau_{y,\text{Tresca}}$ may be found by considering a simple tension test in which $\sigma_1 \neq 0$ and $\sigma_2 = \sigma_3 = 0$. In this case (19.3.18) gives

$$\tau_{\max} = \frac{1}{2}\sigma_1,$$

substitution of which in (19.3.19) gives

$$\frac{1}{2}\sigma_1 \leq \tau_{y,\text{Tresca}}.$$

Denoting the value of the tensile stress σ_1 at yield by Y, we obtain

$$\boxed{\tau_{y,\text{Tresca}} \stackrel{\text{def}}{=} \frac{1}{2}Y.} \tag{19.3.20}$$

The **strength** Y is typically identified with the 0.2% offset **yield strength** in a tension test. The Tresca criterion may therefore alternatively be written as

$$\boxed{\Big(\sigma_1 - \sigma_3\Big) \leq Y.} \tag{19.3.21}$$

REMARKS 19.2

1. Note that the Mises yield condition is expressed in terms of the **equivalent shear stress** $\bar{\tau}$, while the Tresca yield condition is expressed in terms of the **maximum shear stress** $\tau_{\max}$.
2. Since homogeneous simple tension experiments are easier to conduct than homogeneous pure shear experiments, the material parameters $\tau_{y,\text{Mises}}$ and $\tau_{y,\,\text{Tresca}}$ appearing in these two yield conditions are typically evaluated from a simple tension test in terms of the **tensile yield strength** Y according to (19.3.14) and (19.3.20). Thus

$$\frac{\tau_{y,\,\text{Mises}}}{\tau_{y,\,\text{Tresca}}} = \frac{Y/\sqrt{3}}{Y/2} = \frac{2}{\sqrt{3}}. \tag{19.3.22}$$

3. In general three-dimensional formulations of theories of plasticity, there are some mathematical complications associated with theories based on the Tresca yield criterion—here one typically has to solve an eigenvalue problem to calculate the principal stresses, and thereby the maximum shear stress. On the other hand, to calculate the equivalent shear stress or the equivalent tensile stress, one need not perform the eigenvalue calculation for the principal stresses. It is for this reason that the mathematically more tractable theories of plasticity are based on the Mises yield condition.

- *Henceforth, when considering the yield of polycrystalline metallic materials, we shall limit our discussion to the Mises yield condition* (19.3.15).

EXAMPLE 19.2 Yield due to Tresca condition

Consider the one-parameter state of stress from Example 19.1,

$$[\sigma(\zeta)] = \begin{bmatrix} a\zeta & 0 & b\zeta \\ 0 & 0 & 0 \\ b\zeta & 0 & 0 \end{bmatrix}. \tag{19.3.23}$$

To determine the elastic limit of the load parameter based on the Tresca criteria we first need to compute the principal stresses from the characteristic equation for (19.3.23):

$$\det \begin{bmatrix} a\zeta - \lambda & 0 & b\zeta \\ 0 & -\lambda & 0 \\ b\zeta & 0 & -\lambda \end{bmatrix} = (a\zeta - \lambda)\lambda^2 + b^2\zeta^2\lambda = 0. \tag{19.3.24}$$

Solving for λ, gives $\lambda = \{0, a/2 \pm \sqrt{a^2 + 4b^2}\zeta/2\}$. From (19.3.21), one finds

$$(\sigma_1 - \sigma_3) = \sqrt{a^2 + 4b^2}|\zeta| \le Y;$$

hence one has the load limit:

$$|\zeta| \le \frac{Y}{\sqrt{a^2 + 4b^2}}.$$

Observe that for this state of stress, the Tresca criterion is more limiting than the Mises criterion, cf. (19.3.17).

19.3.3 **Coulomb–Mohr yield criterion for cohesive granular materials in compression**

For a cohesive granular material such as sandstone, or a cohesionless granular material such as dry sand—if they may be idealized to be isotropic—the French engineer and physicist Charles-Augustine Coulomb (Coulomb, 1773), as early as 1773, proposed that under *dominantly compressive stress states*, slip along a plane in such a material occurs when the resolved shear stress τ on the plane exceeds a cohesive shear resistance c plus the compressive normal traction σ multiplied by an internal friction coefficient μ (cf., e.g., Timoshenko, 1953, p. 51).

Later, Otto Mohr (Mohr, 1900), with his graphical representation of stress at a point—*the Mohr's-circle*—put Coulomb's failure criterion into a form which allowed for easier utilization in engineering practice. Mohr, using his graphical representation of stresses, proposed a strength theory based on the assumption *that of all planes having the same magnitude of normal stress the weakest one, on which failure is likely to occur, is that with the maximum shearing stress* (cf., e.g., Timoshenko, 1953, p. 287).

The Coulomb–Mohr yield condition involves two material properties:

(a) $c \geq 0$ the **cohesion** of the material (units of stress) which represents a shear resistance of the material; and

(b) μ the **internal friction coefficient** of the material (dimensionless), with $0 \leq \mu < 1$. The internal friction may also be represented by an angle ϕ called the **internal friction angle**, $0 \leq \phi < (\pi/2)$, such that

$$\mu = \tan\phi. \tag{19.3.25}$$

Using these properties and following Mohr and Coulomb's arguments, an involved computation leads to a yield/failure criterion in the following form:

$$\boxed{\tfrac{1}{2}(\sigma_1 - \sigma_3) + \tfrac{1}{2}(\sigma_1 + \sigma_3)\sin\phi - c\cos\phi \leq 0.} \tag{19.3.26}$$

- *Equation (19.3.26) is the widely-used yield/failure criterion due to Coulomb (1773) and Mohr (1900). It is characterized by two material parameters, namely the cohesion c and the angle of internal friction ϕ. This yield/failure criterion is often used to describe the strength characteristics of cohesive granular materials like soils—and often also rocks, concrete, and ceramics—under dominantly **compressive stress states.***[4]

Note that when $\phi = 0$, that is when there is no internal friction, the Coulomb–Mohr criterion (19.3.26) reduces to the Tresca criterion for metals

$$\tfrac{1}{2}(\sigma_1 - \sigma_3) - c \leq 0. \tag{19.3.27}$$

Recalling (19.3.19), the cohesion c for non-frictional materials is the same as the Tresca yield strength in shear, $\tau_{y,\text{Tresca}}$ (cf. eq. (19.3.19) with $c \equiv \tau_{y,\text{Tresca}}$).

19.3.4 The Drucker–Prager yield criterion

This pressure-sensitive yield criterion was proposed for soils by Daniel Drucker and William Prager (Drucker and Prager, 1952). The Drucker–Prager yield function is

$$f(\bar{\tau}, \bar{p}, S) = \bar{\tau} - (S + \alpha\bar{p}),$$

and the corresponding yield condition is

$$\bar{\tau} - (S + \alpha\bar{p}) \leq 0. \tag{19.3.28}$$

Here, S represents the shear yield strength of the material when $\bar{p} = 0$, and $\alpha \geq 0$ represents a *pressure-sensitivity* parameter, which characterizes the increase in the shear yield strength of the material as the pressure increases. The intended validity of this criterion is limited to stress states for which $(S + \alpha\bar{p}) > 0$.

[4] Mohr considered failure in a broad sense—it could be yielding of materials like soils or fracture of materials like sandstone.

REMARKS 19.3

1. Although proposed for soils, this criterion is also widely used for polymeric materials which exhibit pressure-sensitivity of plastic flow.
2. Note that for pressure-sensitive materials, the Drucker–Prager yield condition is phrased in terms of the invariants $\bar{\tau}$ and $\bar{p}$, while the Coulomb–Mohr yield condition is phrased in terms of the maximum and the minimum principal stresses. The former is easier to use in calculations, and various Drucker–Prager approximations to the Coulomb–Mohr yield condition have been discussed in the literature. Two common approximations are based on choosing the parameters (S, α) in the Drucker–Prager yield condition in terms of the parameters (c, ϕ) in the Coulomb–Mohr yield condition according to either

$$\alpha = \frac{6\sin\phi}{\sqrt{3}(3-\sin\phi)} \quad \text{and} \quad S = c \times \frac{6\cos\phi}{\sqrt{3}(3-\sin\phi)}, \tag{19.3.29}$$

or

$$\alpha = \frac{6\sin\phi}{\sqrt{3}(3+\sin\phi)} \quad \text{and} \quad S = c \times \frac{6\cos\phi}{\sqrt{3}(3+\sin\phi)}. \tag{19.3.30}$$

EXAMPLE 19.3 Pressure sensitive yield in a thick-walled sphere

Consider a thick-walled sphere with internal radius a and external radius b subjected to zero external pressure and an internal pressure p_i. If the material of the sphere is governed by the Drucker–Prager criterion with parameters (S, α), both positive-valued, then the yield pressure for the sphere can be found by evaluating (19.3.28) using the solution for the elastic stresses given in (9.1.20):

$$[\sigma] = p_i \frac{1}{b^3/a^3 - 1} \begin{bmatrix} 1 - b^3/r^3 & 0 & 0 \\ 0 & 1 + b^3/2r^3 & 0 \\ 0 & 0 & 1 + b^3/2r^3 \end{bmatrix}. \tag{19.3.31}$$

The state of stress given in (19.3.31) is in the (r, θ, ϕ) coordinate system and is already in a principal stress coordinate system. Thus using $(19.3.11)_{1,2}$,

$$\bar{\tau} = \frac{\left(\dfrac{\sqrt{3}}{2}\dfrac{b^3}{r^3}\right)}{\left(\dfrac{b^3}{a^3} - 1\right)} |p_i| \geq 0, \tag{19.3.32}$$

and

$$\bar{p} = -\frac{p_i}{\left(\dfrac{b^3}{a^3} - 1\right)}. \tag{19.3.33}$$

Substituting these values in (19.3.28) gives

$$\frac{\left(\dfrac{\sqrt{3}}{2}\dfrac{b^3}{r^3}\right)}{\left(\dfrac{b^3}{a^3}-1\right)}|p_i| \le S - \alpha\frac{p_i}{\left(\dfrac{b^3}{a^3}-1\right)}. \tag{19.3.34}$$

Case 1: For $p_i > 0$, so that $\bar{p} < 0$, the condition (19.3.34) requires that

$$\left(\frac{\sqrt{3}}{2}\frac{b^3}{r^3}+\alpha\right)p_i \le S\left(\frac{b^3}{a^3}-1\right). \tag{19.3.35}$$

The largest value of the left-hand side of (19.3.35) occurs at $r = a$, so yielding will first occur at $r = a$ when

$$\boxed{p_i = S\frac{\left(\dfrac{b^3}{a^3}-1\right)}{\left(\dfrac{\sqrt{3}}{2}\dfrac{b^3}{a^3}+\alpha\right)}.} \tag{19.3.36}$$

Case 2: On the other hand for $p_i < 0$, for example if a vacuum is drawn in the sphere, so that $\bar{p} > 0$, (19.3.34) requires that

$$\beta(r)|p_i| \le S\left(\frac{b^3}{a^3}-1\right), \qquad \text{where} \qquad \beta(r) \stackrel{\text{def}}{=} \left(\frac{\sqrt{3}}{2}\frac{b^3}{r^3}-\alpha\right). \tag{19.3.37}$$

In this case we have three subcases that we need to consider:

Case (i): The first case is

$$\beta(r) > 0 \quad \text{for all } r \in [a, b].$$

In this case yielding will first occur at $r = a$ when,

$$\boxed{|p_i| = S\frac{\left(\dfrac{b^3}{a^3}-1\right)}{\left(\dfrac{\sqrt{3}}{2}\dfrac{b^3}{a^3}-\alpha\right)}.} \tag{19.3.38}$$

Case (ii): The second case is

$$\beta(r) < 0 \quad \text{for all } r \in [a, b].$$

In this case, from (19.3.37) we have

$$\text{negative} \times \text{positive} \le \text{positive},$$

which is always true and then the system *never* yields.

Case (iii): The third case is

$$\beta(r) = 0 \quad \text{for some } r = c, \quad a < c < b.$$

Then for $r > c$, (19.3.37) is

$$\text{negative} \times \text{positive} \leq \text{positive}$$

which is always satisfied, and that portion of the sphere does not yield. But for $r < c$, there is the possibility for yield; the critical point occurs at $r = a$. The result is then (19.3.38).

REMARK 19.4

Typical values of α lie in the range $0 \leq \alpha \lesssim 0.66$, while since $a < r < b$ we have

$$\frac{\sqrt{3}}{2}\frac{b^3}{r^3} \geq \frac{\sqrt{3}}{2} = 0.866.$$

Hence for realistic values of the pressure-sensitivity parameter α we expect that

$$\beta(r) \ > 0 \quad \text{for all } r \in [a, b],$$

and the subcase (i) is the only one that we might actually encounter physically should $p_i < 0$.

---

20 One-dimensional plasticity

The theory of linear elasticity furnishes a simple vehicle for the discussion of basic ideas of solid mechanics, and it finds widespread use in modeling the elastic response of engineering components and structures. However, linear elasticity theory can be applied to the description of materials only for relatively small strains, typically $\lesssim 10^{-3}$. Larger deformations in metals lead to plastic flow, permanent set, hysteresis, and other interesting and important phenomena that fall naturally within the purview of plasticity, a topic that we introduce in this chapter.

20.1 Some phenomenological aspects of the elastic-plastic stress-strain response of polycrystalline metals

We begin by discussing some characteristic aspects of the phenomenological elastic-plastic response of metals which are revealed in tension tests on such materials. A stress-strain curve obtained from a simple tension test shows the major features of the elastic-plastic response of a polycrystalline metal. In such an experiment, the length L_0 and cross-sectional area A_0 of a cylindrical specimen are deformed to L and A, respectively. If P denotes the axial force required to effect such a deformation, then the axial *engineering stress* in the specimen is

$$s = \frac{P}{A_0}.$$

In addition, if $\lambda = L/L_0$ denotes the axial stretch, the corresponding axial *engineering strain* is

$$e = \lambda - 1.$$

Figure 20.1 shows a schematic of an engineering-stress versus engineering-strain curve for a metallic specimen. In region OB the stress-strain curve is essentially linear, and reversing the direction of strain from any point in the region OB results in a retracing of the forward straining portion of the stress-strain curve; in this range of small strains, the response of the material is typically idealized to be *linearly elastic*. Beyond the point B, the stress-strain curve deviates from linearity; accordingly, the point B is called the **proportional limit** or **elastic limit.** Upon reversing the direction of strain at any stage of deformation beyond B, say the point C, the stress and strain values do not retrace the forward straining portion of the stress-strain curve; instead, the stress is reduced along an *elastic unloading* curve CD. That is, beyond the proportional limit, unloading to zero stress reduces the strain by an amount called the **elastic strain**, e^{e}, and leaves a permanent **plastic strain**, e^{p}; cf. Fig. 20.1. Another reversal of the strain direction from D (reloading), retraces the unloading curve, and the stress-strain curve turns over to approach the monotonic loading curve, at the point C from which the unloading was initiated.

Continuum Mechanics of Solids. Lallit Anand and Sanjay Govindjee, Oxford University Press (2020).

DOI: 10.1093/oso/9780198864721.001.0001

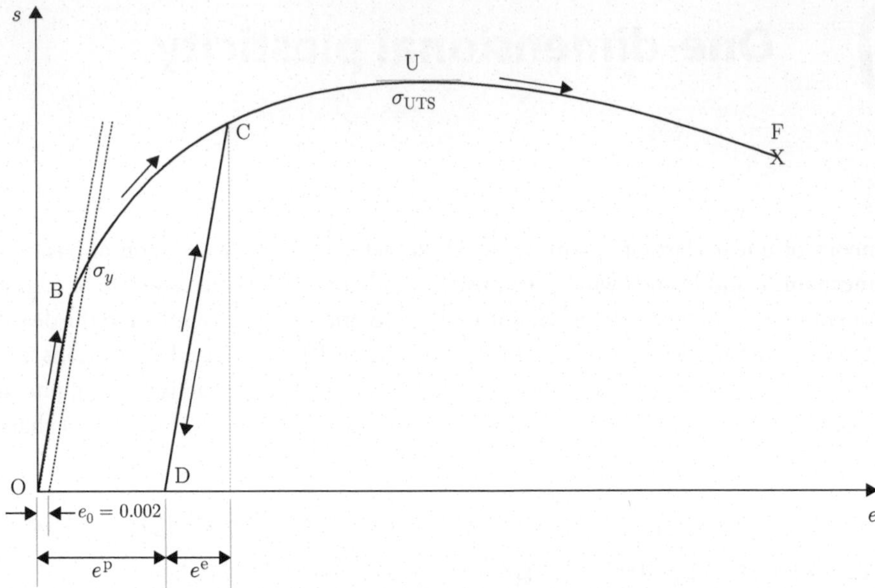

Fig. 20.1 Schematic of an engineering-stress-strain curve for a metallic material.

In theoretical discussions of elastoplasticity, the proportional limit is called the **yield strength** of the material. However, the determination of the onset of deviation from linearity, and hence the proportional limit of the material, is strongly dependent on the sensitivity of the extensometer available to an experimentalist to measure the specimen elongation. In order to circumvent differences in the sensitivity of extensometers available to different experimentalists, it is standard engineering practice to define the yield strength by an offset method: a line parallel to the initial elastic slope is drawn so that it intersects the strain axis at e_0, and the intersection of this line with the stress-strain curve is denoted by σ_y. Usually $e_0 = 0.002$, and in this case σ_y is called the **0.2% offset yield strength** of the material. The 0.2% offset yield strength is more reproducibly measured by different experimentalists than is the proportional limit.

Beyond the point of initial yield, the engineering-stress continues to increase with increasing strain; that is, the specimen is able to withstand a greater axial load despite a reduction in cross-sectional area. This phenomenon is known as **strain-hardening**. At point U on the stress-strain curve the rate of strain-hardening capacity of the material is unable to keep pace with the rate of reduction in the cross-sectional area, and the axial load, and correspondingly the engineering-stress, attains a maximum. The maximum value of the engineering-stress is called the **ultimate tensile strength** of the material, and is denoted by σ_{UTS}. After point U, further deformation occurs with decreasing load (engineering-stress) and a tensile instability phenomenon called **necking** occurs, which typically initiates in a weak cross-section of the specimen. After the neck has initiated, the specimen gauge-length becomes "waisted" in appearance, the axial strain in the specimen is no longer uniform along the gauge-length, and the specimen continues to **neck down** under a complicated and continuously changing triaxial tensile stress state in the necked region, until fracture finally occurs at point F.

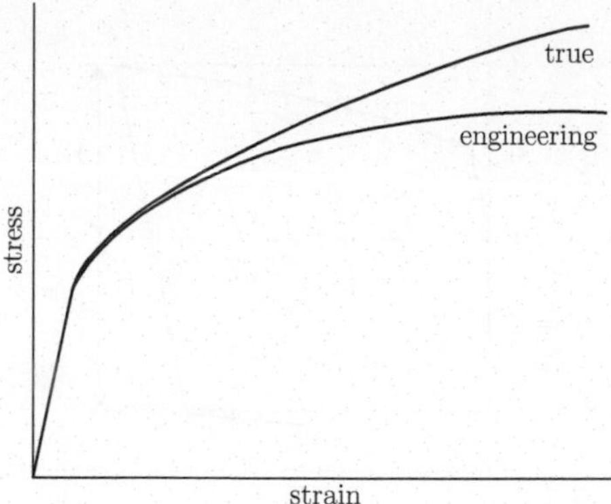

Fig. 20.2 Comparison of a true stress-strain curve with the corresponding engineering stress-strain curve.

The stress-strain response of a material may also be expressed in terms of the *true stress* (i.e., the Cauchy stress), defined by

$$\sigma = \frac{P}{A},$$

and *true strain* (logarithmic, or Hencky, strain), defined by

$$\epsilon = \ln \lambda.$$

The true stress at any point in a tension test may be calculated by taking simultaneous measurements of the load P and current cross-sectional area A of the specimen. However, simultaneous measurements of the axial elongation and the diametrical reduction of a specimen are seldom carried out. Instead, use is made of the experimental observation that plastic flow of metals is essentially **incompressible**—volume change in a tension test is associated only with the elastic response of the material. Thus, for a metallic specimen undergoing plastic deformation that is large in comparison to its elastic response, it is reasonable to assume that the volume of the specimen is conserved, so that $A\,L \approx A_0\,L_0$—provided the deformation is homogeneous, that is for strain levels below the onset of necking at σ_{UTS}. Hence, for any pair of values (s, e), the corresponding pair (σ, ϵ) may be calculated via the relations

$$\sigma = \frac{P}{A} = \frac{P}{A_0}\frac{L}{L_0} = s(1+e), \qquad \epsilon = \ln(1+e).$$

A true stress-strain curve is contrasted with a corresponding engineering stress-strain curve in Fig. 20.2; the differences between the two curves becomes significant after a strain level of a few percent—say 5% or so.

20.1.1 **Isotropic strain-hardening**

The phenomenology of the stress-strain response of metals beyond the elastic limit is quite complicated. Here we discuss a simple idealization of strain-hardening that is frequently used in theories of plasticity. This idealization of actual material response, referred to as **isotropic strain-hardening**, assumes that after reversal of deformation from any level of strain in the plastic regime, the magnitude of the flow stress upon which reverse yielding begins has the

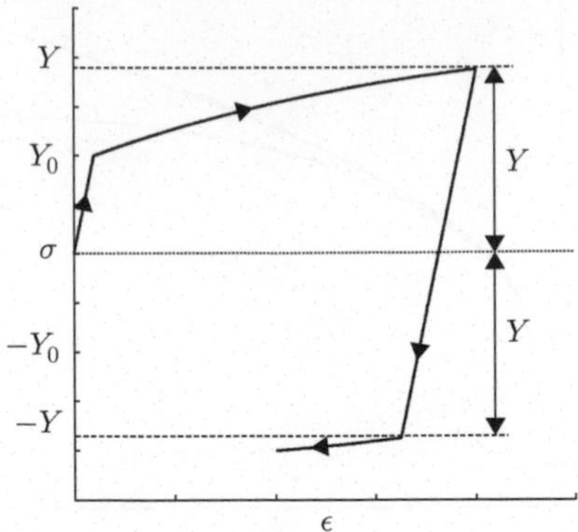

Fig. 20.3 Idealized stress-strain response for an elastic-plastic material with isotropic hardening.

same value as the flow stress from which the unloading was initiated. The stress-strain response corresponding to isotropic strain-hardening is shown schematically in Fig. 20.3.

Let the stress level from which reversed deformation is initiated be denoted by σ_f, and denote the stress at which the stress-strain curve in compression begins to deviate from linearity by $\sigma_r < 0$. The set of endpoints $\{\sigma_r, \sigma_f\}$ of the closed interval $[\sigma_r, \sigma_f]$ is called the **yield set**, and the open interval (σ_r, σ_f) is called the **elastic range**. Let Y, with initial value Y_0, denote the **flow resistance** (a material property) of the material. The initial elastic range then has

$$\sigma_r = -Y_0, \qquad \sigma_f = Y_0,$$

and during subsequent plastic deformation along the hardening curve, the deformation resistance increases from Y_0 to Y due to strain-hardening, and the new elastic range becomes

$$\sigma_r = -Y, \qquad \sigma_f = Y.$$

For such an idealized response of an elastic-plastic material, the magnitude of the stress σ in this elastic range is bounded by the restriction

$$|\sigma| \le Y,$$

generally referred to as a **boundedness inequality**. Plastic flow is possible only when the **yield condition** $|\sigma| = Y$ is satisfied.

The special case corresponding to *no* strain-hardening represents an *elastic-perfectly plastic material.* In this case the boundedness inequality becomes $|\sigma| \le Y_0$, with Y_0 a *constant.*

20.1.2 **Kinematic strain-hardening**

Isotropic strain-hardening is an idealization of the actual hardening behavior of metals. The stress-strain response of metals, and indeed many other engineering materials, under cyclic strains is substantially more complicated than that under monotonic loading. For example, many metals exhibit the *Bauschinger effect*, a phenomenon first observed in metals by Bauschinger (1886). He reported that a metal specimen after receiving a certain amount of axial extension into the plastic range showed a decrease in the magnitude of the flow strength upon subsequent compression.

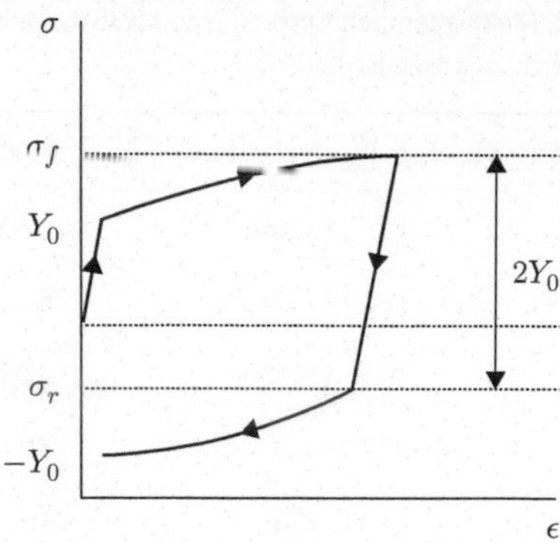

Fig. 20.4 Idealized stress-strain response for an elastic-plastic material with kinematic hardening.

A simple model for strain-hardening that accounts for the Bauschinger effect is referred to as **kinematic hardening.** The stress-strain response for a material with kinematic hardening is shown in Fig. 20.4: the initial flow resistance is Y_0; upon deformation in tension into the plastic regime, the stress increases to σ_f, and, upon reversal of deformation, the material starts to plastically flow at a stress-level $\sigma_r < 0$ and thereafter again continues to harden. The magnitude $|\sigma_r|$ of the stress at which plastic flow recommences upon load reversal is smaller than that at which the reversed loading event was initiated, viz. σ_f, which embodies the *Bauschinger effect*. The asymmetry in the onset of yield upon reversal of loading in such a model arises because of the build-up of an internal stress called the *backstress* and denoted by σ_b, whose magnitude is equal to

$$\sigma_b = \tfrac{1}{2}(\sigma_f + \sigma_r).$$

The term *kinematic hardening* reflects the fact that the center of the *elastic range* in stress-space moves from an initial value of $\sigma = 0$ to a value of $\sigma = \sigma_b$, while the span of the elastic range remains constant at $2Y_0$. This is in contrast to isotropic hardening shown in Fig. 20.3, where the *elastic range* in stress-space stays centered at $\sigma = 0$, while the span of the elastic range expands from $2Y_0$ to $2Y$ as the material strain-hardens. For a material with kinematic hardening as described above, the stress obeys the yield condition

$$|\sigma - \sigma_b| \le Y_0.$$

For many materials the actual strain-hardening behavior—including the Bauschinger effect—may be approximated by a *combination* of isotropic and kinematic hardening.

20.1.3 **Strain-rate and temperature dependence of plastic flow**

Next, physical considerations of the mechanisms of plastic deformation in metals and experimental observations show that plastic flow is both *temperature- and rate-dependent.* Typically,

- the flow resistance increases with increasing strain-rate, and
- it decreases as the temperature increases.

However, the rate-dependence is sufficiently large to merit consideration only at absolute temperatures

$$\vartheta \gtrsim 0.35\,\vartheta_m,$$

Table 20.1 Melting temperatures, ϑ_m, and $0.35\vartheta_m$ values in Kelvin (Celsius) for various metals.

Material	Melting Temp ϑ_m K, (C)	$0.35\,\vartheta_m$ K, (C)
Ti	1941 (1668)	679 (406)
Fe	1809 (1536)	633 (360)
Cu	1356 (1083)	475 (202)
Al	933 (660)	327 (54)
Pb	600 (327)	210 (−63)

where ϑ_m is the melting temperature of the material in degrees absolute. At temperatures

$$\vartheta \lesssim 0.35\,\vartheta_m,$$

the plastic stress-strain response is only slightly rate-sensitive, and in this low homologous temperature regime[1] the plastic stress-strain response of metallic materials is generally assumed to be *rate-independent*.

Table 20.1 shows values of the melting temperatures ϑ_m as well as $0.35\,\vartheta_m$ for some metals in degrees absolute with corresponding centigrade values in parentheses. Thus, the stress-strain response of Ti at room temperature may be idealized as rate-independent, whereas that of Pb cannot, since it will exhibit substantial rate-sensitivity at room temperature.

Attendant manifestations of the strain-rate sensitive stress-strain response of metals are that they exhibit the classical phenomenon of **creep** when subjected a jump-input in stress, and the phenomenon of **stress-relaxation** when subjected to a jump-input in strain. The three major manifestations of time-dependent response of metals: strain-rate sensitive stress response, creep, and stress-relaxation are schematically shown in Fig. 20.5.

In what follows we

- first consider a *one-dimensional theory* of *rate-independent plasticity*, and
- then generalize the one-dimensional theory to the *rate-dependent case.*

20.2 **One-dimensional theory of rate-independent plasticity with isotropic hardening**

One of the central theories of non-linear solid mechanics is a continuum theory describing the elastic-plastic response of metallic materials. Here we develop the central ideas of a rate-independent theory for one-dimensional situations. We do not restrict ourselves to small strains, so that here ϵ represents true or logarithmic strain, and σ the true or Cauchy stress.

[1] The rate sensitivity of the plastic stress-strain response of two materials is (approximately) the same when compared at the same homologous temperature, (ϑ/ϑ_m). Thus the rate sensitivity of the plastic stress-strain response of Ti at 679 K is the about the same as that of Al at 327 K. Note $\vartheta_m = 1941$ K for Ti and $\vartheta_m = 933$ K for Al.

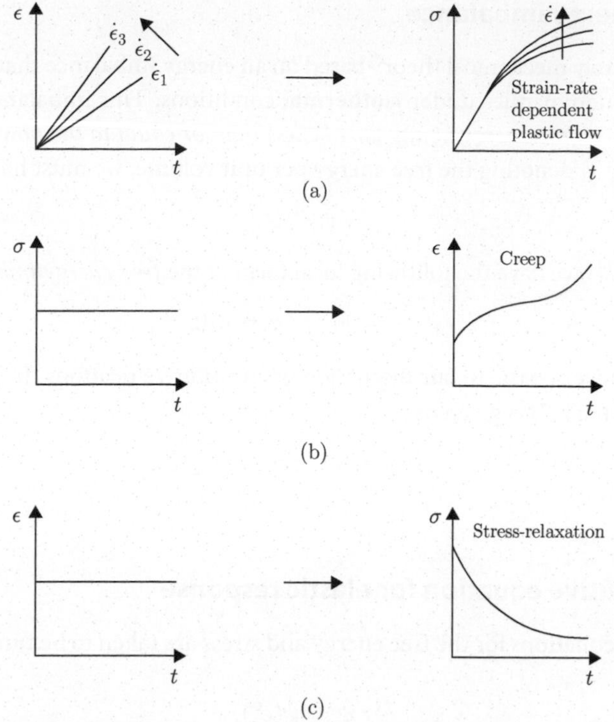

Fig. 20.5 Schematic of: (a) stress response for strain input histories at different strain rates; (b) strain response for a jump-input in stress, creep; (c) stress response for a jump-input in strain, stress-relaxation.

20.2.1 **Kinematics**

We assume that the total strain ϵ may be additively decomposed as

$$\epsilon = \epsilon^{\mathrm{e}} + \epsilon^{\mathrm{p}}, \tag{20.2.1}$$

and call ϵ^{e} and ϵ^{p} the elastic and plastic parts of the strain, respectively. Hence, the strain-rate $\dot{\epsilon}$ also admits the decomposition

$$\dot{\epsilon} = \dot{\epsilon}^{\mathrm{e}} + \dot{\epsilon}^{\mathrm{p}}, \tag{20.2.2}$$

into an elastic part $\dot{\epsilon}^{\mathrm{e}}$, and a plastic part $\dot{\epsilon}^{\mathrm{p}}$.

20.2.2 **Power**

Let σ denote the stress. Then the stress-power per unit volume is

$$\mathcal{P} = \sigma\dot{\epsilon}. \tag{20.2.3}$$

On account of (20.2.2), this too may be decomposed into elastic and plastic parts,

$$\mathcal{P} = \underbrace{\sigma\dot{\epsilon}^{\mathrm{e}}}_{\text{elastic power}} + \underbrace{\sigma\dot{\epsilon}^{\mathrm{p}}}_{\text{plastic power}}. \tag{20.2.4}$$

20.2.3 **Free-energy imbalance**

We consider a purely mechanical theory based on an energy imbalance that represents the first two laws of thermodynamics under **isothermal** conditions. This imbalance requires that *the temporal increase in free energy of any part be less than or equal to the power expended in that part*. That is, with ψ denoting the free energy per unit volume, we must have

$$\dot{\psi} \le \mathcal{P}. \tag{20.2.5}$$

Thus using (20.2.4) we have the following local form of the *free-energy imbalance*:

$$\dot{\psi} - \sigma\dot{\epsilon}^{\mathrm{e}} - \sigma\dot{\epsilon}^{\mathrm{p}} \le 0. \tag{20.2.6}$$

This local imbalance is basic to our discussion of constitutive relations. In particular note that the dissipation (cf. 6.2.7) is given by

$$\mathcal{D} = \sigma\dot{\epsilon}^{\mathrm{e}} + \sigma\dot{\epsilon}^{\mathrm{p}} - \dot{\psi} \ge 0. \tag{20.2.7}$$

20.2.4 **Constitutive equation for elastic response**

The constitutive equations for the free energy and stress are taken to be functions of the elastic strain ϵ^{e} only,

$$\left.\begin{aligned} \psi &= \hat{\psi}(\epsilon^{\mathrm{e}}), \\ \sigma &= \hat{\sigma}(\epsilon^{\mathrm{e}}). \end{aligned}\right\} \tag{20.2.8}$$

The free-energy imbalance (20.2.6) is then equivalent to the requirement that

$$\left[\frac{\partial\hat{\psi}(\epsilon^{\mathrm{e}})}{\partial\epsilon^{\mathrm{e}}} - \hat{\sigma}(\epsilon^{\mathrm{e}})\right]\dot{\epsilon}^{\mathrm{e}} - \sigma\dot{\epsilon}^{\mathrm{p}} \le 0 \tag{20.2.9}$$

hold in all motions of the body. Since this must hold for all motions, in particular it must hold for all elastic motions. Thus the term in the square brackets in (20.2.9) must be zero. This implies that the stress must be given as the derivative of the free energy with respect to the elastic strain,

$$\sigma = \hat{\sigma}(\epsilon^{\mathrm{e}}) = \frac{\partial\hat{\psi}(\epsilon^{\mathrm{e}})}{\partial\epsilon^{\mathrm{e}}}, \tag{20.2.10}$$

and that the dissipation (20.2.7) is given by,

$$\mathcal{D} = \sigma\dot{\epsilon}^{\mathrm{p}} \ge 0. \tag{20.2.11}$$

For metals the elastic strains are typically small, and under these conditions an appropriate free energy function and corresponding relation for stress is

$$\left.\begin{aligned} \psi &= \frac{1}{2}E(\epsilon^{\mathrm{e}})^2, \\ \sigma &= E\,\epsilon^{\mathrm{e}}. \end{aligned}\right\} \tag{20.2.12}$$

Thus, using (20.2.1), the equation for the stress $(20.2.12)_2$ may be written as

$$\sigma = E[\epsilon - \epsilon^{\mathrm{p}}], \tag{20.2.13}$$

and hence

$$\dot{\sigma} = E[\dot{\epsilon} - \dot{\epsilon}^{\mathrm{p}}]. \tag{20.2.14}$$

The basic elastic stress-strain relation (20.2.12) is assumed to hold in all motions of the body, even during plastic flow.

20.2.5 Constitutive equation for the plastic response

We are concerned here in giving an equation that specifies how ϵ^{p} evolves in time, that is an equation for the plastic strain-rate $\dot{\epsilon}^{\mathrm{p}}$. Let

$$n^{\mathrm{p}} \stackrel{\text{def}}{=} \frac{\dot{\epsilon}^{\mathrm{p}}}{|\dot{\epsilon}^{\mathrm{p}}|} \tag{20.2.15}$$

denote the **plastic flow direction,**

$$\dot{\bar{\epsilon}}^{\mathrm{p}} \stackrel{\text{def}}{=} |\dot{\epsilon}^{\mathrm{p}}| \geq 0 \tag{20.2.16}$$

the **equivalent tensile plastic strain-rate**, and

$$\bar{\epsilon}^{\mathrm{p}}(t) = \int_0^t \dot{\bar{\epsilon}}^{\mathrm{p}}(\zeta)\, d\zeta \tag{20.2.17}$$

the **equivalent tensile plastic strain.**

Plastic flow direction

From the dissipation inequality (20.2.11), viz.

$$\mathcal{D} = \sigma\,\dot{\epsilon}^{\mathrm{p}} \geq 0,$$

we note that the "driving force" conjugate to $\dot{\epsilon}^{\mathrm{p}}$ is the stress σ. Accordingly, we assume that the **plastic flow and the stress are co-directional.** That is,[2]

$$n^{\mathrm{p}} = \frac{\dot{\epsilon}^{\mathrm{p}}}{|\dot{\epsilon}^{\mathrm{p}}|} = \frac{\sigma}{|\sigma|} = \mathrm{sign}(\sigma). \tag{20.2.18}$$

Hence

$$\dot{\epsilon}^{\mathrm{p}} = \dot{\bar{\epsilon}}^{\mathrm{p}}\, n^{\mathrm{p}}, \quad \text{with} \quad n^{\mathrm{p}} = \mathrm{sign}(\sigma) \quad \text{and} \quad \dot{\bar{\epsilon}}^{\mathrm{p}} \geq 0. \tag{20.2.19}$$

Note that under this constitutive assumption, the dissipation inequality (20.2.11) reduces to

$$\mathcal{D} = |\sigma|\dot{\bar{\epsilon}}^{\mathrm{p}} \geq 0, \tag{20.2.20}$$

and is trivially satisfied since $|\sigma| \geq 0$ and $\dot{\bar{\epsilon}}^{\mathrm{p}} \geq 0$.

---

[2] $\mathrm{sign}(x) \stackrel{\text{def}}{=} x/|x|$.

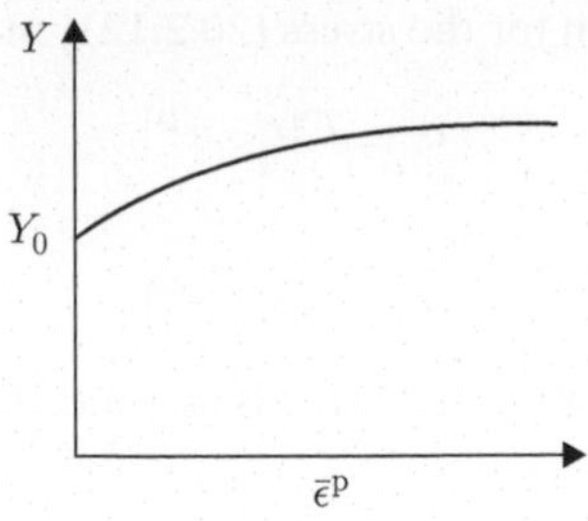

Fig. 20.6 Strain-hardening curve. Variation of the flow strength Y with the equivalent tensile plastic strain $\bar{\epsilon}^{\rm p}$.

Flow strength

Let Y, a positive-valued scalar with dimensions of stress, represent the *resistance to plastic flow* offered by the material; we call it the **flow strength** of the material. We assume that the flow strength is a function of the total plastic strain experienced by the material—independent of its algebraic sign—as represented by the equivalent tensile plastic strain,

$$Y(\bar{\epsilon}^{\rm p}) > 0\,. \tag{20.2.21}$$

Thus the equivalent tensile plastic strain represents a **hardening variable** of the theory, and the function $Y(\bar{\epsilon}^{\rm p})$ characterizes the **strain-hardening** response of the material (Fig. 20.6). The initial value

$$Y_0 \stackrel{\rm def}{=} Y(0) \equiv \sigma_y \tag{20.2.22}$$

represents the initial value of the resistance to plastic flow, and is called the **yield strength** of the material, which is often denoted as σ_y.

Consider the time derivative

$$\dot{\overline{Y(\bar{\epsilon}^{\rm p})}} = H(\bar{\epsilon}^{\rm p})\dot{\bar{\epsilon}}^{\rm p}; \tag{20.2.23}$$

here

$$H(\bar{\epsilon}^{\rm p}) = \frac{dY(\bar{\epsilon}^{\rm p})}{d\bar{\epsilon}^{\rm p}}, \tag{20.2.24}$$

represents the **strain-hardening rate** of the material at a given $\bar{\epsilon}^{\rm p}$, and

- *the material strain hardens or softens accordingly as*

$$H(\bar{\epsilon}^{\rm p}) > 0 \;\text{ or }\; H(\bar{\epsilon}^{\rm p}) < 0. \tag{20.2.25}$$

Yield condition. No-flow conditions. Consistency condition

In the one-dimensional rate-independent theory for isotropic materials, the absolute value of the stress σ cannot be greater than Y; cf. Fig. 20.7, and note that we restrict attention to the isotropic hardening case for now. Accordingly, we assume that there exists a function $f(\sigma, \bar{\epsilon}^{\rm p})$, called the *yield function*, with value

$$f \stackrel{\rm def}{=} |\sigma| - Y(\bar{\epsilon}^{\rm p}) \le 0, \tag{20.2.26}$$

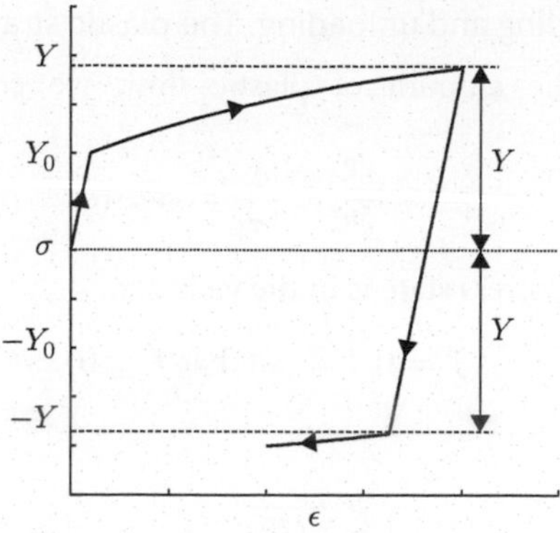

Fig. 20.7 Idealized stress-strain response for an elastic-plastic material with isotropic hardening.

which limits the admissible stresses to lie in the closed interval $[-Y, Y]$. Equation (20.2.26) is called the **yield condition**. The set of endpoints $\{-Y, Y\}$ of the closed interval $[-Y, Y]$ is called the **yield set**, and the *open interval* $(-Y, Y)$ is termed the **elastic range**.

Further, it is assumed that no change in the plastic strain ϵ^p can occur if the stress is in the elastic range. That is,

$$\dot{\epsilon}^p = 0 \qquad \text{whenever} \qquad f < 0. \tag{20.2.27}$$

A change in the plastic strain can only occur if the stress is in the yield set:

$$\dot{\epsilon}^p \neq 0 \quad \text{is possible only if} \quad f = 0. \tag{20.2.28}$$

Consider now a fixed time t and assume that, at that time, $f(t) = 0$, so that the stress state is in the yield set. Then, by (20.2.26), $f(t + \tau) \leq 0$ for all τ and, consequently, $\dot{f}(t) \leq 0$. Thus

$$\boxed{\text{if } f = 0 \text{ then } \dot{f} \leq 0.} \tag{20.2.29}$$

Next, if $f(t) = 0$ and $\dot{f}(t) < 0$, then $f(t + \tau) < 0$ for all sufficiently small $\tau > 0$, so that, by (20.2.27), $\dot{\epsilon}^p(t + \tau) = 0$ for all such τ. Thus it is reasonable to assume $\dot{\epsilon}^p(t) = 0$ in this situation. Expressing this mathematically we have,

$$\text{if } f = 0 \text{ and } \dot{f} < 0, \text{ then } \dot{\epsilon}^p = 0. \tag{20.2.30}$$

The equations (20.2.27) and (20.2.30) combine to form the **no-flow condition**:

$$\boxed{\dot{\epsilon}^p = 0 \text{ if } f < 0 \text{ or if } f = 0 \text{ and } \dot{f} < 0.} \tag{20.2.31}$$

Hence there is no plastic flow when the state of stress is in the elastic range, and there is no plastic flow when the state of stress is in the yield set but is simultaneously moving into the elastic range (a condition known as unloading).

In order for $\dot{\epsilon}^p \neq 0$ at any time, (20.2.28) requires $f = 0$; taken together with (20.2.29) and (20.2.31), this implies that the only possibility for plastic flow is for $f = 0$ *and* $\dot{f} = 0$. This gives the **consistency condition** for plastic flow:

$$\boxed{\text{if } \dot{\epsilon}^p \neq 0, \text{ then } f = 0 \text{ and } \dot{f} = 0.} \tag{20.2.32}$$

That is, plastic flow can only occur if the state of stress *persists* in the yield set.

Conditions describing loading and unloading. The plastic strain-rate $\dot{\epsilon}^{\mathrm{p}}$ in terms of $\dot{\epsilon}$ and σ

In order to determine the amount of plastic flow, we consider the co-directionality assumption:

$$n^{\mathrm{p}} = \frac{\dot{\epsilon}^{\mathrm{p}}}{\dot{\bar{\epsilon}}^{\mathrm{p}}} = \frac{\sigma}{|\sigma|} = \mathrm{sign}\,(\sigma). \tag{20.2.33}$$

Further, we assume that the stress state is in the yield set:

$$f = 0 \qquad \text{(so that } \dot{f} \leq 0\text{)}. \tag{20.2.34}$$

Then,

$$\begin{aligned}
\dot{f} &= \dot{\overline{|\sigma|}} - \dot{\overline{Y(\bar{\epsilon}^{\mathrm{p}})}}, \\
&= \left(\frac{\sigma}{|\sigma|}\right)\dot{\sigma} - H(\bar{\epsilon}^{\mathrm{p}})\dot{\bar{\epsilon}}^{\mathrm{p}}, \\
&= n^{\mathrm{p}}\,\dot{\sigma} - H(\bar{\epsilon}^{\mathrm{p}})\dot{\bar{\epsilon}}^{\mathrm{p}}, \\
&= n^{\mathrm{p}}\,(E[\dot{\epsilon} - \dot{\epsilon}^{\mathrm{p}}]) - H(\bar{\epsilon}^{\mathrm{p}})\dot{\bar{\epsilon}}^{\mathrm{p}}
\end{aligned}$$

or, since $n^{\mathrm{p}}\,\dot{\epsilon}^{\mathrm{p}} = (\dot{\epsilon}^{\mathrm{p}})^2/\dot{\bar{\epsilon}}^{\mathrm{p}} = \dot{\bar{\epsilon}}^{\mathrm{p}}$,

$$\dot{f} = E\,n^{\mathrm{p}}\dot{\epsilon} - [E + H(\bar{\epsilon}^{\mathrm{p}})]\,\dot{\bar{\epsilon}}^{\mathrm{p}}. \tag{20.2.35}$$

To simplify the presentation we will henceforth restrict our attention to the (large) class of materials for which

$$(E + H(\bar{\epsilon}^{\mathrm{p}})) > 0. \tag{20.2.36}$$

Equation (20.2.36) always holds for strain-hardening materials, those with $H(\bar{\epsilon}^{\mathrm{p}}) > 0$ for all $\bar{\epsilon}^{\mathrm{p}}$. In contrast strain-softening materials satisfy $H(\bar{\epsilon}^{\mathrm{p}}) < 0$; in this case the assumption (20.2.36) places a restriction on the maximum amount of allowable softening, as it requires that $H(\bar{\epsilon}^{\mathrm{p}}) > -E$ for all $\bar{\epsilon}^{\mathrm{p}}$.

We now consider conditions which determine whether a material leaves or remains in the yield set when the material point in question is subjected to a loading program characterized by a given strain-rate $\dot{\epsilon}$:

(i) **Elastic unloading** is defined by $n^{\mathrm{p}}\dot{\epsilon} < 0$. In this case, since $\dot{\bar{\epsilon}}^{\mathrm{p}} \geq 0$, (20.2.35) implies that $\dot{f} < 0$, and the no-flow conditions (20.2.31) imply that $\dot{\epsilon}^{\mathrm{p}} = 0$.

(ii) **Plastic loading** is defined by $n^{\mathrm{p}}\dot{\epsilon} > 0$. In this case, if $\dot{\bar{\epsilon}}^{\mathrm{p}} = 0$ then $\dot{f} > 0$, which violates $\dot{f} \leq 0$. Hence $\dot{\bar{\epsilon}}^{\mathrm{p}} > 0$, and since the consistency condition (20.2.32) requires that $\dot{f} = 0$, (20.2.35) gives

$$\dot{\bar{\epsilon}}^{\mathrm{p}} = \left(\frac{E}{E + H(\bar{\epsilon}^{\mathrm{p}})}\right) n^{\mathrm{p}}\dot{\epsilon}. \tag{20.2.37}$$

Thus, since $\dot{\epsilon}^{\mathrm{p}} = \dot{\bar{\epsilon}}^{\mathrm{p}}\,n^{\mathrm{p}}$ and $n^{\mathrm{p}}\,n^{\mathrm{p}} = 1$,

$$\dot{\epsilon}^{\mathrm{p}} = \beta\dot{\epsilon} \neq 0, \tag{20.2.38}$$

where

$$\beta \stackrel{\text{def}}{=} \left(\frac{E}{E + H(\bar{\epsilon}^{\mathrm{p}})} \right). \tag{20.2.39}$$

In this case $\dot{\epsilon}^{\mathrm{p}}$ is fully determined in terms of $\dot{\epsilon}$, the Young's modulus E, and the current strain-hardening rate, or *hardening modulus*, $H(\bar{\epsilon}^{\mathrm{p}})$.

Combining the results of the foregoing discussion, we arrive at an equation for the plastic strain-rate that holds for all time, no matter whether $f = 0$ or $f < 0$:

$$\dot{\epsilon}^{\mathrm{p}} = \begin{cases} 0 & \text{if } f < 0 \quad \text{(behavior in the elastic range),} \\ 0 & \text{if } f = 0 \quad \text{and} \quad n^{\mathrm{p}}\,\dot{\epsilon} < 0 \quad \text{(elastic unloading),} \\ \beta\dot{\epsilon} & \text{if } f = 0 \quad \text{and} \quad n^{\mathrm{p}}\,\dot{\epsilon} > 0 \quad \text{(plastic loading),} \end{cases} \tag{20.2.40}$$

where

$$n^{\mathrm{p}} = \frac{\dot{\epsilon}^{\mathrm{p}}}{\dot{\bar{\epsilon}}^{\mathrm{p}}} = \frac{\sigma}{|\sigma|} = \operatorname{sign}\sigma, \tag{20.2.41}$$

and β is given by (20.2.39).

20.2.6 **Summary of the rate-independent theory with isotropic hardening**

The complete set of constitutive equations consists of:

- A kinematic decomposition

$$\epsilon = \epsilon^{\mathrm{e}} + \epsilon^{\mathrm{p}}, \tag{20.2.42}$$

of the strain ϵ into elastic and plastic strains, ϵ^{e} and ϵ^{p}.

- An elastic stress-strain relation,

$$\sigma = E\epsilon^{\mathrm{e}} = E(\epsilon - \epsilon^{\mathrm{p}}). \tag{20.2.43}$$

- A yield condition,

$$f = |\sigma| - Y(\bar{\epsilon}^{\mathrm{p}}) \le 0, \tag{20.2.44}$$

where f is the yield function, with $Y(\bar{\epsilon}^{\mathrm{p}})$ the flow strength.

- An evolution equation,

$$\dot{\epsilon}^{\mathrm{p}} = \chi\,\beta\dot{\epsilon}, \tag{20.2.45}$$

for the plastic strain. Here, β is the ratio defined by

$$\beta = \frac{E}{E + H(\bar{\epsilon}^{\mathrm{p}})} > 0 \quad \text{(by hypothesis);}$$

χ is a *switching function* defined by

$$\chi = \begin{cases} 0 & \text{if } f < 0, \text{ or if } f = 0 \text{ and } n^{\mathrm{p}}\,\dot{\epsilon} < 0, \qquad \text{where} \qquad n^{\mathrm{p}} = \operatorname{sign}(\sigma), \\ 1 & \text{if } f = 0 \quad \text{and} \quad n^{\mathrm{p}}\,\dot{\epsilon} > 0, \end{cases} \tag{20.2.46}$$

and $H(\bar{\epsilon}^p)$ is the strain-hardening rate defined by

$$H(\bar{\epsilon}^p) = \frac{dY(\bar{\epsilon}^p)}{d\bar{\epsilon}^p}. \qquad (20.2.47)$$

The evolution equation (20.2.45) for ϵ^p is referred to as the **flow rule.**

Note that by (20.2.43)

$$\dot{\sigma} = E[\dot{\epsilon} - \dot{\epsilon}^p],$$

so that using the flow rule (20.2.45) we arrive at *an evolution equation for the stress*:

$$\dot{\sigma} = E[1 - \chi\beta]\dot{\epsilon},$$

which may be written as

$$\dot{\sigma} = E_{\text{tan}}\,\dot{\epsilon}, \qquad (20.2.48)$$

with E_{tan} a *tangent modulus* defined by

$$E_{\text{tan}} = \begin{cases} E & \text{if } \chi = 0, \\ \dfrac{E\,H(\bar{\epsilon}^p)}{E + H(\bar{\epsilon}^p)} & \text{if } \chi = 1. \end{cases} \qquad (20.2.49)$$

20.2.7 Material parameters in the rate-independent theory

To complete the rate-independent constitutive model for a given material, the material properties/functions that need to be determined are:

1. The Young's modulus E.
2. The **flow strength**

   $$Y = Y(\bar{\epsilon}^p) > 0,$$

   with initial value

   $$Y_0 = Y(0) = \sigma_y,$$

   called the **yield strength.**

The yield strength σ_y is determined by the 0.2% plastic strain offset method from a true-stress σ versus true-strain ϵ curve. The complete flow strength curve $Y(\bar{\epsilon}^p)$ is determined as follows:

- Assume that E and the true stress-strain data (σ versus ϵ) have been obtained from a tension or a compression test.
- Then using $\epsilon^p = \epsilon - (\sigma/E)$ the σ versus ϵ data is converted into σ versus ϵ^p data.

  If the data is obtained from a compression test, then convert the data into $|\sigma|$ versus $|\epsilon^p| \equiv \bar{\epsilon}^p$.
- Since $|\sigma| = Y$ during plastic flow, the $|\sigma|$ versus $\bar{\epsilon}^p$ data is **identical** to Y versus $\bar{\epsilon}^p$ data.

Some useful phenomenological forms for the strain-hardening function

1. A widely-used special functional form is the power-law hardening function,

$$Y(\bar{\epsilon}^p) = Y_0 + K(\bar{\epsilon}^p)^n, \tag{20.2.50}$$

with Y_0, K, and n constants.

2. So far we have taken strain-hardening to be characterized directly in the form of a function $Y = Y(\bar{\epsilon}^p)$. At a more fundamental level it is useful to think of Y as *an internal variable* which characterizes the *rate-independent* resistance to plastic flow offered by the material, and presume that Y evolves according to a differential evolution equation of the form

$$\dot{Y} = H(Y)\dot{\bar{\epsilon}}^p \qquad \text{with initial value} \qquad Y(0) = Y_0. \tag{20.2.51}$$

A useful form which fits experimental data for metals reasonably well is

$$\dot{Y} = H\dot{\bar{\epsilon}}^p, \qquad H = H_0\Big(1 - \frac{Y}{Y_s}\Big)^r, \quad Y(0) = Y_0, \tag{20.2.52}$$

with (Y_0, Y_s, H_0, r) constants. The parameter Y_0 represents the initial value of Y, and Y_s represents a *saturation* value at large strains; the parameters H_0 and r control the manner in which Y increases from its initial value Y_0 to its saturation value Y_s.

The integrated form of (20.2.52) is

$$Y(\bar{\epsilon}^p) = Y_s - \left[(Y_s - Y_0)^{(1-r)} + (r-1)\left(\frac{H_0}{(Y_s)^r}\right)\bar{\epsilon}^p\right]^{\dfrac{1}{1-r}} \quad \text{for} \quad r \neq 1. \tag{20.2.53}$$

For $r = 1$ the integrated form of (20.2.52) leads to the classical Voce equation:

$$Y(\bar{\epsilon}^p) = Y_s - (Y_s - Y_0)\exp\left(-\frac{H_0}{Y_s}\bar{\epsilon}^p\right), \tag{20.2.54}$$

which is sometimes known as the exponential saturation hardening model.

EXAMPLE 20.1 Necking in a bar

Consider a cylindrical bar of a material with initial gauge length L_0 and cross-sectional area A_0. The bar is deformed so that gauge length increases to L, the cross-sectional area reduces to A, and the corresponding axial load is P. With $s = P/A_0$ denoting the *engineering* stress, $\sigma = P/A$ denoting the *true* stress, $\lambda = L/L_0$ denoting the axial stretch, and $\epsilon = \ln\lambda$ denoting the *true* or logarithmic strain, consider a non-linear, *incompressible* material, whose stress-strain response under monotonic deformation is described by

$$\sigma = \hat{\sigma}(\epsilon). \tag{20.2.55}$$

Such a "non-linear" material model would approximate the response of a strain-hardening elastic-plastic material under monotonic straining, when the typically small elastic strain is neglected in comparison to the large plastic strain. In terms of the stress σ and the current cross-sectional area A, the load is

$$P = \hat{\sigma}(\epsilon)A, \tag{20.2.56}$$

and for an incompressible material

$$A = A_0\frac{L_0}{L} = \frac{A_0}{\lambda} = A_0 \exp(-\epsilon). \tag{20.2.57}$$

Using eq. (20.2.57) in (20.2.56) and using the definition $s = P/A_0$, we obtain

$$s = \hat{\sigma}(\epsilon) \exp(-\epsilon). \tag{20.2.58}$$

In 1885 Considére argued that a "necking instability" will set in:

- if the engineering stress s versus engineering strain e curve exhibits a maximum (Considére, 1885).

Thus, from (20.2.58),

$$\begin{aligned}\frac{ds}{de} &= \frac{ds}{d\epsilon}\frac{d\epsilon}{de}\\ &= \left[\frac{d\hat{\sigma}(\epsilon)}{d\epsilon}\exp(-\epsilon) - \hat{\sigma}(\epsilon)\exp(-\epsilon)\right]\left[\frac{1}{1+e}\right]\\ &= \left[\frac{d\hat{\sigma}(\epsilon)}{d\epsilon} - \hat{\sigma}(\epsilon)\right]\left[\frac{\exp(-\epsilon)}{1+e}\right];\end{aligned}$$

hence a maximum value of the s versus e curve occurs if

$$\boxed{\frac{d\hat{\sigma}(\epsilon)}{d\epsilon} = \hat{\sigma}(\epsilon).} \tag{20.2.59}$$

That is,

- *necking in an axial bar in tension will occur if the rate of strain-hardening* $d\hat{\sigma}(\epsilon)/d\epsilon$ *decreases to a value equal to the current value of the stress* $\hat{\sigma}(\epsilon)$.

This known as the Considére condition for necking.

Consider a thought experiment in which there exists a geometric imperfection in the bar such that there is a locally reduced cross-section—a cross-section where a "potential neck" might form—and consider such a bar to be subjected to an increment in strain:

- If the material has sufficient strain-hardening capacity so that the increase in the stress-carrying capacity due to strain-hardening is larger than the increase in stress that the locally imperfect region has to withstand because of the decrease of the area, then the bar will respond in a "stable fashion"—the engineering stress s (or the load P) will increase, and the local geometric imperfection will not grow.
- On the other hand if the rate of strain-hardening decreases to a point such that increase in stress-carrying capacity is smaller than the increase in stress due to the reduction in the area, then the bar will be "unstable," the engineering stress s (or the

load P) will decrease, and the neck will grow—that is, the local cross-sectional area will further decrease.

Thus, for a material which shows a gradually decreasing strain-hardening capacity, a "necking instability" in a bar of the material will occur when the rate of strain-hardening $d\sigma/d\epsilon$ is just equal to the current value of the stress σ. If under continued straining the rate of strain-hardening decreases further, then the extra stress in the neck due to the reduction in the cross-sectional area cannot be offset by the increase in stress-carrying capacity due to strain-hardening, and the neck will grow and the engineering stress s (or the load P) will continue to decrease.

For example, the monotonic stress-strain response of a metal undergoing large plastic deformation (negligible elastic strain strains so that $\epsilon = \epsilon^{\mathrm{p}}$) is often fit to a power-law expression of the form

$$\sigma = K\epsilon^n, \tag{20.2.60}$$

where K is called a *strength coefficient* and n the *strain-hardening exponent*. In this case,

$$\frac{d\sigma}{d\epsilon} = nK\epsilon^{n-1} = n\frac{\sigma}{\epsilon},$$

so the Considére condition $\dfrac{d\sigma}{d\epsilon} = \sigma$ is met when

$$\boxed{\epsilon = n.} \tag{20.2.61}$$

20.3 One-dimensional rate-dependent plasticity with isotropic hardening

In what follows, we consider a generalization of the rate-independent theory with isotropic hardening summarized in Sec. 20.2.6 to model rate-dependent plasticity. Rate-dependent plasticity is also known as **viscoplasticity**. Here the kinematical decomposition (20.2.42) and the constitutive equation for the stress (20.2.43) are identical to that in the rate-independent theory. However, the flow rule, that is the evolution equation for ϵ^{p}, is formulated differently in order to model the effect of strain-rate on the response.

As before, from the dissipation inequality (20.2.11) we note that the "driving force" conjugate to $\dot{\epsilon}^{\mathrm{p}}$ is the stress σ, and as before we assume that the *plastic flow and the stress are* **co-directional**. That is,

$$n^{\mathrm{p}} = \frac{\dot{\epsilon}^{\mathrm{p}}}{|\dot{\epsilon}^{\mathrm{p}}|} = \frac{\sigma}{|\sigma|} = \mathrm{sign}(\sigma). \tag{20.3.1}$$

Hence, recalling $\dot{\bar{\epsilon}}^{\mathrm{p}} \stackrel{\mathrm{def}}{=} |\dot{\epsilon}^{\mathrm{p}}|$,

$$\dot{\epsilon}^{\mathrm{p}} = \dot{\bar{\epsilon}}^{\mathrm{p}}\, n^{\mathrm{p}}, \quad \text{with} \quad n^{\mathrm{p}} = \mathrm{sign}(\sigma) \quad \text{and} \quad \dot{\bar{\epsilon}}^{\mathrm{p}} \geq 0. \tag{20.3.2}$$

Note that under this constitutive assumption, the dissipation inequality (20.2.11) reduces to

$$\mathcal{D} = |\sigma|\dot{\bar{\epsilon}}^{\mathrm{p}} \geq 0, \tag{20.3.3}$$

and is trivially satisfied since $|\sigma| \geq 0$ and $\dot{\bar{\epsilon}}^{\mathrm{p}} \geq 0$.

Next, in view of (20.3.3) we assume that whenever there is plastic flow, that is whenever $\dot{\bar{\epsilon}}^{\mathrm{p}} \neq 0$, the magnitude of the stress is constrained to satisfy a rate-dependent expression,

$$|\sigma| = \underbrace{\mathcal{S}(\bar{\epsilon}^{\mathrm{p}}, \dot{\bar{\epsilon}}^{\mathrm{p}})}_{\text{flow strength}}, \tag{20.3.4}$$

where the function $\mathcal{S}(\bar{\epsilon}^{\mathrm{p}}, \dot{\bar{\epsilon}}^{\mathrm{p}}) \geq 0$ represents the **flow strength** of the material at a given $\bar{\epsilon}^{\mathrm{p}}$ and $\dot{\bar{\epsilon}}^{\mathrm{p}}$.

A special form for the strength relation assumes that the dependence of $\mathcal{S}$ on $\bar{\epsilon}^{\mathrm{p}}$ and $\dot{\bar{\epsilon}}^{\mathrm{p}}$ may be written as a separable relation of the type

$$\mathcal{S}(\bar{\epsilon}^{\mathrm{p}}, \dot{\bar{\epsilon}}^{\mathrm{p}}) = \underbrace{Y(\bar{\epsilon}^{\mathrm{p}})}_{\text{rate-independent}} \times \underbrace{g(\dot{\bar{\epsilon}}^{\mathrm{p}})}_{\text{rate-dependent}}, \tag{20.3.5}$$

with

$$Y(\bar{\epsilon}^{\mathrm{p}}) > 0, \tag{20.3.6}$$

a positive-valued scalar with dimensions of stress, and

- $g(\dot{\bar{\epsilon}}^{\mathrm{p}})$ a positive-valued strictly increasing function of $\dot{\bar{\epsilon}}^{\mathrm{p}}$. (20.3.7)

We refer to $Y(\bar{\epsilon}^{\mathrm{p}})$ as the **strain-hardening function**, and to $g(\dot{\bar{\epsilon}}^{\mathrm{p}})$ as the **rate-sensitivity function.**

Next, by (20.3.7) the function $g(\dot{\bar{\epsilon}}^{\mathrm{p}})$ is *invertible*, and the inverse function $\varphi = g^{-1}$ is strictly increasing and hence non-negative for all non-negative arguments. Hence the general relation (20.3.4), with $\mathcal{S}$ given by (20.3.5), may be inverted to give an expression

$$\dot{\bar{\epsilon}}^{\mathrm{p}} = g^{-1}\left(\frac{|\sigma|}{Y(\bar{\epsilon}^{\mathrm{p}})}\right) \equiv \varphi\left(\frac{|\sigma|}{Y(\bar{\epsilon}^{\mathrm{p}})}\right) \geq 0 \tag{20.3.8}$$

for the equivalent plastic strain-rate. Finally, using (20.3.2) and (20.3.8), we obtain

$$\dot{\epsilon}^{\mathrm{p}} = \varphi\left(\frac{|\sigma|}{Y(\bar{\epsilon}^{\mathrm{p}})}\right) \operatorname{sign}(\sigma). \tag{20.3.9}$$

20.3.1 **Power-law rate dependence**

An example of a commonly used rate-sensitivity function is the **power-law function,**

$$g(\dot{\bar{\epsilon}}^{\mathrm{p}}) = \left(\frac{\dot{\bar{\epsilon}}^{\mathrm{p}}}{\dot{\epsilon}_0}\right)^m, \tag{20.3.10}$$

where $m \in (0, 1]$, a constant, is a **rate-sensitivity parameter** and $\dot{\epsilon}_0 > 0$, also a constant, is a reference flow rate. The power-law function satisfies $g(\dot{\bar{\epsilon}}^{\mathrm{p}}) \approx 1$ for $\dot{\bar{\epsilon}}^{\mathrm{p}} \approx \dot{\epsilon}_0$; (20.3.10) is intended to model plastic flows with rates close to $\dot{\epsilon}_0$. In view of (20.3.10), relation (20.3.4) becomes

$$|\sigma| = Y(\bar{\epsilon}^{\mathrm{p}})\left(\frac{\dot{\bar{\epsilon}}^{\mathrm{p}}}{\dot{\epsilon}_0}\right)^m, \tag{20.3.11}$$

and implies that

$$\ln(|\sigma|) = \ln(Y(\bar{\epsilon}^{\mathrm{p}})) + m \ln\left(\frac{\dot{\bar{\epsilon}}^{\mathrm{p}}}{\dot{\epsilon}_0}\right);$$

thus the rate-sensitivity factor m is the slope of the graph of $\ln(|\sigma|)$ versus $\ln(\dot{\bar{\epsilon}}^{\mathrm{p}}/\dot{\epsilon}_0)$.

The power-law function allows one to characterize *nearly rate-independent behavior*, for which m is very small. Indeed, since

$$\lim_{m \to 0} \left(\frac{\dot{\bar{\epsilon}}^{\text{p}}}{\dot{\epsilon}_0} \right)^m = 1,$$

the limit $m \to 0$ in (20.3.11) corresponds to a *rate-independent yield condition*:

$$|\sigma| = Y(\bar{\epsilon}^{\text{p}}). \tag{20.3.12}$$

On the other hand, the limit $m \to 1$ corresponds to a *linearly viscous response*, as one observes in fluids.

Further, for the power-law function (20.3.10), the expressions (20.3.8) and (20.3.9) have the specific form

$$\dot{\bar{\epsilon}}^{\text{p}} = \dot{\epsilon}_0 \left(\frac{|\sigma|}{Y(\bar{\epsilon}^{\text{p}})} \right)^{\frac{1}{m}} \tag{20.3.13}$$

and

$$\dot{\epsilon}^{\text{p}} = \dot{\epsilon}_0 \left(\frac{|\sigma|}{Y(\bar{\epsilon}^{\text{p}})} \right)^{\frac{1}{m}} \text{sign}(\sigma), \tag{20.3.14}$$

when expressed in terms of the inverse of the power-law function.

A plot of the power-law function eqn. (20.3.14) for three different values of the rate-sensitivity parameter m is shown in Fig. 20.8.

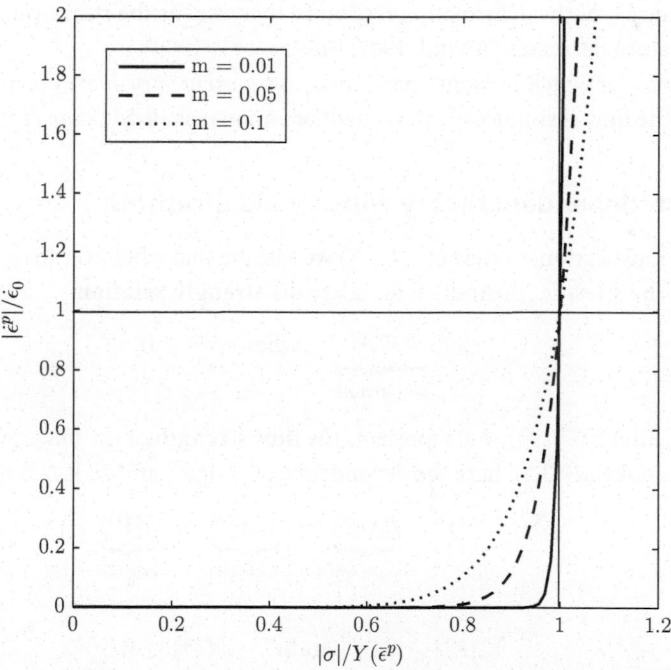

Fig. 20.8 A plot of the power law function (20.3.14) for three different values of the rate-sensitivity parameter m.

20.3.2 **Power-law creep**

Let $\vartheta > 0$ denote a constant temperature. Then for high temperatures $\vartheta \gtrsim 0.5\vartheta_m$, an explicit dependence on ϑ of the viscoplastic strain-rate is accounted for by replacing $\dot{\epsilon}_0$ with

$$\dot{\epsilon}_0 \equiv A \exp\left(-\frac{Q}{R\vartheta}\right), \tag{20.3.15}$$

where A is a pre-exponential factor with units of s^{-1}, Q is an activation energy in units of J/mol, and $R = 8.314$ J/(mol K) is the gas constant. In this case (20.3.14) becomes[3]

$$\dot{\epsilon}^{\mathrm{p}} = A \exp\left(-\frac{Q}{R\vartheta}\right)\left(\frac{|\sigma|}{Y}\right)^{\frac{1}{m}} \mathrm{sign}(\sigma). \tag{20.3.16}$$

Further, the flow resistance Y is taken to evolve according to

$$\begin{aligned} \dot{Y} &= \left[H_0 \left|1 - \frac{Y}{Y_{\mathrm{s}}}\right|^r \mathrm{sign}\left(1 - \frac{Y}{Y_{\mathrm{s}}}\right)\right] |\dot{\epsilon}^{\mathrm{p}}| \quad \text{with initial value} \quad Y_0, \quad \text{and} \\ Y_{\mathrm{s}} &= Y_* \left[\frac{|\dot{\epsilon}^{\mathrm{p}}|}{A \exp\left(-\dfrac{Q}{R\vartheta}\right)}\right]^n, \end{aligned} \tag{20.3.17}$$

where Y_{s} represents the saturation value of Y for a given strain-rate and temperature, and $\{H_0, Y_*, r, n\}$ are strain-hardening parameters. An important feature of the theory is that the saturation value Y_{s} of the deformation resistance increases as the strain-rate $|\dot{\epsilon}^{\mathrm{p}}|$ increases, or as the temperature decreases (Anand, 1982; Brown et al., 1989).

Conditions under which Y evolves are known as *transient or primary* creep, while conditions under which the flow resistance $Y \approx$ constant are known as *steady state creep*.

20.3.3 **A rate-dependent theory with a yield threshold**

As in the previous section, in view of (20.3.3) we assume that whenever there is plastic flow the magnitude of the stress is constrained to satisfy the **strength relation**,

$$|\sigma| = \underbrace{\mathcal{S}(\bar{\epsilon}^{\mathrm{p}}, \dot{\bar{\epsilon}}^{\mathrm{p}})}_{\text{flow strength}} \quad \text{when} \quad \dot{\bar{\epsilon}}^{\mathrm{p}} > 0, \tag{20.3.18}$$

where the function $\mathcal{S}(\bar{\epsilon}^{\mathrm{p}}, \dot{\bar{\epsilon}}^{\mathrm{p}}) \geq 0$ represents the **flow strength** of the material at a given $\bar{\epsilon}^{\mathrm{p}}$ and $\dot{\bar{\epsilon}}^{\mathrm{p}}$. However, unlike (20.3.5), here the dependence of $\mathcal{S}$ on $\bar{\epsilon}^{\mathrm{p}}$ and $\dot{\bar{\epsilon}}^{\mathrm{p}}$ is taken to be given by

$$\mathcal{S}(\bar{\epsilon}^{\mathrm{p}}, \dot{\bar{\epsilon}}^{\mathrm{p}}) = \underbrace{Y_{\mathrm{th}}(\bar{\epsilon}^{\mathrm{p}})}_{\text{rate-independent}} + \underbrace{Y(\bar{\epsilon}^{\mathrm{p}})}_{\text{rate-independent}} \times \underbrace{g(\dot{\bar{\epsilon}}^{\mathrm{p}})}_{\text{rate-dependent}}, \tag{20.3.19}$$

with

$$Y_{\mathrm{th}}(\bar{\epsilon}^{\mathrm{p}}) \geq 0 \quad \text{and} \quad Y(\bar{\epsilon}^{\mathrm{p}}) > 0 \tag{20.3.20}$$

[3] For high temperatures but lower than $\sim 0.5\vartheta_m$, the activation energy is *not constant*, and the temperature dependence is more complicated.

positive-valued scalars with dimensions of stress, and

$$g(\dot{\bar{\epsilon}}^{\mathrm{p}}) \text{ a positive-valued strictly increasing function of } \dot{\bar{\epsilon}}^{\mathrm{p}}. \tag{20.3.21}$$

We refer to:

- $Y_{\mathrm{th}}(\bar{\epsilon}^{\mathrm{p}})$ and $Y(\bar{\epsilon}^{\mathrm{p}})$ as flow resistances, with $Y_{\mathrm{th}}(\bar{\epsilon}^{\mathrm{p}})$ representing a **threshold resistance** to plastic flow;[4] and
- $g(\dot{\bar{\epsilon}}^{\mathrm{p}})$ as the **rate-sensitivity function.**

The strength relation (20.3.18) may then be written as,

$$|\sigma| - Y_{\mathrm{th}}(\bar{\epsilon}^{\mathrm{p}}) = Y(\bar{\epsilon}^{\mathrm{p}})\, g(\dot{\bar{\epsilon}}^{\mathrm{p}}) \quad \text{when} \quad \dot{\bar{\epsilon}}^{\mathrm{p}} > 0. \tag{20.3.22}$$

Since the right-hand side of (20.3.22) is positive-valued, this equation implies that a necessary condition for $\dot{\bar{\epsilon}}^{\mathrm{p}} > 0$ is that,

$$|\sigma| - Y_{\mathrm{th}}(\bar{\epsilon}^{\mathrm{p}}) > 0. \tag{20.3.23}$$

We assume here that this condition is also sufficient for $\dot{\bar{\epsilon}}^{\mathrm{p}} > 0$. This means that plastic flow occurs only when (20.3.23) holds. Equivalently $\dot{\bar{\epsilon}}^{\mathrm{p}} = 0$, and no plastic flow occurs when

$$|\sigma| - Y_{\mathrm{th}}(\bar{\epsilon}^{\mathrm{p}}) < 0. \tag{20.3.24}$$

Relation (20.3.22) may then be inverted to give,

$$\dot{\bar{\epsilon}}^{\mathrm{p}} = \begin{cases} 0 & \text{if } \left(|\sigma| - Y_{\mathrm{th}}(\bar{\epsilon}^{\mathrm{p}})\right) < 0, \\[1ex] \varphi\left(\dfrac{|\sigma| - Y_{\mathrm{th}}(\bar{\epsilon}^{\mathrm{p}})}{Y(\bar{\epsilon}^{\mathrm{p}})}\right) & \text{if } \left(|\sigma| - Y_{\mathrm{th}}(\bar{\epsilon}^{\mathrm{p}})\right) > 0, \end{cases} \tag{20.3.25}$$

where as in the previous section we have written $\varphi(\cdot) \equiv g^{-1}(\cdot)$.

For a simple power-law rate-sensitivity function,

$$g(\dot{\bar{\epsilon}}^{\mathrm{p}}) = \left(\frac{\dot{\bar{\epsilon}}^{\mathrm{p}}}{\dot{\epsilon}_0}\right)^m, \tag{20.3.26}$$

with $m \in (0, 1]$ a constant **rate-sensitivity parameter** and $\dot{\epsilon}_0 > 0$ a constant **reference flow-rate**, the scalar flow equation (20.3.25) may be written as

$$\dot{\bar{\epsilon}}^{\mathrm{p}} = \dot{\epsilon}_0 \left\langle \frac{|\sigma| - Y_{\mathrm{th}}(\bar{\epsilon}^{\mathrm{p}})}{Y(\bar{\epsilon}^{\mathrm{p}})} \right\rangle^{1/m}, \tag{20.3.27}$$

where $\langle\bullet\rangle$ are the Macauley brackets, i.e.,

$$\langle x \rangle = \begin{cases} 0, & x < 0, \\ x, & x \geq 0. \end{cases}$$

Thus, for this rate-sensitive model,

$$\dot{\epsilon}^{\mathrm{p}} = \dot{\epsilon}_0 \left\langle \frac{|\sigma| - Y_{\mathrm{th}}(\bar{\epsilon}^{\mathrm{p}})}{Y(\bar{\epsilon}^{\mathrm{p}})} \right\rangle^{1/m} \operatorname{sign}(\sigma), \tag{20.3.28}$$

---

[4] In the materials science literature, $Y_{\mathrm{th}}(\bar{\epsilon}^{\mathrm{p}})$ is known as the resistance due to the athermal obstacles, and $Y(\bar{\epsilon}^{\mathrm{p}})$ the resistance due to thermally activatable obstacles (Kocks et al., 1975).

and because of the Macauley brackets, the response of such a model is *purely elastic* when $|\sigma| \leq Y_{th}(\bar{\epsilon}^p)$. *Plastic flow only occurs when* $|\sigma| > Y_{th}(\bar{\epsilon}^p)$. That is, $Y_{th}(\bar{\epsilon}^p)$ represents a "yield strength" in this elastic-viscoplastic model.

REMARKS 20.1

1. In the special case $Y_{th} = 0$, plastic flow occurs for all non-zero values of $|\sigma|$, and we recover the model discussed in the previous section.
2. In the special case $Y =$ constant, the flow equation (20.3.28) may be written as

$$\dot{\epsilon}^p = \left\langle \frac{|\sigma| - Y_{th}(\bar{\epsilon}^p)}{\eta^p} \right\rangle^{1/m} \text{sign}(\sigma) \, . \tag{20.3.29}$$

   Here

$$\eta^p \stackrel{\text{def}}{=} \frac{Y}{\dot{\epsilon}_0{}^m}$$

   is a constant, which is sometimes loosely referred to as a "plastic viscosity" parameter — a terminology which is correct only when the rate-sensitivity parameter has a value $m = 1$.
3. A model in which $Y_{th}(\bar{\epsilon}^p) \equiv Y(\bar{\epsilon}^p)$ is also sometimes used in the literature (Cowper and Symonds, 1975).

20.3.4 Summary of the rate-dependent theory with isotropic hardening

To summarize, the constitutive equations in a power-law rate-dependent elastic-viscoplastic theory with a yield threshold and isotropic hardening are given by:

- Elastic-plastic decomposition of ϵ:

$$\epsilon = \epsilon^e + \epsilon^p. \tag{20.3.30}$$

- Free energy function:

$$\psi = \frac{1}{2} E\, (\epsilon^e)^2, \tag{20.3.31}$$

  where $E > 0$ is the Young's modulus.

- Elastic stress-strain relation:

$$\sigma = E[\epsilon^e] = E[\epsilon - \epsilon^p]. \tag{20.3.32}$$

- Flow rule:

$$\dot{\epsilon}^p = \dot{\epsilon}_0 \left\langle \frac{|\sigma| - Y_{th}(\bar{\epsilon}^p)}{Y(\bar{\epsilon}^p)} \right\rangle^{1/m} \text{sign}(\sigma) \, , \tag{20.3.33}$$

where $\dot{\epsilon}_0$ is a reference plastic strain-rate, and $m \in (0, 1]$ is a strain-rate sensitivity parameter. As before,

$$\bar{\epsilon}^{\text{p}}(t) = \int_0^t |\dot{\epsilon}^{\text{p}}(\zeta)|\, d\zeta, \tag{20.3.34}$$

and the positive-valued scalar functions of $\bar{\epsilon}^{\text{p}}$,

$$Y_{\text{th}}(\bar{\epsilon}^{\text{p}}) \geq 0 \qquad \text{and} \qquad Y(\bar{\epsilon}^{\text{p}}) > 0 \tag{20.3.35}$$

represent two flow resistances of the materials, with initial values

$$Y_{\text{th}}(0) = Y_{\text{th}0} \geq 0 \qquad \text{and} \qquad Y(0) = Y_0 > 0. \tag{20.3.36}$$

20.4 One-dimensional rate-independent theory with combined isotropic and kinematic hardening

As discussed in Sec. 20.1.2 many metals display, in addition to isotropic hardening, the Bauschinger effect. The origins of the Bauschinger effect and other related phenomena observed under cyclic straining are attributed to the generation of internal stresses during plastic deformation. The major causes for the generation of such stresses are thought to be due to (i) the plastic-strain incompatibilities at the microstructural level: between grains of different crystallographic orientations in single phase materials, and between grains and hard non-deformable particles in precipitate and dispersion-strengthened materials; and (ii) the polar nature of the dislocation substructures (tangles and cells) which form during forward straining, and partially dissolve upon strain reversals.

There has been an enormous amount of effort over the past 50 years devoted to the development of elastic-plastic constitutive equations for applications to the design of structures under *cyclic straining*. A vast majority of these constitutive models employ a *positive-valued scalar internal variable* to model the isotropic aspects of strain-hardening, and a *strain-like internal variable* to model the directional or "kinematic" aspects of strain-hardening which become pronounced under cyclic conditions. In this section we formulate a one-dimensional rate-independent constitutive theory for combined isotropic and kinematic hardening.

20.4.1 Constitutive equations

The basic equations and the kinematical assumption are as in one-dimensional theory with isotropic hardening, viz.

$$\epsilon = \epsilon^{\text{e}} + \epsilon^{\text{p}}. \tag{20.4.1}$$

Here, we generalize that theory to account for a *free energy associated with plastic flow* resulting in an associated *backstress* and *kinematic hardening*. Specifically, we assume that the free energy has the form

$$\psi = \psi^{\text{e}} + \psi^{\text{p}},$$

with $\psi^{\rm e}$ an elastic energy, and $\psi^{\rm p}$ an energy associated with plastic flow, that is a **defect energy**. As before, we assume that $\dot{\psi}^{\rm e}$ and the elastic power $\sigma\dot{\epsilon}^{\rm e}$ satisfy an *elastic balance*

$$\dot{\psi}^{\rm e} = \sigma\dot{\epsilon}^{\rm e}, \tag{20.4.2}$$

in which case we are left with a *plastic free-energy imbalance*

$$\mathcal{D} = \sigma\dot{\epsilon}^{\rm p} - \dot{\psi}^{\rm p} \geq 0, \tag{20.4.3}$$

which characterizes energy dissipated during plastic flow.

We assume that $\psi^{\rm e}$ and σ are given by the conventional elastic constitutive relations, viz.

$$\psi^{\rm e} = \tfrac{1}{2}E(\epsilon^{\rm e})^2, \qquad \sigma = E\epsilon^{\rm e}, \tag{20.4.4}$$

where $E > 0$ is the Young's modulus of the material. Using (20.4.1), the stress is equivalently given by

$$\sigma = E(\epsilon - \epsilon^{\rm p}). \tag{20.4.5}$$

Then, (20.4.2) is satisfied and we are left with the plastic free-energy imbalance (20.4.3).

Next, as before, let

$$n^{\rm p} = \frac{\dot{\epsilon}^{\rm p}}{|\dot{\epsilon}^{\rm p}|} \tag{20.4.6}$$

denote the **plastic flow direction,**

$$\dot{\bar{\epsilon}}^{\rm p} \stackrel{\rm def}{=} |\dot{\epsilon}^{\rm p}| \geq 0 \tag{20.4.7}$$

the **equivalent tensile plastic strain-rate**, and

$$\bar{\epsilon}^{\rm p}(t) = \int_0^t \dot{\bar{\epsilon}}^{\rm p}(\zeta)\, d\zeta \tag{20.4.8}$$

the **equivalent tensile plastic strain**. Hence,

$$\dot{\epsilon}^{\rm p} = \dot{\bar{\epsilon}}^{\rm p}\, n^{\rm p}, \qquad \text{with} \qquad n^{\rm p} = \frac{\dot{\epsilon}^{\rm p}}{|\dot{\epsilon}^{\rm p}|}, \qquad \text{and} \qquad \dot{\bar{\epsilon}}^{\rm p} \geq 0. \tag{20.4.9}$$

Defect energy

One of the basic ingredients in the choice of constitutive relations describing plastic flow with kinematic hardening is a dimensionless plastic strain-like internal variable,

$$A \geq 0.$$

The field A enters the theory through a constitutive equation for the plastic free energy,

$$\psi^{\rm p} = \hat{\psi}^{\rm p}(A). \tag{20.4.10}$$

For specificity, we assume a simple quadratic form for $\psi^{\rm p}$,

$$\psi^{\rm p} = \frac{1}{2}CA^2, \tag{20.4.11}$$

with $C > 0$ a constant. We assume further that the internal variable A evolves according to the differential equation

$$\dot{A} = \dot{\epsilon}^{\rm p} - \gamma A\dot{\bar{\epsilon}}^{\rm p}, \qquad \text{with initial value} \qquad A = 0, \tag{20.4.12}$$

and $\gamma \geq 0$ a positive constant.

REMARK 20.2

The special case corresponding to $\gamma = 0$ in the evolution (20.4.12) for A is the simplest "kinematic hardening rule", and is variously attributed to Melan (1938), Ishlinskii (1954), and Prager (1956). The case $\gamma > 0$ is the classical *non-linear kinematic hardening* model of Armstrong and Fredrick (1966) with the term $\gamma A \dot{\bar{\epsilon}}^{\mathrm{p}}$ representing a "dynamic recovery" term. The Armstrong–Fredrick model for kinematic hardening is widely used as a simple model to describe cyclic loading phenomena in metals.

From (20.4.11) and (20.4.12) we obtain,

$$\begin{aligned}\dot{\psi}^{\mathrm{p}} &= C A \dot{A}, \\ &= C A \dot{\epsilon}^{\mathrm{p}} - C \gamma A^2 \dot{\bar{\epsilon}}^{\mathrm{p}}, \\ &= \sigma_{\mathrm{back}} \dot{\epsilon}^{\mathrm{p}} - C \gamma A^2 \dot{\bar{\epsilon}}^{\mathrm{p}},\end{aligned} \tag{20.4.13}$$

where we have introduced the notation

$$\sigma_{\mathrm{back}} \stackrel{\mathrm{def}}{=} C\, A, \tag{20.4.14}$$

for a **back stress**—a terminology that shall become clear shortly. In view of (20.4.14), we call C the **back stress modulus.** Note that using (20.4.12) the back stress evolves according to

$$\dot{\sigma}_{\mathrm{back}} = C \dot{\epsilon}^{\mathrm{p}} - \gamma \sigma_{\mathrm{back}} \dot{\bar{\epsilon}}^{\mathrm{p}}. \tag{20.4.15}$$

Substituting $(20.4.13)_3$ in (20.4.3) we obtain,

$$\mathcal{D} = \sigma_{\mathrm{eff}} \dot{\epsilon}^{\mathrm{p}} + C\, \gamma A^2 \dot{\bar{\epsilon}}^{\mathrm{p}} \geq 0, \tag{20.4.16}$$

where we have introduced the notation

$$\sigma_{\mathrm{eff}} \stackrel{\mathrm{def}}{=} \sigma - \sigma_{\mathrm{back}}, \tag{20.4.17}$$

as an **effective stress.** The term $\sigma_{\mathrm{eff}} \dot{\epsilon}^{\mathrm{p}}$ in (20.4.16) may be written as

$$\sigma_{\mathrm{eff}} \dot{\epsilon}^{\mathrm{p}} = \sigma_{\mathrm{eff}} n^{\mathrm{p}} \dot{\bar{\epsilon}}^{\mathrm{p}}. \tag{20.4.18}$$

In view of (20.4.18) the dissipation inequality (20.4.16) may be written as

$$\mathcal{D} = \Big(\sigma_{\mathrm{eff}} n^{\mathrm{p}} + C \gamma A^2 \Big) \dot{\bar{\epsilon}}^{\mathrm{p}} \geq 0. \tag{20.4.19}$$

Plastic flow direction

Guided by the dissipation inequality (20.4.19) we assume that

- *the direction of plastic flow* n^{p} *is* **co-directional** *with the direction of* σ_{eff}*:*

$$n^{\mathrm{p}} = \frac{\sigma_{\mathrm{eff}}}{|\sigma_{\mathrm{eff}}|} \equiv \frac{\sigma - \sigma_{\mathrm{back}}}{|\sigma - \sigma_{\mathrm{back}}|} \qquad \text{whenever} \qquad \dot{\epsilon}^{\mathrm{p}} \neq 0. \tag{20.4.20}$$

Thus, introducing the notation

$$\bar{\sigma}_{\text{eff}} = |\sigma_{\text{eff}}|, \tag{20.4.21}$$

as an **equivalent tensile stress**, under the co-directionality assumption the dissipation (20.4.20) is given by

$$\mathcal{D} = \left(\bar{\sigma}_{\text{eff}} + C\gamma A^2\right)\dot{\bar{\epsilon}}^{\text{p}} \geq 0. \tag{20.4.22}$$

Since all terms appearing in (20.4.22) are either positive or zero, this inequality is trivially satisfied.

Flow resistance

Next, as discussed in the context of (20.2.51) for the theory with isotropic hardening, we introduce a positive-valued stress-dimensioned **flow resistance**,

$$Y(\bar{\epsilon}^{\text{p}}) > 0, \tag{20.4.23}$$

and write

$$H_{\text{iso}}(\bar{\epsilon}^{\text{p}}) \stackrel{\text{def}}{=} \frac{dY(\bar{\epsilon}^{\text{p}})}{d\bar{\epsilon}^{\text{p}}}, \tag{20.4.24}$$

for the rate of change of Y with $\bar{\epsilon}^{\text{p}}$.

Yield function. No-flow conditions. Consistency condition

Next, we introduce a **yield function**

$$f = \bar{\sigma}_{\text{eff}} - Y(\bar{\epsilon}^{\text{p}}), \tag{20.4.25}$$

and assume that the values of $\bar{\sigma}_{\text{eff}}$ are constrained such that,

$$f = \bar{\sigma}_{\text{eff}} - Y(\bar{\epsilon}^{\text{p}}) \leq 0. \tag{20.4.26}$$

Correspondingly, we define the **elastic range** at each $(\bar{\epsilon}^{\text{p}}, A)$ to be the set,

$$\mathcal{E}(\bar{\epsilon}^{\text{p}}, A) \stackrel{\text{def}}{=} \text{the set of all } \sigma \text{ such that } f \leq 0. \tag{20.4.27}$$

Also, we assume that plastic flow is possible when the **yield condition**

$$f = \bar{\sigma}_{\text{eff}} - Y(\bar{\epsilon}^{\text{p}}) = 0 \tag{20.4.28}$$

is satisfied, and define the **yield set** at each $(\bar{\epsilon}^{\text{p}}, A)$ by

$$\mathcal{Y}(\bar{\epsilon}^{\text{p}}, A) \stackrel{\text{def}}{=} \text{the set of all } \sigma \text{ such that } f = 0. \tag{20.4.29}$$

Next, arguing as in our discussion of the theory with isotropic hardening we obtain the **no-flow condition:**

$$\dot{\epsilon}^{\text{p}} = 0 \;\text{ if }\; f < 0 \;\text{ or if }\; f = 0 \;\text{ and }\; \dot{f} < 0. \tag{20.4.30}$$

Hence there is no plastic flow when the state of stress is in the elastic range, and there is no plastic flow when the state of stress is in the yield set but is simultaneously moving into the elastic range.

Also, as in our discussion of the theory with isotropic hardening we obtain the **consistency condition** for plastic flow:

$$\text{if } \dot{\epsilon}^p \neq 0, \text{ then } f = 0 \text{ and } \dot{f} = 0. \tag{20.4.31}$$

That is plastic flow can only occur if the state of stress *persists* in the yield set.

The flow rule

We now show that the no-flow conditions and consistency condition may be used to obtain an equation for $\dot{\epsilon}^p$ in terms of $\dot{\epsilon}$ and σ and σ_{back}. In deriving this relation we use the co-directionality assumption (cf. (20.4.20)):

$$n^p = \frac{\dot{\epsilon}^p}{|\dot{\epsilon}^p|} = \frac{\sigma_{\text{eff}}}{|\sigma_{\text{eff}}|} = \frac{\sigma - \sigma_{\text{back}}}{|\sigma - \sigma_{\text{back}}|}. \tag{20.4.32}$$

First, assume that the stress state is in the yield set,

$$f = 0 \qquad (\text{so that } \dot{f} \leq 0). \tag{20.4.33}$$

Then

$$\begin{aligned}
\dot{f} &= \overline{|\sigma - \sigma_{\text{back}}|}^{\,\cdot} - \overline{Y(\bar{\epsilon}^p)}^{\,\cdot} \\
&= \left(\frac{\sigma - \sigma_{\text{back}}}{|\sigma - \sigma_{\text{back}}|}\right)(\dot{\sigma} - \dot{\sigma}_{\text{back}}) - H_{\text{iso}}(\bar{\epsilon}^p)\dot{\bar{\epsilon}}^p, \\
&= n^p\,\dot{\sigma} - n^p C\,(n^p\dot{\bar{\epsilon}}^p - \gamma A\dot{\bar{\epsilon}}^p) - H_{\text{iso}}(\bar{\epsilon}^p)\dot{\bar{\epsilon}}^p \qquad (\text{by } (20.4.9), (20.4.14), (20.4.32), (20.4.12)), \\
&= n^p\,(E(\dot{\epsilon} - \dot{\bar{\epsilon}}^p n^p)) - [C(1 - \gamma A n^p) + H_{\text{iso}}(\bar{\epsilon}^p)]\dot{\bar{\epsilon}}^p \qquad (\text{by } (20.4.5), (20.4.9)),
\end{aligned}$$

or,

$$\dot{f} = E\,n^p\dot{\epsilon} - [E + H]\,\dot{\bar{\epsilon}}^p, \tag{20.4.34}$$

where we have introduced

$$H \overset{\text{def}}{=} H_{\text{iso}}(\bar{\epsilon}^p) + C\Big(1 - \gamma A n^p\Big) \tag{20.4.35}$$

for an *overall hardening modulus*. As before, we restrict attention to the class of materials for which,

$$(E + H) > 0. \tag{20.4.36}$$

We now deduce conditions that determine whether the material leaves or remains on the yield surface when the material point in question is subjected to a loading program characterized by the strain-rate $\dot{\epsilon}$:

(i) **Elastic unloading** is defined by the condition $n^p\dot{\epsilon} < 0$. In this case (20.4.34) implies that $\dot{f} < 0$, and the no-flow condition (20.4.30) then requires that $\dot{\bar{\epsilon}}^p = 0$.

(ii) **Plastic loading** is defined by the condition $n^p\dot{\epsilon} > 0$. In this case, if $\dot{\bar{\epsilon}}^p = 0$, then $\dot{f} > 0$, which violates $\dot{f} \leq 0$; hence $\dot{\bar{\epsilon}}^p > 0$. The consistency condition (20.4.31) then requires that $\dot{f} = 0$, and (20.4.34) yields

$$\dot{\bar{\epsilon}}^p = \left(\frac{E}{E + H}\right) n^p\dot{\epsilon}. \tag{20.4.37}$$

Thus since, $\dot{\epsilon}^{\rm p} = \dot{\bar{\epsilon}}^{\rm p}\, n^{\rm p}$ and $n^{\rm p} n^{\rm p} = 1$,

$$\dot{\epsilon}^{\rm p} = \beta\dot{\epsilon} \neq 0, \tag{20.4.38}$$

where

$$\beta \stackrel{\rm def}{=} \frac{E}{E+H} > 0. \tag{20.4.39}$$

Combining the results of (i) and (ii) with the condition

$$\dot{\epsilon}^{\rm p} = 0 \qquad \text{if} \qquad f < 0,$$

we arrive at the following (non-smooth) equation for the plastic strain-rate $\dot{\epsilon}^{\rm p}$:

$$\dot{\epsilon}^{\rm p} = \begin{cases} 0 & \text{if } f < 0 \quad \text{(behavior within the elastic range)}, \\ 0 & \text{if } f = 0 \quad \text{and} \quad n^{\rm p}\dot{\epsilon} < 0 \quad \text{(elastic unloading)}, \\ \beta\,\dot{\epsilon} & \text{if } f = 0 \quad \text{and} \quad n^{\rm p}\dot{\epsilon} > 0 \quad \text{(plastic loading)}, \end{cases} \tag{20.4.40}$$

with

$$n^{\rm p} = \frac{\sigma - \sigma_{\rm back}}{|\sigma - \sigma_{\rm back}|} = \operatorname{sign}(\sigma - \sigma_{\rm back}). \tag{20.4.41}$$

20.4.2 Summary of the rate-independent theory with combined isotropic and kinematic hardening

This theory is composed of:

1. A kinematical decomposition:

$$\epsilon = \epsilon^{\rm e} + \epsilon^{\rm p}, \tag{20.4.42}$$

of the strain ϵ into elastic and plastic strains $\epsilon^{\rm e}$ and $\epsilon^{\rm p}$, respectively.

2. A free energy:

$$\psi = \underbrace{\tfrac{1}{2}E(\epsilon^{\rm e})^2}_{\psi^{\rm e}} + \underbrace{\frac{1}{2}\,C\,A^2}_{\psi^{\rm p}}, \tag{20.4.43}$$

Here $E > 0$ is the elastic Young's modulus, and $C > 0$ is a stiffness modulus for the contribution from the plastic-strain-like internal variable A to the free energy.

3. An equation for the stress:

$$\sigma = E\epsilon^{\rm e} = E(\epsilon - \epsilon^{\rm p}). \tag{20.4.44}$$

4. An equation for the back stress:

$$\sigma_{\rm back} = CA. \tag{20.4.45}$$

5. An effective stress, and equivalent tensile stress:

$$\sigma_{\rm eff} \stackrel{\rm def}{=} \sigma - \sigma_{\rm back}, \tag{20.4.46}$$

and

$$\bar{\sigma}_{\rm eff} \stackrel{\rm def}{=} |\sigma_{\rm eff}|. \tag{20.4.47}$$

6. A yield condition:

$$f = \bar{\sigma}_{\rm eff} - Y(\bar{\epsilon}^{\rm p}) \leq 0 \tag{20.4.48}$$

with f the yield function, and $Y(\bar{\epsilon}^{\rm p}) > 0$ an *isotropic flow resistance*.

7. An evolution equation for $\epsilon^{\rm p}$ (the flow rule):

$$\dot{\epsilon}^{\rm p} = \chi\,\beta\dot{\epsilon}. \tag{20.4.49}$$

Here,

$$\beta \stackrel{\text{def}}{=} \frac{E}{E+H} > 0, \tag{20.4.50}$$

with

$$H \stackrel{\text{def}}{=} H_{\rm iso}(\bar{\epsilon}^{\rm p}) + C(1-\gamma A n^{\rm p}) \tag{20.4.51}$$

a hardening modulus, and

$$n^{\rm p} = \text{sign}\,(\sigma - \sigma_{\rm back}). \tag{20.4.52}$$

Also, χ is a switching parameter:

$$\chi = \begin{cases} 0 & \text{if } f<0, \text{ or if } f=0 \text{ and } n^{\rm p}\dot{\epsilon} \le 0, \\ 1 & \text{if } f=0 \quad \text{and} \quad n^{\rm p}\dot{\epsilon} > 0. \end{cases} \tag{20.4.53}$$

8. And evolution equations for the hardening variables:

$$\begin{aligned} &\dot{Y} = H_{\rm iso}(\bar{\epsilon}^{\rm p})\,\dot{\bar{\epsilon}}^{\rm p} \qquad \text{with} \qquad H_{\rm iso}(\bar{\epsilon}^{\rm p}) \stackrel{\text{def}}{=} \frac{dY(\bar{\epsilon}^{\rm p})}{d\bar{\epsilon}^{\rm p}}, \\ &\dot{A} = \dot{\epsilon}^{\rm p} - \gamma\,A\,\dot{\bar{\epsilon}}^{\rm p}, \end{aligned} \tag{20.4.54}$$

where

$$\dot{\bar{\epsilon}}^{\rm p} \stackrel{\text{def}}{=} |\dot{\epsilon}^{\rm p}|, \tag{20.4.55}$$

and

$$\bar{\epsilon}^{\rm p}(t) = \int_0^t \dot{\bar{\epsilon}}^{\rm p}(\zeta)\,d\zeta. \tag{20.4.56}$$

Typical initial conditions presume that the one-dimensional body is initially (at time $t=0$) in a *virgin state* in the sense that

$$\epsilon(x,0) = \epsilon^{\rm p}(x,0) = 0, \qquad Y(x,0) = Y_0, \qquad A(x,0) = 0. \tag{20.4.57}$$

REMARKS 20.3

1. As in (20.2.52) for the theory with only isotropic hardening and no kinematic hardening, a useful form for the evolution equation for the isotropic flow resistance Y is

$$\dot{Y} = H\dot{\bar{\epsilon}}^{\rm p}, \qquad H = H_0\Big(1-\frac{Y}{Y_s}\Big)^r, \quad Y(0) = Y_0, \tag{20.4.58}$$

with (Y_0, Y_s, H_0, r) constants. The parameter Y_0 represents the initial value of Y, and Y_s represents a *saturation* value at large strains; the parameters H_0 and r control the manner in which Y increases from its initial value Y_0 to its saturation value Y_s. The integrated form of (20.4.58) is

REMARKS 20.3 Continued

$$Y(\bar{\epsilon}^{\mathrm{p}}) = Y_s - \left[(Y_s - Y_0)^{(1-r)} + (r-1) \left(\frac{H_0}{(Y_s)^r} \right) \bar{\epsilon}^{\mathrm{p}} \right]^{\dfrac{1}{1-r}} \quad \text{for} \quad r \neq 1. \tag{20.4.59}$$

For $r = 1$ the integrated form of (20.4.58) leads to the classical Voce equation:

$$Y(\bar{\epsilon}^{\mathrm{p}}) = Y_s - (Y_s - Y_0) \exp\left(-\frac{H_0}{Y_s} \bar{\epsilon}^{\mathrm{p}} \right). \tag{20.4.60}$$

2. Note that the dissipation in the theory with combined isotropic and kinematic hardening is

$$\mathcal{D} = \big(Y(\bar{\epsilon}^{\mathrm{p}}) + C\,\gamma A^2\big)\dot{\bar{\epsilon}}^{\mathrm{p}} > 0 \qquad \text{whenever} \qquad \dot{\bar{\epsilon}}^{\mathrm{p}} > 0. \tag{20.4.61}$$

EXAMPLE 20.2 Cyclic loading response

Consider the one-dimensional rate-independent model with combined isotropic and kinematic hardening, where the isotropic flow resistance evolves as defined in (20.4.58). If we wish to determine the response of the model to a cyclic loading, then the standard method for computing the response is to do so using a numerical algorithm. The algorithm of choice for plasticity models is the backward-Euler method. In this scheme the evolution equations are all approximated by the backward-Euler difference formula which is employed inside a predictor-corrector framework; see e.g. Simo and Hughes (1998) for further details. This leads to a system of algebraic equations that are solved incrementally.

If we assume that the loading is specified as strain as a function of time, then the final algorithm can be summarized as:

1. Given: $\{\epsilon_n, \epsilon^{\mathrm{p}}_n, Y_n, q_n\}$, and $\Delta\epsilon_n$ at time t_n
   Calculate: $\{\epsilon_{n+1}, \sigma_{n+1}, \epsilon^{\mathrm{p}}_{n+1}, Y_{n+1}, q_{n+1}\}$, at time $t_{n+1} = t_n + \Delta t$, where for convenience we have used the shorthand notation $q \equiv \sigma_{\mathrm{back}}$.
2. Compute a trial state by assuming that there is no plastic evolution during the load step and test if this indeed was a correct assumption:

$$\left.\begin{aligned}
\sigma^{\mathrm{tr}}_{n+1} &= E(\epsilon_{n+1} - \epsilon^{\mathrm{p}}_n) = \sigma_n + E\Delta\epsilon_n,\\
\xi^{\mathrm{tr}}_{n+1} &= \sigma^{\mathrm{tr}}_{n+1} - q_n,\\
\epsilon^{\mathrm{p,tr}}_{n+1} &= \epsilon^{\mathrm{p}}_n,\\
Y^{\mathrm{tr}}_{n+1} &= Y_n,\\
q^{\mathrm{tr}}_{n+1} &= q_n,\\
f^{\mathrm{tr}}_{n+1} &= |\xi^{\mathrm{tr}}_{n+1}| - Y_n.
\end{aligned}\right\}$$

IF $f^{\mathrm{tr}}_{n+1} \leq 0$ THEN

Elastic step: Set $(\cdots)_{n+1} = (\cdots)^{\mathrm{tr}}_{n+1}$ ELSE

Plastic step:

(a) Solve

$$g(\Delta\bar{\epsilon}^{\mathrm{p}}) = |\bar{\xi}^{\mathrm{tr}}_{\mathrm{n+1}}(\eta)| - \Big(E + (1-\eta)C\Big)\,\Delta\bar{\epsilon}^{\mathrm{p}} - \Big(Y_n + H(Y_n)\Delta\bar{\epsilon}^{\mathrm{p}}\Big) = 0$$

for $\Delta\bar{\epsilon}^{\mathrm{p}}$. Here,

$$\bar{\xi}^{\mathrm{tr}}_{\mathrm{n+1}}(\eta) \stackrel{\mathrm{def}}{=} \xi^{\mathrm{tr}}_{\mathrm{n+1}} + \eta\, q_n, \qquad \eta \stackrel{\mathrm{def}}{=} \frac{\gamma\,\Delta\bar{\epsilon}^{\mathrm{p}}}{1+\gamma\,\Delta\bar{\epsilon}^{\mathrm{p}}}.$$

(b) Update

$$\begin{aligned}
\epsilon^{\mathrm{p}}_{\mathrm{n+1}} &= \epsilon^{\mathrm{p}}_{n} + \Delta\bar{\epsilon}^{\mathrm{p}}\,\mathrm{sign}\,(\bar{\xi}^{\mathrm{tr}}_{\mathrm{n+1}}),\\
\sigma_{\mathrm{n+1}} &= E(\epsilon_{\mathrm{n+1}} - \epsilon^{\mathrm{p}}_{\mathrm{n+1}}),\\
Y_{\mathrm{n+1}} &= Y_n + H(Y_n)\Delta\bar{\epsilon}^{\mathrm{p}},\\
q_{\mathrm{n+1}} &= (1-\eta)\,\left[q_n + C\,\Delta\bar{\epsilon}^{\mathrm{p}}\,\mathrm{sign}\,(\bar{\xi}^{\mathrm{tr}}_{\mathrm{n+1}}(\eta))\right].
\end{aligned}$$

END IF

We can see the range of behaviors possible by considering the following four cases of material parameters.

1. Isotropic hardening only:

$$\begin{aligned}
&E = 200\,\text{GPa},\ Y_0 = 100\,\text{MPa},\\
&H_0 = 1250\ \text{MPa},\ Y_s = 250\,\text{MPa},\ r = 1,\ C = 0\,\text{MPa},\gamma = 0
\end{aligned}$$

2. Kinematic hardening only with no dynamic recovery:

$$E = 200\,\text{GPa},\ Y_0 = 100\,\text{MPa},\ H_0 = 0\,\text{MPa},\ Y_s = 100\,\text{MPa},\ r = 1,\ C = 10\,\text{GPa},\ \gamma = 0$$

3. Kinematic hardening only with dynamic recovery:

$$\begin{aligned}
&E = 200\,\text{GPa},\ Y_0 = 100\,\text{MPa},\\
&H_0 = 0\,\text{MPa},\ Y_s = 100\,\text{MPa},\ r = 1,\ C = 60\,\text{GPa},\gamma = 400
\end{aligned}$$

4. Combined kinematic and isotropic hardening with dynamic recovery:

$$\begin{aligned}
&E = 200\,\text{GPa},\ Y_0 = 100\,\text{MPa},\\
&H_0 = 1250\,\text{MPa},\ Y_s = 250\,\text{MPa},\ r = 1,\ C = 40\,\text{GPa},\gamma = 400
\end{aligned}$$

The numerical calculations will be for a cyclic test at strain rates of $\dot{\epsilon} = \pm 0.001/\text{s}$ between the strain limits of $\epsilon = \pm 0.01$, for twenty load reversals, using a time step of

$$\Delta t = \left(\frac{(Y_0/E)}{|\dot{\epsilon}|}\right).$$

Figures 20.9–20.12 show the resulting stress-strain responses for each of the four cases.

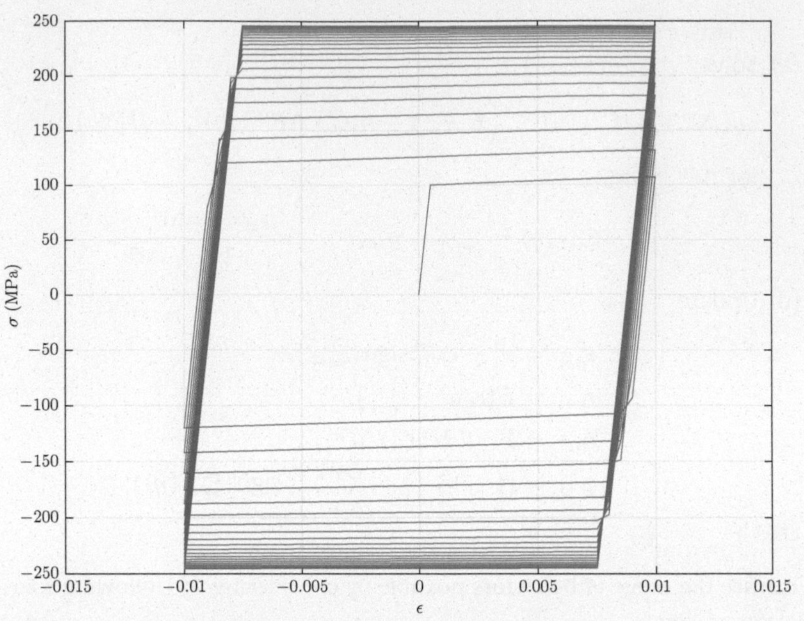

Fig. 20.9 Case 1, only isotropic hardening.

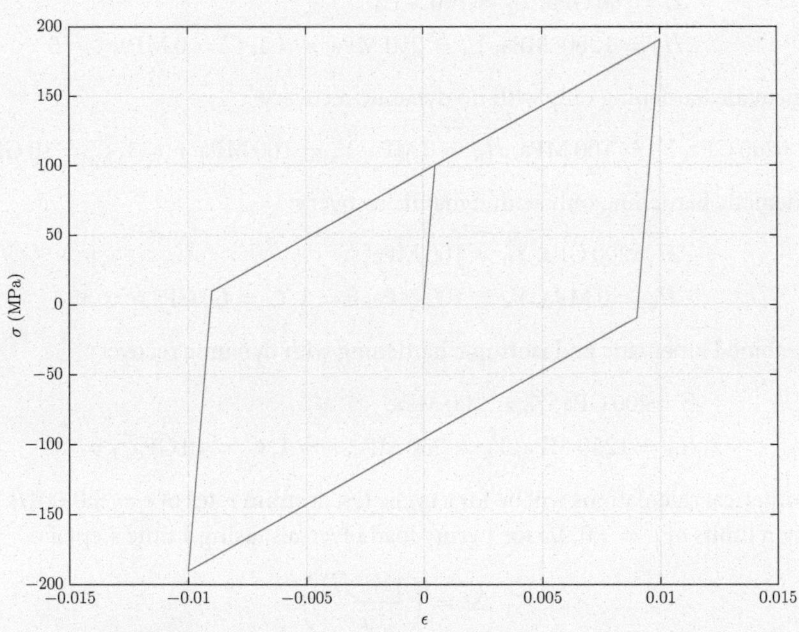

Fig. 20.10 Case 2, only kinematic hardening no dynamic recovery.

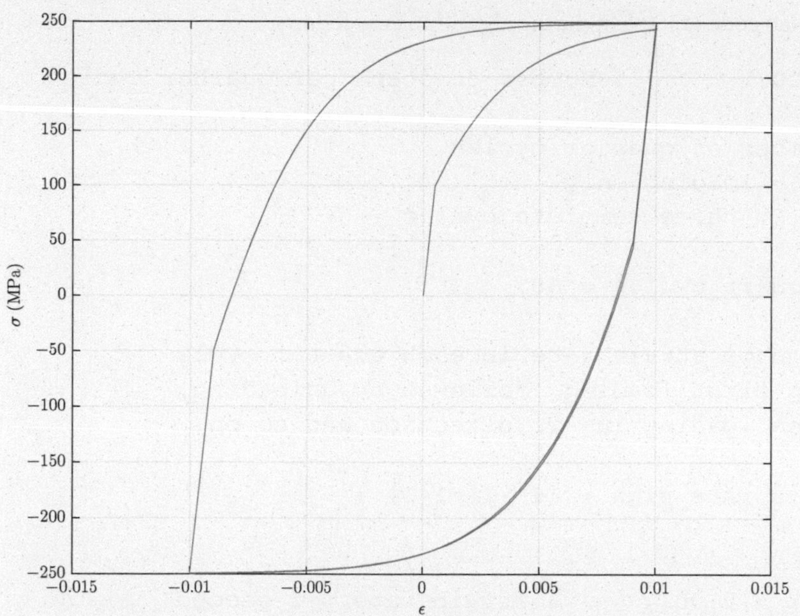

Fig. 20.11 Case 3, only kinematic hardening with dynamic recovery.

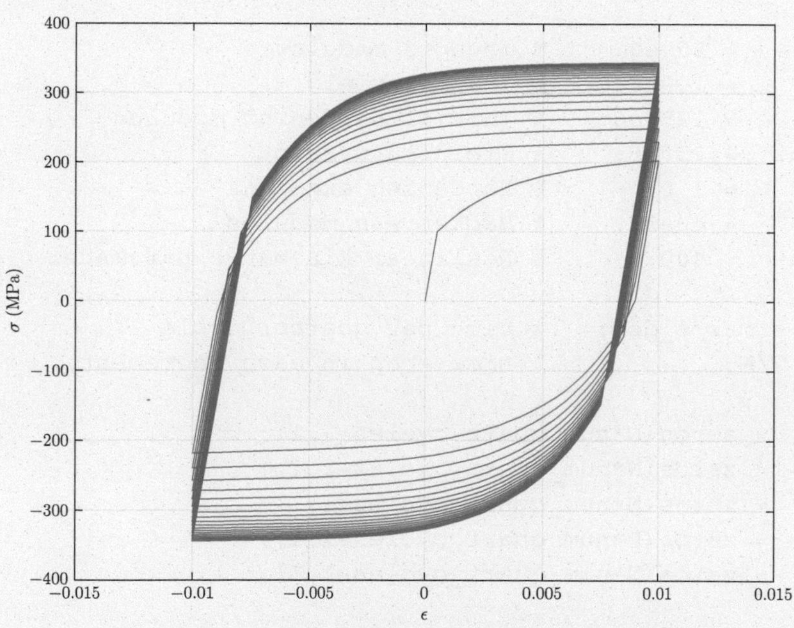

Fig. 20.12 Case 4, combined kinematic and isotropic hardening with dynamic recovery.

Matlab code to implement the algorithm is as follows:

```
N = 20;          % Number of steps per quarter cycle

% Number of quarter cycles
% 1 -> loading only
% 12 -> three complete cycles
%
num_quart_cycles = 80;

% sign of strain rate in each quarter cycle
% e.g first loading, followed by unloading,
% then loading in -ve direction and so on.
%
strain_rate_sign = [1 -1 -1 1]';


e_dot = 0.001;      % strain rate per second
e_max = 0.01;       % absolute value of maximum strain

% Material properties
%
E        = 200e9;    % Young's modulus
Y0       = 100e6;    % Initial yield
h0       = 1250e6;   % Initial hardening modulus
Y_star   = 250e6;    % Yield saturation
r        = 1.0;      % Hardening exponent
C        = 40e9;     % Backstress modulus
gamma    = 400;      % Nonlinear kinematic parameter

T  = e_max/e_dot;    % time per quarter cycle
dt = T/N;            % time step in each increment

e     = zeros(N*num_quart_cycles+1,1);
ep    = zeros(N*num_quart_cycles+1,1);
s     = zeros(N*num_quart_cycles+1,1);
sigma = zeros(N*num_quart_cycles+1,1);
q     = zeros(N*num_quart_cycles+1,1);

% Start the time integration loop
%
n = 1;        % n=1 is the initial condition.
Y(1) = Y0;
```

```
for i=1:num_quart_cycles

    % Strain increment per time step
    ii = mod(i,4);
    if(ii == 0)
      ii = 4;
    end
    de = strain_rate_sign(ii) * e_dot * dt;

    for j=1:N

        % New total strain at the end of this step
        e(n+1) = e(n) + de;

        % Trial stress
        sigma_trial = sigma(n) + E * de;
        xi_trial    = sigma_trial - q(n);
        f_trial     = abs(xi_trial) - Y(n);

        if(f_trial < 0)

            % Elastic step
            %
            sigma(n+1) = sigma_trial;
            Y(n+1)     = Y(n);
            ep(n+1)    = ep(n);
            q(n+1)     = q(n);
        else

            % Solve implicit equation for x=abs(dep)
            x = de/2;  % initial guess
            dx = 100*x;       % a large value

            % Newton Raphson loop
            while (abs(dx) > 1e-4*abs(x))

              eta = gamma*x/(1+gamma*x);
              g = abs(xi_trial + eta*q(n)) - (E + (1-eta)...
              *C)*x - (Y(n) + h0*(1-Y(n)/Y_star)^r * x);

              deta_dx = gamma/(1+gamma*x) - ...
                        gamma*gamma*x/((1+gamma*x)^2);
```

```
                dg        = sign(xi_trial + eta*q(n)) *...
                            deta_dx*q(n) ...
                          - (E + (1-eta)*C) + C*x*deta_dx ...
                          - h0*(1-Y(n)/Y_star)^r;

                % Calculate increment dx.
                dx = -g/dg;
                x = x + dx;
            end
            dep = x;

            eta           = gamma*x/(1+gamma*x);
            xi_bar_trial = xi_trial + eta*q(n);

            % Update ep(n+1), sigma(n+1), q(n+1)
            ep(n+1)      = ep(n) + dep*sign(xi_bar_trial);
            sigma(n+1)  = E*(e(n+1) - ep(n+1));
            Y(n+1)       = Y(n) + h0*(1-Y(n)/Y_star)^r * dep;
            q(n+1)       = (1-eta) * (q(n) + C*dep*sign...
                             (xi_bar_trial));
        end

        n=n+1;
    end
end

plot(e,sigma/1e6,'-');
xlabel('$\epsilon$', 'Interpreter', 'latex','FontSize',16);
ylabel('$\sigma$ (MPa)', 'Interpreter', 'latex',...
 'FontSize',16);
grid on;
```

20.5 **One-dimensional rate-dependent power-law plasticity with combined isotropic and kinematic hardening**

In this section we generalize our discussion of Sec. 20.3.4 to allow for power-law rate-dependent plasticity with a yield threshold and combined isotropic and kinematic hardening. We do not belabor the derivations, but simply summarize the final set of constitutive equations:

1. An elastic-plastic decomposition of ϵ:

$$\epsilon = \epsilon^{\mathrm{e}} + \epsilon^{\mathrm{p}}. \tag{20.5.1}$$

2. A free energy function:

$$\psi = \underbrace{\frac{1}{2} E\,(\epsilon^{\mathrm{e}})^2}_{\psi^{\mathrm{e}}} + \underbrace{\tfrac{1}{2} C A^2}_{\psi^{\mathrm{p}}}, \tag{20.5.2}$$

where $E > 0$ is the elastic Young's modulus, and $C > 0$ is a "back-stress" modulus for the contribution from the plastic-strain-like internal variable to the free energy.

3. An equation for the stress:

$$\sigma = E[\epsilon^{\mathrm{e}}] = E[\epsilon - \epsilon^{\mathrm{p}}]. \tag{20.5.3}$$

4. An equation for the back stress:

$$\sigma_{\mathrm{back}} = CA. \tag{20.5.4}$$

5. An effective stress, equivalent effective stress:

$$\sigma_{\mathrm{eff}} \stackrel{\mathrm{def}}{=} \sigma - \sigma_{\mathrm{back}}, \tag{20.5.5}$$

and

$$\bar{\sigma}_{\mathrm{eff}} \stackrel{\mathrm{def}}{=} |\sigma_{\mathrm{eff}}|. \tag{20.5.6}$$

6. A Flow rule:

$$\dot{\epsilon}^{\mathrm{p}} = \dot{\epsilon}_0 \left\langle \frac{\bar{\sigma}_{\mathrm{eff}} - Y_{\mathrm{th}}(\bar{\epsilon}^{\mathrm{p}})}{Y(\bar{\epsilon}^{\mathrm{p}})} \right\rangle^{1/m} \mathrm{sign}(\sigma_{\mathrm{eff}}). \tag{20.5.7}$$

Here $\dot{\epsilon}_0$ is a reference plastic strain-rate, and $m \in (0, 1]$ is a strain-rate sensitivity parameter. As before,

$$\bar{\epsilon}^{\mathrm{p}}(t) = \int_0^t |\dot{\epsilon}^{\mathrm{p}}(\zeta)|\, d\zeta, \tag{20.5.8}$$

and the positive-valued scalar functions of $\bar{\epsilon}^{\mathrm{p}}$

$$Y_{\mathrm{th}}(\bar{\epsilon}^{\mathrm{p}}) \geq 0 \quad \text{and} \quad Y(\bar{\epsilon}^{\mathrm{p}}) > 0 \tag{20.5.9}$$

represent two flow resistances of the materials.

7. And an evolution equation for the hardening variable A:

$$\dot{A} = \dot{\epsilon}^{\mathrm{p}} - \gamma\, A\, \dot{\bar{\epsilon}}^{\mathrm{p}}, \tag{20.5.10}$$

with $\gamma \geq 0$ a dynamic recovery coefficient.

Typical initial conditions presume that the one-dimensional body is initially at time $t = 0$ in a virgin state, in the sense that

$$\begin{aligned} &\epsilon(x,0) = \epsilon^{\mathrm{p}}(x,0) = 0, \qquad Y_{\mathrm{th}}(x,0) = Y_{\mathrm{th}0} \geq 0, \qquad Y(x,0) = Y_0 > 0, \\ &\text{and} \qquad A(x,0) = 0. \end{aligned} \tag{20.5.11}$$

21 Three-dimensional plasticity with isotropic hardening

21.1 Introduction

In this chapter we consider a three-dimensional plasticity theory for isotropic materials with isotropic hardening, but limit our considerations to *small deformations*, i.e. to strain levels of $\lesssim$ 5% or so. Accordingly, in this and the following sections on small deformation plasticity, we will use the Cauchy stress tensor $\boldsymbol{\sigma}$ and the small deformation strain tensor $\boldsymbol{\epsilon} = (1/2)(\nabla\mathbf{u} + (\nabla\mathbf{u})^{\top})$, as we have previously used in our development of the linear theory of elasticity. We will first formulate a classical rate-independent theory, and then give a simple rate-dependent generalization of this theory. We will further extend the discussion of both the rate-independent and rate-dependent theories to account for kinematic hardening in Chapter 22.

21.2 Basic equations

We take as the starting point the kinematic assumption of the linear theory of elasticity and the balances of momentum. Recall that these equations were derived independent of constitutive relations and are hence applicable to a wide class of materials under the assumption of small deformations. Further, we will utilize the free-energy imbalance for isothermal conditions. Thus, we begin with:

(i) the *strain-displacement relation*

$$\boldsymbol{\epsilon} = \tfrac{1}{2}(\nabla\mathbf{u} + (\nabla\mathbf{u})^{\top}); \tag{21.2.1}$$

(ii) the *momentum balances*

$$\operatorname{div}\boldsymbol{\sigma} + \mathbf{b} = \rho\ddot{\mathbf{u}}, \qquad \boldsymbol{\sigma} = \boldsymbol{\sigma}^{\top}, \tag{21.2.2}$$

with $\boldsymbol{\sigma}$ the *stress*, ρ the *density*, and $\mathbf{b}$ the *body force*; and

(iii) the *free-energy imbalance*

$$\dot{\psi} \le \boldsymbol{\sigma}:\dot{\boldsymbol{\epsilon}}, \tag{21.2.3}$$

with ψ the *free energy* per unit volume.

Continuum Mechanics of Solids. Lallit Anand and Sanjay Govindjee, Oxford University Press (2020).

DOI: 10.1093/oso/9780198864721.001.0001

21.3 **Kinematical assumptions**

Underlying most theories of plasticity is a physical picture that associates with a plastic solid a *microscopic structure*, such as a crystal lattice, that may be stretched and rotated, together with a notion of *defects*, such as dislocations, capable of flowing through that structure. Following our discussion of the one-dimensional theory of plasticity, we introduce the kinematical constitutive assumption that the infinitesimal strain tensor admits the decomposition

$$\boldsymbol{\epsilon} = \boldsymbol{\epsilon}^{\mathrm{e}} + \boldsymbol{\epsilon}^{\mathrm{p}}, \tag{21.3.1}$$

in which:

(i) $\boldsymbol{\epsilon}^{\mathrm{e}}$, the **elastic strain**, represents local strain of the underlying microscopic structure, and

(ii) $\boldsymbol{\epsilon}^{\mathrm{p}}$, the **plastic strain**, represents the local strain of the material due to the formation and motion of dislocations through that structure.

An important general observation is that *the flow of dislocations does not induce changes in volume* (cf., e.g., Bridgman, 1952; Spitzig et al., 1975); *consistent with this, we assume that* $\boldsymbol{\epsilon}^{\mathrm{p}}$ *is deviatoric*, so that

$$\mathrm{tr}\,\boldsymbol{\epsilon}^{\mathrm{p}} = 0. \tag{21.3.2}$$

If in the free-energy imbalance (21.2.3) we account explicitly for the elastic and plastic strains via (21.3.1) we find that

$$\dot{\psi} - \boldsymbol{\sigma}:\dot{\boldsymbol{\epsilon}}^{\mathrm{e}} - \boldsymbol{\sigma}:\dot{\boldsymbol{\epsilon}}^{\mathrm{p}} \le 0. \tag{21.3.3}$$

Further, since $\dot{\boldsymbol{\epsilon}}^{\mathrm{p}}$ is deviatoric, we may conclude that

$$\boldsymbol{\sigma}:\dot{\boldsymbol{\epsilon}}^{\mathrm{p}} = \boldsymbol{\sigma}':\dot{\boldsymbol{\epsilon}}^{\mathrm{p}}, \qquad \text{where} \qquad \boldsymbol{\sigma}' = \boldsymbol{\sigma} - \frac{1}{3}(\mathrm{tr}\,\boldsymbol{\sigma})\mathbf{1}.$$

Hence (21.3.3) becomes

$$\dot{\psi} - \boldsymbol{\sigma}:\dot{\boldsymbol{\epsilon}}^{\mathrm{e}} - \boldsymbol{\sigma}':\dot{\boldsymbol{\epsilon}}^{\mathrm{p}} \le 0. \tag{21.3.4}$$

21.4 **Separability hypothesis**

Most theories of plasticity are based on constitutive relations that separate plastic and elastic response, with the free energy strictly a function of elastic strain.[1] In accord with the assumption of a free energy that is strictly a function of the elastic strain, we find by invoking the Coleman and Noll argument (cf. Sec. 7.1 and Sec. 14.3) that

$$\dot{\psi} = \frac{\partial \psi}{\partial \boldsymbol{\epsilon}^{\mathrm{e}}}:\dot{\boldsymbol{\epsilon}}^{\mathrm{e}} \qquad \text{which leads to} \qquad \boldsymbol{\sigma} = \frac{\partial \psi}{\partial \boldsymbol{\epsilon}^{\mathrm{e}}}, \tag{21.4.1}$$

and consequently the **dissipation inequality**

$$\mathcal{D} = \boldsymbol{\sigma}':\dot{\boldsymbol{\epsilon}}^{\mathrm{p}} \ge 0, \tag{21.4.2}$$

which characterizes the energy dissipated during plastic flow.

[1] However, see Chapter 22 where in order to describe kinematic hardening, we allow for the free energy to also depend on an internal variable which evolves with plastic flow.

In the next few sections we further specify the constitutive assumptions that characterize the elasticity and the plasticity of the material. We assume throughout that *the body is homogeneous and isotropic*. The developments closely parallel those in Chapter 20 for the one-dimensional theory.

21.5 Constitutive characterization of elastic response

As noted, the constitutive equation assumption for the free energy is that it be a function of the elastic strain,

$$\psi = \hat{\psi}(\boldsymbol{\epsilon}^{e}), \tag{21.5.1}$$

which leads to the requirement

$$\boldsymbol{\sigma} = \frac{\partial \hat{\psi}(\boldsymbol{\epsilon}^{e})}{\partial \boldsymbol{\epsilon}^{e}} . \tag{21.5.2}$$

In the elastic-plastic deformation of metals, the elastic strains are typically small, and under these conditions an appropriate free energy function and corresponding relation for stress for isotropic materials are

$$\begin{aligned} \psi &= \mu\,|\boldsymbol{\epsilon}^{e\prime}|^2 + \frac{1}{2}\,\kappa\,(\operatorname{tr}\boldsymbol{\epsilon}^{e})^2, \\ \boldsymbol{\sigma} &= 2\,\mu\boldsymbol{\epsilon}^{e\prime} + \kappa(\operatorname{tr}\boldsymbol{\epsilon}^{e})\,\mathbf{1}, \end{aligned} \tag{21.5.3}$$

where $\mu > 0$ and $\kappa > 0$ are the elastic shear and bulk moduli, respectively. The basic elastic stress-strain relation $(21.5.3)_2$ is assumed to hold in all motions of the body, even during plastic flow.

21.6 Constitutive equations for plastic response

Let

$$\mathbf{n}^{p} = \frac{\dot{\boldsymbol{\epsilon}}^{p}}{|\dot{\boldsymbol{\epsilon}}^{p}|} \tag{21.6.1}$$

denote the **plastic flow direction**, and

$$|\dot{\boldsymbol{\epsilon}}^{p}| \geq 0 \tag{21.6.2}$$

denote the **magnitude of the plastic strain-rate.**

From the dissipation inequality (21.4.2),

$$\mathcal{D} = \boldsymbol{\sigma}' : \dot{\boldsymbol{\epsilon}}^{p} \geq 0, \tag{21.6.3}$$

we note that the "driving force" conjugate to $\dot{\boldsymbol{\epsilon}}^{p}$ is the deviatoric stress $\boldsymbol{\sigma}'$. If we assume that the plastic strain-rate and the deviatoric stress are **co-directional**, that is

$$\mathbf{n}^{p} = \frac{\boldsymbol{\sigma}'}{|\boldsymbol{\sigma}'|}, \tag{21.6.4}$$

then $\mathcal{D} = |\boldsymbol{\sigma}'||\dot{\boldsymbol{\epsilon}}^{p}|$ and hence $\mathcal{D}$ is trivially non-negative. *Henceforth we adopt the assumption* (21.6.4) *of co-directionality*. We note that this assumption is in accord with the early works of Prandtl (1924) and Reuss (1930) on the theory of plasticity. Thus, we assume that

$$\dot{\boldsymbol{\epsilon}}^{\mathrm{p}} = |\dot{\boldsymbol{\epsilon}}^{\mathrm{p}}|\,\mathbf{n}^{\mathrm{p}}, \quad \text{with} \quad \mathbf{n}^{\mathrm{p}} = \frac{\boldsymbol{\sigma}'}{|\boldsymbol{\sigma}'|} \quad \text{and} \quad |\dot{\boldsymbol{\epsilon}}^{\mathrm{p}}| \geq 0. \tag{21.6.5}$$

Recalling the definition of the **Mises equivalent tensile stress,**

$$\begin{aligned}\bar{\sigma} &\stackrel{\text{def}}{=} \sqrt{3/2}\,|\boldsymbol{\sigma}'| = \sqrt{(3/2)\sigma'_{ij}\sigma'_{ij}},\\ &= \left|\left[\frac{1}{2}\left\{(\sigma_{11}-\sigma_{22})^2 + (\sigma_{22}-\sigma_{33})^2 + (\sigma_{33}-\sigma_{11})^2\right\} + 3\left\{\sigma_{12}^2 + \sigma_{23}^2 + \sigma_{31}^2\right\}\right]^{1/2}\right|,\end{aligned} \tag{21.6.6}$$

and introducing an **equivalent tensile plastic strain-rate** defined by

$$\begin{aligned}\dot{\bar{\epsilon}}^{\mathrm{p}} &\stackrel{\text{def}}{=} \sqrt{(2/3)}\,|\dot{\boldsymbol{\epsilon}}^{\mathrm{p}}| = \sqrt{(2/3)\dot{\epsilon}^{\mathrm{p}}_{ij}\dot{\epsilon}^{\mathrm{p}}_{ij}},\\ &= \left|\left[\frac{2}{9}\left\{(\dot{\epsilon}^{\mathrm{p}}_{11}-\dot{\epsilon}^{\mathrm{p}}_{22})^2 + (\dot{\epsilon}^{\mathrm{p}}_{22}-\dot{\epsilon}^{\mathrm{p}}_{33})^2 + (\dot{\epsilon}^{\mathrm{p}}_{33}-\dot{\epsilon}^{\mathrm{p}}_{11})^2\right\} + \frac{4}{3}\left\{(\dot{\epsilon}^{\mathrm{p}}_{12})^2 + (\dot{\epsilon}^{\mathrm{p}}_{23})^2 + (\dot{\epsilon}^{\mathrm{p}}_{31})^2\right\}\right]^{1/2}\right|,\end{aligned} \tag{21.6.7}$$

the flow rule (21.6.5) may be equivalently expressed as

$$\dot{\boldsymbol{\epsilon}}^{\mathrm{p}} = \sqrt{3/2}\,\dot{\bar{\epsilon}}^{\mathrm{p}}\,\mathbf{n}^{\mathrm{p}}, \qquad \text{with} \qquad \mathbf{n}^{\mathrm{p}} = \frac{\boldsymbol{\sigma}'}{|\boldsymbol{\sigma}'|} = \sqrt{3/2}\,\frac{\boldsymbol{\sigma}'}{\bar{\sigma}}. \tag{21.6.8}$$

Note that, under these constitutive assumptions, using (21.6.6), (21.6.7), and (21.6.8), the dissipation inequality (21.6.3) reduces to

$$\mathcal{D} = \bar{\sigma}\dot{\bar{\epsilon}}^{\mathrm{p}} \geq 0, \tag{21.6.9}$$

and is trivially satisfied since, by definition, $\bar{\sigma} \geq 0$ and $\dot{\bar{\epsilon}}^{\mathrm{p}} \geq 0$ (cf. (21.6.6) and (21.6.7)).

For later use we also introduce the **equivalent tensile plastic strain** defined by

$$\bar{\epsilon}^{\mathrm{p}}(t) \stackrel{\text{def}}{=} \int_0^t \dot{\bar{\epsilon}}^{\mathrm{p}}(\zeta)\,d\zeta. \tag{21.6.10}$$

21.6.1 Flow strength

As in the one-dimensional theory with isotropic hardening, let

$$Y(\bar{\epsilon}^{\mathrm{p}}) > 0 \tag{21.6.11}$$

denote a scalar internal variable of the theory, with dimensions of stress, which represents the *resistance to plastic flow* offered by the material. We call Y the **flow strength** of the material. The value of Y is assumed to depend on $\bar{\epsilon}^{\mathrm{p}}$, and it typically increases as $\bar{\epsilon}^{\mathrm{p}}$ increases. Thus the equivalent plastic strain represents a **hardening variable** of the theory, and the function $Y(\bar{\epsilon}^{\mathrm{p}})$ characterizes the **strain-hardening** response of the material. The initial value

$$Y_0 \stackrel{\text{def}}{=} Y(0) \equiv \sigma_y \tag{21.6.12}$$

represents the initial value of the resistance to plastic flow, and is called the **yield strength** of the material, and is often denoted by σ_y. If we consider the rate of change of the flow strength,

$$\dot{\overline{Y(\bar{\epsilon}^{\mathrm{p}})}} = H(\bar{\epsilon}^{\mathrm{p}})\dot{\bar{\epsilon}}^{\mathrm{p}}, \tag{21.6.13}$$

then

$$H(\bar{\epsilon}^{\mathrm{p}}) = \frac{dY(\bar{\epsilon}^{\mathrm{p}})}{d\bar{\epsilon}^{\mathrm{p}}} \tag{21.6.14}$$

represents the **strain-hardening rate** of the material at a given $\bar{\epsilon}^{\mathrm{p}}$. *The material strain-hardens or softens according as*

$$H(\bar{\epsilon}^{\mathrm{p}}) > 0 \text{ or } H(\bar{\epsilon}^{\mathrm{p}}) < 0. \tag{21.6.15}$$

21.6.2 **Mises yield condition. No-flow conditions. Consistency condition**

In the simplest three-dimensional rate-independent theory for isotropic materials, the Mises equivalent tensile stress $\bar{\sigma}$ cannot be greater than Y. Accordingly, we assume that there exists a function $f(\bar{\sigma}, \bar{\epsilon}^{\mathrm{p}})$, called the *yield function*, with value

$$f \overset{\text{def}}{=} \bar{\sigma} - Y(\bar{\epsilon}^{\mathrm{p}}) \leq 0, \tag{21.6.16}$$

which limits the admissible deviatoric stresses $\boldsymbol{\sigma}'$; (21.6.16) is called the **Mises yield condition** (von Mises, 1913).

The spherical "surface"

$$f = \bar{\sigma} - Y(\bar{\epsilon}^{\mathrm{p}}) = 0,$$

with "radius" $Y(\bar{\epsilon}^{\mathrm{p}})$ in the space of symmetric deviatoric tensors is called the **yield surface**, and it is assumed that *plastic flow is possible only when the deviatoric stress $\boldsymbol{\sigma}'$ lies on the yield surface*, and correspondingly, the open ball with radius $Y(\bar{\epsilon}^{\mathrm{p}})$ in the space of symmetric deviatoric tensors is called the **elastic range**, because the plastic flow vanishes there. Thus, we have the requirements that

$$f = 0 \qquad \text{for} \qquad \dot{\boldsymbol{\epsilon}}^{\mathrm{p}} \neq \mathbf{0}, \tag{21.6.17}$$

and

$$\dot{\boldsymbol{\epsilon}}^{\mathrm{p}} = \mathbf{0} \qquad \text{for} \qquad f < 0. \tag{21.6.18}$$

Consider a fixed time t and assume that, at that time, $f(t) = 0$, so that the yield condition is satisfied. Then, by (21.6.16), $f(t+\tau) \leq 0$ for all τ and, consequently, $\dot{f}(t) \leq 0$. Thus,

$$\boxed{\text{if } f = 0, \text{ then } \dot{f} \leq 0.} \tag{21.6.19}$$

Next, if $f(t) = 0$ and $\dot{f}(t) < 0$, then $f(t+\tau) < 0$ for all sufficiently small $\tau > 0$, so that, by (21.6.18), $\dot{\boldsymbol{\epsilon}}^{\mathrm{p}}(t+\tau) = \mathbf{0}$ for all such τ. Hence, $\dot{\boldsymbol{\epsilon}}^{\mathrm{p}}(t) = \mathbf{0}$. Thus,

$$\boxed{\text{if } f = 0 \text{ and } \dot{f} < 0, \text{ then } \dot{\boldsymbol{\epsilon}}^{\mathrm{p}} = \mathbf{0}.} \tag{21.6.20}$$

Equations (21.6.18) and (21.6.20) combine to form the **no-flow condition**:

$$\boxed{\dot{\boldsymbol{\epsilon}}^{\mathrm{p}} = \mathbf{0} \text{ if } f < 0 \text{ or if } f = 0 \text{ and } \dot{f} < 0.} \tag{21.6.21}$$

Next, if $\dot{\boldsymbol{\epsilon}}^{\mathrm{p}} \neq \mathbf{0}$ at a time t, then by (21.6.17)–(21.6.20) it follows that $\dot{f}(t)$ must equal 0. Thus we have the **consistency condition**:

$$\boxed{\text{if } \dot{\boldsymbol{\epsilon}}^{\mathrm{p}} \neq \mathbf{0}, \text{ then } f = 0 \text{ and } \dot{f} = 0.} \tag{21.6.22}$$

REMARK 21.1

The constraint (21.6.16) on the value of the yield function f, the positivity of $\dot{\bar{\epsilon}}^{\mathrm{p}}$, and the first of the no-flow conditions in (21.6.21) are often stated together to read as

$$\boxed{f \leq 0, \qquad \dot{\bar{\epsilon}}^{\mathrm{p}} \geq 0, \qquad \dot{\bar{\epsilon}}^{\mathrm{p}} f = 0,} \tag{21.6.23}$$

and are known as the **Kuhn–Tucker complementarity conditions**. Further, the second of the no-flow conditions in (21.6.21) and the consistency condition (21.6.22) are sometimes combined and alternatively written as

$$\boxed{\dot{\bar{\epsilon}}^{\mathrm{p}} \dot{f} = 0 \text{ when } f = 0,} \tag{21.6.24}$$

and called the *consistency condition.*

We now show that the no-flow and consistency conditions may be used to obtain an equation for $\dot{\boldsymbol{\epsilon}}^{\mathrm{p}}$ in terms of $\dot{\boldsymbol{\epsilon}}$ and $\boldsymbol{\sigma}$. In deriving this relation we recall the co-directionality hypothesis,

$$\mathbf{n}^{\mathrm{p}} = \frac{\dot{\boldsymbol{\epsilon}}^{\mathrm{p}}}{|\dot{\boldsymbol{\epsilon}}^{\mathrm{p}}|} = \frac{\boldsymbol{\sigma}'}{|\boldsymbol{\sigma}'|} = \sqrt{3/2}\,\frac{\boldsymbol{\sigma}'}{\bar{\sigma}}, \tag{21.6.25}$$

which is basic. Assume unless specified otherwise that the yield condition is satisfied:

$$f = 0 \qquad (\text{so that } \dot{f} \leq 0); \tag{21.6.26}$$

cf. (21.6.19). Then[2]

$$\begin{aligned}
\dot{f} &= \overline{\sqrt{3/2}\,|\boldsymbol{\sigma}'|}^{\,\cdot} - \overline{Y(\bar{\epsilon}^{\mathrm{p}})}^{\,\cdot} \\
&= \sqrt{3/2}\,\frac{\boldsymbol{\sigma}'}{|\boldsymbol{\sigma}'|} : \dot{\boldsymbol{\sigma}}' - H(\bar{\epsilon}^{\mathrm{p}})\dot{\bar{\epsilon}}^{\mathrm{p}} && \text{(by (21.6.13))} \\
&= \sqrt{3/2}\,\mathbf{n}^{\mathrm{p}} : \dot{\boldsymbol{\sigma}}' - H(\bar{\epsilon}^{\mathrm{p}})\dot{\bar{\epsilon}}^{\mathrm{p}} && \text{(by (21.6.25))} \\
&= \sqrt{6}\mu\mathbf{n}^{\mathrm{p}} : (\dot{\boldsymbol{\epsilon}}' - \dot{\boldsymbol{\epsilon}}^{\mathrm{p}}) - H(\bar{\epsilon}^{\mathrm{p}})\dot{\bar{\epsilon}}^{\mathrm{p}} && \text{(by (21.5.3))}
\end{aligned}$$

or, since $\mathbf{n}^{\mathrm{p}} : \dot{\boldsymbol{\epsilon}}^{\mathrm{p}} = |\dot{\boldsymbol{\epsilon}}^{\mathrm{p}}| = \sqrt{3/2}\dot{\bar{\epsilon}}^{\mathrm{p}}$ and $\mathbf{n}^{\mathrm{p}}$ is deviatoric,

$$\dot{f} = \sqrt{6}\mu\,\mathbf{n}^{\mathrm{p}} : \dot{\boldsymbol{\epsilon}} - [3\mu + H(\bar{\epsilon}^{\mathrm{p}})]\dot{\bar{\epsilon}}^{\mathrm{p}}. \tag{21.6.27}$$

[2] In the derivation we employ the identity

$$\dot{\overline{|\mathbf{A}|}} = \frac{\mathbf{A}}{|\mathbf{A}|} : \dot{\mathbf{A}},$$

which is derived as follows,

$$\dot{\overline{|\mathbf{A}|^2}} = \dot{\overline{\mathbf{A}:\mathbf{A}}} \quad \Rightarrow \quad 2|\mathbf{A}|\dot{\overline{|\mathbf{A}|}} = 2\mathbf{A}:\dot{\mathbf{A}} \quad \Rightarrow \quad \dot{\overline{|\mathbf{A}|}} = \frac{\mathbf{A}}{|\mathbf{A}|} : \dot{\mathbf{A}}.$$

We henceforth restrict attention to the class of materials for which

$$3\mu + H(\bar{\epsilon}^{\mathrm{p}}) > 0. \tag{21.6.28}$$

The inequality (21.6.28) always holds for strain-hardening materials, for which $H(\bar{\epsilon}^{\mathrm{p}}) > 0$ for all $\bar{\epsilon}^{\mathrm{p}}$. In contrast, strain-softening materials satisfy $H(\bar{\epsilon}^{\mathrm{p}}) < 0$, and (21.6.28) places a restriction on the maximum allowable softening, as it requires that $H(\bar{\epsilon}^{\mathrm{p}}) > -3\mu$ for all $\bar{\epsilon}^{\mathrm{p}}$.

We now deduce conditions that determine whether the material leaves—or remains on—the yield surface when the material point in question is subjected to a loading program characterized by the strain-rate $\dot{\boldsymbol{\epsilon}}$:

(i) **Elastic unloading** is defined by the condition $\mathbf{n}^{\mathrm{p}} : \dot{\boldsymbol{\epsilon}} < 0$. In this case, since $\dot{\bar{\epsilon}}^{\mathrm{p}} \geq 0$, (21.6.27) implies that $\dot{f} < 0$, and the no-flow conditions (21.6.21) then imply that $\dot{\boldsymbol{\epsilon}}^{\mathrm{p}} = \mathbf{0}$.

(ii) **Neutral loading** is defined by the condition $\mathbf{n}^{\mathrm{p}} : \dot{\boldsymbol{\epsilon}} = 0$. If $\dot{\bar{\epsilon}}^{\mathrm{p}} > 0$, then by (21.6.27) $\dot{f} < 0$ and the no-flow conditions (21.6.21) require $\dot{\boldsymbol{\epsilon}}^{\mathrm{p}} = \mathbf{0}$. If on the other hand $\dot{\bar{\epsilon}}^{\mathrm{p}} = 0$ (the only other possibility), then we also have $\dot{\boldsymbol{\epsilon}}^{\mathrm{p}} = \mathbf{0}$ by (21.6.8). Hence in either case, once again, $\dot{\boldsymbol{\epsilon}}^{\mathrm{p}} = \mathbf{0}$ and there is no plastic flow.

(iii) **Plastic loading** is defined by the condition $\mathbf{n}^{\mathrm{p}} : \dot{\boldsymbol{\epsilon}} > 0$. In this case, if $\dot{\bar{\epsilon}}^{\mathrm{p}} = 0$, then $\dot{f} > 0$, which violates $\dot{f} \leq 0$. Hence $\dot{\bar{\epsilon}}^{\mathrm{p}} > 0$, and since the consistency condition (21.6.22) then requires that $\dot{f} = 0$, (21.6.27) yields

$$\dot{\bar{\epsilon}}^{\mathrm{p}} = \left(\frac{\sqrt{6}\mu}{3\mu + H(\bar{\epsilon}^{\mathrm{p}})} \right) \mathbf{n}^{\mathrm{p}} : \dot{\boldsymbol{\epsilon}}. \tag{21.6.29}$$

Thus since, $\dot{\boldsymbol{\epsilon}}^{\mathrm{p}} = |\dot{\boldsymbol{\epsilon}}^{\mathrm{p}}|\mathbf{n}^{\mathrm{p}} = \sqrt{3/2}\,\dot{\bar{\epsilon}}^{\mathrm{p}}\,\mathbf{n}^{\mathrm{p}}$,

$$\dot{\boldsymbol{\epsilon}}^{\mathrm{p}} = \beta(\mathbf{n}^{\mathrm{p}} : \dot{\boldsymbol{\epsilon}})\mathbf{n}^{\mathrm{p}} \neq \mathbf{0}, \tag{21.6.30}$$

where we have introduced the **stiffness ratio**

$$\beta \stackrel{\text{def}}{=} \frac{3\mu}{3\mu + H(\bar{\epsilon}^{\mathrm{p}})} > 0. \tag{21.6.31}$$

21.6.3 **The flow rule**

Combining the results of (i)–(iii) with the condition (21.6.18), we arrive at the following equation for the plastic strain-rate $\dot{\boldsymbol{\epsilon}}^{\mathrm{p}}$:

$$\dot{\boldsymbol{\epsilon}}^{\mathrm{p}} = \begin{cases} \mathbf{0} & \text{if } f < 0 \quad \text{(behavior within the elastic range)}, \\ \mathbf{0} & \text{if } f = 0 \quad \text{and} \quad \mathbf{n}^{\mathrm{p}} : \dot{\boldsymbol{\epsilon}} < 0 \quad \text{(elastic unloading)}, \\ \mathbf{0} & \text{if } f = 0 \quad \text{and} \quad \mathbf{n}^{\mathrm{p}} : \dot{\boldsymbol{\epsilon}} = 0 \quad \text{(neutral loading)}, \\ \beta(\mathbf{n}^{\mathrm{p}} : \dot{\boldsymbol{\epsilon}})\mathbf{n}^{\mathrm{p}} & \text{if } f = 0 \quad \text{and} \quad \mathbf{n}^{\mathrm{p}} : \dot{\boldsymbol{\epsilon}} > 0 \quad \text{(plastic loading)}. \end{cases} \tag{21.6.32}$$

The result (21.6.32) is embodied in the **flow rule**

$$\dot{\boldsymbol{\epsilon}}^{\mathrm{p}} = \chi\beta(\mathbf{n}^{\mathrm{p}} : \dot{\boldsymbol{\epsilon}})\mathbf{n}^{\mathrm{p}}, \qquad \mathbf{n}^{\mathrm{p}} = \sqrt{3/2}\frac{\boldsymbol{\sigma}'}{\bar{\sigma}}, \tag{21.6.33}$$

where

$$\chi = \begin{cases} 0 & \text{if } f < 0, \text{ or if } f = 0 \text{ and } \mathbf{n}^{\mathrm{p}} : \dot{\boldsymbol{\epsilon}} \le 0, \\ 1 & \text{if } f = 0 \quad \text{and} \quad \mathbf{n}^{\mathrm{p}} : \dot{\boldsymbol{\epsilon}} > 0, \end{cases} \tag{21.6.34}$$

is a **switching parameter.**

21.7 Summary

The complete set of constitutive equations for the small deformation **Mises–Hill theory** (von Mises, 1913; Hill, 1950) of rate-independent plasticity with isotropic hardening consist of:

(i) The decomposition

$$\boldsymbol{\epsilon} = \boldsymbol{\epsilon}^{\mathrm{e}} + \boldsymbol{\epsilon}^{\mathrm{p}}, \tag{21.7.1}$$

of the strain $\boldsymbol{\epsilon}$ into elastic and plastic parts $\boldsymbol{\epsilon}^{\mathrm{e}}$ and $\boldsymbol{\epsilon}^{\mathrm{p}}$.

(ii) The elastic stress-strain relation

$$\boldsymbol{\sigma} = 2\mu(\boldsymbol{\epsilon} - \boldsymbol{\epsilon}^{\mathrm{p}}) + (\kappa - (2/3)\mu)(\operatorname{tr}\boldsymbol{\epsilon})\mathbf{1}, \tag{21.7.2}$$

with $\mu > 0$ and $\kappa > 0$ the elastic shear and bulk moduli.

(iii) A yield condition

$$f = \bar{\sigma} - Y(\bar{\epsilon}^{\mathrm{p}}) \le 0, \tag{21.7.3}$$

with

$$\bar{\sigma} = \sqrt{3/2}\,|\boldsymbol{\sigma}'| \tag{21.7.4}$$

the Mises equivalent tensile stress, f the yield function, and $Y(\bar{\epsilon}^{\mathrm{p}})$ the flow strength.

(iv) A system of evolution equations

$$\boxed{\begin{aligned} &\dot{\boldsymbol{\epsilon}}^{\mathrm{p}} = \chi\beta(\bar{\epsilon}^{\mathrm{p}})(\mathbf{n}^{\mathrm{p}} : \dot{\boldsymbol{\epsilon}})\mathbf{n}^{\mathrm{p}}, \quad \mathbf{n}^{\mathrm{p}} = \sqrt{3/2}\,\frac{\boldsymbol{\sigma}'}{\bar{\sigma}}, \\ &\dot{\bar{\epsilon}}^{\mathrm{p}} = \sqrt{2/3}|\dot{\boldsymbol{\epsilon}}^{\mathrm{p}}| \end{aligned}} \tag{21.7.5}$$

for the plastic strain and the equivalent tensile plastic strain, in which

$$\beta(\bar{\epsilon}^{\mathrm{p}}) = \frac{3\mu}{3\mu + H(\bar{\epsilon}^{\mathrm{p}})} > 0$$

is a stiffness ratio,

$$H(\bar{\epsilon}^{\mathrm{p}}) = \frac{dY(\bar{\epsilon}^{\mathrm{p}})}{d\bar{\epsilon}^{\mathrm{p}}} \tag{21.7.6}$$

the hardening modulus, and

$$\chi = \begin{cases} 0 & \text{if } f < 0, \text{ or if } f = 0 \text{ and } \mathbf{n}^{\mathrm{p}} : \dot{\boldsymbol{\epsilon}} \le 0 \\ 1 & \text{if } f = 0 \quad \text{and} \quad \mathbf{n}^{\mathrm{p}} : \dot{\boldsymbol{\epsilon}} > 0 \end{cases} \tag{21.7.7}$$

is a switching parameter.

Constitutive equations of the form (21.7.5) need to be accompanied by initial conditions. Typical initial conditions presume that the body is initially at time $t = 0$, in a virgin state, in the sense that

$$\boldsymbol{\epsilon}(\mathbf{x},0) = \boldsymbol{\epsilon}^{\mathrm{p}}(\mathbf{x},0) = \mathbf{0}, \qquad \text{and} \qquad \bar{\epsilon}^{\mathrm{p}}(\mathbf{x},0) = 0, \tag{21.7.8}$$

so that, $\boldsymbol{\epsilon}^{\mathrm{e}}(\mathbf{x},0) = \mathbf{0}$.

REMARK 21.2

In situations where the state of stress is known *a priori* it is often convenient to write the elastic-plastic constitutive relation in a rate compliance form, that is strain-rate in terms of stress and stress-rate. Combining the relations above, this results in

$$\dot{\boldsymbol{\epsilon}} = \underbrace{\frac{1+\nu}{E}\dot{\boldsymbol{\sigma}} - \frac{\nu}{E}(\mathrm{tr}\dot{\boldsymbol{\sigma}})\mathbf{1}}_{\dot{\boldsymbol{\epsilon}}^{\mathrm{e}}} + \underbrace{(3/2)\dot{\bar{\epsilon}}^{\mathrm{p}}\frac{\boldsymbol{\sigma}'}{\bar{\sigma}}}_{\dot{\boldsymbol{\epsilon}}^{\mathrm{p}}}, \tag{21.7.9}$$

where one notes that $f \le 0$ and $\dot{\bar{\epsilon}}^{\mathrm{p}} f = 0$ must additionally also hold at all times. Since time has no constitutive significance in the rate-independent theory of plasticity, this form of the constitutive relation is often expressed in an incremental form, which in indicial notation reads as

$$d\epsilon_{ij} = \underbrace{\frac{1+\nu}{E}d\sigma_{ij} - \frac{\nu}{E}(d\sigma_{kk})\delta_{ij}}_{d\epsilon^{\mathrm{e}}_{ij}} + \underbrace{(3/2)d\bar{\epsilon}^{\mathrm{p}}\frac{\sigma'_{ij}}{\bar{\sigma}}}_{d\epsilon^{\mathrm{p}}_{ij}}. \tag{21.7.10}$$

EXAMPLE 21.1 Plastically pressurized thin-walled sphere

Consider a thin-walled sphere with an initial radius of $r = 100\,\mathrm{mm}$ and an initial wall-thickness of $t = 1\,\mathrm{mm}$. The sphere is made from an isotropic rate-independent elastic-plastic material which hardens isotropically (power-law hardening), with material properties

$$E = 200\ \text{GPa}, \nu = 0.3, Y_0 = 100\ \text{MPa}, K = 50\ \text{MPa, and } n = 0.25.$$

It is desired to plastically expand the radius of the sphere to $101\,\mathrm{mm}$ by monotonically increasing the internal pressure p in the sphere. We wish to determine the pressure p which will cause this desired expansion, and also the amount of plastic deformation, as measured by $\bar{\epsilon}^{\mathrm{p}}$, in the plastically expanded sphere.

First note that in a spherical thin-walled pressure vessel that the stress state is statically determinate,

$$\sigma_{\theta\theta} = \sigma_{\phi\phi} = \frac{pr}{2t}, \qquad \text{and} \qquad \sigma_{rr} \approx 0. \tag{21.7.11}$$

This stress state in matrix form is

$$[\boldsymbol{\sigma}] = \frac{pr}{t}\begin{bmatrix} 0 & 0 & 0 \\ 0 & 1/2 & 0 \\ 0 & 0 & 1/2 \end{bmatrix}, \tag{21.7.12}$$

and correspondingly the deviatoric stress is

$$[\boldsymbol{\sigma}'] = \frac{pr}{t}\begin{bmatrix} -1/3 & 0 & 0 \\ 0 & 1/6 & 0 \\ 0 & 0 & 1/6 \end{bmatrix}, \tag{21.7.13}$$

and the Mises equivalent tensile stress is

$$\bar{\sigma} = \left|\frac{pr}{2t}\right| = \frac{pr}{2t} \quad (\text{since } p > 0). \tag{21.7.14}$$

Next, using the elastic-plastic constitutive law in compliance form (21.7.9), for the stress state under consideration the total strain-rate is given by

$$[\dot{\boldsymbol{\epsilon}}] = \frac{\dot{p}r}{2Et}\begin{bmatrix} -2\nu & 0 & 0 \\ 0 & 1-\nu & 0 \\ 0 & 0 & 1-\nu \end{bmatrix} + \frac{3}{2}\dot{\bar{\epsilon}}^{\mathrm{p}}\begin{bmatrix} -2/3 & 0 & 0 \\ 0 & 1/3 & 0 \\ 0 & 0 & 1/3 \end{bmatrix}. \tag{21.7.15}$$

Since the loading is monotonic, we can integrate (21.7.15) in time to yield,

$$[\boldsymbol{\epsilon}] = \frac{pr}{2Et}\begin{bmatrix} -2\nu & 0 & 0 \\ 0 & 1-\nu & 0 \\ 0 & 0 & 1-\nu \end{bmatrix} + \frac{3}{2}\bar{\epsilon}^{\mathrm{p}}\begin{bmatrix} -2/3 & 0 & 0 \\ 0 & 1/3 & 0 \\ 0 & 0 & 1/3 \end{bmatrix}. \tag{21.7.16}$$

This gives the hoop strain as

$$\epsilon_{\theta\theta}(\equiv \epsilon_{\phi\phi}) = \frac{(1-\nu)}{E}\frac{pr}{2t} + \frac{1}{2}\bar{\epsilon}^{\mathrm{p}}. \tag{21.7.17}$$

Further, when the material of the pressure vessel has undergone plastic deformation of amount $\bar{\epsilon}^{\mathrm{p}}$, the flow resistance $Y(\bar{\epsilon}^{\mathrm{p}})$ has increased from Y_0 to

$$Y(\bar{\epsilon}^{\mathrm{p}}) = Y_0 + K\,(\bar{\epsilon}^{\mathrm{p}})^n\,. \tag{21.7.18}$$

Since $\bar{\sigma} = Y(\bar{\epsilon}^{\mathrm{p}})$ during plastic flow and $\bar{\sigma}$ is given by (21.7.14), we obtain

$$\frac{pr}{2t} = Y_0 + K\,(\bar{\epsilon}^{\mathrm{p}})^n\,. \tag{21.7.19}$$

Substituting (21.7.19) in (21.7.17) gives

$$\epsilon_{\theta\theta} = \frac{(1-\nu)}{E}\left(Y_0 + K\,(\bar{\epsilon}^{\mathrm{p}})^n\right) + \frac{1}{2}\bar{\epsilon}^{\mathrm{p}}. \tag{21.7.20}$$

When the sphere of initial radius 100 mm has been expanded to radius of 101 mm, the hoop strain is

$$\epsilon_{\theta\theta} = 0.01. \tag{21.7.21}$$

For this value of $\epsilon_{\theta\theta}$, we may solve (21.7.20) numerically for $\bar{\epsilon}^{\rm p}$ to obtain,

$$\bar{\epsilon}^{\rm p} = 0.0192.$$

Finally, using (21.7.19), this value of $\bar{\epsilon}^{\rm p}$, the expanded radius $r = 101$ mm, and a thickness $t = 1$ mm (neglecting the small change in the thickness), the pressure p required to cause the expansion is evaluated from (21.7.19) as

$$p = 2.35\ \text{MPa}. \tag{21.7.22}$$

21.8 Three-dimensional rate-dependent plasticity with isotropic hardening

In what follows, we consider a generalization of our three-dimensional rate-independent theory, summarized in Sec. 21.7, to model rate-dependent plasticity, often called **viscoplasticity**. Here the kinematical decomposition (21.7.1) and the constitutive equation for the stress (21.7.2) are identical to the rate-independent theory. However, the flow rule, that is the evolution equation for $\boldsymbol{\epsilon}^{\rm p}$, is formulated differently.

As before, from the dissipation inequality (21.6.3) we note that the "driving force" conjugate to $\dot{\boldsymbol{\epsilon}}^{\rm p}$ is the deviatoric stress $\boldsymbol{\sigma}'$, and as before we assume that the *plastic flow and the deviatoric stress are* **co-directional**. That is,

$$\mathbf{n}^{\rm p} = \frac{\dot{\boldsymbol{\epsilon}}^{\rm p}}{|\dot{\boldsymbol{\epsilon}}^{\rm p}|} = \frac{\boldsymbol{\sigma}'}{|\boldsymbol{\sigma}'|} = \sqrt{3/2}\,\frac{\boldsymbol{\sigma}'}{\bar{\sigma}}, \qquad \text{with} \qquad \bar{\sigma} = \sqrt{3/2}\,|\boldsymbol{\sigma}'|. \tag{21.8.1}$$

Hence

$$\dot{\boldsymbol{\epsilon}}^{\rm p} = \sqrt{3/2}\,\dot{\bar{\epsilon}}^{\rm p}\,\mathbf{n}^{\rm p}, \quad \text{with} \quad \dot{\bar{\epsilon}}^{\rm p} = \sqrt{2/3}\,|\dot{\boldsymbol{\epsilon}}^{\rm p}| \geq 0. \tag{21.8.2}$$

Note that under this constitutive assumption, the dissipation inequality (21.6.3) reduces to

$$\mathcal{D} = \bar{\sigma}\dot{\bar{\epsilon}}^{\rm p} \geq 0, \tag{21.8.3}$$

and is trivially satisfied.

Next, in view of (21.8.3) we assume that whenever there is plastic flow, that is whenever $\dot{\bar{\epsilon}}^{\rm p} > 0$, the magnitude of the equivalent tensile stress is constrained to satisfy

$$\bar{\sigma} = \underbrace{\mathcal{S}(\bar{\epsilon}^{\rm p}, \dot{\bar{\epsilon}}^{\rm p})}_{\text{flow strength}}, \tag{21.8.4}$$

where the function $\mathcal{S}(\bar{\epsilon}^{\rm p}, \dot{\bar{\epsilon}}^{\rm p}) \geq 0$ represents the **flow strength** of the material at a given $\bar{\epsilon}^{\rm p}$ and $\dot{\bar{\epsilon}}^{\rm p}$. A special simple form for the strength relation assumes that the dependence of $\mathcal{S}$ on $\bar{\epsilon}^{\rm p}$ and $\dot{\bar{\epsilon}}^{\rm p}$ may be written as a separable relation of the type

$$\mathcal{S}(\bar{\epsilon}^{\rm p}, \dot{\bar{\epsilon}}^{\rm p}) = \underbrace{Y(\bar{\epsilon}^{\rm p})}_{\text{rate-independent}} \times \underbrace{g(\dot{\bar{\epsilon}}^{\rm p})}_{\text{rate-dependent}}, \tag{21.8.5}$$

with

$$Y(\bar{\epsilon}^{\mathrm{p}}) > 0 \tag{21.8.6}$$

a positive-valued scalar with dimensions of stress, and

$$g(\dot{\bar{\epsilon}}^{\mathrm{p}}) \text{ a positive-valued strictly increasing function of } \dot{\bar{\epsilon}}^{\mathrm{p}}, \text{ with } g(0) = 0. \tag{21.8.7}$$

We refer to $Y(\bar{\epsilon}^{\mathrm{p}})$ as the **strain-hardening function**, and to $g(\dot{\bar{\epsilon}}^{\mathrm{p}})$ as the **rate-sensitivity function.**

Next, by (21.8.7) the function $g(\dot{\bar{\epsilon}}^{\mathrm{p}})$ is *invertible*, and the inverse function $\varphi = g^{-1}$ is strictly increasing and hence strictly positive for all non-zero arguments; further $\varphi(0) = 0$. Hence the general relation (21.8.4), with $\mathcal{S}$ given by (21.8.5), may be inverted to give an expression

$$\dot{\bar{\epsilon}}^{\mathrm{p}} = g^{-1}\left(\frac{\bar{\sigma}}{Y(\bar{\epsilon}^{\mathrm{p}})}\right) \equiv \varphi\left(\frac{\bar{\sigma}}{Y(\bar{\epsilon}^{\mathrm{p}})}\right) \geq 0 \tag{21.8.8}$$

for the equivalent tensile plastic strain-rate. Finally, using (21.8.2) and (21.8.8), we obtain

$$\dot{\boldsymbol{\epsilon}}^{\mathrm{p}} = \sqrt{3/2}\,\varphi\left(\frac{\bar{\sigma}}{Y(\bar{\epsilon}^{\mathrm{p}})}\right)\mathbf{n}^{\mathrm{p}}. \tag{21.8.9}$$

Thus, in contrast to the rate-independent theory:

- *the plastic strain-rate is non-zero whenever the stress is non-zero: there is no elastic range in which the response of the material is purely elastic, and there are no considerations of a yield condition, a consistency condition, loading/unloading conditions, and so forth.*

21.8.1 **Power-law rate dependence**

An example of a commonly used rate-sensitivity function is the **power-law function**

$$g(\dot{\bar{\epsilon}}^{\mathrm{p}}) = \left(\frac{\dot{\bar{\epsilon}}^{\mathrm{p}}}{\dot{\epsilon}_0}\right)^m, \tag{21.8.10}$$

where $m \in (0, 1]$, a constant, is a **rate-sensitivity parameter** and $\dot{\epsilon}_0 > 0$, also a constant, is the reference flow-rate. The power-law function satisfies $g(\dot{\bar{\epsilon}}^{\mathrm{p}}) \approx 1$ for $\dot{\bar{\epsilon}}^{\mathrm{p}} \approx \dot{\epsilon}_0$. In view of (21.8.10), the relation (21.8.5) becomes

$$\bar{\sigma} = Y(\bar{\epsilon}^{\mathrm{p}})\left(\frac{\dot{\bar{\epsilon}}^{\mathrm{p}}}{\dot{\epsilon}_0}\right)^m, \tag{21.8.11}$$

and this implies that

$$\ln(\bar{\sigma}) = \ln(Y(\bar{\epsilon}^{\mathrm{p}})) + m\ln\left(\frac{\dot{\bar{\epsilon}}^{\mathrm{p}}}{\dot{\epsilon}_0}\right);$$

thus the rate-sensitivity parameter m is the slope of the graph of $\ln(\bar{\sigma})$ versus $\ln(\dot{\bar{\epsilon}}^{\mathrm{p}}/\dot{\epsilon}_0)$.

The power-law function allows one to characterize *nearly rate-independent behavior* in the case that m is very small. Indeed, since

$$\lim_{m\to 0}\left(\frac{\dot{\bar{\epsilon}}^{\mathrm{p}}}{\dot{\epsilon}_0}\right)^m = 1,$$

the limit $m \to 0$ in (21.8.11) corresponds to a *rate-independent yield condition*:

$$\bar{\sigma} = Y(\bar{\epsilon}^{\mathrm{p}}). \tag{21.8.12}$$

On the other hand, the limit $m \to 1$ corresponds to a *linearly viscous response*, as one observes in fluids.

Further, for the power-law function (21.8.10), the expressions (21.8.8) and (21.8.9) have the specific form

$$\dot{\bar{\epsilon}}^{\mathrm{p}} = \dot{\epsilon}_0 \left(\frac{\bar{\sigma}}{Y(\bar{\epsilon}^{\mathrm{p}})} \right)^{\frac{1}{m}} \tag{21.8.13}$$

and

$$\dot{\boldsymbol{\epsilon}}^{\mathrm{p}} = \sqrt{3/2}\,\dot{\epsilon}_0 \left(\frac{\bar{\sigma}}{Y(\bar{\epsilon}^{\mathrm{p}})} \right)^{\frac{1}{m}} \mathbf{n}^{\mathrm{p}}. \tag{21.8.14}$$

21.9 **Summary**

The complete set of constitutive equations for a small deformation **Mises–Hill**-type theory of rate-dependent plasticity with isotropic hardening consist of:

(i) The decomposition

$$\boldsymbol{\epsilon} = \boldsymbol{\epsilon}^{\mathrm{e}} + \boldsymbol{\epsilon}^{\mathrm{p}}, \tag{21.9.1}$$

of the strain $\boldsymbol{\epsilon}$ into elastic and plastic parts $\boldsymbol{\epsilon}^{\mathrm{e}}$ and $\boldsymbol{\epsilon}^{\mathrm{p}}$.

(ii) The elastic stress-strain relation

$$\boldsymbol{\sigma} = 2\mu(\boldsymbol{\epsilon} - \boldsymbol{\epsilon}^{\mathrm{p}}) + (\kappa - (2/3)\mu)(\operatorname{tr}\boldsymbol{\epsilon})\mathbf{1}, \tag{21.9.2}$$

with $\mu > 0$ and $\kappa > 0$ the elastic shear and bulk moduli.

(iii) A system of evolution equations

$$\boxed{\begin{aligned} &\dot{\boldsymbol{\epsilon}}^{\mathrm{p}} = \sqrt{3/2}\,\dot{\bar{\epsilon}}^{\mathrm{p}}\,\mathbf{n}^{\mathrm{p}}, \quad \mathbf{n}^{\mathrm{p}} = \sqrt{3/2}\,\frac{\boldsymbol{\sigma}'}{\bar{\sigma}}, \quad \bar{\sigma} = \sqrt{3/2}\,|\boldsymbol{\sigma}'|, \\ &\dot{\bar{\epsilon}}^{\mathrm{p}} = \varphi\left(\frac{\bar{\sigma}}{Y(\bar{\epsilon}^{\mathrm{p}})} \right), \end{aligned}} \tag{21.9.3}$$

for the plastic strain, and the equivalent tensile plastic strain in which $Y(\bar{\epsilon}^{\mathrm{p}})$ is the *rate-independent* flow strength.

Constitutive equations of the form (21.9.3) need to be accompanied by initial conditions. Typical initial conditions presume that the body is initially, at time $t = 0$, in a virgin state in the sense that

$$\boldsymbol{\epsilon}(\mathbf{x}, 0) = \boldsymbol{\epsilon}^{\mathrm{p}}(\mathbf{x}, 0) = \mathbf{0}, \quad \text{and} \quad \bar{\epsilon}^{\mathrm{p}}(\mathbf{x}, 0) = 0, \tag{21.9.4}$$

so that $\boldsymbol{\epsilon}^{\mathrm{e}}(\mathbf{x}, 0) = \mathbf{0}$.

REMARKS 21.3

1. The major difference from the rate-independent theory of the previous section is that $\dot{\bar{\epsilon}}^{\text{p}}$ is prescribed by a constitutive equation of the form $(21.9.3)_4$, and there is no yield condition, and no loading/unloading conditions. However, if desired, a threshold resistance to viscoplastic flow may be introduced as discussed for the one-dimensional theory in Sec. 20.3.4. Specifically, introducing a threshold resistance $Y_{\text{th}}(\bar{\epsilon}^{\text{p}}) \geq 0$ and adopting a power-law form for the rate-sensitivity function gives,

$$\dot{\bar{\epsilon}}^{\text{p}} = \dot{\epsilon}_0 \left\langle \frac{\bar{\sigma} - Y_{\text{th}}(\bar{\epsilon}^{\text{p}})}{Y(\bar{\epsilon}^{\text{p}})} \right\rangle^{1/m}, \tag{21.9.5}$$

where $\dot{\epsilon}_0 > 0$ is a reference plastic strain-rate, and $m \in (0, 1]$ is a strain-rate sensitivity parameter.

2. An important problem in application of the rate-dependent theory concerns the time integration of the rate equation $(21.9.3)_1$ for ϵ^{p}. This equation is typically *highly non-linear, and mathematically stiff*. With reference to the power-law model (21.8.14) for $\dot{\epsilon}^{\text{p}}$, the stiffness of the equations depends on the strain-rate sensitivity parameter m, and the stiffness increases to infinity as m tends to zero, the rate-independent limit.[3]

21.9.1 **Power-law creep form for high homologous temperatures**

For high temperatures $\vartheta \gtrsim 0.5\vartheta_m$, an explicit dependence on ϑ of the viscoplastic strain-rate is accounted for by replacing $\dot{\epsilon}_0$ by

$$\dot{\epsilon}_0 \equiv A \exp\left(-\frac{Q}{R\vartheta}\right), \tag{21.9.6}$$

where A is a pre-exponential factor with units of s^{-1}, Q is an activation energy in units of J/mol, and $R = 8.314$ J/(mol K) is the gas constant. In this case the scalar flow equation for $\dot{\bar{\epsilon}}^{\text{p}}$ is taken in the widely used *power-law creep* form:

$$\dot{\bar{\epsilon}}^{\text{p}} = A \exp\left(-\frac{Q}{R\vartheta}\right)\left(\frac{\bar{\sigma}}{Y}\right)^{\frac{1}{m}}. \tag{21.9.7}$$

Further the flow resistance Y is taken to evolve according to

$$\dot{Y} = \left[H_0 \left|1 - \frac{Y}{Y_{\text{s}}}\right|^r \text{sign}\left(1 - \frac{Y}{Y_{\text{s}}}\right)\right] \dot{\bar{\epsilon}}^{\text{p}} \quad \text{with initial value} \quad Y_0, \quad \text{and}$$

$$Y_{\text{s}} = Y_* \left[\frac{\dot{\bar{\epsilon}}^{\text{p}}}{A \exp\left(-\frac{Q}{R\vartheta}\right)}\right]^n, \tag{21.9.8}$$

[3] For small values of m special care is required to develop stable constitutive time-integration procedures (cf., e.g., Lush et al., 1989).

where Y_s represents the saturation value of Y for a given strain-rate and temperature, and $\{H_0, Y_*, r, n\}$ are strain-hardening parameters. An important feature of the theory is that the saturation value Y_s of the deformation resistance increases as the strain-rate $\dot{\epsilon}^p$ increases, or as the temperature decreases (Anand, 1982; Brown et al., 1989).

Conditions under which Y evolves are known as *transient or primary* creep, while conditions under which the flow resistance $Y \approx$ constant are known as *steady state creep*.

22 Three-dimensional plasticity with kinematic and isotropic hardening

22.1 Introduction

In this chapter we generalize the rate-independent and rate-dependent theories of plasticity with isotropic hardening of Chapter 21 to also account for kinematic hardening.

22.2 Rate-independent constitutive theory

The basic equations and the kinematical assumption are as in small deformation theory with isotropic hardening, viz.

$$\boldsymbol{\epsilon} = \boldsymbol{\epsilon}^{\mathrm{e}} + \boldsymbol{\epsilon}^{\mathrm{p}}, \qquad \operatorname{tr}\boldsymbol{\epsilon}^{\mathrm{p}} = 0. \tag{22.2.1}$$

Here we generalize that theory to account for a *free energy associated with plastic flow* resulting in an associated *backstress* and *kinematic hardening*. Specifically, we assume that the free energy has the form

$$\psi = \psi^{\mathrm{e}} + \psi^{\mathrm{p}},$$

with ψ^{e} an elastic energy, and ψ^{p} an energy associated with plastic flow, that is a **defect energy**, and we assume that $\dot{\psi}^{\mathrm{e}}$ and the elastic power $\boldsymbol{\sigma}:\dot{\boldsymbol{\epsilon}}^{\mathrm{e}}$ satisfy an *elastic balance*

$$\dot{\psi}^{\mathrm{e}} = \boldsymbol{\sigma}:\dot{\boldsymbol{\epsilon}}^{\mathrm{e}}, \tag{22.2.2}$$

in which case we are left with a *plastic free-energy imbalance*

$$\mathcal{D} = \boldsymbol{\sigma}':\dot{\boldsymbol{\epsilon}}^{\mathrm{p}} - \dot{\psi}^{\mathrm{p}} \geq 0, \tag{22.2.3}$$

which characterizes energy dissipated during plastic flow.

We assume that ψ^{e} and $\boldsymbol{\sigma}$ are given by the conventional elastic constitutive relations viz.,

$$\begin{aligned} \psi^{\mathrm{e}} &= \mu|\boldsymbol{\epsilon}^{\mathrm{e}\prime}|^2 + \tfrac{1}{2}\kappa(\operatorname{tr}\boldsymbol{\epsilon}^{\mathrm{e}})^2, \\ \boldsymbol{\sigma} &= 2\mu\boldsymbol{\epsilon}^{\mathrm{e}} + (\kappa - (2/3)\mu)(\operatorname{tr}\boldsymbol{\epsilon}^{\mathrm{e}})\mathbf{1}, \end{aligned} \tag{22.2.4}$$

or equivalently, using (22.2.1), that the stress is given by

$$\boldsymbol{\sigma} = 2\mu(\boldsymbol{\epsilon} - \boldsymbol{\epsilon}^{\mathrm{p}}) + (\kappa - (2/3)\mu)(\operatorname{tr}\boldsymbol{\epsilon})\mathbf{1}, \tag{22.2.5}$$

and in particular the deviatoric stress is given by

$$\boldsymbol{\sigma}' = 2\mu(\boldsymbol{\epsilon}' - \boldsymbol{\epsilon}^{\mathrm{p}}). \tag{22.2.6}$$

To ensure positive definiteness of the elastic free energy, we assume that $\mu > 0$ and $\kappa > 0$.

Continuum Mechanics of Solids. Lallit Anand and Sanjay Govindjee, Oxford University Press (2020).

DOI: 10.1093/oso/9780198864721.001.0001

Within the present framework, as a basic element, we will introduce a dimensionless internal variable $\mathbf{A}$ that, like ϵ^{p}, is symmetric and deviatoric:

$$\mathbf{A}(\mathbf{X},t), \quad \mathbf{A}=\mathbf{A}^{\top}, \quad \operatorname{tr}\mathbf{A}=0.$$

The internal variable $\mathbf{A}$ will be used to characterize kinematic hardening, and it will enter the theory though a constitutive equation for the plastic free-energy,

$$\psi^{\mathrm{p}}=\hat{\psi}^{\mathrm{p}}(\mathbf{A}), \tag{22.2.7}$$

with $\hat{\psi}^{\mathrm{p}}$ an isotropic function. For specificity we assume a simple quadratic form for ψ^{p},

$$\psi^{\mathrm{p}}=\frac{1}{2}C\,|\mathbf{A}|^2, \tag{22.2.8}$$

with $C>0$ a constant, and assume further that $\mathbf{A}$ evolves according to an evolution equation proposed by Armstrong and Fredrick (1966),

$$\dot{\mathbf{A}}=\dot{\epsilon}^{\mathrm{p}}-\gamma\,\mathbf{A}\dot{\bar{\epsilon}}^{\mathrm{p}}, \tag{22.2.9}$$

with $\gamma\geq 0$ a constant, and

$$\dot{\bar{\epsilon}}^{\mathrm{p}}\stackrel{\text{def}}{=}\sqrt{2/3}\,|\dot{\epsilon}^{\mathrm{p}}|, \tag{22.2.10}$$

the **equivalent tensile plastic strain-rate**. Note that we may write $\dot{\epsilon}^{\mathrm{p}}$ in terms of $\dot{\bar{\epsilon}}^{\mathrm{p}}$ as

$$\dot{\epsilon}^{\mathrm{p}}=\sqrt{3/2}\,\dot{\bar{\epsilon}}^{\mathrm{p}}\,\mathbf{n}^{\mathrm{p}}, \qquad \text{with} \qquad \mathbf{n}^{\mathrm{p}}=\frac{\dot{\epsilon}^{\mathrm{p}}}{|\dot{\epsilon}^{\mathrm{p}}|}. \tag{22.2.11}$$

From (22.2.8) and (22.2.9)

$$\dot{\psi}^{\mathrm{p}}=C\,\mathbf{A}:\dot{\mathbf{A}}=C\,\mathbf{A}:\dot{\epsilon}^{\mathrm{p}}-C\,\gamma|\mathbf{A}|^2\dot{\bar{\epsilon}}^{\mathrm{p}}=\boldsymbol{\sigma}_{\text{back}}:\dot{\epsilon}^{\mathrm{p}}-C\,\gamma|\mathbf{A}|^2\dot{\bar{\epsilon}}^{\mathrm{p}}, \tag{22.2.12}$$

where we have introduced the notation

$$\boldsymbol{\sigma}_{\text{back}}\stackrel{\text{def}}{=}C\,\mathbf{A} \tag{22.2.13}$$

for a **back stress.** In view of (22.2.13), we call C the *back stress modulus*. Also, note that in view of the properties of $\mathbf{A}$, the back stress $\boldsymbol{\sigma}_{\text{back}}$ is symmetric and deviatoric. Further observe that using (22.2.9) the back stress evolves according to

$$\dot{\boldsymbol{\sigma}}_{\text{back}}=C\,\dot{\epsilon}^{\mathrm{p}}-\gamma\boldsymbol{\sigma}_{\text{back}}\dot{\bar{\epsilon}}^{\mathrm{p}}. \tag{22.2.14}$$

Substituting (22.2.12) in (22.2.3) we obtain the dissipation inequality

$$\mathcal{D}=\boldsymbol{\sigma}_{\text{eff}}:\dot{\epsilon}^{\mathrm{p}}+C\,\gamma|\mathbf{A}|^2\dot{\bar{\epsilon}}^{\mathrm{p}}\geq 0, \tag{22.2.15}$$

where we have introduced the notation

$$\boldsymbol{\sigma}_{\text{eff}}\stackrel{\text{def}}{=}\boldsymbol{\sigma}'-\boldsymbol{\sigma}_{\text{back}}, \tag{22.2.16}$$

as an **effective stress** which is also symmetric and deviatoric. Next, the term $\boldsymbol{\sigma}_{\text{eff}}:\dot{\epsilon}^{\mathrm{p}}$ may be written as

$$\boldsymbol{\sigma}_{\text{eff}}:\dot{\epsilon}^{\mathrm{p}}=\left(\sqrt{3/2}\,\boldsymbol{\sigma}_{\text{eff}}:\mathbf{n}^{\mathrm{p}}\right)\dot{\bar{\epsilon}}^{\mathrm{p}}. \tag{22.2.17}$$

In view of (22.2.17) the dissipation inequality (22.2.15) may be written as

$$\mathcal{D} = \left(\sqrt{3/2}\,\boldsymbol{\sigma}_{\text{eff}} : \mathbf{n}^{\text{p}} \, + C\,\gamma|\mathbf{A}|^2\right)\,\dot{\bar{\epsilon}}^{\text{p}} \geq 0. \tag{22.2.18}$$

Guided by this requirement we assume that

- *the direction of plastic flow* $\mathbf{n}^{\text{p}}$ *is* **co-directional** *with the direction of* $\boldsymbol{\sigma}_{\text{eff}}$, that is

$$\mathbf{n}^{\text{p}} = \frac{\boldsymbol{\sigma}_{\text{eff}}}{|\boldsymbol{\sigma}_{\text{eff}}|} \equiv \frac{\boldsymbol{\sigma}' - \boldsymbol{\sigma}_{\text{back}}}{|\boldsymbol{\sigma}' - \boldsymbol{\sigma}_{\text{back}}|}. \tag{22.2.19}$$

Thus, introducing

$$\bar{\sigma}_{\text{eff}} \stackrel{\text{def}}{=} \sqrt{3/2}\,|\boldsymbol{\sigma}_{\text{eff}}|, \tag{22.2.20}$$

as an **equivalent tensile effective stress,** under the co-directionality assumption the dissipation inequality (22.2.18) is given by

$$\mathcal{D} = \left(\bar{\sigma}_{\text{eff}} + C\,\gamma|\mathbf{A}|^2\right)\dot{\bar{\epsilon}}^{\text{p}} \geq 0, \tag{22.2.21}$$

which is always satisfied since we have a product of two terms that are non-negative by definition.

22.2.1 Yield function. Elastic range. Yield set

As before, let

$$\bar{\epsilon}^{\text{p}}(\mathbf{x}, t) = \int_0^t \dot{\bar{\epsilon}}^{\text{p}}(\mathbf{x}, \zeta)\, d\zeta \tag{22.2.22}$$

define the **equivalent tensile plastic strain**, which serves as a monotonically increasing measure of the past history of plastic strain. Next, we introduce a positive-valued **tensile flow resistance**

$$Y(\bar{\epsilon}^{\text{p}}) > 0, \tag{22.2.23}$$

and the derivative

$$H_{\text{iso}}(\bar{\epsilon}^{\text{p}}) \stackrel{\text{def}}{=} \frac{dY(\bar{\epsilon}^{\text{p}})}{d\bar{\epsilon}^{\text{p}}}, \tag{22.2.24}$$

for the **rate of isotropic strain-hardening** of the material at a given $\bar{\epsilon}^{\text{p}}$.

Further, we introduce a **yield function**

$$f = \bar{\sigma}_{\text{eff}} - Y(\bar{\epsilon}^{\text{p}}), \tag{22.2.25}$$

and assume that the values of $\bar{\sigma}_{\text{eff}}$ are constrained by the inequality,

$$f = \bar{\sigma}_{\text{eff}} - Y(\bar{\epsilon}^{\text{p}}) \leq 0. \tag{22.2.26}$$

Next, arguing as in our discussion of the theory with isotropic hardening we obtain the **Kuhn–Tucker complementarity conditions,**

$$f \leq 0, \qquad \dot{\bar{\epsilon}}^{\text{p}} \geq 0, \qquad \dot{\bar{\epsilon}}^{\text{p}} f = 0, \tag{22.2.27}$$

and the **consistency condition,**

$$\dot{\bar{\epsilon}}^{\text{p}} \dot{f} = 0 \quad \text{when} \quad f = 0. \tag{22.2.28}$$

22.2.2 The flow rule

We now show that the Kuhn–Tucker and consistency conditions may be used to obtain an equation for $\dot{\boldsymbol{\epsilon}}^{\rm p}$ in terms of $\dot{\boldsymbol{\epsilon}}$ and $\boldsymbol{\sigma}$ and $\boldsymbol{\sigma}_{\rm back}$. In deriving this relation we recall the co-directionality condition (22.2.19), viz.

$$\mathbf{n}^{\rm p} = \frac{\dot{\boldsymbol{\epsilon}}^{\rm p}}{|\dot{\boldsymbol{\epsilon}}^{\rm p}|} = \frac{\boldsymbol{\sigma}_{\rm eff}}{|\boldsymbol{\sigma}_{\rm eff}|} = \frac{\boldsymbol{\sigma}' - \boldsymbol{\sigma}_{\rm back}}{|\boldsymbol{\sigma}' - \boldsymbol{\sigma}_{\rm back}|}. \tag{22.2.29}$$

Assume unless specified otherwise that the yield condition is satisfied:

$$f = 0 \qquad \text{(so that } \dot{f} \le 0\text{)}. \tag{22.2.30}$$

Then

$$\begin{aligned}
\dot{f} &= \sqrt{3/2}\,\overline{|\boldsymbol{\sigma}' - \boldsymbol{\sigma}_{\rm back}|}^{\,\boldsymbol{\cdot}} - \dot{\overline{Y(\bar{\epsilon}^{\rm p})}},\\
&= \sqrt{3/2}\,\frac{\boldsymbol{\sigma}' - \boldsymbol{\sigma}_{\rm back}}{|\boldsymbol{\sigma}' - \boldsymbol{\sigma}_{\rm back}|} : (\dot{\boldsymbol{\sigma}}' - \dot{\boldsymbol{\sigma}}_{\rm back}) - H_{\rm iso}\dot{\bar{\epsilon}}^{\rm p},\\
&= \sqrt{3/2}\,\mathbf{n}^{\rm p} : \dot{\boldsymbol{\sigma}}' - \sqrt{3/2}\,C\,\mathbf{n}^{\rm p} : \dot{\mathbf{A}} - H_{\rm iso}\dot{\bar{\epsilon}}^{\rm p} \qquad \text{(by (22.2.29), (22.2.13))}\\
&= \sqrt{3/2}\,2\mu\,\mathbf{n}^{\rm p} : (\dot{\boldsymbol{\epsilon}}' - \dot{\boldsymbol{\epsilon}}^{\rm p}) - [C((3/2) - \sqrt{3/2}\,\gamma\,\mathbf{A} : \mathbf{n}^{\rm p}) + H_{\rm iso}]\dot{\bar{\epsilon}}^{\rm p} \quad \text{(by (22.2.6), (22.2.9))}
\end{aligned}$$

or, since $\mathbf{n}^{\rm p}$ is deviatoric,

$$\dot{f} = \sqrt{6}\mu\,\mathbf{n}^{\rm p} : \dot{\boldsymbol{\epsilon}} - [3\mu + H]\,\dot{\bar{\epsilon}}^{\rm p}, \tag{22.2.31}$$

where we have introduced the notation

$$H \stackrel{\rm def}{=} H_{\rm iso} + C\Big((3/2) - \sqrt{3/2}\,\gamma\,\mathbf{A} : \mathbf{n}^{\rm p}\Big), \tag{22.2.32}$$

for an *overall hardening modulus*. We henceforth restrict attention to the class of materials for which

$$g \stackrel{\rm def}{=} 3\mu + H > 0. \tag{22.2.33}$$

We now deduce conditions that determine whether the material leaves or remains on the yield surface when the material point in question is subjected to a loading program characterized by the strain-rate $\dot{\boldsymbol{\epsilon}}$.

(i) **Elastic unloading** is defined by the condition $\mathbf{n}^{\rm p} : \dot{\boldsymbol{\epsilon}} < 0$. In this case (22.2.31) implies that $\dot{f} < 0$, and the consistency condition (22.2.28) then requires that $\dot{\bar{\epsilon}}^{\rm p} = 0$.

(ii) **Neutral loading** is defined by the condition $\mathbf{n}^{\rm p} : \dot{\boldsymbol{\epsilon}} = 0$. In this case $\dot{\bar{\epsilon}}^{\rm p} > 0$ cannot hold, for if it did, then (22.2.31) would imply that $\dot{f} < 0$, which by the consistency condition (22.2.28) once again leads to $\dot{\bar{\epsilon}}^{\rm p} = 0$.

(iii) **Plastic loading** is defined by the condition $\mathbf{n}^{\rm p} : \dot{\boldsymbol{\epsilon}} > 0$. In this case, if $\dot{\bar{\epsilon}}^{\rm p} = 0$, then $\dot{f} > 0$, which violates $\dot{f} \le 0$. Hence $\dot{\bar{\epsilon}}^{\rm p} > 0$, the consistency condition (22.2.28) then requires that $\dot{f} = 0$, and (22.2.31) yields

$$\dot{\bar{\epsilon}}^{\rm p} = \frac{\sqrt{6}\mu}{g}\mathbf{n}^{\rm p} : \dot{\boldsymbol{\epsilon}}. \tag{22.2.34}$$

Thus since, $\dot{\epsilon}^p = \sqrt{3/2}\,\dot{\bar{\epsilon}}^p\,\mathbf{n}^p$,

$$\dot{\epsilon}^p = \beta(\mathbf{n}^p : \dot{\epsilon})\mathbf{n}^p \neq \mathbf{0}, \tag{22.2.35}$$

where we have introduced a stiffness ratio,

$$\beta \stackrel{\text{def}}{=} \frac{3\mu}{3\mu + H} > 0. \tag{22.2.36}$$

Hence

- (22.2.35) *determines* $\dot{\epsilon}^p$ *in terms of* $\dot{\epsilon}, \boldsymbol{\sigma}, \boldsymbol{\sigma}_{\text{back}}$, the elastic shear modulus μ, and *the current hardening modulus* H.

Combining the results of (i)–(iii) with the condition

$$\dot{\epsilon}^p = \mathbf{0} \qquad \text{if} \qquad f < 0,$$

we arrive at the following equation for the plastic strain-rate $\dot{\epsilon}^p$:

$$\dot{\epsilon}^p = \begin{cases} \mathbf{0} & \text{if } f < 0 \quad \text{(behavior within the elastic range)}, \\ \mathbf{0} & \text{if } f = 0 \quad \text{and} \quad \mathbf{n}^p : \dot{\epsilon} < 0 \quad \text{(elastic unloading)}, \\ \mathbf{0} & \text{if } f = 0 \quad \text{and} \quad \mathbf{n}^p : \dot{\epsilon} = 0 \quad \text{(neutral loading)}, \\ \beta\,(\mathbf{n}^p : \dot{\epsilon})\mathbf{n}^p & \text{if } f = 0 \quad \text{and} \quad \mathbf{n}^p : \dot{\epsilon} > 0 \quad \text{(plastic loading)}. \end{cases} \tag{22.2.37}$$

The result (22.2.37) is embodied in the **flow rule**

$$\dot{\epsilon}^p = \chi\,\beta\,(\mathbf{n}^p : \dot{\epsilon})\mathbf{n}^p, \qquad \mathbf{n}^p = \frac{\boldsymbol{\sigma}' - \boldsymbol{\sigma}_{\text{back}}}{|\boldsymbol{\sigma}' - \boldsymbol{\sigma}_{\text{back}}|}, \tag{22.2.38}$$

where

$$\chi = \begin{cases} 0 & \text{if } f < 0, \text{ or if } f = 0 \text{ and } \mathbf{n}^p : \dot{\epsilon} \le 0 \\ 1 & \text{if } f = 0 \quad \text{and} \quad \mathbf{n}^p : \dot{\epsilon} > 0 \end{cases} \tag{22.2.39}$$

is a **switching parameter**.

22.3 Summary of the rate-independent theory with combined isotropic and kinematic hardening

In summary, the developed theory consists of:

1. A free energy:

$$\psi = \underbrace{\mu\,|\epsilon^e|^2 + \tfrac{1}{2}(\kappa - (2/3)\mu)|\operatorname{tr}\epsilon^e|^2}_{\psi^e} + \underbrace{\frac{1}{2}\,C\,|\mathbf{A}|^2}_{\psi^p}. \tag{22.3.1}$$

Here μ and κ are the elastic shear and bulk moduli which satisfy $\mu > 0$ and $\kappa > 0$; and $C > 0$ is a stiffness modulus for the contribution from the internal variable $\mathbf{A}$ to the free energy.

2. An equation for the stress:

$$\boldsymbol{\sigma} = 2\mu(\epsilon - \epsilon^p) + (\kappa - (2/3)\mu)(\operatorname{tr}\epsilon)\mathbf{1}. \tag{22.3.2}$$

3. An equation for the back stress:

$$\boldsymbol{\sigma}_{\text{back}} = C\,\mathbf{A}. \tag{22.3.3}$$

4. An effective stress. Equivalent tensile effective stress:

$$\boldsymbol{\sigma}_{\text{eff}} \stackrel{\text{def}}{=} \boldsymbol{\sigma}' - \boldsymbol{\sigma}_{\text{back}}, \tag{22.3.4}$$

and

$$\bar{\sigma}_{\text{eff}} \stackrel{\text{def}}{=} \sqrt{3/2}\,|\boldsymbol{\sigma}_{\text{eff}}|. \tag{22.3.5}$$

5. A yield condition:

$$f = \bar{\sigma}_{\text{eff}} - Y(\bar{\epsilon}^{\text{p}}) \le 0 \tag{22.3.6}$$

with f the **yield function**, and $Y(\bar{\epsilon}^{\text{p}}) > 0$ the **isotropic flow resistance**.

6. And a system of evolution equations:

$$\boxed{\begin{aligned} &\dot{\boldsymbol{\epsilon}}^{\text{p}} = \chi\,\beta\,(\mathbf{n}^{\text{p}} : \dot{\boldsymbol{\epsilon}})\mathbf{n}^{\text{p}}, \quad \mathbf{n}^{\text{p}} = \frac{\boldsymbol{\sigma}' - \boldsymbol{\sigma}_{\text{back}}}{|\boldsymbol{\sigma}' - \boldsymbol{\sigma}_{\text{back}}|}, \\ &\dot{\bar{\epsilon}}^{\text{p}} = \sqrt{2/3}|\dot{\boldsymbol{\epsilon}}^{\text{p}}|, \\ &\dot{\mathbf{A}} = \dot{\boldsymbol{\epsilon}}^{\text{p}} - \gamma\,\mathbf{A}\,\dot{\bar{\epsilon}}^{\text{p}} \end{aligned}} \tag{22.3.7}$$

for the plastic strain $\boldsymbol{\epsilon}^{\text{p}}$ and the hardening variables $\bar{\epsilon}^{\text{p}}$ and $\mathbf{A}$. Here,

$$\beta = \frac{3\mu}{3\mu + H}, \qquad 3\mu + H > 0,$$

with a hardening modulus

$$H \stackrel{\text{def}}{=} H_{\text{iso}} + C\Big((3/2) - \sqrt{3/2}\,\gamma\,\mathbf{A} : \mathbf{n}^{\text{p}}\Big) \qquad \text{where} \qquad H_{\text{iso}}(\bar{\epsilon}^{\text{p}}) = \frac{dY(\bar{\epsilon}^{\text{p}})}{d\bar{\epsilon}^{\text{p}}}, \tag{22.3.8}$$

and

$$\chi = \begin{cases} 0 & \text{if } f < 0, \text{ or if } f = 0 \text{ and } \mathbf{n}^{\text{p}} : \dot{\boldsymbol{\epsilon}} \le 0 \\ 1 & \text{if } f = 0 \quad \text{and} \quad \mathbf{n}^{\text{p}} : \dot{\boldsymbol{\epsilon}} > 0 \end{cases} \tag{22.3.9}$$

is a switching parameter. Typical initial conditions presume that the body is initially in a virgin state in the sense that

$$\boldsymbol{\epsilon}(\mathbf{x}, 0) = \boldsymbol{\epsilon}^{\text{p}}(\mathbf{x}, 0) = \mathbf{0}, \qquad \bar{\epsilon}^{\text{p}}(\mathbf{x}, 0) = 0, \qquad \mathbf{A}(\mathbf{x}, 0) = \mathbf{0}. \tag{22.3.10}$$

REMARK 22.1

Note that the dissipation in the theory,

$$\mathcal{D} = \big(Y(\bar{\epsilon}^{\text{p}}) + C\,\gamma|\mathbf{A}|^2\big)\dot{\bar{\epsilon}}^{\text{p}} \ge 0, \tag{22.3.11}$$

is always non-negative as required by the second law of thermodynamics.

22.4 **Summary of a rate-dependent theory with combined isotropic and kinematic hardening**

The rate-independent theory summarized in the previous section is straightforwardly generalized by mimicking the arguments in Sec. 21.8 for the rate-dependent theory with isotropic hardening. Such a generalization includes:

1. A free energy:

$$\psi = \underbrace{\mu\,|\boldsymbol{\epsilon}^{\rm e}|^2 + \tfrac{1}{2}(\kappa - (2/3)\mu)|\,{\rm tr}\,\boldsymbol{\epsilon}^{\rm e}|^2}_{\psi^{\rm e}} + \underbrace{\frac{1}{2}\,C\,|\mathbf{A}|^2}_{\psi^{\rm p}}. \tag{22.4.1}$$

   Here μ and κ are the elastic shear and bulk moduli which satisfy $\mu > 0$ and $\kappa > 0$; and $C > 0$ is a stiffness modulus for the contribution from the internal variable $\mathbf{A}$ to the free energy.

2. An equation for the stress:

$$\boldsymbol{\sigma} = 2\mu(\boldsymbol{\epsilon} - \boldsymbol{\epsilon}^{\rm p}) + (\kappa - (2/3)\mu)({\rm tr}\,\boldsymbol{\epsilon})\mathbf{1}. \tag{22.4.2}$$

3. An equation for the back stress:

$$\boldsymbol{\sigma}_{\rm back} = C\,\mathbf{A}. \tag{22.4.3}$$

4. An effective stress. Equivalent tensile effective stress:

$$\boldsymbol{\sigma}_{\rm eff} \stackrel{\rm def}{=} \boldsymbol{\sigma}' - \boldsymbol{\sigma}_{\rm back}, \tag{22.4.4}$$

   and

$$\bar{\sigma}_{\rm eff} \stackrel{\rm def}{=} \sqrt{3/2}\,|\boldsymbol{\sigma}_{\rm eff}|. \tag{22.4.5}$$

5. And a system of evolution equations:

$$\boxed{\begin{aligned} &\dot{\boldsymbol{\epsilon}}^{\rm p} = \sqrt{3/2}\,\dot{\bar{\epsilon}}^{\rm p}\,\mathbf{n}^{\rm p}, \quad \mathbf{n}^{\rm p} = \sqrt{3/2}\,\frac{\boldsymbol{\sigma}_{\rm eff}}{\bar{\sigma}_{\rm eff}}, \\ &\dot{\bar{\epsilon}}^{\rm p} = \varphi\left(\frac{\bar{\sigma}_{\rm eff}}{Y(\bar{\epsilon}^{\rm p})}\right), \\ &\dot{\mathbf{A}} = \dot{\boldsymbol{\epsilon}}^{\rm p} - \gamma\,\mathbf{A}\,\dot{\bar{\epsilon}}^{\rm p} \end{aligned}} \tag{22.4.6}$$

   for the plastic strain $\boldsymbol{\epsilon}^{\rm p}$ and the hardening variables $\bar{\epsilon}^{\rm p}$ and $\mathbf{A}$. Typical initial conditions assume that

$$\boldsymbol{\epsilon}(\mathbf{x},0) = \boldsymbol{\epsilon}^{\rm p}(\mathbf{x},0) = \mathbf{0}, \qquad \bar{\epsilon}^{\rm p}(\mathbf{x},0) = 0, \qquad \mathbf{A}(\mathbf{x},0) = \mathbf{0}. \tag{22.4.7}$$

The major difference from the rate-independent theory of the previous section is that $\dot{\bar{\epsilon}}^{\rm p}$ is prescribed by a constitutive equation of the form $(22.4.6)_2$, and there is no yield condition, and no loading/unloading conditions. However, if desired, a threshold resistance to viscoplastic flow may be introduced as discussed in the one-dimensional theory in Sec. 20.5. Specifically, introducing a threshold resistance $Y_{\text{th}}(\bar{\epsilon}^{\text{p}}) \geq 0$ and adopting a power-law form for the rate-sensitivity function gives,

$$\dot{\bar{\epsilon}}^{\text{p}} = \dot{\epsilon}_0 \left\langle \frac{\bar{\sigma} - Y_{\text{th}}(\bar{\epsilon}^{\text{p}})}{Y(\bar{\epsilon}^{\text{p}})} \right\rangle^{1/m}, \tag{22.4.8}$$

where $\dot{\epsilon}_0 > 0$ is a reference plastic strain-rate, and $m \in (0, 1]$ is a strain-rate sensitivity parameter.

23 Small deformation rate-independent plasticity based on a postulate of maximum dissipation

In this chapter we develop a more general rate-independent theory than that considered previously. The theory is based on a *postulate of maximum dissipation*, which is often credited to von Mises (cf., e.g., Hill, 1950). Below we present the postulate in an axiomatic manner; however, we note that the postulate can also be motivated by considering the material to possess a certain type of stability property, as advocated by Drucker (1959); also cf. Shames and Cozzarelli (1997, §8.11).

23.1 Elastic domain and yield set

We begin by introducing a positive-valued scalar $Y > 0$ with dimensions of stress—called the flow resistance, and consider a *smooth*[1] scalar-valued *yield function*

$$f(\boldsymbol{\sigma}, Y) \tag{23.1.1}$$

of the stress $\boldsymbol{\sigma}$ and Y, and let the boundedness hypothesis for the stress be given by

$$f(\boldsymbol{\sigma}, Y) \le 0. \tag{23.1.2}$$

Correspondingly let

$$\mathcal{E}(Y) = \{\boldsymbol{\sigma} \mid f(\boldsymbol{\sigma}, Y) \le 0\}, \tag{23.1.3}$$

denote the set of stresses in the elastic range for a given Y, and let

$$\mathcal{Y}(Y) = \{\boldsymbol{\sigma} \mid f(\boldsymbol{\sigma}, Y) = 0\} \tag{23.1.4}$$

denote the set of stresses on the yield surface, the yield set.

23.2 Postulate of maximum dissipation and its consequences

The *postulate of maximum dissipation* states that the plastic dissipation

$$\mathcal{D} = \boldsymbol{\sigma} : \dot{\boldsymbol{\epsilon}}^{\mathrm{p}} \ge 0 \tag{23.2.1}$$

caused by the actual stress $\boldsymbol{\sigma}$ for a *given* plastic strain-rate $\dot{\boldsymbol{\epsilon}}^{\mathrm{p}}$, results in *the* **maximum** *possible value for* $\mathcal{D}$ over all stresses $\boldsymbol{\sigma}_*$ in the elastic range $\mathcal{E}(Y)$; that is,

$$\boldsymbol{\sigma} : \dot{\boldsymbol{\epsilon}}^{\mathrm{p}} \ge \boldsymbol{\sigma}_* : \dot{\boldsymbol{\epsilon}}^{\mathrm{p}}, \qquad \forall \boldsymbol{\sigma}_* \in \mathcal{E}(Y). \tag{23.2.2}$$

[1] For the non-smooth case see Simo et al. (1988) and references therein.

Continuum Mechanics of Solids. Lallit Anand and Sanjay Govindjee, Oxford University Press (2020).

DOI: 10.1093/oso/9780198864721.001.0001

The postulate of maximum dissipation may be *equivalently* rephrased as a constrained minimization problem as follows (Simo and Hughes, 1998):

- With $\boldsymbol{\sigma}$ the actual stress, the dissipation,

$$\mathcal{D} = \boldsymbol{\sigma} : \dot{\boldsymbol{\epsilon}}^{\mathrm{p}} \geq 0 \tag{23.2.3}$$

for a *given* plastic strain rate $\dot{\boldsymbol{\epsilon}}^{p}$, represents *the* **minimum** *value of* $-\boldsymbol{\sigma}_* : \dot{\boldsymbol{\epsilon}}^{\mathrm{p}}$ over all stresses $\boldsymbol{\sigma}_* \in \mathcal{E}(Y)$.

We transform the constrained minimization problem above, to an *unconstrained* saddle point problem by introducing λ, a scalar-valued *Lagrange multiplier*, and constructing the Lagrangian function

$$L \stackrel{\text{def}}{=} -\boldsymbol{\sigma}_* : \dot{\boldsymbol{\epsilon}}^{\mathrm{p}} + \lambda\, f(\boldsymbol{\sigma}_*, Y),$$

and seek the stationary point of L with respect to $\boldsymbol{\sigma}_*$ and λ. Here $\dot{\boldsymbol{\epsilon}}^{\mathrm{p}}$ and the state Y are *considered to be known and fixed.*

The solution to this stationarity problem is given by the stress $\boldsymbol{\sigma}$ which satisfies the following **Kuhn–Tucker optimality** conditions (cf. Sec. 23.A):

$$\frac{\partial L}{\partial \boldsymbol{\sigma}_*} = \mathbf{0} \quad \Longrightarrow \quad -\dot{\boldsymbol{\epsilon}}^{\mathrm{p}} + \lambda \frac{\partial f(\boldsymbol{\sigma}, Y)}{\partial \boldsymbol{\sigma}} = \mathbf{0}, \tag{23.2.4}$$

with λ and $\boldsymbol{\sigma}$ subject to the necessary conditions

$$\lambda \geq 0, \quad f(\boldsymbol{\sigma}, Y) \leq 0, \quad \lambda\, f(\boldsymbol{\sigma}, Y) = 0. \tag{23.2.5}$$

Consequences of the maximum dissipation postulate: The maximum dissipation postulate implies the following major results:

(i) A normality flow rule:

$$\dot{\boldsymbol{\epsilon}}^{\mathrm{p}} = \lambda \frac{\partial f(\boldsymbol{\sigma}, Y)}{\partial \boldsymbol{\sigma}}. \tag{23.2.6}$$

This is also called an **associative flow rule** because it is *associated with the yield function* $f(\boldsymbol{\sigma}, Y)$. A schematic diagram of the normality flow rule is shown in Fig. 23.1, depicting the plastic strain-rate to be normal/orthogonal to the yield set.

(ii) The Kuhn–Tucker loading-unloading conditions:

$$\lambda \geq 0, \quad f(\boldsymbol{\sigma}, Y) \leq 0, \quad \lambda\, f(\boldsymbol{\sigma}, Y) = 0. \tag{23.2.7}$$

(iii) The elastic domain

$$\mathcal{E}(Y) = \{\boldsymbol{\sigma} \mid f(\boldsymbol{\sigma}, Y) \leq 0\}$$

is **convex.**

Further, as discussed in Chapter 21 (cf. eq. (21.6.22)), during plastic flow we must satisfy the **consistency condition**

$$\text{if } \lambda > 0, \text{ then } f = 0 \text{ and } \dot{f} = 0, \tag{23.2.8}$$

which may alternatively be written as

$$\lambda \dot{f} = 0 \qquad \text{when} \qquad f = 0. \tag{23.2.9}$$

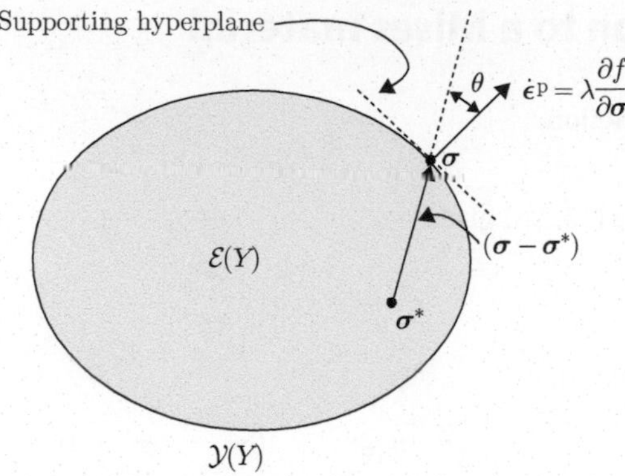

Fig. 23.1 Convex elastic domain $\mathcal{E}(Y)$ and the normality flow rule.

PROOF: (i) and (ii) follow immediately from (23.2.4) and (23.2.5).

To show that the convexity condition (iii) for the elastic domain $\mathcal{E}(Y)$ also follows from the postulate of maximum dissipation, first note that should $\boldsymbol{\sigma} \in \text{int}(\mathcal{E}) \stackrel{\text{def}}{=} \mathcal{E}(Y) \setminus \mathcal{Y}(Y)$, then $\lambda = 0$ by eqn $(23.2.7)_3$ and thus $\dot{\boldsymbol{\epsilon}}^p = 0$, which is the no plastic flow case. Thus necessarily during non-zero plastic flow, $\boldsymbol{\sigma} \in \mathcal{Y}(Y)$ with $\dot{\boldsymbol{\epsilon}}^p$ co-directional with $\partial f(\boldsymbol{\sigma}, Y)/\partial \boldsymbol{\sigma}$, as shown in Fig. 23.1. Consider a stress $\boldsymbol{\sigma}_*$ in the elastic range $\mathcal{E}(Y)$, then the principle of maximum dissipation (23.2.2) requires that

$$(\boldsymbol{\sigma} - \boldsymbol{\sigma}_*) : \dot{\boldsymbol{\epsilon}}^p \geq 0 \qquad \forall \boldsymbol{\sigma}_* \in \mathcal{E}(Y). \tag{23.2.10}$$

If we view the tensors $(\boldsymbol{\sigma} - \boldsymbol{\sigma}_*)$ and $\dot{\boldsymbol{\epsilon}}^p$ as vectors in the space of symmetric tensors, then the angle θ between non-zero "vectors" $(\boldsymbol{\sigma} - \boldsymbol{\sigma}_*)$ and $\dot{\boldsymbol{\epsilon}}^p$ is *defined* by the relation

$$\cos\theta \stackrel{\text{def}}{=} \frac{(\boldsymbol{\sigma} - \boldsymbol{\sigma}_*) : \dot{\boldsymbol{\epsilon}}^p}{|\boldsymbol{\sigma} - \boldsymbol{\sigma}_*|\,|\dot{\boldsymbol{\epsilon}}^p|}. \tag{23.2.11}$$

The principle of maximum dissipation (23.2.10) is thus seen to require that for every interior or boundary point $\boldsymbol{\sigma}_* \in \mathcal{E}(Y)$ we must have $\cos\theta \geq 0$; that is, the angle θ between $(\boldsymbol{\sigma} - \boldsymbol{\sigma}_*)$ and $\dot{\boldsymbol{\epsilon}}^p$ *cannot be obtuse* (cf. Fig. 23.1). Thus all points $\boldsymbol{\sigma}_* \in \mathcal{E}(Y)$ lie to one side of the supporting hyperplane with normal $\dot{\boldsymbol{\epsilon}}^p$ at $\boldsymbol{\sigma}$. This plane is depicted by the dashed line in Fig. 23.1. The side on which $\boldsymbol{\sigma}_*$ must lie is the one opposite to which $\dot{\boldsymbol{\epsilon}}^p$ points. Hence, through **every** point $\boldsymbol{\sigma}$ on the yield surface $\mathcal{Y}(Y)$, there is a plane such that all points $\boldsymbol{\sigma}_* \in \mathcal{E}(Y)$ lie on the same side of the plane. Invoking the *converse supporting hyperplane theorem* (Boyd and Vandenberghe, 2004, Chap. 2) then leads to the conclusion that *the elastic range $\mathcal{E}(Y)$ is a convex set.*

REMARK 23.1

Note that if $f(\cdot)$ is chosen *a priori* convex[2], then $\mathcal{E}$ is convex independent of the maximum dissipation assumption by the *level set theorem*; see e.g. Rockafellar (1970, Theorem 4.6).

---

[2] A function f is convex if and only if $f(\alpha\boldsymbol{\sigma}_1 + (1-\alpha)\boldsymbol{\sigma}_2) \leq \alpha f(\boldsymbol{\sigma}_1) + (1-\alpha) f(\boldsymbol{\sigma}_2)$ for all $\boldsymbol{\sigma}_1, \boldsymbol{\sigma}_2$ in the domain of f and for all $\alpha \in [0, 1]$.

23.3 Application to a Mises material

For the Mises yield function,

$$f = \sqrt{3/2}|\boldsymbol{\sigma}'| - Y, \tag{23.3.1}$$

the normality flow rule (23.2.6) gives

$$\dot{\boldsymbol{\epsilon}}^{\rm p} = \lambda\sqrt{3/2}\frac{\boldsymbol{\sigma}'}{|\boldsymbol{\sigma}'|}. \tag{23.3.2}$$

Equation (23.3.2) implies that

$$\lambda = \sqrt{2/3}\,|\dot{\boldsymbol{\epsilon}}^{\rm p}| \qquad \text{for} \qquad \boldsymbol{\sigma}' \neq \mathbf{0}, \tag{23.3.3}$$

and yields, as a consequence, our previous assumption of **co-directionality**:

$$\frac{\dot{\boldsymbol{\epsilon}}^{\rm p}}{|\dot{\boldsymbol{\epsilon}}^{\rm p}|} = \frac{\boldsymbol{\sigma}'}{|\boldsymbol{\sigma}'|}. \tag{23.3.4}$$

Thus introducing the equivalent tensile stress and the equivalent tensile plastic strain-rate by

$$\bar{\sigma} \stackrel{\rm def}{=} \sqrt{3/2}\,|\boldsymbol{\sigma}'|, \qquad \dot{\bar{\epsilon}}^{\rm p} \stackrel{\rm def}{=} \sqrt{2/3}\,|\dot{\boldsymbol{\epsilon}}^{\rm p}| \equiv \lambda, \tag{23.3.5}$$

the flow rule (23.3.2) may be written as

$$\dot{\boldsymbol{\epsilon}}^{\rm p} = (3/2)\,\dot{\bar{\epsilon}}^{\rm p}\,\frac{\boldsymbol{\sigma}'}{\bar{\sigma}}, \tag{23.3.6}$$

and the complementarity relations become

$$\dot{\bar{\epsilon}}^{\rm p} \geq 0, \quad f \leq 0, \quad \dot{\bar{\epsilon}}^{\rm p} f = 0. \tag{23.3.7}$$

Further, the **consistency condition** (23.2.9) becomes

$$\dot{\bar{\epsilon}}^{\rm p}\dot{f} = 0 \qquad \text{when} \qquad f = 0. \tag{23.3.8}$$

For a Mises material in which the plastic flow rule is based on the normality condition (23.2.6), the resulting final set of constitutive equations are **identical** to those in the theory considered in Chapter 21, in which the flow rule was based on the co-directionality hypothesis (21.6.4).

REMARK 23.2

The postulate of maximum dissipation—while widely used—is not valid for all materials. In particular, for materials for which the yield function depends on $\operatorname{tr}\boldsymbol{\sigma}$, a consequence of the normality flow rule is that there will be an attendant plastic volume change; however, the magnitude of the plastic volume change predicted by the normality flow rule, often *far exceeds* that which is actually measured experimentally. This occurs for example for granular materials, for which it is well-documented that even though the yield function for such materials depends on $\operatorname{tr}\boldsymbol{\sigma}$, plastic flow under fully developed flows is almost incompressible, and that the normality flow rule therefore does not hold for such materials.

Appendices

23.A Kuhn–Tucker optimality conditions

We recall a result from optimization theory (cf., e.g., Strang, 2007, p. 724):

PROBLEM: Find the minimum of a cost (or objective) function

$$C(\mathbf{x}), \quad \mathbf{x} \in \mathbb{R}^n,$$

subject to m inequality constraints

$$(A_i(\mathbf{x}) - b_i) \leq 0, \quad i = 1, \cdots, m.$$

SOLUTION: This problem is transformed from a constrained minimization problem to an unconstrained minimization problem as follows. Introduce m *Lagrange multipliers*

$$\boldsymbol{\lambda} \in \mathbb{R}^m,$$

corresponding to the m constraints, and construct the Lagrangian function

$$L = C(\mathbf{x}) + \sum_{i=1}^{m} \lambda_i \left(A_i(\mathbf{x}) - b_i\right),$$

and seek the unconstrained minimization of L. The unconstrained minimum occurs where

$$\frac{\partial L}{\partial x_j} = \frac{\partial C}{\partial x_j} + \sum_{i=1}^{m} \lambda_i \frac{\partial A}{\partial x_j} = 0, \quad j = 1, \cdots, n, \tag{23.A.1}$$

with $\boldsymbol{\lambda}$ and $\mathbf{x}$ also subject to

$$\lambda_i \geq 0, \quad (A_i(\mathbf{x}) - b_i) \leq 0, \quad \lambda_i(A_i(\mathbf{x}) - b_i) = 0, \quad i = 1, \cdots, m. \tag{23.A.2}$$

Equations (23.A.1) and (23.A.2) are known as the **Kuhn–Tucker optimality conditions** and are the necessary conditions for a minimizer (see e.g. Bertsekas (1995, Chap. 3)).

24 Some classical problems in rate-independent plasticity

24.1 Elastic-plastic torsion of a cylindrical bar

A simple example which displays most of the characteristic features of the elastic-plastic response of materials is provided by the problem of torsion of a cylindrical bar.

24.1.1 Kinematics

Consider a homogeneous circular cylinder of length L and radius R. Because of the cylindrical geometry of the body, it is convenient to use a cylindrical coordinate system with orthonormal base vectors $\{\mathbf{e}_r, \mathbf{e}_\theta, \mathbf{e}_z\}$ and coordinates $\{r, \theta, z\}$. Let the axis of the cylinder coincide with the $\mathbf{e}_z$-direction, and let the end-faces of the cylinder coincide with $z = 0$ and $z = L$; cf. Fig. 24.1.

Pure torsion of the shaft is characterized by the following displacement field:

- no change in length in the axial direction occurs, $u_z = 0$;
- no change in length in the radial direction, $u_r = 0$; and
- the tangential displacement varies as

$$u_\theta = \alpha z r, \tag{24.1.1}$$

where $\alpha =$ constant is the **twist per unit length.**

This state of affairs occurs independent of the material response, elastic or plastic, given appropriately imposed boundary conditions and isotropic material response; cf. Sec. 9.3.2.

The components of the infinitesimal strain tensor in cylindrical coordinates are

$$\begin{aligned}
\epsilon_{rr} &= \frac{\partial u_r}{\partial r}, \qquad \epsilon_{\theta\theta} = \frac{1}{r}\frac{\partial u_\theta}{\partial \theta} + \frac{u_r}{r}, \qquad \epsilon_{zz} = \frac{\partial u_z}{\partial z},\\
\epsilon_{r\theta} &= \frac{1}{2}\left(\frac{1}{r}\frac{\partial u_r}{\partial \theta} + \frac{\partial u_\theta}{\partial r} - \frac{u_\theta}{r}\right) = \epsilon_{\theta r},\\
\epsilon_{\theta z} &= \frac{1}{2}\left(\frac{1}{r}\frac{\partial u_z}{\partial \theta} + \frac{\partial u_\theta}{\partial z}\right) = \epsilon_{z\theta},\\
\epsilon_{zr} &= \frac{1}{2}\left(\frac{\partial u_z}{\partial r} + \frac{\partial u_r}{\partial z}\right) = \epsilon_{rz}.
\end{aligned} \tag{24.1.2}$$

Hence, for the case of pure torsion the only non-vanishing strain component is

$$\epsilon_{\theta z} = \epsilon_{z\theta} = \frac{1}{2}\alpha r. \tag{24.1.3}$$

Continuum Mechanics of Solids. Lallit Anand and Sanjay Govindjee, Oxford University Press (2020).

DOI: 10.1093/oso/9780198864721.001.0001

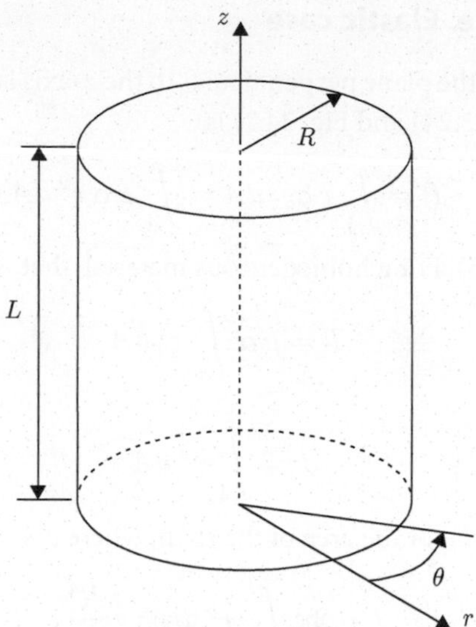

Fig. 24.1 An undeformed cylinder of length L and radius R.

24.1.2 **Elastic constitutive equation**

We begin by reviewing the elastic case. From the elastic constitutive equation

$$\boldsymbol{\sigma} = 2\mu\boldsymbol{\epsilon} + (\kappa - (2/3)\mu)(\operatorname{tr}\boldsymbol{\epsilon})\mathbf{1}, \tag{24.1.4}$$

the only non-zero components of stress are

$$\sigma_{\theta z}(= \sigma_{z\theta}) = 2\mu\epsilon_{\theta z} = \mu\,\alpha\,r, \tag{24.1.5}$$

where μ is the elastic **shear modulus** of the material.

24.1.3 **Equilibrium**

From the general equilibrium equations (in the absence of body forces) in cylindrical coordinates

$$\begin{aligned}
&\frac{\partial \sigma_{rr}}{\partial r} + \frac{1}{r}\frac{\partial \sigma_{r\theta}}{\partial \theta} + \frac{\partial \sigma_{rz}}{\partial z} + \frac{1}{r}(\sigma_{rr} - \sigma_{\theta\theta}) = 0,\\
&\frac{\partial \sigma_{\theta r}}{\partial r} + \frac{1}{r}\frac{\partial \sigma_{\theta\theta}}{\partial \theta} + \frac{\partial \sigma_{\theta z}}{\partial z} + \frac{2}{r}\,\sigma_{\theta r} = 0,\\
&\frac{\partial \sigma_{zr}}{\partial r} + \frac{1}{r}\frac{\partial \sigma_{z\theta}}{\partial \theta} + \frac{\partial \sigma_{zz}}{\partial z} + \frac{\sigma_{zr}}{r} = 0,
\end{aligned} \tag{24.1.6}$$

we note that since $\sigma_{\theta z}$ is independent of θ and z, that the stress field (24.1.5) identically satisfies equilibrium.

24.1.4 Resultant torque: Elastic case

By (24.1.5) the traction on the plane perpendicular to the z-axis is $\mathbf{t} = \boldsymbol{\sigma}\mathbf{e}_z = \sigma_{\theta z}\mathbf{e}_\theta$ and hence the resultant torque (cf. (9.3.24) and Fig. 24.2) is

$$T = \int_A r\,\sigma_{\theta z}\,dA = \int_A \mu\,\alpha\,r^2\,dA, \tag{24.1.7}$$

where we have used (24.1.5). For a homogeneous material, that is μ =constant, (24.1.7) gives

$$T = \mu\,\alpha \int_A r^2\,dA. \tag{24.1.8}$$

Now,

$$J \stackrel{\text{def}}{=} \int_A r^2\,dA \tag{24.1.9}$$

is the *polar moment of inertia* of the area of the shaft, where $dA = rdrd\theta$, and thus

$$J = 2\pi \int_0^R r^3\,dr = \frac{\pi R^4}{2}, \tag{24.1.10}$$

where R is the outer radius of the shaft.[1]

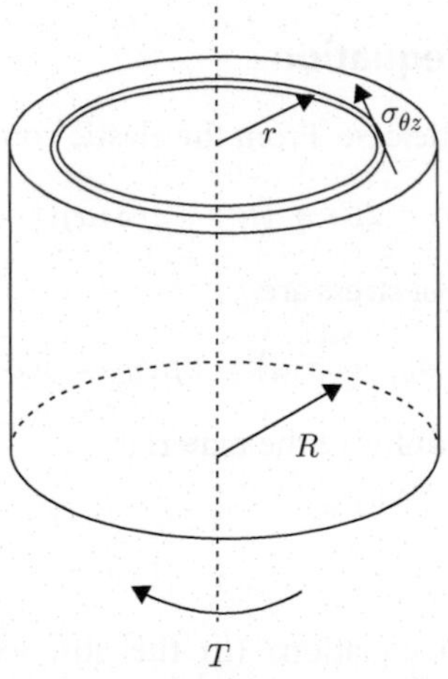

Fig. 24.2 Schematic of forces in torsion.

[1] For a **hollow** circular tube with an inner radius R_i and an outer radius R_o, J is given as the difference in the moments of inertia of solid rods of radius R_o and R_i, respectively:

$$J = \frac{\pi R_\text{o}^4}{2} - \frac{\pi R_\text{i}^4}{2}.$$

For a **thin-walled tube** of nominal radius R and wall-thickness $t \ll R$,

$$J = 2\pi R^3 t.$$

Thus,

$$T = (\mu J)\alpha. \tag{24.1.11}$$

The quantity (μJ) is called the **torsional stiffness** of the shaft, and Eq. (24.1.11) is the basic torque-twist equation.

Shear stress in terms of applied torque and geometry

Since the angle of twist per unit length is

$$\alpha = \frac{T}{\mu J}, \tag{24.1.12}$$

and the shear strain is

$$\epsilon_{\theta z} = \frac{1}{2}\alpha r = \frac{T\,r}{2\mu J}, \tag{24.1.13}$$

the shear stress $\sigma_{\theta z} = 2\mu\epsilon_{\theta z}$ is given by

$$\sigma_{\theta z}(r) = \frac{T\,r}{J}, \tag{24.1.14}$$

which is the basic relation for the shear stress in elastic torsion.

24.1.5 **Total twist of a shaft**

We can integrate the expression (24.1.12) for the twist per unit length to get a **total twist** over a given length of a shaft:

$$\Theta = \int_0^L \alpha\,dz = \int_0^L \frac{T}{\mu J}\,dz. \tag{24.1.15}$$

If T, μ, and J are constant, we get

$$\Theta = \frac{TL}{\mu J}. \tag{24.1.16}$$

- *Equations* (24.1.11), (24.1.14), *and* (24.1.16) *are the basic equations for the design and analysis of homogeneous circular elastic rods in torsion.*

24.1.6 **Elastic-plastic torsion**

We now consider pure torsion of shafts in the elastic-plastic regime. We limit our discussion to an *elastic-perfectly-plastic* material with a constant flow strength Y in tension. We will develop a relation for the torque required to cause the outer fibers of the shaft to begin to deform plastically, as well as a relation for the torque at which the complete section of the shaft might become plastic. We will also derive a procedure to predict the **elastic spring-back** and **residual stresses** in the shaft after unloading.

Onset of yield

The Mises equivalent tensile stress is

$$\bar{\sigma} = \left| \left[\frac{1}{2} \left\{ (\sigma_{rr} - \sigma_{\theta\theta})^2 + (\sigma_{\theta\theta} - \sigma_{zz})^2 + (\sigma_{zz} - \sigma_{rr})^2 \right\} + 3 \left\{ \sigma_{r\theta}^2 + \sigma_{\theta z}^2 + \sigma_{zr}^2 \right\} \right]^{1/2} \right|, \tag{24.1.17}$$

and for pure torsion

$$\bar{\sigma} = \sqrt{3}\,|\sigma_{\theta z}|. \tag{24.1.18}$$

Thus, for a perfectly plastic material with **tensile yield strength** $Y(0) = \sigma_y$, plastic flow is initiated when

$$\bar{\sigma} = \sqrt{3}\,|\sigma_{\theta z}| = \sigma_y, \tag{24.1.19}$$

or when

$$|\sigma_{\theta z}| = \tau_y, \tag{24.1.20}$$

where

$$\tau_y \stackrel{\text{def}}{=} \frac{\sigma_y}{\sqrt{3}} \tag{24.1.21}$$

is the **shear yield strength** of the material. Let

$$\tau \stackrel{\text{def}}{=} \sigma_{\theta z}, \qquad \text{and} \qquad \gamma \stackrel{\text{def}}{=} 2\epsilon_{\theta z} = \alpha r. \tag{24.1.22}$$

The quantity

$$\gamma_y \stackrel{\text{def}}{=} \frac{\tau_y}{\mu} \tag{24.1.23}$$

is the **shear yield strain** of the material. In terms of τ and γ, a schematic of the shear stress-strain curve for an elastic-perfectly-plastic material is shown in Fig. 24.3.

In the elastic range, the non-zero stress in torsion is (cf. (24.1.5))

$$\sigma_{\theta z} = (\mu\alpha)r \tag{24.1.24}$$

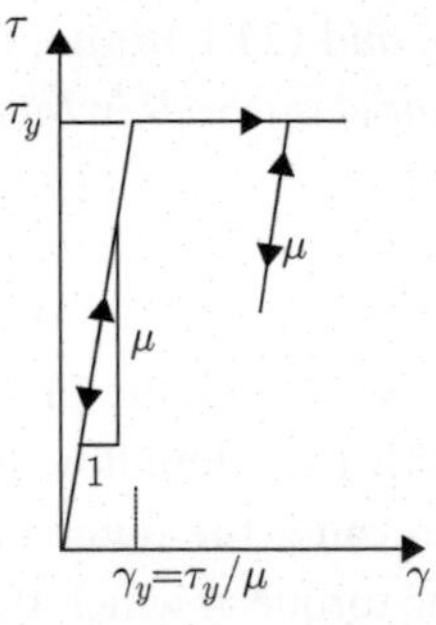

Fig. 24.3 Schematic of an elastic-perfectly-plastic constitutive model in shear.

and is a maximum at $r = R$. This implies that plastic yielding is initiated at the outer radius of the shaft. From (24.1.3), we see that the engineering shear strain

$$\gamma = 2\epsilon_{\theta z} = \alpha r \tag{24.1.25}$$

varies linearly with r for each twist α. Let α_y denote the twist when the strain at the outer fibers $r = R$ first reaches the yield strain γ_y,

$$\gamma_y = \frac{\tau_y}{\mu} = \alpha_y \times R.$$

This gives

$$\alpha_y = \frac{\gamma_y}{R} = \frac{\tau_y}{\mu R}. \tag{24.1.26}$$

With increasing twist $\alpha > \alpha_y$, the strain distribution continues to vary linearly across the radius (cf. 24.1.25); however, the stress τ will not continue to vary linearly across the radius. For an elastic-perfectly-plastic material under monotonically increasing magnitudes of shear strain

$$\tau = \begin{cases} \mu\gamma & \text{if } \gamma \leq \gamma_y, \\ \tau_y & \text{if } \gamma > \gamma_y. \end{cases} \tag{24.1.27}$$

Figure 24.4 schematically shows the stress distribution as a function of radial position for a value of $\alpha > \alpha_y$, when a core of the shaft at $0 \leq r \leq c$ is elastic, while the outer region $c \leq r \leq R$ is deforming plastically.

Torque-twist relation

With $\tau(r)$ denoting the shear stress distribution due to torsion, the resultant torque is

$$T = \int_A r\,\tau(r)\,dA, \tag{24.1.28}$$

and since $dA = rdrd\theta$, and since the shear stress varies only with r,

$$T = 2\pi \int_0^R \tau(r)\,r^2 dr. \tag{24.1.29}$$

Torque-twist relation for $\alpha \leq \alpha_y$

When $\alpha \leq \alpha_y$ the cross-section is deforming elastically, so that by (24.1.11)

$$T = (\mu J)\alpha \qquad \text{for} \qquad \alpha \leq \alpha_y. \tag{24.1.30}$$

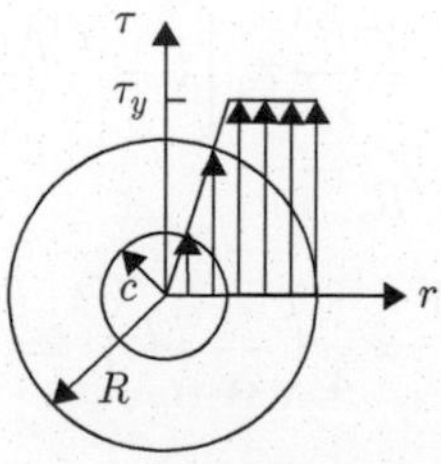

Fig. 24.4 Stress distribution as a function of radial position for $\alpha > \alpha_y$.

The value of the torque at the elastic limit, that is when $\alpha = \alpha_y$, is therefore given by

$$T_y = \mu J \alpha_y = \mu J \frac{\tau_y}{\mu R} = \frac{\tau_y J}{R} = \frac{\pi R^3}{2}\tau_y,$$

where we have used (24.1.26) and (24.1.10). Thus the outer fibers of the shaft begin to yield plastically when the torque T reaches the value

$$\boxed{T_y = \frac{\pi R^3}{2}\tau_y.} \tag{24.1.31}$$

Torque-twist relation for $\alpha > \alpha_y$

Now suppose that α is increased to a value greater than α_y; the shear strain still varies linearly through the radius, but $\gamma \geq \gamma_y$ over a certain region $c \leq r \leq R$; here $c > 0$ is the r-coordinate of the transition between the elastic core and the plastic outer shell; cf. Fig. 24.4. We can solve for c in terms of the twist α and the yield strain γ_y. Since

$$\gamma_y = \alpha c$$

at $r = c$, we obtain

$$c = \frac{\gamma_y}{\alpha} \qquad \text{for } \alpha \geq \alpha_y. \tag{24.1.32}$$

Thus for $\alpha > \alpha_y$ the shear stress distribution through the radius of the shaft is given by

$$\tau = \begin{cases} \mu \alpha r & \text{in } 0 \leq r \leq c, \\ \tau_y & \text{in } c \leq r \leq R, \end{cases} \tag{24.1.33}$$

with c given by (24.1.32). This stress distribution is sketched in Fig. 24.4.

Next, using (24.1.29), the expression for the torque becomes

$$\begin{aligned} T &= 2\pi \left[\int_0^c \mu \alpha r^3\, dr + \int_c^R \tau_y r^2\, dr \right] \\ &= 2\pi \left[\frac{\mu \alpha c^4}{4} + \frac{\tau_y}{3}\left(R^3 - c^3\right) \right] \\ &= 2\pi \left[\frac{\mu \gamma_y c^3}{4} + \frac{\tau_y}{3}\left(R^3 - c^3\right) \right] && (\text{using } \alpha c = \gamma_y) \\ &= 2\pi \left[\frac{\tau_y c^3}{4} + \frac{\tau_y}{3}\left(R^3 - c^3\right) \right] && (\text{using } \tau_y = \mu \gamma_y). \end{aligned}$$

Hence,

$$T = \frac{2\pi R^3}{3}\tau_y \left[1 - \frac{1}{4}\left(\frac{c}{R}\right)^3 \right]. \tag{24.1.34}$$

Next, using $c = \gamma_y/\alpha$ and $\alpha_y = \gamma_y/R$,

$$\frac{1}{4}\left(\frac{c}{R}\right)^3 = \frac{1}{4}\left(\frac{\gamma_y}{\alpha}\frac{\alpha_y}{\gamma_y}\right)^3 = \frac{1}{4}\left(\frac{\alpha_y}{\alpha}\right)^3,$$

and (24.1.34) may be written as

$$\boxed{T = \frac{2\pi R^3}{3}\tau_y \left[1 - \frac{1}{4}\left(\frac{\alpha_y}{\alpha}\right)^3\right].} \tag{24.1.35}$$

Summarizing, the torque-twist relation can be expressed as

$$\boxed{T = \begin{cases} \mu\left(\dfrac{\pi R^4}{2}\right)\alpha & \text{for } \alpha \le \alpha_y \equiv \tau_y/\mu R, \\ \dfrac{2\pi R^3}{3}\tau_y\left[1 - \frac{1}{4}\left(\dfrac{\alpha_y}{\alpha}\right)^3\right] & \text{for } \alpha \ge \alpha_y. \end{cases}} \tag{24.1.36}$$

The *fully plastic limit* occurs as $c \to 0$ (equivalently $\alpha \to \infty$):

$$T_p = \lim_{\alpha\to\infty}\left\{\frac{2\pi R^3}{3}\tau_y\left[1 - \frac{1}{4}\left(\frac{\alpha_y}{\alpha}\right)^3\right]\right\}$$

or

$$\boxed{T_p = \frac{2\pi R^3}{3}\tau_y.} \tag{24.1.37}$$

Note that since $\alpha \to \infty$ implies $\gamma \to \infty$ for any $r > 0$, this limit is purely *formal*, it can never be reached; in reality, other phenomena such as fracture would intervene at substantially lower strain levels.

Using (24.1.31) we note that

$$\boxed{T_p = \frac{4}{3}T_y.} \tag{24.1.38}$$

Thus, the ultimate torque that a shaft can support is greater than the torque that first produces yielding at the outer fibers. *Therefore, inelastic-torsion behavior of shafts provides an additional strength margin that can be effectively used in design.* A sketch of the torque-twist relation for elastic-plastic torsion of a shaft is shown in Fig. 24.5.

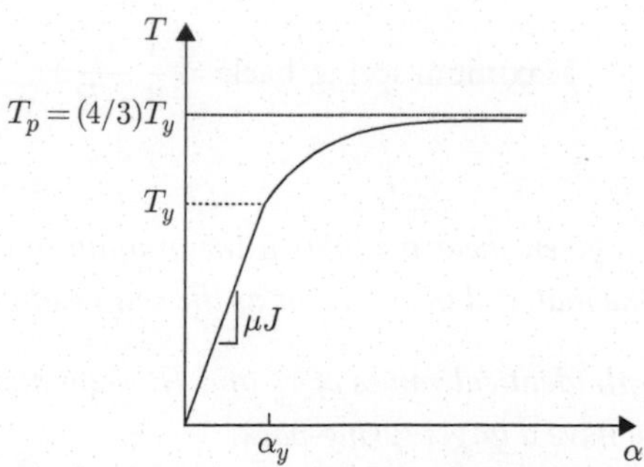

Fig. 24.5 Schematic of torque-twist plot for elastic-plastic torsion.

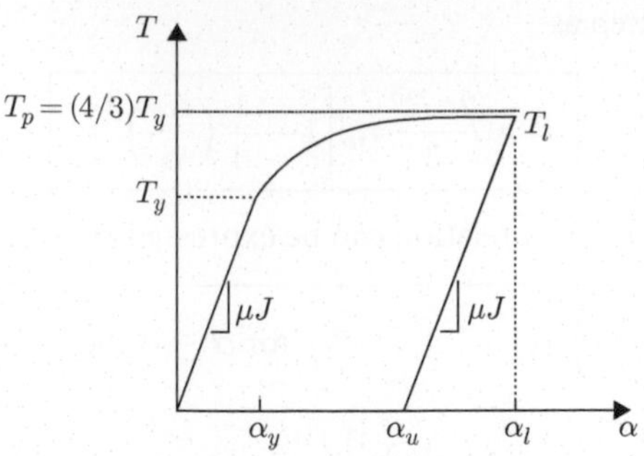

Fig. 24.6 Schematic of elastic unloading from a point (α_l, T_l).

REMARK 24.1

If the material has some strain-hardening capacity, then it has an even greater load-carrying capacity in the plastic range than the non-hardening case considered here.

Spring-back

With reference to Fig. 24.6, let T_l denote the torque in the elastic-plastically loaded state at a twist α_l. Then since the unloading occurs elastically with a slope μJ, the spring-back is defined by

$$\text{Spring-back} = \alpha_l - \alpha_u = \frac{T_l}{\mu J}. \tag{24.1.39}$$

where α_u is the twist per unit length in the unloaded state. The upper limit to the amount of possible spring-back is

$$\text{Maximum spring-back} = \frac{T_p}{\mu J}$$

or using (24.1.10) and (24.1.37)

$$\text{Maximum spring-back} = \frac{4}{3}\frac{\tau_y}{\mu R}. \tag{24.1.40}$$

Thus:

- *For a material with a given shear modulus μ the amount of spring-back increases as the yield strength τ_y of the material increases, or as the shaft radius R decreases.*
- *For two materials with identical values of τ_y and R, the material with lower value of the shear modulus μ will have a larger spring-back.*

Residual stress

- *Residual stresses invariably arise whenever a component has undergone non-homogeneous elastic-plastic deformations.*

Here, we examine the residual stress-state for the simple case of elastic-plastic torsion of the circular shaft under study.

The residual stress state is given by

$$\tau(r)_{\text{unloaded}} = \tau(r)_{\text{loaded}} - \Delta\tau \tag{24.1.41}$$

where

$$\Delta\tau = \mu\Delta\gamma = \mu\left(\alpha_{\text{loaded}} - \alpha_{\text{unloaded}}\right) r. \tag{24.1.42}$$

Since

$$T_l = \mu J \left(\alpha_{\text{loaded}} - \alpha_{\text{unloaded}}\right),$$

relation (24.1.42) gives

$$\Delta\tau = \frac{T_l\, r}{J}, \tag{24.1.43}$$

and hence (24.1.41) becomes

$$\tau(r)_{\text{unloaded}} = \tau(r)_{\text{loaded}} - \frac{T_l\, r}{J}. \tag{24.1.44}$$

To be specific, when T_l is equal to the fully plastic torque $T_p = (2\pi R^3 \tau_y)/3$, that is when

$$T_l = \frac{2\pi R^3 \tau_y}{3},$$

the stress distribution across the shaft is

$$\tau(r)_{\text{loaded}} = \tau_y \quad \text{for} \quad 0 \le r \le R. \tag{24.1.45}$$

Next, since $J = \pi R^4/2$, we have

$$\frac{T_l\, r}{J} = \frac{2\pi R^3 \tau_y}{3} \times r \times \frac{2}{\pi R^4} = \frac{4}{3}\frac{\tau_y\, r}{R}, \tag{24.1.46}$$

and hence

$$\tau(r)_{\text{unloaded}} = \tau_y\left(1 - \frac{4}{3}\frac{r}{R}\right) \quad \text{for} \quad 0 \le r < R. \tag{24.1.47}$$

This unloaded, or **residual stress field** is schematically shown in Fig. 24.7.

The value of the residual stress on the outer surface, $r = R$, has a *negative* value $(-\tau_y/3)$, and it varies linearly with r up to a *positive* value of τ_y at $r = 0$.

- *It is important to note that nowhere does the magnitude of the residual stress state exceed the shear yield strength τ_y.*

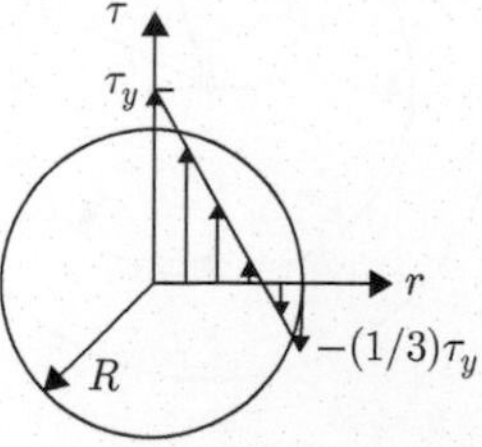

Fig. 24.7 Residual stress field after unloading from T_p.

24.2 **Spherical pressure vessel**

Let the internal and external radii of a thick-walled pressure vessel under an internal pressure p be denoted a and b. The pressure vessel is made from a homogeneous *isotropic* elastic-perfectly-plastic material with a constant yield strength Y. Neglecting inertia and body forces, we study the problem of how the spherical shell responds as the pressure is increased from when the shell is initially elastic, to when it first starts to deform plastically, and finally becomes fully plastic. To solve this problem, we use a spherical coordinate system (r, θ, ϕ) with origin at the center of the spherical pressure vessel, so that the body when under pressure occupies the region $a \le r \le b$ (Fig. 24.8).

For this problem, there are only traction boundary conditions, so any solution to the displacement field that we shall determine will be modulo a rigid displacement. The prescribed tractions are

$$\begin{aligned} \hat{\mathbf{t}} &= p\mathbf{e}_r \quad && \text{on} \quad r = a, \\ \hat{\mathbf{t}} &= \mathbf{0} \quad && \text{on} \quad r = b. \end{aligned} \tag{24.2.1}$$

Since the *outward* unit normal at $r = a$ is $-\mathbf{e}_r$, and that on $r = b$ is $\mathbf{e}_r$, the traction boundary conditions are

$$\begin{aligned} \boldsymbol{\sigma}(-\mathbf{e}_r) &= p\mathbf{e}_r \quad && \text{on} \quad r = a, \\ \boldsymbol{\sigma}(\mathbf{e}_r) &= \mathbf{0} \quad && \text{on} \quad r = b, \end{aligned} \tag{24.2.2}$$

and hence

$$\begin{aligned} \sigma_{rr} &= -p \quad && \text{on} \quad r = a, \\ \sigma_{rr} &= 0 \quad && \text{on} \quad r = b. \end{aligned} \tag{24.2.3}$$

24.2.1 **Elastic analysis**

As in Sec. 9.1, because of the homogeneity and isotropy of the material, and because of the symmetry of the loading, we *assume* that the displacement field $\mathbf{u}$ is directed in the radial direction and depends only on r for both elastic and plastic response:

$$\mathbf{u} = u_r(r)\mathbf{e}_r \quad \text{with} \quad u_\theta = u_\phi = 0. \tag{24.2.4}$$

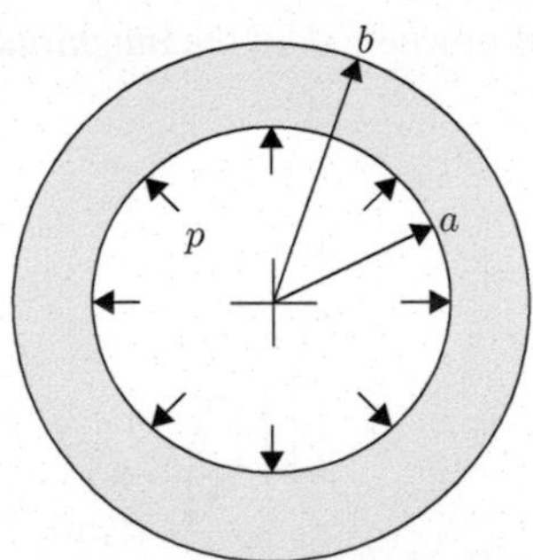

Fig. 24.8 Spherical pressure vessel under an internal pressure p.

Recall from our discussion of a similar problem in the isotropic theory of elasticity, in the absence of body forces the displacement equation of equilibrium is

$$(\lambda + 2\mu)\nabla\, \mathrm{div}\, \mathbf{u} - \mu\, \mathrm{curl}\,\mathrm{curl}\, \mathbf{u} = \mathbf{0}, \tag{24.2.5}$$

where μ and λ are the Lamé moduli. Also, recalling the expression for the curl of a vector field $\mathbf{v}$ in spherical coordinates

$$\mathrm{curl}\, \mathbf{v} = \left\{\frac{1}{r}\frac{\partial v_\phi}{\partial \theta} - \frac{1}{r\sin\theta}\frac{\partial v_\theta}{\partial \phi} + \cot\theta \frac{v_\phi}{r}\right\}\mathbf{e}_r + \left\{\frac{1}{r\sin\theta}\frac{\partial v_r}{\partial \phi} - \frac{\partial v_\phi}{\partial r} - \frac{v_\phi}{r}\right\}\mathbf{e}_\theta + \left\{\frac{\partial v_\theta}{\partial r} - \frac{1}{r}\frac{\partial v_r}{\partial \theta} + \frac{v_\theta}{r}\right\}\mathbf{e}_\phi,$$

we see that for the presumed displacement field (24.2.4) the curl vanishes:

$$\mathrm{curl}\, \mathbf{u} = \mathbf{0}.$$

Using this in (24.2.5), and noting that $(\lambda + 2\mu) \neq 0$, we must have

$$\nabla\, \mathrm{div}\, \mathbf{u} = \mathbf{0}, \tag{24.2.6}$$

and hence $\mathrm{div}\, \mathbf{u}$ is a *constant*. We write

$$\mathrm{div}\, \mathbf{u} = 3A, \tag{24.2.7}$$

where $3A$ is an integration constant, where we have introduced the factor of 3 for later convenience. In a spherical coordinate system the divergence of a vector field $\mathbf{v}$ is given by

$$\mathrm{div}\, \mathbf{v} = \frac{\partial v_r}{\partial r} + \frac{1}{r}\frac{\partial v_\theta}{\partial \theta} + \frac{1}{r\sin\theta}\frac{\partial v_\phi}{\partial \phi} + \frac{\cot\theta}{r}v_\theta + \frac{2v_r}{r}.$$

Hence, for the displacement field (24.2.4) we obtain

$$\mathrm{div}\, \mathbf{u} = \frac{du_r}{dr} + \frac{2u_r}{r}, \tag{24.2.8}$$

use of which in (24.2.7) gives

$$\frac{du_r}{dr} + \frac{2u_r}{r} = 3A. \tag{24.2.9}$$

Multiplying (24.2.9) by r^2 gives

$$r^2\left(\frac{du_r}{dr} + 2\frac{u_r}{r}\right) = 3Ar^2,$$

and noting the identity that

$$\frac{d(r^2 u_r)}{dr} = r^2\left(\frac{du_r}{dr} + 2\frac{u_r}{r}\right), \tag{24.2.10}$$

we obtain

$$\frac{d(r^2 u_r)}{dr} = 3Ar^2. \tag{24.2.11}$$

Integrating (24.2.11) we obtain

$$r^2 u_r = Ar^3 + B,$$

where B is another constant of integration, and hence

$$u_r = Ar + \frac{B}{r^2}. \tag{24.2.12}$$

It remains to find the constants A and B, which are determined using the boundary conditions (24.2.3). To this end, from (24.2.12) the only non-zero strain components are

$$\epsilon_{rr} = \frac{du_r}{dr}, \qquad \epsilon_{\theta\theta} = \epsilon_{\phi\phi} = \frac{u_r}{r}. \tag{24.2.13}$$

Hence, using the constitutive equation for the stress in the absence of plasticity, the radial stress component is

$$\begin{aligned}
\sigma_{rr} &= 2\mu\frac{du_r}{dr} + \lambda\left(\frac{du_r}{dr} + \frac{2u_r}{r}\right) \\
&= (2\mu+\lambda)\frac{du_r}{dr} + 2\lambda\left(\frac{u_r}{r}\right) \\
&= (2\mu+\lambda)\left(A - 2\frac{B}{r^3}\right) + 2\lambda\left(A + \frac{B}{r^3}\right) \qquad \text{(using (24.2.12))} \\
&= (2\mu+3\lambda)A - \frac{4\mu B}{r^3}.
\end{aligned} \tag{24.2.14}$$

Also, the hoop stress components are

$$\begin{aligned}
\sigma_{\theta\theta} = \sigma_{\phi\phi} &= 2\mu\frac{u_r}{r} + \lambda\left(\frac{du_r}{dr} + \frac{2u_r}{r}\right) \\
&= (2\mu+2\lambda)\frac{u_r}{r} + \lambda\left(\frac{du_r}{dr}\right) \\
&= (2\mu+2\lambda)\left(A + \frac{B}{r^3}\right) + \lambda\left(A - 2\frac{B}{r^3}\right) \qquad \text{(using (24.2.12))} \\
&= (2\mu+3\lambda)A + 2\mu\frac{B}{r^3},
\end{aligned} \tag{24.2.15}$$

while the shear stress components vanish,

$$\sigma_{r\theta} = \sigma_{\theta\phi} = \sigma_{\phi r} = 0. \tag{24.2.16}$$

Next, using (24.2.14) and the boundary conditions (24.2.3) we obtain

$$\begin{aligned}
(2\mu+3\lambda)A - \frac{4\mu B}{a^3} &= -p, \\
(2\mu+3\lambda)A - \frac{4\mu B}{b^3} &= 0.
\end{aligned} \tag{24.2.17}$$

Solving (24.2.17) for the unknowns A and B, one obtains

$$A = \frac{1}{(2\mu+3\lambda)}\frac{p}{\left(\left(\dfrac{b}{a}\right)^3 - 1\right)}, \qquad B = \frac{b^3}{4\mu}\frac{p}{\left(\left(\dfrac{b}{a}\right)^3 - 1\right)}. \tag{24.2.18}$$

Finally, substituting for A and B in (24.2.12), (24.2.14), and (24.2.15), we obtain

$$u_r = \frac{p\,r}{\left(\left(\frac{b}{a}\right)^3 - 1\right)} \left[\frac{1}{(2\mu + 3\lambda)} + \frac{1}{4\mu}\left(\frac{b}{r}\right)^3\right]. \tag{24.2.19}$$

$$\sigma_{rr} = -\frac{p}{\left(\left(\frac{b}{a}\right)^3 - 1\right)} \left[\left(\frac{b}{r}\right)^3 - 1\right],$$

$$\sigma_{\theta\theta} = \sigma_{\phi\phi} = \frac{p}{\left(\left(\frac{b}{a}\right)^3 - 1\right)} \left[\tfrac{1}{2}\left(\frac{b}{r}\right)^3 + 1\right], \tag{24.2.20}$$

$$\sigma_{r\theta} = \sigma_{\theta\phi} = \sigma_{\phi r} = 0.$$

Since the shear stresses are all zero, $(24.2.20)_3$, the stress components σ_{rr} and $\sigma_{\theta\theta} = \sigma_{\phi\phi}$ are *principal stresses.*

Using the standard relations

$$2\mu + 3\lambda = 3\kappa = \frac{E}{1 - 2\nu}, \qquad 2\mu = \frac{E}{1+\nu},$$

between the elastic constants, the relation (24.2.19) for the displacement may also be written as

$$u_r = \frac{(p/E)\,r}{\left(\left(\frac{b}{a}\right)^3 - 1\right)} \left[(1 - 2\nu) + \frac{(1+\nu)}{2}\left(\frac{b}{r}\right)^3\right]. \tag{24.2.21}$$

24.2.2 **Elastic-plastic analysis**

Onset of yield

The Mises equivalent tensile stress is

$$\bar{\sigma} = \left|\left[\frac{1}{2}\left\{(\sigma_{rr} - \sigma_{\theta\theta})^2 + (\sigma_{\theta\theta} - \sigma_{\phi\phi})^2 + (\sigma_{\phi\phi} - \sigma_{rr})^2\right\} + 3\left\{\sigma_{r\theta}^2 + \sigma_{\theta\phi}^2 + \sigma_{\phi r}^2\right\}\right]^{1/2}\right|, \tag{24.2.22}$$

and for the problem at hand from (24.2.20)

$$\bar{\sigma} = |\sigma_{\theta\theta} - \sigma_{rr}| = \sigma_{\theta\theta} - \sigma_{rr} = \frac{p}{\left(\left(\frac{b}{a}\right)^3 - 1\right)} \left[1 + \frac{3}{2}\left(\frac{b}{r}\right)^3\right]. \tag{24.2.23}$$

We note that the Mises equivalent tensile stress is maximum at $r = a$ and yielding will therefore commence at the inner radius, when

$$\bar{\sigma} = Y,$$

or when the pressure p reaches a value p_y which satisfies

$$\frac{p_y}{\left(\left(\frac{b}{a}\right)^3 - 1\right)}\left[1 + \frac{3}{2}\left(\frac{b}{a}\right)^3\right] = Y,$$

from which

$$\boxed{p_y = \frac{2}{3} Y \left[1 - \left(\frac{a}{b}\right)^3\right].} \tag{24.2.24}$$

Next, using (24.2.24) in relation (24.2.21) for the displacement u_r we find that at initial yield

$$u_{r(y)} = \frac{Y}{E} r \left[\frac{2}{3}(1 - 2\nu)\left(\frac{a}{b}\right)^3 + \frac{(1+\nu)}{3}\left(\frac{a}{r}\right)^3\right], \tag{24.2.25}$$

and hence the radial displacements of the inner and outer radii of the sphere are

$$u_{r(y)}(a) = \frac{Y}{E} a \left[\frac{2}{3}(1 - 2\nu)\left(\frac{a}{b}\right)^3 + \frac{(1+\nu)}{3}\right] \tag{24.2.26}$$

and

$$u_{r(y)}(b) = \frac{Y}{E} b \left[(1 - \nu)\left(\frac{a}{b}\right)^3\right]. \tag{24.2.27}$$

Partly plastic spherical pressure vessel

With increasing pressure a plastic region spreads into the sphere from the inner surface outwards. Let the radius of the boundary between the elastic and plastic portions of the sphere be denoted by c; cf. Fig. 24.9.

Elastic region $c \le r \le b$:

In the elastic region $c \le r \le b$ the displacement and the stresses are still given by (24.2.12), (24.2.14), and (24.2.15):

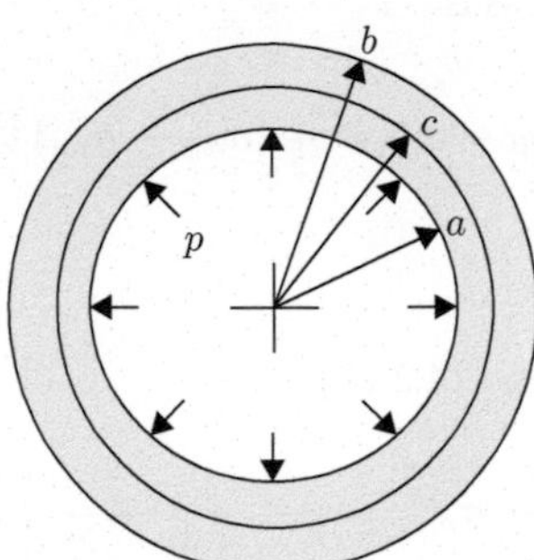

Fig. 24.9 Partly plastic spherical pressure vessel under an internal pressure p. The radius of the boundary between the plastic and elastic portions of the sphere is at $r = c$.

$$\begin{aligned} u_r &= Ar + Br^{-2}, \\ \sigma_{rr} &= (2\mu + 3\lambda)A - 4\mu B r^{-3}, \\ \sigma_{\theta\theta} = \sigma_{\phi\phi} &= (2\mu + 3\lambda)A + 2\mu B r^{-3}, \end{aligned} \tag{24.2.28}$$

but the boundary conditions are now

$$\begin{aligned} \bar{\sigma} = \sigma_{\theta\theta} - \sigma_{rr} &= Y \qquad \text{at } r = c, \\ \sigma_{rr} &= 0 \qquad \text{at } r = b. \end{aligned} \tag{24.2.29}$$

From $(24.2.28)_{2,3}$ and (24.2.29)

$$A = Y\frac{2}{3(2\mu + 3\lambda)} Y \left(\frac{c}{b}\right)^3, \qquad B = \frac{1}{6\mu} Y c^3, \tag{24.2.30}$$

or in terms of E and ν

$$A = \frac{2}{3} Y \frac{(1 - 2\nu)}{E} \left(\frac{c}{b}\right)^3, \qquad B = \frac{1}{3} Y \frac{(1 + \nu)}{E} c^3. \tag{24.2.31}$$

Use of (24.2.31) in (24.2.28) gives that in the elastic domain

$$\left.\begin{aligned} u_r &= \frac{2}{3}\frac{Y}{E}\left(\frac{c}{b}\right)^3 r \left[(1 - 2\nu) + \frac{(1+\nu)}{2}\left(\frac{b}{r}\right)^3\right] \\ \sigma_{rr} &= -\frac{2}{3} Y \left(\frac{c}{b}\right)^3 \left[\left(\frac{b}{r}\right)^3 - 1\right] \\ \sigma_{\theta\theta} = \sigma_{\phi\phi} &= \frac{2}{3} Y \left(\frac{c}{b}\right)^3 \left[\frac{1}{2}\left(\frac{b}{r}\right)^3 + 1\right] \end{aligned}\right\} \quad \text{for} \quad c \le r \le b. \tag{24.2.32}$$

Plastic region $a \le r \le c$:

Next, in the plastic domain $a \le r \le c$, the stress components need to satisfy the yield condition, the equilibrium equation, the boundary condition $\sigma_{rr} = -p$ at $r = a$, and displacement and traction continuity with the elastic solution at $r = c$. From the yield condition $\bar{\sigma} = Y$, with $\bar{\sigma}$ given in (24.2.22),

$$\sigma_{\theta\theta} - \sigma_{rr} = Y. \tag{24.2.33}$$

Now since the only non-vanishing stress components are σ_{rr} and $\sigma_{\theta\theta} = \sigma_{\phi\phi}$, the only non-trivial equilibrium equation is

$$\frac{\partial \sigma_{rr}}{\partial r} + 2\frac{\sigma_{rr} - \sigma_{\theta\theta}}{r} = 0. \tag{24.2.34}$$

Thus, (24.2.33) and (24.2.34) require that

$$\frac{\partial \sigma_{rr}}{\partial r} = 2\frac{Y}{r}, \tag{24.2.35}$$

which upon integration gives

$$\sigma_{rr} = 2Y \ln r + D, \tag{24.2.36}$$

where D is a constant of integration. Since σ_{rr} must be continuous across the elastic-plastic boundary, from (24.2.36) and (24.2.32)$_2$ we must have that at $r = c$

$$2Y \ln c + D = -\frac{2}{3} Y \left(\frac{c}{b}\right)^3 \left[\left(\frac{b}{c}\right)^3 - 1\right],$$

which gives the integration constant D as

$$D = -2Y \ln c - \frac{2}{3} Y \left[1 - \left(\frac{c}{b}\right)^3\right],$$

substitution of which in (24.2.36), together with use of (24.2.33) gives

$$\left.\begin{aligned} \sigma_{rr} &= -\frac{2}{3} Y \left[1 - \left(\frac{c}{b}\right)^3 + 3 \ln\left(\frac{c}{r}\right)\right] \\ \sigma_{\theta\theta} = \sigma_{\phi\phi} &= \frac{2}{3} Y \left[\tfrac{1}{2} + \left(\frac{c}{b}\right)^3 - 3 \ln\left(\frac{c}{r}\right)\right] \end{aligned}\right\} \quad \text{for} \quad a \le r \le c. \tag{24.2.37}$$

The radius c delineating the extent of the plastic domain is obtained by solving

$$\boxed{p = \frac{2}{3} Y \left[1 - \left(\frac{c}{b}\right)^3 + 3 \ln\left(\frac{c}{a}\right)\right],} \tag{24.2.38}$$

which is the requirement at a given pressure p, that the radial stress must satisfy $\sigma_{rr} = -p$ at $r = a$.

In the plastic domain the displacement is determined from the fact that plastic flow is incompressible, and that any volume change is purely elastic:

$$\operatorname{tr} \boldsymbol{\epsilon} = \operatorname{tr} \boldsymbol{\epsilon}^e = \frac{1}{3\kappa} \operatorname{tr} \boldsymbol{\sigma} = \frac{(1-2\nu)}{E} \operatorname{tr} \boldsymbol{\sigma},$$

or that

$$\epsilon_{rr} + \epsilon_{\theta\theta} + \epsilon_{\phi\phi} = \frac{(1-2\nu)}{E} (\sigma_{rr} + \sigma_{\theta\theta} + \sigma_{\phi\phi}).$$

Using

$$\epsilon_{rr} = \frac{du_r}{dr}, \qquad \epsilon_{\theta\theta} = \epsilon_{\phi\phi} = \frac{u_r}{r},$$

and $\sigma_{\theta\theta} = \sigma_{\phi\phi}$, we have that

$$\frac{du_r}{dr} + 2\frac{u_r}{r} = \frac{(1-2\nu)}{E} (\sigma_{rr} + 2\sigma_{\theta\theta}). \tag{24.2.39}$$

Substituting for σ_{rr} and $\sigma_{\theta\theta}$ from (24.2.37) in (24.2.39) we obtain the following differential equation for u_r:

$$\frac{du_r}{dr} + 2\frac{u_r}{r} = 2(1-2\nu)\frac{Y}{E} \left[\left(\frac{c}{b}\right)^3 + 3 \ln\left(\frac{c}{r}\right)\right], \tag{24.2.40}$$

which, using the identity (24.2.10), may be rewritten as

$$\begin{aligned} \frac{d}{dr}\left(r^2 u_r\right) &= 2(1-2\nu)\frac{Y}{E} \left[\left(\frac{c}{b}\right)^3 - 3 \ln\left(\frac{c}{r}\right)\right] r^2 \\ &= 2(1-2\nu)\frac{Y}{E} \left[\left(\frac{c}{b}\right)^3 r^2 - 3 (\ln c) r^2 + 3 (\ln r) r^2\right], \end{aligned}$$

which upon integration gives

$$r^2 u_r = 2(1-2\nu)\frac{Y}{E}\left[\left(\frac{c}{b}\right)^3 \frac{1}{3}r^3 - 3\ln c\frac{1}{3}r^3 + 3(\ln r)\frac{1}{3}r^3 - \frac{1}{3}r^3\right] - F$$

or

$$u_r = -\frac{2}{3}(1-2\nu)\frac{Y}{E}\left[1 - \left(\frac{c}{b}\right)^3 + 3\ln\left(\frac{c}{r}\right)\right] r + Fr^{-2}, \tag{24.2.41}$$

where F is a constant of integration which is determined by the continuity of the displacement at $r = c$. Given (24.2.41) and $(24.2.32)_1$, this continuity requirement is

$$-\frac{2}{3}(1-2\nu)\frac{Y}{E}\left[1-\left(\frac{c}{b}\right)^3\right]c + Fc^{-2} = \frac{2}{3}\frac{Y}{E}\left[(1-2\nu)\left(\frac{c}{b}\right)^3 + \frac{1+\nu}{2}\right]c,$$

which gives

$$F = (1-\nu)\frac{Y}{E}c^3.$$

Substituting for F in (24.2.41) gives

$$\boxed{u_r = \frac{Y}{E}r\left[(1-\nu)\left(\frac{c}{r}\right)^3 - \frac{2}{3}(1-2\nu)\left\{1-\left(\frac{c}{b}\right)^3 + 3\ln\left(\frac{c}{r}\right)\right\}\right] \quad \text{for} \quad a \le r \le c.} \tag{24.2.42}$$

Fully plastic spherical pressure vessel

When $c = b$, the sphere becomes completely plastic. The corresponding pressure using (24.2.38) is given by

$$\boxed{p^* = 2Y\ln\left(\frac{b}{a}\right).} \tag{24.2.43}$$

From (24.2.42), when $c = b$ the displacements at the inner and outer walls of the sphere are

$$\boxed{u_r^*(a) = \frac{Y}{E}a_0\left[(1-\nu)\left(\frac{b}{a}\right)^3 - 2(1-2\nu)\ln\left(\frac{b}{a}\right)\right]} \tag{24.2.44}$$

and

$$\boxed{u_r^*(b) = (1-\nu)\frac{Y}{E}b.} \tag{24.2.45}$$

Note that from (24.2.38)

$$\frac{dp}{dc} = \frac{2Y}{c}\left[1-\left(\frac{c}{b}\right)^3\right], \tag{24.2.46}$$

so that when $c = b$, the pressure (24.2.43) is a *maximum*.

Residual stresses upon unloading

If the sphere is unloaded from a partially plastic state, then the residual stress field is determined by subtracting from $(24.2.32)_{2,3}$ and (24.2.37) the elastic stress distribution (24.2.20) with p given by (24.2.38). The algebra is tedious and we omit it here.

24.3 Cylindrical pressure vessel

Let the internal and external radii of a thick-walled cylindrical pressure vessel be denoted by a and b, respectively. The pressure vessel is constrained so that it is in a state of *plane strain* along its axial direction, and made from a homogeneous isotropic, *incompressible*,[2] elastic-perfectly-plastic material with a constant yield strength Y. Neglecting inertia and body forces, we consider here the problem of how the cylindrical pressure vessel responds as the pressure is increased from when the cylinder is initially elastic, to when it first starts to deform plastically, and finally becomes fully plastic. To solve this problem, we use a cylindrical coordinate system (r, θ, z) with origin at the center of the pressure vessel (Fig. 24.10).

The prescribed tractions are

$$\begin{aligned} \hat{\mathbf{t}} &= p\,\mathbf{e}_r && \text{on} \quad r = a, \\ \hat{\mathbf{t}} &= \mathbf{0} && \text{on} \quad r = b. \end{aligned} \tag{24.3.1}$$

Since the *outward* unit normal at $r = a$ is $-\mathbf{e}_r$, and that on $r = b$ is $\mathbf{e}_r$, the traction boundary conditions are

$$\begin{aligned} \boldsymbol{\sigma}(-\mathbf{e}_r) &= p\,\mathbf{e}_r && \text{on} \quad r = a, \\ \boldsymbol{\sigma}(\mathbf{e}_r) &= \mathbf{0} && \text{on} \quad r = b, \end{aligned} \tag{24.3.2}$$

and hence

$$\begin{aligned} \sigma_{rr} &= -p && \text{on} \quad r = a, \\ \sigma_{rr} &= 0 && \text{on} \quad r = b. \end{aligned} \tag{24.3.3}$$

24.3.1 Elastic analysis

Because of the homogeneity and isotropy of the material, the symmetry of the loading, and plane strain restraint we *assume* that the displacement field $\mathbf{u}$ is directed in the radial direction and depends only on r:

$$\mathbf{u} = u_r(r)\mathbf{e}_r \qquad \text{with} \qquad u_\theta = u_z = 0. \tag{24.3.4}$$

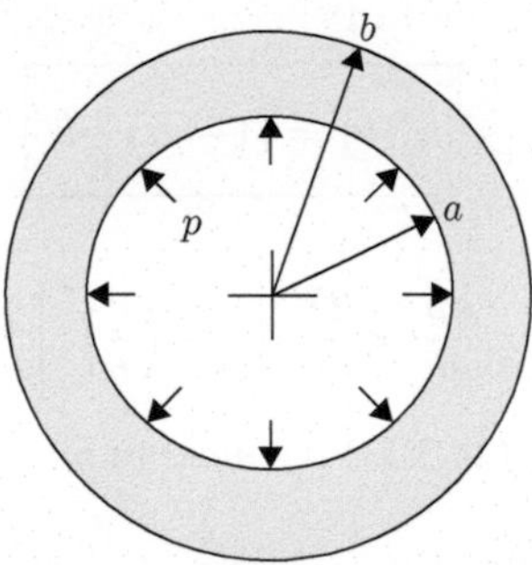

Fig. 24.10 Cross-section of a cylindrical pressure vessel under an internal pressure p.

---

[2] Complete incompressibility greatly simplifies the analysis.

The corresponding non-zero strain components are

$$\epsilon_{rr} = \frac{du_r}{dr}, \qquad \epsilon_{\theta\theta} = \frac{u_r}{r}. \tag{24.3.5}$$

Further, because of incompressibility, $\operatorname{tr}\boldsymbol{\epsilon} = 0$,

$$\frac{du_r}{dr} + \frac{u_r}{r} = 0, \tag{24.3.6}$$

or

$$\frac{du_r}{u_r} = -\frac{dr}{r},$$

integration of which gives

$$u_r = \frac{C}{r}. \tag{24.3.7}$$

Hence, using (24.3.5),

$$\epsilon_{rr} = -\frac{C}{r^2}, \qquad \epsilon_{\theta\theta} = \frac{C}{r^2}. \tag{24.3.8}$$

Observe also that this solution is universal, independent of material response, elastic or plastic; it only depends upon our assumption of incompressibility and plane strain conditions.

To determine the stresses, we first note that we have assumed *complete incompressibility*, which is associated with the situation where the bulk modulus $\kappa \to \infty$. In this case the elastic stress-strain relation reduces to

$$\boldsymbol{\sigma} = -\pi\mathbf{1} + 2\mu\boldsymbol{\epsilon}, \tag{24.3.9}$$

where π is an *indeterminate pressure*—in the sense that it is not determined by the strain $\boldsymbol{\epsilon}$ alone.[3,4] Thus, with the non-zero strain components given by (24.3.8), the non-zero stress components are:

$$\begin{aligned} \sigma_{rr} &= -\pi - 2\mu\frac{C}{r^2}, \\ \sigma_{\theta\theta} &= -\pi + 2\mu\frac{C}{r^2}, \\ \sigma_{zz} &= -\pi. \end{aligned} \tag{24.3.10}$$

Next, satisfaction of equilibrium requires that

$$\frac{d\sigma_{rr}}{dr} + \frac{1}{r}(\sigma_{rr} - \sigma_{\theta\theta}) = 0. \tag{24.3.11}$$

Substitution of $(24.3.10)_{1,2}$ in (24.3.11) gives

$$\frac{d\pi}{dr} = 0 \qquad \Rightarrow \qquad \pi = \text{constant}. \tag{24.3.12}$$

---

[3] The origin of the π term can be motivated by noting that the constitutive response is given by

$$\boldsymbol{\sigma} = 2\mu\boldsymbol{\epsilon} - (2/3)\mu \underbrace{\operatorname{tr}\boldsymbol{\epsilon}}_{=0}\mathbf{1} + \underbrace{\kappa\operatorname{tr}\boldsymbol{\epsilon}}_{-\pi=\infty\cdot 0}\ \mathbf{1}.$$

The product of ∞ and 0 is the indeterminate pressure, $-\pi$.

[4] Note that this indeterminate pressure π is not the same as the pressure p applied to the cylinder.

Further, using the boundary condition $(24.3.3)_2$ at $r = b$ in $(24.3.10)_1$ gives

$$\pi = -2\mu \frac{C}{b^2}, \tag{24.3.13}$$

substitution of which in $(24.3.10)_1$ gives

$$\sigma_{rr} = -\frac{2\mu C}{b^2}\left(\frac{b^2}{r^2} - 1\right). \tag{24.3.14}$$

Next, using boundary condition $(24.3.3)_1$ at $r = a$ in (24.3.14)

$$-p = -\frac{2\mu C}{b^2}\left(\frac{b^2}{a^2} - 1\right), \tag{24.3.15}$$

which gives the integration constant C as

$$C = \frac{p}{\left(\dfrac{b^2}{a^2} - 1\right)}\frac{b^2}{2\mu}. \tag{24.3.16}$$

Also, substitution of (24.3.16) into (24.3.13) gives

$$\pi = -\frac{p}{\left(\dfrac{b^2}{a^2} - 1\right)}. \tag{24.3.17}$$

Thus we see that what we had called the "constitutively indeterminate pressure" is determined in a boundary-value problem by satisfying an appropriate boundary condition.

For convenience, let

$$\bar{p} \stackrel{\text{def}}{=} \frac{p}{\left(\dfrac{b^2}{a^2} - 1\right)} > 0. \tag{24.3.18}$$

Then, substitution of (24.3.18) in (24.3.17) and (24.3.16) gives

$$\pi = -\bar{p}, \qquad \text{and} \qquad C = \frac{\bar{p}b^2}{2\mu}. \tag{24.3.19}$$

Finally substitution of (24.3.19) in (24.3.7) and (24.3.10) gives the elastic solution as

$$\boxed{\left.\begin{aligned} u_r &= \frac{\bar{p}b^2}{2\mu}\frac{1}{r} \\ \sigma_{rr} &= -\bar{p}\left(\frac{b^2}{r^2} - 1\right) \\ \sigma_{\theta\theta} &= \bar{p}\left(\frac{b^2}{r^2} + 1\right) \\ \sigma_{zz} &= \bar{p} \end{aligned}\right\} \quad \text{where} \quad \bar{p} \stackrel{\text{def}}{=} \frac{p}{\left(\dfrac{b^2}{a^2} - 1\right)} > 0.} \tag{24.3.20}$$

24.3.2 **Elastic-plastic analysis**

Onset of yield

The Mises equivalent tensile stress in the cylinder is

$$\bar{\sigma} = \sqrt{(3/2)(\sigma'^2_{rr} + \sigma'^2_{\theta\theta} + \sigma'^2_{zz})}, \tag{24.3.21}$$

and for the problem at hand from (24.3.20),

$$\bar{\sigma} = \sqrt{3}\,\bar{p}\,\frac{b^2}{r^2}. \tag{24.3.22}$$

Thus the equivalent tensile stress is a maximum at $r = a$ and plastic yielding will initiate at $r = a$ when the pressure reaches a value p_y which satisfies

$$\sqrt{3}\,\frac{p_y}{\left(\dfrac{b^2}{a^2} - 1\right)}\,\frac{b^2}{a^2} = Y;$$

hence

$$\boxed{p_y = \frac{Y}{\sqrt{3}}\left(1 - \frac{a^2}{b^2}\right).} \tag{24.3.23}$$

Partly plastic cylindrical pressure vessel

With increasing pressure a plastic region spreads into the cylinder from the inner surface outwards. Let the radius of the boundary between the elastic and plastic portions of the pressure vessel be denoted by c; cf. Fig. 24.11.

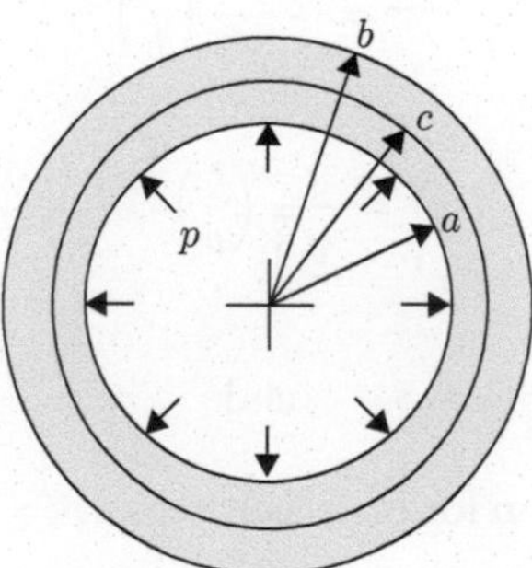

Fig. 24.11 Partly plastic cylindrical pressure vessel under an internal pressure p. The radius of the boundary between the plastic and elastic portions of the cylinder is at $r = c$.

Elastic region $c \le r \le b$:

In the elastic region $c \le r \le b$ the displacement and the stresses are still given by (24.3.7) and (24.3.10):

$$
\begin{aligned}
u_r &= \frac{C}{r},\\
\sigma_{rr} &= -\pi - 2\mu\frac{C}{r^2},\\
\sigma_{\theta\theta} &= -\pi + 2\mu\frac{C}{r^2},\\
\sigma_{zz} &= -\pi
\end{aligned}
\tag{24.3.24}
$$

from which the corresponding equivalent tensile stress is,

$$
\bar{\sigma} = \sqrt{3}\,2\mu\frac{C}{r^2}. \tag{24.3.25}
$$

The boundary conditions needed to determine C and π in this case, however, are now given by,

$$
\begin{aligned}
\bar{\sigma} &= Y \qquad \text{at } r = c,\\
\sigma_{rr} &= 0 \qquad \text{at } r = b.
\end{aligned}
\tag{24.3.26}
$$

From boundary condition $(24.3.26)_1$ and (24.3.25)

$$
C = \frac{Yc^2}{\sqrt{3}\,2\mu}, \tag{24.3.27}
$$

and using boundary condition $(24.3.26)_2$ in $(24.3.24)_2$

$$
\pi = -2\mu\frac{C}{b^2}. \tag{24.3.28}
$$

Hence, using (24.3.27)

$$
\pi = -\frac{Y}{\sqrt{3}}\left(\frac{c}{b}\right)^2. \tag{24.3.29}
$$

For convenience, let

$$
\tilde{p} \stackrel{\text{def}}{=} \frac{Y}{\sqrt{3}}\left(\frac{c}{b}\right)^2 > 0. \tag{24.3.30}
$$

Then

$$
\pi = -\tilde{p} \qquad \text{and} \qquad C = \frac{\tilde{p}b^2}{2\mu}, \tag{24.3.31}
$$

which gives the complete solution for the elastic region $c \le r \le b$ as

$$
\boxed{
\left.
\begin{aligned}
u_r &= \frac{\tilde{p}\,b^2}{2\mu}\frac{1}{r}\\
\sigma_{rr} &= -\tilde{p}\left(\frac{b^2}{r^2} - 1\right)\\
\sigma_{\theta\theta} &= \tilde{p}\left(\frac{b^2}{r^2} + 1\right)\\
\sigma_{zz} &= \tilde{p}
\end{aligned}
\right\}
\qquad \text{where} \qquad \tilde{p} \stackrel{\text{def}}{=} \frac{Y}{\sqrt{3}}\left(\frac{c^2}{b^2}\right) > 0.
}
\tag{24.3.32}
$$

Plastic region $a \leq r \leq c$:

Next, in the plastic domain $a \leq r \leq c$, the displacement u_r must still be of the form

$$u_r = \frac{C}{r}, \qquad (24.3.33)$$

and in principle may have a different value for C in the elastic and plastic regions, *but the displacement must also be continuous at the elastic-plastic boundary.* Hence the value of C in both regions must be the same; that is

$$u_r = \frac{\tilde{p}\, b^2}{2\mu}\frac{1}{r} = \frac{Y}{\sqrt{3}}\frac{1}{2\mu}\frac{c^2}{r} \qquad (24.3.34)$$

must also be valid in the plastic region.

In the plastic domain the stress components need to satisfy the yield condition, the equilibrium equation, and the boundary condition $\sigma_{rr} = -p$ at $r = a$, together with traction and displacement continuity with the elastic solution at $r = c$.

The flow rule requires that

$$\dot{\epsilon}^{\rm p}_{rr} = (3/2)\dot{\bar{\epsilon}}^{\rm p}\sigma'_{rr}/\bar{\sigma}, \quad \dot{\epsilon}^{\rm p}_{\theta\theta} = (3/2)\dot{\bar{\epsilon}}^{\rm p}\sigma'_{\theta\theta}/\bar{\sigma}, \quad \text{and} \quad \dot{\epsilon}^{\rm p}_{zz} = (3/2)\dot{\bar{\epsilon}}^{\rm p}\sigma'_{zz}/\bar{\sigma},$$

and plastic incompressibility demands that

$$\dot{\epsilon}^{\rm p}_{rr} + \dot{\epsilon}^{\rm p}_{\theta\theta} + \dot{\epsilon}^{\rm p}_{zz} = 0.$$

Our restriction to plane strain requires that $\dot{\epsilon}_{zz} = 0$ and overall incompressibility requires the total strain to be deviatoric, implying that $\dot{\epsilon}'_{zz}$ is also zero. From the deviatoric part of the constitutive law in compliance form,

$$\dot{\boldsymbol{\epsilon}}' = \frac{1}{2\mu}\dot{\boldsymbol{\sigma}}' + \frac{3}{2}\dot{\bar{\epsilon}}^{\rm p}\frac{\boldsymbol{\sigma}'}{\bar{\sigma}}, \qquad (24.3.35)$$

we see that

$$\dot{\sigma}'_{zz} = -3\mu\dot{\bar{\epsilon}}^{\rm p}\frac{\sigma'_{zz}}{\bar{\sigma}}. \qquad (24.3.36)$$

The solution to this differential equation is given by

$$\sigma'_{zz} = \begin{cases} A & t < t_y \\ B\exp[-3\mu\bar{\epsilon}^{\rm p}/Y] & t \geq t_y \end{cases} \qquad (24.3.37)$$

where A and B are constants of integration and t_y is the time at which yield starts at a given point in the cylinder. Since we know that $\sigma'_{zz} = 0$ before plastic deformation begins as well as right at the moment that yield starts ($\bar{\epsilon}^{\rm p} = 0$), we see that $A = B = 0$ and that at all times

$$\sigma'_{zz} = 0. \qquad (24.3.38)$$

Hence the plastic incompressibility constraint reduces to $\dot{\epsilon}^{\rm p}_{rr} + \dot{\epsilon}^{\rm p}_{\theta\theta} = 0$. The latter implies that σ'_{rr} and $\sigma'_{\theta\theta}$ have the same *absolute value*, resulting in

$$|\sigma'_{rr}| = |\sigma'_{\theta\theta}| = k. \qquad (24.3.39)$$

Substitution of (24.3.38) and (24.3.39) in the yield condition with

$$\bar{\sigma} = \sqrt{(3/2)(\sigma'^2_{rr} + \sigma'^2_{\theta\theta} + \sigma'^2_{zz})}$$

gives

$$k = \frac{Y}{\sqrt{3}}. \tag{24.3.40}$$

Consider now the equilibrium relation

$$\frac{\partial \sigma_{rr}}{\partial r} + \frac{\sigma_{rr} - \sigma_{\theta\theta}}{r} = 0, \tag{24.3.41}$$

which under (24.3.39) becomes

$$\frac{\partial \sigma_{rr}}{\partial r} - \frac{2Y}{\sqrt{3}r} = 0. \tag{24.3.42}$$

The algebraic sign of the difference of the stresses comes from the physical consideration that $\sigma_{\theta\theta} > \sigma_{rr}$. Solving (24.3.42) and applying the boundary condition $\sigma_{rr}(a) = -p$ gives

$$\sigma_{rr} = -p + \frac{2Y}{\sqrt{3}} \ln\left(\frac{r}{a}\right) \equiv -\pi - (p - \pi) + \frac{2Y}{\sqrt{3}} \ln\left(\frac{r}{a}\right), \tag{24.3.43}$$

where in writing the second of (24.3.43) we have accounted for the arbitrary pressure π for an incompressible material.

Enforcing traction continuity between the elastic zone and plastic zone we find that

$$-\frac{Y}{\sqrt{3}}\frac{c^2}{b^2}\left(\frac{b^2}{c^2} - 1\right) = -p + \frac{2Y}{\sqrt{3}} \ln\left(\frac{c}{a}\right), \tag{24.3.44}$$

which gives an implicit relation for determining the location of the elastic-plastic interface for any given p as

$$p = \frac{Y}{\sqrt{3}}\left[1 - \left(\frac{c}{b}\right)^2 + 2\ln\left(\frac{c}{a}\right)\right]. \tag{24.3.45}$$

Since $\frac{1}{3}\,\mathrm{tr}\,\boldsymbol{\sigma} = -\pi$, we see that the hoop stress in the plastic region takes the form

$$\sigma_{\theta\theta} = -\pi + (p - \pi) - \frac{2Y}{\sqrt{3}} \ln\left(\frac{r}{a}\right), \tag{24.3.46}$$

and the deviatoric stresses are of the form

$$\sigma'_{\theta\theta} = -(p-\pi) + \frac{2Y}{\sqrt{3}} \ln\left(\frac{r}{a}\right), \qquad \sigma'_{\theta\theta} = (p-\pi) - \frac{2Y}{\sqrt{3}} \ln\left(\frac{r}{a}\right), \qquad \sigma'_{zz} = 0. \tag{24.3.47}$$

If we now use $(24.3.47)_2$ together with (24.3.45) in $(24.3.39)_2$, then we find that

$$\pi = -\frac{Y}{\sqrt{3}}\left[\left(\frac{c}{b}\right)^2 - 2\ln\left(\frac{r}{c}\right)\right]. \tag{24.3.48}$$

Finally, gathering the relation (24.3.34) for u_r, and substitution of (24.3.48) in (24.3.43) and (24.3.46) gives the displacement and stress fields in the plastic region of the cylinder as

$$\left.\begin{aligned} u_r &= \frac{Y}{\sqrt{3}}\frac{1}{2\mu}\frac{c^2}{r} \\ \sigma_{rr} &= -\frac{Y}{\sqrt{3}}\left[1-\left(\frac{c}{b}\right)^2+2\ln\frac{c}{r}\right] \\ \sigma_{\theta\theta} &= \frac{Y}{\sqrt{3}}\left[1+\left(\frac{c}{b}\right)^2-2\ln\frac{c}{r}\right] \\ \sigma_{zz} &= \frac{Y}{\sqrt{3}}\left[\left(\frac{c}{b}\right)^2-2\ln\frac{c}{r}\right] \end{aligned}\right\} \quad \text{in} \quad a \le r \le c, \tag{24.3.49}$$

together with the condition locating the elastic-plastic interface:

$$p = \frac{Y}{\sqrt{3}}\left[1-\left(\frac{c}{b}\right)^2+2\ln\frac{c}{a}\right]. \tag{24.3.50}$$

Fully plastic cylindrical pressure vessel

When $c = b$, the cylindrical pressure vessel becomes completely plastic. The corresponding pressure using (24.3.50) is given by

$$\boxed{p^* = \frac{2Y}{\sqrt{3}}\ln\left(\frac{b}{a}\right).} \tag{24.3.51}$$

Note that from (24.3.50)

$$\frac{dp}{dc} = \frac{Y}{\sqrt{3}}\left[-\frac{2c}{b^2}+\frac{2}{c}\right]. \tag{24.3.52}$$

so that when $c = b$, the pressure (24.3.50) is a *maximum*.

25 Rigid-perfectly-plastic materials. Two extremal principles

In engineering applications one often encounters situations in which the plastic strains are very much larger than the elastic strains. Under such circumstances the small elastic strains may be neglected, and the material may be treated as **rigid-plastic**, that is,

$$\boldsymbol{\epsilon} = \boldsymbol{\epsilon}^{\mathrm{p}} \quad \text{and} \quad \dot{\boldsymbol{\epsilon}} = \dot{\boldsymbol{\epsilon}}^{\mathrm{p}}.$$

Further, for sufficiently large plastic strains the flow resistance for many materials does not change significantly with continued plastic flow, and the material may be treated as **perfectly plastic** with a constant plastic flow resistance,

$$Y \equiv \text{constant}.$$

Materials for which these two idealizations hold are called **rigid-perfectly-plastic materials.**

The modeling of materials as rigid-perfectly-plastic facilitates analysis of many problems due to the fact that bodies made from such materials satisfy two extremal principles. In turn, these extremal principles can be used to find upper and lower bounds to the limit load associated with a given boundary-value problem. In this chapter we take up these extremal principles and their exploitation in the methods of *limit analysis*.

25.1 Mixed boundary-value problem for a rigid-perfectly-plastic solid

Let $\mathcal{S}_1$ and $\mathcal{S}_2$ be complementary subsurfaces of the boundary $\partial\mathcal{B}$ of the body $\mathcal{B}$. We consider **boundary conditions** in which the velocity $\mathbf{v} \equiv \dot{\mathbf{u}}$ is specified on $\mathcal{S}_1$ and the surface traction on $\mathcal{S}_2$:

$$\mathbf{v} = \hat{\mathbf{v}} \quad \text{on } \mathcal{S}_1, \qquad \boldsymbol{\sigma}\mathbf{n} = \hat{\mathbf{t}} \quad \text{on } \mathcal{S}_2, \tag{25.1.1}$$

with $\hat{\mathbf{v}}$ and $\hat{\mathbf{t}}$ prescribed functions of $\mathbf{x}$. Then the quasi-static mixed boundary-value problem for a rigid-perfectly-plastic solid may be stated as follows:

Continuum Mechanics of Solids. Lallit Anand and Sanjay Govindjee, Oxford University Press (2020).

DOI: 10.1093/oso/9780198864721.001.0001

Given: The flow resistance Y, a body force field $\mathbf{b}$, and boundary data $\hat{\mathbf{v}}$ and $\hat{\mathbf{t}}$.
Find: A triplet of fields $(\mathbf{v}, \dot{\boldsymbol{\epsilon}}, \boldsymbol{\sigma})$ which satisfy the field equations

$$\left.\begin{aligned} &\dot{\boldsymbol{\epsilon}} = \tfrac{1}{2}(\nabla\mathbf{v} + (\nabla\mathbf{v})^{\top}), \qquad \operatorname{tr}\dot{\boldsymbol{\epsilon}} = 0,\\ &\boldsymbol{\sigma}' = \left(\frac{2Y}{3\dot{\bar{\epsilon}}}\right)\dot{\boldsymbol{\epsilon}} \quad \text{for} \quad \dot{\boldsymbol{\epsilon}} \neq \mathbf{0}, \quad \text{with} \quad \dot{\bar{\epsilon}} \stackrel{\text{def}}{=} \sqrt{2/3}|\dot{\boldsymbol{\epsilon}}| \geq 0,\\ &(\bar{\sigma} - Y) \leq 0, \qquad \dot{\bar{\epsilon}}(\bar{\sigma} - Y) = 0, \qquad \text{with} \qquad \bar{\sigma} \stackrel{\text{def}}{=} \sqrt{3/2}|\boldsymbol{\sigma}'|,\\ &\mathcal{D}(\dot{\boldsymbol{\epsilon}}) = Y\dot{\bar{\epsilon}} > 0, \quad \text{when} \quad \dot{\boldsymbol{\epsilon}} \neq \mathbf{0},\\ &\operatorname{div}\boldsymbol{\sigma} + \mathbf{b} = \mathbf{0}, \end{aligned}\right\} \text{ in } \mathcal{B} \tag{25.1.2}$$

and the boundary conditions (25.1.1); cf. Sec. 21.7 and Remark 21.2. Note also that the superscript indicating plastic strain has been dropped since the total strain in this context is identical to the plastic strain.

25.2 Two extremal principles for rigid-perfectly-plastic materials

In this section we consider two extremal principles for rigid-perfectly-plastic materials.

25.2.1 Extremal principle for power expended by tractions and body forces

This principle is stated in terms of *statically admissible stress fields.* A stress field $\boldsymbol{\sigma}^*$ is said to be **statically admissible** with respect to a traction field $\mathbf{t}^*$ and a body force field $\mathbf{b}^*$ if it

(i) satisfies the equilibrium equation

$$\operatorname{div}\boldsymbol{\sigma}^* + \mathbf{b}^* = \mathbf{0} \quad \text{in} \quad \mathcal{B}; \tag{25.2.1}$$

(ii) satisfies the traction boundary conditions

$$\boldsymbol{\sigma}^*\mathbf{n} = \mathbf{t}^* \quad \text{on} \quad \partial\mathcal{B}; \text{ and} \tag{25.2.2}$$

(iii) does not violate the yield condition

$$f(\boldsymbol{\sigma}^*, Y) = \sqrt{3/2}\,|\boldsymbol{\sigma}^{*\prime}| - Y \leq 0 \quad \text{in} \quad \mathcal{B}. \tag{25.2.3}$$

Let $\boldsymbol{\sigma}^*$ denote a statically admissible stress field, and let $(\mathbf{v}, \dot{\boldsymbol{\epsilon}}, \boldsymbol{\sigma})$ denote the actual velocity field, strain-rate, and stress field in a rigid-perfectly-plastic solid. Then, by assumption (25.2.3) we have that $f(\boldsymbol{\sigma}^*, Y) \leq 0$, and the principle of maximum dissipation requires that (cf. Sec. 23.2)

$$(\boldsymbol{\sigma} - \boldsymbol{\sigma}^*) : \dot{\boldsymbol{\epsilon}} \geq 0. \tag{25.2.4}$$

Integrating this relation over the body gives

$$\int_{\mathcal{B}} (\boldsymbol{\sigma} - \boldsymbol{\sigma}^*) : \dot{\boldsymbol{\epsilon}}\, dv \geq 0. \tag{25.2.5}$$

Next, using the symmetry of the stress difference

$$\boldsymbol{\sigma}^{\mathrm{d}} \stackrel{\text{def}}{=} (\boldsymbol{\sigma} - \boldsymbol{\sigma}^*),$$

the identity

$$\operatorname{div}(\boldsymbol{\sigma}^{\mathrm{d}}\mathbf{v}) = \boldsymbol{\sigma}^{\mathrm{d}} : \nabla\mathbf{v} + \mathbf{v}\cdot\operatorname{div}\boldsymbol{\sigma}^{\mathrm{d}},$$

as well as (25.2.1) and (25.2.2), we have

$$\begin{aligned}
\underbrace{\int_{\mathcal{B}} (\boldsymbol{\sigma} - \boldsymbol{\sigma}^*) : \dot{\boldsymbol{\epsilon}}\, dv}_{\geq 0} &= \int_{\mathcal{B}} \boldsymbol{\sigma}^{\mathrm{d}} : \dot{\boldsymbol{\epsilon}}\, dv = \int_{\mathcal{B}} \boldsymbol{\sigma}^{\mathrm{d}} : \nabla\mathbf{v}\, dv \\
&= \int_{\mathcal{B}} \operatorname{div}(\boldsymbol{\sigma}^{\mathrm{d}}\mathbf{v})\, dv - \int_{\mathcal{B}} \mathbf{v}\cdot\operatorname{div}\boldsymbol{\sigma}^{\mathrm{d}}\, dv \\
&= \int_{\mathcal{B}} \operatorname{div}(\boldsymbol{\sigma}^{\mathrm{d}}\mathbf{v})\, dv + \int_{\mathcal{B}} \mathbf{v}\cdot(\mathbf{b} - \mathbf{b}^*)\, dv \\
&= \int_{\partial\mathcal{B}} (\boldsymbol{\sigma}^{\mathrm{d}}\mathbf{v})\cdot\mathbf{n}\, da + \int_{\mathcal{B}} \mathbf{v}\cdot(\mathbf{b} - \mathbf{b}^*)\, dv \\
&= \int_{\partial\mathcal{B}} (\boldsymbol{\sigma}^{\mathrm{d}}\mathbf{n})\cdot\mathbf{v}\, da + \int_{\mathcal{B}} \mathbf{v}\cdot(\mathbf{b} - \mathbf{b}^*)\, dv \\
&= \left[\int_{\partial\mathcal{B}} (\boldsymbol{\sigma}\mathbf{n})\cdot\mathbf{v}\, da + \int_{\mathcal{B}} \mathbf{b}\cdot\mathbf{v}\, dv\right] - \left[\int_{\partial\mathcal{B}} (\boldsymbol{\sigma}^*\mathbf{n})\cdot\mathbf{v}\, da + \int_{\mathcal{B}} \mathbf{b}^*\cdot\mathbf{v}\, dv\right] \geq 0.
\end{aligned}$$

Noting that the actual tractions are $\mathbf{t} = \boldsymbol{\sigma}\mathbf{n}$ and those associated with the statically admissible field are $\mathbf{t}^* = \boldsymbol{\sigma}^*\mathbf{n}$, we can state the following extremal principle:

EXTREMAL PRINCIPLE FOR POWER EXPENDED BY LOADING SYSTEM: The power expended by the tractions and body forces corresponding to the actual stress field $\boldsymbol{\sigma}$ and the actual velocity field $\mathbf{v}$, is greater than or equal to the power expended by the tractions and body forces corresponding to any statically admissible stress field $\boldsymbol{\sigma}^*$ and the actual velocity field $\mathbf{v}$,

$$\int_{\partial\mathcal{B}} \mathbf{t}\cdot\mathbf{v}\, da + \int_{\mathcal{B}} \mathbf{b}\cdot\mathbf{v}\, dv \geq \int_{\partial\mathcal{B}} \mathbf{t}^*\cdot\mathbf{v} da + \int_{\mathcal{B}} \mathbf{b}^*\cdot\mathbf{v}\, dv. \qquad (25.2.6)$$

REMARK 25.1

It is important to observe that $\boldsymbol{\sigma}^*$ only needs to satisfy (25.2.1)–(25.2.3). In particular there are no requirements that it be associated with any particular strain field or displacement field.

25.2.2 Extremal principle for velocity

This principle is stated in terms of *kinematically admissible velocity fields*. We say that a velocity field $\mathbf{v}^*$ is **kinematically admissible** if it:

(i) satisfies the strain-rate-velocity relation, $\dot{\boldsymbol{\epsilon}}^* = \frac{1}{2}((\nabla\mathbf{v}^*) + (\nabla\mathbf{v}^*)^\top)$;

(ii) gives no volume change, $\operatorname{tr}\dot{\boldsymbol{\epsilon}}^* = 0$; and

(iii) satisfies the velocity boundary condition $\mathbf{v}^* = \hat{\mathbf{v}} \quad \text{on} \quad \mathcal{S}_1$.

For such a kinematically admissible velocity field let

$$\dot{\bar{\epsilon}}^* \stackrel{\text{def}}{=} \sqrt{2/3}\,|\dot{\boldsymbol{\epsilon}}^*| \qquad \text{for } \dot{\boldsymbol{\epsilon}}^* \neq \mathbf{0} \tag{25.2.7}$$

define an equivalent tensile strain-rate, and let

$$\mathcal{D}(\dot{\boldsymbol{\epsilon}}^*) \stackrel{\text{def}}{=} Y\,\dot{\bar{\epsilon}}^* \tag{25.2.8}$$

denote the corresponding dissipation rate per unit volume.

Given a kinematically admissible velocity field $\mathbf{v}^*$, we denote the *internal rate of dissipation* in the body by

$$\mathcal{D}_{\text{int}}(\mathbf{v}^*) \stackrel{\text{def}}{=} \int_{\mathcal{B}} Y\dot{\bar{\epsilon}}^*\,dv, \tag{25.2.9}$$

and denote the *external rate of work* done on the body by

$$\mathcal{W}_{\text{ext}}(\mathbf{v}^*) \stackrel{\text{def}}{=} \int_{\mathcal{B}} \mathbf{b}\cdot\mathbf{v}^*\,da + \int_{\partial\mathcal{B}} \mathbf{t}\cdot\mathbf{v}^*\,da, \tag{25.2.10}$$

where the traction term includes the work of all external tractions, not just those on $\mathcal{S}_2$.

Next, we define a functional

$$\Phi\{\mathbf{v}^*\} = \underbrace{\int_{\mathcal{B}} Y\,\dot{\bar{\epsilon}}^*\,dv}_{\mathcal{D}_{\text{int}}(\mathbf{v}^*)} - \underbrace{\left(\int_{\mathcal{B}} \mathbf{b}\cdot\mathbf{v}^* + \int_{\partial\mathcal{B}} \mathbf{t}\cdot\mathbf{v}^*da\right)}_{\mathcal{W}_{\text{ext}}(\mathbf{v}^*)}, \tag{25.2.11}$$

which for the actual velocity field $\mathbf{v}$ is given by

$$\Phi\{\mathbf{v}\} = \underbrace{\int_{\mathcal{B}} Y\,\dot{\bar{\epsilon}}\,dv}_{\mathcal{D}_{\text{int}}(\mathbf{v})} - \underbrace{\left(\int_{\mathcal{B}} \mathbf{b}\cdot\mathbf{v}da + \int_{\partial\mathcal{B}} \mathbf{t}\cdot\mathbf{v}da\right)}_{\mathcal{W}_{\text{ext}}(\mathbf{v})}. \tag{25.2.12}$$

Extremal principle for velocity: Let $\mathbf{v}$ denote the actual velocity field and let $\mathbf{v}^*$ denote a kinematically admissible velocity field; then

$$\Phi\{\mathbf{v}\} = 0, \tag{25.2.13}$$

and

$$\Phi\{\mathbf{v}^*\} \geq \underbrace{\Phi\{\mathbf{v}\}}_{=0}. \tag{25.2.14}$$

That is, for the actual velocity $\mathbf{v}$ the functional $\Phi\{\mathbf{v}\}$ is equal to zero, and the actual velocity field *minimizes* Φ over all kinematically admissible velocity fields $\mathbf{v}^*$.

PROOF:

(i) $\Phi\{\mathbf{v}\} = 0$ portion of the principle:
Consider the actual solution $(\mathbf{v}, \dot{\epsilon}, \boldsymbol{\sigma})$ to (25.1.2). Using the fact that $\dot{\epsilon}$ is deviatoric, $\boldsymbol{\sigma}$ is symmetric, and the identity $\operatorname{div}(\boldsymbol{\sigma}\mathbf{v}) = \boldsymbol{\sigma} : \nabla\mathbf{v} + \mathbf{v} \cdot \operatorname{div}\boldsymbol{\sigma}$, we have

$$Y\dot{\bar{\epsilon}} = \boldsymbol{\sigma} : \dot{\epsilon} = \boldsymbol{\sigma} : \nabla\mathbf{v} = \operatorname{div}(\boldsymbol{\sigma}\mathbf{v}) - \mathbf{v} \cdot \operatorname{div}\boldsymbol{\sigma}. \tag{25.2.15}$$

Substituting (25.2.15) in (25.2.12) and using the equilibrium equation, $\operatorname{div}\boldsymbol{\sigma} + \mathbf{b} = \mathbf{0}$, we find that

$$\Phi\{\mathbf{v}\} = \int_{\mathcal{B}} \operatorname{div}(\boldsymbol{\sigma}\mathbf{v})\, dv - \int_{\partial\mathcal{B}} \mathbf{t} \cdot \mathbf{v} da. \tag{25.2.16}$$

Next, applying the divergence theorem to the volume integral in (25.2.16), we obtain

$$\Phi\{\mathbf{v}\} = \int_{\partial\mathcal{B}} (\boldsymbol{\sigma}\mathbf{v}) \cdot \mathbf{n}\, da - \int_{\partial\mathcal{B}} \mathbf{t} \cdot \mathbf{v} da. \tag{25.2.17}$$

Noting that

$$\int_{\partial\mathcal{B}} (\boldsymbol{\sigma}\mathbf{v}) \cdot \mathbf{n}\, da = \int_{\partial\mathcal{B}} (\boldsymbol{\sigma}\mathbf{n}) \cdot \mathbf{v}\, da = \int_{\partial\mathcal{B}} \mathbf{t} \cdot \mathbf{v}\, da, \tag{25.2.18}$$

(25.2.17) yields

$$\Phi\{\mathbf{v}\} = 0.$$

(ii) $\Phi\{\mathbf{v}^*\} \geq \Phi\{\mathbf{v}\}$ portion of the principle:
First note that $\boldsymbol{\sigma} : \dot{\epsilon}^* = \boldsymbol{\sigma}' : \dot{\epsilon}^*$, since $\dot{\epsilon}^*$ is deviatoric. This then implies that

$$\boldsymbol{\sigma} : \dot{\epsilon}^* = \boldsymbol{\sigma}' : \dot{\epsilon}^* \leq |\boldsymbol{\sigma}' : \dot{\epsilon}^*| \leq |\boldsymbol{\sigma}'|\, |\dot{\epsilon}^*| = |\boldsymbol{\sigma}'| \sqrt{3/2}\, \dot{\bar{\epsilon}}^*, \tag{25.2.19}$$

where the first inequality follows from the fact that all numbers are less than or equal to their absolute value and the second inequality follows from the Cauchy–Schwarz inequality.[1] Since the yield condition requires $\sqrt{3/2}\,|\boldsymbol{\sigma}'| = \bar{\sigma} \leq Y$, we have

$$\boldsymbol{\sigma} : \dot{\epsilon}^* \leq Y\dot{\bar{\epsilon}}^* \tag{25.2.20}$$

for all kinematically admissible velocity fields.

Integrating (25.2.20) over the whole body gives

$$\int_{\mathcal{B}} Y\dot{\bar{\epsilon}}^* dv - \int_{\mathcal{B}} \boldsymbol{\sigma} : \dot{\epsilon}^* dv \geq 0. \tag{25.2.21}$$

Next, using the symmetry of $\boldsymbol{\sigma}$ and the equation of equilibrium, $\operatorname{div}\boldsymbol{\sigma} + \mathbf{b} = \mathbf{0}$, we obtain

$$\boldsymbol{\sigma} : \dot{\epsilon}^* = \boldsymbol{\sigma} : \nabla\mathbf{v}^* = \operatorname{div}(\boldsymbol{\sigma}\mathbf{v}^*) - \mathbf{v}^* \cdot \operatorname{div}\boldsymbol{\sigma} = \operatorname{div}(\boldsymbol{\sigma}\mathbf{v}^*) + \mathbf{b} \cdot \mathbf{v}^*, \tag{25.2.22}$$

substitution of which in (25.2.21) and use of the divergence theorem yields

$$\int_{\mathcal{B}} Y\dot{\bar{\epsilon}}^* dv - \int_{\partial\mathcal{B}} (\boldsymbol{\sigma}\mathbf{v}^*) \cdot \mathbf{n}\, da - \int_{\mathcal{B}} \mathbf{b} \cdot \mathbf{v}^* dv \geq 0. \tag{25.2.23}$$

---

[1] The Cauchy-Schwarz inequality states that for two tensors $\mathbf{A}$ and $\mathbf{B}$, $|\mathbf{A} : \mathbf{B}| \leq |\mathbf{A}|\,|\mathbf{B}|$. The proof of this inequality follows by noting that $0 \leq (\mathbf{A} - \lambda\mathbf{B}) : (\mathbf{A} - \lambda\mathbf{B})$, where λ is any real number. Expanding the right-hand side we see that $0 \leq \mathbf{A} : \mathbf{A} - \lambda(2\mathbf{A} : \mathbf{B} - \lambda\mathbf{B} : \mathbf{B})$. Picking now a particular λ, viz. $\lambda = \mathbf{A} : \mathbf{B}/\mathbf{B} : \mathbf{B}$, yields $0 \leq \mathbf{A} : \mathbf{A} - (\mathbf{A} : \mathbf{B})^2/\mathbf{B} : \mathbf{B}$ and hence $(\mathbf{A} : \mathbf{B})^2 \leq (\mathbf{A} : \mathbf{A})\,(\mathbf{B} : \mathbf{B})$. Taking square roots on both sides gives the final result: $|\mathbf{A} : \mathbf{B}| \leq |\mathbf{A}|\,|\mathbf{B}|$.

Since,

$$\int_{\partial\mathcal{B}} (\boldsymbol{\sigma}\mathbf{v}^*) \cdot \mathbf{n}\, da = \int_{\partial\mathcal{B}} (\boldsymbol{\sigma}\mathbf{n}) \cdot \mathbf{v}^*\, da = \int_{\partial\mathcal{B}} \mathbf{t} \cdot \mathbf{v}^*\, da, \tag{25.2.24}$$

(25.2.23) becomes

$$\int_{\mathcal{B}} Y \dot{\bar{\epsilon}}^* dv - \int_{\mathcal{B}} \mathbf{b} \cdot \mathbf{v}^* dv - \int_{\partial\mathcal{B}} \mathbf{t} \cdot \mathbf{v}^*\, da \geq 0.$$

Hence we obtain the desired result that

$$\underbrace{\int_{\mathcal{B}} Y \dot{\bar{\epsilon}}^* dv - \int_{\mathcal{B}} \mathbf{b} \cdot \mathbf{v}^* dv - \int_{\partial\mathcal{B}} \mathbf{t} \cdot \mathbf{v}^*\, da}_{\Phi\{\mathbf{v}^*\}} \geq \underbrace{0}_{\Phi\{\mathbf{v}\}}.$$

□

REMARK 25.2

These two extremal principles for rigid-perfectly-plastic materials can be reformulated as **theorems of limit analysis**, which give upper and lower bounds on the "load" under which a body—which may be approximately modeled as rigid-perfectly-plastic—reaches a "limit load." We discuss this next.

25.3 Limit analysis

Consider a plate with an edge-notch loaded as shown in Fig. 25.1. For sufficiently small loads P the part behaves elastically. As the load P is increased, yielding occurs at the notch-tip where the stress concentration is the highest. A further increase in the load increases the size of the plastic zone, but it affects the overall deflection δ of the part only modestly because there is still an elastic region holding the bottom and top halves together, but the load-deflection curve is no longer linear. As the load is increased further, the plastic zone spreads across the remaining ligament and general yield occurs. At this point the load-displacement curve becomes essentially horizontal and the total deformation becomes large. The determination of the load at which the part deforms excessively is of obvious engineering interest. This load is called the **limit load**. Here we have ruled out the possibility that the stress and strain concentrations at the notch tip will cause fracture before the limit load is reached.

In an elastic-plastic calculation, often two of the most interesting results are:

- The critical load at which the component **starts to deform plastically** in some localized region.
- The limit load where the component **starts to deform excessively or "collapses"**.

Since plastic flow must be avoided in many designs of structural components, the critical load at which the component **starts to deform plastically** is all that is often needed. In this case we

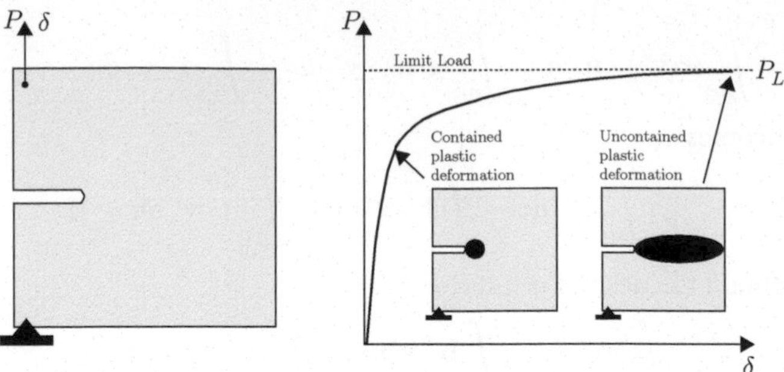

Fig. 25.1 Schematic of the limit load for a notched plate.

do not need to solve the complete plasticity problem; an elastic calculation, together with use of a yield condition, suffices to determine the critical load for the onset of plastic deformation. However, there are situations in which some plasticity in a structure or a component can or must be tolerated, and there are even some situations, such as metal-forming applications, where it is desirable. In these situations one would like to know the **limit load**.

The limit load can be used as a realistic basis for design in the following two common situations:

1. For many structural applications the objective is to prevent excessive deformation. In this case it is sufficient to design to a suitable (large) fraction of the limit load.
2. For metal-forming operations one often wishes to estimate a load which will cause the workpiece to deform plastically to produce the desired shape change. In this case it is sufficient to design for loads just above the limit load of the workpiece.

The question arises as to if one can determine the limit load without having to solve the complete elastic-plastic boundary-value problem. This is the motivation for the approximate procedures of **plastic limit analysis.**

The method of limit analysis is based on two complementary theorems:

- The first, a **lower bound theorem**, is used to calculate a load $P_{\rm lb}$ which is a **guaranteed load capacity under which a system will not undergo excessive deformation.**
- The second, an **upper bound theorem**, is used to calculate a load $P_{\rm ub}$ which is a **guaranteed load capacity which will cause the system to undergo large deformations.**

By using both theorems we can bracket the desired limit load:

$$P_{\rm lb} \leq P_L \leq P_{\rm ub}.$$

If it so happens that $P_{\rm lb} = P_{\rm ub}$ we have an exact limit load; if not, then we at least have some useful information about P_L (Fig. 25.2).

The extremal principles for rigid-perfectly-plastic materials that were discussed in Sec. 25.2, can be reformulated as **theorems of limit analysis**, which give upper and lower bounds on the "load" under which a body, that may be approximately modeled as elastic-perfectly-plastic, reaches a "limit load."

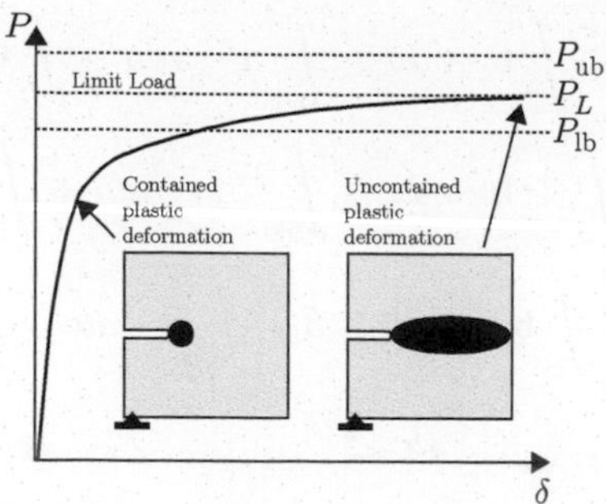

Fig. 25.2 Schematic of the upper bound and lower bound estimates for the limit load for a notched plate.

25.4 Lower bound theorem

Consider a component made from a rigid-perfectly-plastic solid, subjected to a set of tractions of the form $\mathbf{t} = \beta\tilde{\mathbf{t}}$ on $\partial\mathcal{B}$ and a set of body forces of the form $\mathbf{b} = \beta\tilde{\mathbf{b}}$ in $\mathcal{B}$. Here β represents a proportional loading factor and $\tilde{\mathbf{t}}$ and $\tilde{\mathbf{b}}$ represent given loading patterns. We assume that the component collapses when the external loads reach the values

$$\mathbf{t} = \beta_C\,\tilde{\mathbf{t}} \qquad \text{and} \qquad \mathbf{b} = \beta_C\tilde{\mathbf{b}}\,. \tag{25.4.1}$$

We now attempt to find a **lower bound** β_L to the factor β_C by which the external loading can be **proportionally increased** before the solid collapses and the external loads reach their "limit value."

Consider a stress state $\boldsymbol{\sigma}^*$ that is **statically admissible** with respect to a traction field $\mathbf{t}^* = \beta_L\tilde{\mathbf{t}}$ and a body force field $\mathbf{b}^* = \beta_L\tilde{\mathbf{b}}$. Noting that $\boldsymbol{\sigma}^*$ satisfies the yield condition (25.2.3) by assumption, one has the following theorem:

Lower bound theorem: If such a statically admissible stress state can be found, then

$$\beta_L \leq \beta_C, \tag{25.4.2}$$

and the solid will not collapse.

REMARK 25.3

Clearly, the largest value of β_L is of the most practical interest.

Proof: Let $(\mathbf{v}, \boldsymbol{\sigma})$ denote the actual velocity field and stress at collapse: $\beta = \beta_C$, $\mathbf{t} = \beta_C\tilde{\mathbf{t}}$, and $\mathbf{b} = \beta_C\tilde{\mathbf{b}}$. Further let $(\boldsymbol{\sigma}^*, \beta_L)$ denote a statically admissible state where $\mathbf{t}^* = \beta_L\tilde{\mathbf{t}}$ and $\mathbf{b}^* = \beta_L\tilde{\mathbf{b}}$. Now by the extremal theorem (25.2.6) one has that

$$\int_{\partial\mathcal{B}} \mathbf{t}\cdot\mathbf{v}\,da + \int_{\mathcal{B}} \mathbf{b}\cdot\mathbf{v}\,dv \geq \int_{\partial\mathcal{B}} \mathbf{t}^*\cdot\mathbf{v}da + \int_{\mathcal{B}} \mathbf{b}^*\cdot\mathbf{v}\,dv \tag{25.4.3}$$

$$\int_{\partial\mathcal{B}} \beta_C\tilde{\mathbf{t}}\cdot\mathbf{v}\,da + \int_{\mathcal{B}} \beta_C\tilde{\mathbf{b}}\cdot\mathbf{v}\,dv \geq \int_{\partial\mathcal{B}} \beta_L\tilde{\mathbf{t}}\cdot\mathbf{v}da + \int_{\mathcal{B}} \beta_L\tilde{\mathbf{b}}\cdot\mathbf{v}\,dv \tag{25.4.4}$$

$$\beta_C\left[\int_{\partial\mathcal{B}} \tilde{\mathbf{t}}\cdot\mathbf{v}\,da + \int_{\mathcal{B}} \tilde{\mathbf{b}}\cdot\mathbf{v}\,dv\right] \geq \beta_L\left[\int_{\partial\mathcal{B}} \tilde{\mathbf{t}}\cdot\mathbf{v}da + \int_{\mathcal{B}} \tilde{\mathbf{b}}\cdot\mathbf{v}\,dv\right] \tag{25.4.5}$$

$$\beta_C \geq \beta_L\,. \tag{25.4.6}$$

REMARK 25.4

In practice one often does not choose the loading pattern of the statically admissible field to exactly match that of the actual loading. However, the theorem still holds as long as the power of the chosen load pattern for $(\mathbf{t}^*, \mathbf{b}^*)$ equals or exceeds the power of the actual pattern—i.e. the term in the square brackets on the right-hand side of (25.4.5) simply needs to be equal to or greater than the term in the square brackets on the left-hand side of (25.4.5) for the conclusion (25.4.6) to hold true.

25.5 Upper bound theorem

As in the case for the lower bound theorem, consider a component made from a rigid-perfectly-plastic solid, subjected to some known distribution of tractions $\mathbf{t} = \beta\tilde{\mathbf{t}}$ on $\partial\mathcal{B}$ and body forces $\mathbf{b} = \beta\tilde{\mathbf{b}}$ in $\mathcal{B}$, where β is a proportional loading factor and $\tilde{\mathbf{t}}$ and $\tilde{\mathbf{b}}$ are traction and body force loading patterns, respectively. We assume that the component collapses when the proportional loading factor $\beta = \beta_C$:

$$\mathbf{t} = \beta_C\tilde{\mathbf{t}} \qquad \text{and} \qquad \mathbf{b} = \beta_C\tilde{\mathbf{b}}\,. \tag{25.5.1}$$

We now attempt to find an **upper bound** β_U to the factor β_C which will guarantee plastic collapse.

Consider a **kinematically admissible** velocity field $\mathbf{v}^*$. Then the functional defined in (25.2.11) evaluated at the collapse load $\beta = \beta_C$ and the extremal principle for velocities stated in Sec. 25.2.2 give

$$\int_{\mathcal{B}} Y\,\dot{\bar{\epsilon}}^*\,dv - \beta_C\left(\int_{\mathcal{B}} \tilde{\mathbf{b}}\cdot\mathbf{v}^* + \int_{\partial\mathcal{B}} \tilde{\mathbf{t}}\cdot\mathbf{v}^*\,da\right) \geq 0. \tag{25.5.2}$$

Let

$$\beta_U \overset{\text{def}}{=} \frac{\int_{\mathcal{B}} Y\dot{\bar{\epsilon}}^*\,dv}{\int_{\mathcal{B}} \tilde{\mathbf{b}}\cdot\mathbf{v}^*\,dv + \int_{\partial\mathcal{B}} \tilde{\mathbf{t}}\cdot\mathbf{v}^*\,da}, \tag{25.5.3}$$

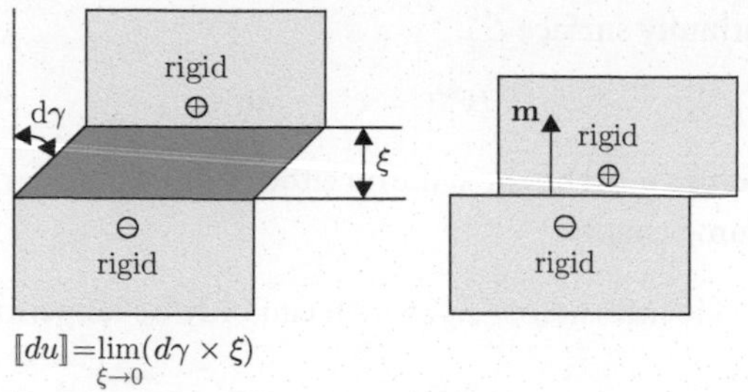

Fig. 25.3 A "velocity discontinuity" as the limit of a simple shear-strain increment in a narrow shear band of thickness ξ in the limit $\xi \to 0$.

then

$$\beta_C \leq \beta_U. \tag{25.5.4}$$

UPPER BOUND THEOREM: An **upper bound** β_U to β_C may be obtained by assuming **any kinematically admissible velocity field**, and computing the ratio of the rate of internal plastic dissipation associated with that velocity field to the rate of external work done by the applied loading pattern against the assumed kinematically admissible velocity field.

REMARK 25.5

Clearly, the smallest value of β_U is of the greatest practical interest.

25.5.1 Upper bound estimates obtained by using block-sliding velocity fields

In engineering practice, upper bound estimates are often obtained using *block-sliding* velocity fields. A sliding displacement increment across a surface does not satisfy the incremental strain-displacement relations, but we consider that such a discontinuous displacement increment across a surface is actually the limit of a simple shear-strain increment in a narrow shear band as depicted in Fig. 25.3.

For the purposes of applying the upper bound theorem, we assume a kinematically admissible velocity field $\mathbf{v}^* = \dot{\mathbf{u}}^*$ at plastic collapse having a finite set of such discontinuities. Each such surface of velocity discontinuity is designated by $\mathcal{S}_d^*$. Let $\mathbf{m}$ be a unit vector normal at some point on such a discontinuity surface, cf. Fig. 25.3, and let $\mathbf{v}^{*\pm}$ and $\boldsymbol{\sigma}^{\pm}$ denote the limiting values of the velocity $\mathbf{v}^*$ and stress $\boldsymbol{\sigma}$ on the two sides of the discontinuity surface. We assume that the following conditions hold:

- We consider kinematically admissible velocity fields which may suffer jump discontinuities in their **tangential components**, but no jumps in their normal components across a velocity discontinuity surface $\mathcal{S}_d^*$,

$$(\mathbf{v}^{*+} - \mathbf{v}^{*-}) \cdot \mathbf{m} = 0. \tag{25.5.5}$$

The normal velocities must be the same on either side of such surfaces in order that there be no plastic volume change.

- The jump in the velocities across $\mathcal{S}_d^*$, which can only be tangential, is denoted by

$$[\![\mathbf{v}^*]\!] \stackrel{\text{def}}{=} \mathbf{v}^{*+} - \mathbf{v}^{*-}, \tag{25.5.6}$$

with jump magnitude

$$[\![v^*]\!] \stackrel{\text{def}}{=} |\mathbf{v}^{*+} - \mathbf{v}^{*-}|. \tag{25.5.7}$$

- In the narrow band of thickness ξ the kinematically admissible velocity is taken to be of the form

$$\mathbf{v}^* = \mathbf{v}^{*-} + [\![\mathbf{v}^*]\!]\frac{\mathbf{m}\cdot\mathbf{x}}{\xi}. \tag{25.5.8}$$

The strain-rate associated with velocity field (25.5.8) is given by

$$\dot{\boldsymbol{\epsilon}}^* = \frac{1}{2\xi}\left([\![\mathbf{v}^*]\!] \otimes \mathbf{m} + \mathbf{m} \otimes [\![\mathbf{v}^*]\!]\right), \tag{25.5.9}$$

which in matrix form, in a coordinate system whose first two directions are aligned with **m** and $[\![\mathbf{v}^*]\!]$, reads

$$[\dot{\boldsymbol{\epsilon}}^*] = [\![v^*]\!] \begin{bmatrix} 0 & 1/2\xi & 0 \\ 1/2\xi & 0 & 0 \\ 0 & 0 & 0 \end{bmatrix}. \tag{25.5.10}$$

Thus the associated equivalent strain-rate $\dot{\bar{\epsilon}}^* = \sqrt{2/3}|\dot{\boldsymbol{\epsilon}}^*| = [\![v^*]\!]/\sqrt{3}\xi$.

For such a block-sliding kinematically admissible velocity field $\mathbf{v}^*$, the rate of internal dissipation over a single discontinuity is given by

$$\mathcal{D}_{\text{int}}(\mathcal{S}_d^*) = \lim_{\xi\to 0}\int_{\mathcal{S}_d^*}\int_0^{\xi} Y\dot{\bar{\epsilon}}^*\, dh\, da = \lim_{\xi\to 0}\int_{\mathcal{S}_d^*} Y\frac{[\![v^*]\!]}{\sqrt{3}\xi}\xi\, da = \int_{\mathcal{S}_d^*} k[\![v^*]\!]\, da = A_{\mathcal{S}_d^*} k[\![v^*]\!], \tag{25.5.11}$$

where h is the thickness coordinate, $k \stackrel{\text{def}}{=} Y/\sqrt{3}$ denotes the **shear yield strength** of the material, and $A_{\mathcal{S}_d^*}$ is the area of $\mathcal{S}_d^*$. Thus,

$$\mathcal{D}_{\text{int}} = \int_{\mathcal{B}} Y\dot{\bar{\epsilon}}^* dv, \tag{25.5.12}$$

may be replaced by

$$\mathcal{D}_{\text{int}} = \sum_{\mathcal{S}_d*} A_{\mathcal{S}_d^*} k\, [\![v^*]\!], \tag{25.5.13}$$

and the upper bound estimate β_U in (25.5.3) becomes

$$\beta_U = \frac{\sum_{\mathcal{S}_d^*} A_{\mathcal{S}_d^*} k\, [\![v^*]\!]}{\int_{\mathcal{B}} \tilde{\mathbf{b}}\cdot\mathbf{v}^* dv + \int_{\partial\mathcal{B}} \tilde{\mathbf{t}}\cdot\mathbf{v}^*\, da}. \tag{25.5.14}$$

25.6 Example problems

25.6.1 Bounds to the limit load for a notched plate

In this simple example problem we shall estimate the limit load for a notched plate of width W and notch length a, subjected to a load P; cf., Fig. 25.4. The plate has a thickness B into the plane of the page. In carrying out the limit analysis, we will rule out the possibility that any stress and strain concentrations at the sharp notch tip will cause fracture before the limit load is reached.

Lower bound

To determine a *lower bound* we neglect body forces, and, with reference to Fig. 25.4 ,we assume a stress field $\boldsymbol{\sigma}^*$ such that there is

- a uniform stress $\sigma^*_{22} \neq 0$, with $\sigma^*_{22} \leq Y$ in region $\boxed{\text{A}}$, and
- all stress components in region $\boxed{\text{B}}$ vanish, $\sigma^*_{ij} = 0$.

This is certainly an unrealistic stress distribution, but it satisfies the conditions of the lower bound theorem. Namely,

- the yield condition,
- the equilibrium equations,

$$\frac{\partial \sigma^*_{11}}{\partial x_1} + \frac{\partial \sigma^*_{12}}{\partial x_2} + \frac{\partial \sigma^*_{13}}{\partial x_3} = 0 \tag{25.6.1}$$

$$\frac{\partial \sigma^*_{21}}{\partial x_1} + \frac{\partial \sigma^*_{22}}{\partial x_2} + \frac{\partial \sigma^*_{23}}{\partial x_3} = 0, \tag{25.6.2}$$

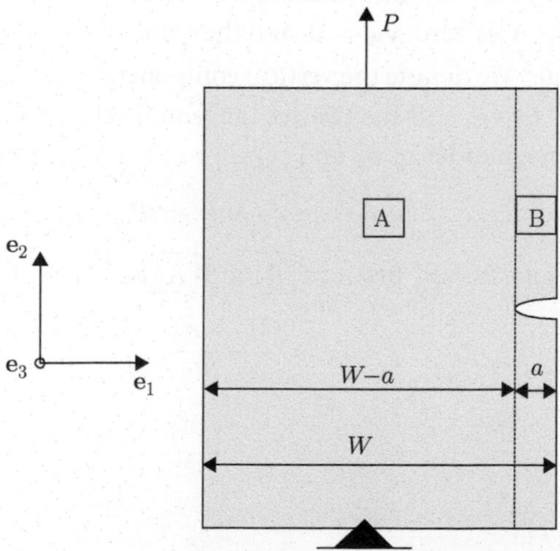

Fig. 25.4 A notched plate of width, W, and notch length, a, subjected to a load, P. The plate has a thickness, B, into the plane of the paper.

- and with
$$P^* = \sigma^*_{22} \times (W - a) \times B, \tag{25.6.3}$$
also the point force traction boundary condition,[2]
$$\mathbf{t}^* = P^* \delta(x_1 - a)\mathbf{e}_2 = \beta_L \delta(x_1 - a)\mathbf{e}_2, \tag{25.6.4}$$
where $x_1 = a$ indicates the location of the applied load and the loading parameter β_L is identified to be P^*.

Thus the lower bound to the collapse loading parameter $\beta_C = P$,
$$\mathbf{t} = P\delta(x_1 - a)\mathbf{e}_2 = \beta_C \delta(x_1 - a)\mathbf{e}_2,$$
is found from the lower bound theorem as
$$P\,|\mathbf{v}_{\text{top}}| \geq P^*\,|\mathbf{v}_{\text{top}}|\,. \tag{25.6.5}$$
Thus P^* is a *lower bound* to the limit load,
$$P_{\text{lb}} = \sigma^*_{22} \times (W - a) \times B. \tag{25.6.6}$$
Further, since *the greatest lower bound is the most useful*, and since $\sigma^*_{22} \leq Y$, we have as the largest, or best, lower bound
$$\boxed{P_{\text{lb}} = Y \times (W - a) \times B.} \tag{25.6.7}$$

Upper bound

Next, as an estimate of an *upper bound* we imagine a **block-sliding velocity** emanating from the tip of the notch through the back face along a plane making some angle θ to the horizontal direction; cf. Fig. 25.5. This assumed velocity field is not realistic because from symmetry one should expect slip on a plane at $-\theta$ to the horizontal as well; but this is not important for use of the upper bound theorem! Any kinematically admissible velocity field suffices.

The velocity of block $\boxed{\text{A}}$ is zero, $\mathbf{v}_A = \mathbf{0}$, and the velocity of the block $\boxed{\text{B}}$ is $\mathbf{v}_B$ inclined at angle θ to the horizontal.[3] We denote the vertical component of the velocity of block $\boxed{\text{B}}$ in the direction of the load P by $\mathbf{v}_P$, and the relative tangential velocity between the two blocks by $(\mathbf{v}_B - \mathbf{v}_A)$. Then, the magnitudes of $\mathbf{v}_P$ and $(\mathbf{v}_B - \mathbf{v}_A)$ are related by
$$|\mathbf{v}_P| = |\mathbf{v}_B - \mathbf{v}_A|\,\sin\theta\,. \tag{25.6.8}$$
Also, the length of the interface S^*_d between $\boxed{\text{B}}$ and $\boxed{\text{A}}$ is
$$l_{BA} = \frac{(W - a)}{\cos\theta}. \tag{25.6.9}$$

[2] The traction boundary conditions are assumed to be satisfied in the sense of St. Venant's Principle.

[3] To de-clutter the notation we drop the superposed $*$ on the symbols for the kinematically admissible velocities. All velocities in our upper bound computations should be understood to be assumed kinematically admissible velocities.

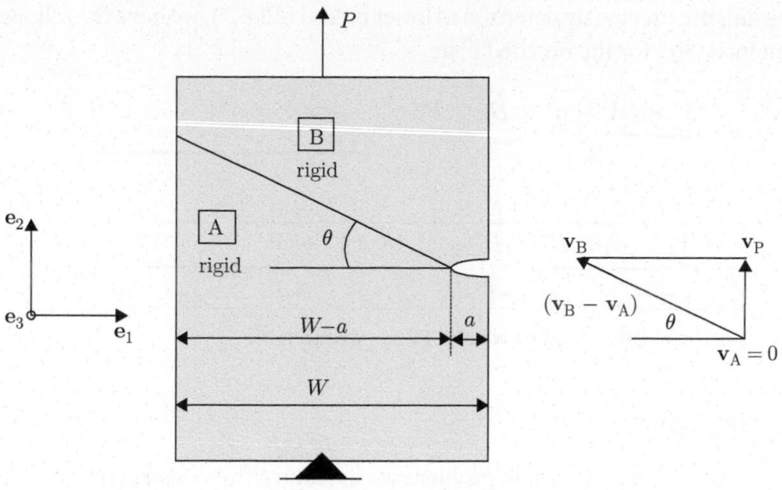

Fig. 25.5 Block-sliding velocity field for an upper bound estimate.

To determine the upper bound we need to calculate the internal dissipation rate corresponding to the assumed velocity field in accordance with (25.5.13):

$$\mathcal{D}_{\text{int}} = (l_{BA} \times B) \times k \times |\mathbf{v}_B - \mathbf{v}_A| = \left(\frac{(W-a)}{\cos\theta} \times B \right) \times k \times |\mathbf{v}_B - \mathbf{v}_A| . \qquad (25.6.10)$$

As before, $k = Y/\sqrt{3}$ is the shear yield strength of the material. Also, the external rate of work corresponding to the assumed velocity field is

$$\mathcal{W}_{\text{ext}} = P \times |\mathbf{v}_P| = P \times |\mathbf{v}_B - \mathbf{v}_A| \sin\theta. \qquad (25.6.11)$$

Thus from (25.5.14)

$$\beta_U = \frac{\mathcal{D}_{\text{int}}}{\mathcal{W}_{\text{ext}}} = \frac{k \times \dfrac{(W-a)}{\cos\theta} \times B \times |\mathbf{v}_B - \mathbf{v}_A|}{P \times |\mathbf{v}_B - \mathbf{v}_A| \sin\theta} = \frac{k}{P} \times \frac{2(W-a)}{\sin 2\theta} \times B, \qquad (25.6.12)$$

or with

$$P_{\text{ub}} \stackrel{\text{def}}{=} \beta_U P,$$

and using the relation, $k = Y/\sqrt{3}$, we have

$$P_{\text{ub}} = \frac{2}{\sqrt{3}} \times Y \times \frac{(W-a)}{\sin 2\theta} \times B.$$

Since we seek the *least* upper bound, and since this occurs for $\theta = \pi/4$, we have the following estimate for an upper bound to the limit load,

$$\boxed{P_{\text{ub}} = \frac{2}{\sqrt{3}} \times Y(W-a) \times B.} \qquad (25.6.13)$$

Finally, using the previously determined lower bound (25.6.7), we have the following bounds on the limit load, P_L, for the notched plate,

$$\underbrace{Y \times (W-a) \times B}_{P_{\rm lb}} \leq P_L \leq \underbrace{\frac{2}{\sqrt{3}} \times Y \times (W-a) \times B}_{P_{\rm ub}},$$

or

$$\boxed{\underbrace{Y \times (W-a) \times B}_{P_{\rm lb}} \leq P_L \leq \underbrace{1.15 \times Y \times (W-a) \times B}_{P_{\rm ub}}.} \tag{25.6.14}$$

We have bounded the limit load to within 15%—wonderful!

25.6.2 **Hodograph**

Before discussing our next example problem we introduce the construction of a **hodograph**, *which is a diagram depicting the velocities and relative velocities for block-sliding velocity fields.* The hodograph eases the computation of the needed quantities in block-sliding estimates for the upper bound.

As a preliminary, recall that in three-dimensional point space the position of a point $\mathbf{x}$ with respect to a rectangular coordinate system with orthonormal base vectors $\{\mathbf{e}_i\}$ and origin $\mathbf{o}$ is specified by the position vector $(\mathbf{x} - \mathbf{o})$, and that of a point $\mathbf{y}$ by the position vector $(\mathbf{y} - \mathbf{o})$. The position of the point $\mathbf{y}$ **relative** to the point $\mathbf{x}$ is given by the **relative position vector** $(\mathbf{y} - \mathbf{x})$; see Fig. 25.6.

Now consider a **two-dimensional** point space whose points represent the velocities of blocks, and whose vectors represent relative velocities. A diagram depicting the velocities and relative velocities for block-sliding in this point space is called a **hodograph.**

The hodograph for two rigid-blocks $\boxed{\rm I}$ and $\boxed{\rm J}$ moving at velocities $\mathbf{v}_I$ and $\mathbf{v}_J$ at angles θ_I and θ_J to the horizontal is shown in Fig. 25.7. Note that the relative velocity between the two blocks $(\mathbf{v}_J - \mathbf{v}_I)$ is necessarily tangential to l_{JI} in order for the velocity field to be kinematically admissible, in particular not to give rise to volume strains.

Next, restricting attention to two dimensional planar problems, the **rate of energy dissipation per unit depth** along a single line of tangential relative velocity discontinuity is

$$\mathcal{D}_{\rm int} = k \times l_{JI} \times |\mathbf{v}_J - \mathbf{v}_I|, \qquad k \stackrel{\rm def}{=} \frac{Y}{\sqrt{3}},$$

where k is the shear flow strength of the material, l_{JI} is the length of discontinuity line between blocks $\boxed{\rm J}$ and $\boxed{\rm I}$, and $|\mathbf{v}_J - \mathbf{v}_I|$ is the absolute value of the relative velocity. For velocity fields

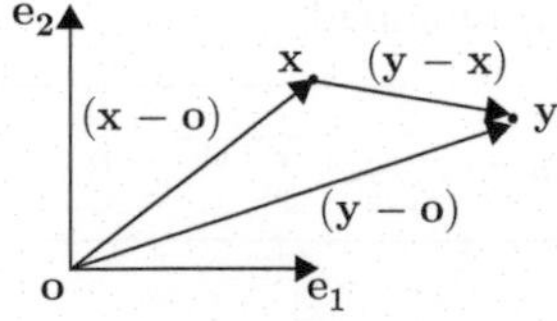

Fig. 25.6 Relative position vector $(\mathbf{y} - \mathbf{x})$ of points $\mathbf{x}$ and $\mathbf{y}$.

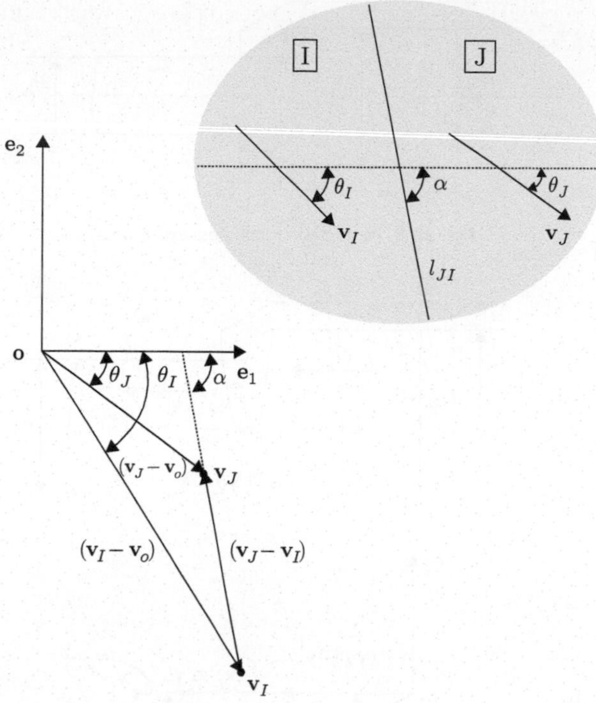

Fig. 25.7 Hodograph showing relative velocity vector $(\mathbf{v}_J - \mathbf{v}_I)$ of velocities $\mathbf{v}_J$ and $\mathbf{v}_I$.

involving more than one line of tangential relative velocity discontinuity, the total rate of energy dissipation per unit depth is

$$\mathcal{D}_{\text{int}} = \sum_{\mathcal{S}_d^*} k \times l_{JI} \times |\mathbf{v}_J - \mathbf{v}_I|.$$

A carefully drawn hodograph facilitates the determination of the terms in this expression.

25.6.3 Plane strain frictionless extrusion

For metal-forming operations where it is required to estimate a load which will cause the workpiece to deform plastically to produce the desired shape change, we need to design for loads which are a suitably small multiple of the limit load of the workpiece. Thus, for forming problems, a determination of the **upper bound** to the limit load is of primary interest.

Here we determine such an upper bound for plane strain, **frictionless** extrusion of a material which may be idealized as a non-hardening rigid plastic material with a **constant** tensile flow strength, Y. A schematic of the extrusion problem is shown in Fig. 25.8. The initial thickness of the workpiece is $2h_i$, and the final thickness is $2h_o$. The die semi-cone angle is α.

Given the symmetry of the problem, it is suffices to consider only one-half of the geometry as depicted in Fig. 25.9(a). To determine the upper bound to the limit load we will adopt a block-sliding velocity field. As a simple velocity field, consider three rigid blocks $\boxed{\text{A}}$, $\boxed{\text{B}}$, and

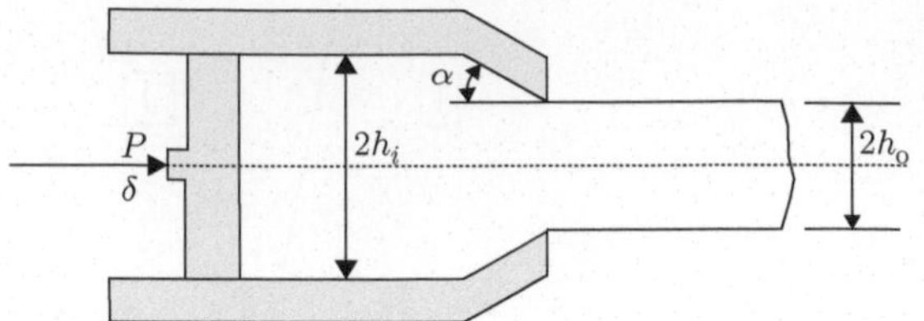

Fig. 25.8 Plane strain frictionless extrusion.

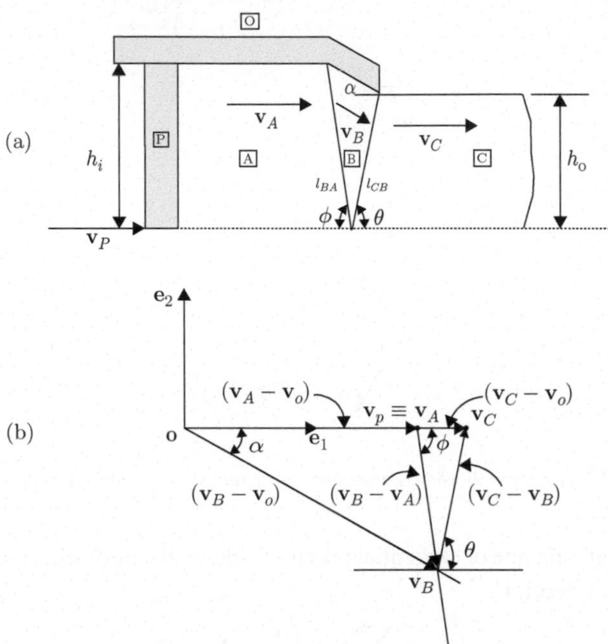

Fig. 25.9 (a) Geometry for the plane strain extrusion. (b) Hodograph.

C, with velocities $\mathbf{v}_A$, $\mathbf{v}_B$, and $\mathbf{v}_C$, **constant** within each block. The movement of block A forces block B downward and to the right, thus forcing block C to move to the right.

In order to satisfy the condition of plastic incompressibility, we must choose the velocity vectors to give no relative normal velocity across each boundary between blocks. That is, *all relative velocity vectors between rigidly moving blocks must be tangential to the boundaries between the respective blocks.* Thus, in the present problem, the relative velocity $(\mathbf{v}_B - \mathbf{v}_A)$ must be tangential to the velocity discontinuity line l_{BA}, and the relative velocity $(\mathbf{v}_C - \mathbf{v}_B)$ must be tangential to the velocity discontinuity line l_{CB}. We assume that line l_{BA} is at angle $(\pi - \phi)$ to the horizontal $\mathbf{e}_1$-direction, and line l_{CB} is at angle θ to the horizontal, as depicted in Fig. 25.9(a). Note that once h_o, h_i, and α are fixed, then only one of the angles θ or ϕ is independent. For example, if θ is fixed, then ϕ is a function of h_o, h_i, α, and θ.

The velocity of the punch $\mathbf{v}_P$ is prescribed and is in the $\mathbf{e}_1$-direction. The velocity of the block A is identical to the punch velocity, $\mathbf{v}_A \equiv \mathbf{v}_P$. The velocity $\mathbf{v}_B$ of block B is in a direction which is at an angle α to the $\mathbf{e}_1$-direction; its magnitude needs to be determined.

The velocity $\mathbf{v}_C$ of block $\boxed{\text{C}}$ is in the $\mathbf{e}_1$-direction; its magnitude needs to be determined. The discontinuity line l_{BA} is at angle $(\pi - \phi)$ to $\mathbf{e}_1$, and its length is $l_{BA} = h_i / \sin\phi$. The magnitude of the relative velocity $|\mathbf{v}_B - \mathbf{v}_A|$ needs to be determined. The discontinuity line l_{CB} is at angle θ to $\mathbf{e}_1$, and its length is $l_{CB} = h_0 / \sin\theta$. The magnitude of the relative velocity $|\mathbf{v}_C - \mathbf{v}_B|$ needs to be determined.

The velocities $\mathbf{v}_B$ and $\mathbf{v}_C$, and the relative velocities $|\mathbf{v}_B - \mathbf{v}_A|$ and $|\mathbf{v}_C - \mathbf{v}_B|$, are determined from a hodograph for the problem. The steps involved in the construction of the hodograph are listed below, and diagrammed in Fig. 25.9(b). The origin $\mathbf{o}$ in the hodograph represents the zero velocity of the extrusion die depicted as block $\boxed{\text{O}}$.

STEPS INVOLVED IN THE CONSTRUCTION OF THE HODOGRAPH:

1. Plot a vector of magnitude $|\mathbf{v}_P - \mathbf{v}_o| \equiv |\mathbf{v}_A - \mathbf{v}_o|$ starting from $\mathbf{o}$ parallel to $\mathbf{e}_1$. This represents the velocities of the punch $\boxed{\text{P}}$ and the block $\boxed{\text{A}}$ relative to the die $\boxed{\text{O}}$.
2. Plot a line emanating from the origin $\mathbf{o}$ at angle $-\alpha$ to $\mathbf{e}_1$. This represents the direction of velocity $(\mathbf{v}_B - \mathbf{v}_o)$ of the block $\boxed{\text{B}}$ relative to that of the die $\boxed{\text{O}}$. The magnitude of $(\mathbf{v}_B - \mathbf{v}_o)$ is not known at this point.
3. Plot a line emanating from the tip of the relative velocity vector $(\mathbf{v}_A - \mathbf{v}_o)$ at angle $-\phi$ to $\mathbf{e}_1$. For plastic incompressibility, the relative velocity vector $(\mathbf{v}_B - \mathbf{v}_A)$ must be parallel to this line. The intersection of the lines from Step 2 and this step give the relative velocities $(\mathbf{v}_B - \mathbf{v}_o)$ and $(\mathbf{v}_B - \mathbf{v}_A)$.
4. Since $\mathbf{v}_C$ is parallel to the $\mathbf{e}_1$ direction, draw a line emanating from the origin $\mathbf{o}$. The magnitude of the relative velocity $(\mathbf{v}_C - \mathbf{v}_o)$ is not known at this point.
5. Since the relative velocity between blocks $\boxed{\text{C}}$ and $\boxed{\text{B}}$ is at angle θ to $\mathbf{e}_1$, plot a line emanating from the tip of $(\mathbf{v}_B - \mathbf{v}_o)$ at an angle θ to $\mathbf{e}_1$. The intersection of this line with the line from Step 4 fixes the magnitudes of the relative velocities $(\mathbf{v}_C - \mathbf{v}_o)$ and $(\mathbf{v}_C - \mathbf{v}_B)$.

An upper bound to the **limit extrusion pressure,** P_e (per unit depth of the workpiece), for the plane strain frictionless extrusion is obtained by equating the external rate of work per unit depth

$$\mathcal{W}_{\text{ext}} = P_{\text{ub}} \times 2h_i \times |\mathbf{v}_P|$$

to the internal rate of work per unit depth

$$\mathcal{D}_{\text{int}} = 2\,(k \times l_{BA} \times |\mathbf{v}_B - \mathbf{v}_A| + k \times l_{CB} \times |\mathbf{v}_C - \mathbf{v}_B|).$$

From the hodograph geometry in Fig. 25.9(a),

$$l_{BA} = \frac{h_i}{\sin\phi}, \quad \text{and } l_{CB} = \frac{h_o}{\sin\theta}.$$

Also, from the hodograph in Fig. 25.9(b),

$$\frac{|\mathbf{v}_B - \mathbf{v}_A|}{\sin\alpha} = \frac{|\mathbf{v}_A - \mathbf{v}_o|}{\sin(\phi - \alpha)} \quad \Longrightarrow \quad |\mathbf{v}_B - \mathbf{v}_A| = |\mathbf{v}_p| \frac{\sin\alpha}{\sin(\phi - \alpha)},$$

and

$$\frac{|\mathbf{v}_C - \mathbf{v}_B|}{\sin\phi} = \frac{|\mathbf{v}_B - \mathbf{v}_A|}{\sin\theta} \quad \Longrightarrow \quad |\mathbf{v}_C - \mathbf{v}_B| = |\mathbf{v}_p| \frac{\sin\alpha}{\sin(\phi - \alpha)} \frac{\sin\phi}{\sin\theta}.$$

Thus

$$P_{\rm ub} = k \times \left[\frac{\sin\alpha}{\sin\phi\,\sin(\phi-\alpha)} + \frac{h_o}{h_i}\frac{\sin\alpha\,\sin\phi\,\sin\phi}{\sin(\phi-\alpha)\,\sin^2\theta} \right]. \tag{25.6.15}$$

The values h_i, h_o, and α are specified by the geometry of the extrusion process to be modeled. With these quantities fixed and a given θ, the angle ϕ is determined by θ, h_i, h_o, and α. Thus, for given h_i, h_o, and α we may calculate the least upper bound for the assumed velocity field by minimizing $P_{\rm ub}$ with respect to θ.

To fix ideas, for $h_o = \frac{1}{2}h_i$ and $\alpha = 45°$, if we assume a simple block sliding velocity field with $\theta = 45°$, then $\phi = 90°$, and (25.6.15) for this case gives,

$$P_{\rm ub} = k \times \left[\frac{1/\sqrt{2}}{1/\sqrt{2}} + \frac{1}{2}\,\frac{(1/\sqrt{2})}{(1/\sqrt{2})(1/2)} \right] = 2k = \frac{2}{\sqrt{3}}Y.$$

That is, an upper bound to the limit pressure for extrusion is

$$\boxed{P_{\rm e} \le 1.15Y.} \tag{25.6.16}$$

The best possible upper bound can be found by minimizing (25.6.15) with respect to θ.

25.6.4 **Plane strain indentation of a semi-infinite solid with a flat punch**

In this example we calculate an upper bound to the load P required for plane strain indentation of a semi-infinite workpiece with a punch of width W; the punch has a thickness B into the plane of the page (Fig. 25.10). This problem is the plane strain analog of a hardness test. The boundary between the punch and the workpiece will be assumed to be well-lubricated so that it may be modeled as *frictionless*. The assumption of a frictionless boundary is not essential, and we will relax it shortly.

We assume the block-sliding velocity field depicted in Fig. 25.11(a). The punch is denoted by block $\boxed{\rm P}$, and the rigid blocks $\boxed{\rm A}$, $\boxed{\rm B}$, and $\boxed{\rm C}$ are assumed to be *equilateral triangles*. Because of the symmetry of the assumed velocity field, only one-half of the punch/workpiece system need be considered. The material below or outside of the triangular regions $\boxed{\rm A}$, $\boxed{\rm B}$, and $\boxed{\rm C}$ is assumed to be rigid and stationary, and it is denoted as block $\boxed{\rm O}$. As the punch moves downward with velocity $\mathbf{v}_P$, shear occurs along discontinuity lines $l_{AO}, l_{BA}, l_{BO}, l_{CB}, l_{CO}$. We assume that **frictionless conditions** prevail along the boundary l_{AP}, between the punch and the workpiece, so that no energy is dissipated along l_{AP}. The absolute velocity in block $\boxed{\rm A}$, $\mathbf{v}_A$, must be parallel to l_{AO}, and has a vertical component equal to $\mathbf{v}_P$. The velocity of $\mathbf{v}_B$ of block $\boxed{\rm B}$ must be parallel to l_{BO}, and finally, the velocity $\mathbf{v}_C$ of block $\boxed{\rm C}$ must be parallel to l_{CO}. The steps involved in the construction of the hodograph are listed below. The origin **o** in the hodograph represents the zero velocity of the stationary part $\boxed{\rm O}$ of the workpiece.

STEPS INVOLVED IN CONSTRUCTING THE HODOGRAPH:

1. Plot a vector of magnitude $|\mathbf{v}_P - \mathbf{v}_o|$ starting from the origin **o** in the $-\mathbf{e}_2$-direction to represent the velocity of the punch.
2. Plot a line emanating from **o** at angle $-60°$ from $\mathbf{e}_1$. This line gives the direction of the relative velocity $(\mathbf{v}_A - \mathbf{v}_o)$.

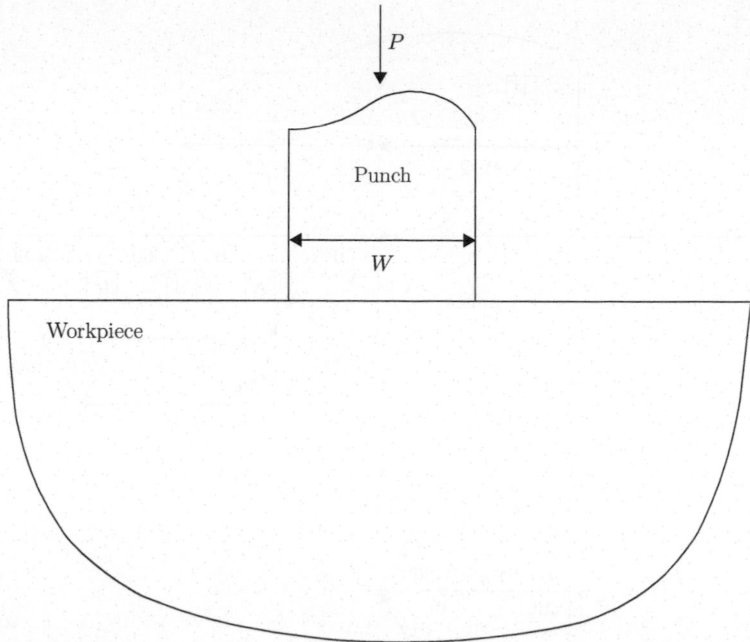

Fig. 25.10 Schematic of a plane strain indentation of a semi-infinite solid with a flat punch.

3. The relative velocity $(\mathbf{v}_A - \mathbf{v}_P)$ must be parallel to the discontinuity line l_{AP}, which is in the $\mathbf{e}_1$-direction. Plot a line emanating from $\mathbf{v}_P$ in the $\mathbf{e}_1$-direction. The intersection of this line and the one drawn in Step 2 fixes the velocity $\mathbf{v}_A$, and therefore also the relative velocities $(\mathbf{v}_A - \mathbf{v}_o)$ and $(\mathbf{v}_A - \mathbf{v}_P)$.
4. The relative velocity $(\mathbf{v}_B - \mathbf{v}_A)$ must be parallel to the discontinuity line l_{BA}, which is at an angle 60° to the $\mathbf{e}_1$-direction. Plot a line emanating from $\mathbf{v}_A$ in this direction.
5. Plot a line emanating from $\mathbf{o}$ in the $\mathbf{e}_1$-direction. This line gives the direction of the relative velocity $(\mathbf{v}_B - \mathbf{v}_o)$. The intersection of this line and the one drawn in Step 4 fixes the velocity $\mathbf{v}_B$, and therefore also the relative velocities $(\mathbf{v}_B - \mathbf{v}_o)$ and $(\mathbf{v}_B - \mathbf{v}_A)$.
6. Plot a line emanating from $\mathbf{o}$ at angle 60° from $\mathbf{e}_1$. This line gives the direction of the relative velocity $(\mathbf{v}_C - \mathbf{v}_o)$.
7. The relative velocity $(\mathbf{v}_C - \mathbf{v}_B)$ must be parallel to the discontinuity line l_{CB}, which is at an angle 120° to the $\mathbf{e}_1$-direction. Plot a line emanating from $\mathbf{v}_B$ in this direction. The intersection of this line and the one drawn in Step 6 fixes the velocity $\mathbf{v}_C$, and therefore also the relative velocities $(\mathbf{v}_C - \mathbf{v}_o)$ and $(\mathbf{v}_C - \mathbf{v}_B)$.

An upper bound to the **limit indentation load**, $P_{\rm l}$, for the plane strain frictionless indentation is obtained by equating the external rate of work,

$$\mathcal{W}_{\rm ext} = P_{\rm ub} \times |\mathbf{v}_P|,$$

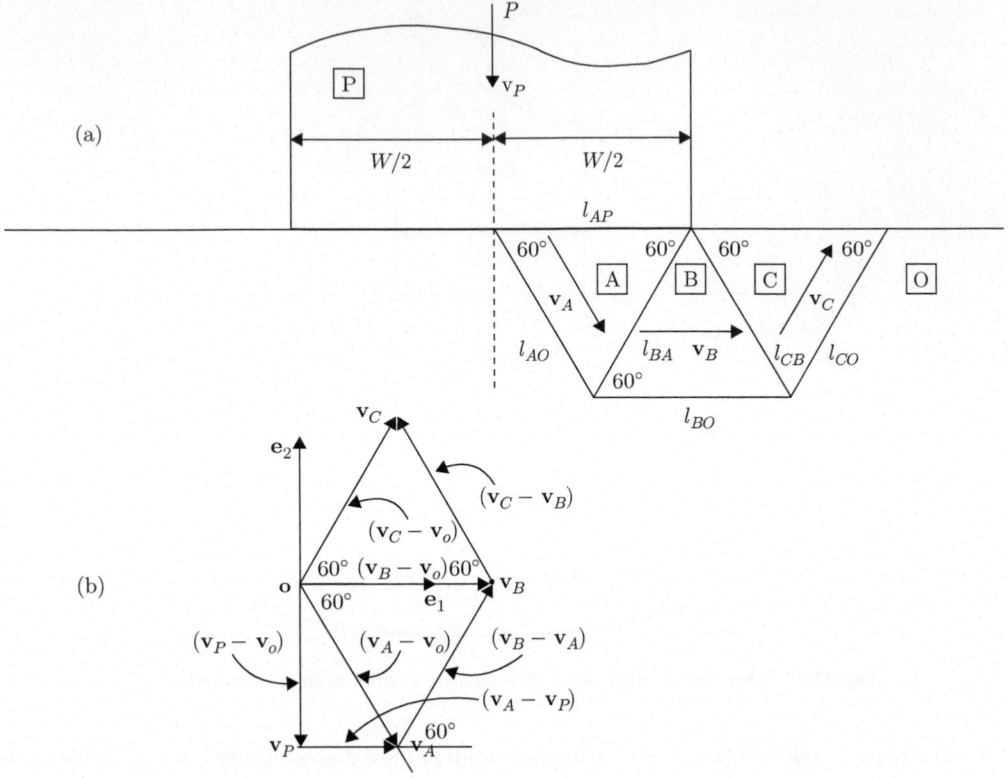

Fig. 25.11 (a) Geometry for the plane strain indentation. (b) Hodograph.

to the internal rate of dissipation

$$\mathcal{D}_{\text{int}} = 2\left(k \times B \times (l_{AO} \times |\mathbf{v}_A - \mathbf{v}_o| + l_{BO} \times |\mathbf{v}_B - \mathbf{v}_o| + l_{CO} \times |\mathbf{v}_C - \mathbf{v}_o| + l_{BA} \times |\mathbf{v}_B - \mathbf{v}_A| + l_{CB} \times |\mathbf{v}_C - \mathbf{v}_B|)\right).$$

Since

$$l_{AO} = l_{BO} = l_{CO} = l_{BA} = l_{CB} = \frac{W}{2},$$

by the geometry of the equilateral triangles, and

$$|\mathbf{v}_A - \mathbf{v}_o| = |\mathbf{v}_B - \mathbf{v}_o| = |\mathbf{v}_C - \mathbf{v}_o| = |\mathbf{v}_B - \mathbf{v}_A| = |\mathbf{v}_C - \mathbf{v}_B| = \frac{|\mathbf{v}_P - \mathbf{v}_o|}{\cos 30^\circ} = \frac{2|\mathbf{v}_P|}{\sqrt{3}}, \tag{25.6.17}$$

we obtain

$$P_{\text{ub}} \times |\mathbf{v}_P| = 2\left[k \times B \times \frac{W}{2} \times \frac{2}{\sqrt{3}}|\mathbf{v}_P| \times 5\right],$$

or

$$P_{\text{ub}} = \frac{10}{\sqrt{3}}k\,(W \times B) = \frac{10}{\sqrt{3}}\frac{Y}{\sqrt{3}}\,(W \times B) = (3.33Y)\,(W \times B).$$

That is, an upper bound to the limit indentation load is

$$\boxed{P_{\rm i} \le (3.33Y)\,(W \times B).} \tag{25.6.17}$$

This upper bound load to the plane strain indentation problem may be used as an estimate of the resistance to indentation or the "hardness", H, of a material. The hardness of a material is defined as the indentation load per unit area,

$$H \overset{\rm def}{=} \frac{P_{\rm i}}{W \times B}.$$

Thus, an upper bound to the hardness of a material is

$$\boxed{H \le 3.33\,Y.} \tag{25.6.18}$$

Accounting for friction between the indenter and the workpiece: It is not essential to assume frictionless conditions between the punch and the workpiece. A conservative assumption is that there is **sticking friction**[4] along l_{AP}. In this case the energy dissipated along l_{AP} is

$$k \times B \times l_{AP} \times |\mathbf{v}_A - \mathbf{v}_P|\,,$$

and since

$$l_{AP} = \frac{W}{2}\,, \text{and } |\mathbf{v}_A - \mathbf{v}_P| = \frac{1}{\sqrt{3}}\,|\mathbf{v}_P|\,,$$

the upper bound is obtained from

$$P_{\rm ub} \times |\mathbf{v}_P| = 2\left[k \times B \times \frac{W}{2} \times \frac{2}{\sqrt{3}}\,|\mathbf{v}_P| \times 5 + k \times B \times \frac{W}{2} \times \frac{1}{\sqrt{3}}\,|\mathbf{v}_P|\right].$$

This gives

$$P_{\rm ub} = \left[\frac{10}{\sqrt{3}} + \frac{1}{\sqrt{3}}\right] k\,(W \times B) = \frac{11}{\sqrt{3}}\,\frac{Y}{\sqrt{3}}\,(W \times B) = (3.67Y)(W \times B),$$

and hence

$$\boxed{P_{\rm i} \le (3.67Y)\,(W \times B),} \tag{25.6.19}$$

together with

$$\boxed{H \le 3.67Y.} \tag{25.6.20}$$

Thus, as expected, indentation of a rough workpiece with a rough punch will require a larger pressure than the frictionless case, by about 10 percent.

[4] Note, sticking friction does *not* imply that the velocities of block $\boxed{\rm A}$ and $\boxed{\rm P}$ are the same, rather they move relative to each other as though they were stuck together and a plastic slip discontinuity appeared between them.

PART IX

Fracture and fatigue

26 Linear elastic fracture mechanics

26.1 Introduction

Fracture mechanics is concerned with failure by cracking of materials under a wide variety of loadings and environments. Applications range from the microscale of materials where crack sizes may be a fraction of a micron, to engineering structures with cracks in the millimeter to centimeter scale, all the way up to earthquake rupture where faults are many kilometers in extent. Examples of problems addressed with methods of fracture mechanics include cracking in ship structures, heat-affected zones of welded bridges and offshore structures, aircraft landing gear, tail and engine attachments, turbine discs and blades, gas transmission pipelines, train rails and wheels, and even artificial heart valves, to name just a few.

As reviewed by Rice (1985), modern developments of fracture mechanics may be traced to the works of Irwin (1948) and Orowan (1948), who reinterpreted and extended the classical work on fracture of brittle materials by Griffith (1921). Irwin's approach in particular brought progress in theoretical solid mechanics, especially on linear elastic stress analysis of cracked bodies, to bear on the practical problems of crack growth testing and structural integrity. The theory of fracture based on Irwin's work is called "linear elastic fracture mechanics"—we discuss it in this chapter.

Some inelasticity is almost always present in the vicinity of a stressed crack-tip. If the zone of inelasticity is sufficiently small—a situation known as "small-scale-yielding"—then solutions from linear elasticity can be used to analyze data from test specimens. This data can, in turn, be used in conjunction with other linear elastic crack solutions to predict failure of cracked structural components.

In essence, the theory of linear elastic fracture mechanics boils down to the application of a fracture criterion which states that initiation of growth of a pre-existing crack of a given size will occur when a parameter K—which characterizes the intensity of stress and strain fields in the vicinity of a crack-tip—reaches a critical value K_c,

$$\boxed{K = K_c,} \tag{26.1.1}$$

while if $K < K_c$ then there is no crack growth. In (26.1.1):

- The quantity K is called the **stress intensity factor**. It is a function of the applied loads, the crack length, and dimensionless groups of geometric parameters.

- The quantity K_c is a **critical stress intensity factor**. It is a material property known as the **fracture toughness**. It measures the resistance of a material to the propagation of a crack.

Continuum Mechanics of Solids. Lallit Anand and Sanjay Govindjee, Oxford University Press (2020).

DOI: 10.1093/oso/9780198864721.001.0001

In the following sections, we examine the foundations of linear elastic fracture mechanics more deeply, noting special results in linear elastic crack-tip stress analysis, consideration of the size and shape of the crack-tip plastic zones, and of the inherent limits of applicability of linear elastic fracture mechanics.

26.2 **Asymptotic crack-tip stress fields. Stress intensity factors**

Consider a sharp crack in a prismatic isotropic linear elastic body. There are three basic loading modes associated with relative crack face displacements for a cracked body. These are described with respect to Fig. 26.1 as: (a) the tensile opening mode, or Mode I; (b) the in-plane sliding mode, or Mode II; and (c) the anti-plane tearing mode, or Mode III.

Let us first examine the nature of the stress fields in the vicinity of the crack-tip. We shall describe these fields in terms of the stress components σ_{ij} with respect to a coordinate system with origin at a point along the crack front. For ease of presentation, we shall identify position in terms of (r, θ)-coordinates of a cylindrical coordinate system; cf. Fig. 26.2. In what follows

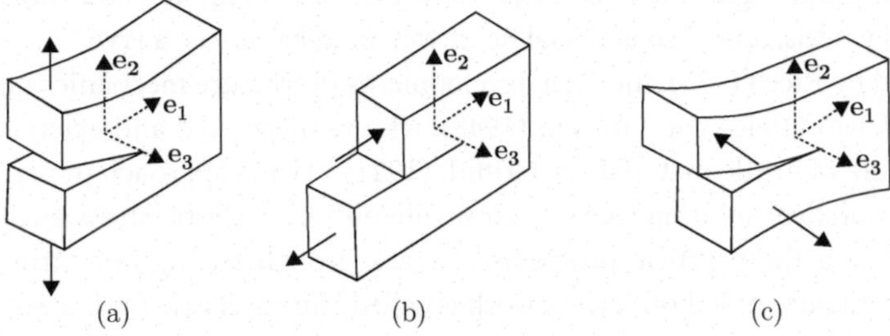

Fig. 26.1 Three basic loading modes for a cracked body: (a) Mode I, tensile opening mode. (b) Mode II, in-plane sliding mode. (c) Mode III, anti-plane tearing mode.

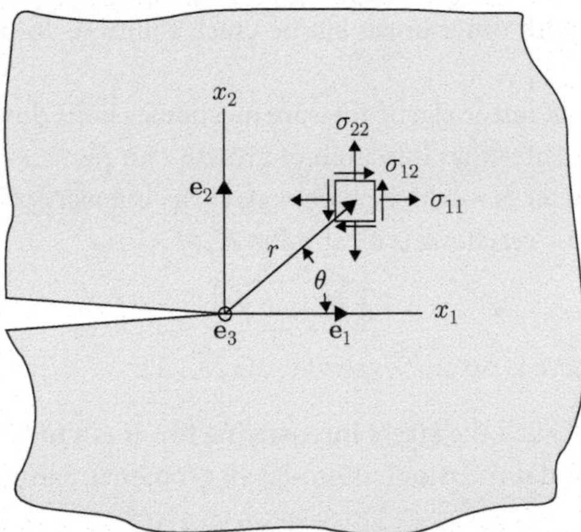

Fig. 26.2 Stress components with respect to a coordinate system with origin on the crack front.

we recall the *asymptotic stress analysis* of Sec. 9.5 which gives us the variation of the components σ_{ij} with position (r, θ) as $r \to 0$, for each of the three basic modes listed above.

1. **Mode I, tensile opening mode**:

$$\sigma_{\alpha\beta} = \frac{K_{\rm I}}{\sqrt{2\pi r}}\, f^{\rm I}_{\alpha\beta}(\theta),$$
$$u_\alpha = u^0_\alpha + \frac{K_{\rm I}}{2\mu}\sqrt{\frac{r}{2\pi}}\, g^{\rm I}_\alpha(\theta;\varkappa), \quad \text{with} \tag{26.2.1}$$
$$\varkappa \overset{\rm def}{=} \begin{cases} 3-4\nu & \text{plane strain} \\ \dfrac{3-\nu}{1+\nu} & \text{plane stress.} \end{cases}$$

The functions $f^{\rm I}_{\alpha\beta}(\theta)$ and $g^{\rm I}_\alpha(\theta;\varkappa)$ are given in (9.6.1) and (9.6.2), from which we note that[1]

$$f^{\rm I}_{22}(\theta = 0) \equiv 1, \tag{26.2.2}$$

so that ahead of the crack

$$\sigma_{22} = \frac{K_{\rm I}}{\sqrt{2\pi r}}. \tag{26.2.3}$$

The factor $K_{\rm I}$ is called the **stress intensity factor** in Mode I. It has units of stress times square root of length, $\text{MPa}\sqrt{\text{m}}$. For plane strain $\sigma_{33} = \nu(\sigma_{rr} + \sigma_{\theta\theta})$ and for plane stress $\sigma_{33} = 0$.

2. **Mode II, in-plane sliding mode**:

$$\sigma_{\alpha\beta} = \frac{K_{\rm II}}{\sqrt{2\pi r}}\, f^{\rm II}_{\alpha\beta}(\theta),$$
$$u_\alpha = u^0_\alpha + \frac{K_{\rm II}}{2\mu}\sqrt{\frac{r}{2\pi}}\, g^{\rm II}_\alpha(\theta;\varkappa), \quad \text{with} \tag{26.2.4}$$
$$\varkappa \overset{\rm def}{=} \begin{cases} 3-4\nu & \text{plane strain} \\ \dfrac{3-\nu}{1+\nu} & \text{plane stress.} \end{cases}$$

The functions $f^{\rm II}_{\alpha\beta}(\theta)$ and $g^{\rm II}_\alpha(\theta;\nu)$ are given in (9.6.3) and (9.6.4), from which we note that

$$f^{\rm II}_{12}(\theta = 0) \equiv 1, \quad \text{also} \quad \sigma_{11} = \sigma_{22} = 0 \quad \text{when} \quad \theta = 0. \tag{26.2.5}$$

For plane strain $\sigma_{33} = \nu(\sigma_{rr} + \sigma_{\theta\theta})$ and for plane stress $\sigma_{33} = 0$.

[1] The qualifiers **plane strain** and **plane stress** refer to the two-dimensional specializations of the general three-dimensional equations of linear elasticity, under which the asymptotic stress fields are derived, cf., Sec. 9.5. **In fracture mechanics the typical situation is of global plane stress.** However, as we shall see later, considerations of the ratio of the thickness of the prismatic body to the size of the plastic zone which develops at the tip of a crack will lead us to describe a state of **local plane plastic strain** at the tip of a crack in a prismatic body which is in a state of **global plane stress.** Further details will be given when we consider the effects of plastic zones at crack-tips.

3. **Mode III, anti-plane tearing mode:**

$$\sigma_{\alpha 3} = \frac{K_{\rm III}}{\sqrt{2\pi r}}\, f_\alpha^{\rm III}(\theta),$$
$$u_3 = u_3^0 + \frac{K_{\rm III}}{2\mu}\sqrt{\frac{r}{2\pi}}\left(4\sin\left(\frac{\theta}{2}\right)\right). \tag{26.2.6}$$

The equations above predict that the magnitude of the stress components increase rapidly as one approaches the crack-tip. Indeed,

- since the non-zero stress components are proportional to $1/\sqrt{r}$, they approach infinity as $r \to 0$! That is, a **mathematical singularity** exists at the crack-tip for the stress (and strain) fields, although the displacements and energies remain bounded.

The non-zero stress components in each of the three modes are proportional to the parameters $K_{\rm I}$, $K_{\rm II}$, and $K_{\rm III}$, respectively,[2] and the remaining terms only give the variation with r and θ. Thus, the **magnitude** of the stress components near the crack-tip can be characterized by giving the values of $K_{\rm I}$, $K_{\rm II}$, and $K_{\rm III}$. It is for this reason, that these quantities are called **the stress intensity factors** for Modes I, II, and III, respectively.

For a general three-dimensional problem the singular stress fields, at any point along the crack edge, will be a linear superposition of Modes I, II, and III, and in any plane problem the crack-tip singularity fields are a linear superposition of Mode I and Mode II. For example

$$\sigma_{\alpha\beta}(r,\theta) = \frac{1}{\sqrt{2\pi r}}\left(K_{\rm I}\, f_{\alpha\beta}^{\rm I}(\theta) + K_{\rm II}\, f_{\alpha\beta}^{\rm II}(\theta)\right),$$
$$u_\alpha(r,\theta) = u_\alpha^0 + \frac{1}{4\mu}\sqrt{\frac{r}{2\pi}}\left(K_{\rm I}\, g_\alpha^{\rm I}(\theta;\varkappa) + K_{\rm II}\, g_\alpha^{\rm II}(\theta;\varkappa)\right), \tag{26.2.7}$$

as $r \to 0$ in plane problems.

The task of stress analysis in linear elastic fracture mechanics (LEFM) is to evaluate $K_{\rm I}$, $K_{\rm II}$, $K_{\rm III}$, and their dependence on geometry, load type, and load amplitude. Many methods for obtaining K-solutions have been developed in the literature. Several hundred solutions, mostly for two-dimensional configurations, are now available. For simple geometries, or where a complex structure can be approximated by a simple model, it may be possible to use the handbooks by Tada et al. (2000) and Murakami (1987). If a solution cannot be obtained directly from a handbook, then a numerical method will need to be employed, as is invariably the case these days for real structures with complex boundary conditions.

Since the three crack-tip loading parameters $(K_{\rm I}, K_{\rm II}, K_{\rm III})$ define the loading imparted to the fracture process zone, we might develop a general expression for fracture initiation under mixed mode conditions (conditions where more than one of the stress intensity factors are non-zero), and this could be phrased as

$$f(K_{\rm I}, K_{\rm II}, K_{\rm III}) = \text{constant},$$

for some material-dependent function f. However,

[2] The subscripts "I", "II", and "III" are Roman numerals, and should therefore be read as "one", "two", and "three".

- in most engineering applications it is observed that globally brittle fracture of structures occurs by a crack propagating in a direction perpendicular to the local maximum principal stress direction.

Accordingly, in what follows,

- we shall develop a fracture criterion in terms of the stress intensity factor K_{I} only.

26.3 **Configuration correction factors**

For the important prototypical case of a finite crack of length $2a$ in a large body subject to a far-field stress $\sigma_{22}^{\infty} \equiv \sigma^{\infty}$, Fig. 26.3, the Mode I stress intensity factor is given by

$$\boxed{K_{\text{I}} = \sigma^{\infty}\sqrt{\pi a}.} \tag{26.3.1}$$

For other geometrical configurations, in which a characteristic crack dimension is a and a characteristic applied tensile stress is σ^{∞}, we will write the corresponding stress intensity factor as

$$\boxed{K_{\text{I}} = Q\sigma^{\infty}\sqrt{\pi a}, \qquad Q \equiv \text{configuration correction factor},} \tag{26.3.2}$$

where

- Q is a *dimensionless* factor needed to account for a geometry different from that of Fig. 26.3. We call Q the **configuration correction factor**. It is usually given in terms of *dimensionless ratios* of relevant geometrical quantities.

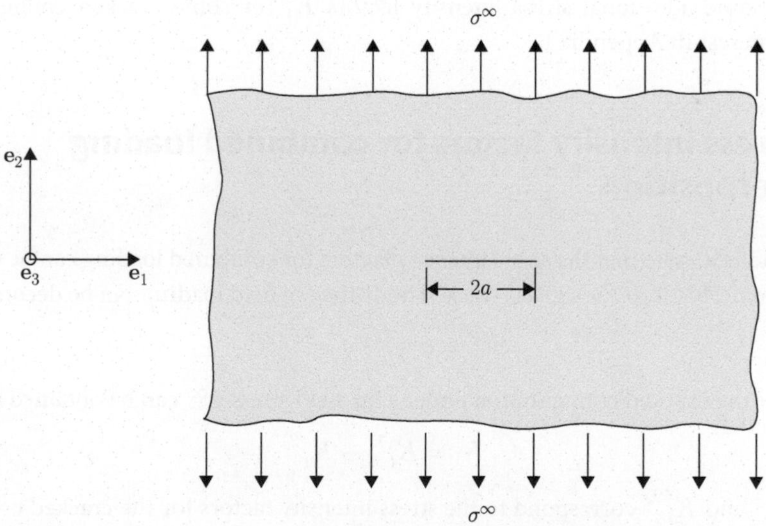

Fig. 26.3 Finite crack of length $2a$ in a large body subject to a far-field stress $\sigma_{22}^{\infty} \equiv \sigma^{\infty}$.

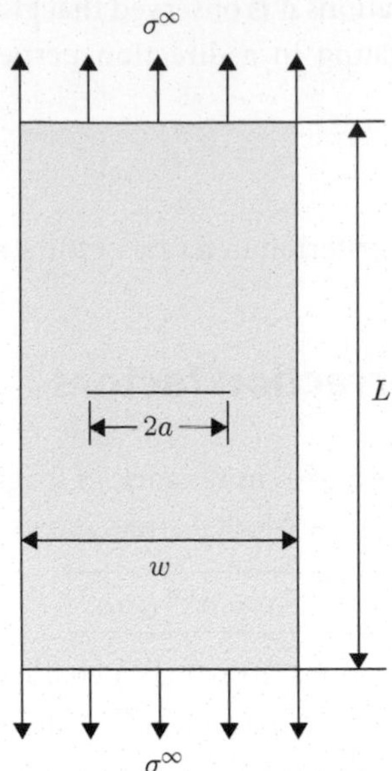

Fig. 26.4 Finite crack of length $2a$ in a long strip $(L > 3w)$ of finite width w subjected to a far-field tensile stress σ^∞.

For example, for a center crack in a long $(L > 3w)$ strip of finite width w, Fig. 26.4, the stress intensity factor is given by

$$\boxed{K_{\mathrm{I}} = Q\sigma^\infty\sqrt{\pi a}, \qquad Q = \hat{Q}\left(\frac{a}{w}\right) \approx \left(\sec\left(\frac{\pi a}{w}\right)\right)^{1/2}.} \tag{26.3.3}$$

We list some additional stress intensity factors K_{I} for some cracked configurations of practical interest in Appendix B.

26.4 **Stress intensity factors for combined loading by superposition**

It is important to note that the stress intensity factors for combined loading can be obtained by *superposition*. That is, if for a given cracked body the far-field loading can be decomposed into

$$\sigma^\infty = \sigma^{\infty\,(1)} + \sigma^{\infty\,(2)},$$

then K_{I} for the cracked configuration under a far-field stress σ^∞ can be obtained as

$$K_{\mathrm{I}} = K_{\mathrm{I}}^{(1)} + K_{\mathrm{I}}^{(2)},$$

where $K_{\mathrm{I}}^{(1)}$ and $K_{\mathrm{I}}^{(2)}$ correspond to the stress intensity factors for the cracked configuration under the far-field stresses $\sigma^{\infty\,(1)}$ and $\sigma^{\infty\,(2)}$, respectively.

26.5 Limits to applicability of K_{I}-solutions

26.5.1 Limit to applicability of K_{I}-solutions because of the asymptotic nature of the K_{I}-stress fields

Recall that the *asymptotic* stress field in the vicinity of a crack-tip under Mode I is given by equations (26.2.1). For the sharp crack of length $2a$ in an infinite body under a far-field tension $\sigma^\infty \equiv \sigma_{22}^\infty$, it is possible to obtain the **complete** solution. The complete solution provides the stress component σ_{22} on the symmetry plane $x_2 = 0$ as (cf., e.g., Hellan, 1984, Appendix B)

$$\sigma_{22}(x_1, 0) = \begin{cases} 0 & \text{if } |x_1| < a, \\ \dfrac{\sigma_{22}^\infty |x_1|}{\sqrt{x_1^2 - a^2}} & \text{if } |x_1| \geq a, \end{cases} \tag{26.5.1}$$

where x_1 is measured from the center of the crack, Fig. 26.5.

Consider a point which is at a distance r ahead of the right crack-tip, with $x_1 = a + r, \theta = 0$. The complete expression (26.5.1) gives

$$\sigma_{22}^{(\text{complete})}(x_1 = a + r,\ 0) = \frac{\sigma_{22}^\infty \sqrt{a}(1 + (r/a))}{\sqrt{2r}\sqrt{1 + \dfrac{r}{2a}}}, \tag{26.5.2}$$

while from the asymptotic solution (26.2.1), for small r we have

$$\sigma_{22}^{(\text{asymptotic})}(x_1 = a + r,\ 0) \to \frac{K_{\mathrm{I}}}{\sqrt{2\pi r}} \qquad \text{as } r \to 0. \tag{26.5.3}$$

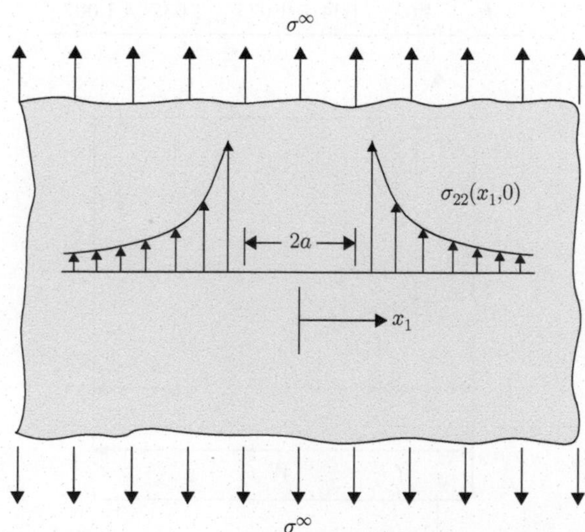

Fig. 26.5 Variation of $\sigma_{22}(x_1, 0)$ with x_1 in an infinite plate with a finite crack of length $2a$ subjected to a far-field tensile stress $\sigma_{22} \equiv \sigma^\infty$.

In the limit $r \to 0$, equations (26.5.2) and (26.5.3) give

$$\sigma_{22}(x_1 = a + r,\ 0) = \frac{\sigma_{22}^{\infty}\sqrt{a}}{\sqrt{2r}} = \frac{K_{\rm I}}{\sqrt{2\pi r}}\,,$$

and thus $K_{\rm I} = \sigma_{22}^{\infty}\sqrt{\pi a}$, as discussed previously.

We now ask the question, how close to the crack-tip must one be before (26.5.3) is an accurate enough approximation to (26.5.2)? One possible answer is based on a comparison of the ratio R of (26.5.2) to (26.5.3):

$$R = \frac{\sigma_{22}^{\rm (complete)}(x_1 = a + r,\ 0)}{\sigma_{22}^{\rm (asymptotic)}(x_1 = a + r,\ 0)} = \frac{1 + (r/a)}{\sqrt{1 + \dfrac{r}{2a}}}.$$

Some tabulated pairs of $\{(r/a), R\}$ are given in Table 26.1, from which we note that it is necessary to be within a radius of $\approx$10% of the crack length before the asymptotic formula in terms of the $K_{\rm I}$-parameter gives this stress component within $\sim$7% of the exact value.

This observation can be generalized to other crack configurations as follows:

- In order for the asymptotic fields based on $K_{\rm I}$ to acceptably approximate the complete elastic fields at a finite distance r from a crack-tip, it is necessary that r be no more distant from the crack-tip than a few percent of other characteristic in-plane dimensions, such

Table 26.1 Distance from crack-tip values of R.

r/a	1	0.5	0.1	0.05	0.01
R	1.63	1.34	1.073	1.037	1.007

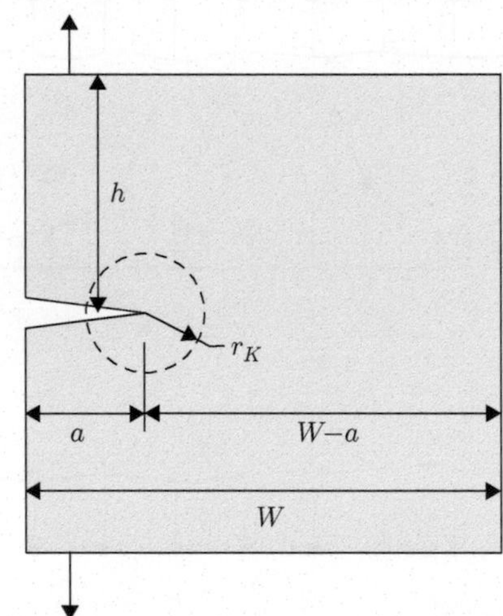

Fig. 26.6 $K_{\rm I}$ dominant region: $\{r \mid r < r_K;\ r_K \ll a,\ (W-a),\ h\}$

as the crack length a, the remaining uncracked ligament $(W - a)$, and the distance h from the crack-tip to a point of application of load; cf. Fig. 26.6.

This limitation on the ability of K_{I} to describe field parameters is purely a feature of the asymptotic nature of our elasticity solution, and ultimately involves only *relative* lengths, e.g., r/a etc. For example, with respect to Fig. 26.6, the asymptotic fields based on K_{I} should acceptably approximate the complete elastic fields at distances up to r_K from the crack-tip, provided $r_K \ll a,\ (W - a),\ h$.

26.5.2 Limit to applicability of K_{I}-solutions because of local inelastic deformation in the vicinity of the crack-tip

Another limitation to the applicability of K_{I} comes from considerations of local inelastic deformation in the vicinity of the crack-tip in response to the large stresses. A simple estimate of the size of the crack-tip plastic zone can be made by considering the stress component σ_{22} ahead of the crack. From (26.2.1),

$$\sigma_{22}(r,\,\theta = 0) = \frac{K_{\mathrm{I}}}{\sqrt{2\pi r}}\,. \tag{26.5.4}$$

This stress component exceeds the tensile yield strength σ_y of a material at points closer to the tip than the distance

$$r_{\mathrm{I}p} \approx \frac{1}{2\pi}\left(\frac{K_{\mathrm{I}}}{\sigma_y}\right)^2. \tag{26.5.5}$$

The length $r_{\mathrm{I}p}$ is called the plastic zone size; cf. Fig. 26.7. This one-dimensional estimate of the plastic zone size neglects many important details in the development of the crack-tip plastic zone, but on dimensional grounds alone provides the correct order of magnitude of the absolute linear dimension of the region at the crack-tip where the assumption of elastic material response would be invalid.

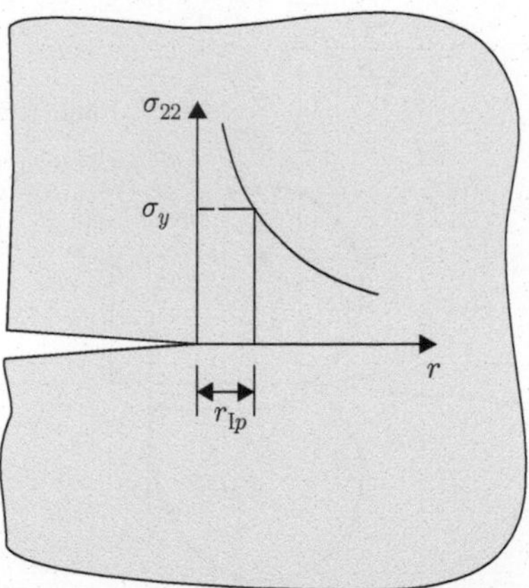

Fig. 26.7 A simple estimate for the plastic zone size.

A somewhat more accurate estimate of the crack-tip plastic zone caused by the multiaxial stress state at the crack-tip can be obtained by using the Mises yield condition

$$\left| \begin{bmatrix} \dfrac{1}{2}\left\{(\sigma_{11}-\sigma_{22})^2+(\sigma_{22}-\sigma_{33})^2+(\sigma_{33}-\sigma_{11})^2\right\} \\ +\,3\left\{\sigma_{12}^2+\sigma_{23}^2+\sigma_{31}^2\right\} \end{bmatrix}^{1/2} \right| \le \sigma_y. \tag{26.5.6}$$

In order to estimate the size and shape of the crack-tip plastic zone we insert the elastically calculated near-tip stress components of (26.2.1) into (26.5.6), and define the plastic zone radius $r_{\mathrm{Ip}}(\theta)$ as the distance from the crack-tip at an angle θ at which one has equality in (26.5.6). This gives the following expressions for the plastic zone radial coordinate as a function of θ:

$$r_{\mathrm{Ip}}(\theta) = \frac{1}{4\pi}\left(\frac{K_{\mathrm{I}}}{\sigma_y}\right)^2 \begin{cases} \left[(3/2)\sin^2\theta + (1-2\nu)^2(1+\cos\theta)\right] & \text{plane strain,} \\ \left[(3/2)\sin^2\theta + (1+\cos\theta)\right] & \text{plane stress.} \end{cases} \tag{26.5.7}$$

The boundaries of the plastic zone for plane stress and plane strain (with $\nu = 0.3$) predicted by (26.5.7) are shown in Fig. 26.8, where the coordinates have been made dimensionless by dividing x_1 and x_2 by $(K_{\mathrm{I}}/\sigma_y)^2$.

From (26.5.7) we may obtain the following estimates for the maximum extent of the plastic zone:

$$r_{\mathrm{Ip,max}} \approx \begin{cases} \dfrac{3}{8\pi}\left(\dfrac{K_{\mathrm{I}}}{\sigma_y}\right)^2 & \text{at } \theta \approx \pm\pi/2 \quad \text{for plane strain } (\nu = 0.3), \\ \dfrac{5}{8\pi}\left(\dfrac{K_{\mathrm{I}}}{\sigma_y}\right)^2 & \text{at } \theta \approx \pm\pi/2 \quad \text{for plane stress.} \end{cases} \tag{26.5.8}$$

Finally, because these estimates do not account for the effects of crack-tip plasticity in causing a redistribution of the stress fields from those calculated elastically, nor for effects of strain-

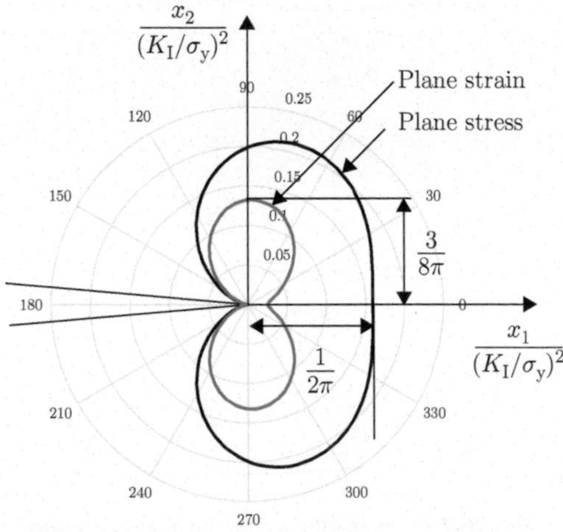

Fig. 26.8 Plastic zones for plane stress and plane strain.

hardening, the plastic zone shapes and sizes are only approximate. However, the qualitative difference in plastic zone shape between plane strain and plane stress is correct. In plane stress, the plastic zone advances directly ahead of the tip, while in plane strain, the major part of the plastic zone extends above and below the crack plane.

For simplicity, in our further discussions we assign the nominal crack-tip plastic zone to be of size

$$\boxed{r_{\rm I\mathit{p}} \equiv \frac{1}{2\pi}\left(\frac{K_{\rm I}}{\sigma_y}\right)^2,} \tag{26.5.9}$$

and we assume that the plastic zone may be approximated as a circular disc of radius $r_{\rm I\mathit{p}}$ centered at the crack-tip.

26.5.3 **Small-scale yielding (ssy)**

For Mode I loading, yielding at the tip of a crack will occur over a region whose maximum dimension is given by $r_{\rm I\mathit{p}}$; cf. (26.5.9). We emphasize that

- linear elastic fracture mechanics is based on the concept of small-scale yielding (ssy).

Small-scale yielding is said to hold when the applied load levels are sufficiently small (less than approximately one-half of the general yielding loads) such that there exists a radius $r = r_K$ about the crack-tip (exterior to the plastic zone), cf. Fig. 26.9, with the following properties:

- $r_{\rm I\mathit{p}} \ll r_K$, so that the stress field at r_K is free of any perturbations due to plasticity.
- $r_K \ll$ {crack length and other relevant in-plane geometric length quantities}, so that the elastic solution at r_K is given accurately in terms of the K_I-solution.

From a practical viewpoint, it is necessary that the plastic zone be small compared to the distance from the crack-tip to any in-plane boundary of a component, such as the distances a, $(W - a)$, and h in Fig. 26.9. A distance of $15 \times r_{\rm I\mathit{p}}$ is generally considered to be sufficient for small-scale yielding conditions to prevail. Hence,

$$\boxed{\text{if}\quad a,\ (W-a),\ h \gtrsim 15 \times \frac{1}{2\pi}\left(\frac{K_{\rm I}}{\sigma_y}\right)^2, \quad \text{then ssy holds,}}$$

and linear elastic fracture mechanics (LEFM) is applicable. The boxed condition must be satisfied for all three of

$$a,\ (W - a),\ \text{and}\ h,$$

otherwise it is possible that the plastic zone might extend to one of the boundaries, and in this case the situation approaches gross yielding prior to fracture, and LEFM will not be applicable.

Thus,

- *under small-scale yielding conditions the asymptotically computed stress field* (26.2.1) *is close to the complete stress field for all material points on $r = r_K$. The stress magnitude for all material points on $r = r_K$ is governed solely by the value of $K_{\rm I}$. That is, there is a one-parameter characterization of the crack-tip region stress field.*

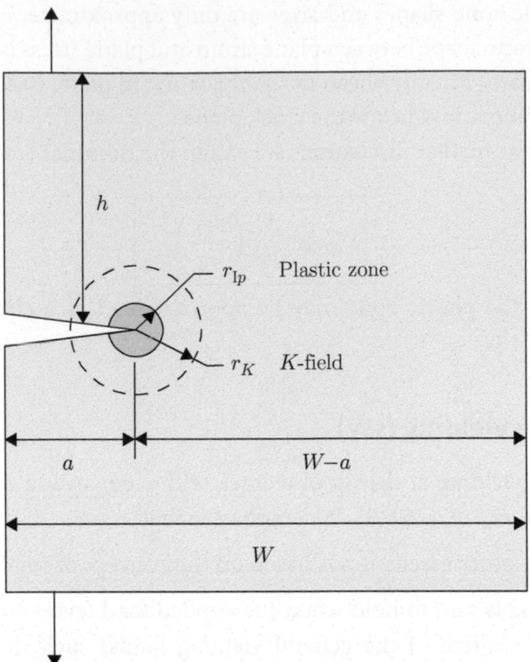

Fig. 26.9 Small-scale yielding for LEFM to be applicable.

26.6 Criterion for initiation of crack extension

Presuming that deviations from linearity occur only over a region that is small compared to geometrical dimensions (*small-scale yielding*), then the linear elastic stress-intensity factor controls the local deformations field. In particular,

- *two bodies with cracks of different size and with different manners of load application, but which are otherwise identical, will have identical near crack-tip deformation fields if the stress intensity factors are equal.*

Hence, under small-scale yielding conditions,

- **whatever** *is occurring at the crack-tip is driven by the elastic stress fields at r_K. Since the value of the stress field for all material points on r_K is governed solely by the value of K_{I}, the specific physical processes of material separation occurring at the crack-tip are driven by K_{I}.*

Accordingly, in the theory of linear elastic fracture mechanics (LEFM) the criterion for crack extension in response to a slowly applied loading of monotonically increasing magnitude is taken as

$$\boxed{K_{\rm I} = K_c,} \tag{26.6.1}$$

where $K_{\rm I}$ is the applied **stress intensity factor**, and K_c is a material and thickness dependent critical value of $K_{\rm I}$ for crack extension, called the **fracture toughness**. The fracture toughness for a given material and component thickness is determined experimentally.

26.7 **Fracture toughness testing**

Fracture toughness values are obtained by testing fatigue-cracked specimens[3] at a given temperature and loading rate.[4] Standard specimen geometries and test procedures used to obtain such data at slow loading rates are given in ASTM E399 (2013).

It is important that ssy conditions prevail in the test and thus certain constraints on minimum specimen dimensions must be satisfied in order to successfully obtain the fracture toughness value from a test which has been conducted. These dimensional requirements can be broadly separated into requirements on (see Fig. 26.9):

1. the in-plane dimensions a, $(W - a)$, h; and
2. the specimen thickness B.

In each case, the geometrical dimensions of a test specimen are compared with a "material" length dimension r_c corresponding to the nominal maximum plastic zone size of equation (26.5.9) evaluated for $K_{\rm I} = K_c$:

$$r_c \equiv \frac{1}{2\pi}\left(\frac{K_c}{\sigma_y}\right)^2. \tag{26.7.1}$$

The length r_c is called the **critical crack-tip plastic zone size**, but this is unknown prior to conducting the test.

Thus the process begins by choosing a specimen, performing a fracture toughness test to determine K_c, then checking that the specimen dimensions are a sufficiently large multiple of the computed r_c from (26.7.1). If they are not, the specimen needs to be resized and the process repeated in order to obtain a valid test.

The first set of conditions on minimum specimen dimensions ensure that the in-plane specimen dimensions are sufficiently large so that small-scale yielding prevailed up to crack-initiation at $K_{\rm I} = K_c$. For operational purposes the ASTM Standard E399 requires

$$\boxed{a,\ (W - a),\ h \gtrsim 15 \times r_c.} \tag{26.7.2}$$

If this requirement is met, then a **valid** K_c value has been obtained in the test.

For a given material,

- the fracture toughness value K_c generally depends on the thickness B of the test specimen.

---

[3] Fatigue cracking is employed in order to prepare cracks of relatively standard sharpnesses.

[4] Note that loading rate is here specified by the time rate of change of the applied stress intensity factor, $dK_{\rm I}/dt$.

- As the thickness B increases, K_c is found to decrease to an asymptotic value which is called K_{Ic}. This is the value that one finds reported in tables of material properties.

Regardless of the value of the thickness B (within reason), we expect that an unconstrained planar elastic body loaded in its plane will globally be in a state of plane stress. That is, the average through-thickness normal stress, as well as the average out-of-plane shear stresses, will vanish. However,

- when plastic deformation occurs at the crack-tip, the notions of "plane stress" or "plane strain" in the plastic zone at the crack-tip are governed by the size of the plastic zone relative to the specimen thickness B.
- For local plane strain in the plastic zone at the crack-tip, it is essential that $r_{\mathrm{Ip}} \ll B$, and
- for local plane stress in the plastic zone at the crack-tip, we must have $r_{\mathrm{Ip}} \gtrsim B$.

This is schematically depicted in Fig. 26.10.

If the plastic zone size r_{Ip} is much smaller than B, then the relatively massive elastic portion of the body surrounding the plastic zone acts to restrain the tendency towards substantial through-thickness thinning. Figure 26.10(c) schematically shows that in this case the dimpling effects at the surfaces of the specimen in the plastic zone will be limited to a small fraction of the thickness B. Thus, if $r_{\mathrm{Ip}} \ll B$, then along a very large part of the crack-tip **the plastic strain field will be essentially one of plane strain.** On the other hand, if the plastic zone size r_{Ip} is equal to or larger than the thickness B, then the elastic portion of the body surrounding the

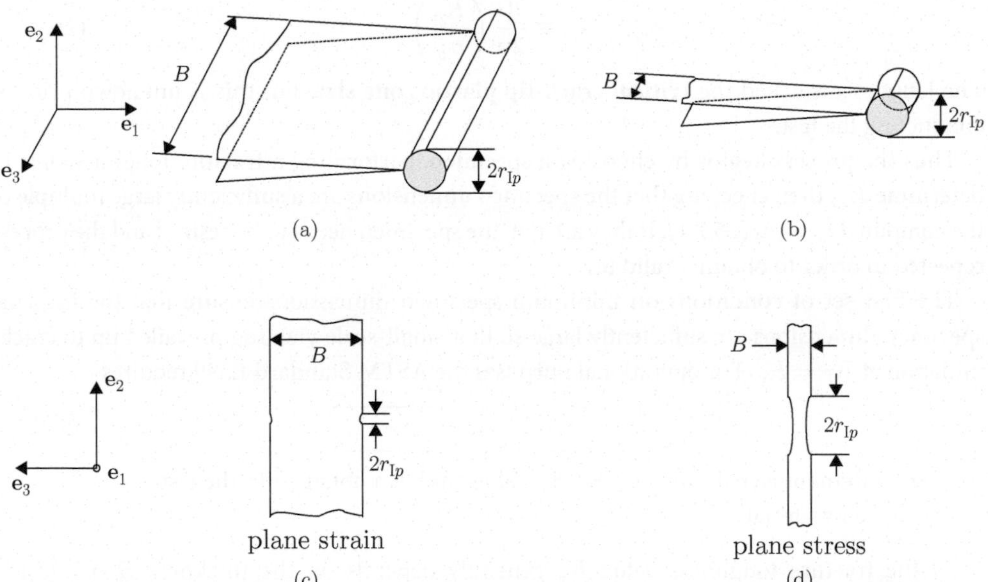

Fig. 26.10 Schematics (not to scale): (a) plane strain crack front conditions, $B \gtrsim 15 \times r_{\mathrm{Ip}}$. (b) plane stress crack front conditions, $r_{\mathrm{Ip}} \gtrsim B$. (c) Effect of specimen thickness on the transverse contraction in the plastic zone near the crack-tip for $B \gtrsim 15 \times r_{\mathrm{Ip}}$. (d) Effect of specimen thickness on the transverse contraction in the plastic zone near the crack-tip for $r_{\mathrm{Ip}} \gtrsim B$.

plastic zone does not provide any restraint against through-thickness thinning, and because of plastic incompressibility, the large in-plane plastic strains produce large negative strains parallel to the crack front. Figure 26.10(d) schematically shows that in this case the dimpling effects at the surfaces of the specimen in the plastic zone will be a large fraction of the thickness B. Thus, if $r_{\mathrm{I}p} \gtrsim B$, **the plastic strain field will be essentially one of plane stress.**

Hence,

- *since the plastic strain field in the vicinity of the crack-tip for large values of the thickness B corresponds essentially to plane plastic strain, K_{Ic} is called the* **plane strain fracture toughness**. Further,
- *since K_{Ic} is* **thickness independent** *it is a* **material property**.

There is a corresponding asymptotic value of the critical crack-tip plastic zone size, r_{Ic}, called the **plane strain critical crack-tip plastic zone size**. It is given by

$$\boxed{r_{\mathrm{Ic}} \equiv \frac{1}{2\pi}\left(\frac{K_{\mathrm{Ic}}}{\sigma_y}\right)^2.} \tag{26.7.3}$$

From the results of many experiments it has been found that the critical thickness B_c at which K_c has decreased to the K_{Ic} asymptote is given by

$$B_c \approx 15 \times r_{\mathrm{Ic}}. \tag{26.7.4}$$

Thus, the second requirement of the ASTM E399 plane strain fracture toughness test is that the thickness B satisfy

$$\boxed{B \gtrsim 15 \times \frac{1}{2\pi}\left(\frac{K_{\mathrm{Ic}}}{\sigma_y}\right)^2.} \tag{26.7.5}$$

If a test is performed on specimens which satisfy the geometric requirement on the in-plane dimensions for small-scale yielding,

$$\boxed{a,\ (W-a),\ h \gtrsim 15 \times r_{\mathrm{Ic}},} \tag{26.7.6}$$

and also the requirement on the thickness, (26.7.5), for plastic plane strain at the crack-tip, then the measured fracture toughness will be a **valid plane strain fracture toughness** K_{Ic}.

EXAMPLE 26.1 LEFM testing and component analysis

1. You are to determine critical loading conditions for fracture of a crack-containing component of non-trivial geometry. The component is made of a homogeneous isotropic material which we will call "Material X". Typically sized components made from Material X fail in a macroscopically *brittle* manner.

   You obtain a large planar coupon of Material X, of thickness t, from which you fabricate a square test specimen containing a small central crack of size $2a = 2$ cm, cf. Fig. 26.11, which is much smaller than the width of the square specimen. In the laboratory, the specimen is loaded by applying a uniform tensile stress σ^{∞} on the top and bottom boundaries. The specimen fractures in a brittle manner when σ^{∞} reaches a value of 200 MPa. Assuming linear elastic fracture mechanics (LEFM), determine a *candidate* value for Material X's Mode I fracture toughness.

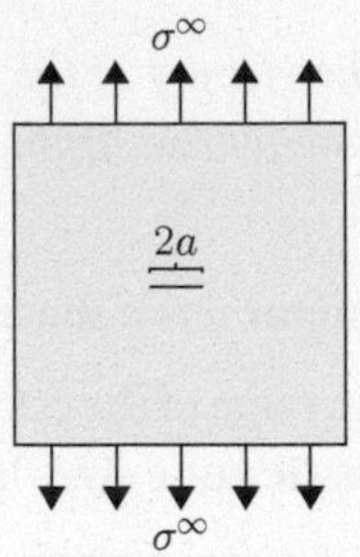

Fig. 26.11 Plate of thickness t with a crack of length $2a$.

2. Suppose that Material X is a metallic alloy having tensile yield strength σ_y. What quantitative bound should apply to σ_y such that the value of fracture toughness determined above is *valid* within the LEFM methodology?
3. In order for the value of the fracture toughness determined above to represent a plane strain fracture toughness value for Material X, what quantitative bound should the specimen thickness, t, satisfy?
4. The through-cracked component of interest has thickness t and a four-blade geometry, as shown schematically in Fig. 26.12. The body contains a centrally located, vertically oriented crack of total length $2a$, as shown, and is loaded by a tensile stress σ applied to the indicated edges of the blades. We wish to predict the critical value of σ for fracture, which we denote as σ_c, at which the component will fracture.

   This component's geometry is modeled using finite elements, including the sharp crack. Linear elastic material properties are assigned, and a numerical simulation is obtained for the case of $\sigma = 1$ MPa. The finite element mesh is extremely refined near the crack-tip, so you may assume that the data extracted from the simulation results, and plotted in Fig. 26.12(b), is accurate. The log-log plot shows the dependence of the stress component σ_{xx} as position is varied along the y-axis (parallel to the crack direction), where the top crack-tip is located at $x = y = 0$.

   (a) Explain the $-1/2$ slope in this plot for small values of y. Use the plot to evaluate the stress intensity factor of the finite element simulation.

   (b) The crack size in the finite element mesh has total length $2a = 24$ mm; use the information given to define and calculate a value for the configuration correction factor, Q, for this crack configuration.

   (c) Predict the critical stress, σ_c, for this component.

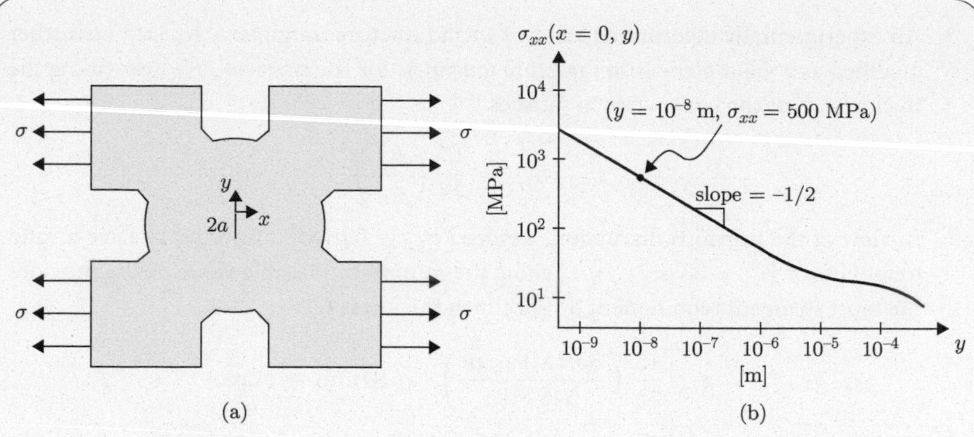

Fig. 26.12 (a) Geometry of actual part. (b) Finite element results for σ_{xx} versus position y at $x = 0$. The top crack-tip is located at $x = y = 0$.

Solution

1. For the crack-configuration in Fig. 26.11 the stress intensity factor is $K_{\rm I} = \sigma^\infty \sqrt{\pi a}$, where a is the half-crack length (= 1 cm). The LEFM methodology assumes that the value of $K_{\rm I}$ at fracture reaches a material-dependent value termed the critical fracture toughness, K_c; that is, fracture is assumed to occur when

$$K_c = K_{\rm I} = \sigma_c^\infty \sqrt{\pi a} = 200\,\text{MPa}\sqrt{\pi \times 10^{-2}\,\text{m}} \approx 35.4\,\text{MPa}\sqrt{\text{m}}.$$

2. The major conditional aspect of LEFM is the requirement of *small-scale yielding* (ssy), which in practice requires that the size of the crack-tip plastic zone at fracture, $r_{\rm Ic}$, be sufficiently small compared to relevant in-plane specimen dimensions. A simple estimate of the plastic zone size, $r_{\rm Ip}$, for any value of applied stress intensity factor, $K_{\rm I}$, is

$$r_{\rm Ip} \approx \frac{1}{2\pi}\left(\frac{K_{\rm I}}{\sigma_y}\right)^2,$$

and in the given configuration, taking a as the (only) relevant in-plane dimension, ssy is typically obtained provided

$$a \geq 15\, r_{\rm Ip} \approx \frac{15}{2\pi}\left(\frac{K_{\rm I}}{\sigma_y}\right)^2.$$

Applying this inequality to the test results and re-arranging using values for a and $K_{\rm I} = K_c$ gives

$$\sigma_y \geq \sqrt{15/2\pi}\,\frac{K_c}{\sqrt{a}} = \sqrt{15/2\pi}\,\frac{35.44\,\text{MPa}\sqrt{\text{m}}}{\sqrt{10^{-2}\,\text{m}}} \approx 548\,\text{MPa}.$$

Note that a value of 548 MPa is only a required lower bound for σ_y, in order that the test described yields a valid K_c. In practice, σ_y may be larger than 548 MPa.

3. An experimentally determined value of a valid fracture toughness, K_c, can be further qualified as a valid *plane strain fracture toughness* for the material, K_{Ic}, providing the thickness, t, of the test specimen satisfies

$$t \gtrsim 15 r_{\mathrm{Ic}} = \frac{15}{2\pi}\left(\frac{K_c}{\sigma_y}\right)^2.$$

In view of the previous discussion, we need $\sigma_y \geq 548\,\mathrm{MPa}$ in order to have a valid toughness of $K_c \approx 35\,\mathrm{MPa}\sqrt{m}$. Taking the minimum possible value for σ_y provides the most stringent requirement on specimen thickness t,

$$t \gtrsim \frac{15}{2\pi}\left(\frac{35.4\mathrm{MPa}\sqrt{\mathrm{m}}}{548\,\mathrm{MPa}}\right)^2 = 10\,\mathrm{mm} \approx 1\,\mathrm{cm}.$$

Note that if the actual yield strength of Material X exceeds 548 MPa, then a valid K_{Ic} could emerge, even when testing specimens with thickness $t < 1\,\mathrm{cm}$.

4. For the cracked four-blade geometry:
   (a) The asymptotic structure of the crack-tip opening stress field a distance $r > 0$ directly ahead of a Mode I crack-tip is given by

   $$\sigma_{xx}(r;\theta = 0) = \frac{K_{\mathrm{I}}}{\sqrt{2\pi r}} + \text{higher order terms},$$

   as $r \to 0$. Here we note that $r = y$ and that the crack opening stress component is σ_{xx}. Taking logarithms on both sides provides

   $$\log(\sigma_{xx}(y)) = \log(K_{\mathrm{I}}/(2\pi)) - \frac{1}{2}\log(y),$$

   with obvious implications for explaining the asymptotic $-1/2$ slope on log-log coordinates in Fig. 26.12(b).

   Matching the given point ($y = 10^{-8}m$; $\sigma_{xx} = 500\,\mathrm{MPa}$) to the asymptotic result of the finite element simulation gives

   $$K_{\mathrm{I}}^{(FEM)} \approx \sigma_{xx}(r)\sqrt{2\pi r} = 500\,\mathrm{MPa}\sqrt{2\pi\, 10^{-8} m} = 0.1253\,\mathrm{MPa}\sqrt{\mathrm{m}}.$$

   (b) A generic formula describing the dependence of the stress intensity factor, K_{I}, on applied loading σ and total crack length $2a$ can be expressed in terms of a dimensionless configuration correction factor Q such that

   $$K_{\mathrm{I}} = Q\sigma\sqrt{\pi a}.$$

   Applying this formula to the results from the finite element solution gives

   $$Q = \frac{K_{\mathrm{I}}}{\sigma\sqrt{\pi a}} = \frac{0.1253\,\mathrm{MPa}\sqrt{\mathrm{m}}}{1\,\mathrm{MPa}\sqrt{0.012\pi\, m}} = 0.646.$$

(c) The critical stress, σ_c can be determined using the value of K_c, the crack size $2a$, and the configuration cotrection factor Q:

$$K_c = \sigma_c Q\sqrt{\pi a} \qquad \Rightarrow \qquad \sigma_c = \frac{K_c}{Q\sqrt{\pi a}} = \frac{35.4\,\text{MPa}\sqrt{\text{m}}}{0.646\sqrt{.012\pi\,\text{m}}} = 281\,\text{MPa}.$$

26.8 **Plane strain fracture toughness data**

Extensive K_{Ic} data for metallic materials are now available (cf., e.g., Matthews, 1973; Hudson and Seward, 1978). Representative values for K_{Ic} (at room temperature under slow loading rates, and in the absence of effects of aggressive environments) for several metals are given in Table 26.2, where the values of corresponding critical plastic zone sizes calculated according to (26.7.3) are also shown.

Table 26.2 Typical values of K_{Ic} for some metallic materials at room temperature. Also listed are values of $r_{\text{Ic}} = \dfrac{1}{2\pi}\left(\dfrac{K_{\text{Ic}}}{\sigma_y}\right)^2$ and $L^* = 16\,r_{\text{Ic}}$.

Material	E, GPa	σ_y, MPa	K_{Ic}, MPa$\sqrt{\text{m}}$	r_{Ic}, mm	L^*, mm
Metals					
Steels					
AISI-1045	210	269	50	5.5	88.0
AISI-1144	210	540	66	2.4	38.4
ASTM A470-8	210	620	60	1.5	24.0
ASTM A533-B	210	483	153	16.0	256.0
ASTM A517-F	210	760	187	9.6	153.6
AISI-4130	210	1090	110	1.6	25.6
AISI-4340	210	1593	75	0.4	6.4
200-Grade Maraging	210	1310	123	1.4	22.4
250-Grade	210	1786	74	0.3	4.8

continued

Table 26.2 Continued

Material	E, GPa	σ_y, MPa	K_{Ic}, MPa$\sqrt{\text{m}}$	r_{Ic}, mm	L^*, mm
Metals					
Aluminum Alloys					
2014-T651	72	415	24	0.5	8.0
2024-T4	72	330	34	1.7	27.2
2219-T37	72	315	41	2.7	43.2
6061-T651	72	275	34	2.4	38.4
7075-T651	72	503	27	0.5	8.0
7039-T651	72	338	32	1.4	22.4
Titanium Alloys					
Ti-6Al-4V	108	1020	50	0.4	6.4
Ti-4Al-4Mo-2Sn-0.5Si	108	945	72	0.9	14.4
Ti-6Al-2Sn-4Zr-6Mo	108	1150	23	0.1	1.6

Also included in Table 26.2 is the crack length $L^* = 2a^*$ which in a configuration corresponding to Fig. 26.3 would cause fracture initiation at an applied stress of $\sigma^\infty = \sigma_y/2$:

$$\frac{\sigma_y}{2}\sqrt{\pi\,(L^*/2)} = K_{\text{Ic}},$$

which gives

$$L^* = 16\left(\frac{1}{2\pi}\left(\frac{K_{\text{Ic}}}{\sigma_y}\right)^2\right) = 16 r_{\text{Ic}}. \tag{26.8.1}$$

Note that L^* is essentially the characteristic length dimension which specimen crack length, remaining ligament, and thickness must exceed in order to obtain a valid K_{Ic} value for the material. The combination of high K_{Ic} and low σ_y leads to relatively large values of the critical plastic zone size, and rather large values of L^* are required before fracture initiation will occur at stress levels equal to one-half the yield stress.

Representative values for K_{Ic} for several polymers and ceramics are given in Table 26.3.

Table 26.3 Typical values of K_{Ic} for some polymers and ceramics materials at room temperature.

Material	E, GPa	K_{Ic}, MPa$\sqrt{\text{m}}$
(Non-metals)		
Polymers		
Epoxies	3	0.3 – 0.5
PS	3.25	0.6 – 2.3
PMMA	3. – 4.	1.2 – 1.7
PC	2.35	2.5 – 3.8
PVC	2.5 – 3.	1.9 – 2.5
PETP	3	3.8 – 6.1
Ceramics		
Soda-Lime Glass	73	0.7
MgO	250	3
Al_2O_3	350	3–5
Al_2O_3, 15% ZrO_2	350	10.
Si_3N_4	310	4–5
SiC	410	3.4

REMARK 26.1

As seen from these tables, values for K_{Ic} range from less than 1 to over 180 MPa$\sqrt{\text{m}}$. At the lower end of this range are brittle materials, which, upon loading remain elastic until they fracture. For these, linear-elastic fracture mechanics work well and the fracture toughness itself is a well-defined property. The theory also works well for relatively brittle materials such as the aluminum and titanium alloys used in the aerospace industry with $r_{Ic} \lesssim 2$mm. These materials require moderate-sized specimens to satisfy the small-scale yielding condition.

At the upper end lie super-tough materials, all of which show substantial plasticity before they fracture. In these super-tough materials when an applied load causes a crack to extend, the size of the plastic zone r_{Ic} often exceeds 10 mm (cf. e.g. A533B in Table 26.2). Such materials require very large specimens to satisfy the small-scale yielding condition—a size approaching the size of a filing cabinet. While such specimens have been tested, they are quite impractical. Further, even if the fracture toughness for such ductile materials is measured, the small-scale-yielding condition limits the use of the fracture criterion $K_I = K_{Ic}$ to large structures containing large cracks. For such materials it becomes necessary to use a "non-linear theory of fracture mechanics" which accounts for large elastic-plastic deformations together with an accounting of microscale damage processes—a topic which is beyond the scope of this book.

27 Energy-based approach to fracture

27.1 Introduction

As an alternate to the LEFM based on

$$K_{\rm I} \le K_{\rm Ic}, \tag{27.1.1}$$

where $K_{\rm I}$ is a **Mode I stress intensity factor** and $K_{\rm Ic}$ is a material property called the **critical stress intensity factor**, we now discuss an energy-based approach to fracture. This approach, which was pioneered by Griffith (1921), is based on the notion of an **energy release rate** denoted by $\mathcal{G}$, and introduces a fracture criterion

$$\mathcal{G} \le \mathcal{G}_c, \tag{27.1.2}$$

where $\mathcal{G}_c$, is a **material property** called the **toughness** or the **critical energy release rate**. We shall discuss the details of such an energy-based approach to fracture in this chapter for elastic materials.

The energy-based approach also allows us to consider fracture of non-linear elastic materials. However, when specialized to consideration of fracture of *linear* isotropic materials, we shall show that the energy release rate $\mathcal{G}$ is given by

$$\mathcal{G} = \frac{1}{\bar{E}}(K_{\rm I}^2 + K_{\rm II}^2) + \frac{1+\nu}{E}K_{\rm III}^2, \tag{27.1.3}$$

where

$$\bar{E} = \begin{cases} \dfrac{E}{(1-\nu^2)} & \text{for plane strain,} \\ E & \text{for plane stress.} \end{cases} \tag{27.1.4}$$

Thus, under pure Mode I loading,

$$\mathcal{G} = \frac{K_{\rm I}^2}{\bar{E}}. \tag{27.1.5}$$

In this case the fracture condition (27.1.2) may be written as in (27.1.1) with

$$K_{\rm Ic} \stackrel{\rm def}{=} \sqrt{\bar{E}\mathcal{G}_c}. \tag{27.1.6}$$

Continuum Mechanics of Solids. Lallit Anand and Sanjay Govindjee, Oxford University Press (2020).

DOI: 10.1093/oso/9780198864721.001.0001

REMARK 27.1

It is largely due to Irwin (1948) that the interpretation of the fracture condition in linear elastic fracture mechanics has shifted from the energy-based condition (27.1.2) to one based on the stress intensity (27.1.1). Mathematically, for isotropic linear elastic materials *under small-scale yielding conditions*, the two approaches are entirely equivalent.

27.2 Energy release rate

27.2.1 Preliminaries

For simplicity we limit our discussion to a two-dimensional body $\mathcal{B}$ of *unit thickness*, which we identify with the region of space it occupies in a fixed reference configuration. We assume that the body contains a sharp, but not necessarily straight, edge crack; cf. Fig. 27.1.

The crack is modeled as a smooth, non-intersecting curve $\mathbf{z}(s)$ parameterized by arc length s, $0 \leq s \leq a$, with $s = 0$ corresponding to the intersection of the crack with the boundary $\partial\mathcal{B}$ of $\mathcal{B}$, and $s = a$ labels the crack-tip, which is allowed to advance with time. For convenience we write

$$\mathcal{C}(a) = \{\mathbf{z}(s) : 0 \leq s \leq a\}$$

for the set of points comprising the crack, $\mathbf{z}_a = \mathbf{z}(a)$ for the *position of the crack-tip*, and

$$\mathbf{e}(a) = \frac{d}{da}\mathbf{z}_a \tag{27.2.1}$$

for the *direction of propagation*. Since s measures arc length, $|d\mathbf{z}/ds| = 1$ and $\mathbf{e}(a)$ is a unit vector. The outward unit normal to the boundary $\partial\mathcal{B}$ of the body is denoted by $\boldsymbol{\nu}$.

We consider an initial crack of length $a > 0$ at $t = 0$, and presume that $da/dt > 0$ for all t in some interval $[0, t_1)$, *so that* a *is a strictly increasing function of* t *for* $0 \leq t < t_1$, or equivalently, for a in an interval $[a_0, a_1)$. Thus,

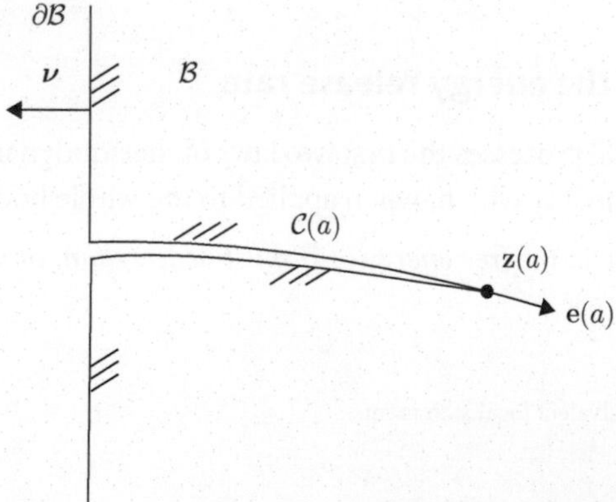

Fig. 27.1 A body $\mathcal{B}$ with an edge crack $\mathcal{C}(a)$ with tip at $\mathbf{z}(a)$, and direction of propagation $\mathbf{e}(a)$. The outward unit normal to the boundary $\partial\mathcal{B}$ of the body is $\boldsymbol{\nu}$.

- if we confine our attention to this interval, we may use "a" as our timescale, and in this sense,
- $\mathbf{e}(a)$ also represents the "velocity" of the crack-tip.

We assume that the body is homogeneous and *hyperelastic*, so that the symmetric Cauchy stress $\boldsymbol{\sigma}$ is given as the derivative with respect to the strain ϵ of a free-energy function $\psi(\epsilon)$ measured per unit volume of the body:

$$\boldsymbol{\sigma} = \frac{\partial \psi(\epsilon)}{\partial \epsilon}, \tag{27.2.2}$$

where

$$\epsilon = \frac{1}{2}(\nabla \mathbf{u} + \nabla \mathbf{u}^{\top}) \tag{27.2.3}$$

is the small strain tensor.

We will need to consider fields $\varphi(\mathbf{x}, a)$ which depend on the parameter a, which is *the length of crack* in the body. In particular, with $\mathbf{u}(\mathbf{x}, a)$ the displacement field in $\mathcal{B}$ at a crack length a, the derivative

$$\frac{\partial \mathbf{u}(\mathbf{x}, a)}{\partial a} \tag{27.2.4}$$

represents a "velocity" of the body associated with crack extension. Further, (27.2.2), (27.2.3), and the chain rule yield

$$\frac{\partial \psi}{\partial a} = \boldsymbol{\sigma} : \nabla \left(\frac{\partial \mathbf{u}}{\partial a} \right). \tag{27.2.5}$$

Here and in what follows, we write

$$\psi(\mathbf{x}, a) \equiv \psi(\epsilon(\mathbf{x}, a)),$$

and

$$\boldsymbol{\sigma}(\mathbf{x}, a) \equiv \boldsymbol{\sigma}(\epsilon(\mathbf{x}, a)),$$

when convenient. We limit our discussion to quasi-static deformations in the absence of body forces; in this case $\boldsymbol{\sigma}(\mathbf{x}, a)$ obeys the equilibrium equation

$$\operatorname{div} \boldsymbol{\sigma} = \mathbf{0}. \tag{27.2.6}$$

27.2.2 Definition of the energy release rate

Recall that for isothermal processes the first two laws of thermodynamics can be collapsed into a single dissipation inequality which, when applied to the whole body, asserts

- *that rate of change in the free energy of $\mathcal{B}$ must be less than, or at most equal to, the power expended on $\mathcal{B}$.*[1]

[1] Cf. Sec. 5.3.1 for the equivalent local statement.

As a crack extends the displacement field $\mathbf{u}(\mathbf{x}, a)$ in the body varies, and the power due to the boundary tractions $\boldsymbol{\sigma\nu}$ on $\partial\mathcal{B}$ is

$$\mathcal{W}_{\text{ext}}(\mathcal{B}) = \int_{\partial\mathcal{B}} (\boldsymbol{\sigma\nu}) \cdot \dot{\mathbf{u}} ds = \left(\int_{\partial\mathcal{B}} (\boldsymbol{\sigma\nu}) \cdot \frac{\partial \mathbf{u}}{\partial a} ds \right) \dot{a}, \tag{27.2.7}$$

while the rate of change of the free energy of the body is

$$\mathcal{W}_{\text{int}}(\mathcal{B}) = \frac{d}{dt} \int_{\mathcal{B}} \psi(\epsilon) dA = \left(\frac{d}{da} \int_{\mathcal{B}} \psi(\epsilon) dA \right) \dot{a}. \tag{27.2.8}$$

Thus the total **dissipation** Φ is

$$\Phi \stackrel{\text{def}}{=} \mathcal{W}_{\text{ext}}(\mathcal{B}) - \mathcal{W}_{\text{int}}(\mathcal{B}) \stackrel{\text{def}}{=} \mathcal{G}\dot{a} \geq 0, \tag{27.2.9}$$

where the quantity

$$\boxed{\mathcal{G} \stackrel{\text{def}}{=} \int_{\partial\mathcal{B}} (\boldsymbol{\sigma\nu}) \cdot \frac{\partial \mathbf{u}}{\partial a} ds - \frac{d}{da} \int_{\mathcal{B}} \psi(\epsilon) dA,} \tag{27.2.10}$$

is called the **energy release rate**. From (27.2.9) we see that the energy release rate $\mathcal{G}$ is the "thermodynamic force" conjugate to $\dot{a}$.

REMARK 27.2

Following common practice in the fracture mechanics literature we have denoted the crack length with a. This notation has required us denote an area element with dA instead of da, as used in the rest of the book.

27.3 Griffith's fracture criterion

We use the thermodynamically motivated definition of an energy release rate $\mathcal{G}$ to formulate a crack-propagation criterion with a threshold. Specifically, we introduce a **material property**,

- $\mathcal{G}_c > 0$, called the **toughness** (or critical energy release rate),

which represents a resistance to crack extension offered by the material, and require that the **fracture condition**,

$$\mathcal{G} = \mathcal{G}_c \quad \text{be satisfied for} \quad \dot{a} > 0, \tag{27.3.1}$$

so that *the dissipation Φ is positive during crack growth.*

Note that (27.3.1) does not fully suffice to characterize fracture of rate-independent materials; an additional assumption, namely that the energy release rate $\mathcal{G}$ cannot be greater than $\mathcal{G}_c$,

$$\mathcal{G} \leq \mathcal{G}_c, \tag{27.3.2}$$

is needed. Thus when $\mathcal{G} < \mathcal{G}_c$ the fracture condition (27.3.1) is not satisfied, so that necessarily $\dot{a} = 0$; therefore,

$$\dot{a} = 0 \quad \text{for} \quad \mathcal{G} < \mathcal{G}_c. \tag{27.3.3}$$

The fracture criterion (27.3.1) is due to Griffith (1921). Griffith's original considerations of fracture were limited to an *ideally brittle material*, for which the fracture energy is the energy required to create two new surfaces as the crack extends; that is, with γ_s the surface energy of the solid,

$$\mathcal{G}_c = 2\gamma_s. \tag{27.3.4}$$

Since Griffith's time the concept of $\mathcal{G}_c$ as a material parameter characterizing the fracture energy of a material has been broadened to include a much wider range of materials. Irwin (1948) and Orowan (1948) argued that (27.3.1) will still apply *under small-scale yielding conditions* if $\mathcal{G}_c$ is reinterpreted as the combined energy per unit area of crack advance going into the formation of new surface area, as well as the plastic deformation and the fracture processes that occur in the small process zone at the crack-tip. It is only under such "small-scale yielding" circumstances that $\mathcal{G}_c$ may be regarded as a material property, independent of the macroscopic geometry of the body and the loading conditions.

Note that the fracture criterion (27.3.1) circumvents the need to consider the details of the physical mechanisms of material separation at the crack-tip. The energy dissipated at the microscopic level is *not included* in the continuum calculation of the energy release rate $\mathcal{G}$.

In order to use the Griffith criterion, we need two essential ingredients:

1. A value of the energy release rate $\mathcal{G}$ for the particular geometrical configuration of the body with a crack, the loading conditions on its boundary, and the elastic characteristics of the material.
2. A value of the critical energy release rate $\mathcal{G}_c$ of the material.

A value of $\mathcal{G}_c$ for a given material can only be obtained through suitable experiments. Values of $\mathcal{G}_c$ can vary widely, from about 1 J/m^2 for separation of atomic planes of brittle materials, to 10^8 J/m^2 or more for ductile materials, for which plastic flow and ductile tearing determine the energy associated with fracture. Approximate values of $\mathcal{G}_c$ for some common materials are:

Glasses: 10 J/m^2, Ceramics: 50 J/m^2, Aluminum alloys: 8–30 kJ/m^2, Mild steels: 100 kJ/m^2.

For most materials the surface energy part of the critical energy release rate is only ≈ 1 J/m^2, and therefore small relative to the other contributions to the dissipation associated with fracture.

27.3.1 **An example**

In engineering applications, instead of distributed boundary tractions, one often considers concentrated boundary forces. Consider a concentrated force of magnitude P acting in the direction of a unit vector $\mathbf{e}$ at a point $\mathbf{x}^*$ on the boundary $\partial\mathcal{B}$ of a body $\mathcal{B}$, such that the applied traction $\mathbf{t} = P\mathbf{e}\,\delta(\mathbf{x} - \mathbf{x}^*)$, and let $\Delta = \mathbf{u} \cdot \mathbf{e}$ be the corresponding component of

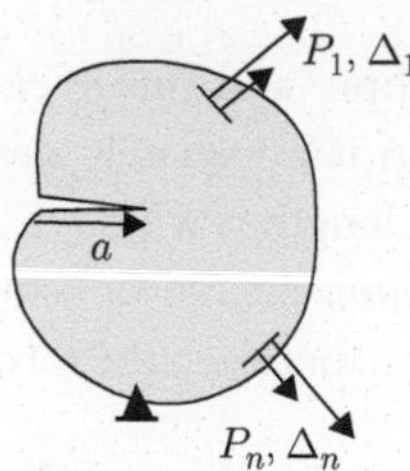

Fig. 27.2 Boundary with discrete loads P_i and associated load-point displacements Δ_i through which P_i do work.

the local displacement vector $\mathbf{u}$ at the point of application of the concentrated force (Fig. 27.2). Considering that $\boldsymbol{\sigma}\boldsymbol{\nu}$ must equal the applied traction for points on the boundary of $\mathcal{B}$, the first term in (27.2.10) reduces to $P\partial\Delta/\partial a$. If we allow for N such instances of concentrated forces P_i with corresponding displacements Δ_i, then (27.2.10) can be written as[2]

$$\boxed{\mathcal{G} = \sum_{i=1}^{N} P_i \frac{d\Delta_i}{da} - \frac{d}{da}\int_{\mathcal{B}} \psi dA.} \tag{27.3.5}$$

EXAMPLE 27.1 Double cantilever beam

As a simple example of Griffith's fracture criterion, consider a beam of height $2h$ with an edge crack of length a extending along its midplane (Fig. 27.3). The crack may be opened symmetrically, either by imposed displacements Δ, or by imposed forces P per unit depth into the plane of the page.

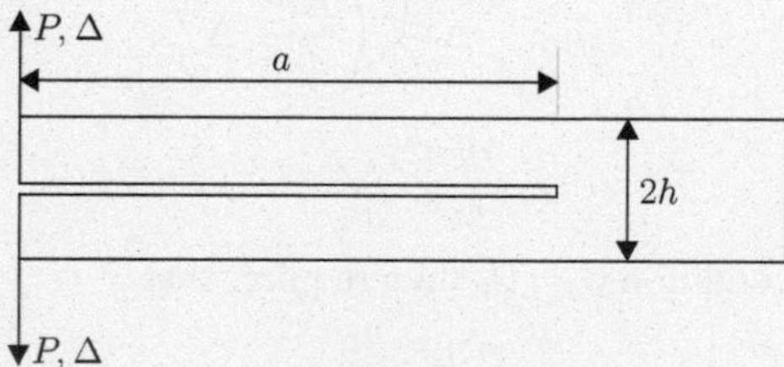

Fig. 27.3 A schematic diagram of a double cantilever beam. The crack of length a opens under the action of either opposed forces P, or opposed displacements Δ of the load points.

[2] Note, we can consider P_i and Δ_i as *generalized forces* and their corresponding *generalized displacements*. For example, two very closely spaced equal and opposite forces form a couple, and by an appropriate limiting process, they can be made to represent a local *concentrated moment* M. The equal and opposite displacements which correspond to the forces in the couple, in the limit, represent a local *rotation* θ. Thus, some of the P_i values may represent moments, and the corresponding Δ_i values may represent rotations whose rotation axis and sense matches the corresponding applied moment.

Assume that the body is isotropic and *linear elastic* with Young's modulus E and Poisson's ratio ν. The deformation is adequately described by assuming that each arm of the body deforms as a beam of length a which is cantilevered at the crack-tip. Such a configuration is referred to as a *double cantilever beam*.

For this geometry, the relationship between the force P and the load point displacement Δ is

$$P = \frac{Eh^3}{4a^3}\,\Delta, \qquad \text{or conversely} \qquad \Delta = \frac{4a^3}{Eh^3}\,P. \tag{27.3.6}$$

The corresponding energy in each arm is $(1/2)P\Delta$, so that for the cracked body which consists of two arms,

$$\int_{\mathcal{B}} \psi dA = P\Delta, \tag{27.3.7}$$

For a double cantilever beam the expression (27.3.5) for the energy release rate reduces to

$$\mathcal{G} = 2P\frac{d\Delta}{da} - \frac{d}{da}(P\Delta), \tag{27.3.8}$$

which gives

$$\mathcal{G} = P\frac{d\Delta}{da} - \Delta\frac{dP}{da}. \tag{27.3.9}$$

Crack initiation under displacement control: First consider the situation when the double cantilever beam is under displacement control. Using the Griffith criterion we determine the value of $\Delta = \Delta_c$ necessary to initiate fracture. For the case of imposed displacements, Δ is a given value independent of a, so $d\Delta/da = 0$, and hence using (27.3.9) and $(27.3.6)_1$,

$$\begin{aligned} \mathcal{G} &= -\Delta\frac{d}{da}\left(\frac{Eh^3}{4a^3}\Delta\right), \\ &= \frac{3Eh^3}{4a^4}\Delta^2. \end{aligned} \tag{27.3.10}$$

The Griffith fracture condition $\mathcal{G} = \mathcal{G}_c$ then requires that

$$\frac{3Eh^3}{4a^4}\Delta_c^2 = \mathcal{G}_c, \tag{27.3.11}$$

which gives

$$\Delta_c = \left(\frac{4\,a^4\,\mathcal{G}_c}{3Eh^3}\right)^{1/2}. \tag{27.3.12}$$

This is the boundary displacement Δ which must be imposed on each arm of the double cantilever specimen to initiate crack growth in a double cantilever beam specimen with an initial crack of length a.

From (27.3.11) we note that if $\mathcal{G}(a) = \mathcal{G}_c$ at a certain crack length a, then $\mathcal{G}(a + \Delta a) < \mathcal{G}_c$ after any increment Δa of crack advance. This implies that Δ_c must be *continually increased for the crack to advance.* Such crack growth is called *stable.* In general, stable crack growth can occur only from states for which

$$\frac{\partial \mathcal{G}(a)}{\partial a} < 0, \tag{27.3.13}$$

where the partial derivative is calculated at fixed boundary displacements.

Crack initiation under load control: Next we consider a load-controlled situation and determine the value $P = P_c$ necessary to initiate fracture. For the load-controlled case, P is a given value independent of a, so $dP/da = 0$, and using (27.3.9) and (27.3.6)$_2$ we obtain

$$\begin{aligned} \mathcal{G} = P\frac{d\Delta}{da} &= P\frac{d}{da}\left(\frac{4a^3}{Eh^3}P\right), \\ &= \frac{12a^2}{Eh^3}P^2. \end{aligned} \tag{27.3.14}$$

Griffith's fracture condition then gives that the critical value $P = P_c$ is determined by

$$\frac{12a^2}{Eh^3}P_c^2 = \mathcal{G}_c, \tag{27.3.15}$$

which gives

$$P_c = \left(\frac{Eh^3\,\mathcal{G}_c}{12a^2}\right)^{1/2}. \tag{27.3.16}$$

This is the force P which must be applied to each arm of the double cantilever beam to initiate crack growth.

From (27.3.15) we note if $\mathcal{G}(a) = \mathcal{G}_c$ at a certain crack length a, then $\mathcal{G}(a + \Delta a) > \mathcal{G}_c$ for any increment Δa of crack advance. That is, the state of incipient fracture under load control is *unstable.* The crack cannot grow under quasi-static equilibrium conditions; inertial and/or material rate effects will become important after the onset of crack growth under load control.

It should be noted that the value of Δ corresponding to the critical force P_c in (27.3.16) is identical to that in (27.3.12); similarly, the value of P corresponding to the critical displacement (27.3.12) is identical to (27.3.16). Hence the force-deflection conditions at the state of incipient fracture for applied displacement and imposed force are the same, but the states themselves are fundamentally different—because the displacement-controlled situation is *stable*, while the load-controlled situation is *unstable.*

27.4 Energy release rate and Eshelby's tensor

Recall the definition (27.2.10) of the energy release rate, viz.

$$\boxed{\mathcal{G} \stackrel{\text{def}}{=} \int_{\partial\mathcal{B}} (\boldsymbol{\sigma}\boldsymbol{\nu}) \cdot \frac{\partial \mathbf{u}}{\partial a}\, ds - \frac{d}{da}\int_{\mathcal{B}} \psi(\epsilon)\, dA.} \tag{27.4.1}$$

In general it is difficult to calculate the energy release rate $\mathcal{G}$ using (27.4.1) because of the difficulty in the computation of the "rate-like" quantities

$$\frac{d}{da}\int_{\mathcal{B}} \psi\, dA \qquad \text{and} \qquad \frac{\partial \mathbf{u}(\mathbf{x}, a)}{\partial a}, \tag{27.4.2}$$

whose computation involves the field $\mathbf{u}(\mathbf{x}, a + da)$ for crack lengths close to a, because when the crack extends the domain of integration $\mathcal{B}$ changes. Instead of (27.4.1) which involves an integral of the free energy of the whole body as well as the power expended by boundary tractions, it can be shown that *the energy release rate may be related to the energy released from, and the power expended on, an infinitesimally small loop moving with the crack-tip.*

Specifically, let $D_\delta(a)$ denote a disc of radius δ centered at—and moving with—the crack-tip, cf. Fig 27.4, with the outward unit normal to ∂D_δ denoted by $\mathbf{n}$. Then $\mathcal{G}$ may be alternatively expressed as

$$\boxed{\mathcal{G} = \lim_{\delta \to 0}\left[\mathbf{e}(a) \cdot \int_{\partial D_\delta} \Big(\psi \mathbf{n} - (\nabla \mathbf{u})^{\top}(\boldsymbol{\sigma}\mathbf{n})\Big)\, ds\right].} \tag{27.4.3}$$

This result is verified in Appendix 27.A. *The importance of this result lies in the observation that $\mathcal{G}$ may be computed from a knowledge of the displacement field $\mathbf{u}(\mathbf{x}, a)$ at a given crack length a* (Gurtin, 1979).

The expression (27.4.3) for the energy release rate $\mathcal{G}$ may be written as

$$\boxed{\mathcal{G} = \lim_{\delta \to 0}\left[\mathbf{e}(a) \cdot \int_{\partial D_\delta} \boldsymbol{\Sigma}\mathbf{n}\, ds\right],} \tag{27.4.4}$$

where

$$\boxed{\boldsymbol{\Sigma} \stackrel{\text{def}}{=} \psi \mathbf{1} - (\nabla \mathbf{u})^{\top}\boldsymbol{\sigma},} \tag{27.4.5}$$

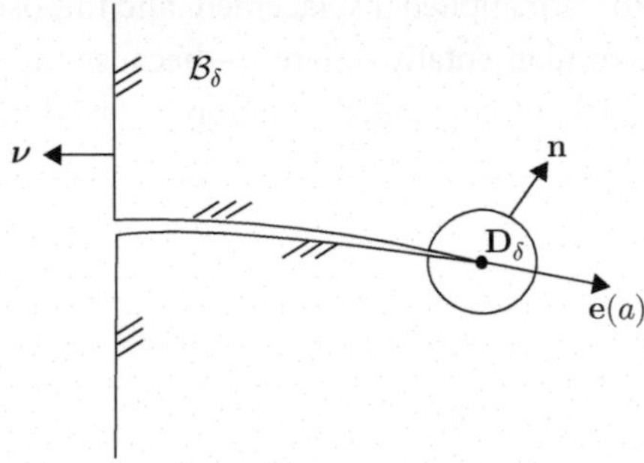

Fig. 27.4 $D_\delta(a)$ is a disc of radius δ centered at–and moving with–the crack-tip. The outward unit normal to ∂D_δ is $\mathbf{n}$, while that to $\partial\mathcal{B}$ is $\boldsymbol{\nu}$. Also, $\mathcal{B}_\delta = \mathcal{B} - D_\delta$.

is known as *Eshelby's energy-momentum tensor*, which was introduced by Eshelby (1951, 1970, 1975) in his discussion of defects in elastic solids. The term $\boldsymbol{\Sigma}\mathbf{n}$ represents a "traction" corresponding to the Eshelby tensor $\boldsymbol{\Sigma}$, and

$$\int_{\partial D_\delta} \boldsymbol{\Sigma}\mathbf{n}\, ds$$

in (27.4.4) represents the **total configurational force** acting on the boundary ∂D_δ of a disc of radius D_δ surrounding the crack-tip. Recall that $\mathbf{e}(a)$ represents the "velocity" of the crack.

- *Thus the term within the brackets on the right-hand side of* (27.4.4) *represents the "power expended" by the total configurational force acting on* $\mathcal{D}_\delta$, *and the energy release rate* $\mathcal{G}$ *is equal to the value of this power expenditure in the limit* $\delta \to 0$.

27.4.1 A conservation integral I

Now consider a region F with boundary ∂F in the body with *no stress singularity*, cf. Fig. 27.5, and consider the integral

$$I \overset{\text{def}}{=} \int_{\partial F} \boldsymbol{\Sigma}\mathbf{n} ds. \tag{27.4.6}$$

Using the divergence theorem in (27.4.6), we find that

$$I = \int_F \operatorname{div} \boldsymbol{\Sigma}\, dA. \tag{27.4.7}$$

Next, we show that the Eshelby energy-momentum tensor $\boldsymbol{\Sigma}$ satisfies the "equilibrium equation",

$$\operatorname{div} \boldsymbol{\Sigma} = \mathbf{0} \qquad \Rightarrow \qquad \operatorname{div}(\psi \mathbf{1} - (\nabla \mathbf{u})^\top \boldsymbol{\sigma}) = \mathbf{0}. \tag{27.4.8}$$

This can be shown by first recalling that,

$$\sigma_{kl} = \frac{\partial \psi}{\partial \epsilon_{kl}} \qquad \text{and} \qquad \frac{\partial \sigma_{kl}}{\partial x_l} = 0.$$

Indeed, since

$$\mathbf{e}_i \cdot (\operatorname{div}(\psi \mathbf{1})) = \frac{\partial}{\partial x_j}(\psi \delta_{ij}) = \frac{\partial \psi}{\partial x_i} = \sigma_{kl}\frac{\partial \epsilon_{kl}}{\partial x_i} = \sigma_{kl}\frac{\partial^2 u_k}{\partial x_l \partial x_i},$$

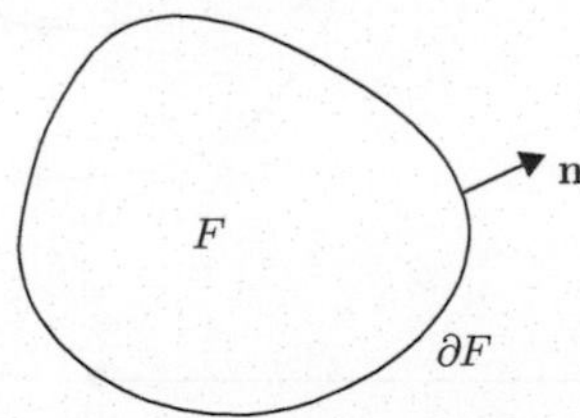

Fig. 27.5 A domain F with boundary ∂F in $\mathcal{B}$.

where we have used the symmetry of σ_{kl}, and since

$$\mathbf{e}_i \cdot (\mathrm{div}((\nabla\mathbf{u})^\top \boldsymbol{\sigma})) = \frac{\partial}{\partial x_l}\,(((\nabla\mathbf{u})^\top\boldsymbol{\sigma})_{il}) = \frac{\partial}{\partial x_l}\left(\frac{\partial u_k}{\partial x_i}\,\sigma_{kl}\right) = \frac{\partial u_k}{\partial x_i}\underbrace{\frac{\partial \sigma_{kl}}{\partial x_l}}_{=0} + \frac{\partial^2 u_k}{\partial x_k \partial x_i}\sigma_{kl}$$

$$= \sigma_{kl}\frac{\partial^2 u_k}{\partial x_k \partial x_i},$$

the "equilibrium equation" (27.4.8) follows.

Thus, on account of (27.4.8), the contour integral,

$$\boxed{I \stackrel{\text{def}}{=} \int_{\partial F} \boldsymbol{\Sigma}\mathbf{n}ds = \int_{\partial F} \left(\psi\mathbf{n} - (\nabla\mathbf{u})^\top\boldsymbol{\sigma}\mathbf{n}\right) ds = 0,} \qquad (27.4.9)$$

and hence I represents a *conservation integral* around a region with *no stress singularities.*

27.5 J-integral

Next, we show that the conservation integral I for a domain *without a stress singularity*, leads us to a definition of Rice's celebrated *path-independent integral*, J, around the crack-tip of a straight crack (Rice, 1968a).

Consider a straight crack, as in Fig. 27.6. Choose a contour Γ_2, and let Γ_1 be another contour which is small enough so that Γ_2 encloses Γ_1, and consider the region F between Γ_2 and Γ_1. Then the stress singularity free region F is bounded by Γ_2 and Γ_1, and the two faces of the crack, so that

$$\partial F = \Gamma_2 + L^+ + \Gamma_1 + L^-.$$

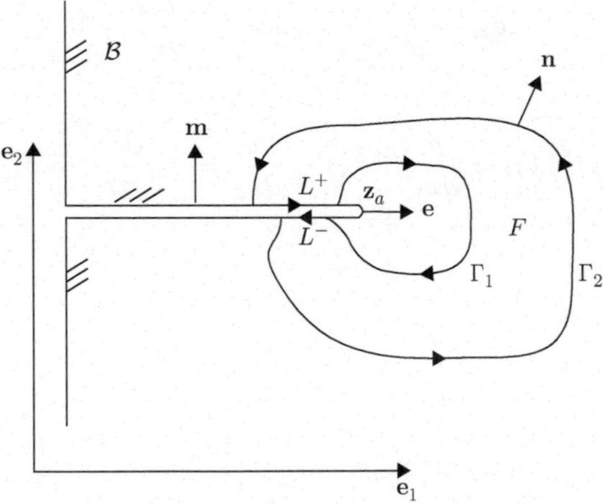

Fig. 27.6 A body $\mathcal{B}$ with a straight edge crack and two contours Γ_1 and Γ_2 surrounding the crack-tip.

On the segments L^+ and L^- of ∂F the upward normal $\mathbf{m}$ to the crack plane is perpendicular to $\mathbf{e}$. Thus,

$$
\begin{aligned}
0 &= I \\
&= \int_{\partial F} \Big(\psi\mathbf{n} - (\nabla\mathbf{u})^{\top}(\boldsymbol{\sigma}\mathbf{n})\Big)\,ds \\
&= \int_{\Gamma_2} \Big(\psi\mathbf{n} - (\nabla\mathbf{u})^{\top}(\boldsymbol{\sigma}\mathbf{n})\Big)\,ds \underbrace{- \int_{\Gamma_1} \Big(\psi\mathbf{n} - (\nabla\mathbf{u})^{\top}(\boldsymbol{\sigma}\mathbf{n})\Big)\,ds}_{(a)} \\
&\quad \underbrace{- \int_{L^+} \Big(\psi\mathbf{m} - (\nabla\mathbf{u})^{\top}(\boldsymbol{\sigma}\mathbf{m})\Big)\,ds}_{(b)} + \int_{L^-} \Big(\psi\mathbf{m} - (\nabla\mathbf{u})^{\top}(\boldsymbol{\sigma}\mathbf{m})\Big)\,ds.
\end{aligned}
$$

Note the algebraic sign in term (a) is negative because we assume that $\mathbf{n}$ points away from the crack on both Γ_1 and Γ_2; the algebraic sign in term (b) is negative because the outward normal on L^+ points down but we have replaced it by $\mathbf{m}$ which points up. Taking the scalar product of this relation with $\mathbf{e}_1$ we obtain,

$$
\begin{aligned}
0 = \mathbf{e}_1 \cdot \int_{\Gamma_2} \Big(\psi\mathbf{n} - (\nabla\mathbf{u})^{\top}(\boldsymbol{\sigma}\mathbf{n})\Big)\,ds - \mathbf{e}_1 \cdot \int_{\Gamma_1} \Big(\psi\mathbf{n} - (\nabla\mathbf{u})^{\top}(\boldsymbol{\sigma}\mathbf{n})\Big)\,ds \\
- \mathbf{e}_1 \cdot \int_{L^+} \Big(\psi\mathbf{m} - (\nabla\mathbf{u})^{\top}(\boldsymbol{\sigma}\mathbf{m})\Big)\,ds + \mathbf{e}_1 \cdot \int_{L^-} \Big(\psi\mathbf{m} - (\nabla\mathbf{u})^{\top}(\boldsymbol{\sigma}\mathbf{m})\Big)\,ds
\end{aligned}
$$

For a straight crack, $\mathbf{e}_1 \cdot \mathbf{m} = 0$, and since the crack faces are traction free, $\boldsymbol{\sigma}\mathbf{m} = \mathbf{0}$, this reduces to

$$
0 = \mathbf{e}_1 \cdot \int_{\Gamma_2} \Big(\psi\mathbf{n} - (\nabla\mathbf{u})^{\top}(\boldsymbol{\sigma}\mathbf{n})\Big)\,ds - \mathbf{e}_1 \cdot \int_{\Gamma_1} \Big(\psi\mathbf{n} - (\nabla\mathbf{u})^{\top}(\boldsymbol{\sigma}\mathbf{n})\Big)\,ds. \tag{27.5.1}
$$

DEFINITION: Let Γ denote any smooth non-intersecting path which begins and ends on the crack faces and which surrounds the tip, and let $\mathbf{n}$ denote the unit normal on Γ directed away from the crack tip. We define

$$
\boxed{J(\Gamma) \stackrel{\text{def}}{=} \mathbf{e}_1 \cdot \int_{\Gamma} \Big(\psi\mathbf{n} - (\nabla\mathbf{u})^{\top}(\boldsymbol{\sigma}\mathbf{n})\Big)\,ds,} \tag{27.5.2}
$$

and call $J(\Gamma)$ the **J-integral for the path** Γ (Rice, 1968a).

Result (27.5.1) shows that

$$
\boxed{J(\Gamma_1) = J(\Gamma_2).} \tag{27.5.3}
$$

That is,

- *$J(\Gamma)$ is independent of the path Γ surrounding the crack tip.*

27.5.1 J-integral and the energy release rate $\mathcal{G}$

Next, consider a disc D_δ of radius δ and a contour Γ surrounding the crack-tip, as shown in Fig. 27.7. From the definition (27.5.2) of the J-integral and the result (27.4.4), we immediately have the important conclusion that the energy release rate $\mathcal{G}$ may be evaluated as,

$$\mathcal{G} = \lim_{\delta \to 0} J(\partial D_\delta). \tag{27.5.4}$$

However, as we have shown in (27.5.3), for a *straight crack* J is independent of the path surrounding the crack-tip. Thus, the energy release rate for a straight crack is also given by,

$$\mathcal{G} = J(\Gamma). \tag{27.5.5}$$

To summarize, for a straight crack,

- *the J-integral is defined as*

$$\boxed{\begin{aligned} J(\Gamma) &\stackrel{\text{def}}{=} \mathbf{e}_1 \cdot \int_\Gamma \Big(\psi \mathbf{n} - (\nabla \mathbf{u})^\top (\boldsymbol{\sigma} \mathbf{n})\Big)\, ds \\ &= \int_\Gamma \left(\psi n_1 - \sigma_{ij} n_j \frac{\partial u_i}{\partial x_1}\right) ds \\ &= \int_\Gamma \left(\psi \delta_{1j} - \sigma_{ij} \frac{\partial u_i}{\partial x_1}\right) n_j ds \\ &= \int_\Gamma \left(\psi dx_2 - \sigma_{ij} n_j \frac{\partial u_i}{\partial x_1}\, ds\right) \end{aligned}} \tag{27.5.6}$$

  for *any* curve Γ surrounding the crack-tip. In writing the last of (27.5.6) we have used the fact that $n_1 = dx_2/ds$.

- *$J(\Gamma)$ is independent of the path Γ surrounding the tip.*
- *The energy release rate is given by,*

$$\boxed{\mathcal{G} = J(\Gamma).} \tag{27.5.7}$$

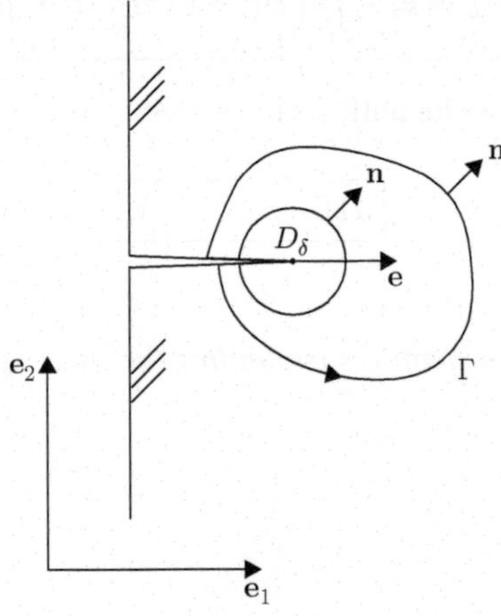

Fig. 27.7 A disc D_δ of radius δ and a contour Γ surrounding the crack-tip.

The utility of J is that we can evaluate it in the far field; but, by taking Γ arbitrarily close to the crack-tip, we can gain information on crack-tip singularity, or on concentrated strain at a notch surface. Thus J may be used for the following purposes:

(i) As an aid in asymptotic stress analysis.

(ii) As a characterizing parameter for crack-tip deformation.

(iii) Use as a crack extension criterion.

EXAMPLE 27.2 Semi-infinite crack in an infinite sheet

The J-integral may be evaluated almost by inspection for the configuration shown in Fig. 27.8 in which there is a semi-infinite flat-surfaced notch in an infinite strip of height h, and loads are applied by clamping the upper and lower surfaces of the strip so that the displacement vector $\mathbf{u}$ is constant on each clamped boundary.

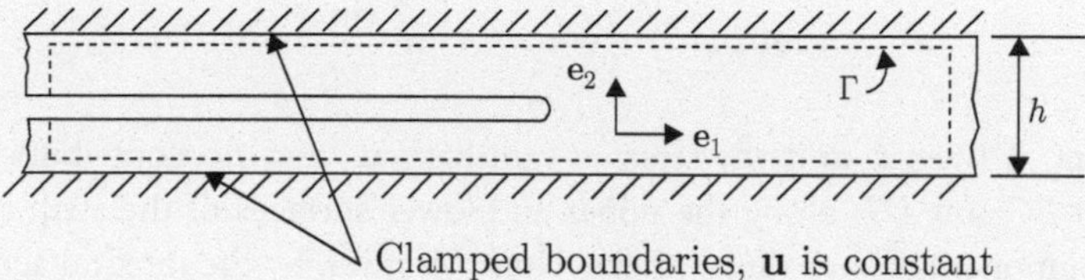

Fig. 27.8 Infinite strip with semi-infinite crack, with constant displacement imposed by clamping boundaries; cf. Rice (1968a).

Take Γ to be the dashed curve shown which stretches out to $x_1 = \pm\infty$. From (27.5.6) we have

$$J(\Gamma) = \int_\Gamma \left(\psi dx_2 - (\boldsymbol{\sigma}\mathbf{n}) \cdot \frac{\partial \mathbf{u}}{\partial x_1}\, ds\right). \tag{27.5.8}$$

There is no contribution to J from the portion of Γ along the clamped boundaries since $dx_2 = 0$ and $(\partial \mathbf{u}/\partial x_1) = \mathbf{0}$ there. Also, at $x_1 = -\infty$, $\psi = 0$, and $(\partial \mathbf{u}/\partial x_1) = \mathbf{0}$. The entire contribution to J comes from the portion of Γ at $x_1 = +\infty$, and since $(\partial \mathbf{u}/\partial x_1) = \mathbf{0}$ there, we have

$$J = \psi_{+\infty}\, h, \tag{27.5.9}$$

where $\psi_{+\infty}$ is the free energy density at $x = +\infty$.

As a concrete case, if one considers a linear elastic incompressible material, then assuming the vertical displacement at $x_1 = +\infty$ is $\hat{u}$, the non-zero strain components will be $\epsilon_{22} = \hat{u}/h$ and $\epsilon_{33} = -\hat{u}/h$, thus $\psi_{+\infty} = \frac{1}{2}\boldsymbol{\epsilon} : \mathbb{C} : \boldsymbol{\epsilon} = (2\mu + \lambda)(\hat{u}/h)^2$, where μ and λ are the Lamé constants.

EXAMPLE 27.3 Infinite strip with a semi-infinite crack with a bending load

As a second similar example, Fig. 27.9 shows an infinite strip with a semi-infinite crack where the loads are applied by couples M per unit thickness on the beam-like arms, so that a state of pure bending results in the arms; that is, all stresses vanish except $\sigma_{11} \equiv \sigma$.

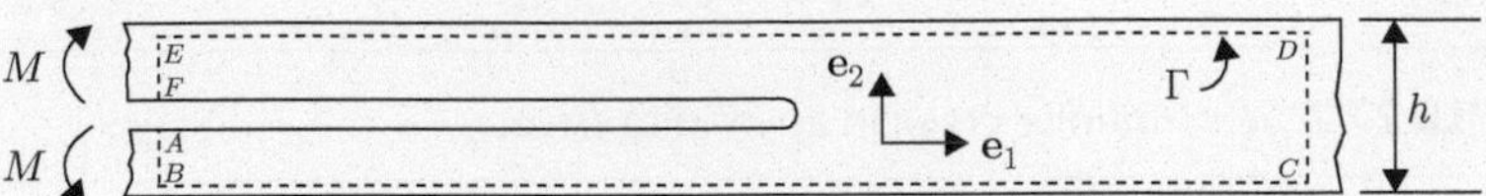

Fig. 27.9 Infinite strip with semi-infinite crack, with pure bending of beam-like arms; cf. Rice (1968a).

For the contour $\Gamma = ABCDEF$ shown by the dashed line, no contribution to

$$J(\Gamma) = \int_{\Gamma} \Big(\psi n_1 - \sigma_{ij} n_j \frac{\partial u_i}{\partial x_1} \Big) ds$$

occurs along CD as ψ and the stresses vanish there, and no contribution occurs for portions of BC and DE along the upper and lower surfaces of the strip as n_1 and the traction components $\sigma_{ij} n_j$ vanish there. Thus J is given by the integral across the beam arms along the segments AB and EF, and their contributions are identical, because of symmetry. Along segment EF we have: (i) $n_1 = -1$; (ii) $\psi = \sigma^2/2E$; (iii) $\sigma_{11} n_1 = -\sigma$ (note the sign); and (iv) $u_{1,1} = \epsilon_{11} = \sigma/E$. Thus

$$J = 2 \int_{EF} \Big(-\frac{\sigma^2}{2E} + \frac{\sigma^2}{E} \Big) ds = \frac{1}{E} \int_{EF} \sigma^2 ds.$$

Using $\sigma = -M\zeta/I$, with $I = (1/12)(h/2)^3$ per unit thickness into the plane of the page, and where ζ is measured from the center-line of the beam arm and $ds = d\zeta$, we have

$$J = \frac{1}{E} \int_{EF} \sigma^2 ds = \frac{M^2}{EI^2} \int_{-h/4}^{h/4} \zeta^2 d\zeta,$$

which gives

$$J = \frac{M^2}{EI}, \tag{27.5.10}$$

measured per unit thickness into the plane of the page.

27.6 Use of the J-integral to calculate the energy release rate $\mathcal{G}$

The J-integral is widely used to calculate the energy release $\mathcal{G}$ using numerical finite element methods because the integrand in (27.5.6) involves the fields in the body at a *fixed crack length*, and since the path of integration needs to only enclose the tip of the crack, it can be chosen to be far away from the tip of the crack, thus avoiding superfine meshes to resolve the singular fields near the crack-tip.

However, line integrals are not easily calculated in a finite element formulation, but area integrals are. This motivates converting the line integral for J to an area integral or a "domain integral". We discuss next a method that is used to perform this conversion.

Consider an annular region A around the crack-tip that encloses no other crack or void as depicted in Fig. 27.10. For the contour Γ surrounding the crack-tip in Fig. 27.10, for later convenience, we write (27.5.6) in the form

$$\mathcal{G} = J = -\int_{\Gamma} \left(\sigma_{ij} \frac{\partial u_i}{\partial x_1} - \psi \delta_{1j} \right) n_j ds = \int_{\Gamma} \Sigma_{1j} n_j ds. \tag{27.6.1}$$

Next, consider the closed curve

$$C = C_0 + C_+ + \Gamma + C_-$$

shown in Fig. 27.10 with outward unit normal $\mathbf{m}$, where C_+ and C_- are traction-free boundaries. Denote the area enclosed by C as A and let

$$q = \hat{q}(x_1, x_2)$$

denote a weighting function over A

- which has a value of unity on Γ and vanishes on C_0.

The function $q(x_1, x_2)$ may be interpreted as a virtual translation of a material point (x_1, x_2) due to a unit extension of the crack in the $\mathbf{e}_1$-direction (Li et al., 1985). Next, consider the integral

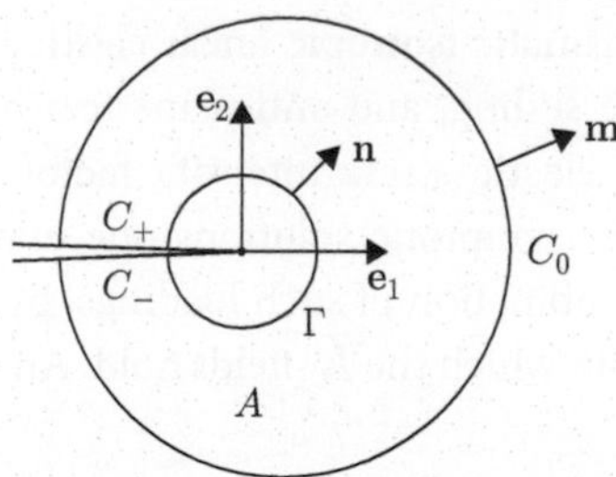

Fig. 27.10 Domain A is enclosed by a closed curve $C = C_0 + C_+ + \Gamma + C_-$, with outward unit normal $\mathbf{m}$.

$$\int_C \Big(\sigma_{ij}\frac{\partial u_i}{\partial x_1} - \psi\delta_{1j}\Big)qm_j ds = \int_{C_0} \Big(\sigma_{ij}\frac{\partial u_i}{\partial x_1} - \psi\delta_{1j}\Big)qm_j ds$$
$$+ \int_{C_+} \Big(\sigma_{ij}\frac{\partial u_i}{\partial x_1} - \psi\delta_{1j}\Big)qm_j ds$$
$$+ \int_{\Gamma} \Big(\sigma_{ij}\frac{\partial u_i}{\partial x_1} - \psi\delta_{1j}\Big)qm_j ds$$
$$+ \int_{C_-} \Big(\sigma_{ij}\frac{\partial u_i}{\partial x_1} - \psi\delta_{1j}\Big)qm_j ds. \tag{27.6.2}$$

The integral on C_0 vanishes because q vanishes there; on Γ, $\mathbf{m} = -\mathbf{n}$ and q is unity. Further, both $\sigma_{ij}m_j = 0$ and $m_1 = 0$ on C_+ and C_-. Thus we have

$$\int_C \Big(\sigma_{ij}\frac{\partial u_i}{\partial x_1} - \psi\delta_{1j}\Big)qm_j ds = \underbrace{-\int_{\Gamma} \Big(\sigma_{ij}\frac{\partial u_i}{\partial x_1} - \psi\delta_{1j}\Big)n_j ds,}_{\equiv \mathcal{G}} \tag{27.6.3}$$

and hence using (27.6.1)

$$\mathcal{G} = \int_C \Big(\sigma_{ij}\frac{\partial u_i}{\partial x_1} - \psi\delta_{1j}\Big)qm_j ds = -\int_C \Sigma_{1j}qm_j ds. \tag{27.6.4}$$

Applying the divergence theorem to (27.6.4) gives

$$\mathcal{G} = -\int_A \frac{\partial}{\partial x_j}(\Sigma_{1j}q)dA = -\int_A \left(\frac{\partial \Sigma_{1j}}{\partial x_j}q + \Sigma_{1j}\frac{\partial q}{\partial x_j}\right)dA. \tag{27.6.5}$$

Since $\dfrac{\partial \Sigma_{1j}}{\partial x_j} = 0$ (cf. (27.4.8)), (27.6.5) reduces to the desired domain integral expression for $\mathcal{G}$ (Li et al., 1985):

$$\boxed{\mathcal{G} = -\int_A \left(\Sigma_{1j}\frac{\partial q}{\partial x_j}\right)dA = \int_A \left(\sigma_{ij}\frac{\partial u_i}{\partial x_1} - \psi\delta_{1j}\right)\frac{\partial q}{\partial x_j}\,dA.} \tag{27.6.6}$$

27.7 Relationship between $\mathcal{G}$, J, and K_{I}, K_{II}, and K_{III}

Consider a sharp crack in a prismatic isotropic linear elastic body. The asymptotic crack-tip solutions for opening, in-plane sliding, and anti-plane tearing modes are given in Sec. 9.5. Let K_{I}, K_{II}, and K_{III} be the elastic stress intensity factors for the three basic crack-tip deformation modes. Using the asymptotic solutions one may evaluate the J-integral for a cracked body subject to any combination of such loadings, by choosing the contour Γ as any path surrounding the crack-tip in which the K-fields hold. An explicit calculation of this leads to (Rice, 1968b, eq. (126)):

$$\boxed{\mathcal{G} = J = \frac{(1-\nu)}{2\mu}(K_{\mathrm{I}}^2 + K_{\mathrm{II}}^2) + \frac{1}{2\mu}K_{\mathrm{III}}^2 \qquad \text{for plane strain.}} \tag{27.7.1}$$

For plane stress we replace ν by $\nu/(1+\nu)$ to obtain

$$\boxed{\mathcal{G} = J = \frac{1}{2\mu(1+\nu)}(K_{\mathrm{I}}^2 + K_{\mathrm{II}}^2) + \frac{1}{2\mu}K_{\mathrm{III}}^2 \qquad \text{for plane stress.}} \tag{27.7.2}$$

Finally, using $E = 2\mu(1+\nu)$ we may rewrite (27.7.1) and (27.7.2) as the following important relation for isotropic linear elastic materials:

$$\boxed{\mathcal{G} = J = \frac{1}{\bar{E}}(K_{\mathrm{I}}^2 + K_{\mathrm{II}}^2) + \frac{1+\nu}{E}K_{\mathrm{III}}^2,} \tag{27.7.3}$$

where

$$\bar{E} = \begin{cases} \dfrac{E}{(1-\nu^2)} & \text{for plane strain,} \\ E & \text{for plane stress.} \end{cases} \tag{27.7.4}$$

In particular for Mode I loading,

$$\boxed{\mathcal{G} = J = \frac{K_{\mathrm{I}}^2}{\bar{E}}.} \tag{27.7.5}$$

In this case the fracture condition

$$\boxed{\mathcal{G} = \mathcal{G}_c} \tag{27.7.6}$$

gives

$$\frac{1}{\bar{E}}K_{\mathrm{I}}^2 = \mathcal{G}_c,$$

which may be written as

$$\boxed{K_{\mathrm{I}} = K_c,} \tag{27.7.7}$$

where

$$\boxed{K_c \stackrel{\text{def}}{=} \sqrt{\bar{E}\mathcal{G}_c}.} \tag{27.7.8}$$

27.8 Use of J as a fracture parameter for elastic-plastic fracture mechanics

So far we have considered the notions of an energy release rate and the J-integral for a hyperelastic material. It is important to see if such ideas can be extended to develop a fracture criterion for materials exhibiting inelastic behavior which is not necessarily confined to a small zone at the crack-tip such as occurs in metals under "small-scale yielding" conditions.

For metallic materials the rate-independent elastic-plastic behavior at low homologous temperatures is best described by an incremental (or flow) theory of plasticity, as discussed in Chapter 21. To date however, there has been no success in formulating a J-integral for the flow theory of plasticity, analogous to the J-integral for non-linear hyperelastic materials. Notwithstanding, as discussed by Hill (1950), under conditions of *continued proportional loading* with *no unloading* anywhere in the body, the equations of the incremental theory of plasticity can be integrated to arrive at a *deformation theory* of plasticity (Hencky, 1924), for which the stress may be determined as the derivative of a potential,[3]

$$\sigma = \frac{\partial W(\epsilon)}{\partial \epsilon}. \tag{27.8.1}$$

Recognizing this, Rice (1968a) used a *deformation theory* of plasticity and the J-integral for a non-linear hyperelastic material (cf. (27.5.6)) to introduce an "elastic-plastic fracture mechanics", which is often used in practice. Rice's (1968a) definition of the J-integral is path-independent and can be used to determine the state near the crack-tip from the solution field far from the crack-tip. Further, it represents a "crack driving force", and

$$J = J_c \tag{27.8.2}$$

can be considered as a criterion for crack extension in an elastic-plastic material under the restrictive conditions of proportional loading. The reader is referred to an exposition on the vast subject of non-linear fracture mechanics by Hutchinson (1979, 1983).

While this theory is quite widely used, fundamental conceptual difficulties exist because the theory relies on the assumption of deformation plasticity, which treats an elastic-plastic material as if it were a "fictitious" non-linear elastic material. This approach is limited because non-proportional loading invariably occurs during fracture of ductile materials; for example unloading processes occur due to the formation of voids in front of the crack-tip, and also due to crack extension. It is well-known that in any situation where a significant amount of non-proportional loading occurs, the J-integral will not be path-independent. Under these circumstances the use of a deformation theory as a surrogate for a bona fide flow theory is not appropriate—a flow theory of plasticity is required for a realistic description of such more general situations. For relatively recent discussions of this issue, and possible alternative considerations based on the use of "material force mechanics", see, e.g., Nguyen et al. (2005) and Simha et al. (2008).

For recent discussions of other approaches for treating inelastic fracture mechanics, see, e.g.,[4]

- Needleman (2014) for developments on the use of *cohesive zone* methods;
- Fries and Belytschko (2010) for developments on the use of the *extended finite element method* (XFEM); and
- Miehe et al. (2010, 2016) for developments on the use of *phase-field* methods.

It should be noted that presence of at least three major approaches to treating inelastic fracture mechanics (and fracture in general) is indicative of the open research nature of this subject.

---

[3] The potential W is not the same as the free energy ψ.

[4] References to the vast earlier literature on these alternative approaches to modeling fracture may be found in the cited references.

Appendices

27.A Equivalence of the two main relations for $\mathcal{G}(a)$

Here we verify (27.4.3); further details regarding the underlying mathematical hypotheses may be found in Gurtin (1979). The objective is to show that

$$\mathcal{G}(a) \stackrel{\text{def}}{=} \int_{\partial\mathcal{B}} (\boldsymbol{\sigma}\boldsymbol{\nu}) \cdot \frac{\partial \mathbf{u}}{\partial a}\, ds - \frac{d}{da}\int_{\mathcal{B}} \psi(\epsilon)\, dA \tag{27.A.1}$$

may be expressed as

$$\mathcal{G}(a) = \lim_{\delta\to 0}\left[\mathbf{e}(a)\cdot \int_{\partial D_\delta} \Big(\psi\mathbf{n} - (\nabla\mathbf{u})^{\top}(\boldsymbol{\sigma}\mathbf{n})\Big)\, ds\right], \tag{27.A.2}$$

where $\mathbf{e}(a)$ is the direction of crack growth; cf. Fig. 27.11. From our discussion in Sec. 27.2.1 we also note that

- we have used the crack length a as our time-scale, and that
- $\mathbf{e}(a)$ also represents the "velocity" of the crack-tip.

We begin by recalling:

Reynold's transport theorem: Let $\mathrm{R}(t)$ denote a bounded region evolving through the reference body $\mathcal{B}$, and let $\mathbf{v}_{\partial\mathrm{R}}(\mathbf{x}, t)$ denote the velocity of the time-dependent boundary $\partial\mathrm{R}(t)$ with outward unit normal $\mathbf{m}$.[5] Then given a scalar function $\varphi(\mathbf{x}, t)$, we have the *transport relation:*

$$\frac{d}{dt}\int_{\mathrm{R}(t)} \varphi(\mathbf{x}, t)\, dv = \int_{\mathrm{R}(t)} \frac{\partial\varphi(\mathbf{x}, t)}{\partial t}\, dv + \int_{\partial\mathrm{R}(t)} \varphi(\mathbf{x}, t)\big(\mathbf{v}_{\partial\mathrm{R}}(\mathbf{x}, t)\cdot\mathbf{m}\big)\, da. \tag{27.A.3}$$

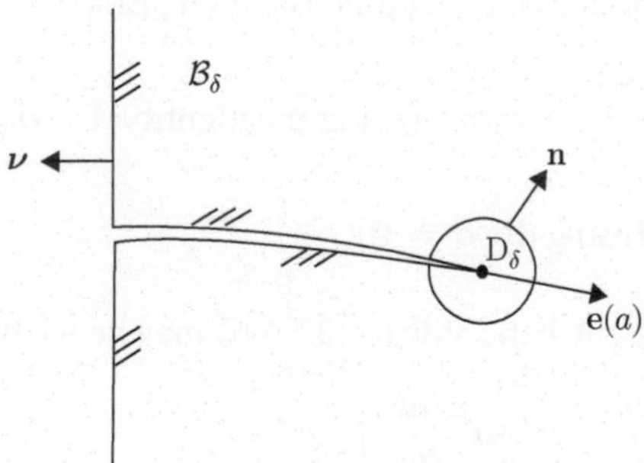

Fig. 27.11 $D_\delta(a)$ is a disc of radius δ centered at, and moving with the crack-tip. The outward unit normal to ∂D_δ is $\mathbf{n}$, while that to $\partial\mathcal{B}$ is $\boldsymbol{\nu}$. Also, $\mathcal{B}_\delta = \mathcal{B} - D_\delta$.

[5] Note $\mathbf{m}$ is a function of $(\mathbf{x}, t)$, which we omit to reduce the visual burden on the reader. Implicitly we also assume that all needed derivatives and integrals are well defined, implying a certain degree of smoothness of $\partial\mathrm{R}$.

Consider first the second term in (27.A.1). We assume that

$$\frac{d}{da}\int_{\mathcal{B}} \psi dA = \lim_{\delta\to 0}\frac{d}{da}\int_{\mathcal{B}_\delta} \psi dA, \tag{27.A.4}$$

where

$$\mathcal{B}_\delta(a) = \mathcal{B} - D_\delta(a),$$

and determine

$$\frac{d}{da}\int_{\mathcal{B}_\delta} \psi dA. \tag{27.A.5}$$

We now apply the transport theorem (27.A.3),[6] with a replacing t. In doing so, we note that

$$\mathbf{v}_{\partial \mathrm{R}}(\mathbf{x}, t)\cdot \mathbf{m} = -\mathbf{e}(a)\cdot\mathbf{n}$$

on ∂D_δ, where $\mathbf{e}(a)$ is the "velocity" of the crack-tip and hence also of the boundary ∂D_δ. The minus sign appears in the last term due to the orientation of $\mathbf{n}$ in Fig. 27.11. The result of applying the theorem, when one assumes that on $\partial\mathcal{B}_\delta - \partial D_\delta$ that either $\mathbf{v}_{\partial R} = \mathbf{0}$ or $\mathbf{e}(a)\cdot\boldsymbol{\nu} = 0$, is

$$\frac{d}{da}\int_{\mathcal{B}_\delta}\psi dA = \int_{\mathcal{B}_\delta}\frac{\partial\psi}{\partial a}\,dA - \mathbf{e}(a)\cdot\int_{\partial D_\delta}\psi\mathbf{n}ds, \tag{27.A.6}$$

where the term

$$-\mathbf{e}(a)\cdot\int_{\partial D_\delta}\psi\mathbf{n}ds$$

represents the energy flow out of $\mathcal{B}_\delta$ across ∂D_δ, which arises because of the motion of D_δ with the tip.

Next, since $\psi(\boldsymbol{\epsilon})$,

$$\begin{aligned}
\frac{\partial\psi}{\partial a} &= \frac{\partial\psi}{\partial\boldsymbol{\epsilon}} : \frac{\partial\boldsymbol{\epsilon}}{\partial a},\\
&= \boldsymbol{\sigma} : \nabla\left(\frac{\partial\mathbf{u}}{\partial a}\right) \quad \text{(because the material is hyperelastic)},\\
&= \operatorname{div}\left(\boldsymbol{\sigma}^{\top}\frac{\partial\mathbf{u}}{\partial a}\right) - \frac{\partial\mathbf{u}}{\partial a}\cdot\operatorname{div}\boldsymbol{\sigma} \quad \text{(using the identity } \operatorname{div}(\boldsymbol{\sigma}^{\top}\mathbf{v}) = \boldsymbol{\sigma}:\nabla\mathbf{v} + \mathbf{v}\cdot\operatorname{div}\boldsymbol{\sigma}),\\
&= \operatorname{div}\left(\boldsymbol{\sigma}^{\top}\frac{\partial\mathbf{u}}{\partial a}\right) \quad \text{(using } \operatorname{div}\boldsymbol{\sigma} = \mathbf{0}).
\end{aligned}$$

Thus the first term on the right-hand side of (27.A.6) may be written as,

$$\begin{aligned}
\int_{\mathcal{B}_\delta}\frac{\partial\psi}{\partial a}\,dA &= \int_{\mathcal{B}_\delta}\operatorname{div}\left(\boldsymbol{\sigma}^{\top}\frac{\partial\mathbf{u}}{\partial a}\right)dA,\\
&= \int_{\partial\mathcal{B}_\delta}\left(\boldsymbol{\sigma}^{\top}\frac{\partial\mathbf{u}}{\partial a}\right)\cdot\boldsymbol{\nu}ds \quad \text{(using the divergence theorem)},\\
&= \int_{\partial\mathcal{B}}\left(\boldsymbol{\sigma}^{\top}\frac{\partial\mathbf{u}}{\partial a}\right)\cdot\boldsymbol{\nu}ds - \int_{\partial D_\delta}\left(\boldsymbol{\sigma}^{\top}\frac{\partial\mathbf{u}}{\partial a}\right)\cdot\mathbf{n}ds,
\end{aligned} \tag{27.A.7}$$

---

6 Restricted to two dimensions.

where the minus sign in the second term arises because of the manner in which $\mathbf{n}$ is directed into $\mathcal{B}_\delta$ around the disc D_δ.[7] Thus

$$\int_{\mathcal{B}_\delta} \frac{\partial \psi}{\partial a}\, dA = \int_{\partial\mathcal{B}} (\boldsymbol{\sigma}\boldsymbol{\nu}) \cdot \frac{\partial \mathbf{u}}{\partial a}\, ds - \int_{\partial D_\delta} (\boldsymbol{\sigma}\mathbf{n}) \cdot \frac{\partial \mathbf{u}}{\partial a}\, ds. \tag{27.A.8}$$

Hence, using (27.A.8) in (27.A.6) we have that (in the limit $\delta \to 0$)

$$\frac{d}{da}\int_{\mathcal{B}_\delta} \psi dA = \int_{\partial\mathcal{B}} (\boldsymbol{\sigma}\boldsymbol{\nu}) \cdot \frac{\partial \mathbf{u}}{\partial a}\, ds - \int_{\partial D_\delta} (\boldsymbol{\sigma}\mathbf{n}) \cdot \frac{\partial \mathbf{u}}{\partial a}\, ds - \mathbf{e}(a) \cdot \int_{\partial D_\delta} \psi \mathbf{n} ds. \tag{27.A.9}$$

Use of (27.A.9) allows us to write (27.A.4) as

$$\frac{d}{da}\int_{\mathcal{B}} \psi dA = \int_{\partial\mathcal{B}} (\boldsymbol{\sigma}\boldsymbol{\nu}) \cdot \frac{\partial \mathbf{u}}{\partial a}\, ds - \lim_{\delta\to 0}\left[\int_{\partial D_\delta} \left(\psi\mathbf{n}\cdot\mathbf{e}(a) + (\boldsymbol{\sigma}\mathbf{n}) \cdot \frac{\partial \mathbf{u}}{\partial a}\right) ds\right]. \tag{27.A.10}$$

Finally, using (27.A.10) in (27.A.1) gives

$$\boxed{\mathcal{G}(a) = \lim_{\delta\to 0}\left[\int_{\partial D_\delta} \left(\psi\mathbf{n}\cdot\mathbf{e}(a) + (\boldsymbol{\sigma}\mathbf{n}) \cdot \frac{\partial \mathbf{u}}{\partial a}\right) ds\right].} \tag{27.A.11}$$

The limit (27.A.11) therefore gives the energy released by the body and absorbed by the moving crack-tip.

Our next result gives an important alternative expression for the power term,

$$(\boldsymbol{\sigma}\mathbf{n}) \cdot \frac{\partial \mathbf{u}}{\partial a},$$

in (27.A.11). Consider the displacement field $\mathbf{u}(\mathbf{x}, a)$ as a function $\mathbf{g}(\mathbf{r}, a)$ of a and the position vector

$$\mathbf{r} = \mathbf{x} - \mathbf{z}_a$$

from the crack-tip. Define

$$\mathbf{v}(\mathbf{x}, a) = \left.\frac{\partial \mathbf{g}(\mathbf{r}, a)}{\partial a}\right|_{\mathbf{r}=\mathbf{x}-\mathbf{z}_a}. \tag{27.A.12}$$

Then

$$\mathbf{u}(\mathbf{x}, a) = \mathbf{g}(\mathbf{x} - \mathbf{z}_a, a),$$

and by (27.2.1) the derivative of $\mathbf{x} - \mathbf{z}_a$ with respect to a holding $\mathbf{x}$ fixed is $-\mathbf{e}(a)$; thus, in view of (27.A.12),

$$\frac{\partial \mathbf{u}(\mathbf{x}, a)}{\partial a} = -\nabla\mathbf{g}(\mathbf{x} - \mathbf{z}_a, a)\mathbf{e}(a) + \mathbf{v}(\mathbf{x}, a).$$

But

$$\nabla\mathbf{u}(\mathbf{x}, a) = \nabla\mathbf{g}((\mathbf{x} - \mathbf{z}_a), a);$$

hence

$$\frac{\partial \mathbf{u}}{\partial a} = -(\nabla\mathbf{u})\mathbf{e} + \mathbf{v}, \tag{27.A.13}$$

[7] The last equality in (27.A.7) is strictly true only in the limit as $\delta \to 0$ but this is our case of interest, so we omit the technical details of including the limit signs.

so that

$$(\boldsymbol{\sigma}\mathbf{n}) \cdot \frac{\partial \mathbf{u}}{\partial a} = -(\boldsymbol{\sigma}\mathbf{n}) \cdot (\nabla \mathbf{u})\mathbf{e} + (\boldsymbol{\sigma}\mathbf{n}) \cdot \mathbf{v}. \tag{27.A.14}$$

Next, since

$$\lim_{\delta\to 0} \int_{\partial D_\delta} (\boldsymbol{\sigma}\mathbf{n}) \cdot \mathbf{v}\, ds = \left[\lim_{\delta\to 0} \int_{D_\delta} \big(\underbrace{\operatorname{div} \boldsymbol{\sigma}}_{=0} \cdot \mathbf{v} + \boldsymbol{\sigma} : \nabla \mathbf{v} \big)\, dA \right] \to 0,$$

we have the following alternative expression for the power term in (27.A.11)

$$\lim_{\delta\to 0} \int_{\partial D_\delta} \boldsymbol{\sigma}\mathbf{n} \cdot \frac{\partial \mathbf{u}}{\partial a}\, ds = -\lim_{\delta\to 0} \mathbf{e} \cdot \int_{\partial D_\delta} (\nabla \mathbf{u})^\top (\boldsymbol{\sigma}\mathbf{n})\, ds. \tag{27.A.15}$$

Substituting (27.A.15) in (27.A.11) gives the desired result (27.A.2)

$$\boxed{\mathcal{G}(a) = \lim_{\delta\to 0} \left[\mathbf{e}(a) \cdot \int_{\partial D_\delta} \Big(\psi \mathbf{n} - (\nabla \mathbf{u})^\top (\boldsymbol{\sigma}\mathbf{n}) \Big)\, ds \right].} \tag{27.A.16}$$

28 Fatigue

28.1 Introduction

The failure of components under the action of repeated fluctuating stresses or strains is called **fatigue failure**. The word "fatigue" was introduced in the mid-1800s in connection with failures of railroad axles which occurred in the then rapidly developing railway industry. It was found that railroad axles, which are subjected to rotating-bending type loads, frequently failed at stress concentrations associated with shoulders between different sections of the axles. Such fatigue failures under cyclic loading conditions appeared to be quite different from failures associated with monotonic testing (Wohler, 1860).

Fracture of a component (or a body) under monotonic conditions may be considered to be an *event*, since it typically occurs rather abruptly. Unlike fracture under monotonic conditions, failure due to fatigue is a *process* which occurs over time in a component which is subjected to fluctuating stresses and strains. Typically, in a component which is subjected to boundary conditions which produce sufficiently large fluctuating stresses and strains in some region of the body, *there is progressive, localized, permanent microstructural change which occurs in that region. This microstructural change may culminate in the initiation of cracks and their subsequent growth to a size which causes final fracture after a sufficient number of stress or strain fluctuations.*

The phrase "permanent microstructural changes" emphasizes the central role of cyclic inelastic deformation in causing irreversible changes in the microstructure.

- *Countless investigations have established that fatigue results from cyclic inelastic deformation in every instance, even though the structure as a whole may be nominally deforming elastically.*

A small inelastic strain excursion applied only once does not cause any substantial changes in the microstructure of materials, but multiple repetitions of very small inelastic strains leads to cumulative damage ending in fatigue failure.

We note that although fatigue is usually associated with metallic materials, it can occur in all engineering materials capable of undergoing inelastic deformation. This includes polymers, and composite materials with plastically deformable phases. Ceramics and inter-metallics can also exhibit fatigue crack nucleation and growth under certain circumstances. However, the irreversible processes in "fatigue" of ceramics are typically not associated with dislocation-based plasticity as in metals, but are related to local micro-cracking, frictional sliding, and particle detachment in the crack-tip process zones in such materials (Ritchie, 1999). In this chapter *we confine our attention to a discussion of fatigue in metallic materials.*

Continuum Mechanics of Solids. Lallit Anand and Sanjay Govindjee, Oxford University Press (2020).

DOI: 10.1093/oso/9780198864721.001.0001

There are two principal methodologies for designing against fatigue failure of components:

(i) *a defect-free approach*, and

(ii) *a defect-tolerant approach*.

In the defect-free approach *no crack-like defects are presumed to pre-exist*. That is, the crack size a is taken to be zero initially. Figure 28.1 shows a schematic of the behavior of the crack length a versus number of cycles of load N for an initially uncracked component. The **number of cycles to fatigue failure** of the component is denoted by N_f. The total number of cycles to failure may be decomposed as

$$N_f = N_i + N_p, \tag{28.1.1}$$

where

- N_i is the number of cycles required to initiate a fatigue crack, and N_p is the number of cycles required to propagate a crack to final fracture after it has initiated.

Although the total fatigue life consists of an initiation life and a propagation life, in the defect-free approach fatigue failure is said to have occurred when a crack has initiated, and since N_p is usually very much smaller than N_i, the fatigue life N_f is approximated as

$$N_f \approx N_i.$$

The initiation life N_i in metals corresponds to the development of a crack which is of a size which is substantially larger than the underlying microstructural grain size. Typically, a fatigue crack is said to have "initiated" when it is readily visible to the naked eye, that is

$$a_i \approx 0.5 \text{ to } 1\,\text{mm}.$$

The defect-free methodology is mostly used to design small components which are *not safety critical*.

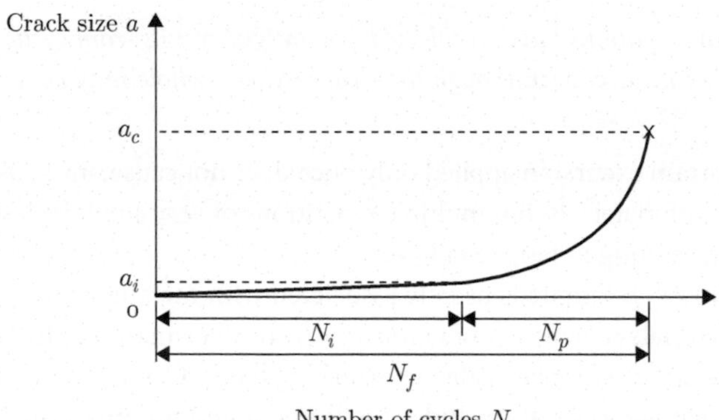

Fig. 28.1 Schematic of crack length a versus number of cycles N in an initially uncracked component.

In contrast to the defect-free methodology, the *defect-tolerant approach* is used in situations where the potential costs of a structural fatigue failure in terms of human life and monetary value is high. The defect-tolerant approach is based on:

1. The assumption that all fabricated components and structures contain a pre-existing population of cracks of an initial size a_i.
   - This initial size is taken to be the crack size a_d which is the largest crack size that can escape detection by non-destructive evaluation (NDE) methods; cf. Fig. 28.2. Also in this approach one assumes that
   $$N_f \approx N_p.$$
2. The requirement that none of these presumed pre-existing cracks be permitted to grow to a critical size during the expected service life of the part or structure. Normally this requires the selection of inspection intervals within the service life.

The major aim of the defect-tolerant approach to fatigue is to reliably predict the growth of pre-existing cracks of specified initial size a_i, shape, location, and orientation in a structure subjected to prescribed cyclic loadings. Provided that this goal can be achieved, then inspection and service intervals can be established such that cracks should be readily detectable well before they have grown to near critical size, a_c.

The defect-tolerant approach is typically used in the design and maintenance of large, fabricated structures such as aircraft, ships, and pressure vessels, where welds are likely sites for initial defects, and the large size of the components may permit substantial subcritical crack growth, so that the enlarged defect can be detected and repaired or replaced well before it reaches a critical dimension.

We discuss the defect-free approach and the defect-tolerant approach for designing against fatigue failures in greater detail in what follows.

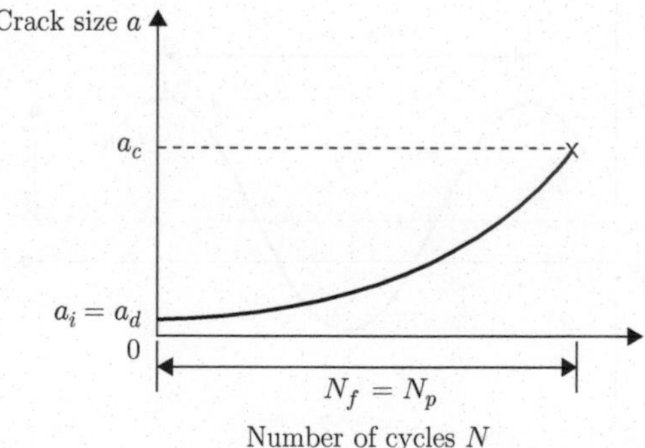

Fig. 28.2 Schematic of crack length a versus number of cycles N in a component with an initial crack size a_i. The initial crack size is taken to be the largest crack size a_d that can escape detection by the NDE technique being employed.

28.2 Defect-free approach

Although the study of the inelastic micromechanisms leading to *initiation of fatigue cracks* in metallic materials has gone on for over a century, and much has been learned (cf. e.g., Suresh, 1998; Pineau et al., 2016), there are many fundamental issues which still remain to be understood and resolved. At present there is no widely accepted continuum-level theory, based on the underlying complex micromechanisms of inelasticity and damage, which is able to predict the initiation of fatigue cracks. Nevertheless there are several important empirical correlations between the number of cycles and magnitudes of applied cyclic stresses or strains which lead to the initiation of fatigue cracks. We discuss some of these basic empirical correlations in this chapter.

28.2.1 S-N curves

The earliest and most common approach in the defect-free methodology for designing against fatigue failure is to use "S-N" curves for a given material. Consider a cylindrical specimen under the action of a time-varying axial stress $\sigma(t)$. With respect to Fig. 28.3, the quantities

$$\Delta\sigma = \sigma_{\max} - \sigma_{\min}, \qquad \sigma_a = \frac{\Delta\sigma}{2}, \qquad \text{and} \qquad \sigma_m = \frac{1}{2}\left(\sigma_{\max} + \sigma_{\min}\right), \tag{28.2.1}$$

are called the **stress range**, the **stress amplitude**, and **mean stress**, respectively.

Consider stress amplitudes σ_a below the tensile strength σ_{UTS} of the material.[1] For a given stress amplitude σ_a the specimen is cycled until a fatigue crack initiates and the number of cycles N_f to initiate such a crack are recorded as a data point (σ_a, N_f). The (σ_a, N_f) data obtained from conducting experiments at various values of σ_a are usually plotted on semi-log

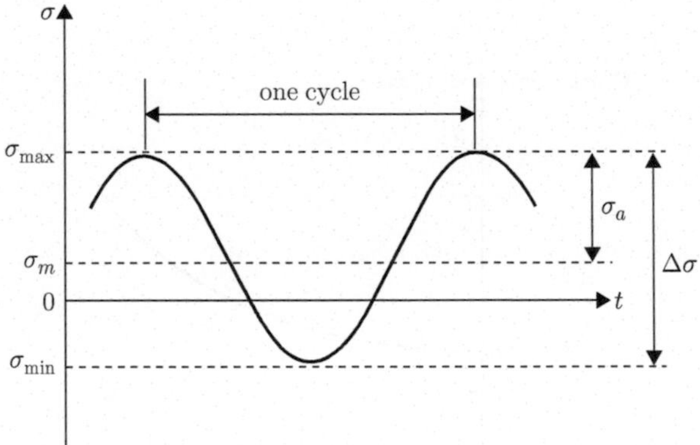

Fig. 28.3 Fatigue testing under constant amplitude stress cycling.

[1] The **tensile strength**, σ_{UTS}, is defined as the maximum value of the stress in an engineering-stress versus engineering-strain curve; cf. Fig. 20.1.

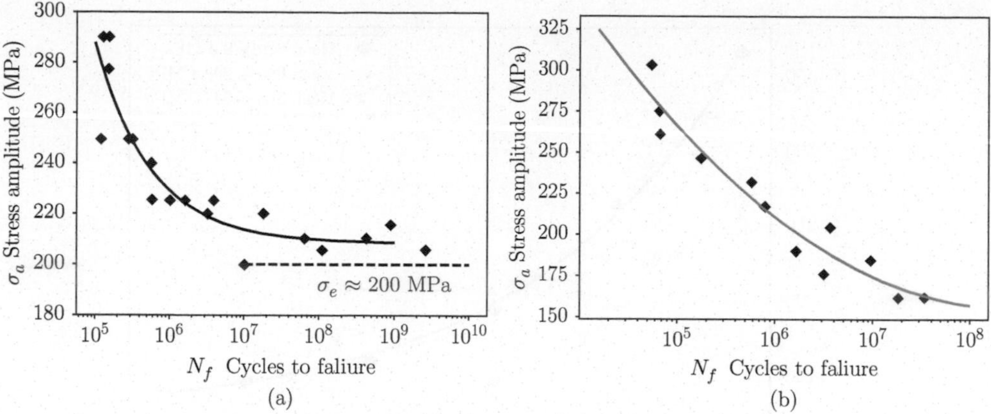

Fig. 28.4 Stress amplitude σ_a versus the number of cycles to failure N_f plotted on a semi-logarithmic scale: (a) For a cast iron; data from Wang et al. (1999, Fig. 1(top)). (b) For an aluminum alloy; data from MacGregor and Grossman (1952, Fig. 7(c)).

scales, and are called **S-N curves.**[2] Figures 28.4 left and right show S-N curves for a spheroidal graphite (SG) cast iron and a 75S-T6 aluminum alloy, respectively.

Note that the S-N curve for the ferrous alloy exhibits a stress amplitude level, denoted as σ_e, below which the material has an "infinite" life. The stress amplitude level σ_e is called the **endurance limit** for the alloy. *Most ferrous alloys exhibit an endurance limit.*

In contrast, the 75S-T6 aluminum alloy, like most other non-ferrous alloys, *does not show an endurance limit.* However, for engineering purposes, a **pseudo-endurance limit** for non-ferrous materials is often defined as the stress amplitude corresponding to a fatigue life of 10^7 cycles.

Thus, if the local stress amplitude σ_a is known (or can be calculated) at a notch where a potential fatigue crack is expected to nucleate, then the number of cycles N_f that it would take to initiate a crack can be read off from an S-N curve for that material. Conversely, if say one desires a certain fatigue life, then the allowable stress amplitude may be determined. To prevent fatigue failure, the local stress amplitude may be controlled to lie below the endurance limit of the material.

EXAMPLE 28.1 Pseudo-endurance limits

Figure 28.5 shows S-N curves obtained from smooth un-notched specimens of 7075-T6 Al at various levels of axial mean stress. The aluminum alloy has an ultimate tensile strength of $\sigma_{UTS} \approx 565$ MPa and a tensile yield strength of $\sigma_y \approx 465$ MPa.

1. Estimate the pseudo-endurance limit for this material at zero mean stress.
2. Estimate the pseudo-endurance limit for this material at a mean stress of 345 MPa.

[2] It is for historical reasons that the stress-amplitude σ_a is denoted by "S".

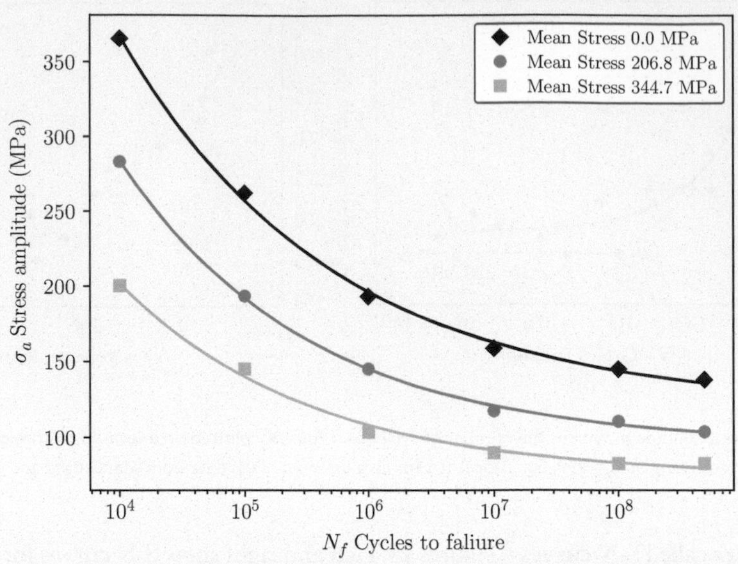

Fig. 28.5 S-N curves for 7075-T6 Al alloy. Effects of mean stress; data from Howell and Miller (1955, Table IV).

Solution:

1. The pseudo-endurance limit for this material at zero mean stress is $\sigma_e \approx 165\,\text{MPa}$.
2. The pseudo-endurance limit for this material at a mean stress of 345 MPa is $\sigma_e \approx 80\,\text{MPa}$. A positive mean stress substantially reduces the endurance limit of this aluminum alloy.

28.2.2 Strain-life approach to design against fatigue failure

Next we turn to another other methodology for defect-free fatigue analysis, viz. the Coffin (1954)–Manson (1953) "strain-life" approach (c.f. e.g., Stephens et al., 2001, and references therein).

The cyclic stress-strain response of metals is usually obtained by cycling cylindrical specimens between certain maximum and minimum axial strain levels, $\epsilon \in [-\epsilon^-, \epsilon^+]$. The stress-strain response observed during cyclic straining is quite different from that observed in monotonic straining, and depending on the initial state of the material and the testing conditions, a material may either cyclically harden, or cyclically soften. However, the cyclic stress amplitude often *saturates* to an essentially constant value after a number of strain reversals. Such stable cyclic behavior of metals can be described in terms of the stress-strain hysteresis loop illustrated in Fig. 28.6.

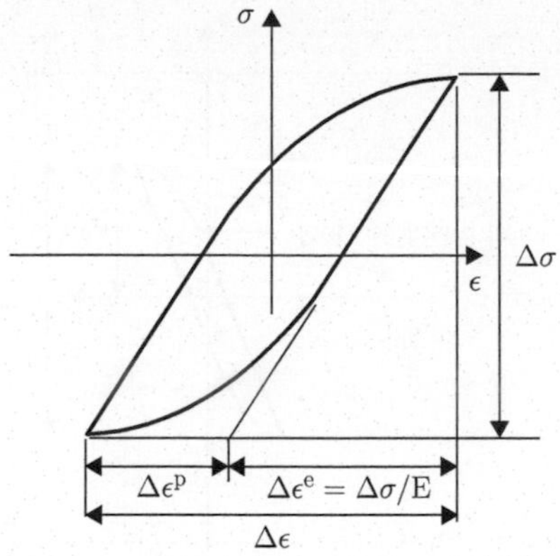

Fig. 28.6 Schematic of a stable stress-strain hysteresis loop.

With respect to this figure, the quantities

$$\Delta\epsilon, \quad \Delta\epsilon^{\mathrm{e}}, \quad \text{and} \quad \Delta\epsilon^{\mathrm{p}}$$

denote the **total strain range, elastic strain range**, and **plastic strain range**, respectively. Clearly,

$$\Delta\epsilon = \Delta\epsilon^{\mathrm{e}} + \Delta\epsilon^{\mathrm{p}}, \quad \text{with} \quad \Delta\epsilon^{\mathrm{e}} = \frac{\Delta\sigma}{E},$$

where $\Delta\sigma$ is the stress range, and E is the Young's modulus. Let

$$\epsilon_a = \frac{\Delta\epsilon}{2}, \quad \epsilon_a^{\mathrm{e}} = \frac{\Delta\epsilon^{\mathrm{e}}}{2}, \quad \text{and} \quad \epsilon_a^{\mathrm{p}} = \frac{\Delta\epsilon^{\mathrm{p}}}{2}$$

denote the **strain amplitude**, the **elastic strain amplitude**, and the **plastic strain amplitude**, respectively. Then

$$\boxed{\epsilon_a = \frac{\sigma_a}{E} + \epsilon_a^{\mathrm{p}}.} \tag{28.2.2}$$

High-cycle fatigue. Basquin's relation

In the high-cycle fatigue regime, i.e. $N_f \gtrsim 10^4$, the stress amplitude σ_a is typically below the macroscopic yield strength σ_y of the material, but because of **microscale plasticity** one still observes small stabilized hysteresis loops; cf. Fig. 28.7. The area within the loop is the energy per unit volume dissipated as plastic work during one cycle. In the high-cycle regime this energy is small, and it decreases as the stress amplitude decreases.

With $2N_f$ denoting **the number of reversals to failure**, Basquin (1910) observed that in the high-cycle fatigue regime the $(\sigma_a, 2N_f)$ data may be approximated by a power-law relation,

$$\boxed{\sigma_a = \sigma_f' \cdot (2N_f)^b.} \tag{28.2.3}$$

The material parameters σ_f' and b are high-cycle fatigue properties of a material. They are called the **fatigue strength coefficient** and the **fatigue strength exponent**, respectively.

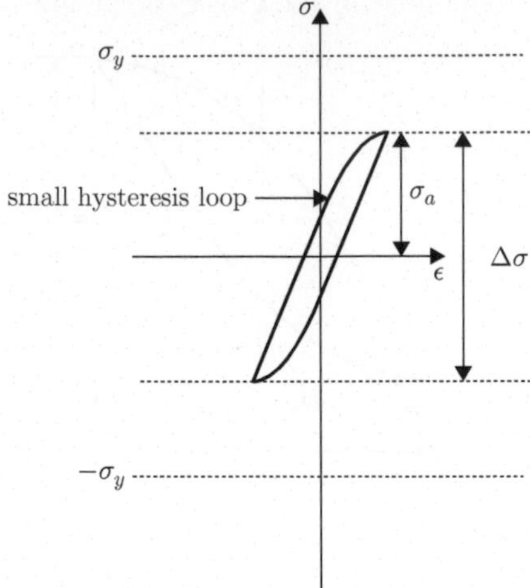

Fig. 28.7 Schematic of a small hysteresis loop produced in the high-cycle regime when the stress amplitude σ_a is typically less than the monotonic yield strength σ_y of the material.

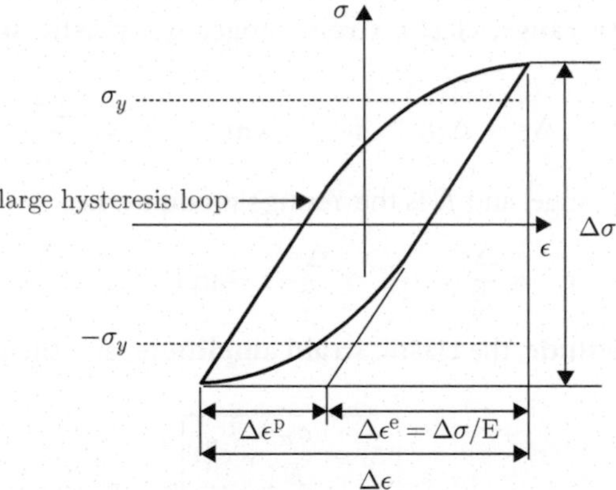

Fig. 28.8 Schematic of a large hysteresis loop produced in the low-cycle regime when the stress amplitude σ_a is typically larger than the monotonic yield strength σ_y of the material.

Low-cycle fatigue. Coffin–Manson relation

If the stress amplitude increases beyond the yield strength σ_y, then the stabilized hysteresis loops become large, cf. Fig. 28.8, and the fatigue life typically decreases below $\approx 10^4$ cycles, which is called the "low-cycle fatigue" regime.

In the mid-1950s, Coffin (1954) and Manson (1953) observed that in the low-cycle fatigue regime the $(\epsilon_a^{\rm p}, 2N_f)$ data may be approximated by a power-law relation,

$$\boxed{\epsilon_a^{\rm p} = \epsilon_f'(2N_f)^{\rm c}.} \tag{28.2.4}$$

The material parameters ϵ_f' and c are the low-cycle fatigue properties for a material. They are called the **fatigue ductility coefficient** and the **fatigue ductility exponent**, respectively.

Strain-life equation for both high-cycle and low-cycle fatigue

One can combine the equation for the cyclic strain amplitude,

$$\epsilon_a = \frac{\sigma_a}{E} + \epsilon_a^{\rm p}, \tag{28.2.5}$$

with the fatigue life relations for the high-cycle regime (28.2.3) and the low-cycle regime (28.2.4), to obtain

$$\boxed{\epsilon_a = \frac{\sigma'_f}{E}(2N_f)^b + \epsilon'_f(2N_f)^c.} \tag{28.2.6}$$

Equation (28.2.6) is the basis for the strain-life approach to defect-free design against fatigue failure. Its use requires separate analysis methods, such as finite element methods, to determine strain amplitude values, ϵ_a, in critical regions of a structural body.

Representative values of fatigue properties $\left(\sigma'_f, b, \epsilon'_f, c\right)$ for some ductile metallic materials are listed in Table 28.1.

The **transition fatigue life**, N_t, is defined as the life at which the corresponding cyclic elastic strain range equals the corresponding cyclic plastic strain range. Dividing (28.2.3) by E to obtain $\epsilon_a^{\rm e}$ and setting this equal to $\epsilon_a^{\rm p}$ in equation (28.2.4), we find that the transition fatigue life is given by the expression

$$N_t = \frac{1}{2}\left(\frac{E\,\epsilon'_f}{\sigma'_f}\right)^{\dfrac{1}{(b-c)}}. \tag{28.2.7}$$

Table 28.1 Representative values of fatigue properties $(\sigma'_f, b, \epsilon'_f, c)$ for some ductile metallic materials.

	Strain-Life Properties			
Material	σ'_f MPa	b	ϵ'_f	c
Steel				
SAE 1020	896	−0.12	0.41	−0.51
SAE 1040	1540	−0.14	0.61	−0.57
Man-Ten	1089	−0.115	0.86	−0.65
RQC-100	938	−0.0648	0.66	−0.69
SAE 4340	1655	−0.076	0.73	−0.62
Aluminum				
2024-T351	1100	−0.124	0.22	−0.59
2024-T4	1015	−0.11	0.21	−0.52
7075-T6	1315	−0.126	0.19	−0.52

At short lives, $N_f < N_t$, plastic strain will predominate and ductility will control the fatigue performance. At long life, $N_f > N_t$, the plastic strain will be far smaller than the elastic strain, and strength will control the fatigue performance.

Mean stress effects on fatigue

The preceding section described relations between stress, plastic strain, and total strain amplitudes and the corresponding fatigue life under conditions of fully reversed loading with zero mean stress; the mean stress σ_m is defined in $(28.2.1)_3$. Mean stress effects on fatigue life are most important at long lives where cyclic plastic straining is small.[3] Many empirical models have been proposed to account for long-life mean stress effects on fatigue. Here we present the simple and widely used assumption that a tensile mean stress, $\sigma_m > 0$ reduces the effective fatigue strength coefficient in (28.2.3), while a compressive mean stress has no effect (Morrow, 1968). The modified form of the Basquin relation then becomes

$$\sigma_a = \begin{cases} \sigma'_f \left(1 - \dfrac{\sigma_m}{\sigma'_f}\right) (2N_f)^b & \sigma_m > 0, \\ \sigma'_f \,(2N_f)^b & \sigma_m \leq 0. \end{cases} \tag{28.2.8}$$

When this equation is inserted into (28.2.6), the governing strain-life equation becomes

$$\boxed{\epsilon_a = \begin{cases} \dfrac{\sigma'_f}{E} \left(1 - \dfrac{\sigma_m}{\sigma'_f}\right) (2N_f)^b + \epsilon'_f \,(2N_f)^c & \sigma_m > 0, \\ \dfrac{\sigma'_f}{E} (2N_f)^b + \epsilon'_f \,(2N_f)^c & \sigma_m \leq 0. \end{cases}} \tag{28.2.9}$$

Cumulative fatigue damage. Miner's rule

The strain-life relations developed in the preceding sections are for **constant amplitude straining** throughout the fatigue life of a component. In order to apply information of this type to the analysis of the fatigue behavior of structural elements which are subjected to other than uniform cyclic straining, it is necessary to develop a formalism for generalization of constant-amplitude life data to variable-amplitude loading. The earliest, and still the most successful, generalizing concept in defect-free fatigue analysis is that of *linear cumulative fatigue damage*, which was first introduced by Palmgren (1924) and by Miner (1945).

Consider a "block" of cyclic strain history, indexed by $i \in [1, M]$, in which n_i cycles of strain amplitude ϵ_{ai} are applied under constant-amplitude conditions. This strain amplitude would result in a fatigue life of N_{fi} cycles. The ith damage increment is then defined by

$$d_i \equiv \frac{n_i}{N_{fi}} \qquad (0 \leq d_i \leq 1). \tag{28.2.10}$$

[3] At low lives, with significant plastic straining, mean stresses quickly relax under strain-controlled limits, or lead to cyclic ratcheting and "run-away" if it is attempted to enforce unequal stress limits.

The assumption of *linear cumulative fatigue damage* states that fatigue failure occurs when

$$\boxed{\sum_{i=1}^{M} d_i = \sum_{i=1}^{M} \frac{n_i}{N_{fi}} = 1, \qquad i = 1, \ldots, M.} \tag{28.2.11}$$

An important limitation of the cumulative damage rule, as presented here, is that there is no explicit accounting for sequence effects on fatigue. However, due to various features of the cyclic stress-strain curve, the gradual change from the monotonic to the cyclic stress-strain curve, and other factors, there can be a block-sequence effect on fatigue life. These effects may be accounted for, to some degree, in computer-based applications of cumulative damage which break arbitrary loading histories into single, sequential reversals, and which simultaneously follow the cyclic stress-strain curve along each segment of the loading history. However, a discussion of these heuristic developments is beyond the scope of this book.

28.3 **Defect-tolerant approach**

In the defect-tolerant approach to fatigue, the structure is *assumed* to have a pre-existing crack of initial size a_i located at the most highly stressed location. The initial size can be

- either the largest pre-existing crack detected by a NDE technique;
- or if no crack has been detected, then the initial crack size is *assigned* to be a_d, where a_d is the minimum crack size which can be reliably detected by the NDE technology employed.

The latter assumption is the more common. The major aim of a defect-tolerant approach to fatigue is to predict reliably the growth of pre-existing cracks of specified initial size a_i, shape, location, and orientation in a structure subjected to prescribed cyclic loading. Provided that this goal can be achieved, then inspection and service intervals can be established so that cracks should be readily detectable well before they have grown to near critical size, a_c. The critical crack size a_c is determined as:

$$\underbrace{Q\,\sigma_{\max}\sqrt{\pi a_c}}_{K_{\mathrm{I}}} = K_{\mathrm{Ic}} \qquad \Rightarrow \qquad a_c = \frac{1}{\pi}\left(\frac{K_{\mathrm{Ic}}}{Q\,\sigma_{\max}}\right)^2.$$

28.3.1 **Fatigue crack growth**

To obtain a fatigue crack growth curve for a particular material, it is necessary to establish reliable fatigue crack growth rate data. Typically, a cracked test specimen is subjected to a constant amplitude cyclic stress range $\Delta\sigma \equiv \sigma_{\max} - \sigma_{\min}$ and two curves: (i) the crack length a versus the number of cycles, and (ii) ΔK_{I} versus the crack length a are obtained, where

$$\Delta K_{\mathrm{I}} = K_{\mathrm{Imax}} - K_{\mathrm{Imin}} = Q\,\sigma_{\max}\sqrt{\pi a} - Q\sigma_{\min}\sqrt{\pi a}, \tag{28.3.1}$$

is the range of the stress intensity factor over one cycle of load, while crack length is essentially constant at value "a". The *crack growth rate* is defined as

$$\frac{da}{dN} \stackrel{\text{def}}{=} \text{slope of crack growth curve at crack length } a$$

$$\equiv \text{crack extension } \Delta a \text{ of a crack length } a \text{ in one cycle.}$$

Thus, from experimentally determined curves of a vs N and knowledge of applied loads and geometry of the test specimen, one can construct da/dN vs a and ΔK_{I} vs a curves. On cross-plotting from these curves to eliminate the variable a, one can construct $\log(da/dN)$ versus $\log(\Delta K_{\mathrm{I}})$ curves.

It was found by Paris (1962) from experimental data, for a given load ratio $R \equiv \sigma_{\min}/\sigma_{\max} = K_{\mathrm{I}\min}/K_{\mathrm{I}\max}$, that the plots of $\log(da/dN)$ versus $\log(\Delta K_{\mathrm{I}})$ obtained from various different specimen-types **superpose on one another to give a single curve for a given material**. Figure 28.9 shows such fatigue crack growth rates over a wide range of stress intensities for ductile pressure vessel A533 steel.

- The fact that such a curve can, to a good approximation, be considered to be a material curve, independent of geometrical factors, is of great practical importance: *The results obtained from simple laboratory specimens can be directly applied to real service conditions, provided the stress intensity factor range in the latter case can be determined.*

A schematic of such a curve is shown in Fig. 28.10. At a fixed R-ratio, the fatigue crack propagation behavior of metallic materials can be divided into three regimes A, B, and C.

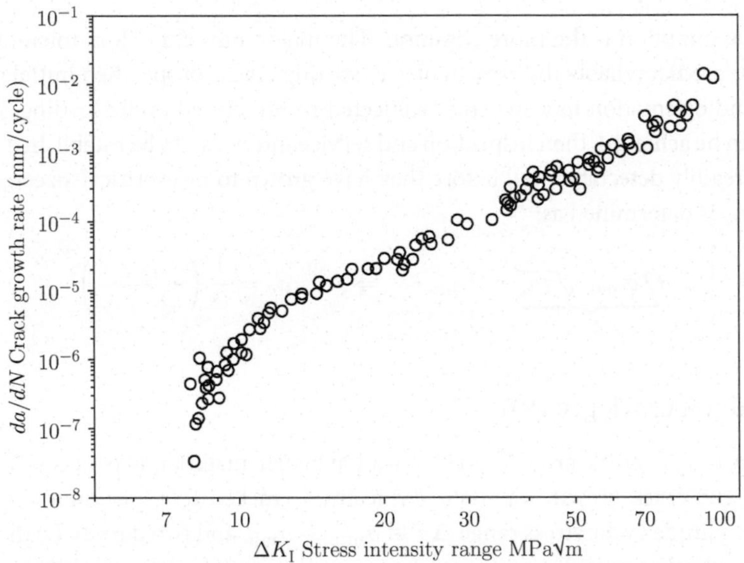

Fig. 28.9 Fatigue crack propagation rate (da/dN) with alternating stress intensity factor (ΔK_{I}) for a ductile pressure vessel steel with $R = 0.1$ at an ambient temperature of 23.9° C; data from (Paris et al., 1972, Fig. 1).

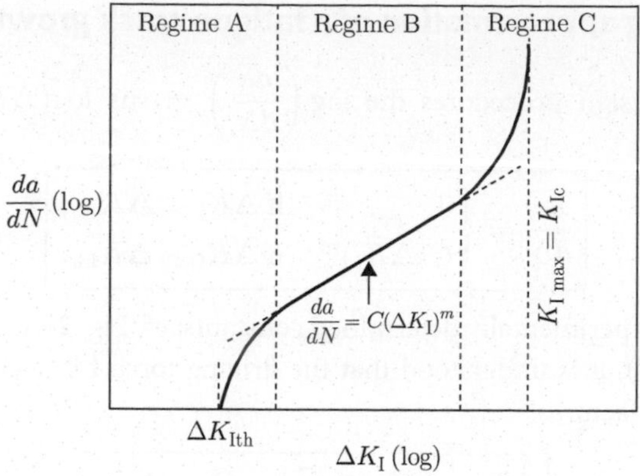

Fig. 28.10 Crack-growth-rate curve, plotting da/dN versus the cyclic stress intensity factor ΔK_{I} in a double-logarithmic plot. There are three characteristic regimes; regime B is described by the Paris law eq. (28.3.2).

Regime A

In this regime the crack growth rate is very low $(da/dN) \lesssim 10^{-9}$ m/cycle, and a crack appears dormant below a fatigue threshold, ΔK_{Ith}. The value of ΔK_{Ith} varies widely, but for many metallic materials it lies in the range $2\ \mathrm{MPa}\sqrt{\mathrm{m}} \lesssim \Delta K_{\mathrm{Ith}} \lesssim\ 10\ \mathrm{MPa}\sqrt{\mathrm{m}}$.

Regime B

In this regime crack growth rate is in the range $10^{-9}\ \lesssim\ (da/dN)\ \lesssim\ 10^{-6}$ m/cycle, and approximately obeys the power-law relation

$$\boxed{\frac{da}{dN} = C(\Delta K_{\mathrm{I}})^m.} \tag{28.3.2}$$

In this equation, C and m are experimentally determined material constants describing the straight-line portion of the $\log da/dN$ vs $\log \Delta K_{\mathrm{I}}$ curve. Over a broad spectrum of engineering alloys, the range of the dimensionless exponent m is $2 \lesssim m \lesssim 7$ with a typical value of $m \approx 4$.

The power law form (28.3.2) of the fatigue crack growth relation was proposed by Paris et al. (1961) and Paris (1962), and is often referred to as the "Paris law".

Regime C

In Regime C the stress levels are high, K_{Imax} approaches K_{Ic}, and the crack growth rates are very high, $(da/dN) \gtrsim 10^{-6}$ m/cycle. Consequently little fatigue crack growth life is involved. Region C has the least importance in most fatigue situations.

28.3.2 Engineering approximation of a fatigue crack growth curve

For crack-tolerant design procedures the $\log\left(\dfrac{da}{dN}\right)$ versus $\log(\Delta K_{\rm I})$ curve is approximated as

$$\boxed{\frac{da}{dN} = \begin{cases} 0 & \text{if } \Delta K_{\rm I} < \Delta K_{\rm Ith}, \\ C(\Delta K_{\rm I})^m & \text{if } \Delta K_{\rm I} \geq \Delta K_{\rm Ith}, \end{cases}} \tag{28.3.3}$$

where C and m are experimentally determined constants; cf. Fig. 28.11.

In applying (28.3.3), it is understood that the driving force for cyclic crack growth is the cyclic stress intensity factor

$$\boxed{\Delta K_{\rm I} = Q\Delta\sigma\sqrt{\pi a}; \qquad Q = \hat{Q}(a).} \tag{28.3.4}$$

This last expression is somewhat dimensionally misleading since Q is dimensionless, while $[a]$ = length. Rather, it is intended to remind us of the possible functional dependence of Q on variable crack length in a structure of fixed geometry (e.g., width w). In general, $Q = \hat{Q}(a/w)$, etc.

REMARK 28.1

In (28.3.3), (da/dN) has units of (m/cycle), and $(\Delta K_{\rm I})^m$ has units of $(\text{MPa}\sqrt{\text{m}})^m$. Hence, the constant

$$C \quad \text{has strange units of} \quad \frac{\text{m/cycle}}{(\text{MPa}\sqrt{\text{m}})^m}\,!$$

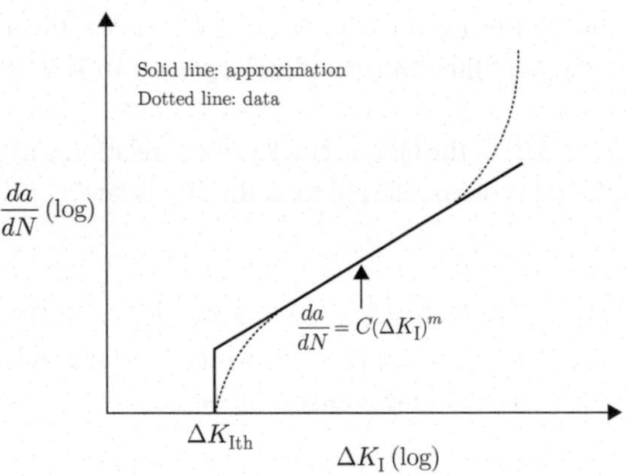

Fig. 28.11 Engineering approximation of a fatigue crack growth curve.

28.3.3 Integration of crack growth equation

By rearranging (28.3.3), we have the differential expression

$$dN = \frac{1}{C}\,\frac{da}{(\Delta K_{\rm I})^m},$$

which can be integrated (on the left with respect to N and on the right with respect to a) as

$$N_{a_i \to a_f} \equiv \int_0^{N_{a_i \to a_f}} dN = \frac{1}{C}\int_{a_i}^{a_f} \frac{da}{\left(\hat{Q}(a)\Delta\sigma\sqrt{\pi a}\right)^m}.$$

In writing the integrated form, we have emphasized that $N_{a_i \to a_f}$ is the number of cycles required to grow a fatigue crack from initial value $a = a_i$ to final crack length $a = a_f$ under the application of a cyclic stress range $\Delta\sigma$ in a material having power-law fatigue crack growth behavior. We have accounted for the dependence of $\Delta K_{\rm I}$ on a by substituting in the cyclic stress intensity factor $\Delta K_{\rm I} = Q\Delta\sigma\sqrt{\pi a}$ under the integral sign.

For constant $\Delta\sigma$,

$$N_{a_i \to a_f} = \frac{1}{C}\,\frac{1}{(\Delta\sigma\sqrt{\pi})^m}\int_{a_i}^{a_f} \frac{da}{\left(\hat{Q}(a)a^{1/2}\right)^m}. \tag{28.3.5}$$

In general, $\hat{Q}(a)$ is a complex function of the crack length, and it is usually necessary to perform the integration numerically. However, if Q is constant, independent of a, then (28.3.5) reduces to

$$N_{a_i \to a_f} = \frac{1}{C}\,\frac{1}{(Q\Delta\sigma\sqrt{\pi})^m}\int_{a_i}^{a_f} a^{-m/2}da. \tag{28.3.6}$$

Thus for constant $\Delta\sigma$ and constant Q, we obtain the following expressions for the number of cycles to grow a crack from its initial size a_i to a final size a_f:

- Case $m \neq 2$:

$$N_f = \frac{2}{(m-2)\,C\,(Q\Delta\sigma\sqrt{\pi})^m}\left[a_i^{(2-m)/2} - a_f^{(2-m)/2}\right] \quad \text{for } m \neq 2. \tag{28.3.7}$$

- Case $m = 2$:

$$N_f = \frac{1}{C}\,\frac{1}{(Q\Delta\sigma\sqrt{\pi})^2}\left[\ln\left(\frac{a_f}{a_i}\right)\right] \quad \text{for } m = 2. \tag{28.3.8}$$

For many additional details regarding fatigue, see Suresh (1998) and Rice et al. (1988).

EXAMPLE 28.2 Fatigue of a cylindrical pressure vessel

A large, capped-cylindrical pressure vessel is being designed for a chemical plant. This is a safety-critical situation where fracture of the pressure vessel during service could result in an explosion which could destroy the chemical plant, release toxic chemicals into the environment, and possibly result in loss of human life.

The radius of the pressure vessel is 2 meters, its length is 10 meters, and the wall thickness is 50 mm. During service the operating conditions will consist of a cyclic pressure history ranging between

$$p_{\text{min}} = 0.5\,\text{MPa and } p_{\text{max}} = 3.5\,\text{MPa}.$$

The material chosen to manufacture the pressure vessel is a high-quality steel with the following properties:

$$\sigma_y = 750\,\text{MPa}, \quad K_{\text{Ic}} = 70\,\text{MPa}\sqrt{\text{m}},$$

and

$$\frac{da}{dN} = C\,(\Delta K_I)^m, \quad \text{with} \quad C = 5 \times 10^{-12}\,\frac{\text{m/cycle}}{(\text{MPa}\sqrt{\text{m}})^m}, \quad \text{and} \quad m = 3.0.$$

1. Due to the safety-critical nature of the pressure vessel, it will be proof-tested with water prior to use. The proof pressure is chosen such that the maximum principal stress in the pressure vessel is equal to $0.60 \times \sigma_y$. What is the proof pressure?
2. The vessel survives the proof test. Based on this outcome presume that there might have been an initial flaw in the pressure vessel, but the stress-intensity caused by the proof-pressure was not sufficient to cause fracture of the pressure vessel. Assuming that a semicircular surface flaw with $Q = 0.65$ could have been present, estimate a value for the initial crack length a_i which escaped detection in the proof-test.
3. Estimate the life of the pressure vessel for the desired cyclic pressure history. The configuration correction factor Q actually varies in this problem, but for simplicity you may take it to remain constant $Q \approx 0.65$ over the fatigue life of the vessel.
4. Periodic inspection levels, N_p, may be defined as follows:

$$N_p = \frac{N_f}{X},$$

where $X \approx 3$ is a typical "safety factor". Would you recommend inspections? If so how often?

For purposes of this preliminary design, do not concern yourself with the effects of welds or other joints.

Solution:

1. Since the ratio of the wall thickness to the mean radius of the pressure vessel is

$$\frac{t}{r} = \frac{0.05}{2} = 0.025 \ll 1,$$

the vessel is thin-walled, and the non-zero stresses may be taken to be given by the standard thin-walled pressure vessel relations

$$\sigma_{\theta\theta} = \frac{pr}{t} \qquad \text{and} \qquad \sigma_{zz} = \frac{pr}{2t}. \tag{28.3.9}$$

The maximum principal stress is clearly the hoop stress $\sigma_{\theta\theta}$. Accordingly, the proof pressure can be determined by setting the maximum principal stress equal to $0.6\sigma_y$,

$$\frac{p_{\text{proof}}\, r}{t} = \sigma_{\text{proof}} = 0.6\sigma_y, \tag{28.3.10}$$

which gives

$$p_{\text{proof}} = \frac{0.6\,\sigma_y\, t}{r} = \frac{0.6 \times 750 \times 0.05}{2} = 11.25\,\text{MPa}. \tag{28.3.11}$$

2. Since the vessel survived the proof test,

$$K_{\text{I,proof}} < K_{\text{Ic}},$$

where

$$K_{\text{I,proof}} = Q\sigma_{\text{proof}}\sqrt{\pi a_i}. \tag{28.3.12}$$

However, in order to estimate a conservatively large value of a_i, we will take

$$K_{\text{I,proof}} = K_{\text{Ic}},$$

which gives

$$a_i = \frac{1}{\pi}\left(\frac{K_{\text{Ic}}}{Q\,\sigma_{\text{proof}}}\right)^2 = \frac{1}{\pi}\left(\frac{70}{0.65 \times 0.6 \times 750}\right)^2 = 18\,\text{mm}. \tag{28.3.13}$$

3. To estimate the fatigue life, we first calculate the maximum and minimum principal stresses to which the presumed initial crack will be subjected,

$$\begin{aligned} \sigma_{\max} &= \frac{p_{\max}\, r}{t} = \frac{3.5 \times 2}{0.05} = 140\,\text{MPa}, \\ \sigma_{\min} &= \frac{p_{\min}\, r}{t} = \frac{0.5 \times 2}{0.05} = 20\,\text{MPa}, \\ \Delta\sigma &= 120\,\text{MPa}, \end{aligned} \tag{28.3.14}$$

and also calculate the critical crack length at which fracture will occur

$$a_c = \frac{1}{\pi}\left(\frac{K_{\text{Ic}}}{Q\,\sigma_{\max}}\right)^2 = \frac{1}{\pi}\left(\frac{70}{0.65 \times 140}\right)^2 = 188\,\text{mm}. \tag{28.3.15}$$

- **Note that** $a_c > t$, which means that a growing fatigue crack will propagate through the thickness of the vessel prior to fracture. Once the crack propagates through the thickness of the wall of vessel, the vessel will leak. This is a desirable condition called "leak before break".

- The life of the vessel before it leaks is determined by $a_f = t$.

For constant $\Delta\sigma$ and Q, and for $m \neq 2$, the integrated form of the fatigue crack growth equation is

$$N_f = \frac{2}{(m-2)\,C\,(Q\Delta\sigma\sqrt{\pi})^m}\left[a_i^{(2-m)/2} - a_f^{(2-m)/2}\right]. \tag{28.3.16}$$

Substituting in appropriate numerical values we obtain

$$N_f = 451,310 \text{ cycles.} \tag{28.3.17}$$

4. It is difficult to inspect large pressure vessels, and the vessel already satisfies the "leak-before-break" criterion. Hence, periodic inspections may not be necessary. However, for a conservative maintenance practice against fatigue failure, if inspection is possible, then for $X = 3$ a possible periodic inspection interval is every

$$N_p = \frac{N_f}{3} = \frac{451,310}{3} \approx 150,000 \text{ cycles.} \tag{28.3.18}$$

PART X

Linear viscoelasticity

29 Linear viscoelasticity

29.1 Introduction

Many structural components undergoing small deformations are reasonably well-described by the stress-strain relation of linear elasticity. Limiting our attention to one-space-dimension, let $\sigma(t)$ denote the stress, and $\epsilon(t)$ the corresponding strain at time t. Recall that for a *linear elastic material* the stress is linearly related to the strain by

$$\sigma(t) = E\epsilon(t), \tag{29.1.1}$$

where E is the *Young's modulus of elasticity*. This constitutive equation for linear elastic materials may also be written in inverted form as

$$\epsilon(t) = J\sigma(t), \qquad \text{where} \qquad J \overset{\text{def}}{=} 1/E \tag{29.1.2}$$

is the *elastic compliance*. Although we have introduced time t as an argument for the strain ϵ, and the stress σ, the stress-strain relation for an elastic material is both *time-independent* and *rate-independent*.

In contrast to elastic materials, a linear viscous fluid obeys the constitutive equation

$$\sigma(t) = \eta\dot{\epsilon}(t), \tag{29.1.3}$$

where η is the *viscosity*, and for Newtonian fluids the stress depends *linearly on the strain-rate*.[1]

In reality the constitutive response of most solid materials deviates from linear elasticity in various ways; a major deviation is when the material exhibits both elastic as well as viscous-like characteristics, and such a response is called **viscoelastic**. Broadly speaking,

- *viscoelastic materials are those for which the relationship between stress and strain depends on time.*

Some phenomena exhibited by viscoelastic materials are:

- if the strain is held constant, the stress decreases with time—**stress relaxation**;
- if the stress is held constant, the strain increases with time—**creep**;
- the stress-strain curve depends on the rate of application of the strain—**strain-rate sensitivity**;

---

[1] Note, viscosity is usually defined in shear, but here for notational convenience, we define it in simple tension-compression.

Continuum Mechanics of Solids. Lallit Anand and Sanjay Govindjee, Oxford University Press (2020).

DOI: 10.1093/oso/9780198864721.001.0001

- if a cyclic stress is applied, then a **phase lag** occurs in the strain response, resulting in **hysteresis**, and leading to a dissipation of energy.

Most engineering materials exhibit some type of viscoelastic response. Although the behavior of common metals such as steels and aluminum alloys at small strains and room temperature does not deviate much from linear elasticity, synthetic engineering polymers and most natural biopolymers display significant viscoelastic effects. In some applications, even a small viscoelastic response can have significant engineering impact. The design and analysis of components made from viscoelastic materials must therefore account for their viscoelastic behavior.

- In what follows we discuss the one-dimensional theory of *linear viscoelasticity*, whose range of validity is for small levels of strain, typically $\epsilon \lesssim 3\%$ to 5%. We will subsequently extend the theory to the three-dimensional setting.

29.2 **Stress-relaxation and creep**

The phenomenon of viscoelasticity is best illustrated by *stress-relaxation* and *creep experiments.* In describing these experiments we will use the mathematical notion of the *Heaviside step function*, $h(t)$, defined by

$$h(t) = \begin{cases} 0 \text{ for } t \le 0, \\ 1 \text{ for } t > 0, \end{cases} \tag{29.2.1}$$

and sketched in Fig. 29.1(a).

We will also need the time-derivative of the Heaviside step function—a function that is zero everywhere except at time $t = 0$. To understand the derivative of $h(t)$ at time zero, one can loosely think of the Heaviside step function as the limiting case of the continuous function $f(t)$ sketched in Fig. 29.1(b); that is,

$$h(t) = \lim_{\tau \to 0} f(t).$$

The derivative of the function $f(t)$ with respect to time t is shown in Fig. 29.2(a). The derivative $\dot{f}(t)$ is zero, except in the interval $[-\tau, \tau]$ where it takes the value $1/2\tau$. Observe that for any value of $\tau > 0$ the area under the graph of $\dot{f}(t)$, the rectangle in Fig. 29.2(a), will be unity. Leaving aside technical details, if we let $\tau \to 0$, then this rectangle will degenerate into a spike, infinitely thin and infinitely high, but still of unit area. It represents a highly singular function $\delta(t) \equiv \dot{h}(t)$, called the *Dirac delta function* and defined by the properties

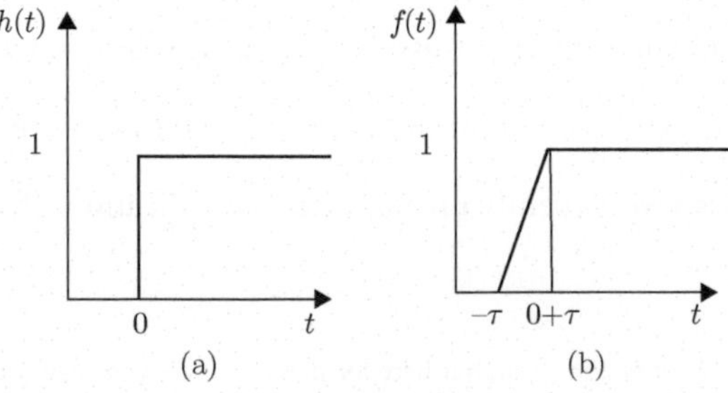

Fig. 29.1 (a) Heaviside step function. (b) Derivation of the Heaviside step function: $h(t) = \lim_{\tau \to 0} f(t)$.

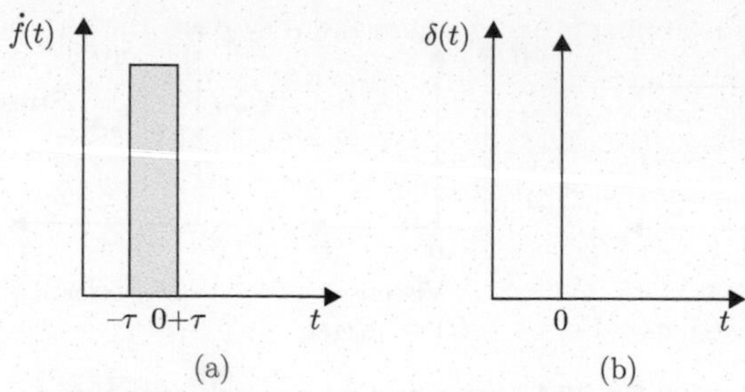

Fig. 29.2 (a) Derivation of the Dirac delta step function: $\delta(t) = \lim_{\tau \to 0} \dot{f}(t)$. (b) Dirac delta function.

$$\delta(t) = \begin{cases} 0 & \text{for } t \neq 0, \\ \infty & \text{for } t = 0, \end{cases} \qquad \text{where} \qquad \int_{-\infty}^{\infty} \delta(t)dt = \int_{0^-}^{0^+} \delta(t)dt = 1, \tag{29.2.2}$$

and for any function $g(t)$, continuous at $t = 0$,

$$\int_{-\infty}^{\infty} g(t)\delta(t)dt = \int_{0^-}^{0^+} g(t)\delta(t)dt = g(0). \tag{29.2.3}$$

The Dirac delta function is sketched in Fig. 29.2(b).

29.2.1 **Stress-relaxation**

Now consider a strain input of the form

$$\epsilon(t) = \epsilon_0 h(t), \qquad \dot{\epsilon}(t) = \epsilon_0\, \delta(t).$$

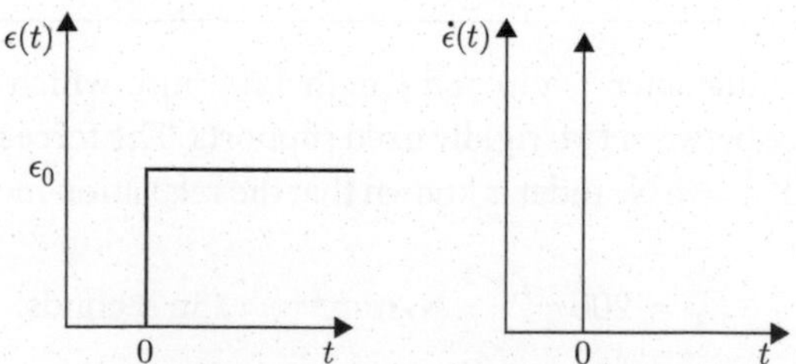

Fig. 29.3 Strain input in a stress-relaxation test.

For the strain input of Fig. 29.3 the stress output corresponding to elastic, viscous, and viscoelastic materials is sketched in Fig. 29.4. Basic features of the stress response for a viscoelastic material are that the largest value of stress $\sigma(0^+)$ is recorded just after applying the strain at $t = 0^+$, with a smooth, steady decay to a lower value $\sigma(\infty)$ for longer times.

For viscoelastic materials, the stress-response function, in response to a *unit* Heaviside strain history is called the **stress-relaxation function**

$$E_r(t) \stackrel{\text{def}}{=} \frac{\sigma(t)}{\epsilon_0},$$

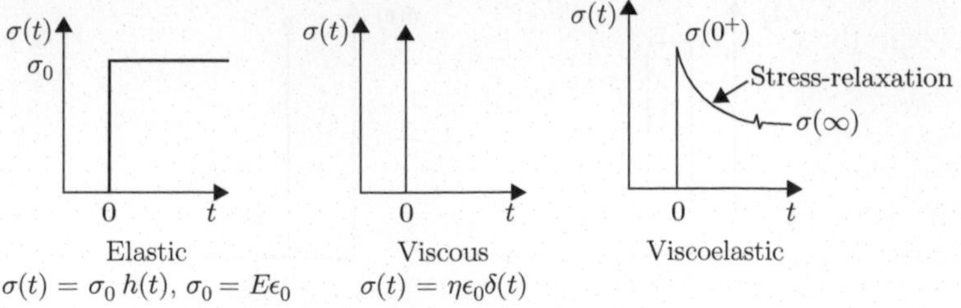

Fig. 29.4 Stress output in a stress-relaxation test.

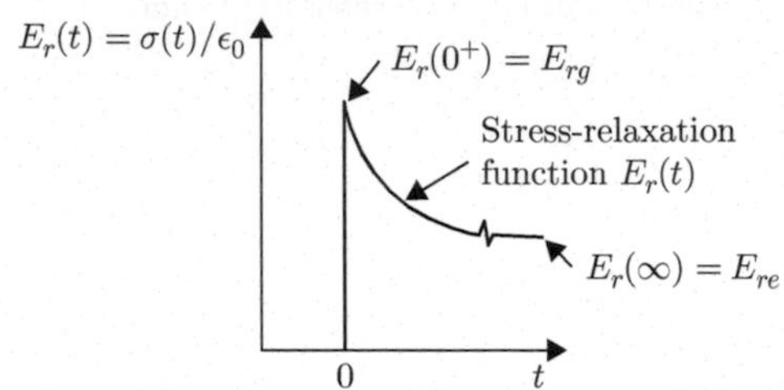

Fig. 29.5 Stress-relaxation function $E_r(t)$ for a viscoelastic material. The value $E_r(0^+)$ is denoted by E_{rg} and called the glassy relaxation modulus, while the value $E_r(\infty)$ is denoted by E_{re} and called the equilibrium relaxation modulus.

and is sketched in Fig. 29.5. The short-time value of this function is called its "glassy" value, $E_r(0^+) \equiv E_{rg}$. The long-time value of this function is called its "equilibrium" value, $E_r(\infty) \equiv E_{re}$.

EXAMPLE 29.1 Stress relaxation in a string

Consider a string of diameter 1 mm and length 1000 mm which is suddenly stretched and clamped in place between two rigidly fixed supports. The force in the string is initially measured to be $F(0^+) = 5\,\text{N}$, and it is known that the relaxation modulus for the material is given by

$$E_r = 200\, e^{-t^{1/8}}\ \text{N/mm}^2 \qquad t \text{ in seconds.} \tag{29.2.4}$$

Determine the initial strain in the string, $\epsilon(0^+)$, and the force in the string after 1 week.

First note that the initial stress in the string is given by

$$\sigma(0^+) = F(0^+)/A = E_{rg}\epsilon(0^+),$$

where $A = \pi 1^2/4 = \pi/4\,\text{mm}^2$ and $E_{rg} = 200\,\text{N/mm}^2$. Thus the initial strain is given by

$$\epsilon(0^+) = \frac{5\,\text{N}}{\pi/4\,\text{mm}^2 \times 200\,\text{N/mm}^2} = 0.032.$$

This implies a distance of 1032 mm between the fixed supports.

The force in the string at a time $t > 0$ is given by

$$F(t) = \sigma(t)\,A = AE_r(t)\epsilon(0^+).$$

So for $t = 1$ week $= 604800$ s,

$$F(604800) = \frac{\pi}{4} \times 200 \times e^{-(604800)^{1/8}} \times 0.032 = 0.026\ \text{N}.$$

29.2.2 **Creep**

Next, consider a stress input of the form

$$\sigma(t) = \sigma_0 h(t),$$

which is sketched in Fig. 29.6. For this stress input the strain output corresponding to elastic, viscous, and viscoelastic materials is sketched in Fig. 29.7. The basic features of the strain output for a viscoelastic material are that the smallest value of strain $\epsilon(0^+)$ is recorded just after applying the stress at $t = 0^+$, with a smooth, steady increase to a higher value $\epsilon(\infty)$ for longer times.

For viscoelastic materials, the strain-response function, in response to a *unit* Heaviside stress history is called the **creep function** or the **creep-compliance function**,

$$J_c(t) \stackrel{\text{def}}{=} \frac{\epsilon(t)}{\sigma_0},$$

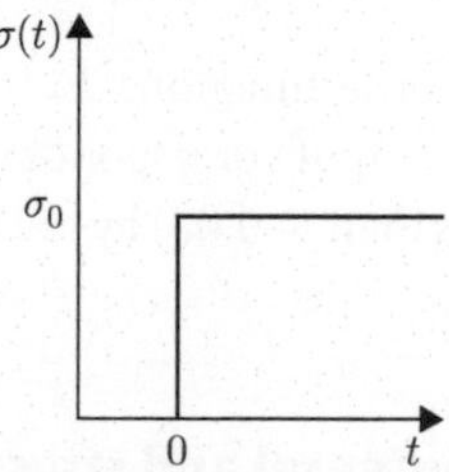

Fig. 29.6 Stress input in a creep test.

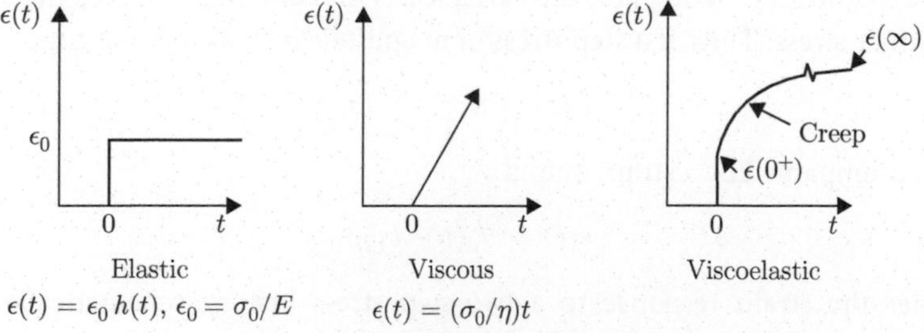

Fig. 29.7 Strain output in a creep test.

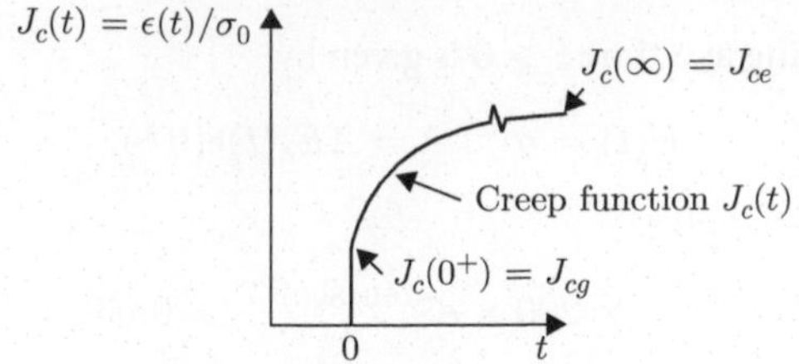

Fig. 29.8 Creep function $J_c(t)$ for a viscoelastic material. The value $J_c(0^+)$ is denoted by J_{cg} and called the glassy creep compliance, while the value $J_c(\infty)$ is denoted by J_{ce} and called the equilibrium creep compliance.

and is sketched in Fig. 29.8. Major features of this function are a small, "glassy" value $J_c(0^+) \equiv J_{cg}$ just after loading, and a steadily increasing value toward a long-time "equilibrium" value, $J_c(\infty) \equiv J_{ce}$.

29.2.3 **Linear viscoelasticity**

- *Viscoelastic behavior is said to be* **linear** *when the stress-relaxation function $E_r(t)$ is independent of the step strain magnitude ϵ_0, and when the creep function $J_c(t)$ is independent of the step stress magnitude σ_0.*

Note that although

$$J_{cg} = 1/E_{rg} \quad \text{and} \quad J_{ce} = 1/E_{re},$$

in general

$$J_c(t) \neq 1/E_r(t).$$

Knowledge of the material response functions $E_r(t)$ or $J_c(t)$ (the two are related—see Sec. 29.5) is sufficient to predict the output corresponding to any input within the *linear range* (typically when the total strain is less than ~ 0.03) by the **Boltzmann superposition principle**, which we describe next.

29.2.4 **Superposition. Creep-integral and stress-relaxation-integral forms of stress-strain relations**

First we note that in $E_r(t)$ and $J_c(t)$ the variable t is the *time elapsed* since the application of the step strain or stress. Thus, if a step stress of magnitude σ_0 was input at time t_1,

$$\sigma(t) = h(t - t_1)\sigma_0,$$

it would be accompanied by a strain output

$$\epsilon(t) = J_c(t - t_1)\sigma_0.$$

Now consider the strain response to a two-step stress history schematically depicted in Fig. 29.9. This stress history may be described mathematically by

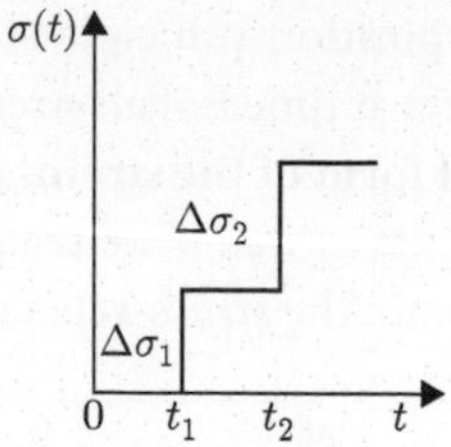

Fig. 29.9 Stress incremented by $\Delta\sigma_1$ at time t_1, and then by $\Delta\sigma_2$ at time t_2.

$$\sigma(t) = h(t - t_1)\Delta\sigma_1 + h(t - t_2)\Delta\sigma_2 \equiv \sum_{i=1}^{2} h(t - t_i)\Delta\sigma_i.$$

The **Boltzmann superposition principle** states that the strain response to this step history in stress is simply

$$\epsilon(t) = \sum_{i=1}^{2} J_c(t - t_i)\Delta\sigma_i.$$

That is, in the linear approximation, the strain is just the sum of the strains corresponding to each step taken separately.[2]

The same arguments apply equally well to histories with an arbitrary number of steps, say N. Thus, if the input stress history is approximated by

$$\sigma(t) = \sum_{i=1}^{N} h(t - t_i)\Delta\sigma_i,$$

the output strain history is

$$\epsilon(t) = \sum_{i=1}^{N} J_c(t - t_i)\Delta\sigma_i.$$

Since it is possible to approximate any physically realizable stress history by an arbitrarily large number of arbitrarily small jumps, passing to the limit $N \to \infty$ in the sums above, the discretized stress history $\sigma(t) = \sum_{i=1}^{N} h(t - t_i)\Delta\sigma_i$, becomes

$$\sigma(t) = \int_{0-}^{t} h(t - \tau)\frac{d\sigma(\tau)}{d\tau}d\tau, \tag{29.2.5}$$

and the discretized output strain history $\epsilon(t) = \sum_{i=1}^{N} J_c(t - t_i)\Delta\sigma_i$, becomes

$$\boxed{\epsilon(t) = \int_{0-}^{t} J_c(t - \tau)\frac{d\sigma(\tau)}{d\tau}d\tau.} \tag{29.2.6}$$

In writing both (29.2.5) and (29.2.6) quiescent conditions have been assumed for all times $t < 0$. We will assume this to hold throughout this chapter.

[2] Coupling effects, depending on both $\Delta\sigma_1$ and $\Delta\sigma_2$ jointly, may occur physically, but in the mathematics they occur only in higher-order non-linear terms.

This integral form of the superposition principle states that the total strain at time t is obtained by superimposing the effect at time t of all stress increments at times $\tau < t$. Relation (29.2.6) is called the **creep-integral form of the strain-stress relation.**

All the preceding remarks hold just as well if we regard the strain history as the input and the stress as the output. We then obtain the **stress-relaxation integral** form of the stress-strain relation:

$$\boxed{\sigma(t) = \int_{0-}^{t} E_r(t-\tau)\frac{d\epsilon(\tau)}{d\tau}d\tau.} \tag{29.2.7}$$

REMARKS 29.1

1. The creep response function $J_c(t)$ and the stress-relaxation response function $E_r(t)$ for a given material are **experimentally measured**; they are properties of the material.
2. The two forms, (29.2.6) and (29.2.7), of the constitutive relation for linear viscoelasticity are equivalent but the inversion of (29.2.6) to (29.2.7) is not trivial.

29.3 **A simple rheological model for linear viscoelastic response**

To get a physical feel for linear viscoelastic behavior it is useful to consider the behavior predicted by *simple analog models* constructed from springs and dashpots representing linear elastic and linear viscous elements, respectively. A classical model, which represents most of the major features of experimentally observed viscoelastic behavior, is called the *standard linear solid* (SLS). It consists of a spring of stiffness E_1, arranged in parallel with an element containing a spring of stiffness E_2 in series with a dashpot with viscosity η, Fig. 29.10.

In this model we have the additive decomposition of the total strain ϵ into an *elastic* part ϵ^{e} and a *viscous* part ϵ^{v}:

$$\epsilon = \epsilon^{\mathrm{e}} + \epsilon^{\mathrm{v}}. \tag{29.3.1}$$

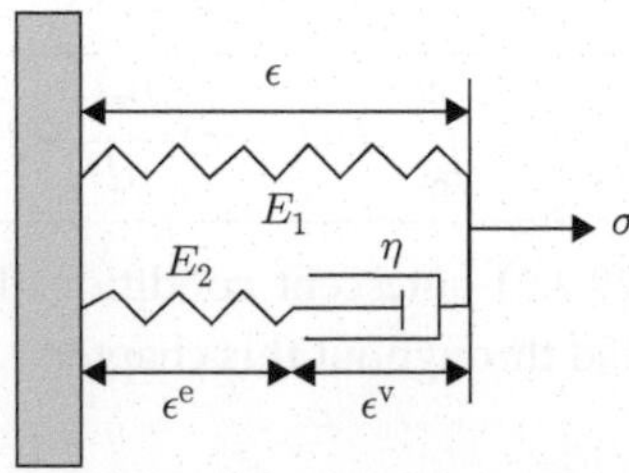

Fig. 29.10 A rheological model for viscoelastic material behavior, called the *standard linear solid.*

As is clear from Fig. 29.10, the total macroscopic stress σ is balanced by the sum of the stresses in springs 1 and 2:

$$\sigma = \underbrace{E_1\,\epsilon}_{\text{stress in spring 1}} + \underbrace{E_2(\epsilon - \epsilon^{\mathrm{v}})}_{\text{stress in spring 2}}\,. \tag{29.3.2}$$

Further, since the dashpot is in series with spring 2, the stress in the dashpot is equal to the stress in spring 2. Since the stress in the dashpot depends linearly on the strain-rate $\dot{\epsilon}^{\mathrm{v}}$ and is given by $\eta\,\dot{\epsilon}^{\mathrm{v}}$, where η is the viscosity of the dashpot, this *internal stress balance* requires that

$$\underbrace{\eta\,\dot{\epsilon}^{\mathrm{v}}}_{\text{stress in dashpot}} = \underbrace{E_2(\epsilon - \epsilon^{\mathrm{v}})}_{\text{stress in spring 2}}\,. \tag{29.3.3}$$

Hence

$$\dot{\epsilon}^{\mathrm{v}} = \frac{1}{\tau_R}(\epsilon - \epsilon^{\mathrm{v}}), \tag{29.3.4}$$

where

$$\tau_R \stackrel{\text{def}}{=} \frac{\eta}{E_2}\,. \tag{29.3.5}$$

Since η has dimensions of (stress $\times$ time), and E_2 has dimensions of stress, the quantity τ_R has dimensions of time, and is called the *relaxation time.*

Summarizing, for this simple rheological model the constitutive equations are

$$\boxed{\begin{aligned} \sigma &= E_1\epsilon + E_2(\epsilon - \epsilon^{\mathrm{v}}),\\ \dot{\epsilon}^{\mathrm{v}} &= \frac{1}{\tau_R}(\epsilon - \epsilon^{\mathrm{v}}), \end{aligned}} \tag{29.3.6}$$

where the quantity ϵ^{v} appears as an *internal variable of the theory* which evolves according to the *flow rule* $(29.3.6)_2$.

We now examine the predictions from these constitutive equations for the phenomena of stress-relaxation and creep.

29.3.1 **Stress-relaxation**

Let $\epsilon = \epsilon_0$ be the *constant* strain imposed in a stress-relaxation experiment. Then equations (29.3.6) become

$$\begin{aligned} \sigma &= E_1\epsilon_0 + E_2(\epsilon_0 - \epsilon^{\mathrm{v}}),\\ \dot{\epsilon}^{\mathrm{v}} &= \frac{1}{\tau_R}(\epsilon_0 - \epsilon^{\mathrm{v}})\,. \end{aligned} \tag{29.3.7}$$

In this case the differential equation $(29.3.7)_2$ is in the standard first-order linear form

$$\frac{d\epsilon^{\mathrm{v}}}{dt} = -\frac{1}{\tau_R}\,\epsilon^{\mathrm{v}} + \frac{\epsilon_0}{\tau_R},$$

and can be solved directly to give

$$\epsilon^{\mathrm{v}} = \epsilon_0\left(1 - \exp\left(-\frac{t}{\tau_R}\right)\right), \tag{29.3.8}$$

where we have used the initial condition

$$\epsilon^{\mathrm{v}}(0) = 0.$$

Substitution of (29.3.8) in $(29.3.7)_1$ gives

$$\sigma(t) = E_1\epsilon_0 + E_2\epsilon_0 \exp\left(-\frac{t}{\tau_R}\right). \tag{29.3.9}$$

Hence, the stress-relaxation function $E_r(t) = \sigma(t)/\epsilon_0$ for the standard linear solid model of viscoelasticity is

$$\boxed{E_r(t) = E_{re} + (E_{rg} - E_{re}) \exp\left(-\frac{t}{\tau_R}\right),} \tag{29.3.10}$$

where the "glassy" and "equilibrium" relaxation moduli are

$$\begin{aligned} E_{rg} &= E_r(0) = E_1 + E_2, \\ E_{re} &= E_r(\infty) = E_1. \end{aligned} \tag{29.3.11}$$

In Fig. 29.11 the function (29.3.10) for the standard linear solid is plotted for $E_1 = 0.5\,\text{GPa}$, $E_2 = 2.5\,\text{GPa}$, and relaxations times of $\tau_R = 5, 10, 20, 40\,\text{s}$. The starting value for E_r is $E_{rg} = E_r(0) = E_1 + E_2 = 3\,\text{GPa}$. For large values of t the relaxation modulus E_r converges to the value $E_{re} = E_r(\infty) = E_1 = 0.5\,\text{GPa}$. The smaller the value of τ_R, the faster the relaxation modulus relaxes to its equilibrium value E_{re}.

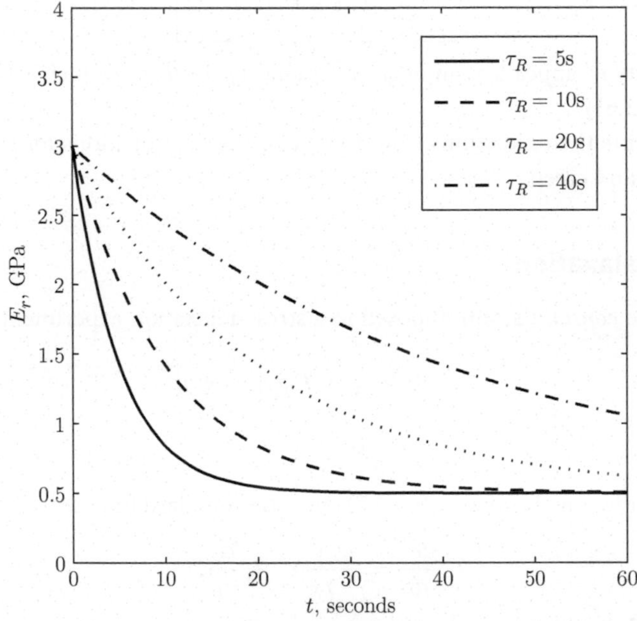

Fig. 29.11 Stress-relaxation function for a standard linear solid, with $E_1 = 0.5\,\text{GPa}$, $E_2 = 2.5\,\text{GPa}$, so that $E_{rg} = 3\,\text{GPa}$, and $E_{re} = 0.5\,\text{GPa}$, for different values of the relaxation time τ_R.

29.3.2 **Creep**

Let $\sigma = \sigma_0$ be the *constant* stress imposed in a creep experiment. In this case, the equation for stress $(29.3.6)_1$ becomes

$$\sigma_0 = E_1\epsilon + \underbrace{E_2(\epsilon - \epsilon^{\mathrm{v}})}_{\text{stress in spring 2}}, \tag{29.3.12}$$

which gives the stress in spring 2 as

$$\underbrace{E_2(\epsilon - \epsilon^{\mathrm{v}})}_{\text{stress in spring 2}} = \sigma_0 - E_1\epsilon. \tag{29.3.13}$$

However, from the *internal stress balance* we also have

$$\underbrace{\eta\dot{\epsilon}^{\mathrm{v}}}_{\text{stress in dashpot}} = \underbrace{E_2(\epsilon - \epsilon^{\mathrm{v}})}_{\text{stress in spring 2}}. \tag{29.3.14}$$

Substituting expression (29.3.13) for the stress in spring 2 in (29.3.14), and rearranging, we obtain

$$\dot{\epsilon}^{\mathrm{v}} = \frac{1}{\eta}(\sigma_0 - E_1\epsilon). \tag{29.3.15}$$

Next, differentiating (29.3.13) we obtain

$$(E_1 + E_2)\dot{\epsilon} = E_2\dot{\epsilon}^{\mathrm{v}}, \tag{29.3.16}$$

and finally substituting (29.3.15) in (29.3.16) we obtain the following differential equation for the strain

$$\frac{d\epsilon}{dt} = -\frac{E_1}{(E_1 + E_2)}\frac{1}{\tau_R}\epsilon + \frac{\sigma_0}{(E_1 + E_2)}\frac{1}{\tau_R} \tag{29.3.17}$$

where, as before $\tau_R = \eta/E_2$ is the *relaxation time.*

Using the initial condition

$$\epsilon(0) = \frac{\sigma_0}{(E_1 + E_2)}, \tag{29.3.18}$$

the differential equation (29.3.17) for strain can be solved to give

$$\epsilon(t) = \frac{\sigma_0}{(E_1 + E_2)}\exp\left(-\frac{E_1}{(E_1 + E_2)}\frac{t}{\tau_R}\right) - \frac{\sigma_0}{E_1}\left(\exp\left(-\frac{E_1}{(E_1 + E_2)}\frac{t}{\tau_R}\right) - 1\right)$$

or, upon rearranging,

$$\epsilon(t) = \sigma_0\left[\frac{1}{E_1} - \left(\frac{1}{E_1} - \frac{1}{(E_1 + E_2)}\right)\exp\left(-\frac{E_1}{(E_1 + E_2)}\frac{t}{\tau_R}\right)\right]. \tag{29.3.19}$$

Hence, the creep function $J_c(t) = \epsilon(t)/\sigma_0$ for the standard linear solid model of viscoelasticity is

$$\boxed{J_c(t) = J_{ce} - (J_{ce} - J_{cg})\exp\left(-\frac{t}{\tau_C}\right),} \tag{29.3.20}$$

where the "glassy" and "equilibrium" creep moduli are

$$J_{cg} = J_c(0) = \frac{1}{(E_1 + E_2)}, \qquad J_{ce} = J_c(\infty) = \frac{1}{E_1}, \tag{29.3.21}$$

and

$$\tau_C \stackrel{\text{def}}{=} \frac{\tau_R(E_1 + E_2)}{E_1} \tag{29.3.22}$$

is the *creep retardation time*. Note that in general $\tau_C \neq \tau_R$.

In Fig. 29.12 the function (29.3.20) for the standard linear solid is plotted for $E_1 = 0.5\,\text{GPa}$, $E_2 = 2.5\,\text{GPa}$, and relaxations times of $\tau_R = 5, 10, 20, 40\,\text{s}$. The starting value for J_c is $J_{cg} = J_c(0) = 1/(E_1 + E_2) = (1/3)\,\text{GPa}^{-1}$. For large values of t the creep function J_c converges to the value $J_{ce} = J_c(\infty) = 1/E_1 = 2.0\,\text{GPa}^{-1}$. The smaller the value of τ_R, the faster the creep compliance approaches to its equilibrium value J_{ce}.

In closing this section,

- *we emphasize that real materials are not made up from springs and dashpots, and even a very large number of them may not be able to represent the behavior of real materials accurately.*
- However, rheological models made up from springs and dashpots are certainly helpful for obtaining a *qualitative* understanding of the behavior of viscoelastic materials, and certain suitably complex rheological models do suggest mathematical forms for $E_r(t)$ and $J_c(t)$ which may be used to fit experimentally obtained data.

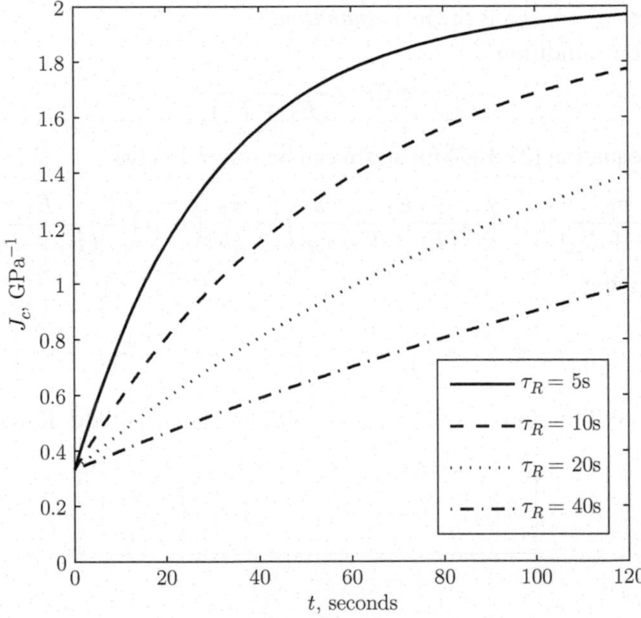

Fig. 29.12 Creep compliance function for a standard linear solid, with $E_1 = 0.5\,\text{GPa}$, $E_2 = 2.5\,\text{GPa}$, and different values of relaxation times τ_R.

29.4 Power-law relaxation functions

The following simple power-law form for the relaxation response,

$$E_r(t) = E_{re} + \frac{E_{rg} - E_{re}}{(1 + t/\tau_0)^n}, \tag{29.4.1}$$

with $\{E_{rg}, E_{re}, \tau_0\}$ positive-valued constants and $n > 0$ is often used to approximate experimentally observed stress relaxation response. As an example, Fig. 29.13 shows a plot for

$$E_{rg} = 10^5 \text{ kPa}, E_{re} = 10^2 \text{ kPa}, \tau_0 = 10^{-4} \text{ s, and } n = 0.35.$$

Here E_{rg} represents the modulus as $t \to 0$, and E_{re} its value as $t \to \infty$, τ_0 a characteristic time which locates the beginning of the transition region, and n the negative slope.

The short- and long-time modulus limits, along with the position of the transition along the log-time axis, and the slope of the mid-section may be readily adjusted through the four material parameters, but it is seldom possible to also represent the proper curvature

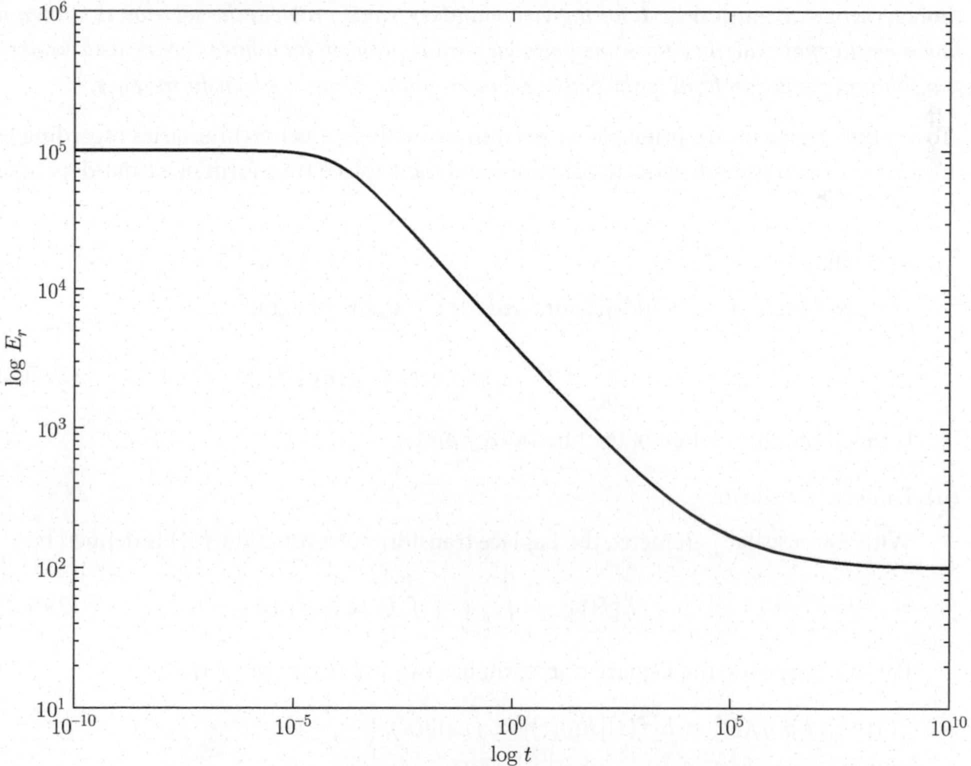

Fig. 29.13 Example of the power-law representation of the relaxation modulus. Time in units of s and modulus in units of kPa.

in the transitions from short- and long-time behavior with this simple power-law relation. Nevertheless, a function of this type can be very useful in capturing the essential features of a problem.

An alternate empirical representation for $E_r(t)$ is,

$$E_r(t) = E_{re} + (E_{rg} - E_{re}) \exp\left[-\left(\frac{t}{\tau_0}\right)^{\beta}\right], \tag{29.4.2}$$

also with four adjustable parameters, $\{E_{rg}, E_{re}, \tau_0, \beta\}$, with $\{E_{rg}, E_{re}, \tau_0\}$ positive-valued constants and $0 < \beta \leq 1$. This relaxation function is termed a *stretched exponential.*[3]

29.5 Correspondence principle

There is an important mathematical observation that helps us solve boundary-value problems involving linear viscoelastic materials. The observation is called *the correspondence principle between linear viscoelasticity and linear elasticity.* This principle reads as follows:

CORRESPONDENCE PRINCIPLE: *If for a given boundary-value problem the solution is known for a linear elastic material, then the same boundary-value problem for a linear viscoelastic material has a solution similar in form to the elastic solution, but in a Laplace transform space.*

To see the origins of this principle we need some mathematical preliminaries regarding the convolution of two time-dependent functions and the Laplace transform of a time-dependent function:

(i) Convolution:

For two functions $f(t)$ and $g(t)$ defined for $t \geq 0$, the integral

$$\int_{0^-}^{t} f(t-\tau)g(\tau)d\tau \equiv (f * g)(t) \tag{29.5.1}$$

is called the *convolution* of the functions f and g.

(ii) Laplace transform:

With s as a scalar parameter, the Laplace transform of a function $f(t)$ is defined by

$$L[f(t)] = \int_{0^-}^{\infty} e^{-st} f(t)\, dt \equiv \bar{f}(s). \tag{29.5.2}$$

For our purposes, the Laplace transform has two important properties:

(a) $L[(f * g)(t)] = L[f(t)]L[g(t)] = \bar{f}(s)\bar{g}(s)$.

(b) $L[\dot{f}(t)] = sL[f(t)] - f(0^-) = s\bar{f}(s) - f(0^-)$.

---

[3] Stretched exponential forms are widely used in polymer physics and functions of this type are called Kohlrausch–Williams–Watts (KWW) functions (Kohlrausch, 1854; Williams and Watts, 1970).

29.5.1 **Correspondence principle in one dimension**

Let us now recall our stress-relaxation integral form of the stress-strain relation for a viscoelastic material (cf. (29.2.7)):

$$\sigma(t) = \int_{0-}^{t} E_r(t-\tau)\frac{d\epsilon(\tau)}{d\tau}d\tau. \tag{29.5.3}$$

Using the definition (29.5.1) of the convolution, this may be written as

$$\sigma(t) = (E_r * \dot{\epsilon})(t). \tag{29.5.4}$$

Taking the Laplace transform of (29.5.4) we obtain

$$\begin{aligned}\bar{\sigma}(s) = L[\sigma(t)] &= L[(E_r * \dot{\epsilon})(t)],\\ &= \bar{E}_r(s)L[\dot{\epsilon}(t)],\\ &= \bar{E}_r(s)\left(s\bar{\epsilon}(s) - \epsilon(0^-)\right),\end{aligned}$$

which implies that for quiescent initial conditions,

$$\boxed{\bar{\sigma}(s) = \bar{E}_r^*(s)\,\bar{\epsilon}(s),} \tag{29.5.5}$$

with

$$\boxed{\bar{E}_r^*(s) = s\bar{E}_r(s).} \tag{29.5.6}$$

This is an important result because the constitutive equation in the Laplace transform domain (parameterization by the transform variable s in place of time t) behaves like an elastic stress-strain relation at each s! This is known as the *correspondence principle* between elastic and viscoelastic behavior.

Thus one procedure to solve a boundary-value problem in linear viscoelasticity is to:

- first find the elastic solution,
- then substitute $\bar{E}_r^*(s)$ for E, and
- finally take the inverse Laplace transform to obtain the viscoelastic time-dependent solution for σ.

REMARK 29.2

The process of computing an inverse Laplace transform is formally defined by the relation

$$f(t) = \frac{1}{2\pi i}\int_{c-i\infty}^{c+i\infty} e^{st}\bar{f}(s)\,ds,$$

where $i = \sqrt{-1}$ is the imaginary unit and c is any real-valued constant to the right of any poles of $\bar{f}(s)$, ensuring that $\int_0^\infty e^{-ct}|f(t)|\,dt < \infty$. To use this relation requires special knowledge of complex analysis; see e.g. Hildebrand (1976, Chap. 11). More typically, Laplace transform inversion is carried out via algebraic means and the use of tables of known transform pairs; see e.g. Kreyszig (1983, Chap. 5).

29.5.2 Connection between $E_r(t)$ and $J_c(t)$ in Laplace transform space

Recall that the creep integral form of the stress-strain relation in linear viscoelasticity is (cf. (29.2.6))

$$\epsilon(t) = \int_{0^-}^{t} J_c(t-\tau)\frac{d\sigma(\tau)}{d\tau}d\tau. \tag{29.5.7}$$

Using the definition (29.5.1) of the convolution, this may be written as

$$\epsilon(t) = (J_c * \dot{\sigma})(t). \tag{29.5.8}$$

Taking the Laplace transform of (29.5.8) we obtain

$$\begin{aligned}\bar{\epsilon}(s) = L[\epsilon(t)] &= L[(J_c * \dot{\sigma})(t)] \\ &= \bar{J}_c(s)L[\dot{\sigma}(t)] \\ &= \bar{J}_c(s)[s\bar{\sigma}(s) - \sigma(0^-)],\end{aligned}$$

or for quiescent initial conditions

$$\boxed{\bar{\epsilon}(s) = \bar{J}_c^*(s)\bar{\sigma}(s),} \tag{29.5.9}$$

where

$$\boxed{\bar{J}_c^*(s) = s\bar{J}_c(s).} \tag{29.5.10}$$

Hence, from (29.5.5) and (29.5.9)

$$\boxed{\bar{E}_r^*(s) \equiv \left(\bar{J}_c^*(s)\right)^{-1},} \tag{29.5.11}$$

or using (29.5.6) and (29.5.10)

$$\boxed{\bar{J}_c(s)\bar{E}_r(s) = \frac{1}{s^2}.} \tag{29.5.12}$$

Thus, for a linear viscoelastic material there is a unique and simple relationship between creep and relaxation behavior in the Laplace transform space.

EXAMPLE 29.2 Extensional response of a viscoelastic rod

Consider a viscoelastic rod of cross-sectional area A that is subjected to a time varying axial load

$$P(t) = [P_o + P_1\sin(\omega t)]h(t).$$

The creep compliance for the rod is known to be

$$J_c(t) = J_{ce} - \Delta J_c e^{-t/\tau_C} \qquad \text{with} \qquad \Delta J_c = J_{ce} - J_{cg}.$$

The time history of the strain in the rod can be found using the correspondence principle. Since the elasticity solution is trivially known to be given by

$$\epsilon = \frac{\sigma}{E},$$

using (29.5.6) and (29.5.12) we can write that for a viscoelastic material in the transform domain

$$\bar{\epsilon}(s) = \frac{\bar{\sigma}(s)}{s\bar{E}_r(s)} = \frac{\bar{\sigma}(s)}{s/(s^2\bar{J}_c(s))} = s\bar{J}_c(s)\bar{\sigma}(s).$$

The necessary Laplace transforms of σ and J_c are easily computed (or found in tables) as:

$$\bar{\sigma}(s) = \frac{1}{A}\left[\frac{P_o}{s} + \frac{P_1\omega}{s^2+\omega^2}\right], \qquad \text{and} \qquad \bar{J}_c(s) = \frac{J_{ce}}{s} - \frac{\Delta J_c}{s+1/\tau_C}.$$

Hence the Laplace transform of the strain is given as

$$\bar{\epsilon}(s) = \frac{s}{A}\left[\frac{J_{ce}}{s} - \frac{\Delta J_c}{s+1/\tau_C}\right]\left[\frac{P_o}{s} + \frac{P_1\omega}{s^2+\omega^2}\right].$$

The inverse Laplace transform can be computed by way of tables or computer algebra systems, yielding

$$\begin{aligned}\epsilon(t) = {} & \frac{J_{ce}}{A}\Big(P_o + P_1\sin(\omega t)\Big) \\ & - \frac{\Delta J_c}{A}\left[\frac{P_1\omega\tau_C\big(\omega\tau_C\sin(\omega t) + \cos(\omega t)\big)}{1+\omega^2\tau_C^2}\right. \\ & \left. + \exp\left(-\frac{t}{\tau_C}\right)\frac{\omega\tau_C\,(\omega\tau_C P_o - P_1) + P_o}{1+\omega^2\tau_C^2}\right].\end{aligned}$$

29.6 Correspondence principles for structural applications

For structural applications involving bending of beams and torsion of shafts, restricted forms of the correspondence principle are useful in practical situations. These principles can be applied directly to simple stress analysis problems in bending of beams and torsion of shafts, without complicating the mathematics by Laplace transforms, convolutions, and the like when the structures are subjected to step loadings in forces or motion.

29.6.1 **Bending of beams made from linear viscoelastic materials**

Before stating versions of the correspondence principle which are applicable to the analysis of beams made from linear viscoelastic materials, we briefly summarize the analysis for bending of beams made from linear elastic materials.

Consider a beam of a prismatic cross-section with its neutral axis aligned with the x-direction of a rectangular Cartesian coordinate system. Let $v(x)$ denote the transverse deflection of the neutral axis under bending; then, for small deflections, the local curvature of the beam is

$$\kappa(x) = \frac{d^2 v(x)}{dx^2}, \tag{29.6.1}$$

and the axial strain is given by

$$\epsilon(x, y) = -\kappa(x)\, y. \tag{29.6.2}$$

These are *kinematical relations*, *independent* of any particular constitutive equation for the material from which the beam is comprised.

For a beam made from a linear elastic material, the constitutive equation for the stress in terms of the strain is

$$\sigma(x, y) = E\, \epsilon(x, y), \tag{29.6.3}$$

with E the Young's modulus. Suppose that the beam has an internal moment distribution $M(x)$. By definition,

$$M(x) = \int_A -\sigma(x, y) y\, dA, \tag{29.6.4}$$

and thus

$$M(x) = (EI)\kappa(x), \qquad \text{with} \qquad I = \int_A y^2\, dA, \tag{29.6.5}$$

where I is the area moment of inertia of the beam cross-section.

Thus, (29.6.2), (29.6.3), and (29.6.5) give that the stress distribution in a linear elastic beam is

$$\sigma(x, y) = -\frac{M(x) y}{I}, \tag{29.6.6}$$

while (29.6.1) and (29.6.5) give the following differential equation

$$\frac{d^2 v(x)}{dx^2} = \frac{M(x)}{EI}, \tag{29.6.7}$$

which can be solved, subject to appropriate boundary conditions, to obtain the transverse deflection of the beam for statically determinate problems; for the general case, including statically indeterminate problems, see e.g. Govindjee (2013).

First correspondence principle for bending of beams made from a linear viscoelastic material

Now consider a beam made from a linear viscoelastic material. First consider a beam problem in which the displacements of selected points are prescribed. As an example we might think of a cantilever beam whose free end is suddenly deflected by a certain amount v_0 and then held in position. Thus given

$$v(x, t) = v(x) h(t), \tag{29.6.8}$$

we inquire as to what are the time-dependent strain and stress distributions $\epsilon(x, y, t)$ and $\sigma(x, y, t)$ in a viscoelastic beam?

Now, the strain distribution that would arise in an elastic beam corresponding to the given application of the deflection of the neutral axis (29.6.8) is

$$\epsilon(x, y, t) = \left[-\left(\frac{d^2v(x)}{dx^2}\right) y\right] h(t). \tag{29.6.9}$$

As a *tentative* solution, we assume that this would also be the strain distribution in a viscoelastic beam.

Now, recall that for a viscoelastic material subjected to a step strain input $\epsilon(t) = \epsilon_0 h(t)$, the stress-relaxation response is given by $\sigma(t) = E_r(t)\epsilon_0$. Thus, in a beam made from a viscoelastic material, we would expect that if it is subjected to a strain distribution (29.6.9), the corresponding stress distribution would be

$$\sigma(x, y, t) = E_r(t)\left(-\frac{d^2v(x)}{dx^2} y\right). \tag{29.6.10}$$

At any time t the stress in a viscoelastic beam would be distributed like the stress in an elastic beam, but with a modulus $E = E_r(t)$, and hence they would satisfy the equilibrium condition. Also, since (29.6.9) satisfies all the kinematic conditions, *we conclude that our tentative solution is the correct solution.* With these observations we are ready to state our first version of the correspondence principle.

Correspondence Principle 1 for bending:[4] *If a statically determinate viscoelastic beam is subjected to forced displacement boundary conditions which are all applied at time $t = 0$ and held constant, then:*

(a) *The strain distribution is time-independent, and the same as that in an elastic beam.*

(b) *The stress distribution is time-dependent, and derived from the elastic solution by replacing the Young's modulus E with the stress-relaxation function $E_r(t)$.*

(c) *The reactions and loadings applied to the beam and the bending moment and shear force distribution along the beam, being consistent with the time-dependent stress field, are also time-dependent, and can be derived from the elastic solution by replacing the Young's modulus E with the stress-relaxation function $E_r(t)$.*

EXAMPLE 29.3 Stress relaxation of a beam with imposed deformation

Consider a simply supported beam of span l, which is subjected to a displacement v_0 at its mid-span $x = l/2$. This displacement is applied at time $t = 0$, and then held constant (Fig. 29.14). What is the reaction force $P(t)$ on the external agency causing the displacement?

[4] This principle's validity has been motivated above by a plausibility argument, however it is rigorously provable using Laplace transform theory for statically determinate beams.

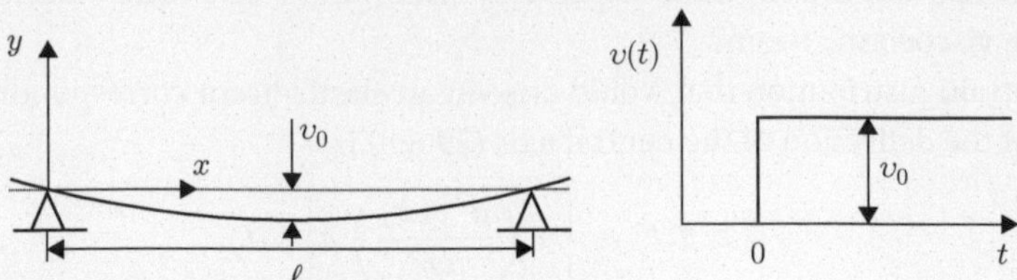

Fig. 29.14 A simply supported beam subjected to a displacement v_0 at its mid-span. The displacement is applied at time $t = 0$, and then held constant.

Linear Elastic Solution:

$$P(t) = \left[\frac{48\, I\, v_0}{l^3}\right]\, Eh(t).$$

Linear Viscoelastic Solution: Using correspondence principle 1, we simply replace the Young's modulus E in the elastic solution by the stress-relaxation function $E_r(t)$,

$$P(t) = \left[\frac{48\, I\, v_0}{l^3}\right]\, E_r(t).$$

Second correspondence principle for bending of beams made from a linear viscoelastic material

Now assume that loads $P(x,t)$ are applied to the beam at time $t = 0$, and then held constant. Since the load $P(x)$ gives rise to moments $M(x)$, we have

$$P(x,t) = P(x)h(t) \Rightarrow M(x,t) = M(x)h(t). \tag{29.6.11}$$

We now inquire as to what are the time-dependent responses

$$\sigma(x,y,t), \quad \epsilon(x,y,t), \quad v(x,t),$$

when the beam is subjected to the step loading (29.6.11), considered above?

Consider a *tentative* solution of the problem of determining $\sigma(x,y,t)$ to be

$$\sigma(x,y,t) = \sigma(x,y)h(t), \tag{29.6.12}$$

where $\sigma(x,y)$ on the right-hand side is the stress that would arise in an elastic beam

$$\sigma(x,y) = -\frac{M(x)y}{I}. \tag{29.6.13}$$

Now, recall that for a viscoelastic material a step stress input $\sigma(t) = \sigma_0 h(t)$ gives a creep response $\epsilon(t) = J_c(t)\sigma_0$. Thus, in a beam made from a viscoelastic material, we would expect that if it is subjected to a stress loading

$$\sigma(x,y,t) = \left(-\frac{M(x)y}{I}\right) h(t), \tag{29.6.14}$$

the corresponding strain would be

$$\epsilon(x, y, t) = J_c(t)\left(-\frac{M(x)y}{I}\right). \tag{29.6.15}$$

At any time t the strains would be distributed like the strains in an elastic beam of modulus $E = 1/J_c(t)$, and hence they would *satisfy all the kinematic conditions of the problem.* Also, the stress distribution (29.6.14) will *satisfy equilibrium. From this we conclude that our tentative solution is the correct solution for a viscoelastic beam.*

Then, recalling (29.6.7) and our observation above, deflection of the neutral axis is obtained by solving

$$\frac{d^2v(x,t)}{dx^2} = \frac{M(x)}{I}J_c(t), \tag{29.6.16}$$

subject to the appropriate boundary conditions. With these observations we are ready to state our second correspondence principle.

Correspondence Principle 2 for bending:[5] *If a statically determinate viscoelastic beam is subjected to loads which are all applied simultaneously at time t = 0, and held constant, then:*

(a) The stress distribution is time-independent, and the same as that in an elastic beam under the same load.

(b) The strain and displacement distributions depend on time, and are derived from the elastic solution by replacing the elastic compliance $1/E$ by the creep compliance $J_c(t)$.

EXAMPLE 29.4 Creep in a simply supported beam

Consider a simply supported beam with a uniformly distributed load p_0 per unit length, which is applied at time $t = 0$ and then held constant (Fig. 29.15). What is the stress distribution in the beam? What is the deflection of the neutral axis?

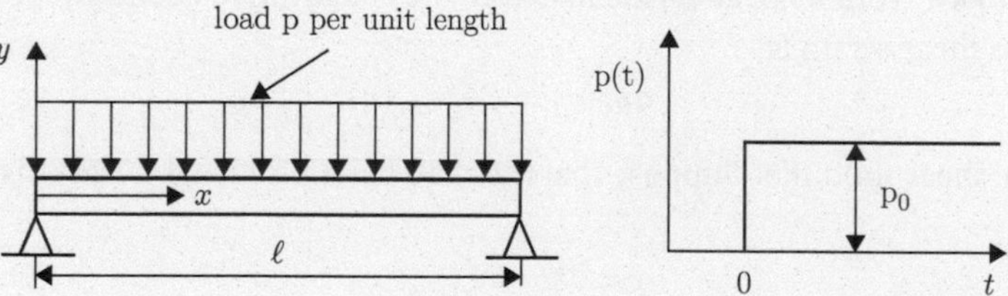

Fig. 29.15 A simply supported beam with a uniformly distributed load p_0 per unit length, which is applied at time $t = 0$ and then held constant.

[5] This principle's validity has been motivated above by a plausibility argument, however it is rigorously provable using Laplace transform theory for statically determinate beams.

Linear Elastic Solution:

$$\sigma(x, y, t) = \left[-\frac{1}{2I} p_0 (lx - x^2) y \right] h(t)$$

$$v(x, t) = \left[-\frac{p_0 x}{24I} \left(l^3 - 2lx^2 + x^3 \right) \right] \frac{1}{E} h(t).$$

Linear Viscoelastic Solution: Using correspondence principle 2, we replace $(1/E)$ in the elastic solution by the creep compliance $J_c(t)$,

$$\sigma(x, y, t) = \left[-\frac{1}{2I} p_0 (lx - x^2) y \right] h(t),$$

$$v(x, t) = \left[-\frac{p_0 x}{24I} \left(l^3 - 2lx^2 + x^3 \right) \right] J_c(t).$$

29.6.2 Torsion of shafts made from linear viscoelastic materials

Before stating versions of the correspondence principle which are applicable to the analysis of torsion of shafts made from linear viscoelastic materials, we briefly summarize the analysis for torsion of shafts made from linear elastic materials.

Consider a cylindrical shaft with its axis aligned with the z-direction of a cylindrical coordinate system. Let α denote the *twist per unit length* of the shaft, then the shear strain distribution in the shaft is given by

$$\epsilon_{\theta z}(r) = \frac{1}{2} \alpha r. \tag{29.6.17}$$

This is a *kinematical relation*, *independent* of any particular constitutive equation for the material from which the shaft is comprised.

For a shaft made from a linear elastic material, the constitutive equation for the shear stress in terms of the shear strain is

$$\sigma_{\theta z}(r) = 2G\, \epsilon_{\theta z}(r), \tag{29.6.18}$$

where G is the shear modulus. Suppose that the shaft has an internal torque distribution $T(z)$. By definition

$$T(z) = \int_A \sigma_{\theta z} r \, dA, \tag{29.6.19}$$

and thus

$$T = (GJ)\alpha, \qquad \text{with} \qquad J = \int_A r^2 \, dA, \tag{29.6.20}$$

where J is the polar area moment of inertia of the shaft cross-section.

Thus, (29.6.17), (29.6.18), and (29.6.20) give that the stress distribution in a linear elastic shaft is

$$\sigma_{\theta z}(r) = \frac{Tr}{J}, \tag{29.6.21}$$

and the angle of twist per unit length is given by

$$\alpha = \frac{T}{GJ}, \tag{29.6.22}$$

which can be integrated to give the twist field

$$\Theta(z) = \Theta(0) + \int_0^z \frac{T}{GJ}\, dz. \tag{29.6.23}$$

For the case of statically indeterminate problems see e.g. Govindjee (2013).

For linear viscoelastic materials, our discussion of the correspondence principles for applied loads and displacements in the case of bending of beams, carries over to the case of torsion of circular shafts and prismatic bars. Here, instead of $E_r(t)$ and $J_c(t)$ that we had introduced in the case of a one-dimensional tensile and compressive situation, we need to introduce

- a *shear-stress-relaxation function* $G_r(t)$, and
- a *shear-strain creep response function* $L_c(t)$.

In terms of the shear-stress relaxation function $G_r(t)$ and the shear-strain creep response function $L_c(t)$ the two correspondence principles for torsion are:[6]

Correspondence Principle 1 for torsion: *If a statically determinate viscoelastic shaft is subjected to an angle of twist which is applied at time* $t = 0$*, and held constant, then:*

(a) The shear-strain distribution is time-independent, and the same as that in an elastic shaft.

(b) The shear-stress distribution and the torque are time-dependent, and derived from the elastic solution by replacing the shear modulus G with shear-stress-relaxation function $G_r(t)$.

Correspondence Principle 2 for torsion: *If a statically determinate viscoelastic shaft is subjected to a torque which is applied at time* $t = 0$*, and held constant, then:*

(a) The shear stress distribution is time-independent, and the same as that in an elastic shaft under the same torque.

(b) The angle of twist depends on time, and is derived from the elastic solution by replacing the shear compliance $1/G$ *by the shear creep compliance* $L_c(t)$.

6 These two principles are rigorously derivable using Laplace transforms for statically determinate problems.

29.7 **Linear viscoelastic response under oscillatory strain and stress**

We have seen that the mechanical behavior of a linear viscoelastic material can be described by a constitutive equation in the equivalent integral forms (cf. (29.2.7) and (29.2.6)):

$$\boxed{\sigma(t) = \int_{0-}^{t} E_r(t-\tau)\frac{d\epsilon(\tau)}{d\tau}d\tau,} \tag{29.7.1}$$

and

$$\boxed{\epsilon(t) = \int_{0-}^{t} J_c(t-\tau)\frac{d\sigma(\tau)}{d\tau}d\tau.} \tag{29.7.2}$$

For a given material, the stress-relaxation response function $E_r(t)$ or the creep response function $J_c(t)$ can be determined *experimentally*, and the relaxation integral form (29.7.1) or the creep integral form (29.7.2) can be used in calculating the response to arbitrary loading histories, of either $\epsilon(\tau)$ or $\sigma(\tau)$.

A particularly common state of loading for viscoelastic materials is that of steady-state oscillatory loads. While the response to such loads can be computed from (29.7.1) or (29.7.2) when the relaxation or creep function is known, it is more common (and instructive) in this context to directly characterize linear viscoelastic materials by subjecting them to oscillatory sinusoidal loads. This characterization of the material response leads to the concepts of storage and loss moduli and storage and loss compliances as equivalent alternates to the relaxation and creep functions, when the steady-state response is what is of interest (Ferry, 1961).

29.7.1 **Oscillatory loads**

Suppose that we apply an oscillatory stress of the form

$$\sigma(t) = \sigma_0 \cos(\omega t), \tag{29.7.3}$$

where σ_0 is the stress amplitude, and ω the angular frequency (radians per second). In linear viscoelastic materials, it is typically found that the resulting strain after the disappearance of transient response terms is of the form

$$\epsilon(t) = \epsilon_0 \cos(\omega t - \delta); \tag{29.7.4}$$

that is, the strain response is an oscillation at the same frequency as the stress, but it **lags behind**—in the sense that it reaches its peak value at a later time—by a phase angle δ; cf. Fig. 29.16. This angle is often referred to as the **loss angle** of the material (for reasons, which will become apparent shortly); it is an important quantity for the characterization of viscoelastic dissipative properties.

Expanding the trigonometric function in the strain response, we get

$$\epsilon(t) = \epsilon_0 \cos(\delta)\cos(\omega t) + \epsilon_0 \sin(\delta)\sin(\omega t). \tag{29.7.5}$$

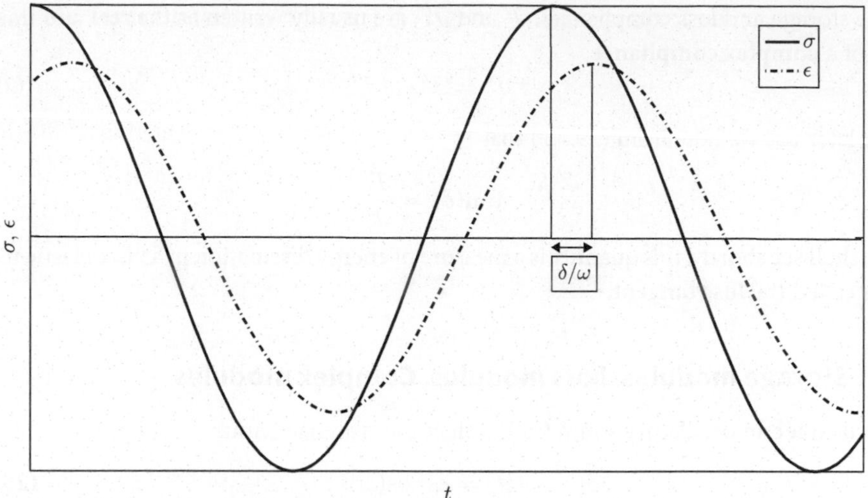

Fig. 29.16 Schematic figure showing an oscillatory stress input $\sigma = \sigma_0 \cos(\omega t)$, and the corresponding steady-state strain output $\epsilon = \epsilon_0 \cos(\omega t - \delta)$ with a phase angle lag δ.

The first term in (29.7.5) is completely in-phase with the stress input (as happens with an ideal elastic material), while the second term is the one contributing to the observed phase-lag. If $\delta = \pi/2$, then the strain response becomes $\epsilon(t) = \epsilon_0 \sin(\omega t)$, and the strain is completely out-of-phase with the stress, sometimes termed *"90-degrees out-of-phase"* with zero output at maximum input and vice versa.

29.7.2 Storage compliance. Loss compliance. Complex compliance

Equation (29.7.5) may be rewritten as

$$\epsilon(t) = \sigma_0 \Big(\frac{\epsilon_0 \cos(\delta)}{\sigma_0} \cos(\omega t) + \frac{\epsilon_0 \sin(\delta)}{\sigma_0} \sin(\omega t) \Big). \tag{29.7.6}$$

Let

$$J' \stackrel{\text{def}}{=} \frac{\epsilon_0}{\sigma_0} \cos(\delta) \qquad \text{and} \qquad J'' \stackrel{\text{def}}{=} \frac{\epsilon_0}{\sigma_0} \sin(\delta), \tag{29.7.7}$$

so that (29.7.6) may be written as

$$\epsilon(t) = \sigma_0 \Big(J' \cos(\omega t) + J'' \sin(\omega t) \Big). \tag{29.7.8}$$

The quantity J' is a measure of how in-phase the strain is with the stress, and is called the **storage compliance**, while J'' is a measure of how out-of-phase the strain is with the stress, and is called the **loss compliance**.[7] Though not immediately apparent J' and J'' are functions of the excitation frequency ω.

---

[7] The oscillatory stress and strain do not have to be in the form of cosine functions, they could also be in the form of sine functions:

$$\sigma(t) = \sigma_0 \sin(\omega t), \qquad \epsilon(t) = \epsilon_0 \sin(\omega t - \delta).$$

The storage and loss compliances, J' and J'', are usually written as the real and imaginary parts of a **complex compliance:**

$$J^* = J' - iJ''. \tag{29.7.9}$$

For later use we note from (29.7.7) that

$$\tan(\delta) = \frac{J''}{J'}. \tag{29.7.10}$$

As we shall see shortly, this quantity is a measure of energy dissipation in a viscoelastic material, and is called the **loss tangent**.

29.7.3 Storage modulus. Loss modulus. Complex modulus

Now consider an oscillatory strain (rather than the stress) as input:

$$\epsilon(t) = \epsilon_0 \cos(\omega t). \tag{29.7.11}$$

In this case we write the stress output as

$$\begin{aligned}\sigma(t) &= \sigma_0 \cos(\omega t + \delta) && (29.7.12)\\ &= \sigma_0 \cos(\delta)\cos(\omega t) - \sigma_0 \sin(\delta)\sin(\omega t), && (29.7.13)\end{aligned}$$

where, for consistency with the previous case, we still consider the strain to lag behind the stress by the phase angle δ. Equation (29.7.13) may be written as

$$\sigma(t) = \epsilon_0 \Big(\frac{\sigma_0 \cos(\delta)}{\epsilon_0} \cos(\omega t) - \frac{\sigma_0 \sin(\delta)}{\epsilon_0} \sin(\omega t)\Big). \tag{29.7.14}$$

Let

$$E' \stackrel{\text{def}}{=} \frac{\sigma_0}{\epsilon_0}\cos(\delta), \qquad E'' \stackrel{\text{def}}{=} \frac{\sigma_0}{\epsilon_0}\sin(\delta), \tag{29.7.15}$$

so that

$$\sigma(t) = \epsilon_0 \Big(E' \cos(\omega t) - E'' \sin(\omega t)\Big). \tag{29.7.16}$$

The quantity E' is a measure of how in-phase the stress is with the strain, and is called the **storage modulus**, while E'' is a measure of how out-of-phase the stress is with the strain, and is called the **loss modulus.**[8]

In this case,

$$\begin{aligned}\epsilon(t) &= \epsilon_0 \sin(\omega t - \delta),\\ &= \epsilon_0 \cos(\delta)\sin(\omega t) - \epsilon_0 \sin(\delta)\cos(\omega t)\\ &= \sigma_0 \Big(J' \sin(\omega t) - J'' \cos(\omega t)\Big),\end{aligned}$$

and again J', the storage compliance, is a measure of how much the strain is "in-phase", and J'' is a measure of how much the strain is "out-of-phase" with the stress input.

[8] Again, the oscillatory strain and stress do not have to be in the form of cosine functions, they could also be in the form of sine functions:

$$\epsilon(t) = \epsilon_0 \sin(\omega t), \qquad \sigma(t) = \sigma_0 \sin(\omega t + \delta).$$

From (29.7.15) we note that the loss tangent may also be expressed in terms of E' and E'' as follows:

$$\tan(\delta) = \frac{E''}{E'} \,. \tag{29.7.17}$$

The storage and loss moduli, E' and E'', are usually written as the real and imaginary parts of a **complex modulus**:

$$E^* = E' + iE''. \tag{29.7.18}$$

From (29.7.9) and (29.7.18) we note that

$$\begin{aligned} J^*E^* &= (J' - iJ'')(E' + iE'') \\ &= J'\,E' + iJ'E'' - iJ''E' + J''E'' \\ &= \cos^2(\delta) + i\cos(\delta)\sin(\delta) - i\sin(\delta)\cos(\delta) + \sin^2(\delta), \end{aligned}$$

or

$$J^*E^* = 1. \tag{29.7.19}$$

29.8 Formulation for oscillatory response using complex numbers

The discussion above can be nicely synthesized using complex numbers. Both cosine and sine oscillations are handled using Euler's formula

$$e^{ix} = \cos(x) + i\sin(x).$$

Thus, for a stress input

$$\sigma(t) = \sigma_0 e^{i\omega t}, \tag{29.8.1}$$

the strain output is

$$\begin{aligned} \epsilon(t) &= \epsilon_0 e^{i(\omega t - \delta)}, \\ &= \epsilon_0\, e^{-i\delta} e^{i\omega t}, \\ &= \epsilon_0\,(\cos(\delta) - i\sin(\delta))e^{i\omega t}, \\ &= \sigma_0 \left[\frac{\epsilon_0}{\sigma_0}(\cos(\delta) - i\sin(\delta))\right] e^{i\omega t}, \\ &= \sigma_0 [J' - iJ'']e^{i\omega t}, \end{aligned}$$

or

$$\boxed{\epsilon(t) = \sigma_0 J^* e^{i\omega t} = J^*\sigma(t).} \tag{29.8.2}$$

In this case,

$$\begin{aligned} \sigma(t) &= \sigma_0 \sin(\omega t + \delta), \\ &= \sigma_0 \cos(\delta)\sin(\omega t) + \sigma_0 \sin(\delta)\cos(\omega t) \\ &= \epsilon_0 \Big(E' \sin(\omega t) + E'' \cos(\omega t)\Big), \end{aligned}$$

and again E', the storage modulus, is a measure of how much the stress is "in-phase", and E'' is a measure of how much the stress is "out-of-phase" with the strain input.

Recall that the creep compliance $J_c(t)$ was defined as the creep response to a unit step in stress:

$$J_c(t) \stackrel{\text{def}}{=} \frac{\epsilon(t)}{\sigma_0}.$$

In an entirely analogous manner, the complex compliance can be defined as the ratio of the steady-state strain response (i.e. ignoring short-time transient terms) to a sinusoidal stress input:

$$\boxed{J^*(\omega) \stackrel{\text{def}}{=} \frac{\epsilon(t)}{\sigma_0 e^{i\omega t}}.} \tag{29.8.3}$$

Similarly, for a strain input

$$\epsilon(t) = \epsilon_0 e^{i\omega t}, \tag{29.8.4}$$

the stress output is

$$\begin{aligned}
\sigma(t) &= \sigma_0 e^{i(\omega t + \delta)}, \\
&= \sigma_0\, e^{i\delta} e^{i\omega t}, \\
&= \sigma_0\, (\cos(\delta) + i \sin(\delta)) e^{i\omega t}, \\
&= \epsilon_0 \left[\frac{\sigma_0}{\epsilon_0}(\cos(\delta) + i \sin(\delta))\right] e^{i\omega t}, \\
&= \epsilon_0 [E' + iE''] e^{i\omega t},
\end{aligned}$$

or

$$\boxed{\sigma(t) = \epsilon_0 E^* e^{i\omega t} = E^* \epsilon(t).} \tag{29.8.5}$$

Recall that the stress-relaxation modulus $E_r(t)$ was defined as the stress response to a unit step in strain:

$$E_r(t) \stackrel{\text{def}}{=} \frac{\sigma(t)}{\epsilon_0}.$$

In an entirely analogous manner, the complex modulus can be defined as the ratio of the steady-state stress response (i.e. ignoring short-time transient terms) to a sinusoidal strain input:

$$\boxed{E^*(\omega) \stackrel{\text{def}}{=} \frac{\sigma(t)}{\epsilon_0 e^{i\omega t}}.} \tag{29.8.6}$$

- *Note that the complex compliance $J^*(\omega)$ and the complex modulus $E^*(\omega)$ are functions of the frequency ω, and hence so also is the loss tangent* $\tan(\delta(\omega))$.

29.8.1 **Energy dissipation under oscillatory conditions**

The stress power per unit volume is

$$\mathcal{P} = \sigma\dot{\epsilon}, \tag{29.8.7}$$

and the work done per unit volume over a time span $[0, t]$ is

$$W = \int_0^t \sigma\dot{\epsilon}\, dt. \tag{29.8.8}$$

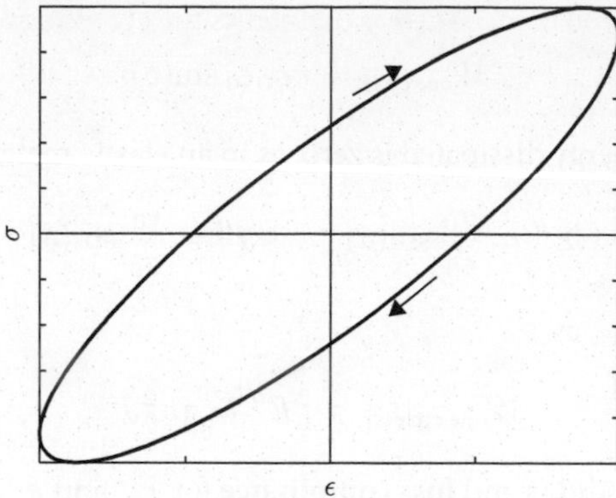

Fig. 29.17 Schematic figure showing a stress-strain hysteresis loop corresponding to an oscillatory strain input $\epsilon = \epsilon_0 \sin(\omega t)$, and the corresponding stress output $\sigma = \sigma_0 \sin(\omega t + \delta)$.

Now, consider a sinusoidal strain input

$$\epsilon(t) = \epsilon_0 \sin(\omega t), \tag{29.8.9}$$

and the corresponding stress output for a viscoelastic material

$$\sigma(t) = \sigma_0 \sin(\omega t + \delta). \tag{29.8.10}$$

A schematic of the stress-strain hysteresis loop resulting from these sinusoidal strain and stress histories is shown in Fig. 29.17.

The energy lost in *one cycle* through internal friction is given by the area of the hysteresis loop:

$$W_{\text{one cycle}} = \int_0^T \sigma \dot{\epsilon}\, dt\,, \tag{29.8.11}$$

where

$$T \stackrel{\text{def}}{=} \frac{2\pi}{\omega} \tag{29.8.12}$$

is the **period of the oscillation**.

Substituting the expression (29.8.10) for the stress, and the time derivative of (29.8.9) for the strain-rate, we get

$$\begin{aligned}
W_{\text{one cycle}} &= \omega \sigma_0 \epsilon_0 \int_0^T \sin(\omega t + \delta) \cos(\omega t)\, dt \\
&= \omega \sigma_0 \epsilon_0 \int_0^T \sin(\omega t) \cos(\omega t) \cos(\delta) + \cos^2(\omega t) \sin(\delta)\, dt \\
&= \omega \sigma_0 \epsilon_0 \int_0^T \frac{1}{2} \sin(2\omega t) \cos(\delta) + \frac{1}{2} [1 + \cos(2\omega t)] \sin(\delta)\, dt \\
&= \frac{1}{2} \omega \sigma_0 \epsilon_0\, T \sin(\delta),
\end{aligned}$$

or

$$W_{\text{one cycle}} = \pi\sigma_0\epsilon_0 \sin(\delta). \tag{29.8.13}$$

Thus, when $\delta = 0$ the energy dissipated is zero, as in an elastic material. Recalling that

$$E'' = \frac{\sigma_0}{\epsilon_0}\sin(\delta), \qquad J'' = \frac{\epsilon_0}{\sigma_0}\sin(\delta)$$

we can also write $W_{\text{one cycle}}$ as

$$W_{\text{one cycle}} = \pi\epsilon_0^2 E'' = \pi\sigma_0^2 J''\,, \tag{29.8.14}$$

hence the names **loss modulus** and **loss compliance** for E'' and J'', respectively.

Note, the energy stored after one complete strain cycle is zero because the material has been brought back to its original configuration. To understand the relation of the stored energy to the dissipated energy we instead integrate over a 1/4 cycle, $\frac{T}{4} = \pi/(2\omega)$. In this case:

$$\begin{aligned}
W_{\text{quarter cycle}} &= \int_0^{\pi/(2\omega)} \sigma\dot{\epsilon}\,dt = \omega\sigma_0\epsilon_0 \int_0^{\pi/(2\omega)} \sin(\omega t + \delta)\cos(\omega t)\,dt \\
&= \omega\sigma_0\epsilon_0 \left[\underbrace{\int_0^{\pi/(2\omega)} \sin(\omega t)\cos(\omega t)\cos(\delta)\,dt}_{\text{stored energy}} + \underbrace{\int_0^{\pi/(2\omega)} \cos^2(\omega t)\sin(\delta)\,dt}_{\text{dissipated energy}}\right] \\
&= \omega\sigma_0\epsilon_0 \left[\underbrace{\int_0^{\pi/(2\omega)} \frac{1}{2}\sin(2\omega t)\cos(\delta)\,dt}_{\text{stored energy}} + \underbrace{\int_0^{\pi/(2\omega)} \frac{1}{2}(1+\cos(2\omega t))\sin(\delta)\,dt}_{\text{dissipated energy}}\right] \\
&= \omega\sigma_0\epsilon_0 \left[\underbrace{\left(-\frac{1}{4\omega}\cos(2\omega t)\cos(\delta)\right)_0^{\pi/(2\omega)}}_{\text{stored energy}}\right. \\
&\qquad \left. + \underbrace{\left(\frac{1}{2}(t + \frac{1}{2\omega}\sin(2\omega t))\sin(\delta)\right)_0^{\pi/(2\omega)}}_{\text{dissipated energy}}\right] \\
&= \omega\sigma_0\epsilon_0 \left[\underbrace{\left(\frac{1}{2\omega}\cos(\delta)\right)}_{\text{stored energy}} + \underbrace{\left(\frac{\pi}{4\omega}\sin(\delta)\right)}_{\text{dissipated energy}}\right]
\end{aligned}$$

or

$$W_{\text{quarter cycle}} = \underbrace{\left(\frac{\sigma_0\epsilon_0}{2}\cos(\delta)\right)}_{\text{stored energy}} + \underbrace{\left(\frac{\sigma_0\epsilon_0\pi}{4}\sin(\delta)\right)}_{\text{dissipated energy}}, \tag{29.8.15}$$

where the first term on the right represents the energy stored in a quarter cycle, while the second term represents the energy dissipated in a quarter cycle.

The **damping capacity** of a viscoelastic material is defined by

$$\boxed{\text{damping capacity} \stackrel{\text{def}}{=} \frac{\text{dissipated energy in quarter cycle}}{\text{stored energy in quarter cycle}} = \frac{\pi}{2}\tan(\delta).} \tag{29.8.16}$$

Thus the damping capacity of a linearly viscoelastic material depends only on the phase or loss angle δ, which we recall is frequency-dependent.

The quantity

$$\boxed{\tan(\delta(\omega)) = \frac{E''(\omega)}{E'(\omega)} = \frac{J''(\omega)}{J'(\omega)}}$$

is referred to by a variety of names:

$$\tan(\delta) \equiv \text{loss tangent} \equiv \text{internal friction} \equiv \text{mechanical damping}.$$

It may be considered a fundamental measure of damping in a linear viscoelastic material. Typical values of $\tan(\delta)$ for a few materials at room temperature and various frequencies are shown in Table 29.1.

Table 29.1 Loss tangent for selected materials at room temperature and selected frequencies.

Material	Frequency ($\nu = \omega/2\pi$)	Loss tangent ($\tan\delta$)
Sapphire	30kHz	5×10^{-9}
Silicon	20 kHz	3×10^{-8}
Cu-31%Zn–Brass	6 kHz	9×10^{-5}
310 Stainless Steel	1 kHz	0.001
Aluminum	1 Hz	0.001
Nitinol (55Ni-45Ti)	1 kHz	0.028
PMMA	1 kHz	0.1

29.9 More on complex variable representation of linear viscoelasticity

Consider an input strain history of the form

$$\epsilon(t) = \epsilon^* \exp(i\omega t), \qquad t \in [0, \infty), \tag{29.9.1}$$

where ϵ^* is a time-independent complex strain amplitude; $\epsilon^* = |\epsilon^*|e^{i\angle\epsilon^*}$ in polar form showing its magnitude and angle. Assume that the output stress history has the form

$$\sigma(t) = \sigma^* \exp(i\omega t), \qquad t \in [0, \infty), \tag{29.9.2}$$

where $\sigma^* = |\sigma^*|e^{i\angle\sigma^*}$ is a time-independent complex stress amplitude in polar form.

As a material model for linear viscoelasticity we consider a constitutive equation of the form

$$\sigma^* = E^*(\omega)\epsilon^*, \tag{29.9.3}$$

where $E^*(\omega)$ is a frequency-dependent *complex modulus*.

Let $|E^*(\omega)|$ denote the magnitude of the complex modulus $E^*(\omega)$, and $\delta(\omega) = \angle E^*(\omega)$ the phase angle—both real-valued—so that $E^*(\omega)$ has the polar form (see Fig. 29.18)

$$E^*(\omega) = |E^*(\omega)| \exp(i\delta(\omega)). \tag{29.9.4}$$

Thus, the complex modulus consists of a real part and an imaginary part,

$$E^*(\omega) = E'(\omega) + iE''(\omega) \tag{29.9.5}$$

with

$$E'(\omega) = |E^*(\omega)| \cos\delta(\omega), \quad E''(\omega) = |E^*(\omega)| \sin\delta(\omega), \quad \text{and} \quad \tan\delta(\omega) = \frac{E''(\omega)}{E'(\omega)}. \tag{29.9.6}$$

Further, substituting (29.9.4) and (29.9.3) in (29.9.2) gives

$$\sigma(t) = [|E^*(\omega)| \exp(i\delta(\omega))]\, \epsilon(t), \tag{29.9.7}$$

which shows that the stress is phase-shifted by $\delta(\omega)$ relative to the strain.

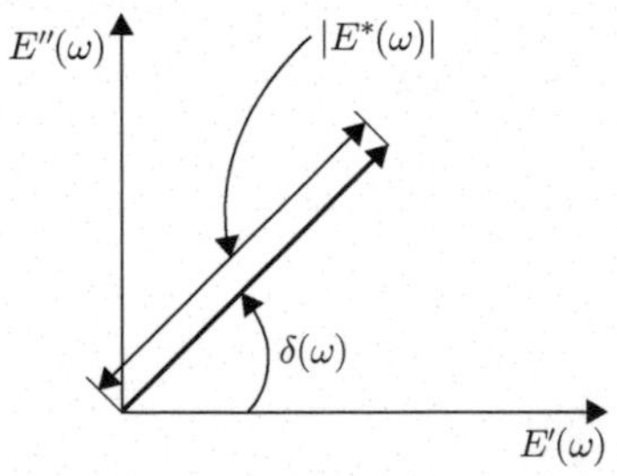

Fig. 29.18 Polar form of the complex modulus.

EXAMPLE 29.5 Storage and loss modulus for the Kelvin-Voigt solid

The Kelvin–Voigt model of a viscoelastic material consists of a spring in parallel with a dashpot. Thus the stress can be expressed as

$$\sigma(t) = E\epsilon(t) + \eta\dot{\epsilon}(t),$$

where E and η are material constants. The storage and loss moduli for this model can be found be adopting complex notation. In particular

$$\sigma^* \exp(i\omega t) = E\epsilon^* \exp(i\omega t) + \eta i\omega\epsilon^* \exp(i\omega t).$$

Canceling the exponential terms one finds that

$$\sigma^* = \underbrace{[E + i\omega\eta]}_{E^*}\,\epsilon^*\,.$$

Introducing $\tau_R = \eta/E$, we can express the complex modulus as

$$E^* = E\,[1 + i\tau_R\omega]$$

and identify the storage and loss moduli as

$$E'(\omega) = \text{Re}\{E^*\} = E,$$
$$E''(\omega) = \text{Im}\{E^*\} = E\tau_R\omega.$$

29.9.1 $E'(\omega)$, $E''(\omega)$, and $\tan\delta(\omega)$ for the standard linear solid

Consider the standard linear solid depicted as in Fig. 29.19. Here, $E^{(0)}$ and $(E^{(1)}, \eta^{(1)})$ are the constitutive moduli in the two branches of the standard linear solid, with $\tau_R^{(1)} = \eta^{(1)}/E^{(1)}$ the relaxation time for the branch with the Maxwell element.[9] For the standard linear solid the constitutive equations are

$$\boxed{\begin{aligned} &\sigma(t) = E^{(0)}\epsilon(t) + E^{(1)}(\epsilon(t) - \epsilon^{v(1)}(t)),\\ &\dot{\epsilon}^{v(1)}(t) = \frac{1}{\tau_R^{(1)}}(\epsilon(t) - \epsilon^{v(1)}(t)), \qquad \tau_R^{(1)} \overset{\text{def}}{=} \frac{\eta^{(1)}}{E^{(1)}}, \end{aligned}} \tag{29.9.8}$$

where the quantity $\epsilon^{v(1)}$ appears as an *internal variable of the theory* which evolves according to the *flow rule* (29.9.8)$_2$.

[9] We use the superscripts (1) for the properties $E^{(1)}$ and $\eta^{(1)}$ of the spring and dashpot in the Maxwell element, because shortly we will extend this model in Example 29.6 to a model called the "generalized Maxwell model" which contains several Maxwell elements in parallel.

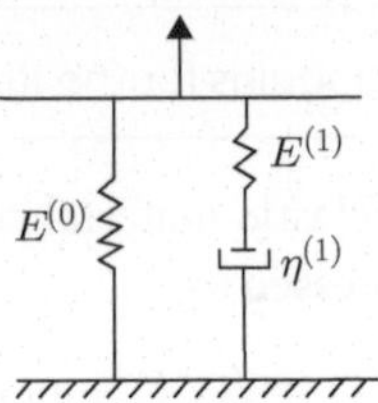

Fig. 29.19 Standard linear solid.

The model relates the three time-dependent functions,

$$\epsilon(t), \quad \sigma(t), \quad \text{and} \quad \epsilon^{v(1)}(t).$$

Consider a cyclic test with an input sinusoidal strain of the form

$$\epsilon(t) = \epsilon^* \exp(i\omega t).$$

We wish to solve for $\sigma(t)$ for the standard linear solid, and hence determine $E'(\omega)$, $E''(\omega)$, and $\tan\delta(\omega)$ for this material model. The solution in general depends on the initial condition. However, after the transients damp out the solution for $\sigma(t)$ will become sinusoidal, with some phase-shift relative to the input strain history. It is this steady state that interests us. Accordingly, we look for a *steady-state* solution of the form

$$\begin{aligned} \sigma(t) &= \sigma^* \exp(i\omega t), \\ \epsilon^{v(1)}(t) &= \epsilon^{v(1)*} \exp(i\omega t). \end{aligned} \tag{29.9.9}$$

In terms of the complex amplitudes the model (29.9.8) becomes

$$\begin{aligned} \sigma^* &= E^{(0)}\epsilon^* + E^{(1)}(\epsilon^* - \epsilon^{v(1)*}) \\ (i\omega)\epsilon^{v(1)*} &= \frac{1}{\tau_R^{(1)}}(\epsilon^* - \epsilon^{v(1)*}), \end{aligned} \tag{29.9.10}$$

which is a pair of *algebraic equations*. Equation $(29.9.10)_2$ gives

$$\epsilon^{v(1)*} = \left(\frac{1}{1 + (i\tau_R^{(1)}\omega)}\right)\epsilon^*, \tag{29.9.11}$$

substitution of which in $(29.9.10)_1$ gives

$$\sigma^* = \left(E^{(0)} + \left(\frac{E^{(1)}(i\tau_R^{(1)}\omega)}{1 + (i\tau_R^{(1)}\omega)}\right)\right)\epsilon^*. \tag{29.9.12}$$

Recalling the definition (29.9.3) for the complex modulus $E^*(\omega)$, we have

$$\boxed{E^*(\omega) \stackrel{\text{def}}{=} E^{(0)} + \frac{E^{(1)}(i\tau_R^{(1)}\omega)}{1 + (i\tau_R^{(1)}\omega)}.} \tag{29.9.13}$$

We may rewrite (29.9.13) as

$$\boxed{E^*(\omega) \overset{\text{def}}{=} \underbrace{\left[E^{(0)} + \frac{E^{(1)}(\tau_R^{(1)}\omega)^2}{1+(\tau_R^{(1)}\omega)^2}\right]}_{E'(\omega)} + i \underbrace{\left[\frac{E^{(1)}(\tau_R^{(1)}\omega)}{1+(\tau_R^{(1)}\omega)^2}\right]}_{E''(\omega)}.} \tag{29.9.14}$$

Recall that for a standard linear solid, the "glassy" and "equilibrium" relaxation moduli are given by

$$\boxed{E_{rg} \overset{\text{def}}{=} E^{(0)} + E^{(1)}, \qquad E_{re} \overset{\text{def}}{=} E^{(0)}.} \tag{29.9.15}$$

Hence,

$$\boxed{\begin{aligned} E^*(\omega) &= E'(\omega) + i\,E''(\omega) \\ E'(\omega) &\overset{\text{def}}{=} E_{re} + \frac{(E_{rg}-E_{re})(\tau_R^{(1)}\omega)^2}{1+(\tau_R^{(1)}\omega)^2} = \frac{E_{re} + E_{rg}(\tau_R^{(1)}\omega)^2}{1+(\tau_R^{(1)}\omega)^2} \\ E''(\omega) &\overset{\text{def}}{=} \frac{(E_{rg}-E_{re})(\tau_R^{(1)}\omega)}{1+(\tau_R^{(1)}\omega)^2} \\ \tan\delta(\omega) &= \frac{(E_{rg}-E_{re})(\tau_R^{(1)}\omega)}{E_{re} + E_{rg}(\tau_R^{(1)}\omega)^2}, \end{aligned}} \tag{29.9.16}$$

- *Observe the frequency dependency of the storage modulus E', the loss modulus E'', and* $\tan(\delta)$.

An important dimensionless number for viscoelastic materials is the **Deborah number** which is defined as

$$\text{De} \overset{\text{def}}{=} \frac{\text{relaxation time}}{\text{timescale of experiment}} \equiv \tau_R^{(1)}\omega. \tag{29.9.17}$$

Figure 29.20 shows a plot of the storage modulus E', loss modulus E'', and $\tan(\delta)$ as functions of the Deborah number $\tau_R\omega$ for a standard linear solid, for the following material properties:

$$E_{re} = 1\,\text{MPa}, \qquad E_{rg} = 1\,\text{GPa}, \qquad \text{and} \qquad \tau_R^{(1)} = 1\,\text{s}.$$

Thus for the standard linear solid we note that:

- When the loading frequency is low relative to the relaxation time, that is when the Deborah number is small, $\tau_R^{(1)}\omega \ll 1$, the storage modulus is the same as the relaxed modulus $E' = E_{re}$.
- When the loading frequency is high relative to the relaxation time, that is when the Deborah number is large, $\tau_R^{(1)}\omega \gg 1$, the storage modulus is the same as the glassy or unrelaxed modulus $E' = E_{rg}$.
- The loss modulus E'' has a maximum at some intermediate frequency, in the vicinity of $\tau_R^{(1)}\omega \approx 1$, as also does $\tan(\delta)$.

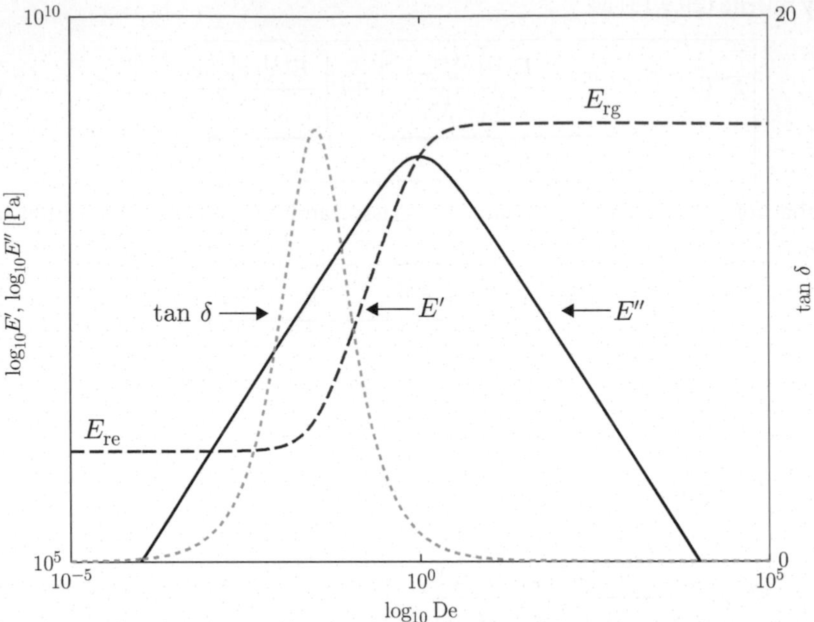

Fig. 29.20 Storage modulus E', loss modulus E'', and $\tan(\delta)$ as functions of the Deborah number $\mathrm{De} = \tau_R\omega$ for a standard linear solid.

Such a frequency-dependent variation of the storage modulus E', loss modulus E'', and $\tan(\delta)$ is common to most polymeric materials. At low Deborah numbers a polymer is said to behave rubber-like and has a low storage modulus $E' = E_{re}$, which is largely independent of frequency. At high Deborah numbers a polymer is glassy with a storage modulus $E' = E_{rg}$, which is again largely independent of frequency. At intermediate frequencies a polymer behaves as a viscoelastic solid, and its storage modulus increases with increasing frequency. The loss modulus and E'' and $\tan(\delta)$ are near zero at low and high Deborah numbers, and peak in the range of intermediate Deborah numbers.

EXAMPLE 29.6 Generalized Maxwell model in one-dimension

(1) Consider the generalization of the standard linear solid shown in Fig. 29.21, which consists of a linear spring of modulus $E^{(0)}$ in parallel with several Maxwell elements, with elastic constants, viscosities, and relaxation times

$$E^{(\alpha)}, \qquad \eta^{(\alpha)}, \qquad \tau_R^{(\alpha)} \stackrel{\text{def}}{=} \frac{\eta^{(\alpha)}}{E^{(\alpha)}}, \qquad \alpha = 1, \ldots, M,$$

and derive the constitutive equation for $\sigma(t)$ in terms of an input strain history $\epsilon(t)$.

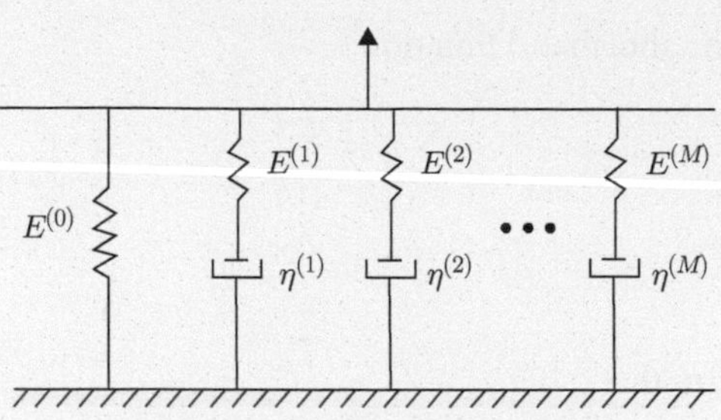

Fig. 29.21 Generalized Maxwell model.

(2) In a stress-relaxation test a constant strain ϵ_0 is applied at time $t = 0$ and thereafter held constant. Derive the stress relaxation modulus

$$E_r(t) = \frac{\sigma(t)}{\epsilon_0} \tag{29.9.18}$$

for the rheological model shown in Fig. 29.21, in terms of the properties $E^{(0)}$, $E^{(\alpha)}$, and $\tau_R^{(\alpha)}$. What is the equilibrium value $E_{re} \stackrel{\text{def}}{=} E_r(\infty)$ for the generalized Maxwell model?

(3) Consider an input strain history of the form

$$\epsilon(t) = \epsilon^* \exp(i\omega t), \tag{29.9.19}$$

where ϵ^* is a time-independent complex strain amplitude. Solve for a steady-state solution for $\sigma(t)$ in the form

$$\sigma(t) = \sigma^* \exp(i\omega t), \tag{29.9.20}$$

where σ^* is a complex stress related to ϵ^* by a constitutive equation of the form

$$\sigma^* = E^*(\omega)\epsilon^*, \tag{29.9.21}$$

where $E^*(\omega)$ is a frequency-dependent *complex modulus*, which consists of a real part and an imaginary part:

$$E^*(\omega) = E'(\omega) + iE''(\omega). \tag{29.9.22}$$

Show that for the generalized Maxwell model

$$\boxed{\begin{aligned} E'(\omega) &= E^{(0)} + \sum_\alpha \frac{E^{(\alpha)}(\omega\tau_R^{(\alpha)})^2}{1+(\omega\tau_R^{(\alpha)})^2}, \\ E''(\omega) &= \sum_\alpha \frac{E^{(\alpha)}(\omega\tau_R^{(\alpha)})}{1+(\omega\tau_R^{(\alpha)})^2}, \\ \tan\delta(\omega) &= \frac{E''(\omega)}{E'(\omega)}, \end{aligned}} \tag{29.9.23}$$

where we have used the shorthand notation

$$\sum_{\alpha} \equiv \sum_{\alpha=1}^{M}.$$

Solution

(1) Kinematics requires that

$$\epsilon(t) = \epsilon^{(0)}(t) = \epsilon^{(\alpha)}(t), \tag{29.9.24}$$

and equilibrium requires that

$$\sigma(t) = \sigma^{(0)}(t) + \sum_{\alpha} \sigma^{(\alpha)}(t). \tag{29.9.25}$$

The stress in branch (0) is given by the constitutive equation

$$\sigma^{(0)}(t) = E^{(0)}\epsilon^{(0)}(t), \tag{29.9.26}$$

while the stress in the αth Maxwell element is given by

$$\left.\begin{aligned} \sigma^{(\alpha)}(t) &= E^{(\alpha)}\left(\epsilon^{(\alpha)}(t) - \epsilon^{v(\alpha)}(t)\right), \\ \dot{\epsilon}^{v(\alpha)}(t) &= \frac{1}{\tau_R^{(\alpha)}}\left(\epsilon^{(\alpha)}(t) - \epsilon^{v(\alpha)}(t)\right). \end{aligned}\right\} \tag{29.9.27}$$

Substituting (29.9.26) and (29.9.27) in (29.9.25), and using (29.9.24) we obtain,

$$\boxed{\begin{aligned} \sigma(t) &= E^{(0)}\epsilon^{(0)}(t) + \sum_{\alpha} E^{(\alpha)}\left(\epsilon(t) - \epsilon^{v(\alpha)}(t)\right), \\ \dot{\epsilon}^{v(\alpha)}(t) &= \frac{1}{\tau_R^{(\alpha)}}\left(\epsilon(t) - \epsilon^{v(\alpha)}(t)\right). \end{aligned}} \tag{29.9.28}$$

(2) For a constant input strain history,

$$\epsilon(t) \equiv \epsilon_0 \qquad \text{for} \qquad t > 0,$$

(29.9.28) reduces to

$$\left.\begin{aligned} \sigma(t) &= E^{(0)}\epsilon_0 + \sum_{\alpha} E^{(\alpha)}\left(\epsilon_0 - \epsilon^{v(\alpha)}(t)\right), \\ \dot{\epsilon}^{v(\alpha)}(t) &= -\frac{\epsilon^{v(\alpha)}(t)}{\tau_R^{(\alpha)}} + \frac{\epsilon_0}{\tau_R^{(\alpha)}}. \end{aligned}\right\} \tag{29.9.29}$$

Equations $(29.9.29)_2$ for each α may be integrated, subject to the initial conditions

$$\epsilon^{v(\alpha)}(0^+) = 0,$$

to give

$$\epsilon^{v(\alpha)}(t) = \epsilon_0 \left(1 - \exp\left(\frac{-t}{\tau_R^{(\alpha)}}\right)\right), \tag{29.9.30}$$

substitution of which in $(29.9.29)_1$, and factoring out ϵ_0, gives the stress relaxation modulus as

$$\boxed{E_r(t) = \frac{\sigma(t)}{\epsilon_0} = E^{(0)} + \sum_\alpha E^{(\alpha)} \exp\left(-\frac{t}{\tau_R^{(\alpha)}}\right).} \tag{29.9.31}$$

This form of the stress-relaxation function is called a **Prony series**. From (29.9.31), we see that at long time, $t \to \infty$, the relaxation modulus is simply

$$\boxed{E_r(\infty) = E^{(0)}.} \tag{29.9.32}$$

(3) For a cyclic input, we seek a solution of the form

$$\begin{aligned} \epsilon(t) &= \epsilon^* \exp(i\omega t), \\ \sigma(t) &= \sigma^* \exp(i\omega t), \\ \epsilon^{v(\alpha)}(t) &= \epsilon^{v(\alpha)*} \exp(i\omega t). \end{aligned} \tag{29.9.33}$$

In terms of the complex amplitudes the model (29.9.28) becomes

$$\begin{aligned} \sigma^* &= E^{(0)}\epsilon^* + \sum_\alpha E^{(\alpha)}(\epsilon^* - \epsilon^{v(\alpha)*}), \\ (i\omega)\epsilon^{v(\alpha)*} &= \frac{1}{\tau_R^{(\alpha)}}(\epsilon^* - \epsilon^{v(\alpha)*}), \end{aligned} \tag{29.9.34}$$

which is a set of $M+1$ *algebraic equations*. Equation $(29.9.34)_2$ gives

$$\epsilon^{v(\alpha)*} = \left(\frac{1}{1 + (i\omega\tau_R^{(\alpha)})}\right)\epsilon^*, \tag{29.9.35}$$

substitution of which in $(29.9.34)_1$ gives

$$\sigma^* = \left(E^{(0)} + \sum_\alpha \frac{E^{(\alpha)}(i\omega\tau_R^{(\alpha)})}{1 + (i\omega\tau_R^{(\alpha)})}\right)\epsilon^*. \tag{29.9.36}$$

Recalling the definition (29.9.21) for the complex modulus $E^*(\omega)$, we have

$$\boxed{E^*(\omega) \stackrel{\text{def}}{=} E^{(0)} + \sum_\alpha \frac{E^{(\alpha)}(i\omega\tau_R^{(\alpha)})}{1 + (i\omega\tau_R^{(\alpha)})}.} \tag{29.9.37}$$

Since,

$$\frac{E^{(\alpha)}(i\omega\tau_R^{(\alpha)})}{1+(i\omega\tau_R^{(\alpha)})} = \left(\frac{E^{(\alpha)}(i\omega\tau_R^{(\alpha)})}{1+(i\omega\tau_R^{(\alpha)})}\right) \times \left(\frac{1-(i\omega\tau_R^{(\alpha)})}{1-(i\omega\tau_R^{(\alpha)})}\right)$$

$$= \left(\frac{E^{(\alpha)}(i\omega\tau_R^{(\alpha)}) + E^{(\alpha)}(\omega\tau_R^{(\alpha)})^2}{1+(\omega\tau_R^{(\alpha)})^2}\right)$$

$$= \frac{E^{(\alpha)}(\omega\tau_R^{(\alpha)})^2}{1+(\omega\tau_R^{(\alpha)})^2} + i\,\frac{E^{(\alpha)}(\omega\tau_R^{(\alpha)})}{1+(\omega\tau_R^{(\alpha)})^2},$$

we may rewrite (29.9.37) as

$$E^*(\omega) \stackrel{\text{def}}{=} \underbrace{\left[E^{(0)} + \sum_\alpha \frac{E^{(\alpha)}(\omega\tau_R^{(\alpha)})^2}{1+(\omega\tau_R^{(\alpha)})^2}\right]}_{E'(\omega)} + i\underbrace{\left[\sum_\alpha \frac{E^{(\alpha)}(\omega\tau_R^{(\alpha)})}{1+(\omega\tau_R^{(\alpha)})^2}\right]}_{E''(\omega)}, \tag{29.9.38}$$

which gives

$$E'(\omega) = E^{(0)} + \sum_\alpha \frac{E^{(\alpha)}(\omega\tau_R^{(\alpha)})^2}{1+(\omega\tau_R^{(\alpha)})^2},$$

$$E''(\omega) = \sum_\alpha \frac{E^{(\alpha)}(\omega\tau_R^{(\alpha)})}{1+(\omega\tau_R^{(\alpha)})^2}, \tag{29.9.39}$$

$$\tan\delta(\omega) = \frac{E''(\omega)}{E'(\omega)}.$$

29.10 Time-integration procedure for one-dimensional constitutive equations for linear viscoelasticity based on the generalized Maxwell model

For a linear viscoelastic material the response, using the generalized Maxwell model, is given by

$$\sigma(t) = \int_{0^-}^{t} E_r(t-\tau)\frac{d\epsilon(\tau)}{d\tau}\,d\tau \qquad \text{with} \qquad E_r(t) = E^{(0)} + \sum_\alpha E^{(\alpha)} \exp\left(-\frac{t}{\tau_R^{(\alpha)}}\right), \tag{29.10.1}$$

where $E_r(t)$ defines the relaxation function based on the generalized Maxwell material. Hence,

$$
\begin{aligned}
\sigma(t) &= \int_{0-}^{t} E_r(t-\tau)\frac{d\epsilon(\tau)}{d\tau}\,d\tau, \\
&= E^{(0)}\epsilon(t) + \sum_{\alpha} \underbrace{\int_{0-}^{t} E^{(\alpha)} \exp\left(-\frac{t-\tau}{\tau_R^{(\alpha)}}\right)\frac{d\epsilon(\tau)}{d\tau}\,d\tau}_{g^{(\alpha)}(t)}, \\
&= E^{(0)}\epsilon(t) + \sum_{\alpha} g^{(\alpha)}(t).
\end{aligned}
\tag{29.10.2}
$$

In this section we formulate a numerical time-integration procedure to calculate the history variables

$$
g^{(\alpha)}(t) \stackrel{\text{def}}{=} \int_{0-}^{t} E^{(\alpha)} \exp\left(-\frac{t-\tau}{\tau_R^{(\alpha)}}\right)\frac{d\epsilon(\tau)}{d\tau}\,d\tau \tag{29.10.3}
$$

in $(29.10.2)_3$ for a deformation-driven problem in a time interval $[0, T]$. This time-integration procedure is known as the *Herrmann–Peterson recursion relation* and is based on the work of Herrmann and Peterson (1968) and Taylor et al. (1970).

Our objective is to develop a time-integration method that allows us to compute $g^{(\alpha)}(t_{n+1})$ assuming $g^{(\alpha)}(t_n)$ is known and $t_{n+1} = t_n + \Delta t$, $\Delta t > 0$. Noting that

$$
\exp\left(-\frac{t_{n+1}}{\tau_R^{(\alpha)}}\right) = \exp\left(-\frac{t_n + \Delta t}{\tau_R^{(\alpha)}}\right) = \exp\left(-\frac{t_n}{\tau_R^{(\alpha)}}\right)\exp\left(-\frac{\Delta t}{\tau_R^{(\alpha)}}\right), \tag{29.10.4}
$$

and separating the deformation history into a period $0 \leq \tau \leq t_n$ for which the result is presumed to be known, and into the current unknown time step $t_n \leq \tau \leq t_{n+1}$, yields

$$
\begin{aligned}
g^{(\alpha)}(t_{n+1}) &= E^{(\alpha)} \int_{0-}^{t_{n+1}} \exp\left(-\frac{t_{n+1}-\tau}{\tau_R^{(\alpha)}}\right)\frac{d\epsilon(\tau)}{d\tau}\,d\tau \\
&= E^{(\alpha)} \int_{0-}^{t_n} \exp\left(-\frac{t_{n+1}-\tau}{\tau_R^{(\alpha)}}\right)\frac{d\epsilon(\tau)}{d\tau}\,d\tau + E^{(\alpha)} \int_{t_n}^{t_{n+1}} \exp\left(-\frac{t_{n+1}-\tau}{\tau_R^{(\alpha)}}\right) \\
&\qquad \frac{d\epsilon(\tau)}{d\tau}\,d\tau \\
&= \exp\left(-\frac{\Delta t}{\tau_R^{(\alpha)}}\right) E^{(\alpha)} \int_{0}^{t_n} \exp\left(-\frac{t_n-\tau}{\tau_R^{(\alpha)}}\right)\frac{d\epsilon(\tau)}{d\tau}\,d\tau \\
&\quad + E^{(\alpha)} \int_{t_n}^{t_{n+1}} \exp\left(-\frac{t_{n+1}-\tau}{\tau_R^{(\alpha)}}\right)\frac{d\epsilon(\tau)}{d\tau}\,d\tau \\
&= \exp\left(-\frac{\Delta t}{\tau_R^{(\alpha)}}\right) g^{(\alpha)}(t_n) + E^{(\alpha)} \int_{t_n}^{t_{n+1}} \exp\left(-\frac{t_{n+1}-\tau}{\tau_R^{(\alpha)}}\right)\frac{d\epsilon(\tau)}{d\tau}\,d\tau.
\end{aligned}
\tag{29.10.5}
$$

Observing the definition of the derivative,

$$\frac{d\epsilon(\tau)}{d\tau} = \lim_{\Delta\tau\to 0} \frac{\Delta\epsilon(\tau)}{\Delta\tau} = \lim_{\Delta t\to 0} \frac{\epsilon(t_{n+1}) - \epsilon(t_n)}{\Delta t}, \tag{29.10.6}$$

allows for the approximate expression

$$\begin{aligned} g^{(\alpha)}(t_{n+1}) &\approx \exp\left(-\frac{\Delta t}{\tau_R^{(\alpha)}}\right) g^{(\alpha)}(t_n) \\ &\quad + E^{(\alpha)} \left(\frac{\epsilon(t_{n+1}) - \epsilon(t_n)}{\Delta t}\right) \int_{t_n}^{t_{n+1}} \exp\left(-\frac{t_{n+1} - \tau}{\tau_R^{(\alpha)}}\right) d\tau. \end{aligned} \tag{29.10.7}$$

Now,

$$\int \exp\left(-\frac{a-x}{b}\right) dx = b \exp\left(\frac{x-a}{b}\right) + \text{const},$$

hence

$$\begin{aligned} \int_{t_n}^{t_{n+1}} \exp\left(-\frac{t_{n+1} - \tau}{\tau_R^{(\alpha)}}\right) d\tau &= \left[\tau_R^{(\alpha)} \exp\left(\frac{\tau - t_{n+1}}{\tau_R^{(\alpha)}}\right)\right]_{t_n}^{t_{n+1}} \\ &= \tau_R^{(\alpha)} \left[\exp\left(\frac{t_{n+1} - t_{n+1}}{\tau_R^{(\alpha)}}\right) - \exp\left(\frac{t_n - t_{n+1}}{\tau_R^{(\alpha)}}\right)\right] \\ &= \tau_R^{(\alpha)} \left[1 - \exp\left(-\frac{\Delta t}{\tau_R^{(\alpha)}}\right)\right], \end{aligned} \tag{29.10.8}$$

use of which in (29.10.7), gives the *Herrmann–Peterson recursion relation*:

$$\boxed{\begin{aligned} \sigma(t_{n+1}) &= E^{(0)}\epsilon(t_{n+1}) + \sum_{\alpha} g^{(\alpha)}(t_{n+1}) \\ g^{(\alpha)}(t_{n+1}) &= \exp\left(-\frac{\Delta t}{\tau_R^{(\alpha)}}\right) g^{(\alpha)}(t_n) + E^{(\alpha)} \frac{\tau_R^{(\alpha)}}{\Delta t} \left[1 - \exp\left(-\frac{\Delta t}{\tau_R^{(\alpha)}}\right)\right] \Delta\epsilon, \end{aligned}} \tag{29.10.9}$$

where $g^{(\alpha)}(t_n)$ is the value of the αth convolution integral at time $t = t_n$, $\Delta t = t_{n+1} - t_n$, and $\Delta\epsilon = \epsilon(t_{n+1}) - \epsilon(t_n)$.

This strain-driven integration algorithm is unconditionally stable for small and large time steps and it is second-order accurate. It is easily generalized to three dimensions for isotropic linear viscoelastic constitutive equations, and is widely used in finite element implementations of linear viscoelasticity.

EXAMPLE 29.7 Numerical computation of transient response in a viscoelastic material

Consider a standard linear solid in one dimension with $E^{(0)} = 1$ MPa, $E^{(1)} = 1$ MPa, and $\eta = 1$ MPa s. Taking the strain input as

$$\epsilon(t) = h(t)\sin(\omega t) \quad \text{with} \quad \omega = 1 \text{ rad/s},$$

we can use the Herrmann–Peterson recursion relations (cf. (29.10.9)) to compute and plot the stress response in any time interval, say, $T = [0, 6\pi]$ s.

For the given strain history and material properties, the resulting stress and strain histories and cross-plot are computed to be as in Fig. 29.22.

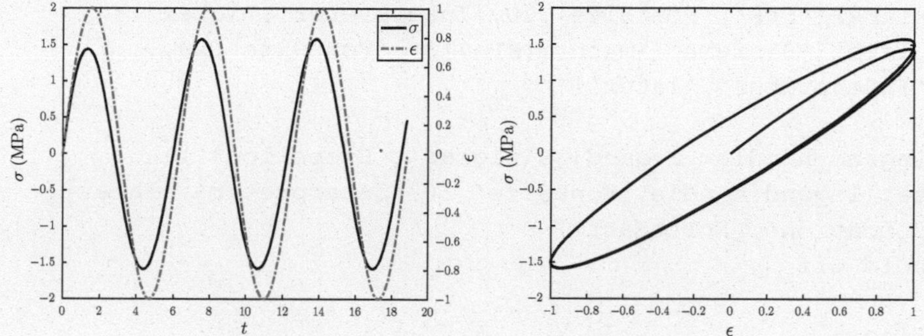

Fig. 29.22 Stress and strain time histories and cross-plot.

A simple Matlab program to generate the response:

```
tot_time    = 6*pi;           % total time
N           = 1000;           % number of increments
dt          = tot_time/N;     % increment in time
time        = zeros(N,1);     % time
epsilon     = zeros(N,1);     % strain
sigma       = zeros(N,1);     % stress
g1          = zeros(N,1);     % contribution to stress
omega       = 1;              % frequency rads/sec

for i=2:N
    time(i)  = time(i-1)+dt;
    if (time(i) <= 0.)
       epsilon(i)  =  0.;
    else
       epsilon(i)   =  sin(omega*time(i));
    end

    de = epsilon(i) - epsilon(i-1);
```

```
    g1(i) = exp(-dt/1)*g1(i-1) + (1/dt)*(1- exp(-dt/1))*de;
    sigma(i) = epsilon(i) + g1(i);
end

figure
yyaxis right;
plot(time,epsilon,'k-.','LineWidth',2);
ylabel('$\epsilon$','FontSize',20,'Interpreter','latex');
hold on;
yyaxis left;
plot(time,sigma,'k-','LineWidth',2);
xlabel('$t$','FontSize',20,'Interpreter','latex');
ylabel('$\sigma (\mathrm{MPa})$','FontSize',20,...
'Interpreter','latex');

legend_handle= legend('$\sigma$','$\epsilon$');
set(legend_handle,'FontSize',20,'Interpreter','latex',...
'Location','NorthEast')
hold off

figure
plot(epsilon,sigma,'k-','LineWidth',2);
xlabel('$\epsilon$','FontSize',20,'Interpreter','latex');
ylabel('$\sigma (\mathrm{MPa})$','FontSize',20,...
'Interpreter','latex');
```

29.11 **Three-dimensional constitutive equation for linear viscoelasticity**

Recall the one-dimensional stress-relaxation-integral form of the stress-strain relations (cf. (29.2.7)):

$$\sigma(t) = \int_{0-}^{t} E_r(t-\tau)\frac{d\epsilon(\tau)}{d\tau}d\tau. \tag{29.11.1}$$

This stress-strain relation is generalized to three dimensions below; we limit our attention to *isotropic* materials.

Recall that for a linear elastic material the stress-strain relation may be written as

$$\boldsymbol{\sigma} = \mathbb{C}\boldsymbol{\epsilon}, \tag{29.11.2}$$

where the fourth-order elasticity tensor $\mathbb{C}$ has the component representation[10]

$$C_{ijkl} = 2G\left[\frac{1}{2}(\delta_{ik}\delta_{jl} + \delta_{il}\delta_{jk}) - \frac{1}{3}\delta_{ij}\delta_{kl}\right] + K\delta_{ij}\delta_{kl}\,. \tag{29.11.3}$$

For a linear viscoelastic material the corresponding stress-strain relation is

$$\boxed{\boldsymbol{\sigma}(t) = \int_{0-}^{t} \mathbb{G}(t-\tau) : \frac{d\boldsymbol{\epsilon}(\tau)}{d\tau}\,d\tau.} \tag{29.11.4}$$

where $\mathbb{G}(t)$ is the fourth-order **stress-relaxation tensor** with components

$$G_{ijkl}(t) \stackrel{\text{def}}{=} 2G_r(t)\left[\frac{1}{2}(\delta_{ik}\delta_{jl} + \delta_{il}\delta_{jk}) - \frac{1}{3}\delta_{ij}\delta_{kl}\right] + K_r(t)\delta_{ij}\delta_{kl}, \tag{29.11.5}$$

where $G_r(t)$ is the **stress-relaxation shear modulus** and $K_r(t)$ is the **stress-relaxation bulk modulus.**

Experimental investigations have shown that for many polymeric materials, viscoelastic behavior is mainly related to the isochoric part of the deformation, and that the volume dilatation may be considered as being purely elastic,

$$K_r(t) = K \equiv \text{constant}.$$

This results in considerably simplified expressions for the stress-strain relations for isotropic linear viscoelastic materials, viz.

$$\boxed{\sigma_{ij}(t) = \int_{0-}^{t} 2G_r(t-\tau)\frac{d\epsilon'_{ij}(\tau)}{d\tau}\,d\tau + K\,\epsilon_{kk}(t)\delta_{ij}.} \tag{29.11.6}$$

In direct notation, this stress-strain relation reads as

$$\boxed{\boldsymbol{\sigma}(t) = \int_{0-}^{t} 2G_r(t-\tau)\frac{d\boldsymbol{\epsilon}'(\tau)}{d\tau}\,d\tau + K\ \operatorname{tr}\boldsymbol{\epsilon}(t)\mathbf{1}.} \tag{29.11.7}$$

REMARK 29.3

A widely used model for $G_r(t)$ in the constitutive equation (29.11.7), is based on the generalized Maxwell model shown in Fig. 29.23, which consists of a linear spring of modulus $G^{(0)}$, in parallel with several Maxwell elements, with elastic constants, viscosities, and relaxation times

$$G^{(\alpha)}, \qquad \eta^{(\alpha)}, \qquad \tau_R^{(\alpha)} \stackrel{\text{def}}{=} \frac{\eta^{(\alpha)}}{G^{(\alpha)}}, \qquad \alpha = 1, \ldots, M.$$

---

[10] In our previous discussion of linear elasticity we had denoted the shear and bulk modulus by μ and κ, respectively. Here we denote these moduli by G and K. Both notations (μ, κ) and (G, K) for these moduli are widely used in the literature, with (G, K) being more common in the viscoelasticity literature.

REMARK 29.3 Continued

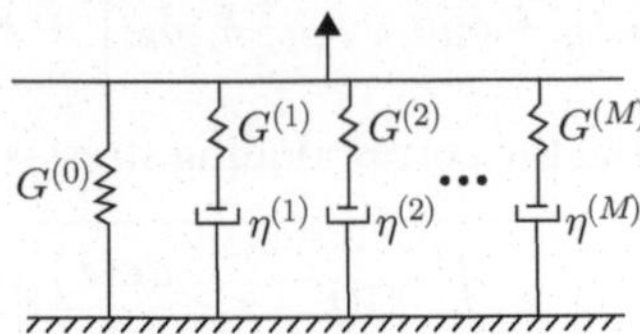

Fig. 29.23 Generalized Maxwell model in terms of shear moduli and viscosities.

Mimicking the derivations leading to (29.9.31) leads to the following **Prony series** form for the stress-relaxation modulus in shear:

$$G_r(t) = G^{(0)} + \sum_{\alpha} G^{(\alpha)} \exp\left(-\frac{t}{\tau_R^{(\alpha)}}\right). \tag{29.11.8}$$

29.11.1 Boundary-value problem for isotropic linear viscoelasticity

The three-dimensional initial-value boundary-value problem for a viscoelastic body can be formulated as follows. First assume quiescent conditions up to time $t = 0$, such that

$$\left.\begin{aligned} \mathbf{u}(\mathbf{x}, t) &= \mathbf{0} \\ \boldsymbol{\epsilon}(\mathbf{x}, t) &= \mathbf{0} \\ \boldsymbol{\sigma}(\mathbf{x}, t) &= \mathbf{0} \end{aligned}\right\} \qquad \text{for all } \mathbf{x} \in \mathcal{B} \text{ and all } t \leq 0. \tag{29.11.9}$$

Then, neglecting inertial effects, the basic equations of the theory are

$$\left.\begin{aligned} \boldsymbol{\epsilon} &= \frac{1}{2}\left(\nabla\mathbf{u} + (\nabla\mathbf{u})^{\top}\right) \\ \operatorname{div}\boldsymbol{\sigma} + \mathbf{b} &= \mathbf{0} \\ \boldsymbol{\sigma}(t) &= \int_{0-}^{t} 2G_r(t-\tau)\frac{d\boldsymbol{\epsilon}'(\tau)}{d\tau}d\tau + K \operatorname{tr}\boldsymbol{\epsilon}(t)\mathbf{1} \end{aligned}\right\} \qquad \text{for all } \mathbf{x} \in \mathcal{B} \text{ and all } t > 0. \tag{29.11.10}$$

Further, the boundary conditions—with $\mathcal{S}_1$ and $\mathcal{S}_2$ complementary subsurfaces of the boundary $\partial\mathcal{B}$—are:

$$\begin{aligned} \mathbf{u}(\mathbf{x}, t) &= \underbrace{\hat{\mathbf{u}}(\mathbf{x}, t)}_{\text{prescribed}}, \qquad \text{for } \mathbf{x} \in \mathcal{S}_1 \text{ and all } t > 0, \\ \boldsymbol{\sigma}(\mathbf{x}, t)\mathbf{n}(\mathbf{x}) &= \underbrace{\hat{\mathbf{t}}(\mathbf{x}, t)}_{\text{prescribed}}, \qquad \text{for } \mathbf{x} \in \mathcal{S}_2 \text{ and all } t > 0. \end{aligned} \tag{29.11.11}$$

29.11.2 Correspondence principle in three dimensions

Recall that the Laplace transform of a scalar, vector, or tensor-valued function $\varphi(\mathbf{x}, t)$ is

$$\bar{\varphi}(\mathbf{x}, s) = \int_{0-}^{\infty} e^{-st}\varphi(\mathbf{x}, t)dt. \tag{29.11.12}$$

For conditions under which we can take the derivative inside the integral sign, we have

$$\nabla\bar{\varphi}(\mathbf{x}, s) = \int_{0-}^{\infty} e^{-st}\nabla\varphi(\mathbf{x}, t)dt\,. \tag{29.11.13}$$

Hence taking the Laplace transform of all relevant quantities in (29.11.10) and (29.11.11), we have

$$\left.\begin{aligned} \bar{\boldsymbol{\epsilon}} &= \frac{1}{2}\left(\nabla\bar{\mathbf{u}} + (\nabla\bar{\mathbf{u}})^{\top}\right) \\ \operatorname{div}\bar{\boldsymbol{\sigma}} + \bar{\mathbf{b}} &= \mathbf{0} \\ \bar{\boldsymbol{\sigma}} &= \bar{\mathbb{C}}:\bar{\boldsymbol{\epsilon}}, \qquad \text{where} \qquad \bar{\mathbb{C}} \overset{\text{def}}{=} s\bar{\mathbb{G}}(s) \end{aligned}\right\} \quad \text{in } \mathcal{B}, \tag{29.11.14}$$

and the boundary conditions are

$$\begin{aligned} \bar{\mathbf{u}}(s) &= \underbrace{\bar{\hat{\mathbf{u}}}(\mathbf{x}, s)}_{\text{prescribed}} \qquad \text{for } \mathbf{x} \in \mathcal{S}_1, \\ \bar{\boldsymbol{\sigma}}(\mathbf{x}, s)\mathbf{n}(\mathbf{x}) &= \underbrace{\bar{\hat{\mathbf{t}}}(\mathbf{x}, s)}_{\text{prescribed}} \qquad \text{for } \mathbf{x} \in \mathcal{S}_2. \end{aligned} \tag{29.11.15}$$

Thus a problem with a linear viscoelastic material is simply a problem in linear elasticity, with s as a parameter. Hence, a solution procedure is to find the elastic solution, substitute $\bar{\mathbb{C}}(s) \equiv s\bar{\mathbb{G}}(s)$ for $\mathbb{C}$, and take the inverse Laplace transform.

This observation is called *the correspondence principle between linear viscoelasticity and linear elasticity.* In essence, the correspondence principle states that:

- If for a given boundary-value problem the solution to the problem is known for a linear elastic material, then the same boundary-value problem for a linear viscoelastic material has a solution similar in form to the elastic solution, but in a Laplace transform space with an appropriate substitution for the elastic moduli.

EXAMPLE 29.8 Interconversion between $E_r(t)$ and $G_r(t)$ for the standard linear solid

The determination of $G_r(t)$ can be performed via shear experiments such as the torsion of a thin-walled tubular specimen. However, it is more common to measure $E_r(t)$ and then convert the result to $G_r(t)$. Let us assume a standard linear solid for $G_r(t)$ with elastic bulk response and use the correspondence principle to find the relation between $E_r(t)$ and $G_r(t)$.

We begin by noting that the overall stress-strain relation is given by

$$\boldsymbol{\sigma}(t) = \int_{0-}^{t} 2G_r(t-\tau)\frac{d\boldsymbol{\epsilon}'(\tau)}{d\tau}d\tau + K \operatorname{tr}\boldsymbol{\epsilon}(t)\mathbf{1}.$$

If we compute the Laplace transform of this relation, we find that

$$\bar{\boldsymbol{\sigma}}(s) = \left[2s\bar{G}_r(s)\mathbb{I}^{\text{symdev}} + 3K\mathbb{I}^{\text{vol}}\right] : \bar{\boldsymbol{\epsilon}}(s), \tag{29.11.16}$$

where s is the transform variable.

Consider now the experiment used to measure $E_r(t)$. In the standard experiment a strain in say the 1-direction is applied in a step manner to a rod, $\epsilon_{11} = \epsilon_o h(t)$. In the transverse directions the stress is zero $\sigma_{22}(t) = \sigma_{33}(t) \equiv \sigma_T(t) = 0$, with the strains $\epsilon_{22}(t) = \epsilon_{33}(t) \equiv \epsilon_T(t)$ unknown. The measured response $\sigma_{11}(t)$ is divided by ϵ_o to give $E_r(t)$.

Note first that

$$\epsilon'_{11} = \epsilon_{11} - \frac{1}{3}(\epsilon_{11} + 2\epsilon_T) = \frac{2}{3}(\epsilon_{11} - \epsilon_T)$$
$$\bar{\epsilon}'_{11} = \frac{2}{3}(\epsilon_o\bar{h} - \bar{\epsilon}_T)$$

and

$$\epsilon'_T = \epsilon_T - \frac{1}{3}(\epsilon_{11} + 2\epsilon_T) = \frac{1}{3}(\epsilon_T - \epsilon_{11})$$
$$\bar{\epsilon}'_T = \frac{1}{3}(\bar{\epsilon}_T - \epsilon_o\bar{h}).$$

If we now evaluate (29.11.16) for this strain state we find

$$\bar{\sigma}_{11} = 2s\bar{G}_r\frac{2}{3}(\epsilon_o\bar{h} - \bar{\epsilon}_T) + K(\epsilon_o\bar{h} + 2\bar{\epsilon}_T), \tag{29.11.17}$$

$$\bar{\sigma}_T = 2s\bar{G}_r\frac{1}{3}(\bar{\epsilon}_T - \epsilon_o\bar{h}) + K(\epsilon_o\bar{h} + 2\bar{\epsilon}_T) = 0. \tag{29.11.18}$$

Equation (29.11.18) can be solved for $\bar{\epsilon}_T$ to give,

$$\bar{\epsilon}_T = \frac{\frac{2}{3}s\bar{G}_r - K}{\frac{2}{3}s\bar{G}_r + 2K}\epsilon_o\bar{h}.$$

This can be substituted into (29.11.17) to give the Laplace transform of the stress in the 1-direction as

$$\bar{\sigma}_{11} = \frac{9Ks\bar{G}_r}{3K + s\bar{G}_r}\epsilon_o\bar{h}.$$

Since $\bar{h} = 1/s$, we find that the Laplace transform of the relaxation modulus $E_r(t)$ is given by

$$\bar{E}_r = \frac{9K\bar{G}_r}{3K + s\bar{G}_r}. \tag{29.11.19}$$

For the standard linear solid $G_r(t) = G_{re} + (G_{rg} - G_{re})\exp[-t/\tau_{RG}]$, where τ_{RG} is the relaxation time in shear. Inserting into (29.11.19) and computing the inverse

Laplace transform (using either the method of partial fractions together with basic Laplace transform tables or using a computer algebra system) one finds that

$$E_r(t) = E_{re} + (E_{rg} - E_{re}) \exp[-t/\tau_{RE}], \tag{29.11.20}$$

where

$$E_{rg} = \frac{9KG_{rg}}{G_{rg} + 3K}, \qquad E_{re} = \frac{9KG_{re}}{G_{re} + 3K}, \qquad \tau_{RE} = \tau_{RG}\frac{G_{rg} + 3K}{G_{re} + 3K}.$$

Here, τ_{RE} is the relaxation time in tension.

Often, in experiments what is actually measured are the parameters of (29.11.20), viz. $\{E_{rg}, E_{re}, \tau_{RE}\}$. From such experimentally measured quantities the corresponding quantities in shear, viz. $\{G_{rg}, G_{re}, \tau_{RG}\}$, are

$$G_{rg} = \frac{3KE_{rg}}{9K - E_{rg}}, \qquad G_{re} = \frac{3KE_{re}}{9K - E_{re}}, \qquad \tau_{RG} = \tau_{RE}\frac{9K - E_{rg}}{9K - E_{re}}.$$

From the results of this problem, when working with materials being modeled as standard linear solids with respect to their deviatoric behavior and elastic with respect to their volumetric behavior, it is possible to interconvert between the relaxation functions measured in shear to those measured in tension. Observe that the glassy and equilibrium moduli convert using the standard formulae for elastic solids. However, the relaxation times in shear and tension differ; they become the same only in the limit $K \to \infty$, that is as the material becomes more incompressible.

29.12 **Temperature dependence. Dynamic mechanical analysis (DMA)**

Temperature is one of the most important environmental variables to affect the response of materials. For polymeric materials in engineering this is particularly so because normal use conditions are relatively close to a characteristic temperature for such materials called the *glass-transition temperature*.

Dynamic mechanical analysis (DMA) is a widely used experimental technique for measuring this thermal dependency of the viscoelastic properties of a material. A frequency, typically 1 Hz, and an amplitude of oscillatory stress are selected and maintained as constants during a temperature sweep in which a heating rate between 1 and 2°C per minute is selected, and the temperature is raised from a desired starting temperature to an endpoint.[11] The most common graphical presentation involves plots of the storage modulus E', the loss modulus E'', and $\tan(\delta)$ as a function of temperature.

[11] See standard ASTM D4065-12 (2012) for details.

29.12.1 Representative DMA results for amorphous polymers

Figure 29.24 shows a typical DMA result for polycarbonate (PC), an **amorphous thermoplastic**. The full-scale plot begins at $-60\,^\circ$C and ends at 175 °C. It can be seen that there is little change in the storage modulus E' between the initial temperature and 140 °C. However, between 140 and 160 °C the storage modulus drops by over two orders of magnitude, and in this temperature range the material loses its usefulness as a structural material. This reasonably abrupt change in physical properties is associated with the onset of short-range molecular motion known as the **glass transition.**

Figure 29.25 shows the region of the glass transition in greater detail. We see that the loss modulus E'' rises to a maximum as the storage modulus E' is in its most rapid rate of descent. *The temperature at the peak of the loss modulus is a common definition of the* **glass-transition temperature** T_g. However, note that the "glass transition" is not a sharp event at a given temperature, but occurs over a temperature range of approximately 20 °C for PC. Since $\tan(\delta) = E''/E'$, the $\tan(\delta)$ curve follows the loss modulus curve closely. At low temperatures leading up to the glass transition $\tan(\delta)$ is well below 0.1. The rapid rise in the $\tan(\delta)$ curve coincides with rapid drop in E' and the rapid increase in E''. Above 150 °C the $\tan(\delta)$ curve rises rapidly, and reaches a peak above 1.2. Once the glass transition is complete, the loss modulus E'' drops back to a level close to the pre-transition values. However, because of the drastic decline in the value of E', the $\tan(\delta)$ value does not decline significantly.

The pattern of the variation of E', E'', and $\tan(\delta)$ observed here for polycarbonate, is typical of all amorphous polymeric materials. The key difference lies in the value of T_g, and the value of E' below T_g. As an additional example, Fig. 29.26 shows a DMA plot for another amorphous polymer poly(methyl methacrylate) (PMMA). Due to its more complex molecular structure the glass transition for PMMA occurs over a wider temperature range than for PC.

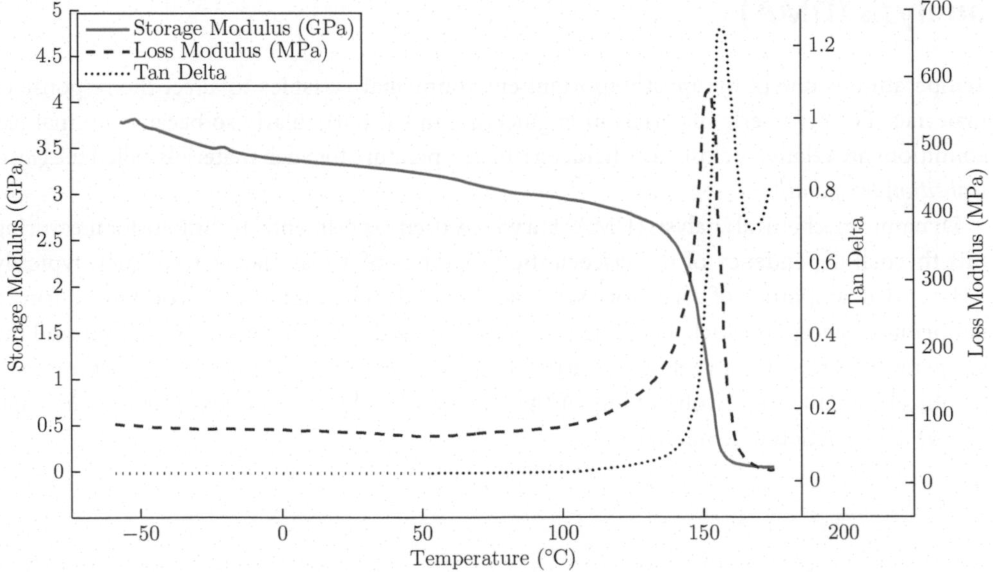

Fig. 29.24 Storage and loss properties for a polycarbonate (PC) (GE Lexan 141R): E', E'', and $\tan(\delta)$ versus temperature.

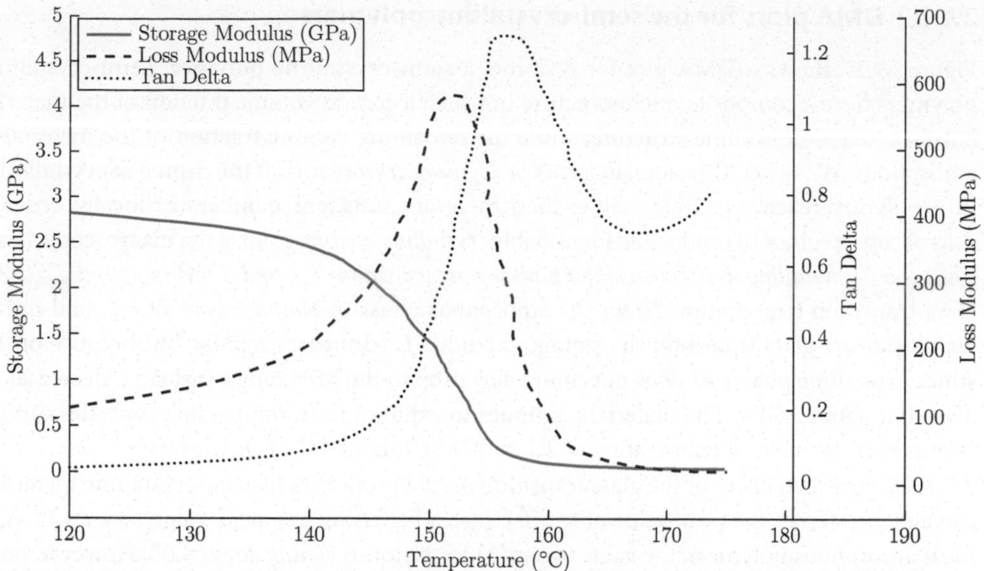

Fig. 29.25 Expanded plot of storage and loss properties for a polycarbonate (PC) (GE Lexan 141R) near T_g.

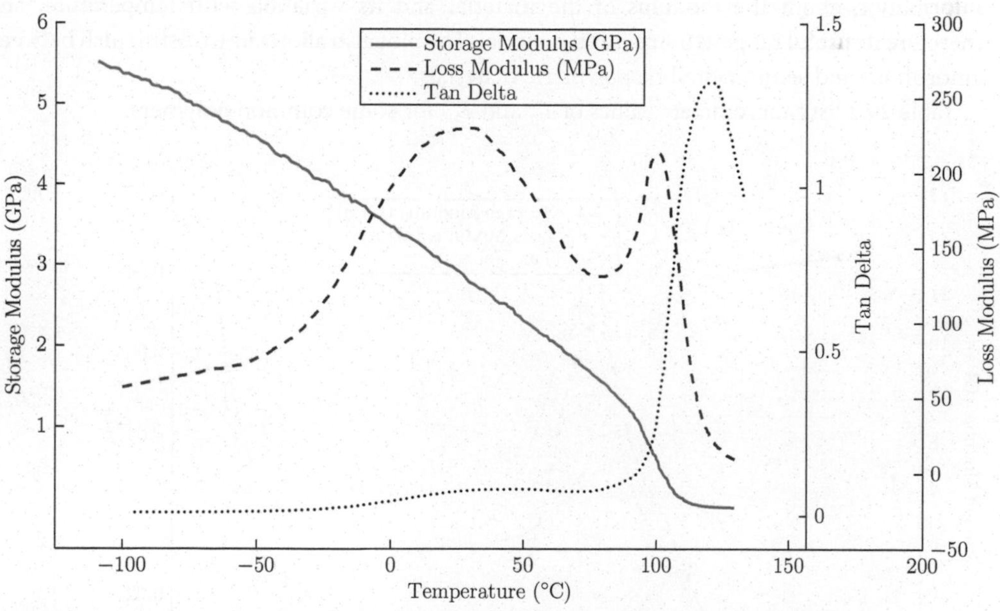

Fig. 29.26 Storage and loss properties for a poly(methyl methacrylate) (PMMA): E', E'', and $\tan(\delta)$ versus temperature.

29.12.2 **DMA plots for the semi-crystalline polymers**

Figure 29.27 shows a DMA plot for Nylon-6, a **semi-crystalline polymer**. Semi-crystalline polymers have a composite microstructure in which a certain volume fraction of the material has an ordered crystalline structure, while the remaining volume fraction of the material is amorphous. We speak, therefore, in terms of *degree of crystallinity*. If the degree of crystallinity in a polymer reaches $\approx 30 - 35\%$, then there are sufficient numbers of locally ordered microscopic regions to produce an identifiable *crystalline melting point* at the macroscopic level. *Thus, semi-crystalline materials exhibit both a glass transition T_g, and a melting point T_m.* The glass-transition temperature T_g for the amorphous phase of Nylon-6 is $\approx 50\,°\mathrm{C}$, and in the vicinity of the glass transition the storage modulus E' decreases rapidly, but because of the stiffer crystalline phase, E' does not completely drop to the MPa range; instead it decreases to a level of about 5 GPa. The material continues to exhibit useful solid-state properties until it approaches the melting temperature of $T_m \approx 215\,°\mathrm{C}$, about $145\,°\mathrm{C}$ higher than T_g.

The diminished effect of the glass transition on the properties of semi-crystalline materials can also be seen in the peak value of $\tan(\delta)$. Instead of rising to a peak value above 1.0, as in most amorphous polymers, the value of $\tan(\delta)$ for Nylon-6 is only about 0.05. However, once the semicrystalline material approaches the melting point, the value of $\tan(\delta)$ increases rapidly as the material changes from a viscoelastic solid to a viscous fluid.

From an engineering point of view, the most useful and accessible information available from a DMA test is the plot of storage modulus E' versus temperature. It gives us quantitative information about the modulus of the material and its variation with temperature, and therefore its useful range as a structural material. The plot also allows us to distinguish between amorphous and semi-crystalline polymeric materials.

Table 29.2 lists approximate values of T_g and T_m for some common polymers.

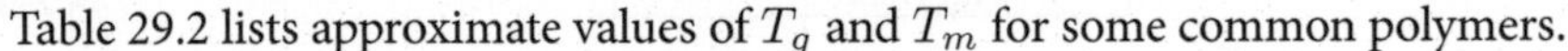

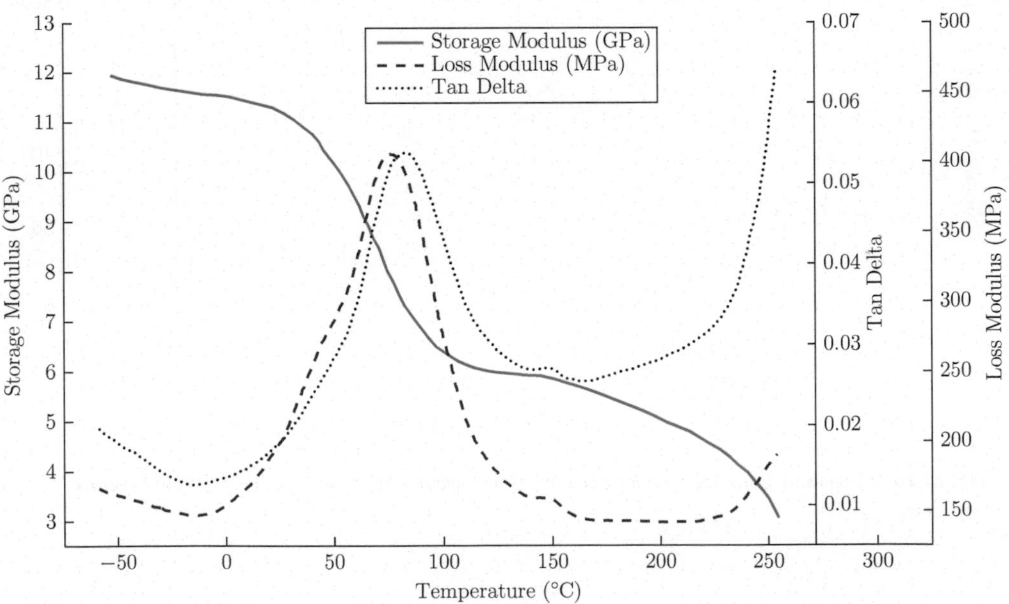

Fig. 29.27 Storage and loss properties for a Nylon-6: E', E'', and $\tan(\delta)$ versus temperature.

Table 29.2 T_g and T_m of some polymers.

Polymer	T_g,°C	T_m,°C
High density Polyethylene, HDPE	−90	137
Low density Polyethylene, LDPE	−110	115
Polypropylene, PP	−18	176
Polymethyl methacrylate, PMMA	105	
Polyvinyl chloride, PVC	87	212
Polytetrafluoro ethylene, PTFE	126	327
Polystyrene, PS	100	
Polyethylene terephthalate, PET	69	265
Polyamide (Nylon), PA	50	215
Polyoxy methylene (Delrin), POM	−87	175
Polycarbonate, PC	150	
Polyimide, PI	280–330	
Polyamide-imide, PAI	277–289	
Polyphenylene sulfide, PPS	85	285
Polysulfone, PSU	193	
Polyether ether ketone, PEEK	143	334

29.13 Effect of temperature on $E_r(t)$ and $J_c(t)$. Time-temperature equivalence

The viscoelastic functions $E_r(t)$and $J_c(t)$ depend not only on time t, but also depend strongly on the temperature T:

$$E_r(t,T), \qquad J_c(t,T).$$

As an example, consider the stress-relaxation function. Suppose that a series of stress-relaxation experiments have been performed at different temperatures, and the results plotted as $E_r(t,T)$ versus $\log_{10} t$, for each temperature, as shown schematically in Fig. 29.28. For a wide variety of polymers it is possible to shift the stress-relaxation curves at different temperatures *horizontally*, so that they *superimpose* on each other. If this is true for a given polymer, then the material obeys *time-temperature equivalence*,[12] and the material is called *thermo-rheologically simple*.

[12] Time-temperature equivalence is also often referred to as *time-temperature superposition*.

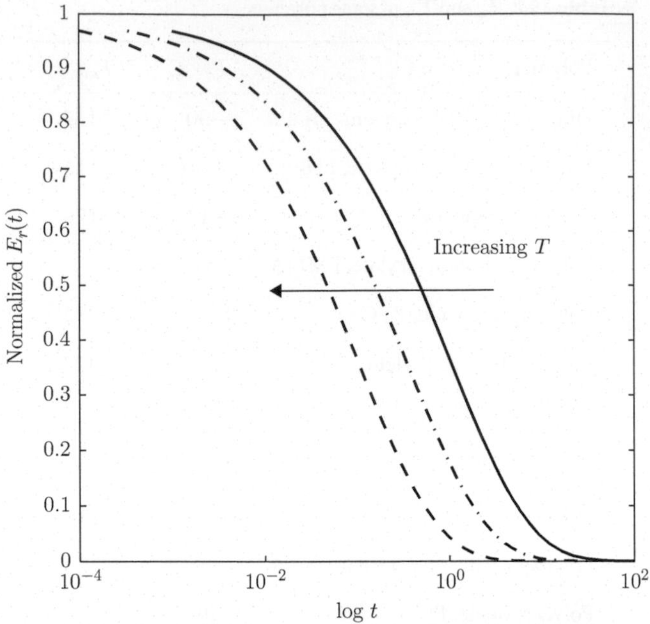

Fig. 29.28 Schematic diagram showing the variation of the relaxation modulus $E_r(t)$ with temperature, for thermo-rheologically simple materials.

Such a time-temperature dependence is evidence that temperature changes are effectively either accelerating or retarding the dominant viscoelastic processes occurring in the material.

Suppose that the response curve at a reference temperature T_{ref} is $E_r(t, T_{\text{ref}})$. Then, to superimpose the response curve at a temperature $T \neq T_{\text{ref}}$ onto the response curve at T_{ref}, one needs to shift the response curve for T horizontally, on a logarithmic scale by a factor $\log_{10} a(T, T_{\text{ref}})$, where $a(T, T_{\text{ref}})$, a property of the material, is called the **shift factor**, defined as

$$a(T, T_{\text{ref}}) \stackrel{\text{def}}{=} \frac{t(T)}{t(T_{\text{ref}})}. \qquad (29.13.1)$$

- If $T > T_{\text{ref}}$, since it takes less time at higher temperatures for the modulus to drop to the same level as it would at a lower temperature, the required shift has a value $a(T, T_{\text{ref}}) < 1$.
- Likewise, if $T < T_{\text{ref}}$ then the shift factor has a value $a(T, T_{\text{ref}}) > 1$.

Hence, for thermo-rheologically simple materials the relaxation function may be written as

$$E_r(t, T) = E_r(t(T_{\text{ref}}), T_{\text{ref}}), \qquad \text{with} \qquad t(T_{\text{ref}}) = \frac{t(T)}{a(T, T_{\text{ref}})}.$$

Fig. 29.29(a) shows representative relaxation data for the amorphous polymer PMMA for several temperatures. The curve constructed by time-temperature superposition is called the *master curve*. The following procedure is used. A temperature called the reference temperature is chosen, for which there is no shift. The master curve is generated by horizontally shifting the experimental curves for temperatures other than the reference temperature, until all the

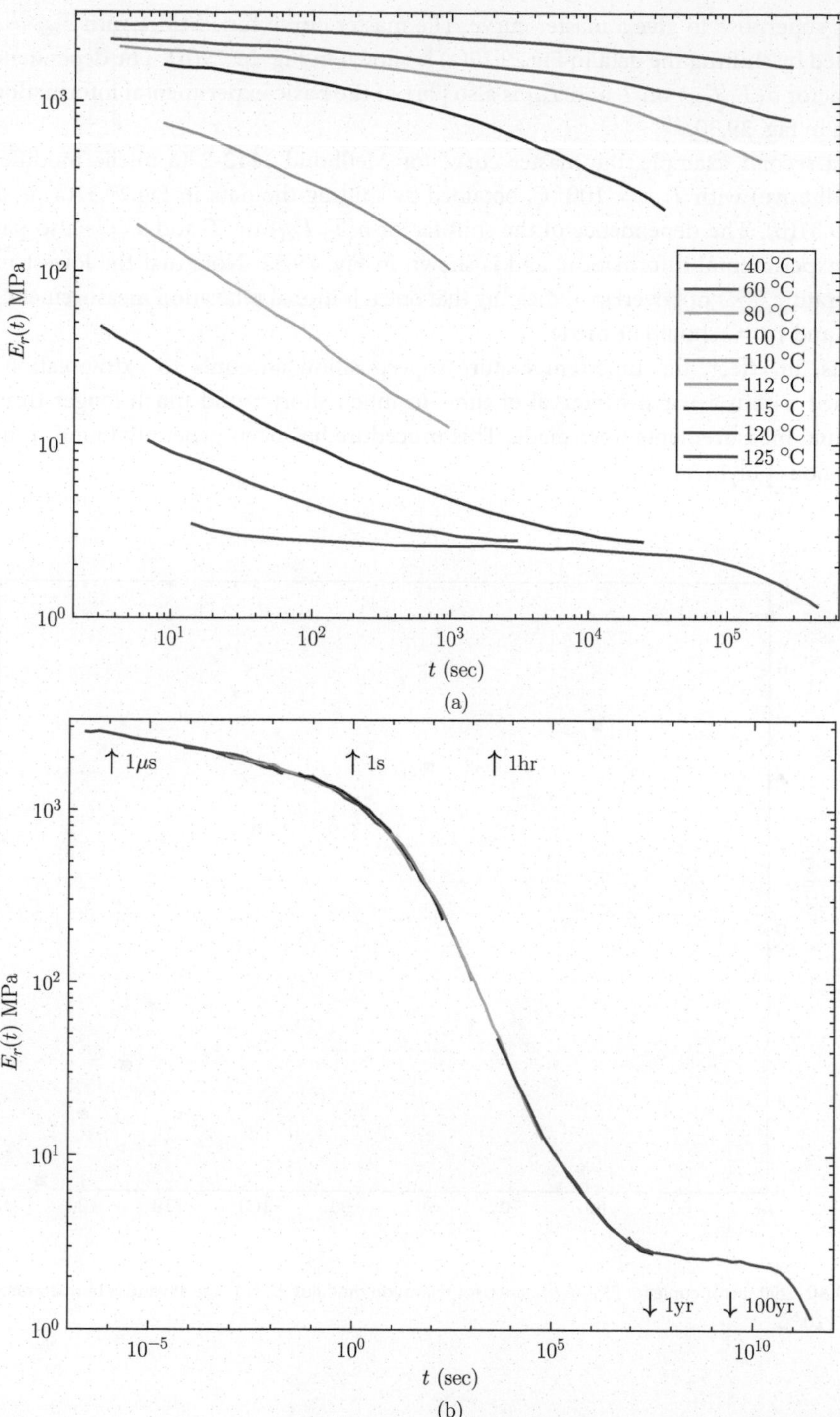

Fig. 29.29 (a) Relaxation modulus $E_r(t)$ for PMMA at various temperatures (data adapted from McLoughlin and Tobolsky, 1952). (b) Master curve for PMMA with $T_{\text{ref}} = 110\,°\text{C}$.

curves superpose to give a master curve. The master curve for PMMA with $T_{\text{ref}} = 110\,^\circ\text{C}$, obtained by shifting the data in Fig.29.29(a) is shown in Fig. 29.29(b). The dependence of the shift factor $a(T, T_{\text{ref}})$ on T and T_{ref} is also part of the basic experimental information, and is shown in Fig. 29.30.

As a second example, the master curve for Metlbond 1113-2 (a nitrile modified epoxy film adhesive) with $T_{\text{ref}} = 100\,^\circ\text{C}$, obtained by shifting the data in Fig.29.31(a) is shown in Fig. 29.31(b). The dependence of the shift factor $a(T, T_{\text{ref}})$ on T and T_{ref} is also part of the basic experimental information, and is shown in Fig. 29.32. Note that the lowest measured temperature does not overlap, indicating that one additional relaxation measurement between $57\,^\circ\text{C}$ and $70\,^\circ\text{C}$ should be made.

Thus, in effect, the time-temperature superposition amounts to extrapolation of data obtained within a narrow interval of time, to much shorter and much longer times where no actual measurements were made. This procedure has been generally found to hold with amorphous polymers.[13]

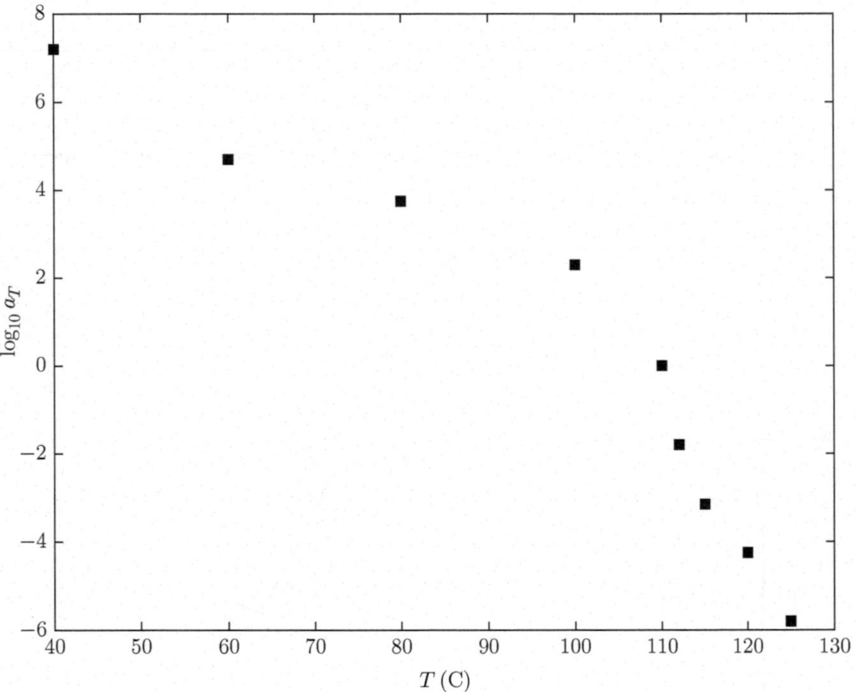

Fig. 29.30 Shift factor curve for PMMA obtained for relaxation modulus $E_r(t)$ data at various temperatures. $T_{\text{ref}} = 110\,^\circ\text{C}$.

[13] Superposition is generally not possible for semi-crystalline polymers.

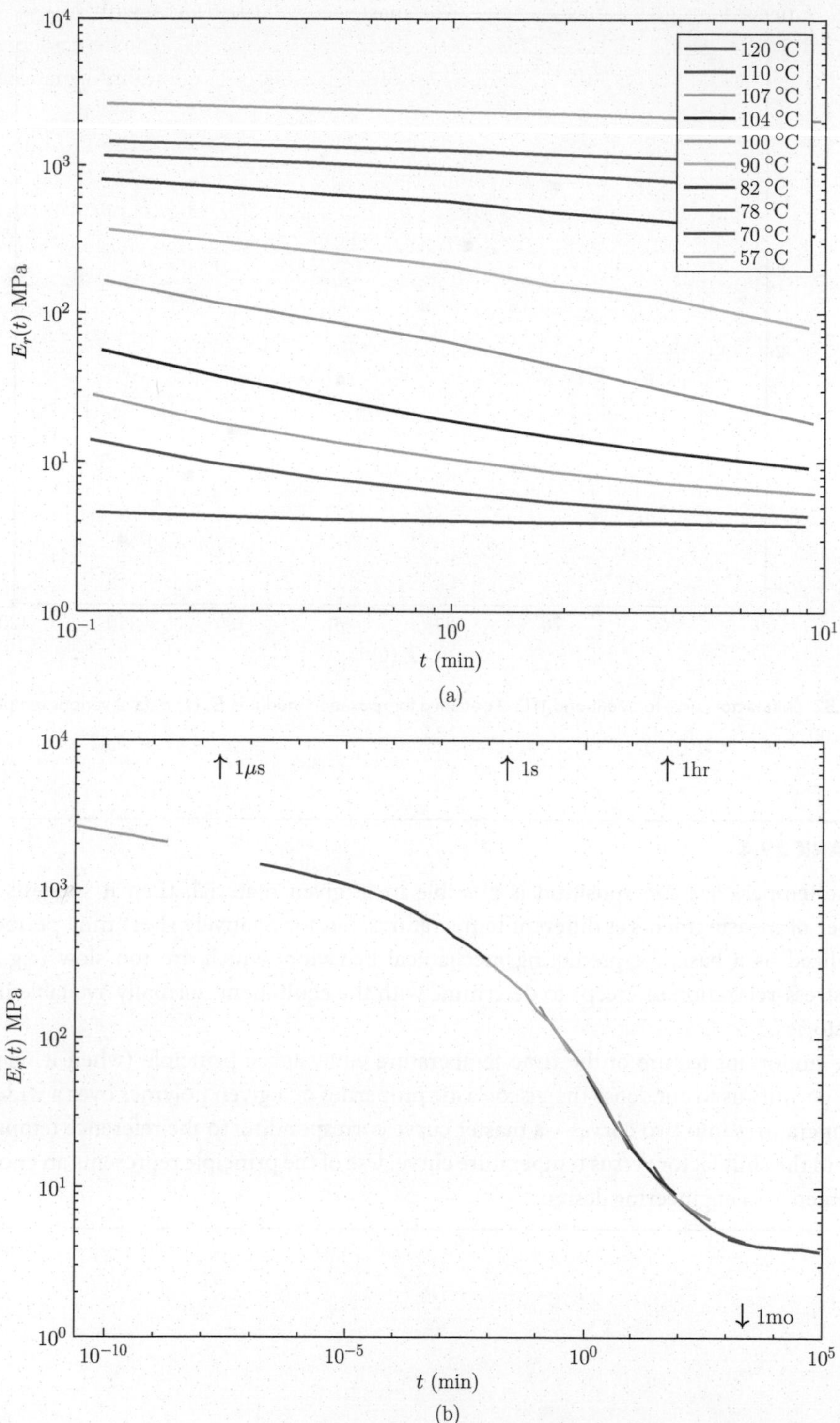

Fig. 29.31 (a) Relaxation modulus $E_r(t)$ for Metlbond 1113-2 at various temperatures (data adapted from Renieri, 1976). (b) Master curve for Metlbond 1113-2 with $T_{\text{ref}} = 100\,°\text{C}$.

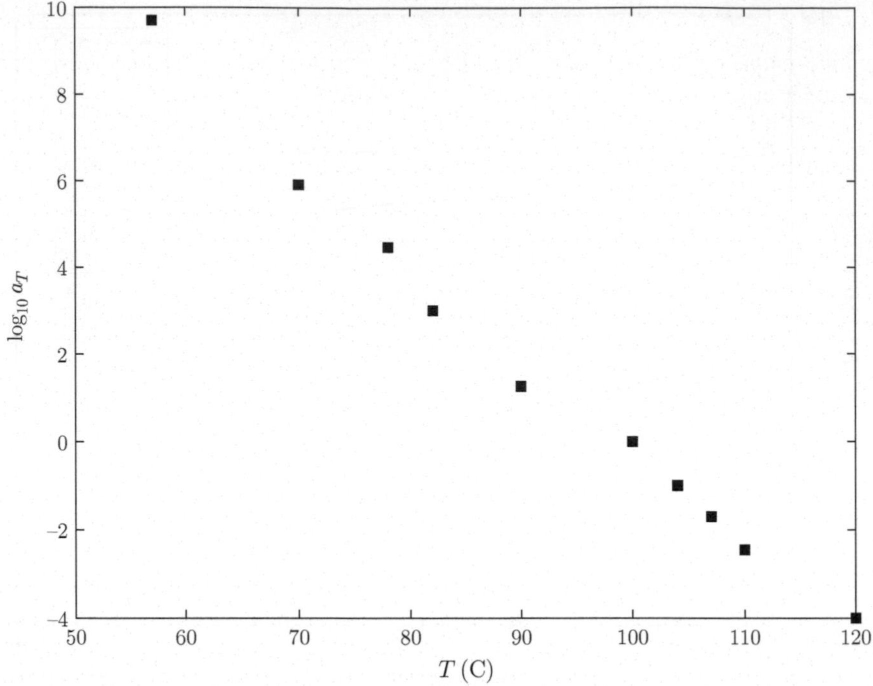

Fig. 29.32 Shift factor curve for Metlbond 1113-2 obtained for relaxation modulus $E_r(t)$ data at various temperatures. $T_{\text{ref}} = 100\,°$C.

REMARK 29.4

If time-temperature superposition is possible for a given material, then it suggests that a number of measurements at different temperatures, but for relatively short time periods, can be utilized as a basis for predicting mechanical behaviors which are too slow (e.g., long-term stress-relaxation or creep) to determine with the equipment normally available in most laboratories.

The important feature of the time-temperature equivalence principle (when it works), is that it permits us to condense the viscoelastic properties of a given polymer over a wide range of temperatures into two curves—a master curve corresponding to the reference temperature T_{ref}, and the shift factor versus temperature curve. Use of the principle represents an enormous convenience in engineering design.

29.13.1 **Shift factor. Williams-Landel-Ferry (WLF) equation**

For an *amorphous polymer*, Williams, Landel, and Ferry found that a plot of $\log_{10} a(T, T_{\text{ref}})$ versus temperature T falls on a single curve, which may be approximated by

$$\log_{10} a(T, T_{\text{ref}}) = \frac{-C_1(T - T_{\text{ref}})}{(C_2 + T - T_{\text{ref}})}, \tag{29.13.2}$$

where C_1 and C_2 are constants for a given polymer. Equation (29.13.2) is called the WLF equation. Further, they observed that the behavior of different polymers is sufficiently similar that if the reference temperature T_{ref} is taken as the glass-transition temperature T_g of an amorphous polymer, then *C_1 and C_2 are "universal constants" which hold for almost all amorphous polymers irrespective of chemical composition*:

$$C_1 = 17.44, \qquad C_2 = 51.6°\text{C}, \qquad \text{when} \qquad T_{\text{ref}} = T_g \text{ in } °\text{C}. \tag{29.13.3}$$

The typical temperature range in which the WLF equation works well is between T_g and $T_g + 100\,°\text{C}$. Below T_g, the WLF equation deviates significantly from test data.

PART XI

Finite elasticity

30 Finite elasticity

This chapter discusses the classical theory of finite deformations of elastic solids under isothermal conditions. This theory is of central importance in continuum mechanics. In addition to being important in its own right, it is also important because it helps delineate the central steps in the construction of sound constitutive theories for finite deformation of solids. We begin, in the next section, with a brief review of some kinematical relations which we have previously discussed in detail in Chapter 3.

30.1 Brief review of kinematical relations

We consider a homogeneous body $\mathcal{B}$ identified with the region of space it occupies in a fixed *reference configuration*, and denote by $\mathbf{X}$ an arbitrary material point of $\mathcal{B}$. A *motion* of $\mathcal{B}$ is then a smooth one-to-one mapping

$$\mathbf{x} = \chi(\mathbf{X}, t)\,, \tag{30.1.1}$$

which gives the place $\mathbf{x}$ occupied by the material point $\mathbf{X}$ at time t in the *deformed body* $\mathcal{B}_t$ (Fig. 30.1).

Corresponding to the motion (30.1.1), the *deformation gradient*, *velocity*, and *acceleration* are given by

$$\mathbf{F} = \nabla\chi \equiv \frac{\partial\chi(\mathbf{X}, t)}{\partial\mathbf{X}}, \qquad \dot{\chi} = \frac{\partial\chi(\mathbf{X}, t)}{\partial t}, \qquad \text{and} \qquad \ddot{\chi} = \frac{\partial^2\chi(\mathbf{X}, t)}{\partial t^2}, \tag{30.1.2}$$

respectively.

The relation between an elemental volume element dv_{R} in the reference body $\mathcal{B}$ and the corresponding volume element dv in the deformed body $\mathcal{B}_t$ is

$$dv = J dv_{\mathrm{R}}, \quad \text{with} \quad J = \det\mathbf{F}. \tag{30.1.3}$$

We restrict our attention to the physically relevant case,

$$J = \det\mathbf{F} > 0. \tag{30.1.4}$$

Further, the relation between an elemental oriented area element $\mathbf{n}_{\mathrm{R}} da_{\mathrm{R}}$ in the reference body $\mathcal{B}$ and its image $\mathbf{n}\, da$ in the deformed body $\mathcal{B}_t$ is given by Nanson's relation

$$\mathbf{n}\, da = J\mathbf{F}^{-\top}\mathbf{n}_{\mathrm{R}} da_{\mathrm{R}}. \tag{30.1.5}$$

Basic to our discussion of constitutive equations for elastic solids is the *polar decomposition*

$$\mathbf{F} = \mathbf{R}\mathbf{U} = \mathbf{V}\mathbf{R}, \tag{30.1.6}$$

Continuum Mechanics of Solids. Lallit Anand and Sanjay Govindjee, Oxford University Press (2020).

DOI: 10.1093/oso/9780198864721.001.0001

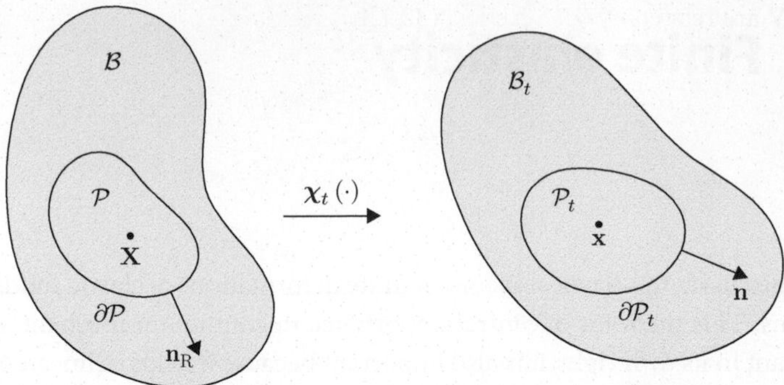

Fig. 30.1 Reference undeformed configuration $\mathcal{B}$ and a current deformed configuration $\mathcal{B}_t$ of a body.

of the *deformation gradient* into a *proper orthogonal tensor* or *rotation* **R**, together with a *symmetric and positive definite right stretch tensor* **U**, and another *symmetric and positive-definite left stretch tensor* **V**, satisfying

$$\mathbf{U} = \sqrt{\mathbf{F}^\top\mathbf{F}} \qquad \text{and} \qquad \mathbf{V} = \sqrt{\mathbf{F}\mathbf{F}^\top}. \tag{30.1.7}$$

The tensors

$$\mathbf{C} = \mathbf{U}^2 = \mathbf{F}^\top\mathbf{F} \qquad \text{and} \qquad \mathbf{B} = \mathbf{V}^2 = \mathbf{F}\mathbf{F}^\top, \tag{30.1.8}$$

respectively, are referred to as the *right* and *left Cauchy–Green tensors*. Important consequences of (30.1.8) and the polar decomposition (30.1.6) are the relations

$$\mathbf{V} = \mathbf{R}\mathbf{U}\mathbf{R}^\top \qquad \text{and} \qquad \mathbf{B} = \mathbf{R}\mathbf{C}\mathbf{R}^\top. \tag{30.1.9}$$

Being symmetric and positive definite, **U** and **V** admit spectral representations of the form

$$\begin{aligned} \mathbf{U} &= \sum_{i=1}^{3} \lambda_i \mathbf{r}_i \otimes \mathbf{r}_i, \\ \mathbf{V} &= \sum_{i=1}^{3} \lambda_i \mathbf{l}_i \otimes \mathbf{l}_i, \end{aligned} \tag{30.1.10}$$

where

(i) the non-negative **principal stretches** λ_1, λ_2, and λ_3 are the eigenvalues of **U** and, by (30.1.9), also of **V**;

(ii) the orthonormal **right principal directions** $\mathbf{r}_1$, $\mathbf{r}_2$, and $\mathbf{r}_3$ are the eigenvectors of **U**,

$$\mathbf{U}\mathbf{r}_i = \lambda_i \mathbf{r}_i \qquad \text{(no sum on } i\text{)}; \tag{30.1.11}$$

(iii) the orthonormal **left principal directions** $\mathbf{l}_1$, $\mathbf{l}_2$, and $\mathbf{l}_3$ are the eigenvectors of **V**,

$$\mathbf{V}\mathbf{l}_i = \lambda_i \mathbf{l}_i \qquad \text{(no sum on } i\text{)}. \tag{30.1.12}$$

Since $\mathbf{U}$ and $\mathbf{V}$ are related by $\mathbf{V} = \mathbf{R}\mathbf{U}\mathbf{R}^{\top}$, (30.1.10) yields

$$\sum_{i=1}^{3} \lambda_i \mathbf{R}\mathbf{r}_i \otimes \mathbf{R}\mathbf{r}_i = \sum_{i=1}^{3} \lambda_i \mathbf{l}_i \otimes \mathbf{l}_i,$$

and therefore implies that the principal directions are related via

$$\mathbf{l}_i = \mathbf{R}\mathbf{r}_i, \qquad i = 1, 2, 3, \tag{30.1.13}$$

and that,

$$\mathbf{R} = \mathbf{l}_i \otimes \mathbf{r}_i. \tag{30.1.14}$$

The tensors $\mathbf{C}$ and $\mathbf{B}$ have the following forms when expressed in terms of principal stretches and directions:

$$\begin{aligned} \mathbf{C} &= \sum_{i=1}^{3} \lambda_i^2\, \mathbf{r}_i \otimes \mathbf{r}_i, \\ \mathbf{B} &= \sum_{i=1}^{3} \lambda_i^2\, \mathbf{l}_i \otimes \mathbf{l}_i. \end{aligned} \tag{30.1.15}$$

Further, since $\mathbf{F} = \mathbf{R}\mathbf{U}$,

$$\mathbf{F} = \sum_{i=1}^{3} \lambda_i \mathbf{l}_i \otimes \mathbf{r}_i. \tag{30.1.16}$$

30.2 Balance of linear and angular momentum expressed spatially

Recall from our earlier discussion in Sec. 4.3 that balance of linear and angular momentum implies that there exists a tensor field $\boldsymbol{\sigma}$ in the *deformed body* $\mathcal{B}_t$, called the **Cauchy stress**, which has the following properties (cf. Sec. 4.3.4):

(i) For each unit vector $\mathbf{n}$ the traction $\mathbf{t}(\mathbf{n})$ is given by

$$\mathbf{t}(\mathbf{n}) = \boldsymbol{\sigma}\mathbf{n}, \qquad t_i(\mathbf{n}) = \sigma_{ij} n_j; \tag{30.2.1}$$

(ii) $\boldsymbol{\sigma}$ is symmetric,

$$\boldsymbol{\sigma} = \boldsymbol{\sigma}^{\top}, \qquad \sigma_{ij} = \sigma_{ji}; \tag{30.2.2}$$

and

(iii) $\boldsymbol{\sigma}$ satisfies the local equation of motion

$$\operatorname{div} \boldsymbol{\sigma} + \mathbf{b} = \rho \dot{\mathbf{v}}, \qquad \sigma_{ij,j} + b_i = \rho \dot{v}_i, \tag{30.2.3}$$

where $\dot{\mathbf{v}}(\mathbf{x}, t) = \ddot{\chi}(\underbrace{\chi^{-1}(\mathbf{x}, t)}_{\mathbf{X}}, t)$ is the spatial description of the acceleration $\ddot{\chi}(\mathbf{X}, t)$.

When working with solids the use of a purely spatial description can be problematic. On the other hand, solids typically possess stress-free reference configurations with respect to which one may measure strain and develop constitutive equations. For that reason we now reformulate the basic balance laws for balance of linear and angular momentum in a referential setting.

30.3 First Piola stress. Force and moment balances expressed referentially

With $\boldsymbol{\sigma}$ the symmetric Cauchy stress, the contact force on a part $\mathcal{P}_t \subset \mathcal{B}_t$ in the deformed body is given by

$$\int_{\partial\mathcal{P}_t} \boldsymbol{\sigma}\mathbf{n}\,da.$$

Using Nanson's relation (30.1.5) for the transformation of oriented area elements, this force can be expressed as an integral over the surface $\partial\mathcal{P}$, defined implicitly by $\partial\mathcal{P}_t = \chi_t(\partial\mathcal{P})$:

$$\int_{\partial\mathcal{P}_t} \boldsymbol{\sigma}\mathbf{n}\,da = \int_{\partial\mathcal{P}} J\,\boldsymbol{\sigma}\,\mathbf{F}^{-\top}\mathbf{n}_{\mathrm{R}}\,da_{\mathrm{R}}. \tag{30.3.1}$$

If we define

$$\boxed{\mathbf{P} \stackrel{\text{def}}{=} J\boldsymbol{\sigma}\mathbf{F}^{-\top},} \tag{30.3.2}$$

then (30.3.1) may be written as

$$\int_{\partial\mathcal{P}} \mathbf{P}\mathbf{n}_{\mathrm{R}}\,da_{\mathrm{R}} = \int_{\partial\mathcal{P}_t} \boldsymbol{\sigma}\mathbf{n}\,da. \tag{30.3.3}$$

- Thus, $\mathbf{P}$ represents the stress measured per unit area in the reference body; $\mathbf{P}$ is referred to as the **first Piola stress.**[1] Note that in general the first Piola stress is **not symmetric.**

Next, we define

$$\mathbf{b}_{\mathrm{R}} = J\mathbf{b}, \tag{30.3.4}$$

so that $\mathbf{b}_{\mathrm{R}}$ represents the conventional body force measured per unit volume in the reference body. Thus, by (30.1.3),

$$\int_{\mathcal{P}_t} \mathbf{b}\,dv = \int_{\mathcal{P}} \mathbf{b}_{\mathrm{R}}\,dv_{\mathrm{R}}. \tag{30.3.5}$$

Since $\ddot{\chi}$ represents the referential description of $\dot{\mathbf{v}}$, we conclude using $\rho_{\mathrm{R}} = \rho J$ that

$$\overline{\int_{\mathcal{P}_t} \rho\mathbf{v}\,dv} = \int_{\mathcal{P}_t} \rho\dot{\mathbf{v}}\,dv = \int_{\mathcal{P}} \rho_{\mathrm{R}}\ddot{\chi}\,dv_{\mathrm{R}} = \overline{\int_{\mathcal{P}} \rho_{\mathrm{R}}\dot{\chi}\,dv_{\mathrm{R}}}^{\,\cdot}. \tag{30.3.6}$$

[1] Shortly we will define a **second Piola stress.** Also note that **P** is often called the "first Piola–Kirchhoff" stress in the literature.

Using (30.3.3), (30.3.5), and (30.3.6), the balance of linear momentum in the deformed body, viz.

$$\int_{\partial\mathcal{P}_t} \boldsymbol{\sigma}\mathbf{n}\, da + \int_{\mathcal{P}_t} \mathbf{b}\, dv = \overline{\int_{\mathcal{P}_t} \rho\dot{\mathbf{v}}\, dv}^{\,\boldsymbol{\cdot}}, \tag{30.3.7}$$

when expressed referentially, yields the following **balance of linear momentum**:

$$\boxed{\int_{\partial\mathcal{P}} \mathbf{P}\mathbf{n}_{\mathrm{R}}\, da_{\mathrm{R}} + \int_{\mathcal{P}} \mathbf{b}_{\mathrm{R}}\, dv_{\mathrm{R}} = \overline{\int_{\mathcal{P}} \rho_{\mathrm{R}}\dot{\boldsymbol{\chi}}\, dv_{\mathrm{R}}}^{\,\boldsymbol{\cdot}}.} \tag{30.3.8}$$

Next, since $\mathcal{P}$ is a part of the reference body $\mathcal{B}$, and using Div to represent the referential divergence operator, the divergence theorem yields

$$\int_{\partial\mathcal{P}} \mathbf{P}\mathbf{n}_{\mathrm{R}}\, da_{\mathrm{R}} = \int_{\mathcal{P}} \operatorname{Div}\mathbf{P}\, dv_{\mathrm{R}},$$

and (30.3.8) implies that

$$\int_{\mathcal{P}} (\operatorname{Div}\mathbf{P} + \mathbf{b}_{\mathrm{R}} - \rho_{\mathrm{R}}\ddot{\boldsymbol{\chi}})\, dv_{\mathrm{R}} = \mathbf{0}.$$

Since the material region $\mathcal{P}$ is arbitrary, we arrive at the referential relation expressing the local form of **balance of linear momentum**:

$$\boxed{\operatorname{Div}\mathbf{P} + \mathbf{b}_{\mathrm{R}} = \rho_{\mathrm{R}}\ddot{\boldsymbol{\chi}}, \qquad \frac{\partial P_{ij}}{\partial X_j} + b_{\mathrm{R}i} = \rho_{\mathrm{R}}\ddot{\chi}_i.} \tag{30.3.9}$$

Further, since balance of angular momentum requires that the Cauchy stress $\boldsymbol{\sigma}$ be symmetric, the first Piola stress $\mathbf{P}$, defined in (30.3.2), must satisfy

$$\boxed{\mathbf{P}\mathbf{F}^{\top} = \mathbf{F}\mathbf{P}^{\top}, \qquad P_{ik}F_{jk} = F_{ik}P_{jk}.} \tag{30.3.10}$$

Equation (30.3.10) represents the local **angular moment balance**, expressed referentially.

EXAMPLE 30.1 Piola stress

Consider a bar with reference configuration as shown in Fig. 30.2(a) that is deformed in two steps, first by stretching it in the $\mathbf{e}_2$-direction by an amount λ, which is accompanied by a uniform transverse stretch ν in the $\mathbf{e}_1$- and $\mathbf{e}_3$-directions:

$$\mathbf{U} = \nu\mathbf{e}_1 \otimes \mathbf{e}_1 + \lambda\mathbf{e}_2 \otimes \mathbf{e}_2 + \nu\mathbf{e}_3 \otimes \mathbf{e}_3,$$

followed by a rotation about the $\mathbf{e}_3$-axis

$$\mathbf{R} = \mathbf{e}_3 \otimes \mathbf{e}_3 + (\mathbf{e}_1 \otimes \mathbf{e}_1 + \mathbf{e}_2 \otimes \mathbf{e}_2)\cos\theta - (\mathbf{e}_1 \otimes \mathbf{e}_2 - \mathbf{e}_2 \otimes \mathbf{e}_1)\sin\theta$$

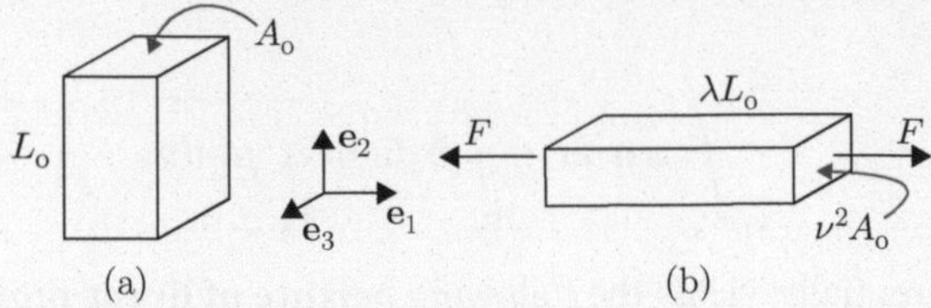

Fig. 30.2 Reference (a) and deformed (b) configurations of an elastic bar.

by an angle $\theta = -\pi/2$, so that

$$\mathbf{R} = \mathbf{e}_3 \otimes \mathbf{e}_3 + \mathbf{e}_1 \otimes \mathbf{e}_2 - \mathbf{e}_2 \otimes \mathbf{e}_1,$$

and hence that the overall deformation gradient is given by

$$\mathbf{F} = \mathbf{RU} = \lambda \mathbf{e}_1 \otimes \mathbf{e}_2 + \nu(\mathbf{e}_3 \otimes \mathbf{e}_3 - \mathbf{e}_2 \otimes \mathbf{e}_1). \tag{30.3.11}$$

The final configuration of the bar is shown in Fig. 30.2(b).

It is assumed that the deformation is caused by a force F applied along the main axis of the bar and that the stress is uniform and of the form

$$\boldsymbol{\sigma} = \left(\frac{F}{\nu^2 A_0}\right) \mathbf{e}_1 \otimes \mathbf{e}_1. \tag{30.3.12}$$

Observe that the traction on any surface with normal $\mathbf{e}_1$ in the deformed body is given by

$$\mathbf{t} = \boldsymbol{\sigma}\mathbf{e}_1 = \left(\frac{F}{\nu^2 A_o}\right) \mathbf{e}_1. \tag{30.3.13}$$

We can find the first Piola stress by noting that

$$\mathbf{F}^{-\top} = \frac{1}{\lambda}\mathbf{e}_1 \otimes \mathbf{e}_2 + \frac{1}{\nu}(\mathbf{e}_3 \otimes \mathbf{e}_3 - \mathbf{e}_2 \otimes \mathbf{e}_1), \tag{30.3.14}$$

and that $J = \lambda\nu^2$. Then the first Piola stress is given by

$$\begin{aligned}
\mathbf{P} &= J\boldsymbol{\sigma}\mathbf{F}^{-\top}, \\
&= (\lambda\nu^2)\left(\frac{F}{\nu^2 A_0}\right)(\mathbf{e}_1 \otimes \mathbf{e}_1)\left(\frac{1}{\lambda}\mathbf{e}_1 \otimes \mathbf{e}_2 + \frac{1}{\nu}(\mathbf{e}_3 \otimes \mathbf{e}_3 - \mathbf{e}_2 \otimes \mathbf{e}_1)\right), \\
&= \frac{F}{A_0}\mathbf{e}_1 \otimes \mathbf{e}_2.
\end{aligned}$$

Operationally, $\mathbf{P}$ acts on surface normals relative to the undeformed configuration of the body. In particular the loaded end of the bar is oriented in the 2-direction in the undeformed configuration. Thus

$$\mathbf{P}\mathbf{e}_2 = \frac{F}{A_0}\mathbf{e}_1, \tag{30.3.15}$$

which is a traction where the magnitude is given as force per unit undeformed area (i.e. engineering stress) and its orientation matches the actual traction orientation in the deformed body—the $\mathbf{e}_1$-direction.

30.4 **Change of observer. Change of frame**

An *observer* $\mathcal{O}$ is assumed to be equipped with the means of measuring physical quantities, and in particular is equipped to measure: (a) the relative positions of points in observed space $\mathcal{E}$, and (b) the progress of time in $\mathbb{R}$ in their *frame of reference*. Thus an observer $\mathcal{O}$ records an event to occur at a place $\mathbf{x} \in \mathcal{E}$ and time $t \in \mathbb{R}$. Consider now a second observer $\mathcal{O}^*$ with their own frame of reference, and who is possibly in relative motion with respect to the first observer. We assume that the clocks carried by the two observers read the same time, and we say that the two observers are *equivalent* if at any time t they agree about: (i) the orientations of relative positions in observed space, and (ii) the distance between arbitrary pairs of points in observed space. Then the general relation between the recorded spatial location $\mathbf{x}$ of an event by observer $\mathcal{O}$ and the location $\mathbf{x}^*$ of the same event by observer $\mathcal{O}^*$, is given by a mapping $\mathcal{F}$ which transforms spatial points $\mathbf{x}$ to spatial points,

$$\begin{aligned} \mathbf{x}^* &= \mathcal{F}(\mathbf{x}), \\ &\stackrel{\text{def}}{=} \mathbf{y}(t) + \mathbf{Q}(t)(\mathbf{x} - \mathbf{o}), \end{aligned} \tag{30.4.1}$$

where $\mathbf{Q}(t)$ is a **rotation**, $\mathbf{y}(t)$ a **spatial point**, and $\mathbf{o}$ the spatial origin for observer $\mathcal{O}$. We refer to $\mathbf{Q}$ as the **frame rotation**, and note that $\mathcal{F}$ has the form of a rigid transformation of observed space to itself. The mapping $\mathcal{F}$ is known as a "change of frame" or equivalently as a "change of observer"—we use the two phrases interchangeably.

Since constitutive equations which describe the *internal interactions* between the parts of a body should be independent of an observer and their particular frame of reference, use of the notion of a change in observer or a change in frame has important implications for the forms that constitutive relations of a body can take (Truesdell and Noll, 1965). To delineate the restrictions placed on the constitutive equations of finite elasticity by such a requirement, we first detail below how kinematic quantities transform under a change in observer, and then consider how stresses should transform. This will then lead to the criteria that constitutive laws in finite elasticity must satisfy.[2]

30.4.1 **Transformation rules for kinematic fields**

Since $\mathbf{x} = \chi(\mathbf{X}, t)$ is a spatial point, but $\mathbf{X}$ is a material point in $\mathcal{B}$ and hence invariant under a change of observer,[3] the motion χ transforms according to

$$\chi^*(\mathbf{X}, t) = \mathbf{y}(t) + \mathbf{Q}(t)(\chi(\mathbf{X}, t) - \mathbf{o}). \tag{30.4.2}$$

Taking the gradient of this relation with respect to $\mathbf{X}$ yields a transformation law

$$\mathbf{F}^*(\mathbf{X}, t) = \mathbf{Q}(t)\mathbf{F}(\mathbf{X}, t) \tag{30.4.3}$$

for the deformation gradient.

---

[2] For a more in-depth discussion of the material of this section see, for example, Gurtin et al. (2010, Chapters 20, 21, and 36).

[3] We assume tacitly that the reference configuration $\mathcal{B}$ is unaffected by a change of frame.

Consider next the tensor fields $\mathbf{U}$ and $\mathbf{C}$. Since $\mathbf{C} = \mathbf{U}^2 = \mathbf{F}^\top\mathbf{F}$, by (30.4.3),

$$\mathbf{C}^* = (\mathbf{QF})^\top\mathbf{QF} = \mathbf{F}^\top\mathbf{F} = \mathbf{C},$$
$$\mathbf{U}^* = \sqrt{\mathbf{C}^*} = \sqrt{\mathbf{C}} = \mathbf{U}.$$

The fields $\mathbf{C}$ *and* $\mathbf{U}$ *are thus invariant to a change in observer.*

Next, since $\mathbf{U}$ is invariant, $\mathbf{F}^* = \mathbf{QF}$, and $\mathbf{R} = \mathbf{F}\mathbf{U}^{-1}$, it follows that $\mathbf{R}^* = \mathbf{QR}$. In summary, we have the following change of observer **transformation laws** for kinematic fields:

$$\begin{aligned}
\mathbf{F}^* &= \mathbf{QF},\\
\mathbf{R}^* &= \mathbf{QR},\\
\mathbf{U}^* &= \mathbf{U},\\
\mathbf{C}^* &= \mathbf{C},\\
\mathbf{V}^* &= \mathbf{QVQ}^\top,\\
\mathbf{B}^* &= \mathbf{QBQ}^\top.
\end{aligned} \tag{30.4.4}$$

30.4.2 Transformation rules for the Cauchy stress and the Piola stress

Consider the Cauchy stress $\boldsymbol{\sigma}$ at a particular point and time, and recall that the mapping

$$\mathbf{t} = \boldsymbol{\sigma}\mathbf{n} \tag{30.4.5}$$

carries spatial vectors $\mathbf{n}$ (unit normals) to spatial vectors $\mathbf{t}$ (tractions). It is reasonable to assume that a second (equivalent) observer $\mathcal{O}^*$ will observe these vectors as having the same magnitude but with an orientation rotated by the frame rotation $\mathbf{Q}$; thus we assume that the normal $\mathbf{t}$ and $\mathbf{n}$ transform according to

$$\mathbf{t}^* = \mathbf{Qt} \quad \text{and} \quad \mathbf{n}^* = \mathbf{Qn}. \tag{30.4.6}$$

If $\boldsymbol{\sigma}^*$ denotes the Cauchy stress in the new frame, then $\mathbf{t}^* = \boldsymbol{\sigma}^*\mathbf{n}^*$. Thus $\mathbf{Qt} = \boldsymbol{\sigma}^*\mathbf{Qn}$, or

$$\mathbf{t} = (\mathbf{Q}^\top\boldsymbol{\sigma}^*\mathbf{Q})\mathbf{n}. \tag{30.4.7}$$

Combining (30.4.5) with (30.4.7) gives

$$(\boldsymbol{\sigma} - \mathbf{Q}^\top\boldsymbol{\sigma}^*\mathbf{Q})\mathbf{n} = \mathbf{0};$$

and since this holds for all $\mathbf{n}$, assumptions (30.4.6) imply that the Cauchy stress $\boldsymbol{\sigma}$ transforms under a change in frame as:

$$\boldsymbol{\sigma}^* = \mathbf{Q}\boldsymbol{\sigma}\mathbf{Q}^\top. \tag{30.4.8}$$

Further since $J^* = J$ and $\mathbf{F}^* = \mathbf{QF}$, the Piola stress $\mathbf{P}$ correspondingly transforms according to

$$\begin{aligned}
\mathbf{P}^* &= J^*\boldsymbol{\sigma}^*(\mathbf{F}^*)^{-\top}\\
&= J\,\mathbf{Q}\boldsymbol{\sigma}\mathbf{Q}^\top(\mathbf{QF})^{-\top}\\
&= \mathbf{Q}(J\boldsymbol{\sigma}\,\mathbf{Q}^\top\mathbf{Q}^{-\top}\mathbf{F}^{-\top})\\
&= \mathbf{Q}\,\underbrace{(J\boldsymbol{\sigma}\mathbf{F}^{-\top})}_{\mathbf{P}},
\end{aligned}$$

or

$$\mathbf{P}^* = \mathbf{QP}. \tag{30.4.9}$$

30.4.3 Form invariance of constitutive equations. Material-frame-indifference

Since constitutive equations which describe the *internal interactions* between the parts of a body should not depend on an external frame of reference which is used to describe them, as a general assumption in continuum mechanics it is required that:

- *all constitutive equations be invariant in form under changes of observers.*

Specifically, if a variable Φ is related to list of variables Λ through a constitutive response function $\hat{\Phi}(\cdot)$, then so also are Φ^* and Λ^* for all changes in observers (cf., e.g. Gurtin et al., 2010, Chapter 36):

$$\text{if} \quad \Phi = \hat{\Phi}(\Lambda) \quad \text{then} \quad \Phi^* = \hat{\Phi}(\Lambda^*). \tag{30.4.10}$$

That is, all observers use the same constitutive response function $\hat{\Phi}(\cdot)$. This hypothesis is referred to as the principle of **material-frame-indifference**, and is used to restrict constitutive relations (Truesdell and Noll, 1965).

In Sec. 30.6 we will ensure that the constitutive relations that we propose to use for finite elasticity satisfy this principle.

30.5 Free-energy imbalance under isothermal conditions

As we have done with the balance of linear and angular momentum, it is also useful to express the free-energy imbalance in terms of the first Piola stress. In particular, considering the developments of Sec. 5.3 for isothermal processes, the first two laws collapse into a single dissipation inequality which asserts that temporal changes in free energy of a part $\mathcal{P}$ be not greater than the power expended on $\mathcal{P}$. Thus, let

- ψ_R denote the free energy density per unit reference volume.

Then the *free-energy imbalance under isothermal conditions* requires that for each part $\mathcal{P}$ of $\mathcal{B}$,

$$\overline{\int_{\mathcal{P}} \psi_R \, dv_R}^{\,\cdot} \le \underbrace{\int_{\partial\mathcal{P}} \mathbf{P}\mathbf{n}_R \cdot \dot{\boldsymbol{\chi}} \, da_R + \int_{\mathcal{P}} (\mathbf{b}_R - \rho_R \ddot{\boldsymbol{\chi}}) \cdot \dot{\boldsymbol{\chi}} \, dv_R}_{\text{Power expended on } \mathcal{P}}. \tag{30.5.1}$$

Converting the integral over the boundary $\partial\mathcal{P}$ in (30.5.1) to a volume integral we obtain

$$\overline{\int_{\mathcal{P}} \psi_R \, dv_R}^{\,\cdot} \le \int_{\mathcal{P}} \mathbf{P} : \dot{\mathbf{F}} \, dv_R + \int_{\mathcal{P}} (\operatorname{Div} \mathbf{P} + \mathbf{b}_R - \rho_R \ddot{\boldsymbol{\chi}}) \cdot \dot{\boldsymbol{\chi}} \, dv_R. \tag{30.5.2}$$

Using the balance of linear momentum (30.3.9) and noting that $\mathcal{P}$ is arbitrary, we have the following *local free-energy imbalance* under isothermal conditions:

$$\dot{\psi}_R - \mathbf{P} : \dot{\mathbf{F}} \le 0. \tag{30.5.3}$$

Following Sec. 5.3.1, we also note that the dissipation per unit reference volume per unit time is given by

$$\mathcal{D}_{\mathrm{R}} = \mathbf{P} : \dot{\mathbf{F}} - \dot{\psi}_{\mathrm{R}} \geq 0. \tag{30.5.4}$$

30.5.1 The second Piola stress

Next, we introduce a new stress measure which allows us to express the free-energy imbalance (30.5.3) in a form more amenable to applications. Recall from (30.3.2) that the first Piola stress is defined by

$$\mathbf{P} = J\boldsymbol{\sigma}\mathbf{F}^{-\top}. \tag{30.5.5}$$

We now introduce a new stress measure

$$\boxed{\mathbf{S} \stackrel{\text{def}}{=} \mathbf{F}^{-1}\mathbf{P} = J\,\mathbf{F}^{-1}\boldsymbol{\sigma}\mathbf{F}^{-\top},} \tag{30.5.6}$$

called the **second Piola stress.** Note that since $\boldsymbol{\sigma}$ is symmetric, the second Piola stress is also **symmetric.**

Next, we also write the local free-energy imbalance (30.5.3) in terms of the second Piola stress. Specifically the stress power $\mathbf{P} : \dot{\mathbf{F}}$ may be written as:

$$\begin{aligned}
\mathbf{P} : \dot{\mathbf{F}} &= \mathbf{FS} : \dot{\mathbf{F}} && (30.5.7)\\
&= \mathbf{S} : \mathbf{F}^{\top}\dot{\mathbf{F}} && (30.5.8)\\
&= \frac{1}{2}\mathbf{S} : \left(\mathbf{F}^{\top}\dot{\mathbf{F}} + \dot{\mathbf{F}}^{\top}\mathbf{F}\right) \qquad \text{(since } \mathbf{S} \text{ is symmetric)} && (30.5.9)\\
&= \frac{1}{2}\mathbf{S} : \dot{\mathbf{C}}, && (30.5.10)
\end{aligned}$$

where in the last step the identity $\dot{\mathbf{C}} = \overline{\mathbf{F}^{\top}\mathbf{F}}^{\boldsymbol{\cdot}} = \mathbf{F}^{\top}\dot{\mathbf{F}} + \dot{\mathbf{F}}^{\top}\mathbf{F}$ has been used. Then, use of (30.5.10) in the free-energy imbalance (30.5.3) yields

$$\dot{\psi}_{\mathrm{R}} - \frac{1}{2}\mathbf{S} : \dot{\mathbf{C}} \leq 0, \tag{30.5.11}$$

and the dissipation rate per unit reference volume becomes

$$\mathcal{D}_{\mathrm{R}} = \frac{1}{2}\mathbf{S} : \dot{\mathbf{C}} - \dot{\psi}_{\mathrm{R}} \geq 0. \tag{30.5.12}$$

REMARKS 30.1

1. Just as with $\mathbf{C}$, the second Piola stress is also invariant under a change of observer; i.e., using (30.4.3) and (30.4.9),

$$\mathbf{S}^{*} = \mathbf{F}^{*-1}\mathbf{P}^{*} = \mathbf{F}^{-1}\mathbf{Q}^{\top}\mathbf{Q}\mathbf{P} = \mathbf{S}. \tag{30.5.13}$$

2. It is reasonable to assume that two equivalent observers will measure the same free-energy density field over $\mathcal{B}$ for the same motion; i.e.,

$$\psi_{\mathrm{R}}^{*} = \psi_{\mathrm{R}}, \qquad \text{thus the free energy is also invariant.} \tag{30.5.14}$$

REMARKS 30.1 Continued

3. Further, using the principle (30.4.10) we assume that the functional forms for a constitutive equation for the free-energy are the same for all observers. That is, if for example the free-energy is taken to depend on the deformation gradient, then

$$\psi_R = \hat{\psi}_R(\mathbf{F}) \qquad \text{and} \qquad \psi_R^* = \hat{\psi}_R(\mathbf{F}^*). \tag{30.5.15}$$

Note also, (30.5.14) and (30.5.15) imply that

$$\hat{\psi}_R(\mathbf{QF}) = \hat{\psi}_R(\mathbf{F}) \qquad \forall \mathbf{Q}. \tag{30.5.16}$$

To ensure that (30.5.16) is always satisfied we can assume that $\hat{\psi}_R$ depends on its argument $\mathbf{F}$ only through the combination $\mathbf{F}^\top\mathbf{F}(\equiv \mathbf{C})$. Formally, we write

$$\psi_R = \hat{\psi}_R(\mathbf{F}) = \bar{\psi}_R(\mathbf{C}). \tag{30.5.17}$$

30.6 Constitutive theory

Since elastic materials are assumed to have a constitutive response that only depends on the current kinematic state of the material (cf. Sec. 7.1), we assume the following special set of constitutive equations for ψ_R and $\mathbf{S}$:

$$\left.\begin{aligned} \psi_R &= \hat{\psi}_R(\mathbf{F}) = \bar{\psi}_R(\mathbf{C}) \\ \mathbf{S} &= \bar{\mathbf{S}}(\mathbf{C}), \end{aligned}\right\} \tag{30.6.1}$$

where in writing $(30.6.1)_1$ we wish to make clear that dependence of the free energy on the deformation gradient $\mathbf{F}$ is taken to appear only through the right Cauchy–Green tensor $\mathbf{C} = \mathbf{F}^\top\mathbf{F}$ such that the invariance assumption (30.5.16) is automatically satisfied. Further, note that since $\mathbf{C}$ and $\mathbf{S}$ are invariant under a change in frame the constitutive equation $(30.6.1)_2$ for the second Piola stress is also invariant under a change in frame.

30.6.1 Thermodynamic restrictions

Recalling also that elastic materials do not dissipate energy and since

$$\overline{\bar{\psi}_R(\mathbf{C})}^{\boldsymbol{\cdot}} = \frac{\partial \bar{\psi}_R(\mathbf{C})}{\partial \mathbf{C}} : \dot{\mathbf{C}}, \tag{30.6.2}$$

satisfaction of (30.5.12) with $\mathcal{D}_R = 0$ requires that the constitutive equations (30.6.1) satisfy

$$\left(\frac{1}{2}\bar{\mathbf{S}}(\mathbf{C}) - \frac{\partial \bar{\psi}_R(\mathbf{C})}{\partial \mathbf{C}}\right) : \dot{\mathbf{C}} = 0. \tag{30.6.3}$$

Since this expression must be satisfied at all $\mathbf{C}$ and for all $\dot{\mathbf{C}}$, we see that[4]

- *the free energy determines the second Piola stress via the* **stress relation**:

$$\boxed{\mathbf{S} = \bar{\mathbf{S}}(\mathbf{C}) = 2\,\frac{\partial \bar{\psi}_{\mathrm{R}}(\mathbf{C})}{\partial \mathbf{C}}\,.} \tag{30.6.4}$$

Next, $(30.5.6)_1$ and (30.6.4) imply that the first Piola stress is given by a constitutive equation of the form

$$\boxed{\mathbf{P} = 2\mathbf{F}\frac{\partial \bar{\psi}_{\mathrm{R}}(\mathbf{C})}{\partial \mathbf{C}}.} \tag{30.6.5}$$

Similarly, $(30.5.6)_2$ and (30.6.4) imply that the Cauchy stress is given by a constitutive equation of the form

$$\boxed{\boldsymbol{\sigma} = 2J^{-1}\mathbf{F}\,\frac{\partial \bar{\psi}_{\mathrm{R}}(\mathbf{C})}{\partial \mathbf{C}}\mathbf{F}^{\top}.} \tag{30.6.6}$$

Materials consistent with (30.6.4), (30.6.5), and (30.6.6) are commonly termed **hyperelastic**.

30.7 Summary of basic equations. Initial/boundary-value problems

30.7.1 Basic field equations

The basic field equations describing the motion of an elastic body consist of the kinematical relations defining the deformation gradient and right Cauchy–Green tensor, the relation determining the Piola stress, and the local balance of linear and angular momentum:

$$\left.\begin{aligned}
\mathbf{F} = \nabla \boldsymbol{\chi}, \qquad \mathbf{C} = \mathbf{F}^{\top}\mathbf{F},\\
\mathbf{P} = 2\mathbf{F}\,\frac{\partial \bar{\psi}_{\mathrm{R}}(\mathbf{C})}{\partial \mathbf{C}},\\
\operatorname{Div}\mathbf{P} + \mathbf{b}_{\mathrm{R}} = \rho_{\mathrm{R}}\ddot{\boldsymbol{\chi}},\\
\mathbf{P}\mathbf{F}^{\top} = \mathbf{F}\mathbf{P}^{\top}.
\end{aligned}\right\} \tag{30.7.1}$$

These equations hold at all material points $\mathbf{X} \in \mathcal{B}$. In the quasi-static setting $(30.7.1)_4$ is replaced by

$$\operatorname{Div}\mathbf{P} + \mathbf{b}_{\mathrm{R}} = \mathbf{0}. \tag{30.7.2}$$

[4] This logical argument is generally known as the *Coleman–Noll procedure* (Coleman and Noll, 1963); cf. Sec. 7.1.

30.7.2 A typical initial/boundary-value problem

Let $\mathcal{S}_1$ and $\mathcal{S}_2$ be mutually exclusive complementary subsurfaces of the boundary $\partial\mathcal{B}$, so that $\mathcal{S}_1 \cup \mathcal{S}_2 = \partial\mathcal{B}$. As possible boundary conditions one might specify the motion on $\mathcal{S}_1$ and the surface traction on $\mathcal{S}_2$:

$$\left.\begin{aligned} \chi &= \hat{\chi} \quad \text{a prescribed function on } \mathcal{S}_1 \text{ for all } t \geq 0, \\ \mathbf{P}\mathbf{n}_{\mathrm{R}} &= \hat{\mathbf{t}}_{\mathrm{R}} \quad \text{a prescribed function on } \mathcal{S}_2 \text{ for all } t \geq 0. \end{aligned}\right\} \tag{30.7.3}$$

Standard *initial conditions* involve a specification of the initial deformation and the velocity:

$$\chi(\mathbf{X}, 0) = \chi_0(\mathbf{X}), \qquad \dot{\chi}(\mathbf{X}, 0) = \mathbf{v}_0(\mathbf{X}), \tag{30.7.4}$$

with χ_0 and $\mathbf{v}_0$ prescribed functions on $\mathcal{B}$.

The *initial/boundary-value problem* corresponding to the prescribed data

$$\{\mathbf{b}_{\mathrm{R}}, \rho_{\mathrm{R}}, \hat{\chi}, \hat{\mathbf{t}}_{\mathrm{R}}, \chi_0, \mathbf{v}_0\}$$

then consists in finding a motion $\chi(\mathbf{X}, t)$—defined for $\mathbf{X}$ in $\mathcal{B}$ and $t \geq 0$—that satisfies the field equations (30.7.1) for $\mathbf{X}$ in $\mathcal{B}$ and $t \geq 0$, the boundary conditions (30.7.3) on $\partial\mathcal{B}$ for $t \geq 0$, and the initial conditions (30.7.4) for $\mathbf{X}$ in $\mathcal{B}$.

Also of importance are quasi-static boundary-value problems involving solutions of the basic field equations and boundary conditions. Since initial conditions are irrelevant to such problems, the data consists of $\{\mathbf{b}_{\mathrm{R}}, \hat{\chi}, \hat{\mathbf{t}}_{\mathrm{R}}\}$ and the corresponding boundary-value problem consists in finding a deformation χ satisfying the field equations (30.7.1)—with the linear momentum balance replaced by the equilibrium balance (30.7.2)—and the natural time-independent analogs of the boundary conditions (30.7.3).

30.8 Material symmetry

Within the constitutive framework under discussion, a material **symmetry transformation** is defined as a *rotation of the reference configuration* that leaves the pointwise response to deformation unaltered. Because of the underlying thermodynamic structure, the energetic response to deformation determines the stress response, a result that allows us to define material symmetry in terms of the response function for the free energy (cf. Sec. 7.2 for a discussion of the linear case).

Choose an arbitrary point $\mathbf{Y}$ in $\mathcal{B}$, and let $\chi_{\mathbf{F}}$ denote a **homogeneous deformation** with deformation gradient $\mathbf{F}$:

$$\chi_{\mathbf{F}}(\mathbf{X}) = \mathbf{Y} + \mathbf{F}(\mathbf{X} - \mathbf{Y}). \tag{30.8.1}$$

Then, given a rotation $\mathbf{Q}$, consider the following thought experiments:

- **Experiment** 1. Deform $\mathcal{B}$ with the homogeneous deformation $\chi_{\mathbf{F}}$. In this case the deformation gradient is $\mathbf{F}$, and the free energy is

$$\psi_{\mathrm{R}1} = \hat{\psi}_{\mathrm{R}}(\mathbf{F}) = \bar{\psi}_{\mathrm{R}}(\mathbf{C}). \tag{30.8.2}$$

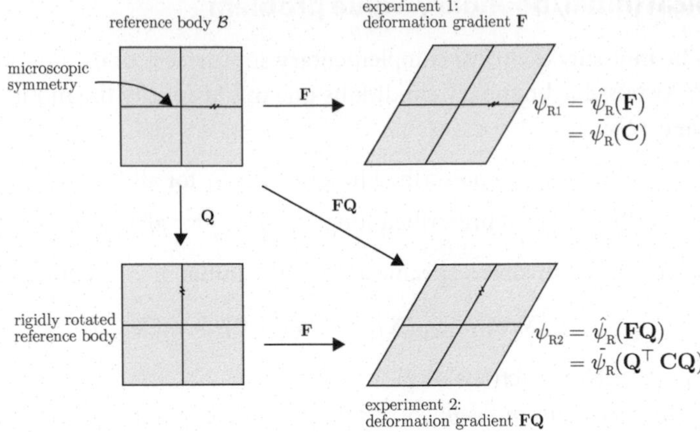

Fig. 30.3 Experiment 1: a homogeneous deformation with deformation gradient $\mathbf{F}$ and corresponding free energy $\psi_{R1} = \hat{\psi}(\mathbf{F}) = \bar{\psi}(\mathbf{C})$. Experiment 2: two successive homogeneous deformations with deformation gradients $\mathbf{Q}$ and $\mathbf{F}$ resulting in a free-energy $\psi_{R2} = \hat{\psi}_R(\mathbf{FQ}) = \bar{\psi}(\mathbf{Q}^\top\mathbf{CQ})$.

- **Experiment** 2. First transform $\mathcal{B}$ with a rotation $\mathbf{Q}$ via the homogeneous deformation

$$\chi_{\mathbf{Q}}(\mathbf{X}) = \mathbf{Y} + \mathbf{Q}(\mathbf{X} - \mathbf{Y}),$$

and then deform the rotated body with the homogeneous deformation $\chi_{\mathbf{F}}$ as in experiment 1 (Fig. 30.3). In this experiment the composite deformation gradient relative to the original reference body is

$$\mathbf{F}_2 = \nabla\chi_{\mathbf{F}}(\chi_{\mathbf{Q}}(\mathbf{X})) = \mathbf{FQ}. \tag{30.8.3}$$

Then

$$\mathbf{C}_2 = \mathbf{F}_2^\top\mathbf{F}_2 = \mathbf{Q}^\top\mathbf{F}^\top\mathbf{FQ} = \mathbf{Q}^\top\mathbf{CQ}, \tag{30.8.4}$$

and the free energy in the second experiment is given by

$$\psi_{R2} = \hat{\psi}_R(\mathbf{FQ}) = \bar{\psi}_R(\mathbf{Q}^\top\mathbf{CQ}). \tag{30.8.5}$$

In general, we would not expect ψ_{R1} to equal ψ_{R2}; however, if for a given rotation $\mathbf{Q}$, $\psi_{R1} = \psi_{R2}$ for *every* $\mathbf{F}$, then $\mathbf{Q}$ is referred to as a material symmetry transformation or simply as a symmetry transformation. Thus, by (30.8.2) and (30.8.5), a **symmetry transformation** is a rotation $\mathbf{Q}$ such that

$$\hat{\psi}_R(\mathbf{F}) = \hat{\psi}_R(\mathbf{FQ}) \tag{30.8.6}$$

for every deformation gradient $\mathbf{F}$; or equivalently,

$$\bar{\psi}_R(\mathbf{C}) = \bar{\psi}_R(\mathbf{Q}^\top\mathbf{CQ}) \tag{30.8.7}$$

for every symmetric, positive definite tensor $\mathbf{C}$.

30.9 **Isotropy**

We say that a material is **isotropic** if every rotation is a material symmetry transformation.

- We assume, henceforth that the body $\mathcal{B}$ under consideration is composed of an *isotropic material*, so that

$$\bar{\psi}_{\text{R}}(\mathbf{Q}^{\top}\mathbf{C}\mathbf{Q}) = \bar{\psi}_{\text{R}}(\mathbf{C}) \tag{30.9.1}$$

for all rotations $\mathbf{Q}$ and all symmetric, positive definite tensors $\mathbf{C}$.

Note that (30.9.1) holds trivially when $\mathbf{Q}$ is replaced by $-\mathbf{Q}$, and therefore the phrase "for all rotations" may be replaced by "for all orthogonal tensors." A scalar or tensor function which is invariant under the full orthogonal group is referred to as an *isotropic function*; thus

- $\bar{\psi}_{\text{R}}$ is an *isotropic function.*

Next, since (30.9.1) holds for all rotations $\mathbf{Q}$, we may, without loss in generality, choose $\mathbf{Q} = \mathbf{R}^{\top}$, where $\mathbf{R}$ is the rotation in the polar decomposition of the deformation gradient: $\mathbf{F} = \mathbf{R}\mathbf{U} = \mathbf{V}\mathbf{R}$. Recalling the relation

$$\mathbf{B} = \mathbf{R}\mathbf{C}\mathbf{R}^{\top} \tag{30.9.2}$$

between the right and left Cauchy–Green tensors, we conclude that

$$\bar{\psi}_{\text{R}}(\mathbf{C}) = \bar{\psi}_{\text{R}}(\mathbf{B}), \tag{30.9.3}$$

and hence for isotropic materials we may also express the free energy in terms of $\mathbf{B}$:

$$\psi_{\text{R}} = \bar{\psi}_{\text{R}}(\mathbf{B}).$$

From (30.6.1) and (30.5.10) it follows that

$$\dot{\psi}_{\text{R}} = \frac{1}{2}\mathbf{S}:\dot{\mathbf{C}} = \mathbf{P}:\dot{\mathbf{F}},$$

and hence

$$\frac{\partial\bar{\psi}_{\text{R}}(\mathbf{B})}{\partial\mathbf{B}}:\dot{\mathbf{B}} = \mathbf{P}:\dot{\mathbf{F}}.$$

Further,

$$\begin{aligned}\dot{\mathbf{B}} &= \overline{\mathbf{F}\mathbf{F}^{\top}}^{\boldsymbol{\cdot}}\\ &= (\dot{\mathbf{F}}\mathbf{F}^{\top} + \mathbf{F}\dot{\mathbf{F}}^{\top})\\ &= 2\,\text{sym}(\dot{\mathbf{F}}\mathbf{F}^{\top}),\end{aligned}$$

and, since

$$\frac{\partial\bar{\psi}_{\text{R}}(\mathbf{B})}{\partial\mathbf{B}} \text{ is symmetric,}$$

it follows that

$$\begin{aligned}\frac{\partial\bar{\psi}_{\text{R}}(\mathbf{B})}{\partial\mathbf{B}}:\dot{\mathbf{B}} &= 2\frac{\partial\bar{\psi}_{\text{R}}(\mathbf{B})}{\partial\mathbf{B}}:(\dot{\mathbf{F}}\mathbf{F}^{\top})\\ &= 2\left[\frac{\partial\bar{\psi}_{\text{R}}(\mathbf{B})}{\partial\mathbf{B}}\mathbf{F}\right]:\dot{\mathbf{F}};\end{aligned}$$

therefore

$$\left[2\frac{\partial\bar{\psi}_{\mathrm{R}}(\mathbf{B})}{\partial\mathbf{B}}\mathbf{F}-\mathbf{P}\right]:\dot{\mathbf{F}}=0,$$

and since $\dot{\mathbf{F}}$ is arbitrary, we are led to the relation

$$\boxed{\mathbf{P}=2\frac{\partial\bar{\psi}_{\mathrm{R}}(\mathbf{B})}{\partial\mathbf{B}}\mathbf{F}.} \tag{30.9.4}$$

Finally, since $\boldsymbol{\sigma}=J^{-1}\mathbf{P}\mathbf{F}^{\top}$, we obtain an expression giving the Cauchy stress as a function of $\mathbf{B}$ only:

$$\boxed{\boldsymbol{\sigma}=2J^{-1}\frac{\partial\bar{\psi}_{\mathrm{R}}(\mathbf{B})}{\partial\mathbf{B}}\mathbf{B}} \tag{30.9.5}$$

(recall that $J=\sqrt{\det\mathbf{B}}$). Since $\bar{\psi}_{\mathrm{R}}$ is an isotropic function, an important consequence of (30.9.5) is that $\mathbf{B}$ and $\boldsymbol{\sigma}$ share the same eigenvectors (Gurtin, 1981, Sec. 37) and hence $\boldsymbol{\sigma}$ *commutes with* $\mathbf{B}$:

$$\mathbf{B}\boldsymbol{\sigma}=\boldsymbol{\sigma}\mathbf{B}. \tag{30.9.6}$$

30.10 **Isotropic scalar functions**

Isotropic functions are known to have very particular forms. As noted above, a scalar function $\hat{g}(\mathbf{A})$ is **isotropic** if, for each tensor $\mathbf{A}$,

$$\hat{g}(\mathbf{A})=\hat{g}(\mathbf{Q}\mathbf{A}\mathbf{Q}^{\top})$$

for all rotations $\mathbf{Q}$. Examples of isotropic scalar functions are:

(i) the trace and determinant functions;

(ii) the principal invariants $I_1(\mathbf{A})$, $I_2(\mathbf{A})$, and $I_3(\mathbf{A})$, defined as

$$\begin{aligned} I_1(\mathbf{A})&=\operatorname{tr}\mathbf{A},\\ I_2(\mathbf{A})&=\frac{1}{2}\left[\operatorname{tr}^2(\mathbf{A})-\operatorname{tr}(\mathbf{A}^2)\right],\\ I_3(\mathbf{A})&=\det\mathbf{A}. \end{aligned} \tag{30.10.1}$$

For a tensor $\mathbf{A}$ the invariants (30.10.1) can be computed from the eigenvalues ω_i of $\mathbf{A}$ via the relations

$$\begin{aligned} I_1(\mathbf{A})&=\omega_1+\omega_2+\omega_3,\\ I_2(\mathbf{A})&=\omega_1\omega_2+\omega_2\omega_3+\omega_3\omega_1,\\ I_3(\mathbf{A})&=\omega_1\omega_2\omega_3. \end{aligned} \tag{30.10.2}$$

For convenience we write $\mathcal{I}_{\mathbf{A}}$ for the list of principal invariants:

$$\mathcal{I}_{\mathbf{A}}\stackrel{\text{def}}{=}(I_1(\mathbf{A}),I_2(\mathbf{A}),I_3(\mathbf{A})). \tag{30.10.3}$$

Representation Theorem for Isotropic Scalar Functions *A scalar function*

$$\hat{g} : \mathcal{A} \to \mathbb{R} \qquad (\mathcal{A} \subset \mathrm{Sym})$$

is isotropic if and only if there is a scalar function $\tilde{g}$ *such that*

$$\hat{g}(\mathbf{A}) = \tilde{g}\,(\mathcal{I}_{\mathbf{A}}) \tag{30.10.4}$$

for every $\mathbf{A}$ *in* $\mathcal{A}$ (the domain of $\hat{g}$ and a subset of the set of symmetric tensors).[5]

30.11 **Free energy for isotropic materials expressed in terms of invariants**

Using the representation relation (30.10.4), an isotropic scalar function of a symmetric tensor $\mathbf{B}$ may be expressed as a function of the principal invariants $I_1(\mathbf{B})$, $I_2(\mathbf{B})$, and $I_3(\mathbf{B})$ of $\mathbf{B}$. The constitutive relation for the free energy of an isotropic elastic solid may therefore be written in the form

$$\psi_{\mathrm{R}} = \tilde{\psi}_{\mathrm{R}}(\mathcal{I}_{\mathbf{B}}), \tag{30.11.1}$$

where

$$\mathcal{I}_{\mathbf{B}} \stackrel{\text{def}}{=} (I_1(\mathbf{B}), I_2(\mathbf{B}), I_3(\mathbf{B})). \tag{30.11.2}$$

Then, by (30.9.5),

$$\boldsymbol{\sigma} = \frac{2}{\sqrt{I_3}} \left[\frac{\partial \tilde{\psi}_{\mathrm{R}}(\mathcal{I}_{\mathbf{B}})}{\partial \mathbf{B}}\right] \mathbf{B}. \tag{30.11.3}$$

It is straightforward to establish the following identities (cf. Sec. 30.A):

$$\begin{aligned} \frac{\partial I_1(\mathbf{B})}{\partial \mathbf{B}} &= \mathbf{1}, \\ \frac{\partial I_2(\mathbf{B})}{\partial \mathbf{B}} &= I_1(\mathbf{B})\mathbf{1} - \mathbf{B}, \\ \frac{\partial I_3(\mathbf{B})}{\partial \mathbf{B}} &= I_3(\mathbf{B})\mathbf{B}^{-1}. \end{aligned} \tag{30.11.4}$$

Using the identities (30.11.4) we arrive at the following expression for the Cauchy stress for an isotropic material:

$$\boxed{\boldsymbol{\sigma} = \frac{2}{\sqrt{I_3}} \left[I_3 \frac{\partial \tilde{\psi}_{\mathrm{R}}(\mathcal{I}_{\mathbf{B}})}{\partial I_3} \mathbf{1} + \left(\frac{\partial \tilde{\psi}_{\mathrm{R}}(\mathcal{I}_{\mathbf{B}})}{\partial I_1} + I_1 \frac{\partial \tilde{\psi}_{\mathrm{R}}(\mathcal{I}_{\mathbf{B}})}{\partial I_2} \right) \mathbf{B} - \frac{\partial \tilde{\psi}_{\mathrm{R}}(\mathcal{I}_{\mathbf{B}})}{\partial I_2} \mathbf{B}^2 \right].} \tag{30.11.5}$$

An alternative to (30.11.5) follows from the Cayley–Hamilton equation (cf. (1.2.46)):

$$\mathbf{B}^3 - I_1(\mathbf{B})\mathbf{B}^2 + I_2(\mathbf{B})\mathbf{B} - I_3(\mathbf{B})\mathbf{1} = \mathbf{0}. \tag{30.11.6}$$

Multiplying (30.11.6) by $\mathbf{B}^{-1}$ yields

$$\mathbf{B}^2 = I_1\mathbf{B} - I_2\mathbf{1} + I_3\mathbf{B}^{-1}, \tag{30.11.7}$$

[5] Cf., e.g., Truesdell and Noll (1965, p. 28) or Gurtin (1981, Sec. 37).

and using this relation to eliminate $\mathbf{B}^2$ from (30.11.5), we find that

$$\boxed{\boldsymbol{\sigma} = \beta_0(\mathcal{I}_{\mathbf{B}})\mathbf{1} + \beta_1(\mathcal{I}_{\mathbf{B}})\mathbf{B} + \beta_2(\mathcal{I}_{\mathbf{B}})\mathbf{B}^{-1},} \tag{30.11.8}$$

with

$$\begin{aligned}
\beta_0(\mathcal{I}_{\mathbf{B}}) &= \frac{2}{\sqrt{I_3}}\left(I_2\frac{\partial\tilde{\psi}_{\mathrm{R}}(\mathcal{I}_{\mathbf{B}})}{\partial I_2} + I_3\frac{\partial\tilde{\psi}_{\mathrm{R}}(\mathcal{I}_{\mathbf{B}})}{\partial I_3}\right),\\
\beta_1(\mathcal{I}_{\mathbf{B}}) &= \frac{2}{\sqrt{I_3}}\frac{\partial\tilde{\psi}_{\mathrm{R}}(\mathcal{I}_{\mathbf{B}})}{\partial I_1},\\
\beta_2(\mathcal{I}_{\mathbf{B}}) &= -2\sqrt{I_3}\,\frac{\partial\tilde{\psi}_{\mathrm{R}}(\mathcal{I}_{\mathbf{B}})}{\partial I_2}.
\end{aligned} \tag{30.11.9}$$

30.12 Free energy for isotropic materials expressed in terms of principal stretches

The spectral representations of the left stretch tensor $\mathbf{V}$ and the left Cauchy–Green tensor $\mathbf{B} = \mathbf{V}^2$ are

$$\mathbf{V} = \sum_{i=1}^{3} \lambda_i \mathbf{l}_i \otimes \mathbf{l}_i, \qquad \mathbf{B} = \sum_{i=1}^{3} \omega_i \mathbf{l}_i \otimes \mathbf{l}_i, \qquad \omega_i = \lambda_i^2, \tag{30.12.1}$$

where $(\lambda_1, \lambda_2, \lambda_3)$ are the positive-valued principal stretches, and $(\mathbf{l}_1, \mathbf{l}_2, \mathbf{l}_3)$ are the orthonormal principal directions. The principal invariants $I_1(\mathbf{B})$, $I_2(\mathbf{B})$, and $I_3(\mathbf{B})$ of $\mathbf{B}$ may be expressed in terms of the principal stretches as

$$\begin{aligned}
I_1(\mathbf{B}) &= \lambda_1^2 + \lambda_2^2 + \lambda_3^2,\\
I_2(\mathbf{B}) &= \lambda_1^2\lambda_2^2 + \lambda_2^2\lambda_3^2 + \lambda_3^2\lambda_1^2,\\
I_3(\mathbf{B}) &= \lambda_1^2\lambda_2^2\lambda_3^2.
\end{aligned} \tag{30.12.2}$$

Using (30.12.2) in (30.11.1), to express the free energy in terms of the principal stretches:

$$\psi = \tilde{\psi}_{\mathrm{R}}(\mathcal{I}_{\mathbf{B}}) = \breve{\psi}_{\mathrm{R}}(\lambda_1, \lambda_2, \lambda_3). \tag{30.12.3}$$

Taking into account $(30.12.1)_3$, then by the chain rule and (30.11.3), the Cauchy stress is given by

$$\begin{aligned}
\boldsymbol{\sigma} &= \frac{2}{\lambda_1\lambda_2\lambda_3}\left[\frac{\partial\breve{\psi}_{\mathrm{R}}(\lambda_1, \lambda_2, \lambda_3)}{\partial\mathbf{B}}\right]\mathbf{B}\\
&= \frac{2}{\lambda_1\lambda_2\lambda_3}\left[\sum_{i=1}^{3}\frac{\partial\breve{\psi}_{\mathrm{R}}(\lambda_1, \lambda_2, \lambda_3)}{\partial\lambda_i}\frac{\partial\lambda_i}{\partial\mathbf{B}}\right]\mathbf{B}\\
&= \frac{1}{\lambda_1\lambda_2\lambda_3}\left[\sum_{i=1}^{3}\frac{1}{\lambda_i}\frac{\partial\breve{\psi}_{\mathrm{R}}(\lambda_1, \lambda_2, \lambda_3)}{\partial\lambda_i}\frac{\partial\omega_i}{\partial\mathbf{B}}\right]\mathbf{B}.
\end{aligned} \tag{30.12.4}$$

When the squared principal stretches ω_i are distinct, the ω_i and the principal directions $\mathbf{l}_i$ may be considered as functions of $\mathbf{B}$, and in this case it may be shown that

$$\frac{\partial \omega_i}{\partial \mathbf{B}} = \mathbf{l}_i \otimes \mathbf{l}_i. \tag{30.12.5}$$

Granted (30.12.5), (30.12.1) and (30.12.4) imply that

$$\boxed{\boldsymbol{\sigma} = \frac{1}{\lambda_1\lambda_2\lambda_3} \sum_{i=1}^{3} \lambda_i \frac{\partial \breve{\psi}_{\mathrm{R}}(\lambda_1, \lambda_2, \lambda_3)}{\partial \lambda_i} \mathbf{l}_i \otimes \mathbf{l}_i.} \tag{30.12.6}$$

Finally, since

$$\mathbf{F}^{-\top} = \sum_{i=1}^{3} \lambda_i^{-1} \mathbf{l}_i \otimes \mathbf{r}_i,$$

the Piola stress $\mathbf{P} = J\boldsymbol{\sigma}\mathbf{F}^{-\top}$ is given by

$$\boxed{\mathbf{P} = \sum_{i=1}^{3} \frac{\partial \breve{\psi}_{\mathrm{R}}(\lambda_1, \lambda_2, \lambda_3)}{\partial \lambda_i} \mathbf{l}_i \otimes \mathbf{r}_i.} \tag{30.12.7}$$

A consequence of (30.12.6) is that the principal values of the Cauchy stress are given by

$$\boxed{\sigma_i = \frac{\lambda_i}{(\lambda_1\lambda_2\lambda_3)} \frac{\partial \breve{\psi}_{\mathrm{R}}(\lambda_1, \lambda_2, \lambda_3)}{\partial \lambda_i} \quad \text{(no sum)}.} \tag{30.12.8}$$

Further, assuming $\mathbf{R} \equiv \mathbf{1}$, so that the deformation is a *pure homogeneous strain*, the Piola stress $\mathbf{P}$ is symmetric with principal values given by

$$\boxed{s_i = \frac{\partial \breve{\psi}_{\mathrm{R}}(\lambda_1, \lambda_2, \lambda_3)}{\partial \lambda_i}.} \tag{30.12.9}$$

Equations (30.12.8) and (30.12.9) in terms of principal stresses are widely used in comparing theory with experiment for isotropic materials.[6]

30.13 **Incompressibility**

Material constraints are constitutive assumptions that limit the class of constitutive processes a body may undergo. Here we consider the constraint of incompressibility.

An **incompressible material** is one for which only isochoric motions are possible. This constraint is expressed mathematically as

$$J = \det \mathbf{F} \equiv 1. \tag{30.13.1}$$

[6] For the cases where the principal stretches are not all distinct, see e.g. Ogden (1984, Sec. 6.1.4).

This constraint ensures that each part $\mathcal{P} \subset \mathcal{B}$ deforms to a spatial region $\mathcal{P}_t$ with the same volume; viz. $\text{vol}(\mathcal{P}_t) = \text{vol}(\mathcal{P})$ for all $\mathcal{P} \subset \mathcal{B}$. An immediate consequence is that

$$\det \mathbf{C} \equiv 1 \qquad \text{and} \qquad \det \mathbf{B} \equiv 1. \tag{30.13.2}$$

In what follows, it will be useful to note that

$$0 = \overline{\det \mathbf{C}}^{\boldsymbol{\cdot}} = (\det \mathbf{C})\mathbf{C}^{-1} : \dot{\mathbf{C}}$$

for an incompressible material. The constraint $(30.13.2)_1$ is thus equivalent to

$$\mathbf{C}^{-1} : \dot{\mathbf{C}} = 0. \tag{30.13.3}$$

30.14 Incompressible elastic materials

The constitutive assumption for elastic materials (30.6.1) together with the thermodynamic requirement that elastic materials have zero dissipation leads to

$$\left(\frac{1}{2}\bar{\mathbf{S}}(\mathbf{C}) - \frac{\partial \bar{\psi}_{\mathrm{R}}(\mathbf{C})}{\partial \mathbf{C}} \right) : \dot{\mathbf{C}} = 0. \tag{30.14.1}$$

Now, for incompressible materials $\mathbf{C}$ is restricted to the set of positive definite tensors with determinant equal to unity, and $\dot{\mathbf{C}}$ satisfies $\mathbf{C}^{-1} : \dot{\mathbf{C}} = 0$. Thus, unlike with (30.6.3), we cannot conclude that the quantity in the parenthesis in (30.14.1) must be equal to zero. Rather we can only conclude that it is a scalar multiple of $\mathbf{C}^{-1}$. Without loss of generality we assume the scalar multiplier to be $-(1/2)p$, such that

$$\frac{1}{2}\bar{\mathbf{S}}(\mathbf{C}) - \frac{\partial \bar{\psi}_{\mathrm{R}}(\mathbf{C})}{\partial \mathbf{C}} = -\frac{1}{2}p\mathbf{C}^{-1}, \tag{30.14.2}$$

where p is an arbitrary scalar. Thus the second Piola stress is given by

$$\boxed{\mathbf{S} = \bar{\mathbf{S}}(\mathbf{C}) = -p\mathbf{C}^{-1} + 2\frac{\partial \bar{\psi}_{\mathrm{R}}(\mathbf{C})}{\partial \mathbf{C}}.} \tag{30.14.3}$$

Equivalently, recalling that $\mathbf{P} = \mathbf{FS}$ and $\boldsymbol{\sigma} = J^{-1}\mathbf{PF}^{\top}$,

$$\boxed{\begin{aligned} \mathbf{P} &= -p\mathbf{F}^{-\top} + 2\mathbf{F}\frac{\partial \bar{\psi}_{\mathrm{R}}(\mathbf{C})}{\partial \mathbf{C}}, \\ \boldsymbol{\sigma} &= -p\mathbf{1} + 2\mathbf{F}\frac{\partial \bar{\psi}_{\mathrm{R}}(\mathbf{C})}{\partial \mathbf{C}}\mathbf{F}^{\top}, \end{aligned}} \tag{30.14.4}$$

in which p remains an *arbitrary scalar field*; note, $J = 1$ in the incompressible case.

30.14.1 Free energy for incompressible isotropic elastic bodies

Assume that the body is both incompressible and isotropic. Then, bearing in mind (30.13.1), we have

$$\bar{\psi}_R(\mathbf{B}) = \tilde{\psi}_R(\mathcal{I}_\mathbf{B}), \tag{30.14.5}$$

where

$$\mathcal{I}_\mathbf{B} \stackrel{\text{def}}{=} (I_1(\mathbf{B}), I_2(\mathbf{B})) \tag{30.14.6}$$

denotes the list of non-trivial principal invariants. Under these circumstances the Cauchy stress is given by

$$\boldsymbol{\sigma} = -p\mathbf{1} + 2\frac{\partial\tilde{\psi}_R(\mathcal{I}_\mathbf{B})}{\partial\mathbf{B}}\mathbf{B}, \tag{30.14.7}$$

and the counterpart of the stress relation (30.11.5) is

$$\boxed{\boldsymbol{\sigma} = -p\mathbf{1} + (\beta_1(\mathcal{I}_\mathbf{B}) - I_1\beta_2(\mathcal{I}_\mathbf{B}))\,\mathbf{B} + \beta_2(\mathcal{I}_\mathbf{B})\mathbf{B}^2,} \tag{30.14.8}$$

and that of (30.11.8) is

$$\boxed{\boldsymbol{\sigma} = -p\mathbf{1} + \beta_1(\mathcal{I}_\mathbf{B})\mathbf{B} + \beta_2(\mathcal{I}_\mathbf{B})\mathbf{B}^{-1},} \tag{30.14.9}$$

with

$$\begin{aligned} \beta_1(\mathcal{I}_\mathbf{B}) &= 2\frac{\partial\tilde{\psi}_R(\mathcal{I}_\mathbf{B})}{\partial I_1}, \\ \beta_2(\mathcal{I}_\mathbf{B}) &= -2\frac{\partial\tilde{\psi}_R(\mathcal{I}_\mathbf{B})}{\partial I_2}. \end{aligned} \tag{30.14.10}$$

When the free energy is expressed in terms of principal stretches,

$$\psi_R = \breve{\psi}_R(\lambda_1, \lambda_2, \lambda_3),$$

the constraint of incompressibility reads

$$\lambda_1\lambda_2\lambda_3 = 1,$$

and for λ_i distinct, the counterparts of expressions (30.12.6) and (30.12.7) are

$$\boxed{\begin{aligned} \mathbf{P} &= \sum_{i=1}^{3}\left(\frac{\partial\breve{\psi}_R(\lambda_1, \lambda_2, \lambda_3)}{\partial\lambda_i} - \frac{p}{\lambda_i}\right)\mathbf{l}_i \otimes \mathbf{r}_i, \\ \boldsymbol{\sigma} &= \sum_{i=1}^{3}\left(\lambda_i\frac{\partial\breve{\psi}_R(\lambda_1, \lambda_2, \lambda_3)}{\partial\lambda_i} - p\right)\mathbf{l}_i \otimes \mathbf{l}_i, \end{aligned}} \tag{30.14.11}$$

with $\mathbf{l}_i$ and $\mathbf{r}_i$ being the left and right principal directions.

A consequence of $(30.14.11)_2$ is that the principal values of the Cauchy stress are given by

$$\boxed{\sigma_i = \lambda_i\frac{\partial\breve{\psi}_R(\lambda_1, \lambda_2, \lambda_3)}{\partial\lambda_i} - p \quad \text{(no sum).}} \tag{30.14.12}$$

Further, assuming $\mathbf{R} \equiv \mathbf{1}$, so that the deformation is a *pure homogeneous strain*, the Piola stress $\mathbf{P}$ is symmetric with principal values given by

$$\boxed{s_i = \frac{\partial \breve{\psi}_{\mathrm{R}}(\lambda_1, \lambda_2, \lambda_3)}{\partial \lambda_i} - p\lambda_i^{-1}.} \tag{30.14.13}$$

EXAMPLE 30.2 Biaxial stretching of a thin incompressible elastic sheet

The biaxial stretching of a thin sheet of an incompressible material can be used to directly measure the derivatives of the free energy function, viz. β_1 and β_2 (Rivlin and Saunders, 1951).

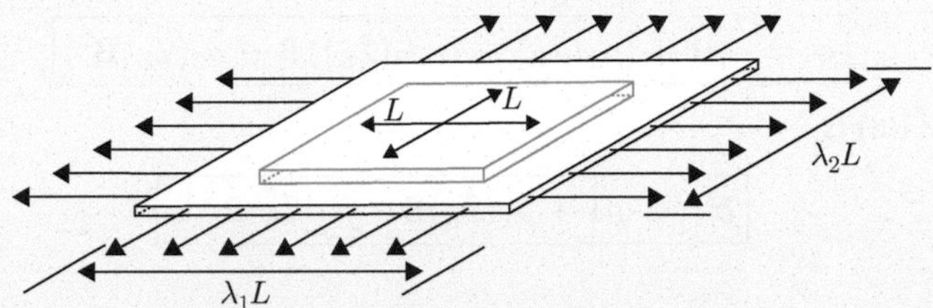

Fig. 30.4 Biaxial stretching of an elastic sheet.

The imposed deformation gradient in the biaxial stretching (cf. Fig. 30.4) of an incompressible sheet of material is given by

$$\mathbf{F} = \lambda_1 \mathbf{e}_1 \otimes \mathbf{e}_1 + \lambda_2 \mathbf{e}_2 \otimes \mathbf{e}_2 + (\lambda_1 \lambda_2)^{-1} \mathbf{e}_3 \otimes \mathbf{e}_3, \tag{30.14.14}$$

where λ_1 and λ_2 are experimentally controlled values of the in-plane stretching of the sheet. In the experiment the stresses σ_{11} and σ_{22} are measured and thus also known. From (30.14.9) we have

$$\sigma_{11} = -p + \beta_1(\mathcal{I}_{\mathbf{B}})\lambda_1^2 + \beta_2(\mathcal{I}_{\mathbf{B}})\lambda_1^{-2} \tag{30.14.15}$$

$$\sigma_{22} = -p + \beta_1(\mathcal{I}_{\mathbf{B}})\lambda_2^2 + \beta_2(\mathcal{I}_{\mathbf{B}})\lambda_2^{-2} \tag{30.14.16}$$

$$0 = -p + \beta_1(\mathcal{I}_{\mathbf{B}})\lambda_1^{-2}\lambda_2^{-2} + \beta_2(\mathcal{I}_{\mathbf{B}})\lambda_1^2\lambda_2^2. \tag{30.14.17}$$

where (30.14.17) follows in the biaxial test as the surface of the sheet is free of load. Since the deformation is homogeneous (and the stretches are constant), one can use (30.14.17) to solve for the unknown scalar field as

$$p = \beta_1(\mathcal{I}_{\mathbf{B}})\lambda_1^{-2}\lambda_2^{-2} + \beta_2(\mathcal{I}_{\mathbf{B}})\lambda_1^2\lambda_2^2. \tag{30.14.18}$$

Substituting back into (30.14.15) and (30.14.16), we find

$$\sigma_{11} = -\beta_1(\mathcal{I}_{\mathbf{B}})\lambda_1^{-2}\lambda_2^{-2} - \beta_2(\mathcal{I}_{\mathbf{B}})\lambda_1^2\lambda_2^2 + \beta_1(\mathcal{I}_{\mathbf{B}})\lambda_1^2 + \beta_2(\mathcal{I}_{\mathbf{B}})\lambda_1^{-2} \tag{30.14.19}$$

$$= (\lambda_1^2 - \lambda_1^{-2}\lambda_2^{-2})\beta_1(\mathcal{I}_{\mathbf{B}}) + (\lambda_1^{-2} - \lambda_1^2\lambda_2^2)\beta_2(\mathcal{I}_{\mathbf{B}}) \tag{30.14.20}$$

$$\sigma_{22} = -\beta_1(\mathcal{I}_{\mathbf{B}})\lambda_1^{-2}\lambda_2^{-2} - \beta_2(\mathcal{I}_{\mathbf{B}})\lambda_1^2\lambda_2^2 + \beta_1(\mathcal{I}_{\mathbf{B}})\lambda_2^2 + \beta_2(\mathcal{I}_{\mathbf{B}})\lambda_2^{-2} \tag{30.14.21}$$

$$= (\lambda_2^2 - \lambda_1^{-2}\lambda_2^{-2})\beta_1(\mathcal{I}_{\mathbf{B}}) + (\lambda_2^{-2} - \lambda_1^2\lambda_2^2)\beta_2(\mathcal{I}_{\mathbf{B}}). \tag{30.14.22}$$

Equations (30.14.20) and (30.14.22) constitute two linear equations that allow one to determine $\beta_1(\mathcal{I}_{\mathbf{B}})$ and $\beta_2(\mathcal{I}_{\mathbf{B}})$ in terms of the experimentally controlled and measured values $(\lambda_1, \lambda_2, \sigma_{11}, \sigma_{22})$:

$$\beta_1(\mathcal{I}_{\mathbf{B}}) = \frac{1}{\lambda_1^2 - \lambda_2^2}\left[\frac{\lambda_1^2\sigma_{11}}{\lambda_1^2 - \lambda_1^{-2}\lambda_2^{-2}} - \frac{\lambda_2^2\sigma_{22}}{\lambda_2^2 - \lambda_1^{-2}\lambda_2^{-2}}\right] \tag{30.14.23}$$

$$\beta_2(\mathcal{I}_{\mathbf{B}}) = \frac{1}{\lambda_1^2 - \lambda_2^2}\left[-\frac{\sigma_{11}}{\lambda_1^2 - \lambda_1^{-2}\lambda_2^{-2}} + \frac{\sigma_{22}}{\lambda_2^2 - \lambda_1^{-2}\lambda_2^{-2}}\right]. \tag{30.14.24}$$

In this way, by controlling (λ_1, λ_2) and measuring $(\sigma_{11}, \sigma_{22})$, one can map out the functional dependency of $\beta_1(\mathcal{I}_{\mathbf{B}})$ and $\beta_2(\mathcal{I}_{\mathbf{B}})$ on the invariants of $\mathbf{B}$.

30.14.2 **Simple shear of a homogeneous, isotropic, incompressible elastic body**

In this section we discuss the problem of simple shear of a homogeneous, isotropic, incompressible elastic body. This problem demonstrates a central feature of the large deformation theory of isotropic elasticity: *it is impossible to produce simple shear by applying a shear stress alone.* Let $\mathcal{B}$ be a homogeneous, isotropic, incompressible body in the shape of a cube. Consider a homogeneous deformation defined in Cartesian components by

$$x_1 = X_1 + \gamma X_2, \qquad x_2 = X_2, \qquad x_3 = X_3, \tag{30.14.25}$$

where γ is the *amount of shear*. The matrix of the deformation gradient $\mathbf{F}$ corresponding to (30.14.25) is

$$[\mathbf{F}] = \begin{bmatrix} 1 & \gamma & 0 \\ 0 & 1 & 0 \\ 0 & 0 & 1 \end{bmatrix}, \tag{30.14.26}$$

and the matrices of the left Cauchy–Green tensor $\mathbf{B} = \mathbf{F}\mathbf{F}^{\top}$ and its inverse are

$$[\mathbf{B}] = \begin{bmatrix} 1+\gamma^2 & \gamma & 0 \\ \gamma & 1 & 0 \\ 0 & 0 & 1 \end{bmatrix} \quad \text{and} \quad [\mathbf{B}]^{-1} = \begin{bmatrix} 1 & -\gamma & 0 \\ -\gamma & 1+\gamma^2 & 0 \\ 0 & 0 & 1 \end{bmatrix}. \tag{30.14.27}$$

Also,

$$\det(\mathbf{B} - \omega\mathbf{1}) = -\omega^3 + (3+\gamma^2)\omega^2 - (3+\gamma^2)\omega + 1,$$

and the list of principal invariants is

$$I_1(\mathbf{B}) = 3 + \gamma^2, \qquad I_2(\mathbf{B}) = 3 + \gamma^2, \qquad I_3(\mathbf{B}) = 1. \tag{30.14.28}$$

The material response functions β_1, and β_2 (cf. (30.14.10)) which are functions of the invariants of $\mathbf{B}$, are therefore functions of γ^2 alone. For convenience we write,

$$\beta_1(\gamma^2) \equiv \beta_1(3+\gamma^2, 3+\gamma^2),$$
$$\beta_2(\gamma^2) \equiv \beta_2(3+\gamma^2, 3+\gamma^2).$$

Using (30.14.9), (30.14.10), and the kinematical relations discussed above, we find that for an isotropic, incompressible elastic body

$$\begin{bmatrix}\sigma_{11} & \sigma_{12} & \sigma_{13}\\ \sigma_{21} & \sigma_{22} & \sigma_{23}\\ \sigma_{31} & \sigma_{32} & \sigma_{33}\end{bmatrix} = -p\begin{bmatrix}1 & 0 & 0\\ 0 & 1 & 0\\ 0 & 0 & 1\end{bmatrix} + \beta_1(\gamma^2)\begin{bmatrix}1+\gamma^2 & \gamma & 0\\ \gamma & 1 & 0\\ 0 & 0 & 1\end{bmatrix} + \beta_2(\gamma^2)\begin{bmatrix}1 & -\gamma & 0\\ -\gamma & 1+\gamma^2 & 0\\ 0 & 0 & 1\end{bmatrix}. \tag{30.14.29}$$

Thus,

$$\begin{aligned}
\sigma_{11} &= -p + (1+\gamma^2)\beta_1(\gamma^2) + \beta_2(\gamma^2),\\
\sigma_{22} &= -p + \beta_1(\gamma^2) + (1+\gamma^2)\beta_2(\gamma^2),\\
\sigma_{33} &= -p + \beta_1(\gamma^2) + \beta_2(\gamma^2),\\
\sigma_{12} &= \mu(\gamma^2)\gamma,\\
\sigma_{13} &= \sigma_{23} = 0,
\end{aligned} \tag{30.14.30}$$

where the non-linear shear modulus $\mu(\gamma^2)$ entering the shear stress σ_{12} is defined by

$$\mu(\gamma^2) = \beta_1(\gamma^2) - \beta_2(\gamma^2). \tag{30.14.31}$$

Next, on physical grounds we expect that σ_{12} is positive when γ is positive; that is, we expect that

$$\mu(\gamma^2) > 0, \tag{30.14.32}$$

an inequality which is satisfied if

$$\beta_1(\gamma^2) > \beta_2(\gamma^2). \tag{30.14.33}$$

Assuming the 3-direction faces of the cube to be traction-free, we have that $\sigma_{33} = 0$. Thus from $(30.14.30)_3$ we see that the arbitrary scalar field p is given by

$$p = \beta_1(\gamma^2) + \beta_2(\gamma^2). \tag{30.14.34}$$

With this choice, we obtain

$$\begin{aligned}
\sigma_{11} &= \beta_1(\gamma^2)\gamma^2,\\
\sigma_{22} &= \beta_2(\gamma^2)\gamma^2,\\
\sigma_{12} &= \mu(\gamma^2)\gamma,
\end{aligned} \tag{30.14.35}$$

and

$$\sigma_{13} = \sigma_{23} = \sigma_{33} = 0. \tag{30.14.36}$$

In the linear theory of elasticity, the normal stresses σ_{11}, and σ_{22} vanish in simple shear. On the other hand, in the non-linear theory a shear stress alone does not suffice to induce simple shear: additional normal stresses σ_{11} and σ_{22} are required. For these normal stresses to vanish, both β_1 and β_2 would have to vanish, and this, in turn, would imply that $\mu = 0$. Thus, if

$$\mu(\gamma^2) \neq 0 \quad \text{for} \quad \gamma \neq 0, \tag{30.14.37}$$

which is a physically reasonable assumption, then

- *it is impossible to produce simple shear by applying shear stresses alone.*

Further, (30.14.31) and (30.14.35) imply that

$$\boxed{\sigma_{11} - \sigma_{22} = \gamma\sigma_{12}.} \tag{30.14.38}$$

This relation is independent of the material response functions β_1 and β_2; therefore:

- *equation* (30.14.38) *is satisfied by every isotropic elastic body in simple shear.*[7]

If experimental measurements are inconsistent with this relation, then the material being studied is not an isotropic elastic material.

A consequence of (30.14.31) and (30.14.38) is that

$$\mu(\gamma^2) \neq 0 \quad \text{implies that} \quad \sigma_{11} \neq \sigma_{22}. \tag{30.14.39}$$

Further from (30.14.10) we see that if (as is reasonable to assume)

$$\frac{\partial\tilde{\psi}_{\text{R}}(\mathcal{I}_{\mathbf{B}})}{\partial I_1} > 0 \quad \text{and} \quad \frac{\partial\tilde{\psi}_{\text{R}}(\mathcal{I}_{\mathbf{B}})}{\partial I_2} > 0,$$

then

$$\beta_1(\gamma^2) > 0 \quad \text{and} \quad \beta_2(\gamma^2) < 0,$$

and from $(30.14.35)_{1,2}$, we see that the signs of the normal stresses are

$$\sigma_{11} > 0 \quad \text{and} \quad \sigma_{22} < 0, \tag{30.14.40}$$

and that these signs are unchanged when the sign of shear strain is reversed.

Appendices

30.A **Derivatives of the invariants of a tensor**

In this subsection we verify the computation of the derivatives of the principal invariants of a tensor $\mathbf{A}$:

$$I_1(\mathbf{A}) = \operatorname{tr}\mathbf{A}, \qquad I_2(\mathbf{A}) = \frac{1}{2}[(\operatorname{tr}\mathbf{A})^2 - \operatorname{tr}(\mathbf{A}^2)], \qquad I_3(\mathbf{A}) = \det\mathbf{A}.$$

1. Using the definition (2.1.2) of the derivative of a scalar, vector, or tensor field, the derivative of $I_1(\mathbf{A}) = \operatorname{tr}\mathbf{A}$ is found by noting that the following relation holds for all $\mathbf{H}$:

$$\left(\frac{\partial I_1(\mathbf{A})}{\partial\mathbf{A}}\right):\mathbf{H} = \frac{d}{d\alpha}\left[I_1(\mathbf{A} + \alpha\mathbf{H})\right]\big|_{\alpha=0} = \operatorname{tr}\mathbf{H} = \mathbf{1}:\mathbf{H}.$$

[7] This result is true not only for incompressible, but also for compressible elastic materials (cf., Gurtin et al., 2010, Chapter 51).

So

$$\frac{\partial I_1(\mathbf{A})}{\partial \mathbf{A}} = \mathbf{1}.$$

2. Next, we calculate the derivative of $I_2(\mathbf{A}) = \frac{1}{2}\left[(\operatorname{tr}\mathbf{A})^2 - \operatorname{tr}(\mathbf{A}^2)\right]$. Since,

$$\begin{aligned}
I_2(\mathbf{A} + \alpha\mathbf{H}) &= \frac{1}{2}\left[(\operatorname{tr}(\mathbf{A} + \alpha\mathbf{H}))^2 - \operatorname{tr}((\mathbf{A} + \alpha\mathbf{H})^2)\right] \\
&= \frac{1}{2}\left[(\operatorname{tr}\mathbf{A} + \alpha \operatorname{tr}\mathbf{H})^2 - \operatorname{tr}((\mathbf{A} + \alpha\mathbf{H})(\mathbf{A} + \alpha\mathbf{H}))\right] \\
&= \frac{1}{2}\left[(\operatorname{tr}\mathbf{A})^2 + 2\alpha \operatorname{tr}\mathbf{A} \operatorname{tr}\mathbf{H} + \alpha^2(\operatorname{tr}\mathbf{H})^2 - \operatorname{tr}(\mathbf{A}^2) + \alpha \operatorname{tr}(\mathbf{A}\mathbf{H} + \mathbf{H}\mathbf{A})\right. \\
&\quad \left. + \alpha^2 \operatorname{tr}(\mathbf{H}^2)\right],
\end{aligned}$$

we have for all $\mathbf{H}$

$$\begin{aligned}
\left(\frac{\partial I_2(\mathbf{A})}{\partial \mathbf{A}}\right) : \mathbf{H} &= \frac{d}{d\alpha}\left[I_2(\mathbf{A} + \alpha\mathbf{H})\right]\big|_{\alpha=0} = (\operatorname{tr}\mathbf{A})(\operatorname{tr}\mathbf{H}) - \operatorname{tr}(\mathbf{A}\mathbf{H}) \\
&= ((\operatorname{tr}\mathbf{A})\mathbf{1} - \mathbf{A}^{\top}) : \mathbf{H}.
\end{aligned}$$

So

$$\frac{\partial I_2(\mathbf{A})}{\partial \mathbf{A}} = (\operatorname{tr}\mathbf{A})\mathbf{1} - \mathbf{A}^{\top}.$$

3. Recall that the determinant of the tensor $(\mathbf{A} - \omega\mathbf{1})$ admits the representation

$$\det(\mathbf{A} - \omega\mathbf{1}) = -\omega^3 + I_1(\mathbf{A})\omega^2 - I_2(\mathbf{A})\omega + I_3(\mathbf{A}).$$

Then, with $\omega = -1$

$$\det(\mathbf{1} + \mathbf{A}) = 1 + \operatorname{tr}(\mathbf{A}) + o(|\mathbf{A}|).$$

Thus, for $\mathbf{A}$ invertible and $\mathbf{H}$ any tensor,

$$\begin{aligned}
\det(\mathbf{A} + \alpha\mathbf{H}) &= \det[(\mathbf{1} + \alpha\mathbf{H}\mathbf{A}^{-1})\mathbf{A}] \\
&= (\det\mathbf{A}) \det(\mathbf{1} + \alpha\mathbf{H}\mathbf{A}^{-1}) \\
&= (\det\mathbf{A})[1 + \alpha \operatorname{tr}(\mathbf{H}\mathbf{A}^{-1}) + o(|\alpha\mathbf{H})|] \\
&= \det(\mathbf{A}) + \alpha(\det\mathbf{A}) \operatorname{tr}(\mathbf{H}\mathbf{A}^{-1}) + o(|\alpha\mathbf{H}|).
\end{aligned}$$

Therefore,

$$\left(\frac{\partial \det(\mathbf{A})}{\partial \mathbf{A}}\right) : \mathbf{H} = \frac{d}{d\alpha}\left[I_3(\mathbf{A} + \alpha\mathbf{H})\right]\big|_{\alpha=0} = (\det\mathbf{A}) \operatorname{tr}(\mathbf{H}\mathbf{A}^{-1}) = \left((\det\mathbf{A})\mathbf{A}^{-\top}\right) : \mathbf{H}.$$

Hence

$$\frac{\partial \det(\mathbf{A})}{\partial \mathbf{A}} = (\det\mathbf{A})\mathbf{A}^{-\top}.$$

Thus, for the left Cauchy–Green tensor $\mathbf{B}$ which is symmetric, these formulae give the derivatives listed in (30.11.4).

31 Finite elasticity of elastomeric materials

31.1 Introduction

The major macroscopic physical characteristic of an elastomeric (or rubber-like) material is its ability to sustain large (essentially) reversible strains under the action of small stresses. A typical engineering stress, s, versus axial stretch, λ, curve for a vulcanized natural rubber in simple tension is shown in Fig. 31.1.

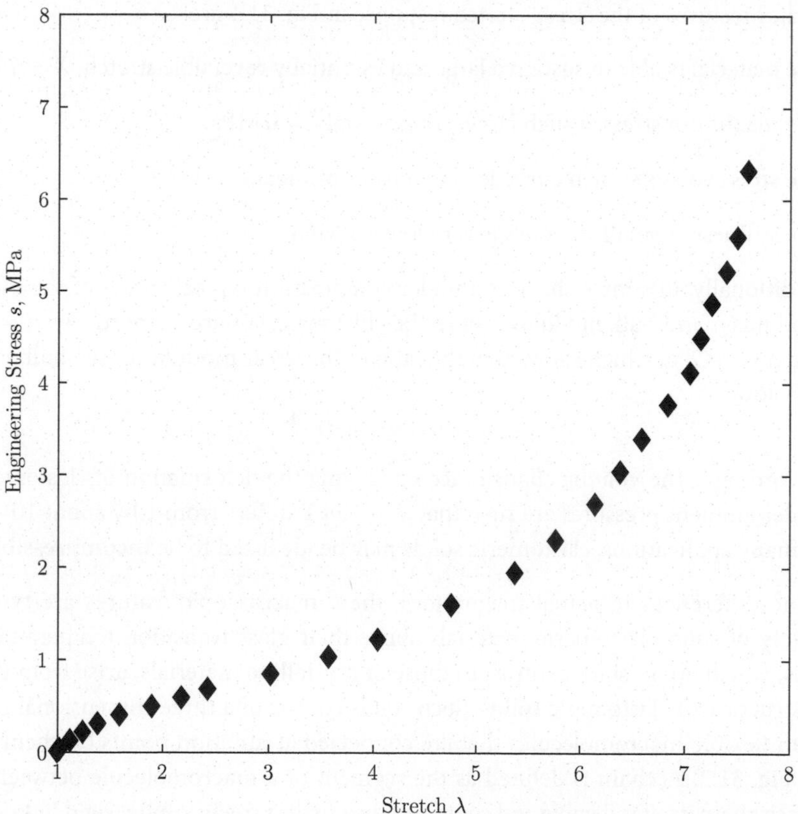

Fig. 31.1 A typical engineering stress-versus-stretch curve for vulcanized natural rubber in simple tension. Data from Treloar (1944).

Continuum Mechanics of Solids. Lallit Anand and Sanjay Govindjee, Oxford University Press (2020).

DOI: 10.1093/oso/9780198864721.001.0001

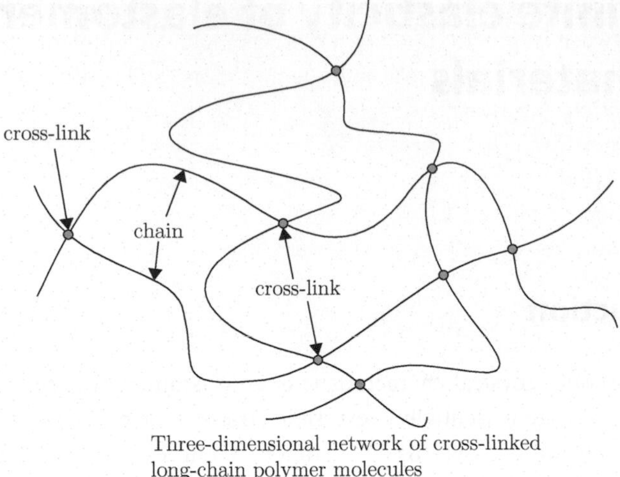

Fig. 31.2 Schematic of a chemically cross-linked elastomeric material.

The major features of the stress-stretch response in Fig. 31.1 are:

- The material is able to sustain a large and essentially reversible stretch, $\lambda \sim 7$.[1]
- The maximum stress at such large strains is only ~ 6 MPa.
- The stress versus stretch curve is markedly non-linear.
- The stiffness at small stretches is very low $\sim$1 MPa.
- Additionally, in simple shear and under compression typical values of the initial shear modulus μ and bulk modulus κ of rubber-like materials are, respectively, $\mu \sim 0.5$ MPa, and $\kappa \sim 2$ GPa, which shows that the ratio of the shear modulus to the bulk modulus is very low,[2]

  $$(\mu/\kappa) \sim 10^{-4}.$$

  Accordingly, the volume changes accompanying the deformation of elastomeric solids under ambient pressures are such that $J = \det \mathbf{F}$ differs from 1 by about 10^{-4}. Hence, in many applications, elastomeric solids may be idealized to be **incompressible.**

With some differences in actual magnitudes, these macroscopic features are typical for a wide variety of solid elastomeric materials above their glass transition temperatures. These properties, which are in stark contrast to those of crystalline materials, arise from the unique internal structure of elastomeric solids. Such solids consist of a three-dimensional network of long-chain flexible macromolecules that are connected at junction points by chemical cross-links, cf. Fig. 31.2. A chain is defined as the segment of a macromolecule between junction points. Each chain itself is flexible and contains more-or-less freely jointed rigid links composed

---

[1] Experimental data often shows a small amount of *hysteresis*, but we shall ignore this in our ensuing discussion.

[2] These values compare with $\mu = 82$ GPa and $\kappa = 158$ GPa for steel, where $\mu/\kappa = 0.52$—a typical value for a metal.

of repeating chemical elements. Because of the large number of repeating units in a chain, for a given distance between its junction points, a chain can take on many internal arrangements of the rigid links without the need for straining chemical bonds. This property allows for the easy and large deformation of elastomeric materials wherein the deformation-induced free energy changes occur primarily due to changes in the so-called *configurational entropy* of the solid, and not due to changes in the internal energy of the solid. This is directly opposite to what happens when one deforms a crystalline solid, where deformation-induced changes in free energy arise due to changes in internal energy (from the straining of chemical bonds) and entropy changes are minimal.

The literature on special free energy functions for isotropic incompressible materials which have been proposed to model elastomeric materials is *vast*; for reviews see Boyce and Arruda (2000), Marckmann and Verron (2006), and Beda (2014). In this chapter we discuss a few prominent free energy functions which have found significant use in practice.

31.2 Neo-Hookean free-energy function

The simplest model for the non-linear elastic response of an isotropic incompressible elastomeric or rubber-like material is provided by the **neo-Hookean** free energy function proposed by Treloar (1943) on the basis of Gaussian statistics and a molecular network theory (cf., Mark and Erman, 2007):

$$\boxed{\begin{aligned}\psi_{\mathrm{R}} &= \frac{\mu_0}{2}(I_1 - 3)\\[1ex] &= \frac{\mu_0}{2}(\lambda_1^2 + \lambda_2^2 + \lambda_3^2 - 3),\end{aligned}} \tag{31.2.1}$$

with the principal stretches $(\lambda_1, \lambda_2, \lambda_3)$ satisfying the incompressibility constraint

$$\lambda_1 \lambda_2 \lambda_3 = 1, \tag{31.2.2}$$

and with $\mu_0 > 0$ a constant. We refer to an incompressible, isotropic, hyperelastic body with free energy of this form as a **neo-Hookean material**. The principal values of the Cauchy stress for a neo-Hookean material are then given by

$$\boxed{\sigma_i = \mu_0 \lambda_i^2 - p.} \tag{31.2.3}$$

In this case, in simple shear the shear stress is given by

$$\sigma_{12} = \mu_0 \gamma, \tag{31.2.4}$$

so that the constant $\mu_0 > 0$ represents the ground state rubbery **shear modulus** of the neo-Hookean material.

The first systematic series of experiments on the mechanical properties of rubber-like solids were carried out by Treloar (1944). The primary objective of Treloar's experiments was to determine the range of validity of the neo-Hookean free energy (31.2.1). This simple form of the free energy provides an adequate first approximation to the behavior of rubber-like solids, and is generally regarded as a valid prototype for this class of materials. It has the advantage of being easy to treat mathematically, so that solutions to many problems have been obtained by its use.

31.3 Mooney–Rivlin free-energy function

Another simple free energy which depends not only on the first invariant I_1 but also on the second invariant

$$I_2 = \lambda_2^2\lambda_3^2 + \lambda_3^2\lambda_1^2 + \lambda_1^2\lambda_2^2 \equiv \lambda_1^{-2} + \lambda_2^{-2} + \lambda_3^{-2}, \tag{31.3.1}$$

is provided by the **Mooney–Rivlin** (Mooney, 1940; Rivlin, 1948; Rivlin and Saunders, 1951) free energy function

$$\boxed{\begin{aligned}\psi_R &= \frac{C_1}{2}(I_1 - 3) + \frac{C_2}{2}(I_2 - 3)\\ &= \frac{C_1}{2}(\lambda_1^2 + \lambda_2^2 + \lambda_3^2 - 3) + \frac{C_2}{2}(\lambda_1^{-2} + \lambda_2^{-2} + \lambda_3^{-2} - 3),\end{aligned}} \tag{31.3.2}$$

with $C_1 > 0$ and $C_2 > 0$ constants. We refer to an incompressible, isotropic, hyperelastic body with free energy of this form as a **Mooney–Rivlin material**. The principal values of the Cauchy stress for a Mooney–Rivlin material are given by

$$\boxed{\sigma_i = (C_1\lambda_i^2 - C_2\lambda_i^{-2}) - p.} \tag{31.3.3}$$

From our discussion of simple shear in Sec. 30.14.2 we see that for a Mooney–Rivlin material in simple shear the shear stress is given by

$$\sigma_{12} = \mu_0\gamma \qquad \text{with a shear modulus} \quad \mu_0 \stackrel{\text{def}}{=} C_1 + C_2 > 0, \tag{31.3.4}$$

and the normal stresses are

$$\sigma_{11} = C_1\gamma^2 \quad \text{and} \quad \sigma_{22} = -C_2\gamma^2. \tag{31.3.5}$$

Since $C_1 > 0$ the normal stress σ_{11} is *tensile*, and since $C_2 > 0$, the normal stress σ_{22} is *compressive*.

The free energy (31.3.2) was proposed by Mooney (1940) as the most general form of a free energy function which yields a linear relationship between shear stress σ_{12} and shear strain γ in simple shear. With a suitable choice of the constants C_1 and C_2, the Mooney–Rivlin form of the free energy function gives a marginally better fit to the experimental data for vulcanized natural rubber in simple tension, pure-shear, and equi-biaxial tension, than the neo-Hookean form. However, since $C_2 = 0$ for the neo-Hookean form, the normal stress $\sigma_{22} = 0$ for a neo-Hookean material in contradiction to experimental evidence for some elastomers (Rivlin and Saunders, 1951; Gent and Rivlin, 1952), which is a limitation of the neo-Hookean free energy.

EXAMPLE 31.1 Inflation of a Mooney–Rivlin balloon

Consider an incompressible elastic thin-walled spherical balloon of inner radius R and wall thickness T. When the balloon is subjected to an internal pressure it will inflate. Common experience with novelty balloons indicates that the internal pressure p_i required to inflate the balloon will increase with very little expansion of the radius of the balloon until a critical pressure is reached, at which point the balloon will rapidly expand.

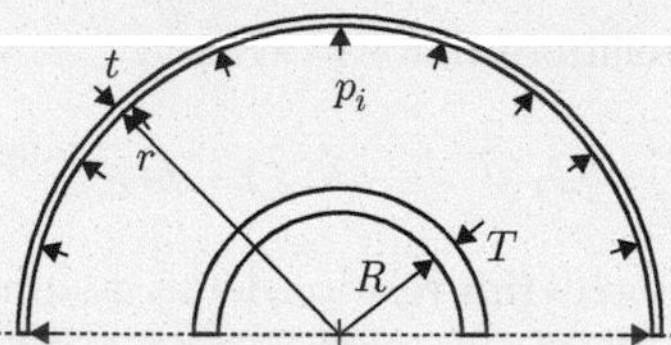

Fig. 31.3 Schematic cross-sectional geometry of an inflating thin-walled spherical balloon.

A schematic of such a balloon is shown in Fig. 31.3, where R and T are the reference dimensions of the balloon and r and t are the radius and thickness when subjected to an inflation pressure p_i. If the material of the balloon is an incompressible Mooney–Rivlin material, then the stress in the material will be of the form

$$\boldsymbol{\sigma} = -p\mathbf{1} + C_1\mathbf{B} - C_2\mathbf{B}^{-1}. \tag{31.3.6}$$

Further, symmetry tells us that in the plane of the material, the stress will be a constant (by the thin-walled assumption):

$$\sigma = \frac{p_i r}{2t}. \tag{31.3.7}$$

Also making the standard thin-walled assumption of plane stress, the through-thickness stress can be approximated to be zero.

Along every great circle of the sphere, the stretch of the material (in the plane of the balloon) is $\lambda = r/R$. Thus at every point on the sphere, all tangential directions to the surface of the sphere have the same stretch λ and any such two mutually orthogonal directions serve as principal directions for **B**. The third principal direction is the cross product of these two directions and thus the radial direction at each point. Using these directions as a basis to represent the left Cauchy–Green tensor we have

$$[\mathbf{B}] = \begin{bmatrix} \lambda^2 & 0 & 0 \\ 0 & \lambda^2 & 0 \\ 0 & 0 & \lambda^{-4} \end{bmatrix} \tag{31.3.8}$$

and in the same coordinate frame, from (31.3.6),

$$\begin{bmatrix} \sigma & 0 & 0 \\ 0 & \sigma & 0 \\ 0 & 0 & 0 \end{bmatrix} = -p \begin{bmatrix} 1 & 0 & 0 \\ 0 & 1 & 0 \\ 0 & 0 & 1 \end{bmatrix} + C_1 \begin{bmatrix} \lambda^2 & 0 & 0 \\ 0 & \lambda^2 & 0 \\ 0 & 0 & \lambda^{-4} \end{bmatrix} - C_2 \begin{bmatrix} \lambda^{-2} & 0 & 0 \\ 0 & \lambda^{-2} & 0 \\ 0 & 0 & \lambda^4 \end{bmatrix}. \tag{31.3.9}$$

Since the deformation is homogeneous (and the stretches are constant), the third of these equations can be used to determine the unknown scalar field p as

$$-p = -C_1\lambda^{-4} + C_2\lambda^4. \tag{31.3.10}$$

Substituting this back into (31.3.9) gives the in-plane stress as

$$\sigma = C_1(\lambda^2 - \lambda^{-4}) - C_2(\lambda^{-2} - \lambda^4). \tag{31.3.11}$$

Employing (31.3.7) and the relations that $r = \lambda R$ and $t = \lambda^{-2}T$, we find that

$$p_i = \frac{2T}{R}\left[C_1(\lambda^{-1} - \lambda^{-7}) - C_2(\lambda^{-5} - \lambda)\right]. \tag{31.3.12}$$

To understand the prediction of this relation, let us assume a balloon with $R = 10$ mm, $T = 0.5$ mm, and a shear modulus $\mu = 0.25$ MPa. If we define $\zeta = C_2/C_1$, then we can rewrite (31.3.12) as

$$p_i = \frac{2T}{R}\frac{\mu}{1+\zeta}\left[(\lambda^{-1} - \lambda^{-7}) - \zeta(\lambda^{-5} - \lambda)\right]. \tag{31.3.13}$$

A plot of (31.3.13) is shown in Fig. 31.4 for various values of ζ. Observe that the response shows a clear pressure maximum. If the pressure is maintained, then a very rapid expansion of the balloon will occur. Also note that in the neo-Hookean case, $\zeta = 0$, the pressure monotonically decreases after the peak value. For $\zeta > 0$, the curves eventually rise again as is more commonly experienced.

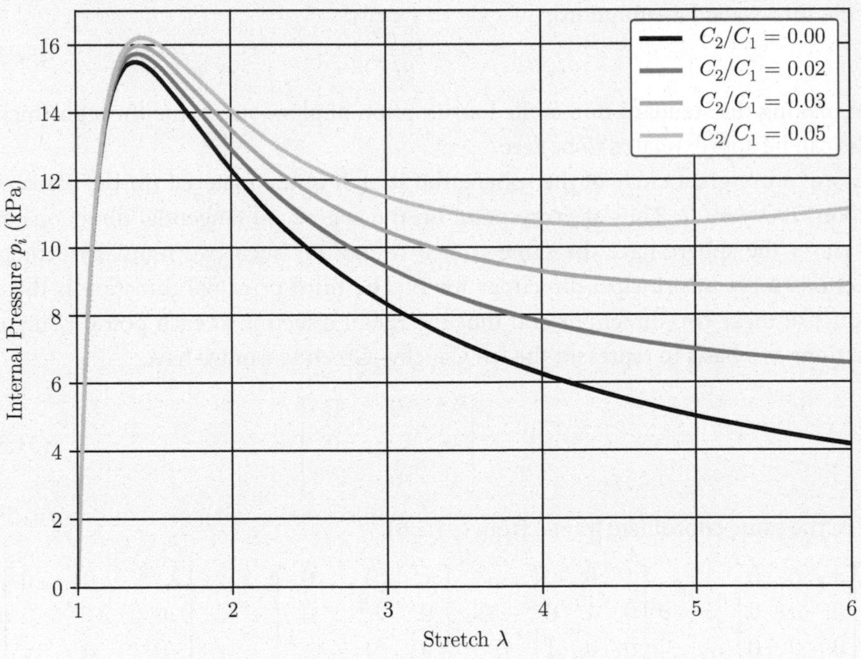

Fig. 31.4 Pressure versus stretch response for a Mooney–Rivlin balloon.

31.3.1 **The Mooney–Rivlin free energy expressed in an alternate fashion**

We may express the Mooney–Rivlin free energy (31.3.2) alternatively as

$$\psi_{\mathrm{R}} = \sum_{r=1}^{2} \frac{\mu_r}{\alpha_r} (\lambda_2^{\alpha_r} + \lambda_1^{\alpha_r} + \lambda_3^{\alpha_r} - 3), \tag{31.3.14}$$

in which

$$\mu_1 \equiv C_1 > 0, \qquad \alpha_1 = 2 \qquad \text{and} \qquad \mu_2 \equiv -C_2 < 0, \qquad \alpha_2 = -2. \tag{31.3.15}$$

These constants satisfy the requirement

$$\sum_{r=1}^{2} \mu_r \alpha_r = \mu_1 \alpha_1 + \mu_2 \alpha_2 = 2\mu_0 > 0, \tag{31.3.16}$$

where μ_0 is the shear modulus of the material in the undeformed stress-free reference configuration. Further,

$$\mu_r \alpha_r > 0 \quad \text{for each } r = 1, 2. \tag{31.3.17}$$

Using (30.14.12) the principal Cauchy stresses are given by

$$\sigma_i = \sum_{r=1}^{2} \mu_r \lambda_i^{\alpha_r} - p, \qquad i = 1, 2, 3. \tag{31.3.18}$$

31.4 **Ogden free-energy function**

As a generalization of the form (31.3.14) for the Mooney–Rivlin free energy in terms of principal stretches $(\lambda_1, \lambda_2, \lambda_3)$, Ogden (1972a; 1986) has proposed a free energy function of the form

$$\psi_{\mathrm{R}} = \sum_{r=1}^{M} \frac{\mu_r}{\alpha_r} (\lambda_2^{\alpha_r} + \lambda_1^{\alpha_r} + \lambda_3^{\alpha_r} - 3), \tag{31.4.1}$$

which contains M terms, and in which μ_r and α_r are both material constants, with the latter not necessarily being integers. For consistency with the classical neo-Hookean theory, these constants must satisfy the requirement

$$\sum_{r=1}^{M} \mu_r \alpha_r = 2\mu_0, \tag{31.4.2}$$

where μ_0 is the ground state shear modulus of the material in the undeformed stress-free reference configuration. Further, considerations of physically realistic response and material stability lead to the inequalities

$$\mu_r \alpha_r > 0 \quad \text{for each } r = 1, \ldots, M \quad \text{(no sum over } r\text{)}. \tag{31.4.3}$$

Using (30.14.12) the principal Cauchy stresses are given by

$$\sigma_i = \sum_{r=1}^{M} \mu_r \lambda_i^{\alpha_r} - p, \qquad i = 1, 2, 3. \tag{31.4.4}$$

Table 31.1 Representative values from Ogden (1972a) of the material parameters in the neo-Hookean, Mooney-Rivlin, and Ogden free energies for a vulcanized natural rubber studied by Treloar (1944).

Model	α_r	μ_r, MPa
Neo-Hookean, $M = 1$	$\alpha_1 = 2$	$\mu_1 (\equiv \mu_0) = 0.382$
Mooney–Rivlin, $M = 2$	$\alpha_1 = 2$	$\mu_1 = 0.334$
	$\alpha_2 = -2$	$\mu_2 = -0.048$
Ogden, $M = 3$	$\alpha_1 = 1.3$	$\mu_1 = 0.6176$
	$\alpha_2 = 5.0$	$\mu_2 = 0.0012$
	$\alpha_3 = -2.0$	$\mu_3 = -0.0098$

For $M = 1$, and particular values of material constants

$$\mu_1 \equiv \mu_0, \qquad \alpha_1 = 2,$$

the Ogden model reduces to the neo-Hookean model, and for $M = 2$, and particular values of material constants

$$\mu_1 = C_1, \quad \alpha_1 = 2, \quad \text{and} \quad \mu_2 = -C_2, \quad \alpha_2 = -2,$$

the Ogden model reduces to the Mooney–Rivlin model.

The Ogden model is most commonly used with $M = 3$ or $M = 4$, and the predictions of such a model are generally quite satisfactory. Ogden (1972a) showed that he could obtain good agreement between his model and experimental data from simple tension, pure-shear and equi-biaxial tension tests for an elastomer. Table 31.1 lists values for the six material parameters for $M = 3$, estimated from the experiments by Treloar (1944) on a vulcanized natural rubber.[3]

31.5 Arruda–Boyce free-energy function

An alternate popular model for elastomers at high levels of deformation is the Arruda–Boyce model (Arruda and Boyce, 1993). The **Arruda–Boyce free energy** takes as its main kinematic variable the *root-mean-square stretch*, i.e. the average stretch in the three principal directions:

$$\bar{\lambda} \stackrel{\text{def}}{=} \sqrt{\frac{\lambda_1^2 + \lambda_2^2 + \lambda_3^2}{3}} = \sqrt{\frac{I_1(\mathbf{C})}{3}} = \sqrt{\frac{I_1(\mathbf{B})}{3}}. \tag{31.5.1}$$

[3] For a discussion of issues related to material parameter determination for this model see Ogden et al. (2004).

We call $\bar{\lambda}$ the **effective stretch**. In terms of this kinematic variable the free energy is taken as

$$\psi_{\mathrm{R}} = \hat{\psi}_{\mathrm{R}}(\bar{\lambda}) \tag{31.5.2}$$

$$= \mu_0\, \lambda_L^2 \left[\left(\frac{\bar{\lambda}}{\lambda_L} \right) \beta + \ln\left(\frac{\beta}{\sinh \beta} \right) \right] \qquad \text{with} \qquad \beta = \mathcal{L}^{-1}\left(\frac{\bar{\lambda}}{\lambda_L} \right), \tag{31.5.3}$$

where $\mathcal{L}^{-1}$ is the function inverse to the Langevin function $\mathcal{L}$, which is defined as

$$\mathcal{L}(\zeta) \stackrel{\text{def}}{=} \coth\zeta - \zeta^{-1}. \tag{31.5.4}$$

This functional form for the free energy ψ_{R} involves two material parameters:

- μ_0, the ground state *shear modulus*, and
- λ_L, called the *network locking stretch.*

This seemingly complex form is motivated by statistical mechanical considerations of the change in entropy of individual polymer chains at large degrees of extension.

Then, as discussed in Sec. 30.14.1, the principal values of the Piola stress **P** (for the case when $\mathbf{R} = \mathbf{1}$) and the Cauchy stress $\boldsymbol{\sigma}$ are given by

$$s_i = \frac{\partial \hat{\psi}_{\mathrm{R}}(\bar{\lambda})}{\partial \lambda_i} - p\lambda_i^{-1}, \tag{31.5.5}$$

and

$$\sigma_i = \lambda_i \frac{\partial \hat{\psi}_{\mathrm{R}}(\bar{\lambda})}{\partial \lambda_i} - p \quad \text{(no sum)}, \tag{31.5.6}$$

respectively. Thus, since

$$\frac{\partial \hat{\psi}_{\mathrm{R}}(\bar{\lambda})}{\partial \lambda_i} = \frac{\partial \hat{\psi}_{\mathrm{R}}(\bar{\lambda})}{\partial \bar{\lambda}} \frac{\partial \bar{\lambda}}{\partial \lambda_i} = \frac{\partial \hat{\psi}_{\mathrm{R}}(\bar{\lambda})}{\partial \bar{\lambda}} \frac{1}{\sqrt{3}} \frac{1}{2} \frac{2\lambda_i}{\sqrt{\lambda_1^2 + \lambda_2^2 + \lambda_3^2}} = \left(\frac{1}{3\bar{\lambda}} \frac{\partial \hat{\psi}_{\mathrm{R}}(\bar{\lambda})}{\partial \bar{\lambda}} \right) \lambda_i,$$

we may write the derivative $\dfrac{\partial \hat{\psi}_{\mathrm{R}}(\bar{\lambda})}{\partial \lambda_i}$ as

$$\frac{\partial \hat{\psi}_{\mathrm{R}}(\bar{\lambda})}{\partial \lambda_i} = \mu\, \lambda_i, \tag{31.5.7}$$

where we have introduced a **generalized shear modulus**

$$\boxed{\mu \stackrel{\text{def}}{=} \frac{1}{3\bar{\lambda}} \frac{\partial \hat{\psi}_{\mathrm{R}}(\bar{\lambda})}{\partial \bar{\lambda}},} \tag{31.5.8}$$

which depends on the effective stretch $\bar{\lambda}$.

Hence, using (31.5.7) and (31.5.8), the equation for the principal Piola stresses (31.5.5) may be written as

$$\boxed{s_i = \mu\lambda_i - p\lambda_i^{-1},} \tag{31.5.9}$$

while that for the Cauchy stresses (31.5.6) may be written as

$$\boxed{\sigma_i = \mu\lambda_i^2 - p,} \tag{31.5.10}$$

and the Cauchy stress is thus given by

$$\boxed{\boldsymbol{\sigma} = -p\mathbf{1} + \mu\mathbf{B}.} \tag{31.5.11}$$

For the Arruda–Boyce free energy (31.5.3),

$$\frac{\partial\hat{\psi}_{\mathrm{R}}(\bar{\lambda})}{\partial\bar{\lambda}} = \mu_0\lambda_L\mathcal{L}^{-1}\left(\frac{\bar{\lambda}}{\lambda_L}\right),$$

use of which in (31.5.8) gives the generalized shear modulus,

$$\boxed{\mu = \mu_0\left(\frac{\lambda_L}{3\bar{\lambda}}\right)\mathcal{L}^{-1}\left(\frac{\bar{\lambda}}{\lambda_L}\right).} \tag{31.5.12}$$

Note that since $\mathcal{L}^{-1}(\bar{\lambda}/\lambda_L) \to \infty$ as $\bar{\lambda} \to \lambda_L$, the generalized modulus $\mu \to \infty$ as $\bar{\lambda} \to \lambda_L$, a situation which reflects the limited chain-extensibility of elastomeric materials. Note also that as $\lambda_L \to \infty$, $\mu \to \mu_0$ and we recover the neo-Hookean result.

Arruda and Boyce (1993) have shown the utility of the constitutive model (31.5.11) with μ given by (31.5.12) in representing the classical stress-strain data of Treloar (1944) for vulcanized natural rubber under uniaxial tension, pure-shear, and equi-biaxial extension, as well as their own stress-strain data for simple compression and plane strain compression on silicone, gum, and neoprene rubbers. As an example, Fig. 31.5 shows a fit of the Arruda–Boyce

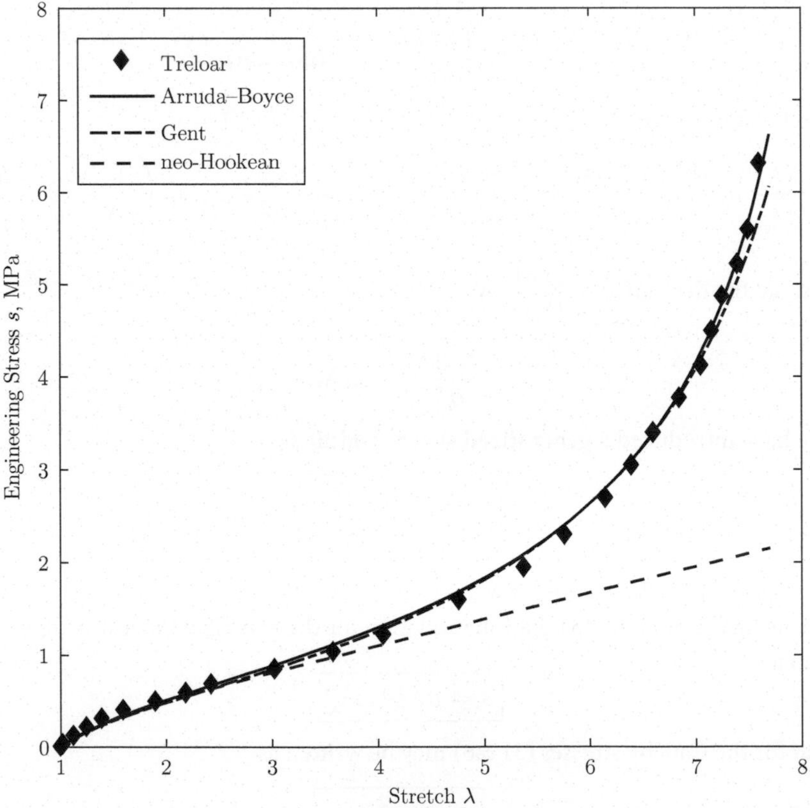

Fig. 31.5 Fit of the neo-Hookean, Arruda–Boyce, and Gent models to the engineering stress versus stretch curve for a vulcanized natural rubber in simple tension. Experimental data from Treloar (1944).

model to the experimental data of Treloar (1944) for a vulcanized natural rubber in simple tension at room temperature. The material parameters used to obtain the fit are

$$\mu_0 = 0.28\,\text{MPa}, \quad \lambda_L = 5.12. \tag{31.5.13}$$

Figure 31.5 also shows a fit of the neo-Hookean model to Treloar's data with

$$\mu_0 = 0.28\,\text{MPa}. \tag{31.5.14}$$

As expected, the neo-Hookean model cannot represent the stiffening response of this elastomer at very large stretches, but is able to adequately represent the data to a moderate stretch level of $\lambda \approx 3$.

REMARK 31.1

In generating the plot shown in Fig. 31.5 we have used the following approximant to $\mathcal{L}^{-1}(\zeta)$,

$$\boxed{\mathcal{L}^{-1}(\zeta) \approx \zeta\,\frac{3-\zeta^2}{1-\zeta^2} \qquad \text{with} \qquad \zeta \stackrel{\text{def}}{=} \frac{\bar{\lambda}}{\lambda_L},} \tag{31.5.15}$$

which is its $[3/2]$ Padé approximation (Cohen, 1991). Use of this approximant to $\mathcal{L}^{-1}(\zeta)$ considerably simplifies numerical calculations.

31.6 **Gent free-energy function**

As a final example of a model that incorporates the finite extensibility of the polymer chains comprising an elastomeric network, we consider a model proposed by Gent (1996, 1999) in the form of the free energy function

$$\psi_{\text{R}} = -\frac{\mu_0}{2}\, I_m \ln\left(1 - \frac{I_1 - 3}{I_m}\right), \tag{31.6.1}$$

with two material parameters

$$\mu_0 > 0 \qquad \text{and} \qquad I_m > 0,$$

in which I_m is a stiffening parameter which sets a bound on the maximum possible value of I_1; that is

$$I_1 < 3 + I_m.$$

For the Gent free energy the Cauchy stress is

$$\boxed{\boldsymbol{\sigma} = -p\mathbf{1} + \mu\mathbf{B},} \tag{31.6.2}$$

with the generalized shear modulus from (31.6.1) given by

$$\boxed{\mu \stackrel{\text{def}}{=} \mu_0 \left(\frac{I_m}{I_m - (I_1 - 3)}\right).} \tag{31.6.3}$$

Note that in the limit of $I_m \to \infty$ we obtain that

$$\mu \equiv \mu_0 \tag{31.6.4}$$

which is the classical neo-Hookean limit. Also, in the limit $I_1 \to 3 + I_m$ we obtain that

$$\mu \to \infty, \tag{31.6.5}$$

which reflects the limited chain-extensibility and rapid strain-hardening observed in experiments.

Figure 31.5 also shows a fit of the Gent model to the experimental data of Treloar (1944). The material parameters used to obtain the fit are

$$\mu_0 = 0.27\,\text{MPa}, \quad I_m = 86. \tag{31.6.6}$$

The predictions from the Gent and the Arruda–Boyce models are almost *indistinguishable*, except at the largest values of stretch. Relative to the Arruda–Boyce model, the Gent model has the advantage of mathematical simplicity, which allows for detailed analysis and explicit solution of particular boundary-value problems.

EXAMPLE 31.2 Classic torsion of a round bar

As an example of the behavior of the Gent material, consider a round bar of length L_0 and radius R_0. Let us assume that the bar's deformation is given by the classical deformation of a bar in torsion, viz. plane sections remain plane and rotate by an amount that increases linearly from one end of the bar to the other. Assuming a coordinate frame with origin at the center of the bar (at the lower end), the motion is given by

$$x_1 = \sqrt{(X_1)^2 + (X_2)^2}\cos\left(\tan^{-1}(X_2/X_1) + \alpha X_3\right), \tag{31.6.7}$$

$$x_2 = \sqrt{(X_1)^2 + (X_2)^2}\sin\left(\tan^{-1}(X_2/X_1) + \alpha X_3\right), \tag{31.6.8}$$

$$x_3 = X_3, \tag{31.6.9}$$

where α is a given constant, representing the twist per unit length. The given motion indicates that a cross-section at height X_3 rotates (rigidly) by an amount αX_3.

The left Cauchy–Green deformation tensor for this motion is conveniently written in polar coordinates as

$$[\mathbf{B}] = \begin{bmatrix} 1 & 0 & 0 \\ 0 & 1 + (r\alpha)^2 & r\alpha \\ 0 & r\alpha & 1 \end{bmatrix}, \tag{31.6.10}$$

where $r = \sqrt{(x_1)^2 + (x_2)^2}$. Thus the Cauchy stress is given by

$$\boldsymbol{\sigma} = -p\mathbf{1} + \mu_0 \frac{I_m}{I_m - (r\alpha)^2}\mathbf{B}. \tag{31.6.11}$$

In polar coordinates,

$$[\boldsymbol{\sigma}] = -p\begin{bmatrix} 1 & 0 & 0 \\ 0 & 1 & 0 \\ 0 & 0 & 1 \end{bmatrix} + \mu_0 \frac{I_m}{I_m - (r\alpha)^2}\begin{bmatrix} 1 & 0 & 0 \\ 0 & 1+(r\alpha)^2 & r\alpha \\ 0 & r\alpha & 1 \end{bmatrix}. \tag{31.6.12}$$

The unknown scalar field p can be found from the equilibrium equations. In polar coordinates, with the assumed kinematics, the relevant equilibrium equation is the one for radial force balance:

$$\frac{\partial \sigma_{rr}}{\partial r} + \frac{1}{r}\frac{\partial \sigma_{\theta\theta}}{\partial \theta} + \frac{\partial \sigma_{rz}}{\partial z} + \frac{1}{r}(\sigma_{rr} - \sigma_{\theta\theta}) = 0. \tag{31.6.13}$$

Substituting from (31.6.12) gives

$$\frac{dp}{dr} = \frac{d}{dr}\left[\mu_0 \frac{I_m}{I_m - (r\alpha)^2}\right] - r\alpha^2 \mu_0 \frac{I_m}{I_m - (r\alpha)^2}. \tag{31.6.14}$$

The lateral traction free boundary condition,

$$\boldsymbol{\sigma}\mathbf{e}_r = \mathbf{0}, \tag{31.6.15}$$

implies

$$p(R_0) = \mu_0 \frac{I_m}{I_m - (R_0\alpha)^2}, \tag{31.6.16}$$

which can be used as a boundary condition to integrate (31.6.14):

$$p(r) = \mu_0 \frac{I_m}{I_m - (r\alpha)^2} + \frac{1}{2} I_m \mu_0 \ln\left(\frac{I_m - (r\alpha)^2}{I_m - (R_0\alpha)^2}\right). \tag{31.6.17}$$

It is interesting to note that the traction on the end of the rod involves both a tangential component as well as an axial component:

$$\mathbf{t}_{\text{end}} = \boldsymbol{\sigma}\mathbf{e}_z = \sigma_{\theta z}\mathbf{e}_\theta + \sigma_{zz}\mathbf{e}_z \tag{31.6.18}$$

$$\sigma_{\theta z} = \mu_0 \frac{I_m r\alpha}{I_m - (r\alpha)^2} \tag{31.6.19}$$

$$\sigma_{zz} = -\frac{1}{2} I_m \mu_0 \ln\left(\frac{I_m - (r\alpha)^2}{I_m - (R_0\alpha)^2}\right). \tag{31.6.20}$$

Thus the imposition of the classical torsion deformation results in a reaction torque of magnitude

$$T = \int_A \sigma_{\theta z} r \, dr d\theta = \frac{\pi \mu_0 I_m}{\alpha^2}\left[\sqrt{I_m} \ln\left(\frac{\sqrt{I_m} + \alpha R_0}{\sqrt{I_m} - \alpha R_0}\right) - 2\alpha R_0\right], \tag{31.6.21}$$

and an axial force of magnitude

$$N = \int_A \sigma_{zz} \, dr d\theta = -2\pi I_m \mu_0 \left[\frac{\sqrt{I_m}\tanh^{-1}(R_0\alpha/\sqrt{I_m})}{\alpha} - R_0\right]. \tag{31.6.22}$$

The resultant torque, T, and the normal force, N, as a function of the twist per unit length α, for $\mu_0 = 0.27$ MPa and $I_m = 86$, are shown in Fig. 31.6. Observe that a compressive normal force is required to impose the pure torsion deformation. If a compressive force is not applied to a torsion specimen to maintain pure torsion, then the specimen will elongate—which is known as the "Poynting effect" (Poynting, 1909).

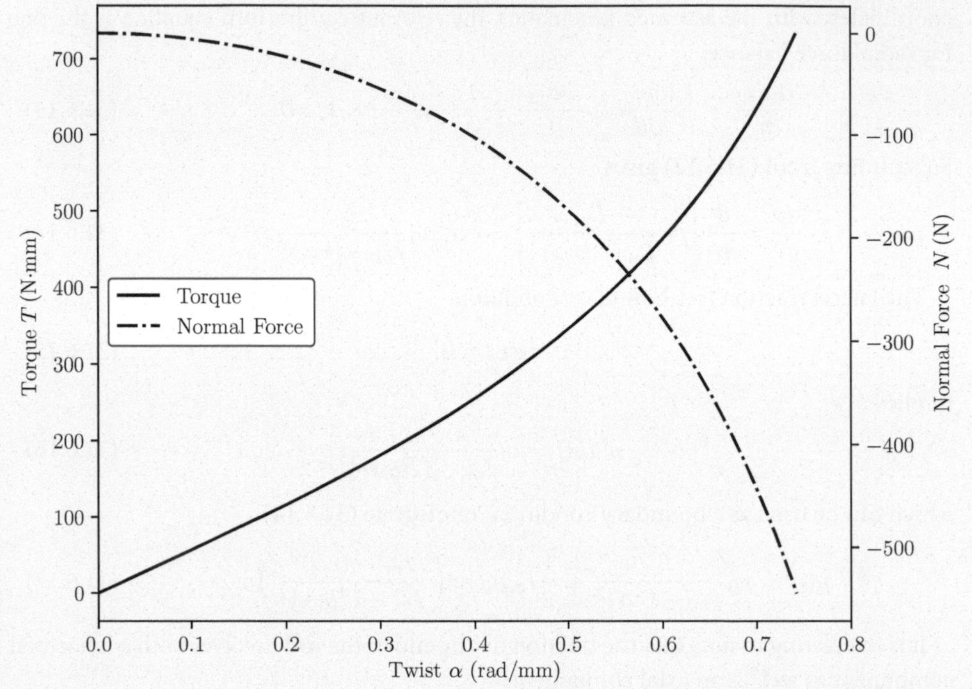

Fig. 31.6 Torque and normal force required to twist a circular rod ($L_0 = 50$ mm, $R_0 = 10$ mm) made of a Gent material.

31.6.1 I_2-dependent extension of the Gent free-energy function

Before closing this section, we note that unlike the Mooney–Rivlin free energy (31.3.2) which depends on both I_1 and I_2, the Gent free energy function depends only on I_1. A dependence on I_2 is easily accounted for by modifying the Gent free energy by using a previous proposal by Gent and Thomas (1958), to read as (Pucci and Saccomandi, 2002):

$$\psi_{\text{R}} = -\frac{\mu_0}{2} I_m \ln\left(1 - \frac{I_1 - 3}{I_m}\right) + C_2 \ln\left(\frac{I_2}{3}\right), \tag{31.6.23}$$

where $C_2 > 0$ is an additional constant reflecting the I_2-dependence. Ogden et al. (2004) have shown that this simple three-parameter free energy is useful for representing many aspects of the large deformation non-linear response of elastomeric materials. Representative values reported by these authors for the vulcanized natural rubber studied experimentally by Treloar (1944) are

$$\mu_0 = 0.242\,\text{MPa}, \quad I_m = 77.93, \quad C_2 = 0.102\,\text{MPa}.$$

31.7 Slightly compressible isotropic elastomeric materials

No material is completely incompressible. Thus, for reasons of more accurate modeling, as well as for computational considerations, it is sometimes desirable to account for the slight compressibility of elastomers. A theory for slightly compressible elastic materials can be developed based on a multiplicative decomposition

$$\mathbf{F} = \mathbf{F}^v \bar{\mathbf{F}}, \tag{31.7.1}$$

of the deformation gradient $\mathbf{F}$ into volumetric and isochoric factors $\mathbf{F}^v$ and $\bar{\mathbf{F}}$—as proposed by Flory (1961), where

$$\mathbf{F}^v = J^{1/3}\mathbf{1} \qquad \text{and} \qquad \bar{\mathbf{F}} = J^{-1/3}\mathbf{F}. \tag{31.7.2}$$

Decomposition (31.7.1) and the representations (31.7.2) lead to a multiplicative decomposition of the left Cauchy–Green tensor in the form:

$$\left.\begin{aligned} \mathbf{B} &= \mathbf{B}^v \bar{\mathbf{B}}, \\ \mathbf{B}^v &= J^{2/3}\mathbf{1}, \\ \bar{\mathbf{B}} &= J^{-2/3}\mathbf{B}. \end{aligned}\right. \tag{31.7.3}$$

Then, presuming that the free energy of a slightly compressible isotropic elastic solid can be expressed in separable isochoric and volumetric contributions,

$$\psi_{\text{R}} = \underbrace{\bar{\psi}_{\text{R}}(\bar{I}_1, \bar{I}_2)}_{\text{isochoric}} + \underbrace{\psi^v_{\text{R}}(J)}_{\text{volumetric}}, \tag{31.7.4}$$

where

$$\bar{I}_1 = I_1(\bar{\mathbf{B}}) \text{ and } \bar{I}_2 = I_2(\bar{\mathbf{B}})$$

are the first and second invariants of $\bar{\mathbf{B}}$, the Cauchy stress may be obtained after some effort (cf., Gurtin et al., 2010, Chapter 55) as

$$\boldsymbol{\sigma} = 2J^{-1}\text{dev}\left[\frac{\partial \bar{\psi}_{\text{R}}(\bar{I}_1, \bar{I}_2)}{\partial \bar{I}_1}\bar{\mathbf{B}} + \frac{\partial \bar{\psi}_{\text{R}}(\bar{I}_1, \bar{I}_2)}{\partial \bar{I}_2}(\bar{I}_1\mathbf{1} - \bar{\mathbf{B}})\bar{\mathbf{B}}\right] - p\mathbf{1}, \tag{31.7.5}$$

where $\text{dev}\mathbf{A} \stackrel{\text{def}}{=} \mathbf{A} - (1/3)(\text{tr}\mathbf{A})\mathbf{1}$ is the deviatoric operator, and

$$p \stackrel{\text{def}}{=} -(1/3)\,\text{tr}\,\boldsymbol{\sigma} = -\frac{\partial \psi^v_{\text{R}}(J)}{\partial J}, \tag{31.7.6}$$

is a pressure.

Table 31.2 Sample volumetric energy expressions and their derivatives.

Model	$\psi_{\text{R}}^{v}(J)$	$\dfrac{\partial \psi_{\text{R}}^{v}}{\partial J}$
1	$\frac{1}{2}\kappa(J-1)^2$	$\kappa(J-1)$
2	$\frac{1}{2}\kappa\,(\ln J)^2$	$\kappa\frac{\ln J}{J}$
3	$\frac{1}{\alpha^2}\kappa\,(\cosh(\alpha(J-1))-1)$	$\kappa\frac{1}{\alpha}\sinh(\alpha(J-1))$
4	$\frac{1}{4}\kappa\,(J^2-1-2\ln J)$	$\kappa\frac{1}{2}\left(J-\frac{1}{J}\right)$
5	$\frac{1}{2\alpha}\kappa\exp\left(\alpha\,(\ln J)^2\right)$	$\kappa\frac{\ln J}{J}\exp\left(\alpha\,(\ln J)^2\right)$

31.7.1 **Models for the volumetric energy**

As with the incompressible free energy, the literature contains a wide array of options for the volumetric energy term $\psi_{\text{R}}^{v}(J)$. A few common models are given in Table 31.2, in which κ is the ground state bulk modulus in all the models, and α is an additional dimensionless parameter.

REMARKS 31.2

1. Model 1 is very common and easy to work with. It provides a linear pressure–volume relation. However, the material may be compressed to zero volume ($J = 0$) with finite energy, which is not physically realistic, and this can cause difficulties in numerical finite element computations. Nonetheless, for materials with large bulk modulus at modest pressures, it works well.
2. Model 2 repairs the problem with Model 1 in that it requires infinite energy to compress the material to zero volume. However at high dilations ($J \to \infty$) the accompanying pressure goes to zero ($p \to 0$) instead of ($p \to -\infty$). Thus caution should be exercised when using this model in the presence of large positive dilation.
3. Model 3 was proposed in Bischoff et al. (2001) to better model observed pressure–volume non-linearities, which it does nicely. However, like Model 1, it allows the material to be compressed to zero volume with finite energy.
4. Model 4 was proposed in Simo and Taylor (1991) and is a particularization of a more general model due to Ogden (1972b). Like Model 2, it requires infinite energy to compress material to zero volume. However, unlike Model 2, here $J \to \infty$ implies $p \to -\infty$. These two features make it quite suitable for numerical finite element computations.
5. Lastly Model 5 (see e.g. Montella et al., 2016) has the same limiting behaviors as Model 4, but can model higher degrees of non-linearity in the pressure–volume response. As an example,

REMARKS 31.2 Continued

Fig. 31.7 shows the fit of this model to the elastomeric pressure volume data reported by Adams and Gibson (1930). The fits for their three materials (labeled here as A, B, and C) are made with the parameters listed in Table 31.3 (after extending the data by non-linear extrapolation to include a point at $p = 0$). For all three materials the agreement is seen to be quite good. Similar good results for this data can also be achieved with Model 3. However, the other models are much less accurate over the full range of this data set.

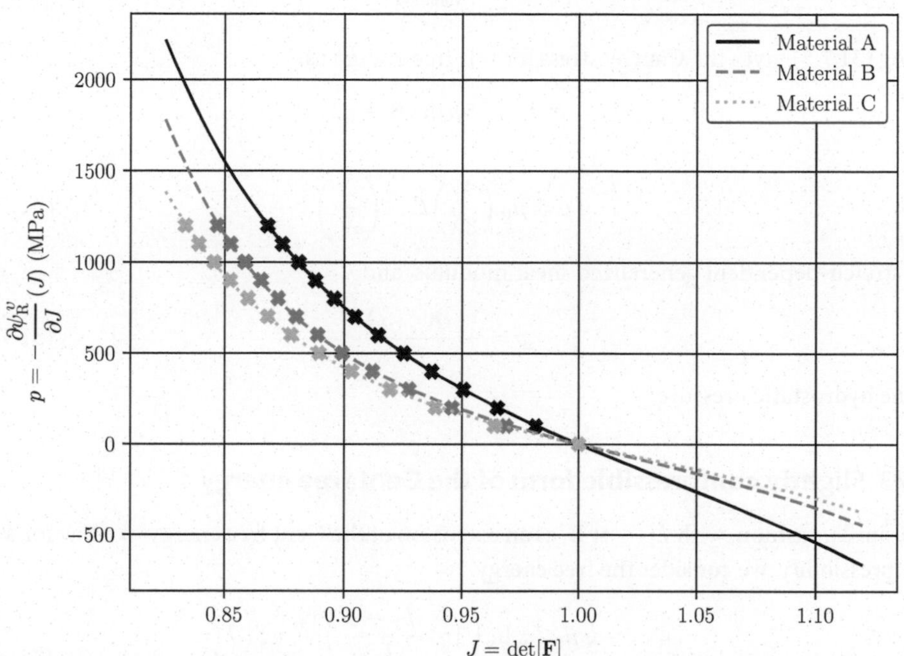

Fig. 31.7 Fit of the pressure volume data of Adams and Gibson (1930) (the X markers) using Model 5 (the continuous curves).

Table 31.3 Parameters for fitting Model 5 to the data of Adams and Gibson (1930).

Material A	$\kappa = 5.43$ GPa	$\alpha = 15.2$
Material B	$\kappa = 3.31$ GPa	$\alpha = 22.6$
Material C	$\kappa = 2.90$ GPa	$\alpha = 19.4$

31.7.2 **Slightly compressible form of the Arruda-Boyce free energy**

Redefine the effective stretch as

$$\bar{\lambda} \stackrel{\text{def}}{=} \sqrt{\frac{I_1(\bar{\mathbf{B}})}{3}}, \tag{31.7.7}$$

so that it represents an **effective distortional stretch**. Then, as an extension of the Arruda–Boyce free energy to allow for a slight compressibility, we consider the free energy (Anand, 1996)

$$\psi_{\mathrm{R}} = \underbrace{\mu_0\,\lambda_{\mathrm{L}}^2\left[\left(\frac{\bar{\lambda}}{\lambda_{\mathrm{L}}}\right)\beta + \ln\left(\frac{\beta}{\sinh\beta}\right)\right]}_{\text{isochoric}} + \underbrace{\psi_{\mathrm{R}}^{v}(J)}_{\text{volumetric}}, \quad \text{where} \quad \beta = \mathcal{L}^{-1}\left(\frac{\bar{\lambda}}{\lambda_{\mathrm{L}}}\right). \tag{31.7.8}$$

Using (31.7.5) gives the Cauchy stress for this free energy as,

$$\boldsymbol{\sigma} = J^{-1}\,\mu\,\mathrm{dev}\bar{\mathbf{B}} - p\mathbf{1}, \tag{31.7.9}$$

with

$$\mu = \mu_0\left(\frac{\lambda_{\mathrm{L}}}{3\bar{\lambda}}\right)\mathcal{L}^{-1}\left(\frac{\bar{\lambda}}{\lambda_{\mathrm{L}}}\right) \tag{31.7.10}$$

the stretch-dependent generalized shear modulus, and

$$p = -\frac{\partial\psi_{\mathrm{R}}^{v}(J)}{\partial J} \tag{31.7.11}$$

as the hydrostatic pressure.

31.7.3 **Slightly compressible form of the Gent free energy**

In a similar fashion, with $\bar{I}_1 = \mathrm{tr}\,\bar{\mathbf{B}}$, as an extension of the Gent free energy to allow for a slight compressibility, we consider the free energy

$$\psi_{\mathrm{R}} = \underbrace{-\tfrac{1}{2}\,\mu_0\,I_m \ln\left(1 - \frac{\bar{I}_1 - 3}{I_m}\right)}_{\text{isochoric}} + \underbrace{\psi_{\mathrm{R}}^{v}(J)}_{\text{volumetric}}. \tag{31.7.12}$$

Using (31.7.5) gives the Cauchy stress as,

$$\boldsymbol{\sigma} = J^{-1}\,\mu\,\mathrm{dev}\bar{\mathbf{B}} - p\mathbf{1}, \tag{31.7.13}$$

with

$$\mu \stackrel{\text{def}}{=} \mu_0\left(\frac{I_m}{I_m - (\bar{I}_1 - 3)}\right) \tag{31.7.14}$$

the $\bar{I}_1$-dependent generalized shear modulus, and

$$p = -\frac{\partial\psi_{\mathrm{R}}^{v}(J)}{\partial J} \tag{31.7.15}$$

as the hydrostatic pressure.

PART XII
Appendices

Appendix A Cylindrical and spherical coordinate systems

A.1 Introduction

The general equations of continuum mechanics are best expressed in either the direct notation, or in rectangular Cartesian coordinates. However, for the solution of particular problems it is often preferable to employ other coordinate systems. The most commonly encountered coordinate systems are

1. The **cylindrical coordinate** system which is good for solids that are symmetric around an axis.
2. The **spherical coordinate** system which is used when there is symmetry about a point.

In what follows we derive some important vector and tensor identities which will help us solve problems formulated in these coordinate systems.

A.2 Cylindrical coordinates

Cylindrical coordinates (r, θ, z) are related to rectangular coordinates (x_1, x_2, x_3) by

$$\begin{aligned} r &= [x_1^2 + x_2^2]^{\frac{1}{2}}, \quad r \geq 0, \\ \theta &= \tan^{-1}(x_2/x_1), \quad 0 \leq \theta \leq 2\pi, \\ z &= x_3, \quad -\infty < z < \infty, \end{aligned} \tag{A.2.1}$$

$$x_1 = r\cos\theta, \quad x_2 = r\sin\theta, \quad x_3 = z. \tag{A.2.2}$$

The orthonormal base vectors in the cylindrical coordinate system are directed in the radial, tangential, and axial directions as illustrated in Fig. A.1, and denoted by $\{\mathbf{e}_r, \mathbf{e}_\theta, \mathbf{e}_z\}$.

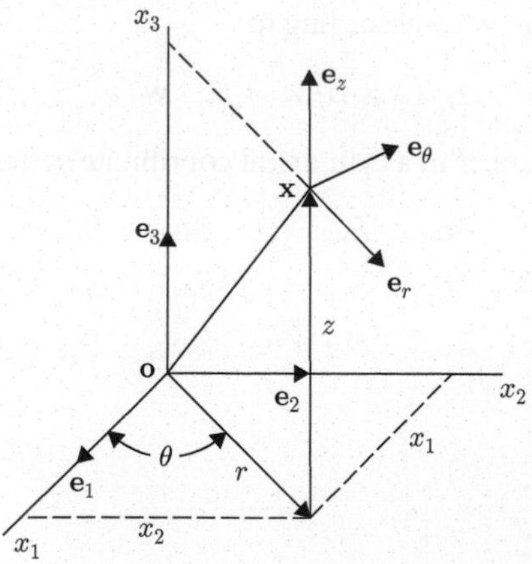

Fig. A.1 Cylindrical coordinate system.

Continuum Mechanics of Solids. Lallit Anand and Sanjay Govindjee, Oxford University Press (2020).

DOI: 10.1093/oso/9780198864721.001.0001

The position vector $\mathbf{r} = \mathbf{x} - \mathbf{o}$ is given by

$$\mathbf{r} = r\cos\theta\mathbf{e}_1 + r\sin\theta\,\mathbf{e}_2 + z\,\mathbf{e}_3, \tag{A.2.3}$$

and with respect to a cylindrical coordinate system a differential element $\mathrm{d}\mathbf{x}$ at $\mathbf{x}$ is given by

$$\mathrm{d}\mathbf{x} = \mathrm{d}r\,\mathbf{e}_r + (r\mathrm{d}\theta)\,\mathbf{e}_\theta + \mathrm{d}z\,\mathbf{e}_z. \tag{A.2.4}$$

Thus, since $\mathrm{d}\mathbf{r} = \mathrm{d}\mathbf{x}$,

$$\mathbf{e}_r = \frac{\partial\mathbf{r}}{\partial r}, \qquad \mathbf{e}_\theta = \frac{1}{r}\frac{\partial\mathbf{r}}{\partial\theta}, \qquad \mathbf{e}_z = \frac{\partial\mathbf{r}}{\partial z}.$$

Using this and (A.2.3) we obtain that $\{\mathbf{e}_r, \mathbf{e}_\theta, \mathbf{e}_z\}$ are related to the orthonormal base vectors $\{\mathbf{e}_1, \mathbf{e}_2, \mathbf{e}_3\}$ in the rectangular system by

$$\begin{aligned}\mathbf{e}_r &= \cos\theta\,\mathbf{e}_1 + \sin\theta\,\mathbf{e}_2,\\ \mathbf{e}_\theta &= -\sin\theta\,\mathbf{e}_1 + \cos\theta\,\mathbf{e}_2,\\ \mathbf{e}_z &= \mathbf{e}_3.\end{aligned} \tag{A.2.5}$$

From (A.2.5),

$$\frac{\partial\mathbf{e}_r}{\partial\theta} = \mathbf{e}_\theta, \qquad \frac{\partial\mathbf{e}_\theta}{\partial\theta} = -\mathbf{e}_r, \qquad \frac{\partial\mathbf{e}_z}{\partial\theta} = \mathbf{0}. \tag{A.2.6}$$

Components of vectors and tensors

Let $\mathbf{v}$ be a vector with components

$$v_r = \mathbf{v}\cdot\mathbf{e}_r, \qquad v_\theta = \mathbf{v}\cdot\mathbf{e}_\theta, \qquad v_z = \mathbf{v}\cdot\mathbf{e}_z, \tag{A.2.7}$$

with respect to $\{\mathbf{e}_r, \mathbf{e}_\theta, \mathbf{e}_z\}$, then

$$\mathbf{v} = v_r\,\mathbf{e}_r + v_\theta\,\mathbf{e}_\theta + v_z\,\mathbf{e}_z. \tag{A.2.8}$$

Recall that a tensor $\mathbf{S}$ is a linear map that assigns to each vector $\mathbf{u}$ a vector $\mathbf{v}$ according to

$$\mathbf{v} = \mathbf{S}\mathbf{u}.$$

Also the **tensor product** of two vectors $\mathbf{a}$ and $\mathbf{b}$ is defined as the tensor $\mathbf{a}\otimes\mathbf{b}$ that assigns to each vector $\mathbf{w}$ the vector $(\mathbf{b}\cdot\mathbf{w})\,\mathbf{a}$ according to

$$(\mathbf{a}\otimes\mathbf{b})\,\mathbf{w} = (\mathbf{b}\cdot\mathbf{w})\,\mathbf{a}.$$

The components of a tensor $\mathbf{S}$ in a cylindrical coordinate system are defined by

$$\begin{aligned}S_{rr} &= \mathbf{e}_r\cdot\mathbf{S}\mathbf{e}_r, & S_{r\theta} &= \mathbf{e}_r\cdot\mathbf{S}\mathbf{e}_\theta, & S_{rz} &= \mathbf{e}_r\cdot\mathbf{S}\mathbf{e}_z,\\ S_{\theta r} &= \mathbf{e}_\theta\cdot\mathbf{S}\mathbf{e}_r, & S_{\theta\theta} &= \mathbf{e}_\theta\cdot\mathbf{S}\mathbf{e}_\theta, & S_{\theta z} &= \mathbf{e}_\theta\cdot\mathbf{S}\mathbf{e}_z,\\ S_{zr} &= \mathbf{e}_z\cdot\mathbf{S}\mathbf{e}_r, & S_{z\theta} &= \mathbf{e}_z\cdot\mathbf{S}\mathbf{e}_\theta, & S_{zz} &= \mathbf{e}_z\cdot\mathbf{S}\mathbf{e}_z.\end{aligned}$$

Hence,

$$\begin{aligned}\mathbf{S} = & S_{rr}\mathbf{e}_r \otimes \mathbf{e}_r + S_{r\theta}\mathbf{e}_r \otimes \mathbf{e}_\theta + S_{rz}\mathbf{e}_r \otimes \mathbf{e}_z \\ & + S_{\theta r}\mathbf{e}_\theta \otimes \mathbf{e}_r + S_{\theta\theta}\mathbf{e}_\theta \otimes \mathbf{e}_\theta + S_{\theta z}\mathbf{e}_\theta \otimes \mathbf{e}_z \\ & + S_{zr}\mathbf{e}_z \otimes \mathbf{e}_r + S_{z\theta}\mathbf{e}_z \otimes \mathbf{e}_\theta + S_{zz}\mathbf{e}_z \otimes \mathbf{e}_z,\end{aligned} \tag{A.2.9}$$

and the matrix of the components of S with respect to $\{\mathbf{e}_r, \mathbf{e}_\theta, \mathbf{e}_z\}$ is

$$[\mathbf{S}] = \begin{bmatrix} S_{rr} & S_{r\theta} & S_{rz} \\ S_{\theta r} & S_{\theta\theta} & S_{\theta z} \\ S_{zr} & S_{z\theta} & S_{zz} \end{bmatrix}.$$

Next, we consider some vector and tensor identities in cylindrical coordinates.

Gradient of a scalar field

We start with the gradient of a scalar field ψ, which is a vector quantity denoted by $\nabla\psi$. Recall that the derivative of ψ in a direction $\mathbf{u}$ at a point $\mathbf{x}$ is

$$\nabla\psi(\mathbf{x}) \cdot \mathbf{u} = \lim_{\Delta s \to 0} \frac{\psi(\mathbf{x} + \Delta s\, \mathbf{u}) - \psi(\mathbf{x})}{\Delta s},$$

where Δs is the arc length in the direction $\mathbf{u}$. That is, the scalar product of $\nabla\psi$ with $\mathbf{u}$ gives the rate of change of ψ in that direction. Choosing $\mathbf{u}$ to be successively along the $\mathbf{e}_r$, $\mathbf{e}_\theta$, and $\mathbf{e}_z$ directions, and recalling that the arc lengths in these directions are Δr, $r\Delta\theta$, and Δz, respectively, we obtain

$$\nabla\psi \cdot \mathbf{e}_r = \frac{\partial\psi}{\partial r}, \quad \nabla\psi \cdot \mathbf{e}_\theta = \frac{1}{r}\frac{\partial\psi}{\partial\theta}, \quad \nabla\psi \cdot \mathbf{e}_z = \frac{\partial\psi}{\partial z}.$$

Hence,

$$\boxed{\nabla\psi = \frac{\partial\psi}{\partial r}\mathbf{e}_r + \frac{1}{r}\frac{\partial\psi}{\partial\theta}\mathbf{e}_\theta + \frac{\partial\psi}{\partial z}\mathbf{e}_z.} \tag{A.2.10}$$

This can be written in operational form as

$$\nabla\psi = \left(\mathbf{e}_r\frac{\partial}{\partial r} + \mathbf{e}_\theta\frac{1}{r}\frac{\partial}{\partial\theta} + \mathbf{e}_z\frac{\partial}{\partial z}\right)\psi,$$

where the vector[1]

$$\nabla = \left(\mathbf{e}_r\frac{\partial}{\partial r} + \mathbf{e}_\theta\frac{1}{r}\frac{\partial}{\partial\theta} + \mathbf{e}_z\frac{\partial}{\partial z}\right) \tag{A.2.11}$$

denotes the **del operator** in cylindrical coordinates.

---

[1] This vector has components that are *operations not numbers*! It is illegal but very useful.

Gradient of a vector field

The gradient of $\mathbf{v}$, denoted by $\nabla\mathbf{v}$ is a second-order tensor defined by[2]

$$\nabla\mathbf{v} = [\nabla \otimes \mathbf{v}]^{\top}. \tag{A.2.12}$$

Hence, using (A.2.6), (A.2.11), and (A.2.8),

$$\begin{aligned}
\nabla\mathbf{v} &= \left[\left(\mathbf{e}_r\frac{\partial}{\partial r} + \mathbf{e}_\theta\frac{1}{r}\frac{\partial}{\partial\theta} + \mathbf{e}_z\frac{\partial}{\partial z}\right)\otimes(v_r\,\mathbf{e}_r + v_\theta\,\mathbf{e}_\theta + v_z\,\mathbf{e}_z)\right]^{\top}\\
&= \left[\frac{\partial v_r}{\partial r}\mathbf{e}_r\otimes\mathbf{e}_r + \frac{\partial v_\theta}{\partial r}\mathbf{e}_r\otimes\mathbf{e}_\theta + \frac{\partial v_z}{\partial r}\mathbf{e}_r\otimes\mathbf{e}_z\right.\\
&\quad + \frac{1}{r}\frac{\partial v_r}{\partial\theta}\mathbf{e}_\theta\otimes\mathbf{e}_r + \frac{v_r}{r}\mathbf{e}_\theta\otimes\mathbf{e}_\theta + \frac{1}{r}\frac{\partial v_\theta}{\partial\theta}\mathbf{e}_\theta\otimes\mathbf{e}_\theta - \frac{v_\theta}{r}\mathbf{e}_\theta\otimes\mathbf{e}_r + \frac{1}{r}\frac{\partial v_z}{\partial\theta}\mathbf{e}_\theta\otimes\mathbf{e}_z\\
&\quad \left. + \frac{\partial v_r}{\partial z}\mathbf{e}_z\otimes\mathbf{e}_r + \frac{\partial v_\theta}{\partial z}\mathbf{e}_z\otimes\mathbf{e}_\theta + \frac{\partial v_z}{\partial z}\mathbf{e}_z\otimes\mathbf{e}_z\right]^{\top},
\end{aligned}$$

or

$$\boxed{\begin{aligned}
\nabla\mathbf{v} &= \frac{\partial v_r}{\partial r}\mathbf{e}_r\otimes\mathbf{e}_r + \left(\frac{1}{r}\frac{\partial v_r}{\partial\theta} - \frac{v_\theta}{r}\right)\mathbf{e}_r\otimes\mathbf{e}_\theta + \frac{\partial v_r}{\partial z}\mathbf{e}_r\otimes\mathbf{e}_z\\
&\quad + \frac{\partial v_\theta}{\partial r}\mathbf{e}_\theta\otimes\mathbf{e}_r + \left(\frac{1}{r}\frac{\partial v_\theta}{\partial\theta} + \frac{v_r}{r}\right)\mathbf{e}_\theta\otimes\mathbf{e}_\theta + \frac{\partial v_\theta}{\partial z}\mathbf{e}_\theta\otimes\mathbf{e}_z\\
&\quad + \frac{\partial v_z}{\partial r}\mathbf{e}_z\otimes\mathbf{e}_r + \frac{1}{r}\frac{\partial v_z}{\partial\theta}\mathbf{e}_z\otimes\mathbf{e}_\theta + \frac{\partial v_z}{\partial z}\mathbf{e}_z\otimes\mathbf{e}_z,
\end{aligned}} \tag{A.2.13}$$

and the matrix of components of $\nabla\mathbf{v}$ with respect to $\{\mathbf{e}_r, \mathbf{e}_\theta, \mathbf{e}_z\}$ is

$$\boxed{[\nabla\mathbf{v}] = \begin{bmatrix}
\dfrac{\partial v_r}{\partial r} & \dfrac{1}{r}\dfrac{\partial v_r}{\partial\theta} - \dfrac{v_\theta}{r} & \dfrac{\partial v_r}{\partial z}\\[2ex]
\dfrac{\partial v_\theta}{\partial r} & \dfrac{1}{r}\dfrac{\partial v_\theta}{\partial\theta} + \dfrac{v_r}{r} & \dfrac{\partial v_\theta}{\partial z}\\[2ex]
\dfrac{\partial v_z}{\partial r} & \dfrac{1}{r}\dfrac{\partial v_z}{\partial\theta} & \dfrac{\partial v_z}{\partial z}
\end{bmatrix}.} \tag{A.2.14}$$

Divergence of a vector field

The divergence of $\mathbf{v}$, denoted by $\operatorname{div}\mathbf{v}$, is the scalar field defined as

$$\operatorname{div}\mathbf{v} = \operatorname{tr}\nabla\mathbf{v} = \nabla\cdot\mathbf{v}.$$

Hence, using (A.2.6),

$$\operatorname{div}\mathbf{v} = \left(\mathbf{e}_r\frac{\partial}{\partial r} + \mathbf{e}_\theta\frac{1}{r}\frac{\partial}{\partial\theta} + \mathbf{e}_z\frac{\partial}{\partial z}\right)\cdot(v_r\,\mathbf{e}_r + v_\theta\,\mathbf{e}_\theta + v_z\,\mathbf{e}_z),$$

[2] This is easily verified in Cartesian components: $(\nabla\mathbf{v})_{ij} = v_{i,j}$ and $((\nabla\otimes\mathbf{v})^{\top})_{ij} = v_{i,j}$.

or

$$\boxed{\operatorname{div}\mathbf{v} = \frac{\partial v_r}{\partial r} + \frac{1}{r}\frac{\partial v_\theta}{\partial \theta} + \frac{\partial v_z}{\partial z} + \frac{v_r}{r}.} \tag{A.2.15}$$

Divergence of a tensor field

The divergence of a tensor field $\mathbf{S}$, denoted by div $\mathbf{S}$, is a vector field defined by

$$(\operatorname{div}\mathbf{S})\cdot\mathbf{a} = \operatorname{div}(\mathbf{S}^\top\mathbf{a}),$$

for every constant vector a.

Using (A.2.9),

$$\mathbf{S}^\top\mathbf{a} = (\mathbf{e}_r\cdot\mathbf{a})\,\mathbf{s}_r + (\mathbf{e}_\theta\cdot\mathbf{a})\,\mathbf{s}_\theta + (\mathbf{e}_z\cdot\mathbf{a})\,\mathbf{s}_z,$$

where

$$\begin{aligned}
\mathbf{s}_r &= S_{rr}\,\mathbf{e}_r + S_{r\theta}\,\mathbf{e}_\theta + S_{rz}\,\mathbf{e}_z,\\
\mathbf{s}_\theta &= S_{\theta r}\,\mathbf{e}_r + S_{\theta\theta}\,\mathbf{e}_\theta + S_{\theta z}\,\mathbf{e}_z,\\
\mathbf{s}_z &= S_{zr}\,\mathbf{e}_r + S_{z\theta}\,\mathbf{e}_\theta + S_{zz}\,\mathbf{e}_z.
\end{aligned}$$

Hence,

$$\begin{aligned}
(\mathbf{S}^\top\mathbf{a})_r &= S_{rr}\,(\mathbf{e}_r\cdot\mathbf{a}) + S_{\theta r}\,(\mathbf{e}_\theta\cdot\mathbf{a}) + S_{zr}\,(\mathbf{e}_z\cdot\mathbf{a}),\\
(\mathbf{S}^\top\mathbf{a})_\theta &= S_{r\theta}\,(\mathbf{e}_r\cdot\mathbf{a}) + S_{\theta\theta}\,(\mathbf{e}_\theta\cdot\mathbf{a}) + S_{z\theta}\,(\mathbf{e}_z\cdot\mathbf{a}),\\
(\mathbf{S}^\top\mathbf{a})_z &= S_{rz}\,(\mathbf{e}_r\cdot\mathbf{a}) + S_{\theta z}\,(\mathbf{e}_\theta\cdot\mathbf{a}) + S_{zz}\,(\mathbf{e}_z\cdot\mathbf{a}).
\end{aligned}$$

Then, upon using (A.2.15)

$$\begin{aligned}
(\operatorname{div}\mathbf{S})\cdot\mathbf{a} = &\frac{\partial}{\partial r}\left[S_{rr}\,(\mathbf{e}_r\cdot\mathbf{a}) + S_{\theta r}\,(\mathbf{e}_\theta\cdot\mathbf{a}) + S_{zr}\,(\mathbf{e}_z\cdot\mathbf{a})\right]\\
&+ \frac{1}{r}\frac{\partial}{\partial\theta}\left[S_{r\theta}\,(\mathbf{e}_r\cdot\mathbf{a}) + S_{\theta\theta}\,(\mathbf{e}_\theta\cdot\mathbf{a}) + S_{z\theta}\,(\mathbf{e}_z\cdot\mathbf{a})\right]\\
&+ \frac{\partial}{\partial z}\left[S_{rz}\,(\mathbf{e}_r\cdot\mathbf{a}) + S_{\theta z}\,(\mathbf{e}_\theta\cdot\mathbf{a}) + S_{zz}\,(\mathbf{e}_z\cdot\mathbf{a})\right]\\
&+ \frac{1}{r}\left[S_{rr}\,(\mathbf{e}_r\cdot\mathbf{a}) + S_{\theta r}\,(\mathbf{e}_\theta\cdot\mathbf{a}) + S_{zr}\,(\mathbf{e}_z\cdot\mathbf{a})\right],
\end{aligned}$$

or

$$\begin{aligned}
(\operatorname{div}\mathbf{S})\cdot\mathbf{a} = &\frac{\partial S_{rr}}{\partial r}\,(\mathbf{e}_r\cdot\mathbf{a}) + \frac{\partial S_{\theta r}}{\partial r}\,(\mathbf{e}_\theta\cdot\mathbf{a}) + \frac{\partial S_{zr}}{\partial r}\,(\mathbf{e}_z\cdot\mathbf{a})\\
&+ \frac{S_{rr}}{r}\,(\mathbf{e}_r\cdot\mathbf{a}) + \frac{S_{\theta r}}{r}\,(\mathbf{e}_\theta\cdot\mathbf{a}) + \frac{S_{zr}}{r}\,(\mathbf{e}_z\cdot\mathbf{a})\\
&+ \frac{\partial S_{rz}}{\partial z}\,(\mathbf{e}_r\cdot\mathbf{a}) + \frac{\partial S_{\theta z}}{\partial z}\,(\mathbf{e}_\theta\cdot\mathbf{a}) + \frac{\partial S_{zz}}{\partial z}\,(\mathbf{e}_z\cdot\mathbf{a})\\
&+ \frac{1}{r}\left[\frac{\partial S_{r\theta}}{\partial\theta}(\mathbf{e}_r\cdot\mathbf{a}) + S_{r\theta}(\mathbf{e}_\theta\cdot\mathbf{a})\right.\\
&\left.+ \frac{\partial S_{\theta\theta}}{\partial\theta}(\mathbf{e}_\theta\cdot\mathbf{a}) - S_{\theta\theta}(\mathbf{e}_r\cdot\mathbf{a}) + \frac{\partial S_{z\theta}}{\partial\theta}(\mathbf{e}_z\cdot\mathbf{a})\right],
\end{aligned}$$

from which

$$
\boxed{
\begin{aligned}
(\operatorname{div}\mathbf{S})_r &= \frac{\partial S_{rr}}{\partial r} + \frac{1}{r}\frac{\partial S_{r\theta}}{\partial \theta} + \frac{\partial S_{rz}}{\partial z} + \frac{1}{r}(S_{rr} - S_{\theta\theta}),\\
(\operatorname{div}\mathbf{S})_\theta &= \frac{\partial S_{\theta r}}{\partial r} + \frac{1}{r}\frac{\partial S_{\theta\theta}}{\partial \theta} + \frac{\partial S_{\theta z}}{\partial z} + \frac{1}{r}(S_{\theta r} + S_{r\theta}),\\
(\operatorname{div}\mathbf{S})_z &= \frac{\partial S_{zr}}{\partial r} + \frac{1}{r}\frac{\partial S_{z\theta}}{\partial \theta} + \frac{\partial S_{zz}}{\partial z} + \frac{S_{zr}}{r}.
\end{aligned}
}
\tag{A.2.16}
$$

Curl of a vector field

The curl of $\mathbf{v}$, denoted by $\operatorname{curl}\mathbf{v}$, is a vector field defined as

$$\operatorname{curl}\mathbf{v} = \nabla \times \mathbf{v}.$$

Hence,

$$\operatorname{curl}\mathbf{v} = \left(\mathbf{e}_r \frac{\partial}{\partial r} + \mathbf{e}_\theta \frac{1}{r}\frac{\partial}{\partial \theta} + \mathbf{e}_z \frac{\partial}{\partial z}\right) \times (v_r\,\mathbf{e}_r + v_\theta\,\mathbf{e}_\theta + v_z\,\mathbf{e}_z),$$

which upon noting that

$$
\begin{aligned}
\mathbf{e}_r \times \mathbf{e}_r &= \mathbf{0}, \quad & \mathbf{e}_\theta \times \mathbf{e}_z &= -\mathbf{e}_z \times \mathbf{e}_\theta = \mathbf{e}_r,\\
\mathbf{e}_\theta \times \mathbf{e}_\theta &= \mathbf{0}, \quad & \mathbf{e}_z \times \mathbf{e}_r &= -\mathbf{e}_r \times \mathbf{e}_z = \mathbf{e}_\theta,\\
\mathbf{e}_z \times \mathbf{e}_z &= \mathbf{0}, \quad & \mathbf{e}_r \times \mathbf{e}_\theta &= -\mathbf{e}_\theta \times \mathbf{e}_r = \mathbf{e}_z,
\end{aligned}
$$

gives

$$
\boxed{
\operatorname{curl}\mathbf{v} = \left\{\frac{1}{r}\frac{\partial v_z}{\partial \theta} - \frac{\partial v_\theta}{\partial z}\right\}\mathbf{e}_r + \left\{\frac{\partial v_r}{\partial z} - \frac{\partial v_z}{\partial r}\right\}\mathbf{e}_\theta + \left\{\frac{\partial v_\theta}{\partial r} - \frac{1}{r}\frac{\partial v_r}{\partial \theta} + \frac{v_\theta}{r}\right\}\mathbf{e}_z.
}
\tag{A.2.17}
$$

Laplacian of a scalar field

The Laplacian of a scalar ψ, denoted by $\triangle\,\psi$, is a scalar field defined as

$$\triangle\,\psi = \operatorname{div}\nabla\psi.$$

Thus, using (A.2.10) and (A.2.15)

$$
\boxed{
\triangle\,\psi = \frac{\partial^2 \psi}{\partial r^2} + \frac{1}{r^2}\frac{\partial^2 \psi}{\partial \theta^2} + \frac{\partial^2 \psi}{\partial z^2} + \frac{1}{r}\frac{\partial \psi}{\partial r}.
}
\tag{A.2.18}
$$

Laplacian of a vector field

The Laplacian of a vector $\mathbf{v}$, denoted by $\triangle\,\mathbf{v}$, is a vector field defined as

$$\triangle\,\mathbf{v} = \operatorname{div}\nabla\mathbf{v}.$$

From (A.2.16) and (A.2.14),

$$\boxed{\triangle \mathbf{v} = \left(\triangle\, v_r - \frac{2}{r^2}\frac{\partial v_\theta}{\partial \theta} - \frac{1}{r^2}v_r\right)\mathbf{e}_r + \left(\triangle\, v_\theta + \frac{2}{r^2}\frac{\partial v_r}{\partial \theta} - \frac{1}{r^2}v_\theta\right)\mathbf{e}_\theta + (\triangle v_z)\mathbf{e}_z.} \tag{A.2.19}$$

Infinitesimal strain tensor

Let

$$\mathbf{u} = u_r\mathbf{e}_r + u_\theta\mathbf{e}_\theta + u_z\mathbf{e}_z$$

denote the components of the displacement vector with respect to $\{\mathbf{e}_r, \mathbf{e}_\theta, \mathbf{e}_z\}$. Then (A.2.14) gives the components of the displacement gradient tensor $\mathbf{H} \equiv \nabla\mathbf{u}$ in the cylindrical coordinate system, and hence the symmetric infinitesimal strain tensor $\boldsymbol{\epsilon}$ defined by

$$\boldsymbol{\epsilon} = \frac{1}{2}\left[\mathbf{H} + \mathbf{H}^\top\right]$$

has the following components in the cylindrical coordinate system

$$\boxed{\begin{aligned}
\epsilon_{rr} &= \frac{\partial u_r}{\partial r},\\
\epsilon_{\theta\theta} &= \frac{1}{r}\frac{\partial u_\theta}{\partial \theta} + \frac{u_r}{r},\\
\epsilon_{zz} &= \frac{\partial u_z}{\partial z},\\
\epsilon_{r\theta} &= \frac{1}{2}\left(\frac{1}{r}\frac{\partial u_r}{\partial \theta} + \frac{\partial u_\theta}{\partial r} - \frac{u_\theta}{r}\right) = \epsilon_{\theta r},\\
\epsilon_{\theta z} &= \frac{1}{2}\left(\frac{1}{r}\frac{\partial u_z}{\partial \theta} + \frac{\partial u_\theta}{\partial z}\right) = \epsilon_{z\theta},\\
\epsilon_{zr} &= \frac{1}{2}\left(\frac{\partial u_z}{\partial r} + \frac{\partial u_r}{\partial z}\right) = \epsilon_{rz}.
\end{aligned}} \tag{A.2.20}$$

Equation of motion

The matrix of the components of the symmetric stress tensor $\boldsymbol{\sigma}$ with respect to $\{\mathbf{e}_r, \mathbf{e}_\theta, \mathbf{e}_z\}$ is

$$[\boldsymbol{\sigma}] = \begin{bmatrix} \sigma_{rr} & \sigma_{r\theta} & \sigma_{rz} \\ \sigma_{\theta r} & \sigma_{\theta\theta} & \sigma_{\theta z} \\ \sigma_{zr} & \sigma_{z\theta} & \sigma_{zz} \end{bmatrix}.$$

Let (b_r, b_θ, b_z) and $(\ddot{u}_r, \ddot{u}_\theta, \ddot{u}_z)$ denote the components of the body force $\mathbf{b}$ and the acceleration $\ddot{\mathbf{u}}$ in a cylindrical coordinate system. Then, using (A.2.16) the equation of motion,

$$\operatorname{div}\boldsymbol{\sigma} + \mathbf{b} = \rho\ddot{\mathbf{u}},$$

where ρ is the mass density, may be written as the following three equations in the cylindrical coordinate system

$$\boxed{\begin{aligned}
&\frac{\partial \sigma_{rr}}{\partial r} + \frac{1}{r}\frac{\partial \sigma_{r\theta}}{\partial \theta} + \frac{\partial \sigma_{rz}}{\partial z} + \frac{1}{r}(\sigma_{rr} - \sigma_{\theta\theta}) + b_r = \rho \ddot{u}_r,\\
&\frac{\partial \sigma_{\theta r}}{\partial r} + \frac{1}{r}\frac{\partial \sigma_{\theta\theta}}{\partial \theta} + \frac{\partial \sigma_{\theta z}}{\partial z} + \frac{2}{r}\sigma_{\theta r} + b_\theta = \rho \ddot{u}_\theta,\\
&\frac{\partial \sigma_{zr}}{\partial r} + \frac{1}{r}\frac{\partial \sigma_{z\theta}}{\partial \theta} + \frac{\partial \sigma_{zz}}{\partial z} + \frac{\sigma_{zr}}{r} + b_z = \rho \ddot{u}_z.
\end{aligned}} \tag{A.2.21}$$

A.3 Spherical coordinates

Spherical coordinates (r, θ, ϕ) are related to rectangular coordinates (x_1, x_2, x_3) by

$$\begin{aligned}
r &= [x_1^2 + x_2^2 + x_3^2]^{\frac{1}{2}}, \quad r \geq 0,\\
\theta &= \cos^{-1} \frac{x_3}{[x_1^2 + x_2^2 + x_3^2]^{\frac{1}{2}}}, \quad 0 \leq \theta \leq \pi,\\
\phi &= \tan^{-1}(x_2/x_1), \quad 0 \leq \phi < 2\pi,
\end{aligned} \tag{A.3.1}$$

$$x_1 = r \sin\theta \cos\phi, \quad x_2 = r \sin\theta \sin\phi, \quad x_3 = r\cos\theta. \tag{A.3.2}$$

The orthonormal base vectors $\{\mathbf{e}_r, \mathbf{e}_\theta, \mathbf{e}_\phi\}$ in the spherical coordinate system are illustrated in Fig. A.2. The position vector $\mathbf{r} = \mathbf{x} - \mathbf{o}$ is given by

$$\mathbf{r} = r \sin\theta \cos\phi \, \mathbf{e}_1 + r\sin\theta \sin\phi \, \mathbf{e}_2 + r\cos\theta \, \mathbf{e}_3, \tag{A.3.3}$$

and with respect to a spherical coordinate system a differential element $d\mathbf{x}$ at $\mathbf{x}$ is given by

$$d\mathbf{x} = dr\, \mathbf{e}_r + (r d\theta)\, \mathbf{e}_\theta + (r \sin\theta d\phi)\, \mathbf{e}_\phi. \tag{A.3.4}$$

Thus, since $d\mathbf{r} = d\mathbf{x}$,

$$\mathbf{e}_r = \frac{\partial \mathbf{r}}{\partial r}, \quad \mathbf{e}_\theta = \frac{1}{r}\frac{\partial \mathbf{r}}{\partial \theta}, \quad \mathbf{e}_\phi = \frac{1}{r\sin\theta}\frac{\partial \mathbf{r}}{\partial \phi}.$$

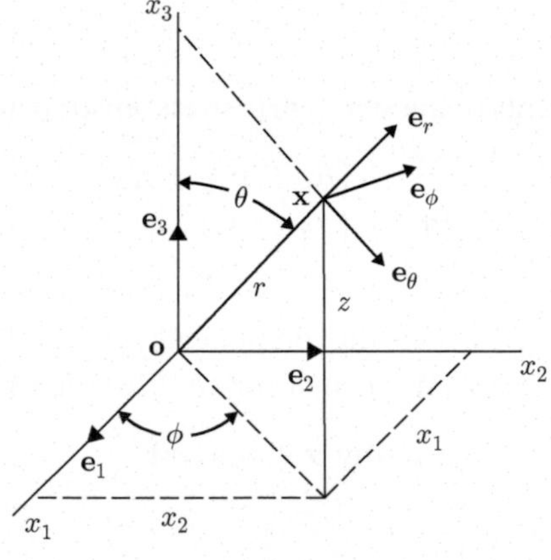

Fig. A.2 Spherical coordinate system.

Using this and (A.3.3) we obtain that $\{\mathbf{e}_r, \mathbf{e}_\theta, \mathbf{e}_\phi\}$ are related to the orthonormal base vectors $\{\mathbf{e}_1, \mathbf{e}_2, \mathbf{e}_3\}$ of the rectangular system by

$$\begin{aligned}\mathbf{e}_r &= \sin\theta\cos\phi\,\mathbf{e}_1 + \sin\theta\sin\phi\,\mathbf{e}_2 + \cos\theta\mathbf{e}_3,\\ \mathbf{e}_\theta &= \cos\theta\cos\phi\,\mathbf{e}_1 + \cos\theta\sin\phi\,\mathbf{e}_2 - \sin\theta\,\mathbf{e}_3,\\ \mathbf{e}_\phi &= -\sin\phi\,\mathbf{e}_1 + \cos\phi\,\mathbf{e}_2.\end{aligned} \tag{A.3.5}$$

From (A.3.5) we obtain the differential relations

$$\begin{aligned}&\frac{\partial\mathbf{e}_r}{\partial r} = \frac{\partial\mathbf{e}_\theta}{\partial r} = \frac{\partial\mathbf{e}_\phi}{\partial r} = \mathbf{0},\\ &\frac{\partial\mathbf{e}_r}{\partial\theta} = \mathbf{e}_\theta, \quad \frac{\partial\mathbf{e}_\theta}{\partial\theta} = -\mathbf{e}_r, \quad \frac{\partial\mathbf{e}_\phi}{\partial\theta} = \mathbf{0},\\ &\frac{\partial\mathbf{e}_r}{\partial\phi} = \sin\theta\,\mathbf{e}_\phi, \quad \frac{\partial\mathbf{e}_\theta}{\partial\phi} = \cos\theta\,\mathbf{e}_\phi, \quad \frac{\partial\mathbf{e}_\phi}{\partial\phi} = -\left(\sin\theta\,\mathbf{e}_r + \cos\theta\,\mathbf{e}_\theta\right).\end{aligned} \tag{A.3.6}$$

Components of vectors and tensors

In the spherical coordinate system a vector $\mathbf{v}$ has the representation

$$\mathbf{v} = v_r\,\mathbf{e}_r + v_\theta\,\mathbf{e}_\theta + v_z\,\mathbf{e}_z, \tag{A.3.7}$$

where $\{v_r, v_\theta, v_\phi\}$ are the components of $\mathbf{v}$ with respect to $\{\mathbf{e}_r, \mathbf{e}_\theta, \mathbf{e}_\phi\}$, and a tensor $\mathbf{S}$ has the representation

$$\begin{aligned}\mathbf{S} = &S_{rr}\mathbf{e}_r\otimes\mathbf{e}_r + S_{r\theta}\mathbf{e}_r\otimes\mathbf{e}_\theta + S_{r\phi}\mathbf{e}_r\otimes\mathbf{e}_\phi\\ &+ S_{\theta r}\mathbf{e}_\theta\otimes\mathbf{e}_r + S_{\theta\theta}\mathbf{e}_\theta\otimes\mathbf{e}_\theta + S_{\theta\phi}\mathbf{e}_\theta\otimes\mathbf{e}_\phi\\ &+ S_{\phi r}\mathbf{e}_\phi\otimes\mathbf{e}_r + S_{\phi\theta}\mathbf{e}_\phi\otimes\mathbf{e}_\theta + S_{\phi\phi}\mathbf{e}_\phi\otimes\mathbf{e}_\phi,\end{aligned} \tag{A.3.8}$$

and the matrix of the components of $\mathbf{S}$ with respect to $\{\mathbf{e}_r, \mathbf{e}_\theta, \mathbf{e}_\phi\}$ is

$$[\mathbf{S}] = \begin{bmatrix} S_{rr} & S_{r\theta} & S_{r\phi}\\ S_{\theta r} & S_{\theta\theta} & S_{\theta\phi}\\ S_{\phi r} & S_{\phi\theta} & S_{\phi\phi}\end{bmatrix}.$$

Gradient of a scalar field

Here,

$$\boxed{\nabla\psi = \frac{\partial\psi}{\partial r}\mathbf{e}_r + \frac{1}{r}\frac{\partial\psi}{\partial\theta}\mathbf{e}_\theta + \frac{1}{r\sin\theta}\frac{\partial\psi}{\partial\phi}\mathbf{e}_\phi.} \tag{A.3.9}$$

This can be written in operational form as

$$\nabla\psi = \left(\mathbf{e}_r\frac{\partial}{\partial r} + \mathbf{e}_\theta\frac{1}{r}\frac{\partial}{\partial\theta} + \mathbf{e}_\phi\frac{1}{r\sin\theta}\frac{\partial}{\partial\phi}\right)\psi,$$

where the vector

$$\nabla = \left(\mathbf{e}_r\frac{\partial}{\partial r} + \mathbf{e}_\theta\frac{1}{r}\frac{\partial}{\partial\theta} + \mathbf{e}_\phi\frac{1}{r\sin\theta}\frac{\partial}{\partial\phi}\right) \tag{A.3.10}$$

denotes the **del operator** in spherical coordinates.

Gradient of a vector field

The gradient of $\mathbf{v}$, denoted by $\nabla\mathbf{v}$ is a second-order tensor defined by

$$\nabla\mathbf{v} = [\nabla \otimes \mathbf{v}]^{\top}. \tag{A.3.11}$$

Using (A.3.10), (A.3.7), and (A.3.6),

$$\begin{aligned}
\nabla\mathbf{v} &= \left[\left(\mathbf{e}_r \frac{\partial}{\partial r} + \mathbf{e}_\theta \frac{1}{r}\frac{\partial}{\partial \theta} + \mathbf{e}_\phi \frac{1}{r\sin\theta}\frac{\partial}{\partial \phi}\right) \otimes (v_r\,\mathbf{e}_r + v_\theta\,\mathbf{e}_\theta + v_\phi\,\mathbf{e}_\phi)\right]^{\top}\\
&= \left[\frac{\partial v_r}{\partial r}\mathbf{e}_r \otimes \mathbf{e}_r + \frac{\partial v_\theta}{\partial r}\mathbf{e}_r \otimes \mathbf{e}_\theta + \frac{\partial v_\phi}{\partial r}\mathbf{e}_r \otimes \mathbf{e}_\phi\right.\\
&\quad + \frac{1}{r}\frac{\partial v_r}{\partial \theta}\mathbf{e}_\theta \otimes \mathbf{e}_r + \frac{v_r}{r}\mathbf{e}_\theta \otimes \mathbf{e}_\theta + \frac{1}{r}\frac{\partial v_\theta}{\partial \theta}\mathbf{e}_\theta \otimes \mathbf{e}_\theta - \frac{v_\theta}{r}\mathbf{e}_\theta \otimes \mathbf{e}_r + \frac{1}{r}\frac{\partial v_\phi}{\partial \theta}\mathbf{e}_\theta \otimes \mathbf{e}_\phi\\
&\quad + \frac{1}{r\sin\theta}\frac{\partial v_r}{\partial \phi}\mathbf{e}_\phi \otimes \mathbf{e}_r + \frac{v_\phi}{r}\mathbf{e}_\phi \otimes \mathbf{e}_\phi\\
&\quad + \frac{1}{r\sin\theta}\frac{\partial v_\theta}{\partial \phi}\mathbf{e}_\phi \otimes \mathbf{e}_\theta + \frac{v_\theta\cot\theta}{r}\mathbf{e}_\phi \otimes \mathbf{e}_\phi\\
&\quad \left. + \frac{1}{r\sin\theta}\frac{\partial v_\phi}{\partial \phi}\mathbf{e}_\phi \otimes \mathbf{e}_\phi - \frac{v_\phi}{r}\mathbf{e}_\phi \otimes \mathbf{e}_r - \frac{v_\phi\cot\theta}{r}\mathbf{e}_\phi \otimes \mathbf{e}_\theta\right]^{\top},
\end{aligned}$$

or

$$\begin{aligned}
\nabla\mathbf{v} &= \frac{\partial v_r}{\partial r}\mathbf{e}_r \otimes \mathbf{e}_r + \left(\frac{1}{r}\frac{\partial v_r}{\partial \theta} - \frac{v_\theta}{r}\right)\mathbf{e}_r \otimes \mathbf{e}_\theta + \left(\frac{1}{r\sin\theta}\frac{\partial v_r}{\partial \phi} - \frac{v_\phi}{r}\right)\mathbf{e}_r \otimes \mathbf{e}_\phi\\
&\quad + \frac{\partial v_\theta}{\partial r}\mathbf{e}_\theta \otimes \mathbf{e}_r + \left(\frac{1}{r}\frac{\partial v_\theta}{\partial \theta} + \frac{v_r}{r}\right)\mathbf{e}_\theta \otimes \mathbf{e}_\theta + \left(\frac{1}{r\sin\theta}\frac{\partial v_\theta}{\partial \phi} - \frac{v_\phi\cot\theta}{r}\right)\mathbf{e}_\theta \otimes \mathbf{e}_\phi\\
&\quad + \frac{\partial v_\phi}{\partial r}\mathbf{e}_\phi \otimes \mathbf{e}_r + \frac{1}{r}\frac{\partial v_\phi}{\partial \theta}\mathbf{e}_\phi \otimes \mathbf{e}_\theta + \left(\frac{1}{r\sin\theta}\frac{\partial v_\phi}{\partial \phi} + \frac{v_\theta\cot\theta}{r} + \frac{v_r}{r}\right)\mathbf{e}_\phi \otimes \mathbf{e}_\phi,
\end{aligned} \tag{A.3.12}$$

and the matrix of components of $\nabla\mathbf{v}$ with respect to $\{\mathbf{e}_r, \mathbf{e}_\theta, \mathbf{e}_\phi\}$ is

$$[\nabla\mathbf{v}] = \begin{bmatrix}
\dfrac{\partial v_r}{\partial r} & \left(\dfrac{1}{r}\dfrac{\partial v_r}{\partial \theta} - \dfrac{v_\theta}{r}\right) & \left(\dfrac{1}{r\sin\theta}\dfrac{\partial v_r}{\partial \phi} - \dfrac{v_\phi}{r}\right)\\[2ex]
\dfrac{\partial v_\theta}{\partial r} & \left(\dfrac{1}{r}\dfrac{\partial v_\theta}{\partial \theta} + \dfrac{v_r}{r}\right) & \left(\dfrac{1}{r\sin\theta}\dfrac{\partial v_\theta}{\partial \phi} - \dfrac{\cot\theta}{r}v_\phi\right)\\[2ex]
\dfrac{\partial v_\phi}{\partial r} & \dfrac{1}{r}\dfrac{\partial v_\phi}{\partial \theta} & \left(\dfrac{1}{r\sin\theta}\dfrac{\partial v_\phi}{\partial \phi} + \dfrac{\cot\theta}{r}v_\theta + \dfrac{v_r}{r}\right)
\end{bmatrix}. \tag{A.3.13}$$

Divergence of a vector field

The divergence of $\mathbf{v}$, denoted by div $\mathbf{v}$, is the scalar field defined as

$$\operatorname{div}\mathbf{v} = \operatorname{tr}\nabla\mathbf{v} = \nabla\cdot\mathbf{v}.$$

From (A.3.13),

$$\boxed{\operatorname{div}\mathbf{v} = \frac{\partial v_r}{\partial r} + \frac{1}{r}\frac{\partial v_\theta}{\partial \theta} + \frac{1}{r\sin\theta}\frac{\partial v_\phi}{\partial \phi} + \frac{\cot\theta}{r}v_\theta + \frac{2v_r}{r}.} \tag{A.3.14}$$

Divergence of a tensor field

The divergence of a tensor field $\mathbf{S}$, denoted by div $\mathbf{S}$, is a vector field defined by

$$(\operatorname{div}\mathbf{S})\cdot\mathbf{a} = \operatorname{div}(\mathbf{S}^\top\mathbf{a}),$$

for every constant vector $\mathbf{a}$.

Using (A.3.8),

$$\mathbf{S}^\top\mathbf{a} = (\mathbf{e}_r\cdot\mathbf{a})\,\mathbf{s}_r + (\mathbf{e}_\theta\cdot\mathbf{a})\,\mathbf{s}_\theta + (\mathbf{e}_\phi\cdot\mathbf{a})\,\mathbf{s}_\phi,$$

where

$$\begin{aligned}
\mathbf{s}_r &= S_{rr}\,\mathbf{e}_r + S_{r\theta}\,\mathbf{e}_\theta + S_{r\phi}\,\mathbf{e}_\phi,\\
\mathbf{s}_\theta &= S_{\theta r}\,\mathbf{e}_r + S_{\theta\theta}\,\mathbf{e}_\theta + S_{\theta\phi}\,\mathbf{e}_\phi,\\
\mathbf{s}_\phi &= S_{\phi r}\,\mathbf{e}_r + S_{\phi\theta}\,\mathbf{e}_\theta + S_{\phi\phi}\,\mathbf{e}_\phi.
\end{aligned}$$

Hence,

$$\begin{aligned}
(\mathbf{S}^\top\mathbf{a})_r &= S_{rr}\,(\mathbf{e}_r\cdot\mathbf{a}) + S_{\theta r}\,(\mathbf{e}_\theta\cdot\mathbf{a}) + S_{\phi r}\,(\mathbf{e}_\phi\cdot\mathbf{a}),\\
(\mathbf{S}^\top\mathbf{a})_\theta &= S_{r\theta}\,(\mathbf{e}_r\cdot\mathbf{a}) + S_{\theta\theta}\,(\mathbf{e}_\theta\cdot\mathbf{a}) + S_{\phi\theta}\,(\mathbf{e}_\phi\cdot\mathbf{a}),\\
(\mathbf{S}^\top\mathbf{a})_\phi &= S_{r\phi}\,(\mathbf{e}_r\cdot\mathbf{a}) + S_{\theta\phi}\,(\mathbf{e}_\theta\cdot\mathbf{a}) + S_{\phi\phi}\,(\mathbf{e}_\phi\cdot\mathbf{a}).
\end{aligned}$$

Then, upon using (A.3.14)

$$\begin{aligned}
(\operatorname{div}\mathbf{S})\cdot\mathbf{a} = {}&\frac{\partial}{\partial r}\left[S_{rr}\,(\mathbf{e}_r\cdot\mathbf{a}) + S_{\theta r}\,(\mathbf{e}_\theta\cdot\mathbf{a}) + S_{\phi r}\,(\mathbf{e}_\phi\cdot\mathbf{a})\right]\\
&+\frac{1}{r}\frac{\partial}{\partial\theta}\left[S_{r\theta}\,(\mathbf{e}_r\cdot\mathbf{a}) + S_{\theta\theta}\,(\mathbf{e}_\theta\cdot\mathbf{a}) + S_{\phi\theta}\,(\mathbf{e}_\phi\cdot\mathbf{a})\right]\\
&+\frac{1}{r\sin\theta}\frac{\partial}{\partial\phi}\left[S_{r\phi}\,(\mathbf{e}_r\cdot\mathbf{a}) + S_{\theta\phi}\,(\mathbf{e}_\theta\cdot\mathbf{a}) + S_{\phi\phi}\,(\mathbf{e}_\phi\cdot\mathbf{a})\right]\\
&+\frac{\cot\theta}{r}\left[S_{r\theta}\,(\mathbf{e}_r\cdot\mathbf{a}) + S_{\theta\theta}\,(\mathbf{e}_\theta\cdot\mathbf{a}) + S_{\phi\theta}\,(\mathbf{e}_\phi\cdot\mathbf{a})\right]\\
&+\frac{2}{r}\left[S_{rr}\,(\mathbf{e}_r\cdot\mathbf{a}) + S_{\theta r}\,(\mathbf{e}_\theta\cdot\mathbf{a}) + S_{\phi r}\,(\mathbf{e}_\phi\cdot\mathbf{a})\right],
\end{aligned}$$

or

$$
\begin{aligned}
(\operatorname{div}\mathbf{S})\cdot\mathbf{a} = {} & \frac{\partial S_{rr}}{\partial r}(\mathbf{e}_r\cdot\mathbf{a}) + \frac{\partial S_{\theta r}}{\partial r}(\mathbf{e}_\theta\cdot\mathbf{a}) + \frac{\partial S_{\phi r}}{\partial r}(\mathbf{e}_\phi\cdot\mathbf{a}) \\
& + \frac{1}{r}\left[\frac{\partial S_{r\theta}}{\partial\theta}(\mathbf{e}_r\cdot\mathbf{a}) + S_{r\theta}(\mathbf{e}_\theta\cdot\mathbf{a})\right. \\
& \left. + \frac{\partial S_{\theta\theta}}{\partial\theta}(\mathbf{e}_\theta\cdot\mathbf{a}) - S_{\theta\theta}(\mathbf{e}_r\cdot\mathbf{a}) + \frac{\partial S_{\phi\theta}}{\partial\theta}(\mathbf{e}_\phi\cdot\mathbf{a})\right] \\
& + \frac{1}{r\sin\theta}\left[\frac{\partial S_{r\phi}}{\partial\phi}(\mathbf{e}_r\cdot\mathbf{a}) + S_{r\phi}\sin\theta(\mathbf{e}_\phi\cdot\mathbf{a})\right. \\
& + \frac{\partial S_{\theta\phi}}{\partial\phi}(\mathbf{e}_\theta\cdot\mathbf{a}) + S_{\theta\phi}\cos\theta(\mathbf{e}_\phi\cdot\mathbf{a}) \\
& \left. + \frac{\partial S_{\phi\phi}}{\partial\phi}(\mathbf{e}_\phi\cdot\mathbf{a}) - S_{\phi\phi}\sin\theta(\mathbf{e}_r\cdot\mathbf{a}) - S_{\phi\phi}\cos\theta(\mathbf{e}_\theta\cdot\mathbf{a})\right] \\
& + \frac{\cot\theta}{r}\left[S_{r\theta}(\mathbf{e}_r\cdot\mathbf{a}) + S_{\theta\theta}(\mathbf{e}_\theta\cdot\mathbf{a}) + S_{\phi\theta}(\mathbf{e}_\phi\cdot\mathbf{a})\right] \\
& + \frac{2}{r}\left[S_{rr}(\mathbf{e}_r\cdot\mathbf{a}) + S_{\theta r}(\mathbf{e}_\theta\cdot\mathbf{a}) + S_{\phi r}(\mathbf{e}_\phi\cdot\mathbf{a})\right],
\end{aligned}
$$

from which

$$
\boxed{
\begin{aligned}
(\operatorname{div}\mathbf{S})_r &= \frac{\partial S_{rr}}{\partial r} + \frac{1}{r}\frac{\partial S_{r\theta}}{\partial\theta} + \frac{1}{r\sin\theta}\frac{\partial S_{r\phi}}{\partial\phi} \\
&\quad + \frac{1}{r}\left(2\,S_{rr} - S_{\theta\theta} - S_{\phi\phi} + \cot\theta\,S_{r\theta}\right), \\
(\operatorname{div}\mathbf{S})_\theta &= \frac{\partial S_{\theta r}}{\partial r} + \frac{1}{r}\frac{\partial S_{\theta\theta}}{\partial\theta} + \frac{1}{r\sin\theta}\frac{\partial S_{\theta\phi}}{\partial\phi} \\
&\quad + \frac{1}{r}\left\{2\,S_{\theta r} + S_{r\theta} + \cot\theta\,(S_{\theta\theta} - S_{\phi\phi})\right\}, \\
(\operatorname{div}\mathbf{S})_\phi &= \frac{\partial S_{\phi r}}{\partial r} + \frac{1}{r}\frac{\partial S_{\phi\theta}}{\partial\theta} + \frac{1}{r\sin\theta}\frac{\partial S_{\phi\phi}}{\partial\phi} \\
&\quad + \frac{1}{r}\left\{2\,S_{\phi r} + S_{r\phi} + \cot\theta\,(S_{\phi\theta} + S_{\theta\phi})\right\}.
\end{aligned}
}
\tag{A.3.15}
$$

Curl of a vector field

The curl of $\mathbf{v}$, denoted by $\operatorname{curl}\mathbf{v}$, is a vector field defined as

$$\operatorname{curl}\mathbf{v} = \nabla\times\mathbf{v}.$$

Hence,

$$\operatorname{curl}\mathbf{v} = \left(\mathbf{e}_r\frac{\partial}{\partial r} + \mathbf{e}_\theta\frac{1}{r}\frac{\partial}{\partial\theta} + \mathbf{e}_\phi\frac{1}{r\sin\theta}\frac{\partial}{\partial\phi}\right)\times(v_r\,\mathbf{e}_r + v_\theta\,\mathbf{e}_\theta + v_\phi\,\mathbf{e}_\phi),$$

which upon noting that

$$
\begin{aligned}
&\mathbf{e}_r\times\mathbf{e}_r = \mathbf{0}, \quad \mathbf{e}_\theta\times\mathbf{e}_\phi = -\mathbf{e}_\phi\times\mathbf{e}_\theta = \mathbf{e}_r, \\
&\mathbf{e}_\theta\times\mathbf{e}_\theta = \mathbf{0}, \quad \mathbf{e}_\phi\times\mathbf{e}_r = -\mathbf{e}_r\times\mathbf{e}_\phi = \mathbf{e}_\theta, \\
&\mathbf{e}_\phi\times\mathbf{e}_\phi = \mathbf{0}, \quad \mathbf{e}_r\times\mathbf{e}_\theta = -\mathbf{e}_\theta\times\mathbf{e}_r = \mathbf{e}_\phi,
\end{aligned}
$$

gives

$$\operatorname{curl}\mathbf{v} = \left\{\frac{1}{r}\frac{\partial v_\phi}{\partial \theta} - \frac{1}{r\sin\theta}\frac{\partial v_\theta}{\partial \phi} + \cot\theta\frac{v_\phi}{r}\right\}\mathbf{e}_r + \left\{\frac{1}{r\sin\theta}\frac{\partial v_r}{\partial \phi} - \frac{\partial v_\phi}{\partial r} - \frac{v_\phi}{r}\right\}\mathbf{e}_\theta + \left\{\frac{\partial v_\theta}{\partial r} - \frac{1}{r}\frac{\partial v_r}{\partial \theta} + \frac{v_\theta}{r}\right\}\mathbf{e}_\phi. \tag{A.3.16}$$

Laplacian of a scalar field

The Laplacian of ψ, denoted by $\triangle\psi$, is a scalar field defined as

$$\triangle\psi = \operatorname{div}\nabla\psi.$$

Thus, using (A.3.9) and (A.3.14),

$$\triangle\psi = \frac{\partial^2\psi}{\partial r^2} + \frac{1}{r^2}\frac{\partial^2\psi}{\partial\theta^2} + \frac{1}{(r\sin\theta)^2}\frac{\partial^2\psi}{\partial\phi^2} + \frac{2}{r}\frac{\partial\psi}{\partial r} + \frac{\cot\theta}{r^2}\frac{\partial\psi}{\partial\theta}. \tag{A.3.17}$$

Laplacian of a vector field

The Laplacian of $\mathbf{v}$, denoted by $\triangle\mathbf{v}$, is a vector field defined as

$$\triangle\mathbf{v} = \operatorname{div}\nabla\mathbf{v}.$$

From (A.3.15) and (A.3.13),

$$\begin{aligned}
(\triangle\mathbf{v})_r &= \triangle v_r - \frac{2}{r^2}\Big(v_r + v_\theta\cot\theta + \frac{\partial v_\theta}{\partial\theta} + \frac{1}{\sin\theta}\frac{\partial v_\phi}{\partial\phi}\Big),\\
(\triangle\mathbf{v})_\theta &= \triangle v_\theta + \frac{1}{r^2}\Big(2\frac{\partial v_r}{\partial\theta} - \frac{1}{\sin^2\theta}v_\theta - 2\frac{\cos\theta}{\sin^2\theta}\frac{\partial v_\phi}{\partial\phi}\Big),\\
(\triangle\mathbf{v})_\phi &= \triangle v_\phi + \frac{1}{r^2\sin\theta}\Big(2\frac{\partial v_r}{\partial\phi} + 2\cot\theta\frac{\partial v_\theta}{\partial\phi} - \frac{1}{\sin\theta}v_\phi\Big).
\end{aligned} \tag{A.3.18}$$

Infinitesimal strain tensor

Let

$$\mathbf{u} = u_r\mathbf{e}_r + u_\theta\mathbf{e}_\theta + u_\phi\mathbf{e}_\phi,$$

denote the components of the displacement vector with respect to $\{\mathbf{e}_r, \mathbf{e}_\theta, \mathbf{e}_\phi\}$. Then (A.3.13) gives the components of the displacement gradient tensor $\mathbf{H} \equiv \nabla\mathbf{u}$ in the spherical coordinate system, and hence the symmetric infinitesimal strain tensor $\boldsymbol{\epsilon}$ defined by

$$\boldsymbol{\epsilon} = \frac{1}{2}[\mathbf{H} + \mathbf{H}^{\top}],$$

has the following components in the spherical coordinate system:

$$
\begin{aligned}
\epsilon_{rr} &= \frac{\partial u_r}{\partial r}, \\
\epsilon_{\theta\theta} &= \frac{1}{r}\frac{\partial u_\theta}{\partial \theta} + \frac{u_r}{r}, \\
\epsilon_{\phi\phi} &= \left(\frac{1}{r\sin\theta}\frac{\partial u_\phi}{\partial \phi} + \frac{\cot\theta}{r}u_\theta + \frac{u_r}{r}\right), \\
\epsilon_{r\theta} &= \frac{1}{2}\left(\frac{1}{r}\frac{\partial u_r}{\partial \theta} + \frac{\partial u_\theta}{\partial r} - \frac{u_\theta}{r}\right) = \epsilon_{\theta r}, \\
\epsilon_{\theta\phi} &= \frac{1}{2}\left(\frac{1}{r\sin\theta}\frac{\partial u_\theta}{\partial \phi} + \frac{1}{r}\frac{\partial u_\phi}{\partial \theta} - \frac{\cot\theta}{r}u_\phi\right) = \epsilon_{\phi\theta}, \\
\epsilon_{\phi r} &= \frac{1}{2}\left(\frac{\partial u_\phi}{\partial r} + \frac{1}{r\sin\theta}\frac{\partial u_r}{\partial \phi} - \frac{u_\phi}{r}\right) = \epsilon_{r\phi}.
\end{aligned}
\tag{A.3.19}
$$

Equation of motion

The matrix of the components of the symmetric stress tensor $\boldsymbol{\sigma}$ with respect to $\{\mathbf{e}_r, \mathbf{e}_\theta, \mathbf{e}_z\}$ is

$$
[\boldsymbol{\sigma}] = \begin{bmatrix} \sigma_{rr} & \sigma_{r\theta} & \sigma_{r\phi} \\ \sigma_{\theta r} & \sigma_{\theta\theta} & \sigma_{\theta\phi} \\ \sigma_{\phi r} & \sigma_{\phi\theta} & \sigma_{\phi\phi} \end{bmatrix}.
$$

Let (b_r, b_θ, b_ϕ) and $(\ddot{u}_r, \ddot{u}_\theta, \ddot{u}_\phi)$ denote the components of the body force $\mathbf{b}$ and the acceleration $\ddot{\mathbf{u}}$ in a spherical coordinate system. Then using (A.3.15) the equation of motion

$$
\operatorname{div}\boldsymbol{\sigma} + \mathbf{b} = \rho\ddot{\mathbf{u}},
$$

where ρ is the mass density, may be written as the following three equations in the spherical coordinate system:

$$
\begin{aligned}
&\frac{\partial \sigma_{rr}}{\partial r} + \frac{1}{r}\frac{\partial \sigma_{r\theta}}{\partial \theta} + \frac{1}{r\sin\theta}\frac{\partial \sigma_{r\phi}}{\partial \phi} + \frac{1}{r}\left(2\,\sigma_{rr} - \sigma_{\theta\theta} - \sigma_{\phi\phi} + \cot\theta\,\sigma_{r\theta}\right) + b_r = \rho\ddot{u}_r, \\
&\frac{\partial \sigma_{\theta r}}{\partial r} + \frac{1}{r}\frac{\partial \sigma_{\theta\theta}}{\partial \theta} + \frac{1}{r\sin\theta}\frac{\partial \sigma_{\theta\phi}}{\partial \phi} + \frac{1}{r}\left\{3\,\sigma_{\theta r} + \cot\theta\,(\sigma_{\theta\theta} - \sigma_{\phi\phi})\right\} + b_\theta = \rho\ddot{u}_\theta, \\
&\frac{\partial \sigma_{\phi r}}{\partial r} + \frac{1}{r}\frac{\partial \sigma_{\phi\theta}}{\partial \theta} + \frac{1}{r\sin\theta}\frac{\partial \sigma_{\phi\phi}}{\partial \phi} + \frac{1}{r}\left\{3\,\sigma_{\phi r} + 2\,\cot\theta\sigma_{\phi\theta}\right\} + b_\phi = \rho\ddot{u}_\phi.
\end{aligned}
\tag{A.3.20}
$$

Appendix B Stress intensity factors for some crack configurations

1. **Finite crack of length $2a$ in an infinitely large body subject to far-field stresses $\sigma_{22}^{\infty} = \sigma^{\infty}$**

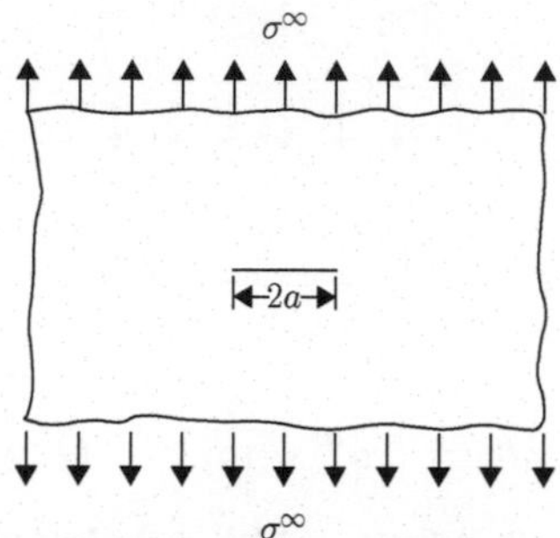

Fig. B.1 Finite crack of length $2a$ in an infinitely large body subject to far-field stresses $\sigma_{22}^{\infty} = \sigma^{\infty}$.

The stress intensity factor is

$$K_{\mathrm{I}} = \sigma^{\infty}\sqrt{\pi a}.$$

For other geometrical configurations, in which a characteristic crack dimension is a and a characteristic applied tensile stress is σ^{∞}, we will write the corresponding stress intensity factor as

$$\boxed{K_{\mathrm{I}} = Q\sigma^{\infty}\sqrt{\pi a}, \qquad Q \equiv \text{configuration correction factor.}}$$

2. **Center-crack in a long $(L > 3w)$ strip of finite width (w)**

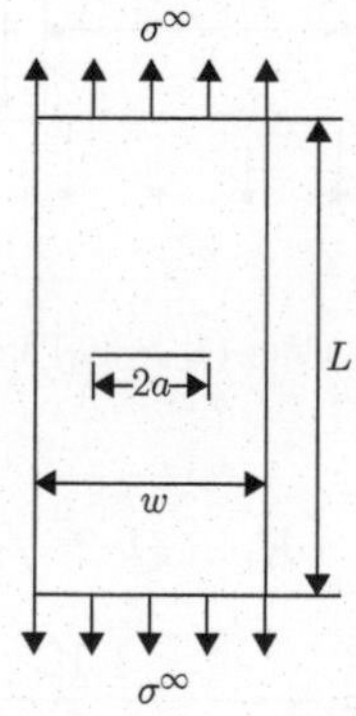

Fig. B.2 Center-crack in a long $(L > 3w)$ strip of finite width (w).

$$K_{\mathrm{I}} = Q\sigma^{\infty}\sqrt{\pi a}, \qquad Q = \hat{Q}\left(\frac{a}{w}\right) \approx \left\{\sec\left(\frac{\pi a}{w}\right)\right\}^{1/2}.$$

Continuum Mechanics of Solids. Lallit Anand and Sanjay Govindjee, Oxford University Press (2020).

DOI: 10.1093/oso/9780198864721.001.0001

3. **Edge-crack in a semi-infinite body**

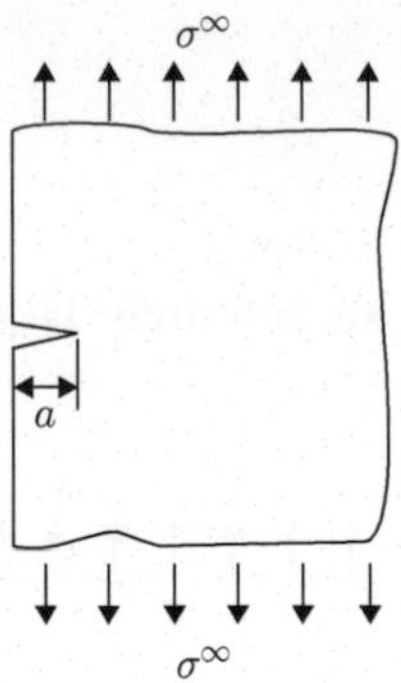

Fig. B.3 Edge-crack in a semi-infinite body.

$$K_{\rm I} = Q\sigma^\infty\sqrt{\pi a}, \qquad Q \approx 1.12$$

4. **Edge-crack in a long $(L > 3w)$ finite-width strip in tension**

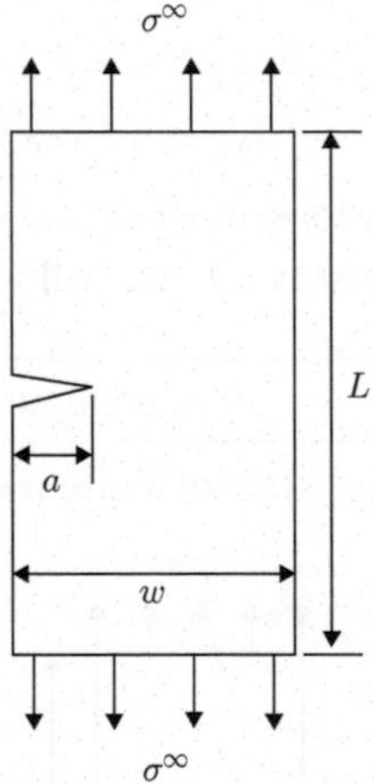

Fig. B.4 Edge-crack in a long $(L > 3w)$ finite-width strip in tension.

$$K_{\rm I} = Q\sigma^\infty\sqrt{\pi a}\,,$$

$$Q = \hat{Q}\left(\frac{a}{w}\right) \approx \frac{1.12}{\left[1 - 0.7\left(\frac{a}{w}\right)^{1.5}\right]^{3.25}}, \quad \frac{a}{w} \leq 0.65.$$

Also, the following slightly more cumbersome formula is more accurate for larger a/w:

$$Q = \hat{Q}\left(\frac{a}{w}\right) \approx 0.265 \times \left(1 - \frac{a}{w}\right)^4 + \frac{[0.857 + 0.265\,a/w]}{(1 - a/w)^{3/2}}.$$

5. **Edge-crack in a long ($L > 3w$) finite-width strip of thickness B subject to a pure bending moment M**

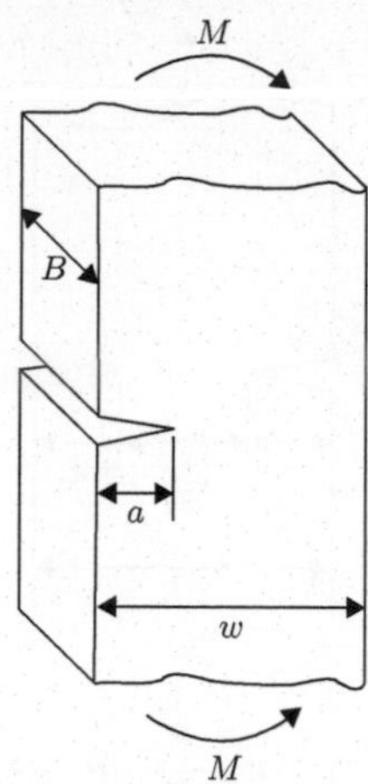

Fig. B.5 Edge-crack in a long ($L > 3w$) finite-width strip of thickness B subject to a pure bending moment M.

$$K_{\mathrm{I}} = Q\,\sigma_{\mathrm{nom}}\,\sqrt{\pi a}\,, \qquad \sigma_{\mathrm{nom}} = \frac{6M}{Bw^2},$$

$$Q \approx \frac{1.12}{\left[1 - \left(\dfrac{a}{w}\right)^{1.82}\right]^{1.285}} - \sin\left(\frac{\pi}{2}\frac{a}{w}\right); \quad \frac{a}{w} \leq 0.7\,.$$

Deep crack approximation for $a/w > 0.7$:

$$K_{\mathrm{I}} \approx (4M)/Bc^{3/2}, \quad \text{where} \quad c \equiv w - a.$$

6. **Symmetric double-edge-cracked strip in tension**

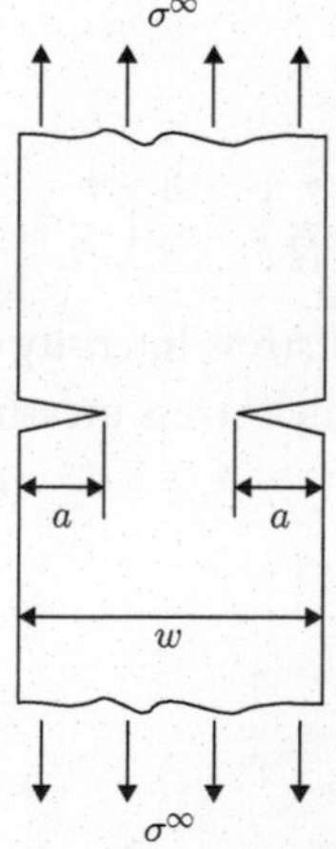

Fig. B.6 Symmetric double-edge-cracked strip in tension.

$$K_{\mathrm{I}} = Q\,\sigma^{\infty}\,\sqrt{\pi a}\,,$$

$$Q = \hat{Q}(a/w) \approx \frac{\tan\left(\dfrac{\pi a}{2w}\right)}{\left(\dfrac{\pi a}{2w}\right)} \times \left[1 + 0.122\cos^4\left(\frac{\pi a}{2w}\right)\right].$$

7. **Externally circumferentially cracked rod in tension**

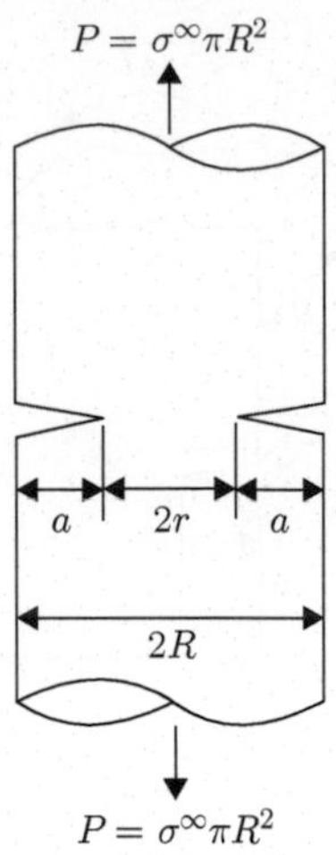

Fig. B.7 Externally circumferentially cracked rod in tension.

$$K_{\rm I} = Q\,\sigma^{\infty}\,\sqrt{\pi a}\,,$$

with

$$Q = \hat{Q}\left(\frac{a}{R}\right) \approx \frac{1.12}{\left[1 - \left(\frac{a}{R}\right)^{1.47}\right]^{1.2}}, \quad \frac{a}{R} \le 0.7\,.$$

Also, in terms of the net section stress,

$$K_{\rm I} = Q_{\rm net}\,\sigma_{\rm net}\,\sqrt{\pi a}; \qquad \sigma_{\rm net} = \frac{P}{\pi r^2},$$

with

$$\begin{aligned} Q_{\rm net} &= \hat{Q}(r/R) \\ &\approx \frac{1}{2}\left(\frac{r}{R}\right)^{1/2}\left[1 + \frac{1}{2}\left(\frac{r}{R}\right) + \frac{3}{8}\left(\frac{r}{R}\right)^2 - 0.363\left(\frac{r}{R}\right)^3 + 0.731\left(\frac{r}{R}\right)^4\right]. \end{aligned}$$

Note that when alternate forms of stress intensity calibration are given, it is important to remember "which Q goes with which stress measure."

8. **Single crack at edge of hole in an infinite plane body in uniaxial tension**

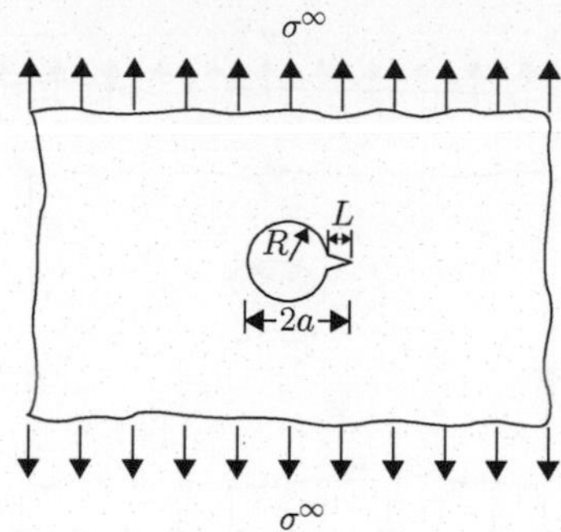

Fig. B.8 Single crack at edge of hole in an infinite plane body in uniaxial tension.

$$K_{\rm I} = Q\sigma^\infty\sqrt{\pi a}\,,$$

$$Q \approx \left[\left\{1 - \sin\left(\frac{\pi R}{a}\right)\right\}\left\{1 - \left(\frac{R}{a}\right)^{16.6}\right\}^{0.9} + \left\{\sin\left(\frac{\pi R}{a}\right)\right\} 1.45\left(\frac{R}{a}\right)^{1/3}\right]^{1/2},$$

for $0 \leq R/a \leq 1$.

9. **Equal double cracks at edges of a hole in an infinite plane body in uniaxial tension**

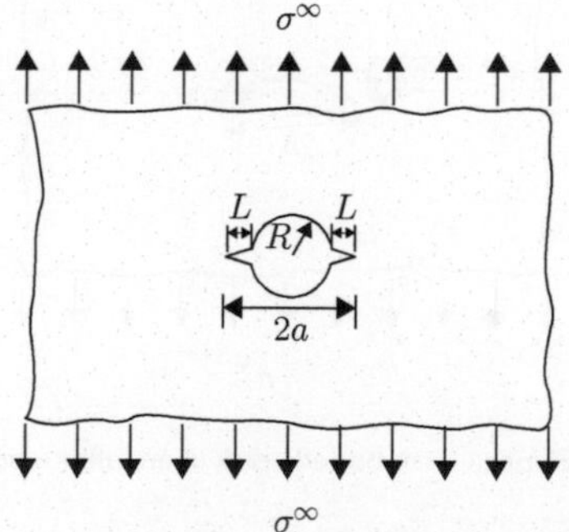

Fig. B.9 Equal double cracks at edges of a hole in an infinite plane body in uniaxial tension.

$$K_{\rm I} = Q\sigma^\infty\sqrt{\pi a}\,,$$

$$Q \approx \left[\left\{1 - \sin\left(\frac{\pi R}{a}\right)\right\}\left\{1 - \left(\frac{R}{a}\right)^{5}\right\}^{0.78} + \left\{\sin\left(\frac{\pi R}{a}\right)\right\}^{1.23}\left(\frac{R}{a}\right)^{0.19}\right]^{1/2},$$

for $0 \leq R/a \leq 1$. Note that for $L/R \gtrsim 0.2$

$$K_{\rm I} \approx \sigma^\infty\sqrt{\pi a}, \text{ with } 2a = 2L + 2R.$$

10. **Penny-shaped buried crack in an infinite body in tension**

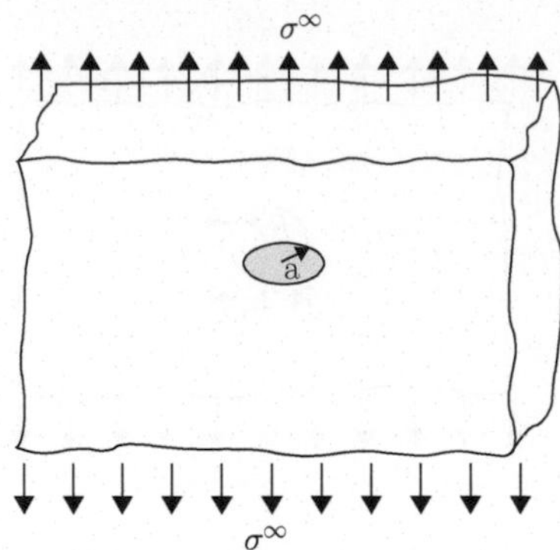

Fig. B.10 Penny-shaped buried crack in an infinite body in tension.

$$K_{\mathrm{I}} = Q\sigma^\infty \sqrt{\pi a}, \qquad Q = \frac{2}{\pi}.$$

11. **Elliptical plan buried crack in an infinite body in tension**

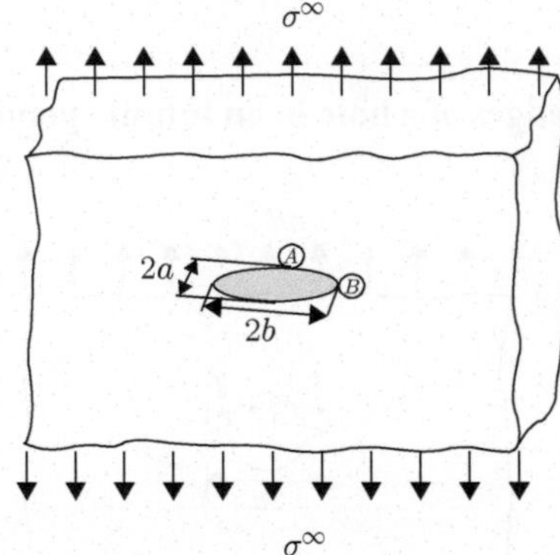

Fig. B.11 Elliptical plan buried crack in an infinite body in tension.

$$K_{\mathrm{I}A} = Q_A\,\sigma^\infty \sqrt{\pi a},$$
$$Q_A = \Phi^{-1} \approx \left\{1 - 0.619\left(\frac{a}{b}\right)\right\}^{1/2},$$
$$\Phi \equiv \int_0^{\pi/2} \sqrt{1 - k^2 \sin^2\phi}\; d\phi,$$
$$k^2 = 1 - (a^2/b^2),$$
$$K_{\mathrm{I}B} = Q_B\,\sigma^\infty \sqrt{\pi b},$$
$$Q_B = \left(\frac{a}{b}\right) Q_A.$$

For $a/b < 1$, the maximum stress intensity factor occurs at A:

$K_{\mathrm{I\ max}} = K_{\mathrm{I}A}$; For $a = b$, $Q_B = Q_A = (2/\pi)$; $K_{\mathrm{I}A}(b \to \infty) = \sigma^\infty \sqrt{\pi a}$.

12. **Semi-elliptical plan surface crack on a semi-infinite body in tension**

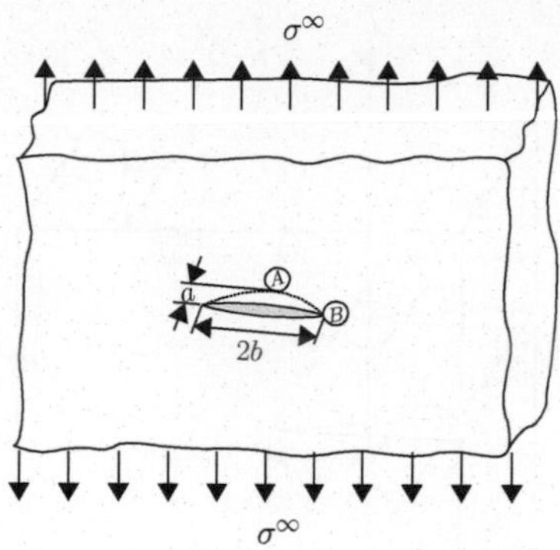

Fig. B.12 Semi-elliptical plan surface crack on a semi-infinite body in tension.

$$K_{IA} = Q_A \sigma^{\infty} \sqrt{\pi a},$$

$$Q_A \approx \left\{1 + 0.12\left(1 - \frac{a}{b}\right)\right\}\left\{1 - 0.619\left(\frac{a}{b}\right)\right\}^{1/2},$$

$$K_{IB} = Q_B \sigma^{\infty} \sqrt{\pi b},$$

$$Q_B \approx \left(\frac{a}{b}\right) Q_A.$$

13. **Standard ASTM three-point-bend test specimen**

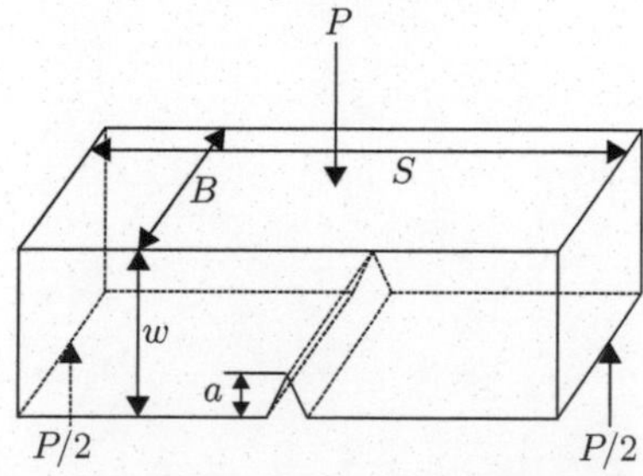

Fig. B.13 Standard ASTM three-point-bend test specimen.

$$K_{\mathrm{I}} = \left\{PS/\left(Bw^{3/2}\right)\right\} \cdot f(a/w),$$

where

$$f(a/w) = \frac{3(a/w)^{1/2}\left[1.99 - (a/w)(1 - a/w) \times (2.15 - 3.93a/w + 2.7a^2/w^2)\right]}{2(1 + 2a/w)(1 - a/w)^{3/2}}.$$

14. **Standard ASTM compact-tension-specimen**

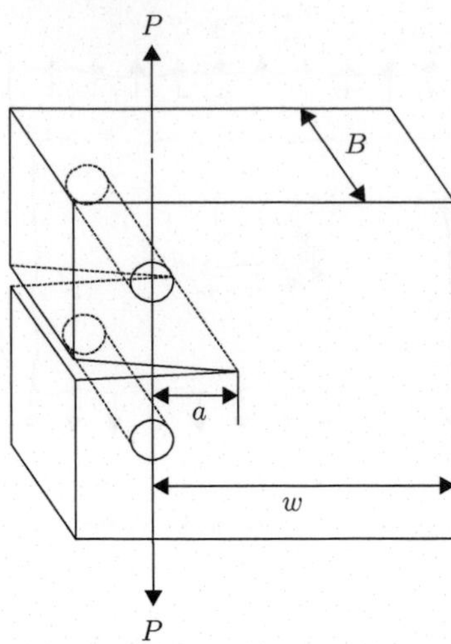

Fig. B.14 Standard ASTM compact-tension-specimen.

$$K_{\mathrm{I}} = \left\{P/\left(Bw^{1/2}\right)\right\} \cdot f(a/w),$$

$$f(a/w) \approx \frac{(2 + a/w)\left(0.886 + 4.64a/w - 13.32a^2/w^2 + 14.72a^3/w^3 - 5.6a^4/w^4\right)}{(1 - a/w)^{3/2}}.$$

Bibliography

L. H. Adams and R. E. Gibson. The compressibility of rubber. *Journal of the Washington Academy of Sciences*, 20:213–23, 1930.

G. B. Airy. On the strains in the interior of beams. *Philosophical Transactions of the Royal Society*, 153:49–79, 1863.

T. Alfrey Jr., E. F. Gurnee, and W. G. Lloyd. Diffusion in glassy polymers. *Journal of Polymer Science Part C: Polymer Symposia*, 12:249–61, 1966.

L. Anand. Constitutive equations for the rate-dependent deformation of metals at elevated temperatures. *ASME Journal of Engineering Materials and Technology*, 104:12–17, 1982.

L. Anand. A constitutive model for compressible elastomeric solids. *Computational Mechanics*, 18 (5):339–55, 1996.

L. Anand. 2014 Drucker Medal Paper: A derivation of the theory of linear poroelasticity from chemoelasticity. *ASME Journal of Applied Mechanics*, 82:111005-(1-11), 2015.

L. Anand and S. Govindjee. *Example Problems for Continuum Mechanics of Solids*. 2020.

D. Annaratone. *Engineering Heat Transfer*. Springer-Verlag, 2010.

P. J. Armstrong and C. O. Fredrick. *A mathematical representation of the multiaxial Bauschinger effect*, 1966. CEGB report No. RD/B/N 73.

E. M. Arruda and M. C. Boyce. A three-dimensional constitutive model for the large stretch behavior of rubber elastic materials. *Journal of the Mechanics and Physics of Solids*, 41:389–412, 1993.

ASTM D4065. *Standard practice for Plastics: Dynamic Mechanical Properties: Determination and Report of Procedures*. American Society for Testing and Materials, West Conshohocken (PA), USA, 2012.

ASTM E399. *Standard Test Method for Linear-Elastic Plane-Strain Fracture Toughness K_{Ic} of Metallic Materials*. American Society for Testing and Materials, West Conshohocken (PA), USA, 2013.

B. A. Auld. *Acoustic Fields and Waves in Solids*, Volume I. Krieger Publishing Co., 1990.

A.-J.-C. Barré de Saint-Venant. *De la torsion des prismes : avec des considérations sur leur flexion ainsi que sur l'équilibre des solides élastiques en général et des formules pratiques pour le calcul de leur résistance à divers efforts s'exerçant simultanément*, Volume 14 of *Des mémoires présentés par divers savants*. Académie des sciences, Paris, 1855.

O. H. Basquin. The exponential law of endurance tests. *Proceedings of the American Society for Testing and Materials*, 10:625–30, 1910.

J. Bauschinger. Über die Veränderung der Position der Elastizitätsgrenze des Eisens und Stahls durch Strecken und Quetschen und durch Erwärmen und Abkühlen und durch oftmals wiederholte Beanspruchungen (On the change of the elastic limit and hardness of iron and steels through extension and compression, through heating and cooling, and through cycling). *Mitteilung aus dem Mechanisch-technischen Laboratorium der Königlichen polytechnischen Hochschule in München*, 13:1–115, 1886.

T. Beda. An approach for hyperelastic model-building and parameters estimation a review of constitutive models. *European Polymer Journal*, 50:97–108, 2014.

D. P. Bertsekas. *Nonlinear Programming*. Athena Scientific, Belmont, MA, 1995.

M. A. Biot. Le problem̃e de la consolidation des matiéres argileuses sous une charge. *Annales de la Societe Scientifique de Bruxelles*, B55:110–13, 1935.

M. A. Biot. General theory of three-dimensional consolidation. *Journal of Applied Physics*, 12:155–64, 1941.

M. A. Biot and D. G. Willis. The elastic coefficients of the theory of consolidation. *Journal of Applied Mechanics*, 24:594–601, 1957.

J. E. Bischoff, E. M. Arruda, and K. Grosh. A new constitutive model for the compressibility of elastomers at finite deformations. *Rubber Chemistry and Technology*, 74:541–59, 2001.

E. Bohn, T. Eckl, M. Kamlah, and R. M. McMeeking. A model for lithium diffusion and stress generation in an intercalation storage particle with phase change. *Journal of the Electrochemical Society*, 160:A1638–A1652, 2013.

B. A. Boley and J. H. Weiner. *Theory of thermal stresses*. Wiley, 1960.

R. M. Bowen. Compressible porous media models using the theory of mixtures. *International Journal of Engineering Science*, 20:697–735, 1982.

M. C. Boyce and E. M. Arruda. Constitutive models of rubber elasticity: a review. *Rubber Chemistry and Technology*, 73:504–23, 2000.

S. Boyd and L. Vandenberghe. *Convex Optimization*. Cambridge University Press, 2004.

P. W. Bridgman. *Studies in large plastic flow and fracture with special emphasis on the effects of hydrostatic pressure*. McGraw-Hill, 1952.

S. B. Brown, K. H. Kim, and L. Anand. An internal variable constitutive model for hot working of metals. *International Journal of Plasticity*, 5:95–130, 1989.

J. Casey. On infinitesimal deformation measures. *Journal of Elasticity*, 28:257–69, 1992.

A. H. D. Cheng. *Poroelasticity*. Springer, 2016.

S. A. Chester and L. Anand. A coupled theory of fluid permeation and large deformations for elastomeric materials. *Journal of the Mechanics and Physics of Solids*, 58:1879–1906, 2010.

S. A. Chester and L. Anand. A thermo-mechanically coupled theory for fluid permeation in elastomeric materials: Application to thermally responsive gels. *Journal of the Mechanics and Physics of Solids*, 59:1978–2006, 2011.

S. A. Chester, C. V. Di Leo, and L. Anand. A finite element implementation of a coupled diffusion-deformation theory for elastomeric gels. *International Journal of Solids and Structures*, 52:1–18, 2015.

J. Christensen and J. Newman. A mathematical model of stress generation and fracture in lithium manganese oxide. *Journal of the Electrochemical Society*, 153:A1019–A1030, 2005.

L. F. Coffin. A study of the effects of cyclic thermal stresses on a ductile metal. *Transactions ASME*, 76:931–50, 1954.

A. Cohen. A Pade approximant to the inverse Langevin function. *Rheologica Acta*, 30:270–3, 1991.

B. D. Coleman and W. Noll. The thermodynamics of elastic materials with heat conduction and viscosity. *Archive for Rational Mechanics and Analysis*, 13:167–78, 1963.

M. Considére. Memoire sur lémploi du fer et de lácier dans les constructions. *Annales des Ponts et Chauseés*, 9:574–605, 1885.

C. A. Coulomb. *Essai sur une application des règles de Maximis & Minimis à quelques Problèmes de Statique, relatifs à l' Architecture*, Volume 7 of *Mémories de Mathématique et de Physique*, pages 343–82. Académie Royale des Sciences, 1773.

O. Coussy. *Mechanics of Porous Media*. John Wiley and Sons Inc., Chichester, 1995.

S. C. Cowin. Bone poroelasticity. *Journal of Biomechanics*, 32:217–302, 1999.

G. R. Cowper and P. S. Symonds. Strain-hardening and strain-rate effects in the impact loading of cantilever beams. Technical Report C11-28, Division of Applied Mathematics, Brown University, September 1975.

J. Curie and P. Curie. Dévelopment par compression de l'életricité polaire dans les cristaux hémièdres à faces inclinées. *Bulletin de Minéralogie*, 3:90–3, 1880.

B. Dacorogna. *Direct Methods in the Calculus of Variations*. Springer-Verlag, 1989.

H. Darcy. *Les Fontaines de la Ville de Dijon*. Victor Dalmont, Paris, 1856.

R. DeHoff. *Thermodynamics in Materials Science*. CRC Press, Boca Raton, 2006.

E. Detournay and A. H. D. Cheng. Fundamentals of Poroelasticity. In J. A. Hudson and C. Fairhurst, editors, *Comprehensive Rock Engineering: Principles Practices and Projects*, pages 113–71. Pergamnon Press, 1993.

P. A. Domenico and F. W. Schwartz. *Physical and Chemical Hydrogeology*. John Wiley & Sons, 1990.

D. C. Drucker. A definition of a stable inelastic material. *ASME Journal of Applied Mechanics*, 26:101–95, 1959.

D. C. Drucker and W. Prager. Soil mechanics and plastic analysis or limit design. *Quarterly of Applied Mathematics*, 10:157–65, 1952.

F. P. Duda, A. C. Souza, and E. Fried. A theory for species migration in finitely strained solid with application to polymer network swelling. *Journal of the Mechanics and Physics of Solids*, 58:515–29, 2010.

J. D. Eshelby. The force on an elastic singularity. *Philosophical Transactions*, A224:87–112, 1951.

J. D. Eshelby. Energy relations and the energy-momentum tensor in continuum mechanics. *Inelastic behavior of solids*, pages 321–35. McGraw-Hill, 1970.

J. D. Eshelby. The elastic energy-momentum tensor. *Journal of Elasticity*, 5:321–35, 1975.

J. D. Ferry. *Viscoelastic Properties of Polymers*. John Wiley & Sons, New York, 1961.

A. Fick. Über Diffusion. *Poggendorff's Annalen der Physik und Chemie*, 94:59–86, 1855.

J. Fish and T. Belytschko. *A First Course in Finite Elements*. Wiley, 2007.

P. J. Flory. Thermodynamic relations for high elastic materials. *Transactions of the Faraday Society*, 57:829–38, 1961.

L. Freund and S. Suresh. *Thin Film Materials: Stress, Defect Formation and Surface Evolution*. Cambridge University Press, Cambridge, 2004.

T.-P. Fries and T. Belytschko. The extended/generalized finite element method: An overview of the method and its applications. *International Journal for Numerical Methods in Engineering*, 84:253–304, 2010.

A. N. Gent. A new constitutive relation for rubber. *Rubber Chemistry and Technology*, 69:59–61, 1996.

A. N. Gent. Elastic instabilities of inflated rubber shells. *Rubber Chemistry and Technology*, 72:263–8, 1999.

A. N. Gent and A. G. Thomas. Forms for the stored (strain) energy function for vulcanized rubber. *Journal of Polymer Science*, 28:625–37, 1958.

A. N. Gent and R. S. Rivlin. Experiments on the mechanics of rubber I: Eversion of a tube. *Proceedings of the Physical Society*, B65:118–21, 1952.

P. Germain. The method of virtual power in continuum mechanics. Part 2: microstructure. *SIAM Journal of Applied Mathematics*, 25:556–75, 1973.

J. W. Gibbs. On the equilibrium of heterogeneous substances. *Transactions of the Connecticut Academy of Arts and Sciences*, III:108–248, 1878.

S. Govindjee. *Engineering Mechanics of Deformable Solids: A Presentation with Exercises*. Oxford University Press, Oxford UK, 2013.

S. Govindjee and J. Simo. Coupled stress–diffusion: Case II. *Journal of the Mechanics and Physics of Solids*, 41:863–87, 1993.

V. Govorukha, M. Kamlah, V. Loboda, and Y. Lapusta. *Fracture mechanics of piezoelectric solids with interface cracks*. Springer, New York, 2017.

A. A. Griffith. The phenomenon of rupture and flow in solids. *Philosophical Transactions of The Royal Society of London*, A221:163–98, 1921.

Y. Guéguen, L. Dormieux, and M. Boutéca. Fundamentals of poromechanics. In Y. Guéguen and M. Boutéca, editors, *Mechanics of fluid-saturated porous materials. International Geophysics Series 89*. Elsevier Academic Press, 2004.

M. E. Gurtin. *The Linear Theory of Elasticity*, Volume VIa/2 of *Handbook of Physics*. Springer-Verlag, 1972.

M. E. Gurtin. On the energy release rate in quasi-static elastic crack propagation. *Journal of Elasticity*, 9:187–95, 1979.

M. E. Gurtin. *An Introduction to Continuum Mechanics*. Academic Press, 1981.

M. E. Gurtin. A gradient theory of single-crystal viscoplasticity that accounts for geometrically necessary dislocations. *Journal of the Mechanics and Physics of Solids*, 50:5–32, 2002.

M. E. Gurtin, E. Fried, and L. Anand. *The Mechanics and Thermodynamics of Continua*. Cambridge University Press, Cambridge, 2010.

K. Hellan. *Introduction to fracture mechanics*. McGraw-Hill, 1984.

H. Hencky. Zur Theorie plastischer Deformationen und der hierdurch im Material hervorgerufenen Nachspannungen. *Zeitschrift für angewandte Mathematik und Mechanik*, 4:323–34, 1924.

L. R. Herrmann and E. F. Peterson. A numerical procedure for visco-elastic stress analysis. In *Proceedings of the Seventh Meeting of ICRPG Mechanical Behavior Working Group*, 1968.

H. Hertz. Über die Berührung fester elastischer Körper. (On the contact of elastic solids) (For English trans. see misc. papers by Hertz, Jones, and Schott, Macmillan. London, 1896). *Journal für die reine und angewandte Mathematik*, 92:156–71, 1882.

F. B. Hildebrand. *Advanced Calculus for Applications*. Prentice-Hall, 2nd edition, 1976.

R. Hill. *The Mathematical Theory of Plasticity*. Oxford University Press, 1950.

W. Hong, X. Zhao, J. Zhou, and Z. Suo. A theory of coupled diffusion and large deformation in polymeric gel. *Journal of the Mechanics and Physics of Solids*, 56:1779–93, 2008.

C. O. Horgan. Recent developments concerning Saint-Venant's principle: A second update. *Applied Mechanics Reviews*, 49:S101–S111, 1996.

F. M. Howell and J. L. Miller. Axial-stress fatigue strengths of several structural aluminum alloys. In *Proceedings of the Fifty-Eighth Annual Meeting of the American Society for Testing and Materials*, pages 955–68, 1955.

M. T. Huber. Specific work of strain as a measure of material effort (English translation). *Archives of Mechanics*, 56:173–190, 1904. (2004 translation by Anna Strek, Czasopismo Techniczne, XXII, 1904, Lwów, Organ Towarzystwa Politechnicznego we Lwowie).

C. M. Hudson and S. K. Seward. Compendium of sources of fracture toughness and fatigue crack growth data for metallic alloys. *International Journal of Fracture*, 14:R151–R184, 1978.

T. J. R. Hughes. *The Finite Element Method: Linear Static and Dynamic Finite Element Analysis*. Dover Publications, 2000.

J. W. Hutchinson. *Nonlinear fracture mechanics*. Technical report, Technical University of Denmark, 1979.

J. W. Hutchinson. Fundamentals of the phenomenological theory of nonlinear fracture mechanics. *ASME Journal of Applied Mechanics*, 50:1042–51, 1983.

G. R. Irwin. Fracture dynamics, 1948. In *Fracturing of Metals*. ASM Symposium (Trans. ASM 40A), Cleveland.

A. Y. Ishlinskii. General theory of plasticity with linear hardening. *Ukrainskij Matematicheskij Zhurnal*, 6:314–25, 1954.

J. D. Jackson. *Classical Electrodynamics*. John Wiley & Sons, 1999.

B. Jaffe, W. R. Cook, and H. Jaffe. *Piezoelectric Ceramics*. Academic Press London, New York, 1971.

K. L. Johnson. *Contact Mechanics*. Cambridge University Press, 1985.

U. F. Kocks, A. S. Argon, and M. F. Ashby. Thermodynamics and kinetics of slip. In *Progress in Material Science*. Pergamon Press, London, 1975.

R. Kohlrausch. Theorie des elektrischen Rückstandes in der Leidener Flasche. *Annalen der Physik*, 167:179–14, 1854.

A. Kovetz. *Electromagnetic Theory*. Oxford University Press, 2000.

E. Kreyszig. *Advanced Engineering Mathematics*. John Wiley & Sons, 5th edition, 1983.

F. C. Larche and J. W. Cahn. The interactions of composition and stress in crystalline solids. *Acta Metallurgica*, 33:331–57, 1985.

F. Li, C. F. Shih, and A. Needleman. A comparison of methods for calculating energy release rates. *Engineering Fracture Mechanics*, 21:405–21, 1985.

J. C. M. Li. Physical chemistry of some microstructural phenomena. *Metallurgical Transactions A*, 9:1353–80, 1978.

J. H. Lienhard IV and J. H. Lienhard V. *A Heat Transfer Textbook*. Phlogiston Press, 2007. Downloadable at: http://web.mit.edu/lienhard/www/ahtt.html.

A. E. H. Love. *A Treatise on the Mathematical Theory of Elasticity*. Cambridge University Press, 1927.

A. M. Lush, G. Weber, and L. Anand. An implicit time-integration procedure for a set of internal variable constitutive equations for isotropic elasto-viscoplasticity. *International Journal of Plasticity*, 5:521–9, 1989.

C. W. MacGregor and N. Grossman. Effects of cyclic loading on mechanical behavior of 24S-T4 and 75S-T6 aluminum alloys and SAE 4130 steel. Technical Report NACA-TN-2812, National Advisory Committee for Aeronautics, 1952.

S. S. Manson. Behavior of materials under conditions of thermal stress. In *Heat Transfer Symposium*, pages 9–75. University of Michigan Engineering Research Institute, 1953.

G. Marckmann and E. Verron. Comparison of hyperelastic models for rubber-like materials. *Rubber Chemistry and Technology*, 79:835–58, 2006.

J. E. Mark and B. Erman. *Rubberlike Elasticity: A Molecular Primer*. Cambridge University Press, 2nd edition, 2007.

W. T. Matthews. Plane strain fracture toughness (K_{Ic}) data handbook for metals. Technical Report AMMRC MS73-6, Watertown, MA, 1973.

J. R. McLoughlin and A. V. Tobolsky. The viscoelastic behavior of polymethyl methacrylate. *Journal of Colloid Science*, 7:555–68, 1952.

J. McNamee and R. E. Gibson. Displacement functions and linear transforms applied to diffusion through porous elastic media. *Quarterly Journal of Mechanics and Applied Mathematics*, 13:98–111, 1960.

A. H. Meitzler, editor. *IEEE Standard on Piezoelectricity*. The Institute of Electrical and Electronics Engineers, Inc., 345 East 47th Street, New York, NY 10017, USA, 1988.

E. Melan. Zur plastiztät des räumlichen Kontinuums. *Ingenieur-Archiv Gesellschaft fürAngewandte Mathematik und Mecahnik*, 9:116–26, 1938.

J. H. Michell. On the direct determination of stress in an elastic solid, with application to the theory of plates. *Proceedings of the London Math Society*, 31:100–124, 1899.

C. Miehe, M. Hofacker, and F. Welschinger. A phase field model for rate-independent crack propagation: Robust algorithmic implementation based on operator splits. *Computer Methods in Applied Mechanics and Engineering*, 199:2765–78, 2010.

C. Miehe, S. Teichtmeister, and F. Aldakheel. Phase-field modelling of ductile fracture: a variational gradient-extended plasticity-damage theory and its micromorphic regularization. *Philosophical Transactions of the Royal Society A: Mathematical, Physical and Engineering Sciences*, 374: 20150170, 2016.

M. A. Miner. Cumulative damage in fatigue. *Journal of Applied Mechanics*, 12:A159, 1945.

R. v. Mises. Mechanik der festen Körper im plastisch-deformablen Zustand. *Nachrichten der königlichen Gesellschaft der Wissenschaften zu Göttingen, Mathematisch-Physikalische Klasse*, pages 582–92, 1913.

O. Mohr. Welche Umstände bedingen die Elastizitätsgrenze und den Bruch eines Materiales? *Zeitschrift des Vereines Deutscher Ingenieure*, 44:1524–30, 1900.

G. Montella, S. Govindjee, and P. Neff. The exponentiated Hencky strain energy in modeling tire derived material for moderately large deformations. *ASME Journal of Engineering Materials and Technology*, 138:031008–(1–12), 2016.

M. Mooney. A theory of large elastic deformation. *Journal of Applied Physics*, 11:582–92, 1940.

J. Morrow. Fatigue properties of metals. In J. A. Graham, J. F. Millan, and F. J. Appl, editors, Fatigue Design Handbook, pages 21–30. SAE, 1968.

Y. Murakami. *Stress intensity factors handbook*. Pergamon, 3rd edition, 1987.

N. Murayama, K. Nakamura, H. Obara, and M. Segawa. The strong piezoelectricity in polyvinylidene fluoride (PVDF). *Ultrasonics*, 14(1):15–24, 1976.

A. Needleman. Some issues in cohesive surface modeling. *Procedia IUTAM*, 10:221–46, 2014.

T. D. Nguyen, S. Govindjee, P. A. Klein, and H. Gao. A material force method for inelastic fracture mechanics. *Journal of the Mechanics and Physics of Solids*, 53:91–121, 2005.

R. W. Ogden. Large deformation isotropic elasticity: On the correlation of theory and experiment for incompressible rubberlike solids. *Proceedings of the Royal Society London*, A326:565–84, 1972a.

R. W. Ogden. Large deformation isotropic elasticity: On the correlation of theory and experiment for compressible rubberlike solids. *Proceedings of the Royal Society London*, A328:567–83, 1972b.

R. W. Ogden. *Non-linear Elastic Deformations*. Ellis Horwood Limited, 1984.

R. W. Ogden. Recent advances in the phenomenological theory of rubber elasticity. *Rubber Chemistry and Technology*, 59:361–83, 1986.

R. W. Ogden, G. Saccomandi, and I. Sigura. Fitting hyperelastic models to experimental datas. *Computational Mechanics*, 34:484–502, 2004.

E. Orowan. Fracture and strength of solids. *Reports on Progress in Physics*, 12:185–232, 1948.

A. G. Palmgren. Die Lebensdauer von Kugellagern. *Zeitschrift des Vereins Deutscher Ingenieure*, 68: 339–41, 1924.

P. C. Paris. *The growth of cracks due to variations in load*. PhD thesis, Lehigh University, 1962.

P. C. Paris, M. P. Gomez, and W. E. Anderson. A rational analytic theory of fatigue. *The Trend in Engineering. Alumni Newsletter of College of Engineering. University of Washington*, 13:9–14, 1961.

P. C. Paris, R. J. Bucci, E. T. Wessel, W. G. Clark, and T. R. Mager. Extensive study of low fatigue crack growth rates in A533 and A508 steels. In H. T. Corten and J. P. Gallagher, editors, *Stress Analysis and Growth of Cracks: Proceedings of the 1971 National Symposium on Fracture Mechanics: Part 1*, pages 141–176. ASTM International, West Conshohocken, PA, 1972.

A. Pineau, D. L. McDowell, E. P. Busso, and S. D. Antolovich. Failure of metals II: Fatigue. *Acta Materialia*, 107:484–507, 2016.

J. H. Poynting. On pressure perpendicular to the shear planes in finite pure shears, and on the lengthening of loaded wires when twisted. *Proceedings of the Royal Society London*, 82: 546–59, 1909.

W. Prager. A new method of analyzing stresses and strains in workhardening plastic solids. *Journal of Applied Mechanics*, 23:493–6, 1956.

L. Prandtl. Spannungsverteilung in plastischen Körpern. In C. B. Biezeno and J. M. Burgers, editors, *Proceedings of the First International Congress for Applied Mechanics, Delft* 1924, pages 43–54. Technische Boekhandel en Drukkerij J. Waltman, Jr., Delft, 1925.

S. Prussin. Generation and distribution of dislocations by solute diffusion. *Journal of Applied Physics*, 32:1876–81, 1961.

E. Pucci and G. Saccomandi. A note on the Gent model for rubber-like materials. *Rubber Chemistry and Technology*, 75:839–51, 2002.

R. Purkayastha and R. M. McMeeking. A linearized model for lithium ion batteries and maps for their performance and failure. *ASME Journal of Applied Mechanics*, 79:031021, 2012.

W. J. M. Rankine. II. On the stability of loose earth. *Philosophical Transactions of the Royal Society of London*, 147:9–27, 1857.

M. P. Renieri. *Rate and time dependent behavior of structural adhesives*. PhD thesis, Virginia Polytechnic Institute and State University, Blacksburg, VA, April 1976.

A. Reuss. Berückistigung der elastischen Formänderungen in der Plastizitätstheorie. *Zeitschrift für angewandte Mathematik und Mechanik*, 10:266–74, 1930.

J. R. Rice. A path independent integral and the approximate analysis of strain concentration by notches and cracks. *ASME Journal of Applied Mechanics*, 35:379–86, 1968a.

J. R. Rice. *Mathematical analysis in the mechanics of fracture*. In H. Liebowitz, editor, Mathematical Foundations, volume 2 of *Fracture an Advanced Treatise*, pages 191–311. Academic Press, New York, 1968b.

J. R. Rice. Fracture mechanics. In J. R. Rice, editor, Solid Mechanics Research Trends and Opportunities, volume 38 of *Applied Mechanics Reviews*, pages 1271–1275. American Society of Mechanical Engineers, 1985.

J. R. Rice. Elasticity of fluid-infilitrated porous solids (poroelasticity). http://esag.harvard.edu/rice/e2-Poroelasticity.pdf, 1998.

J. R. Rice and M. P. Cleary. Some basic stress diffusion solutions for fluid saturated elastic porous media with compressible constituents. *Reviews of Geophysics and Space Physics*, 14:227–41, 1976.

R. C. Rice, B. N. Leis, and D. V. Nelson, editors. *Fatigue Design Handbook*. Society of Automotive Engineers, Inc., 1988.

R. O. Ritchie. Mechanisms of fatigue-crack propagation in ductile and brittle solids. *International Journal of Fracture*, 100:55–83, 1999.

R. S. Rivlin. Large elastic deformations of isotropic materials. IV. Further developments of the general theory. *Philosophical Transactions of the Royal Society of London. Series A*, 241: 379–97, 1948.

R. S. Rivlin and D. W. Saunders. Large elastic deformations of isotropic materials. VII. Experiments on the deformation of rubber. *Philosophical Transactions of the Royal Society of London. Series A*, 243:251–88, 1951.

R. T. Rockafellar. *Convex Analysis*. Princeton University Press, 1970.

J. W. Rudnicki. Coupled deformation-diffusion effects in the mechanics of faulting and geomaterials. *Applied Mechanics Reviews*, 54:483–502, 2001.

M. Sadd. *Elasticity: Theory, Applications, and Numerics*. Elsevier, 2014.

I. H. Shames and F. A. Cozzarelli. *Elastic and Inelastic Stress Analysis*. Taylor & Francis, 1997.

N. K. Simha, F. D. Fischer, G. X. Shan, C. R. Chen, and O. Kolednik. J-integral and crack driving force in elastic-plastic materials. *Journal of the Mechanics and Physics of Solids*, 56:2876–95, 2008.

G. Simmons and H. Wang. *Single crystal elastic constants and calculated polycrystal properties*. MIT Press, 1971.

J. C. Simo and T. J. R. Hughes. *Computational Inelasticity*. Springer, 1998.

J. C. Simo and R. L. Taylor. Quasi-incompressible finite elasticity in principal stretches. Continuum basis and numerical algorithms. *Computer Methods in Applied Mechanics and Engineering*, 85: 273–310, 1991.

J. C. Simo, J. G. Kennedy, and S. Govindjee. Non-smooth multisurface plasticity and viscoplasticity. loading/unloading conditions and numerical algorithms. *International Journal for Numerical Methods in Engineering*, 26:2161–85, 1988.

D. R. Smith. *Variational Methods in Optimization*. Dover Publications, 1998.

W. A. Spitzig, R. J. Sober, and O. Richmond. Pressure-dependence of yielding and associated volume expansion in tempered martensite. *Acta Metallurgica*, 23:885–93, 1975.

R. I. Stephens, A. Fatemi, R. R. Stephens, and H. O. Fuchs. *Metal Fatigue in Engineering*. Wiley, 2nd edition 2001.

G. G. Stoney. The tension of metallic films deposited by electrolysis. *Proceedings of the Royal Society (London)*, A2:172–175, 1909.

G. Strang. *Computational Science and Engineering*. Wellesley-Cambridge Press, 2007.

S. Suresh. *Fatigue of materials*. Cambridge University Press, Cambridge, UK, second edition, 1998.

H. Tada, P. C. Paris, and G. R. Irwin. *The Stress Analysis of Cracks Handbook*. ASME, third edition, 2000.

R. L. Taylor, K. S. Pister, and G. L. Goudreau. Thermomechanical analysis of viscoelastic solids. *International Journal for Numerical Methods in Engineering*, 2:45–9, 1970.

K. Terzaghi. Die Berechnung der Durchlassigkeitsziffer des Tones aus dem Verlauf der hydrodynamischen Spannungserscheinungen. In *Sitzung Berichte*, Volume 132, pages 105–124. Akadamie der Wissenschaften, Wien Mathematische-Naturwissenschaftliche Klasse, 1923.

K. Terzaghi and O. K. Froölicj. *Theories der Setzung von Tonschicheten*. Franz Deuticke, Leipzig, Wien, 1936.

H. F. Tiersten. *Linear piezoelectric plate vibrations*. Plenum Press, New York, 1969.

S. P. Timoshenko. *History of strength of materials*. McGraw-Hill, 1953.

R. A. Toupin. Saint-Venant's principle. *Archives of Rational Mechanics and Analysis*, 18:83–96, 1965.

L. R. G. Treloar. The elasticity of a network of long-chain molecules. II. *Transactions of the Faraday Society*, 39:241–6, 1943.

L. R. G. Treloar. Stress-strain data for vulcanized rubber under various types of deformation. *Transactions of the Faraday Society*, 40:59–70, 1944.

H. Tresca. Mémoire sur l' écoulement des corps solides soumis à de fortes pressions. *Comptes Rendus de l'Académie des Sciences*, 59:754–8, 1864.

J. L. Troutman. *Variational Calculus with Elementary Convexity*. Springer-Verlag, 1983.

C. Truesdell and W. Noll. The non-linear field theories of mechanics. *Handbuch der Physik*, III, 1965.

W. Voigt. *Lehrbuch der Kristallphysik*. B. G. Teubner, Berlin, 1910.

H. F. Wang. *Theory of Linear Poroelasticity with Applications to Geomechanics and Hydrogeology*. Princeton University Press, 2000.

Q. Y. Wang, J. Y. Berard, A. Dubarre, G. Baudry, S. Rathery, and C. Bathias. Gigacycle fatigue of ferrous alloys. *Fatigue & Fracture of Engineering Materials & Structures*, 22:667–72, 1999.

W. Weibull. A statistical theory of the strength of materials. Handlingar Nr 151, Ingeniörsvetenskapsakademiens, Stockholm, 1939.

G. Williams and D. C. Watts. Non-symmetrical dielectric relaxation behaviour arising from a simple empirical decay function. *Transactions of the Faraday Society*, 66:80–5, 1970.

M. L. Williams. Stress singularities resulting from various boundary conditions in angular corners of plates in extension. *Journal of Applied Mechanics*, 19:526–8, 1952.

M. L. Williams. On the stress distribution at the base of a stationary crack. *Journal of Applied Mechanics*, 24:109–14, 1957.

A. Wöhler. Versuche zur Ermittlung der auf die Eisenbahnwagenachsen einwirkenden Kräfte und die Widerstandsfähigkeit des Wagen-Achsen. *Zeitschrift für Bauwesen*, 10:583–616, 1860.

A. Yavari. Compatibility equations of nonlinear elasticity for non-simply-connected bodies. *Archive for Rational Mechanics and Analysis*, 209:237–53, 2013.

X. C. Zhang, W. Shyy, and A. M. Sastry. Numerical simulation of intercalation-induced stress in Li-ion battery electrode particles. *Journal of the Electrochemical Society*, 154:A910–A916, 2007.

O. C. Zienkiewicz, R. L. Taylor, and J. Z. Zhu. *The Finite Element Method: Its Basis and Fundamentals*. Elsevier, Oxford UK, 7th edition, 2013.

Index